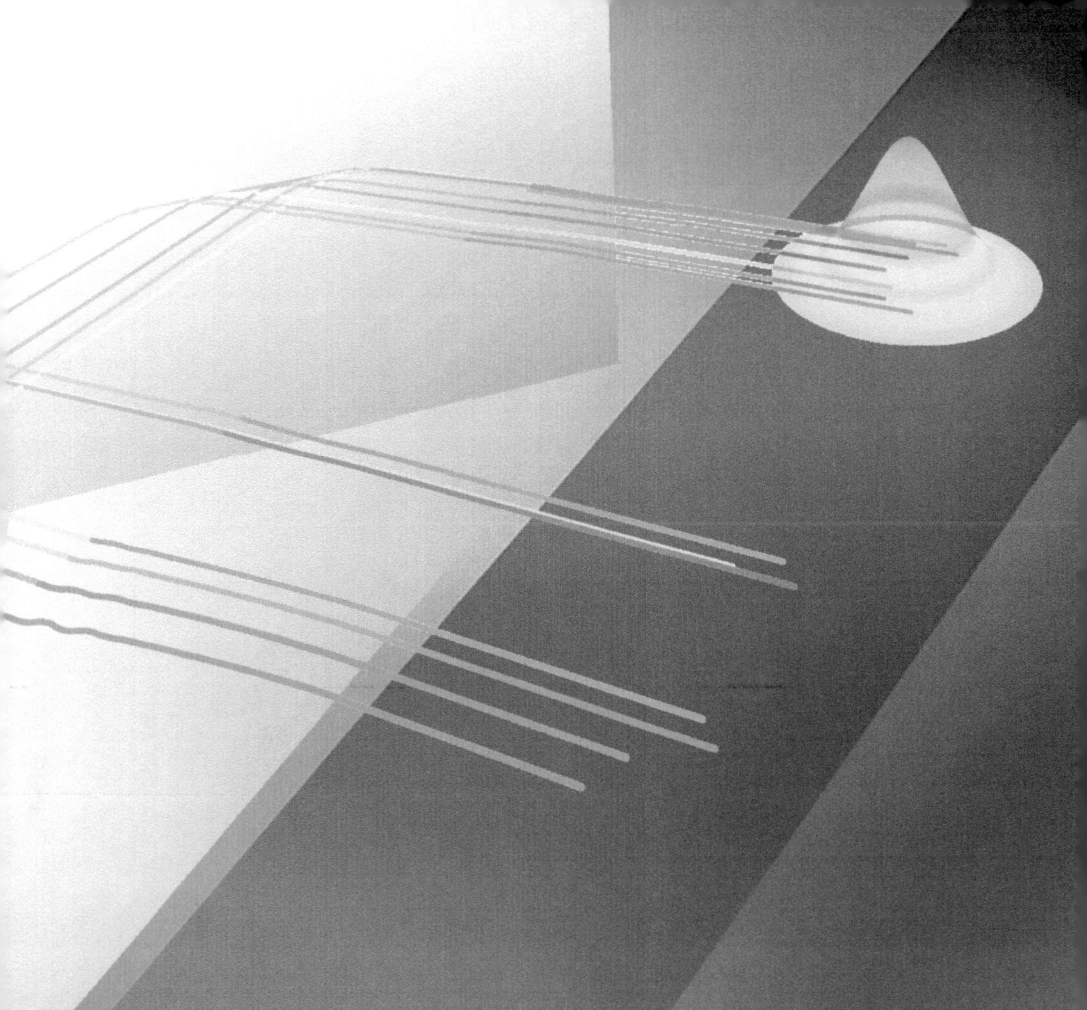

Applied Bohmian Mechanics

Applied Bohmian Mechanics

From Nanoscale Systems to Cosmology

Second Edition

Xavier Oriols
Jordi Mompart

JENNY STANFORD
PUBLISHING

Published by

Jenny Stanford Publishing Pte. Ltd.
Level 34, Centennial Tower
3 Temasek Avenue
Singapore 039190

Email: editorial@jennystanford.com
Web: www.jennystanford.com

British Library Cataloguing-in-Publication Data
A catalogue record for this book is available from the British Library.

Applied Bohmian Mechanics: From Nanoscale Systems to Cosmology (Second Edition)

ISBN 978-981-4800-10-5 (Hardcover)
ISBN 978-0-429-29474-7 (eBook)

Contents

Foreword to the First Edition

Quantum theory, born at the beginning of the twentieth century, represents one of the biggest revolutions ever performed in science. Led by Niels Bohr, the reputed physicist Werner Heisenberg and others devoted their careers both to formulate quantum mechanics in a consistent way, constructing what is known as the Copenhagen or orthodox formulation of quantum mechanics, and to extend it to other realms such as thermodynamics, solid-state physics, relativity, particle physics, and quantum field theory, to cite only a few. Thus, the orthodox formulation of quantum mechanics became the standard formulation to understand the quantum world. In parallel, Louis de Broglie and David Bohm showed that there is an alternative formulation of quantum mechanics, nowadays named Bohmian mechanics, that addresses the problem from a different perspective and provides exactly the same results as the orthodox formulation. However, Bohmian mechanics has been almost ignored by the scientific community until now.

During the second half of the twentieth century, a significant number of physicists dedicated strong efforts to look for practical applications of quantum mechanics. For example, the development of the laser brought a new scenario for precision experiments with an extremely high control on the atomic manipulation. Thus, quantum mechanics evolved from a theory focused on the fundamental principles of nature into an engineering discipline directly involved in the timely needs of our society. One enlightening example in this evolution of the use of quantum mechanics is the emerging field of quantum information science. Can Bohmian mechanics help in *applied* quantum physics?

I have to admit that before reading this book, I only had a vague knowledge of Bohmian mechanics. Now, after having the pleasure of reading it, the persistent question that bothered John S. Bell appears also in my mind: Why has Bohmian mechanics been so ignored among the scientific community when the most *devastating* criticism against it is just that *we are all too busy with our own work to spend time on something that doesn't seem likely to help us make progress with our real problems*? In fact, this book shows that even this criticism is not at all evident. It is mainly a consequence of the very few efforts that have been done to explore the possible utility of the door opened by Louis de Broglie almost a century ago.

Contrarily to others, one of the most original and attractive features of this book is the description of Bohmian mechanics from an engineering point of view. It is time to convert all the physical and mathematical ideas developed by Louis de Broglie, David Bohm, John S. Bell, and many others into *applied* tools for thinking, computing, and understanding quantum phenomena. Not as the only way, but as a complementary and useful alternative. In my opinion, this is the main message of this book.

In addition, since the progress in scientific and engineering research feeds from unexplored routes, I hope that this book will be very welcomed by the scientific community. Students and researchers have a new door to pass through (without closing the others) for *playing* with the Schrödinger and continuity equations in terms of waves and particles when addressing their particular quantum problems.

March 2012

Ignacio Cirac
Garching

Preface to the Second Edition

Quantum engineering, a giant with feet of clay

More than five years after the first edition of our book on applied Bohmian mechanics, our original motivations for writing it are still present. Certainly, today, a lot of publicity about the abilities of the Bohmian theory is still needed among the scientific community. In fact, over these last few years, many research programs have been devoted to the so-called quantum engineering with the goal of developing new materials, new sensors, or new computing strategies based on pure quantum phenomena. Thus, this book can be understood as a promotional presentation on how the Bohmian theory, among others, can help in the design and development of such applications. However, Bohmian mechanics is not a mere computational tool in terms of quantum trajectories, but a complete and ontological theory that provides a consistent explanation on how nature works. In this regard, this book can also be seen as a useful exercise to sincerely question our present understanding of the physical laws that govern the quantum world. After a bit of reflection on this point, many of us will probably conclude that our knowledge about the fundamentals of the quantum world are much more immature and imprecise than what we previously thought. In this sense, we want this book to be a warning on the risks of constructing the new and exciting discipline of quantum engineering as a *giant with feet of clay*.

The beginning of the twentieth century saw the first quantum revolution where novel and original theories were developed to understand unexpected nonclassical phenomena. What determines the structure of the periodic table? Why are some materials metals and some dielectrics, while others behave like semiconductors?

Nowadays, having established answers to these basic questions, a second quantum revolution is starting to take place, focusing on actively capitalizing on our quantum knowledge to alter the face of the physical world, developing a myriad of new quantum technologies. The difference between these two quantum revolutions is just the difference between science and engineering. The first revolution tried to properly understand our physical surroundings, the natural objects around us, while the second one intends to manipulate these surroundings to our own benefit. This is the typical evolution of most scientific disciplines. When scientific knowledge is mature enough, and the necessary technological means are available, engineers can use this knowledge for practical applications.

It is a common belief in our society that quantum theory, after more than a century, is ready to take a leap towards the engineering field. We have certainly outstanding technological means to manipulate quantum systems, even individual atoms, at the nanometer and femtosecond scales. Therefore, many national or international research organizations are focusing their programs towards the effective development of quantum technologies, trying to ensure that money spent on science has a direct impact on our society and its challenges for a better life. This is indeed a legitimate and compelling goal. However, is the quantum theory mature enough to blindly jump from science towards engineering? The pressure from society (in terms of research programs, grants, citing indices, etc.) is so effective that it forces most of the scientists to forget about the maturity of the quantum theory and just focus on (what really matters) the fast development of practical applications in the new and exciting field of quantum engineering.

We argue that the development of quantum engineering cannot be done at the price of forgetting the need for a deeper understanding of the physical laws governing the quantum world. One of the *forgotten* discussions by the new generation of quantum engineers is the *measurement problem*, which remains inside the backbone of the quantum theory. The measurement problem is manifested in the orthodox theory by its failure in explaining which physical interactions among particles constitute a measurement and which do not. In fact, there are many more examples of the immaturity of our quantum knowledge. Our inability to properly describe many-body systems due to their exponential complexity (the so-called many-body problem) makes that most of our understanding is based on a *puerile* single-particle description. We do not have a clear physical picture on the quantum-to-classical transition. What makes a quantum system to behave classically in some circumstances? The fact that there are several quantum theories which are empirically equivalent but radically different at the ontological level is a clear evidence of our bad understanding. The Copenhagen theory is the most extensively investigated and presently the one with more support among the scientific community. Others include spontaneous collapse theories or the many-worlds theory. The one studied in this book, Bohmian mechanics, provides a description of quantum phenomena by particles choreographed by the wave function. In general, neither of these theories is more mature than the orthodox one, but they remove the need of an observer, which relaxes some of the difficulties to understand the measurement at a quantum level. Some of these alternative theories are not free of problems, including the quantum-to-classical transition and the many-body problem, while others still need to be dealt with. We do not mean to imply that these alternative theories are better (how does one quantify *better* here?), but that there is still a lot of work needed to certify that our comprehension of the quantum world is unproblematic.

Let us try to exemplify the risks of developing quantum engineering alone without worrying on its fundamentals. In the orthodox theory, every time we make a measurement a random process occurs. But, as we do not really know what makes a physical interaction to be a measurement, we really do not know the origin of such randomness. In the Bohmian theory, for example, this randomness comes from an uncertainty in the initial position of the particles. With further efforts to clarify the quantum theories, we can perhaps achieve a better understanding of the origin of quantum randomness and then, the exciting new building of application developed along the new discipline of quantum cryptography, based on the unavoidable presence of such intrinsic quantum randomness, will simple melt as a giant with feet of clay. The reader can argue that there are a lot of scientific works supporting the actual status of quantum cryptography. Perhaps this particular warning is completely unfounded and quantum cryptography will certainly remain as robust as we know today.

But, perhaps not. It is enlightening to remember here the theorem that John von Neumann stated in 1932 about the impossibility of explaining quantum mechanics with hidden variables (such as quantum trajectories). This theorem remained an unquestionable truth, and part of the essence of the quantum world, until David Bohm (with an explanation of quantum phenomena in terms of waves and particles) showed that the theorem was wrong (as its own preliminary assumption precludes the existence of Bohmian trajectories). The curious spectacle is not that John von Neumann (an outstanding scientist in many disciplines) made a mistake in a theorem, but that the community (with the exception of Grete Hermann in 1935 that was totally ignored) blindly accepted the theorem for almost half a century.

There are many more examples which certify that our understanding of the quantum world is still immature. The wave function, the basic element in most theories, can be prepared for instance by forcing the quantum system into its ground state, but it cannot be directly measured in a single experiment. The wave function can be measured through a weak protocol, but also Bohmian velocities can be measured though such protocol. We do not even know what is really the wave function at the most fundamental level: A law? A field? A probability transporter.

In summary, the quantum world is so complicated that one century has not been enough for the scientific community to clearly elaborate an unproblematic description of the laws of quantum mechanics. We are not arguing here that research on quantum engineering needs to stop. Just the contrary. The development of quantum engineering and the research on the foundations of the quantum theory has to evolve intimately connected to benefice from each other in achieving better practical applications and a deeper understanding of the quantum world. Otherwise, we will build a giant with feet of clay. We hope that the present book can be viewed as a modest contribution in both directions.

March 2019

Xavier Oriols
Jordi Mompart
Cerdanyola del Vallès, Barcelona

Preface to the First Edition

New cutting-edge ideas come from outside of the main stream

Most of our collective activities are regulated by other people who decide whether they are well done or not. One has to learn some arbitrary symbols to write understandable messages or to read those from others. Human rules over collective activities govern the evolution of our culture. On the contrary, natural systems, from atoms to galaxies, evolve independently of the human rules. We cannot modify physical laws. We can only try to understand them. Nature itself judges, through experiments, whether a plausible explanation for some natural phenomena is correct or incorrect. Nevertheless, in forefront research where the unknowns start to become understandable, the new knowledge is still unstable, somehow immature. It is supported by few experimental evidences, or the evidences are still subjected to different interpretations. Certainly, novel research grows up closely tied to the economical, sociological, or historical circumstances of the involved researchers. A period of time is needed in order to distil new knowledge, separating pure scientific arguments from cultural influences.

The past and the present status of Bohmian mechanics cannot be understood without these cultural considerations. The Bohmian formalism was proposed by Louis de Broglie even before the standard, that is, Copenhagen, explanation of quantum phenomena was established. Bohmian mechanics provides an explanation of quantum phenomena in terms of point particles guided by waves. One object cannot be a wave and a particle simultaneously, but two can, especially if one of the objects is a wave and the other is a particle. Unfortunately, Louis de Broglie himself abandoned these ideas. Later, in the fifties, David Bohm clarified the meaning and applications of this original explanation of quantum phenomena. Bohmian mechanics agrees with all non-relativistic quantum experiments done up to now. However, it remains almost ignored by most of the scientific community. In our opinion, there are no scientific arguments to support its marginal status but only cultural reasons. One of the motivations for writing this book is helping in the maturing process that the scientific community needs about Bohmian mechanics.

Certainly, the distilling process of Bohmian mechanics is being quite slow. Anyone interested enough to walk this causal road of quantum mechanics can be easily confused by many misleading signposts that have been raised in the scientific literature, not only by its detractors, but, unfortunately, very often, also by some of its advocates. Nowadays, following opinions from other reputed physicists (we are easily persuaded by those scientists with authority) is far from being a proper scientific strategy to get our own opinion about Bohmian mechanics.

In any case, since the mathematical structure of Bohmian mechanics is quite simple, it can be easily learned by anyone with only a basic knowledge of classical and quantum mechanics who makes the necessary effort to build his own scientific opinion based on logical deductions, free from cultural influences. The introductory chapter of this book, including a thorough list of exercises and easily programmable algorithms, provides a reasonable and objective source of information in order to achieve this later goal, even for undergraduate students.

Curiously, the fact that Bohmian mechanics is ignored and remains mainly unexplored is an attractive feature for some adventurous scientists. They know that very often *new cutting-edge ideas come from outside of the main stream* and find in Bohmian mechanics a useful tool in their research activity. On the one hand, it provides an explanation of quantum mechanics, in terms of trajectories, that results to be very useful in explaining the dynamics of quantum systems, being thus also a source of inspiration to look for novel quantum phenomena. On the other hand, since it provides an alternative mathematical formulation, Bohmian mechanics offers new computational tools to explore physical scenarios that presently are computationally inaccessible, such as many-particle solutions of the Schrödinger equation. In addition, Bohmian mechanics sheds light on the limits and extensions of our present understanding of quantum mechanics towards other paradigms such as relativity or cosmology, where the internal structure of Bohmian mechanics in terms of well-defined trajectories is very attractive. With all these previous motivations in mind, this book provides ten chapters (apart from the introduction in the first chapter) with practical examples showing how Bohmian mechanics helps us in our daily research activities.

Obviously, there are other books focused on Bohmian mechanics. However, many of them are devoted to the foundations of quantum mechanics emphasizing the difficulties or limitations of the Copenhagen interpretation for providing an ontological description of our world. On the contrary, this book is not focus on the foundations of quantum mechanics, but on the discussion about the practical application of the ideas of de Broglie and Bohm to understand and compute the quantum world. Several examples of such practical applications written by leading experts in different fields, with an extensive updated bibliography, are provided here. The book, in general, is addressed to students in physics, chemistry, electrical engineering, applied mathematics, and nanotechnology, as well as to both theoretical and experimental researchers who seek new computational and interpretative tools for their everyday research activity. We hope that the newcomers to this causal explanation of quantum mechanics will use Bohmian mechanics in their research activity so that Bohmian mechanics will become more and more popular for the broad scientific community. If so, we expect that, in the near future, Bohmian mechanics will be taught regularly at the Universities, not as the unique and revolutionary way of understanding quantum phenomena, but as an additional and useful interpretation of all quantum phenomena in terms of quantum trajectories. In fact, Bohmian mechanics has the ability of removing most of the mysteries of the Copenhagen interpretations and, somehow, simplifying (or demystifying) quantum mechanics. We will be very glad if this book can contribute to shorten the time needed to achieve all these goals.

Finally, we want to acknowledge many different people who have allowed us to embark on and successfully finish this book project. First of all, we want to thank Alfonso Alarcón and Albert Benseny, who became involved in the book project from the very beginning, as two additional editors. We also want to thank the rest of the authors of the book for accepting our invitation to participate in this project and writing their chapters according to the general spirit of the book. Due to page limitations, only nine examples of practical applications of Bohmian mechanics in forefront research activity are presented in this book. Therefore, we want to apologize to many other researchers who could have certainly been also included in the book. We also want to express our gratitude to Pan Stanford Publishing for accepting our book project and for its kind attention during the publishing process.

March 2012

Xavier Oriols
Jordi Mompart
Cerdanyola del Vallès, Barcelona

Introduction

The beginning of the twentieth century brought surprising nonclassical phenomena. Max Planck's explanation of the black-body radiation [1], the work of Albert Einstein on the photoelectric effect [2] and the Niels Bohr's model to account for the electron orbits around the nuclei [3] established what is now known as the *old quantum theory*. To describe and explain these effects, phenomenological models and theories were first developed, without rigorous justification. In order to provide a complete explanation for the underlying physics of such new nonclassical phenomena, physicists were forced to abandon classical mechanics to develop novel, abstract and imaginative formalisms.

In 1924, Louis de Broglie suggested in his doctoral thesis that matter, apart from its intrinsic particle-like behavior, could exhibit also a wave-like one [4]. Three years later he proposed an interpretation of quantum phenomena based on nonclassical trajectories guided by a *wave field* [5]. This was the origin of the pilot-wave formulation of quantum mechanics that we will refer as Bohmian mechanics to account for the following work of David Bohm [6, 7]. In the Bohm formulation, an individual quantum system is formed by a point particle and a guiding wave. Contemporaneously, Max Born and Werner Heisenberg, in the course of their collaboration in Copenhagen with Niels Bohr, provided an original formulation of quantum mechanics without the need of trajectories [8, 9]. This was the origin of the so-called Copenhagen interpretation of quantum phenomena and, since it is the most accepted formulation, it is basically the only one explained at most universities. Thus, it is also known as the *orthodox* formulation of quantum mechanics. In the Copenhagen interpretation, an individual quantum system exhibits its wave or its particle nature depending on the experimental arrangement.

The present status of Bohmian mechanics among the scientific community is quite marginal (the quantum chemistry community is an encouraging exception). Most researchers do not know about it or believe that is not fully correct. There are others that know that quantum phenomena can be interpreted in terms of trajectories, but they think that this formalism cannot be useful in their daily research activity. Finally, there are few researchers, the authors of this book among them, who think that Bohmian mechanics is a useful tool to make progress in frontline research fields involving quantum phenomena.

The main (nonscientific) reason why still many researchers believe that there is something wrong with Bohmian mechanics can be illustrated with Hans Christian Andersen's tale "The Emperor's New Clothes." Two swindlers promise the Emperor the finest clothes that, as they tell him, are invisible to anyone who is unfit for their position. The Emperor cannot see the (nonexisting) clothes, but pretends that he can for fear of appearing stupid. The rest of the people do the same. Advocates of the Copenhagen interpretation have attempted to produce *impossibility proofs* in order to demonstrate that Bohmian mechanics is incompatible with quantum phenomena [10]. Most researchers, who are not aware of the incorrectness of such *proofs*, might conclude that there is some controversy with the Bohmian formulation of quantum mechanics and they prefer not to support it, for fear of appearing discordant. At the end of the tale, during the course of a procession, a small child cries out "The Emperor is Naked!" In the tale of quantum mechanics, David Bohm [6, 7] and John Bell [11] were the first to exclaim to the scientific community "Bohmian mechanics is a correct interpretation of quantum phenomena whose results exactly coincides with the predictions of the orthodox interpretation!".

What Is a Quantum Theory?

Albert Einstein, in the paper entitled "Physics and reality" [12], pointed out the possibility of living in a bizarre world without comprehensible explanations for natural phenomena. He wrote: "The fact that [the world] is comprehensible is a miracle." Similarly, Eugene Wigner wrote: "The unreasonable effectiveness of mathematics in the natural science .... is a wonderful gift which we neither understand nor deserve" [13]. Both reflections were inspired by the previous work of the German philosopher Immanuel Kant who wrote the very same idea almost two centuries before: "The eternal mystery of the world is its comprehensibility." Fortunately, it seems that we live in a comprehensible world.

Kant divided scientific knowledge into three categories: appearance, reality and theory. Appearance is the content of our sensory experience of natural phenomena, which is the empirical outcome of an experiment. Reality is what lies behind all natural phenomena. A theory is a human model that tries to mirror both appearance and reality. A useful theory might predict the outcome of an experiment in a laboratory or the observation of a phenomenon in Nature. Empiricists believe on experimental outcomes (what Kant called appearance) and refuse to speculate about a deeper reality. On the other hand, realists believe that good physical theories explain, or at least provide clues about, the reality of our comprehensible world. Most researchers are a combination of both stereotypes, with variable proportions.

As all human creations, there are successful and unsuccessful theories. When in 1864 James Clerk Maxwell conjectured that light was an electromagnetic vibration, it was believed that all waves had to vibrate in some medium. The medium in which light presumably travels was named *luminiferous ether*. For almost a century eminent scientists believed in this concept blindly. Nowadays, the *luminiferous ether* plays no role at all in modern physical theories [14]. The atomicity of matter is an example of a very successful theory. It was introduced by the British chemist John Dalton in 1808 to explain why some chemical substances need to combine in some fixed ratios. During one century it was thought that atoms were a crazy idea. Marcelin Berthelot said "who [has] ever seen a gas molecule or an atom?," expressing the disdain that many chemists felt for the unseen atoms, which were inaccessible to experiments [14]. Even their defenders saw little hope of ever directly verifying the atomic hypothesis. Nowadays, the fact that everything is made of atoms is one of the most precious knowledge that we get on how Nature works [15], and their images are even routinely seen in the screens of scanning tunneling microscopes [16].

A quantum theory is a human explanation of quantum phenomena. All quantum theories have associated their own intangible reality. The so-called ontology of the theory. For example, the ontology of the Bohmian theory is very simple: everything is formed by point particles guided ("choreographed") by waves. The different quantum theories available today (Copenhagen, Bohmian, many worlds, spontaneous collapse, etc.) are indeed inspired by radically different realities, but all of them provide the same empirical predictions on quantum phenomena. In Kant's words, all of them provide the same explanation of the appearance of our world. As we repeatedly stress, up to now, in spite of many attempts, there is no experimental evidence that can discern between Bohmian and Copenhagen realities (ontologies).

In fact, for practical applications, even wrong theories can be very useful. Most natural phenomena that affect our ordinary life can be exclusively explained in terms of classical mechanics. However, today, we know that the reality behind the classical theory is wrong because it does not provide accurate predictions for some natural phenomena, like relativistic (with particles with high velocities) or quantum (atomistic dimensions) experiments. Surprisingly, the fact that the classical theory is a wrong theory does not demerit its extraordinary utility and our confidence on its predictions within its range of validity.[a] The same is true for most physical theories at a practical level. Even if we could demonstrate in the future that either the Copenhagen or the Bohmian theories is wrong (or both), the practical utility of these theories in their range of validity would not dismiss.

---

[a]We take *classical* planes expecting that they will follow a deterministic trajectory, e.g., from Barcelona to Paris. However, we know that quantum uncertainty precludes us to affirm that there is only one possible trajectory for the fly departing from Barcelona. Even after doing our best to fix the initial conditions of the physical degrees of the plane, there is still an unavoidable quantum randomness implying that several trajectories are possible. Of course, the differences between trajectories are so small at a macroscopic level that the pilot can easily certify that we will arrive to Paris.

How Bohmian Mechanics Helps?

Although there is no experimental evidence against Bohmian mechanics, many researchers believe that Bohmian mechanics is not a useful tool to do research. In the words of Steven Weinberg, in a private exchange of letters with Sheldon Goldstein [17], "In any case, the basic reason for not paying attention to the Bohm approach is not some sort of ideological rigidity, but much simpler—it is just that we are all too busy with our own work to spend time on something that doesn't seem likely to help us make progress with our real problems."

The history of science seems to give credit to Weinberg's sentence. In spite of the controversies that have always been associated with the Copenhagen interpretation since its birth a century ago, its mathematical and computational machinery has enabled physicists, chemists and (quantum) engineers to calculate and predict the outcome of a vast number of experiments, while the contribution of Bohmian mechanics during the same period is much less significative. The differences are due to the fact that Bohmian mechanics remains mainly unexplored.

Contrarily to Weinberg's opinion, we believe that Bohmian mechanics *can help us make progress with our real problems*. There are, at least, three clear reasons why one could be interested in studying quantum problems with Bohmian mechanics:

(1) **Bohmian *explaining***: Even when the Copenhagen mathematical machinery is used to compute observable results, the Bohmian interpretation often offers different interpretational tools. We can find descriptions of electron dynamics such as *an electron crosses a resonant tunneling barrier and interacts with another electron inside the well.* However, according to the orthodox theory, we can only talk about the properties of an electron (for example, its position) when we measure it. Thus, an electron crossing a tunneling region is not rigorously supported within orthodox quantum mechanics, but it is within the Bohmian picture. Thus, in contrast to the Copenhagen theory, Bohmian mechanics allows for an easy visualization of quantum phenomena in terms of trajectories that has important

demystifying or clarifying consequences. In fact, Bohmian mechanics allows for an unambiguous[a] description of measured and unmeasured properties of particles (an electron crossing a tunneling barrier is a description of unmeasured properties). Bohmian mechanics provides a single-event description of the experiment, while Copenhagen quantum mechanics accounts for its statistical or ensemble explanation. We will present several examples in Chapters 2 and 3 emphasizing all these points.

(2) **Bohmian *computing*:** Although the predictions of the Bohmian interpretation reproduce the ones of the orthodox formulation of quantum mechanics, its mathematical formalism is different. In some systems, the Bohmian equations might provide better computational tools than the ones obtained from the orthodox machinery, resulting in a reduction of the computational time, an increase in the number of degrees of freedom directly simulated, etc. We will see examples of these computational issues in quantum chemistry in Chapters 4 and 5, as well as in quantum electron transport in Chapter 6.

(3) **Bohmian *thinking*:** From a more fundamental point of view, alternative formulations of quantum mechanics can provide alternative routes to look for the limits and possible extensions of the quantum theory. In particular, Chapter 7 presents the route to connect Bohmian mechanics with geometrical optics and beyond opening the way to apply the powerful computational tools of quantum mechanics to classical optics, and even to electromagnetism. The natural extension of Bohmian mechanics to the relativistic regime and to quantum field theory are presented in Chapter 8, while Chapter 9 and Chapter 10 discusses its application to cosmology.

The fact that all measurable results of the orthodox quantum mechanics can be exactly reproduced with Bohmian mechanics

---

[a]About the ambiguity of the orthodox explanation of quantum mechanics and the unambiguity of Bohmian mechanics, J. Bell wrote [11] (page 111): "I will try to interest you in the de Broglie–Bohm version of non-relativistic quantum mechanics. It is, in my opinion, very instructive. It is experimentally equivalent to the usual version insofar as the latter is unambiguous."

(and vice versa) is the relevant point that completely justifies why Bohmian mechanics can be used for ***explaining*** or ***computing*** different quantum phenomena in physics, chemistry, electrical engineering, applied mathematics, nanotechnology, etc. In the scientific literature, the **Bohmian *computing*** technique to find the trajectories (without directly computing the wave function) is also known as a *syntactic* technique, while the **Bohmian *explaining*** technique (where the wave function is directly computed first) is referred as the *analytic* technique [18]. Furthermore, the fact that Bohmian mechanics is a theory without observers is an attractive feature for those researchers interested in ***thinking*** about the limits or extensions of the quantum theory.

In order to convince the reader about the practical utility of Bohmian mechanics for ***explaining***, ***computing*** or ***thinking***, we will not present elaborated mathematical developments or philosophical discussions, but provide practical examples. Apart from the first chapter, devoted to an overview of Bohmian mechanics, the book is divided into nine additional chapters with several examples on the practical application of Bohmian mechanics to different research fields, ranging from atomic systems to cosmology. These examples will clearly show that the previous quotation by Weinberg does not have to be always true.

On the Name "Bohmian Mechanics"

Any possible newcomer to Bohmian mechanics can certainly be quite confused and disoriented by the large list of names and slightly different explanations of the original ideas of de Broglie and Bohm that are present in the scientific literature. Different researchers use different names. Certainly, this is an indication that the theory is still not correctly settled down among the scientific community.

In his original works [4, 5], de Broglie used the term *pilot-wave theory* [19], to emphasize the fact that wave fields guided the motion of point particles. After de Broglie abandoned his theory, Bohm rediscovered it in the seminal papers entitled "A Suggested Interpretation of the Quantum Theory in Terms of Hidden Variables"

[6, 7]. The term *hidden variables,*[a] referring to the positions of the particles, was perhaps pertinent in 1952, in the context of the *impossibility proofs* [10]. Nowadays, these words might seem inappropriate because they suggest something metaphysical on the trajectories.[b]

To give credit to both de Broglie and Bohm, some researchers refer to their works as the *de Broglie–Bohm theory* [22].[b] Some reputed researchers argue that de Broglie and Bohm did not provide the same exact presentation of the theory [19, 23]. While de Broglie presented a first order development of the quantum trajectories (integrated from the velocity), Bohm himself did a second order (integrated from the acceleration) emphasizing the role of the quantum potential. The differences between both approaches appear when one considers initial ensembles of trajectories which are not in quantum equilibrium.[c] Except for this issue, which will not be addressed in this book, both approaches are identical.

Many researchers prefer to use the name *Bohmian mechanics* [24]. It is perhaps the most popular name. We know directly from his

---

[a]Note that the term *hidden variables* can also refer to other (local and nonlocal) formulations of quantum mechanics.

[b]Sometimes it is argued that the name *hidden variables* is because Bohmian trajectories cannot be measured directly. However, what is not directly measured in experiments is the (complex) wave function amplitude, while the final positions of particles can be directly measured, for example, by the imprint they leave on a screen. John S. Bell wrote [11] (page 201): "Absurdly, such theories are known as 'hidden variable' theories. Absurdly, for there it is not in the wave function that one finds an image of the visible world, and the results of experiments, but in the complementary 'hidden'(!) variables."

[b]In fact, even de Broglie and Bohm were not the original names of the scientists' families. Louis de Broglie's family, which included dukes, princes, ambassadors and marshals of France, changed their original Italian name Broglia to de Broglie when they established in France in the seventeenth century [20]. David Bohm's father, Shmuel Düm, was born in the Hungarian town of Munkács and, was sent to America when he was young. Upon landing at Ellis Island, he was told by an immigration official that his name, Düm would mean "stupid" in English. The official himself decided to change the name to Bohm [21].

[c]Quantum equilibrium assumes that the initial positions and velocities of Bohmian trajectories are defined compatible with the initial wave function. Then the trajectories computed from Bohm's or de Broglie's formulations will become identical. However, one can select completely arbitrary initial positions from the (first-order) de Broglie explanation and arbitrary initial velocities and positions from the (second-order) Bohm work (see Section 1.2.6).

alive collaborators, Basile Hiley and David Peat [25], that this name irritated David Bohm and he said about its own work, "It's Bohmian nonmechanics." He argued that the *quantum potential* is a nonlocal potential that depends on the relative shape of the wave function and thus it is completely different from other mechanical (such as the gravitational or the electrostatic) potentials which decrease with distance. See this particular discussion in the last chapters of Bohm and Hiley's book entitled *The Undivided Universe: An Ontological Interpretation of Quantum Theory* [26]. He preferred the names *causal* or *ontological interpretation* of quantum mechanics [22, 26]. The latter names emphasize the foundational aspects of its formulation of quantum mechanics.

Finally, another very common term is *quantum hydrodynamics* [18] that underlines the fact that Bohmian trajectories provide a mathematical relationship between the Schrödinger equation and fluid dynamics. In fact, this name is more appropriate when one refers to the Madelung theory [27], which is considered as a precursor of Bohm's work (see Section 1.1.8).

From all these different names, we choose *Bohmian mechanics* because it is short and clearly specifies what we are referring to. It has the inconvenience of not giving credit to the initial work of Louis de Broglie. Although it might be argued that Bohm merely reinterpreted the prior work of de Broglie, we think that he was the first scientist to genuinely understand its significance and implications. As we mentioned, Bohm himself disliked this name. However, as any work of art, the explanation of the quantum phenomena done in the 1952 Bohm's paper does not completely belong to the author,[a] but has become part of our scientific heritage. It has happened many times during the history of science that the mathematical equation developed by a scientist contains much more physical substance than what he/she imagined at the beginning. In any case, we understand Bohmian mechanics as a generic name that includes all those works inspired from the original ideas of Bohm and de Broglie. In Fig. 1, we plot the numbers of citations per year for

---

[a]For example, Erwin Schrödinger, talking about quantum theory, wrote: "I don't like it, and I'm sorry I ever had anything to do with it", but his opinion did not influence the great applicability of his famous equation in the orthodox theory.

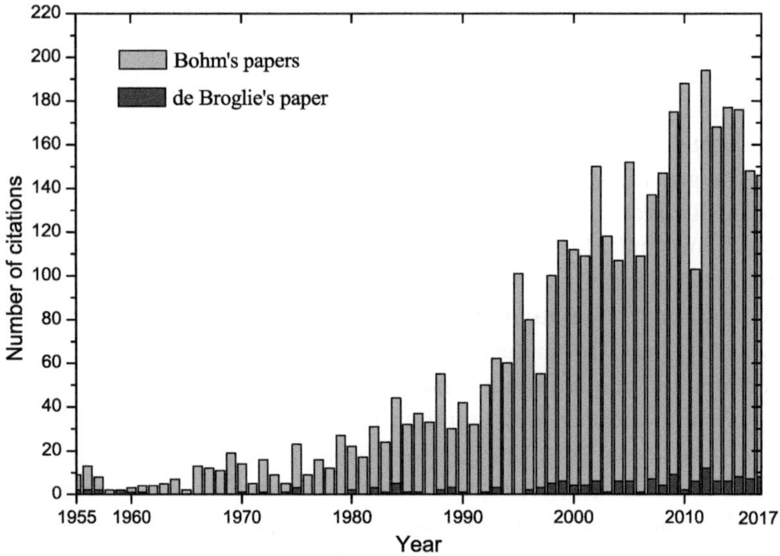

Figure 1 Number of citations per year for (a) the two 1952 David Bohm's papers entitled "A Suggested Interpretation of the Quantum Theory in Terms of Hidden Variables" [6, 7] and (b) Louis de Broglie's paper "La mécanique ondulatorie et la structure atomique de la matière et du rayonnement" [5]. Data retrieved from ISI Web of Knowledge [28] in December 2017.

the 1952 Bohm's seminal papers [6, 7], certifying the exponentially growing influence of these papers, which is not the case for the original work of de Broglie [5].

On the Book Contents

The book contains ten chapters. The first chapter provides an accessible introduction to Bohmian mechanics. The remaining chapters present practical examples of the applicability of Bohmian mechanics. Let us start mentioning the cover of this second edition of the book. It represents the wave and particle nature of electrons according to the Bohmian theory. In particular, we see the Bohmian trajectories of an electron which suffers Klein tunneling when impinging on a triangular potential barrier of a graphene structure. The wave packet of the electron corresponding to the bispinor solution of the Dirac equation (electron with positive and negative energies) is also plotted.

Chapter 1 is the longest one and it is entitled "Overview of Bohmian mechanics." It is written by Xavier Oriols and Jordi Mompart, the editors of the book, both from the Universitat Autònoma de Barcelona, Spain. This chapter is intended to be an introduction to any newcomer interested in Bohmian mechanics. Only basic concepts of classical and quantum mechanics are assumed. The chapter is divided into four different sections. First, the historical development of Bohmian mechanics is explained. Then, Bohmian mechanics for single particle and for many particle systems (with spin and entanglement discussions) are presented. Finally, the topic of Bohmian measurement is addressed. The chapter also contains a list of solved problems and easily implementable codes for computation of Bohmian trajectories.

Chapter 2 is entitled "Hydrogen Photoionization with Strong Lasers." It is written by Albert Benseny from the Okinawa Institute of Science and Technology in Japan; Antonio Picón and Luis Plaja from the Universidad de Salamanca, Spain; Jordi Mompart from the Universitat Autònoma de Barcelona, Spain and Luis Roso from the CLPU, the Laser Center for Ultrashort and Ultraintense Pulses, in Salamanca. They discuss the dynamics of a single hydrogen atom interacting with a strong laser. In particular, the Bohmian trajectories of these electrons represent an interesting illustrating view, with new calculation methods (i.e., **Bohmian *computing***), of both the above threshold ionization and the harmonic generation spectra problems. They do also present a full three-dimensional model to discuss the dynamics of Bohmian trajectories when the light beam and the hydrogen atom exchange spin and orbital angular momenta. The chapter does also provide a practical example on how Bohmian mechanics is computed, with an analytical (i.e., **Bohmian *explaining***) procedure, when full (scalar and vector potentials) electromagnetic fields are considered.

The title of **Chapter 3** is "Atomtronics: Coherent Control of Atomic Flow via Adiabatic Passage" and it is written by Albert Benseny from the Okinawa Institute of Science and Technology in Japan; Joan Bagudà, Xavier Oriols, and Jordi Mompart from the Universitat Autònoma de Barcelona, Spain; and Gerhard Birkl from the Institut für Angewandte Physik, from the Technische Universität Darmstadt in Germany. Here, it is discussed an efficient and robust technique to coherently transport a single neutral atom, a single hole, or even a Bose–Einstein condensate between the two extreme traps of the triple-well potential. The dynamical evolution of this system with the direct integration of the Schrödinger equation presents a very counterintuitive effect: by slowing down the total time duration of the transport process it is possible to achieve atomic transport between the two extreme traps with a very small (almost negligible) probability to populate the middle trap. The analytical (i.e., **Bohmian *explaining***) solution of this problem with Bohmian trajectories enlightens the role of the *particle conservation law* in quantum systems showing that the negligible particle presence is due to a sudden particle acceleration yielding, in fact, ultra-high atomic velocities. The Bohmian contribution opens the discussion about the possible detection of these high velocities or the need for a relativistic formulation to accurately describe such a simple quantum system.

Chapter 4, entitled "Bohmian Pathways into Chemistry: A Brief Overview," is prepared by Ángel S. Sanz, from the Universidad Complutense de Madrid, Spain, and deals with the issue of how the **Bohmian *computing*** abilities have been explored and exploited in chemistry over decades. Interestingly, contrary to physics, Bohmian mechanics has always found a better accommodation and acceptance within different areas of chemistry, where the pedagogical advantages mentioned by John Bell have been widely recognized. Because providing an exhaustive account on the applications (both as a problem solver and as a computational tool) where Bohmian mechanics has been of relevance within chemistry would exceed the scope of the chapter, it has been prepared in a way that may serve the reader as a guide to acquire a general perspective (or impression) on how this trajectory-based quantum approach has permeated the different traditional levels or pathways to approach the problems of interest in chemistry.

Chapter 5, whose title is "Adaptive Quantum Monte Carlo Approach States for High-Dimensional Systems," is written by Eric R. Bittner, Donald J. Kouri, Sean Derrickson, from the University of Houston; and Jeremy B. Maddox, from the Western Kentucky University, in USA. They provide one particular example on the success of Bohmian mechanics in the chemistry community. In this chapter, the authors explain their **Bohmian *computing*** development for knowing the *ab initio* quantum mechanical structure, energetics and thermodynamics of multiatoms systems. They use a variational approach that finds the quantum ground sate (or even excited states at finite temperature) using a statistical modeling approach for determining the best estimate of a quantum potential for a multidimensional system.

Chapter 6 is entitled "Nanoelectronics: Quantum Electron Transport." It is written by Enrique Colomés, Devashish Pandey, Alfonso Alarcón and Xavier Oriols from the Universitat Autònoma de Barcelona, Spain; Zhen Zhan from the Wuhan University, in China; Guillem Albareda from Max Planck Institute for the Structure and Dynamics of Matter in Germany and Fabio Lorenzo Traversa from University of California, in USA. The authors explain the ability of their own many-particle **Bohmian *computing*** algorithm to understand and model nanoscale electron devices. In particular, it is shown that the application of Bohmian mechanics to electron transport in open systems (with interchange of particles and energies) leads to a quantum Monte Carlo algorithm, where randomness appears because of the uncertainties in the number of electrons, their energies and the initial positions of (Bohmian) trajectories. A general, versatile and time-dependent 3D electron transport simulator for nanoelectronic devices, named BITLLES (Bohmian Interacting Transport for nonequiLibrium eLEctronic Structures), is presented showing its ability for a full prediction (DC, AC, fluctuations) of the electrical characteristics of any nanoelectronic device. The BITLLES simulator is also applied to graphene structures (by solving the Dirac equation) as reflected in the cover of this book.

Chapter 7, entitled "Beyond the Eikonal Approximation in Classical Optics and Quantum Physics," is written by Adriano Orefice, Raffaele Giovanelli and Domenico Ditto from the Università degli Studi di Milano, Italy. It is devoted to discuss how **Bohmian *thinking*** can also help in optics, exploring the fact that the time-independent Schrödinger equation is strictly analogous to the Helmholtz equation appearing in classical wave theory. Starting from this equation they obtain indeed, without any omission or approximation, a Hamiltonian set of ray-tracing equations providing (in stationary media) the exact description in term of rays of a family of wave phenomena (such as diffraction and interference) much wider than that allowed by standard geometrical optics, which is contained as a simple limiting case. They show in particular that classical ray trajectories are ruled by a wave potential presenting the same mathematical structure and physical role of Bohm's quantum potential, and that the same equations of motion obtained for classical rays hold, in suitable dimensionless form, for quantum particle dynamics, leading to analogous trajectories and reducing to classical dynamics in the absence of such a potential.

Chapter 8, entitled "Relativistic Quantum Mechanics and Quantum Field Theory," is written by Hrvoje Nikolić from the Rudjer Bošković Institute, Croatia. This chapter presents a clear example on how a **Bohmian *thinking*** on superluminal velocities and nonlocal interactions helps in extending the quantum theory towards relativity and quantum field theory. A relativistic covariant formulation of relativistic quantum mechanics of fixed number of particles (with or without spin) is presented, based on many-time wave functions and on an interpretation of probabilities in the spacetime. These results are used to formulate the Bohmian interpretation of relativistic quantum mechanics in a manifestly relativistic covariant form and are also generalized to quantum field theory. The corresponding Bohmian interpretation of quantum field theory describes an infinite number of particle trajectories. Even though the particle trajectories are continuous, the appearance of creation and destruction of a finite number of particles results from quantum theory of measurements describing entanglement with particle detectors.

Chapter 9, whose title is "Subquantum Accelerating Universe," is written by Pedro F. González-Díaz from the Instituto de Física Fundamental, Consejo Superior de Investigaciones Científicas, Spain and and Alberto Rozas-Fernández from the Instituto de Astrofísica e Ciências do Espaço, in Portugal. Contrarily to the general belief, quantum mechanics does not only govern microscopic systems, but it has influence also on the cosmological domain. However, the extension of the Copenhagen version of quantum mechanics to cosmology is not free from conceptual difficulties: the probabilistic interpretation of the wave function of the whole universe is somehow misleading because we cannot make statistical "measurements" of different realizations of our universe. This chapter deals with two new cosmological models describing the accelerating universe in the spatially flat case. Also in this chapter there is a discussion on the quantum cosmic models that result from the existence of a nonzero entropy of entanglement. In such a realm, they obtain new cosmic solutions for any arbitrary number of spatial dimensions, studying the stability of these solutions, as well as the emergence of gravitational waves in the realm of the most general models.

Finally, **Chapter 10** entitled "Bohmian Quantum Gravity and Cosmology," is written by Nelson Pinto-Neto from the Centro Brasileiro de Pesquisas Físicas, in Brazil, and by Ward Struyve from the Ludwig-Maximilians-Universität München, in Germany. This chapter is another enlightening example on the utility of **Bohmian *thinking*** concerning the nature of space-time and mass in physical theories. The authors discuss how many conceptual problems that appear in a description of gravity in quantum mechanical terms, such as the measurement problem and the problem of time, can be overcome by adopting a Bohmian perspective. In addition to solving conceptual problems, the authors show that **Bohmian *computing*** in quantum cosmology gives new types of semi-classical approximations to quantum gravity, and approximations for quantum perturbations moving in a quantum background.

References

1. M. Planck, *On the Law of Distribution of Energy in the Normal Spectrum*, Annalen der Physik, **4**, 553 (1901).
2. A. Einstein, *Über einen die Erzeugung und Verwandlung des Lichtes betreffenden heuristischen Gesichtspunkt*, Annalen der Physik, **17**, 132 (1905).
3. N. Bohr, *On the Constitution of Atoms and Molecules, Part I*, Philosophical Magazine, **26** 1 (1913); N. Bohr, *On the Constitution of Atoms and Molecules, Part II Systems Containing Only a Single Nucleus*, Philosophical Magazine, **26**, 476 (1913); N. Bohr, *On the Constitution of Atoms and*

Molecules, Part III Systems Containing Several Nuclei, Philosophical Magazine, **26**, 857 (1913).

4. L. de Broglie, *Recherches sur la théorie des quantas*, Annalen de Physique, **3**, 22 (1925).
5. L. de Broglie, *La mécanique ondulatorie et la structure atomique de la matière et du rayonnement*, Journal de Physique et du Radium, **8**, 225 (1927).
6. D. Bohm, *A Suggested Interpretation of the Quantum Theory in Terms of "Hidden" Variables I*, Physical Review, **85**, 166 (1952).
7. D. Bohm, *A Suggested Interpretation of the Quantum Theory in Terms of "Hidden" Variables II*, Physical Review, **85**, 180, (1952).
8. M. Born, *Zur Quantenmechanik der Stovorgänge*, Zeitschrift für Physik, **37**, 863 (1926).
9. W. Heisenberg, *Über quantentheoretishe Umdeutung kinematisher und mechanischer Beziehungen*, Zeitschrift für Physik, **33**, 879 (1925); English translation in Ref. [10].
10. J. von Neumann, *Mathematische Grundlagen der Quantenmechanik* (Springer Verlag, Berlin, 1932); English translation by: R. T. Beyer, *Mathematical Foundations of Quantum Mechanics* (Princeton University Press, Princeton, 1955).
11. J. S. Bell, *Speakable and Unspeakable in Quantum Mechanics* (Cambridge University Press, Cambridge, 1987).
12. A. Einstein, *Physics and reality*, Journal of the Franklin Institute, **221**(3), 349 (1936).
13. E. Wigner, *The Unreasonable Effectiveness of Mathematics in the Natural Sciences*, Communications on Pure and Applied Mathematics, **13**, 1 (1960).
14. N. Herbert, *Quantum reality* (Anchor Books, New York, 1984).
15. "If, in some cataclysm, all of scientific knowledge were to be destroyed, and only one sentence passed on to the next generations of creatures, what statement would contain the most information in the fewest words? I believe it is the atomic hypothesis (or the atomic fact, or whatever you wish to call it) that all things are made of atoms—little particles that move around in perpetual motion, attracting each other when they are a little distance apart, but repelling upon being squeezed into one another. In that one sentence, you will see, there is an enormous amount of information about the world, if just a little imagination and thinking are applied." R. P. Feynman, R. B. Leighton and M. Sands, *Feynman Lectures on Physics* (Addison-Wesley, Caltech USA, 1965).

16. G. Binnig, H. Rohrer, C. Gerber, and E.Weibel, *Surface Studies by Scanning Tunneling Microscopy*, Physical Review Letters, **49**, 57 (1982).
17. Private exchange of letters between S. Goldstein and S. Weinberg; http://www.mathematik.uni-muenchen.de/~bohmmech/BohmHome/weingold.htm
18. R. E. Wyatt *Quantum Dynamics with Trajectories: Introduction to Quantum Hydrodynamics* (Springer, 2005).
19. A. Valentini, *Pilot-Wave Theory: An Alternative Approach to Modern Physics* (Cambridge University Press, Cambridge, 2006).
20. G. Bacciagaluppi and A. Valentini, *Quantum Theory at the Cross-roads: Reconsidering the 1927 Solvay Conference* (Cambridge University Press, Cambridge, 2009).
21. D. Peat, *Infinite Potential: The Life and Times of David Bohm* (Helix Books, Addison-Wesley, 1997).
22. P. R. Holland, *The Quantum Theory of Motion: An account of the de Broglie-Bohm Causal Interpretation of Quantum mechanics* (Cambridge University Press, Cambridge, 1993).
23. S. W. Saunders et al., *Everett and his Critics* (Oxford University Press, 2009) and arXiv: quant-ph/0811081.
24. J. T. Cushing, A. Fine and S. Goldstein, *Bohmian Mechanics and Quantum Theory: An Appraisal* (Kluwer Academic, 1996).
25. Private communication with Basile Hiley and David Peat.
26. D. Bohm and B. J. Hiley, *The Uundivided Universe: An Ontological Intepretation of Quantum Theory* (Routledge & Kegan Paul, London, 1993).
27. E. Madelung, *Quantentheorie in hydrodynamischer Form*, Zeitschrift für Physik, **40**, 322 (1926).
28. http://www.isiknowledge.com/

Chapter 1

Overview of Bohmian Mechanics

Xavier Oriols[a] and Jordi Mompart[b]
[a]*Departament d'Enginyeria Electrònica, Universitat Autònoma de Barcelona, E-08193 Bellaterra, Spain*
[b]*Departament de Física, Universitat Autònoma de Barcelona, E-08193 Bellaterra, Spain*
xavier.oriols@uab.cat; jordi.mompart@uab.cat

This chapter provides a fully comprehensive overview of the Bohmian formulation of quantum phenomena. It starts with a historical review of the difficulties found by Louis de Broglie, David Bohm and John Bell to convince the scientific community about the validity and utility of Bohmian mechanics. Then, a formal explanation of Bohmian mechanics for nonrelativistic single-particle quantum systems is presented. The generalization to many-particle systems, where correlations play an important role, is explained. After that, the measurement process in Bohmian mechanics is discussed. It is emphasized that Bohmian mechanics exactly reproduces the mean value and temporal and spatial correlations obtained from the standard, i.e., 'orthodox', formulation. The ontological characteristics of the Bohmian theory provide a description of measurements in a natural way, without the need of introducing operators for the wave function collapse. Several solved problems are presented at the end of the chapter giving additional mathematical support to

Applied Bohmian Mechanics: From Nanoscale Systems to Cosmology (Second Edition)
Edited by Xavier Oriols and Jordi Mompart

ISBN 978-981-4800-10-5 (Hardcover), 978-0-429-29474-7 (eBook)
www.jennystanford.com

some particular issues. A detailed description of computational algorithms to obtain Bohmian trajectories from the numerical solution of the Schrödinger or the Hamilton–Jacobi equations are presented in an appendix. The motivation of this chapter is twofold. First, as a didactic introduction of the Bohmian formalism which is used in the subsequent chapters. Second, as a self-contained summary for any newcomer interested in using Bohmian mechanics in their daily research activity.

1.1 Historical Development of Bohmian Mechanics

In general, the history of quantum mechanics is explained in textbooks as a chronicle where each step follows naturally from the preceding one. However, it was exactly the opposite. The development of quantum mechanics was a zigzagging route full of misunderstandings and personal disputes. It was a painful history, where scientists were forced to abandon well-established classical concepts and to explore new and imaginative routes. Most of the new routes went nowhere. Others were simply abandoned. Some of the explored routes were successful in providing new mathematical formalisms capable of predicting experiments at the atomic scale. Even such successful routes were painful enough, so relevant scientists, such as Albert Einstein or Erwin Schrödinger, decided not to support them. In this section we will briefly explain the history of one of these routes: Bohmian mechanics. It was first proposed by Louis de Broglie [1], who abandoned it soon afterward, and rediscovered by David Bohm [2, 3] many years later, and it has been ignored by most of the scientific community since then. We will discuss the historical development of Bohmian mechanics to understand its present status. Also, we will introduce the basic mathematical aspects of the theory, while the formal and rigorous structure will be presented in the subsequent sections.

1.1.1 Particles and Waves

The quantum theory revolves around the notions of particles and waves. In classical physics, the concept of a *particle* is very useful for the description of many natural phenomena. A particle is directly related to a *trajectory*[a] $\vec{r}_i[t]$ that defines its position as a continuous function of time, usually found as a solution of a set of differential equations. For example, the planets can be considered particles orbiting around the sun, whose orbits are determined by the classical Newton gravitational laws.

In classical mechanics, it is natural to think that the total number of particles (e.g., planets in the solar system) is conserved, and the particle trajectories must be continuous in time: if a particle goes from one place to another, then, it has to go through all the trajectory positions between these two places. This condition can be summarized with a *local* conservation law:

$$\frac{\partial \rho(\vec{r}, t)}{\partial t} + \vec{\nabla}\vec{j}(\vec{r}, t) = 0 \tag{1.1}$$

where $\rho(\vec{r}, t)$ is the density of particles and $\vec{j}(\vec{r}, t)$ is the particle current density. For an ensemble of point particles at positions $\vec{r}_i[t]$ with velocites $\vec{v}_i[t]$, it follows that $\rho(\vec{r}, t) = \sum \delta(\vec{r} - \vec{r}_i[t])$ and $\vec{j}(\vec{r}, t) = \sum \vec{v}_i[t]\delta(\vec{r} - \vec{r}_i[t])$, with $\delta(\vec{r})$ being the Dirac delta function, satisfying Eq. (1.1). We have used the property that $\partial\delta(\vec{r} - \vec{r}_i[t])/\partial t = -\vec{\nabla}\delta(\vec{r} - \vec{r}_i[t])d\vec{r}_i[t]/dt$.

However, the total number of planets in the solar system could be conserved in another (quite different) way. A phenomenon where a planet disappearing (instantaneously) from its orbit and appearing (instantaneously) at another point far away from its original location would certainly conserve the number of planets but it would violate Eq. (1.1). We must then think of Eq. (1.1) as a law for the local conservation of particles.

Fields, and particularly waves, also appear in many explanations of physical phenomena. The concept of a field was initially introduced to deal with the interaction of distant particles. For example, there is an interaction between the electrons in an emitting radio antenna at the top of a mountain and those in the receiving antenna at home. Such interaction can be explained through the use of an electromagnetic field. Electrons in the transmitter generate

---

[a]In order to avoid confusion, let us emphasize that in the orthodox formulation of quantum mechanics, the concept of a particle is not directly related to the concept of a trajectory. For example, the electron is a particle but there is not trajectory in the orthodox ontology, as we will discuss later.

an electromagnetic field, a radio wave, that propagates through the atmosphere and *arrives* at our antenna, affecting its electrons. Finally, a loudspeaker transforms the electron motion into music at home.

The simplest example of a wave is the so-called plane wave:

$$\psi(\vec{r}, t) = e^{i(\omega t - \vec{k}\cdot\vec{r})} \tag{1.2}$$

where the angular frequency ω and the wave vector $\vec{k}$ refer respectively to its temporal and spatial behavior. In particular, the angular frequency ω specifies when the temporal behavior of such wave is repeated. The value of $\psi(\vec{r_1}, t_1)$ at position $\vec{r_1}$ and time t_1 is identical to $\psi(\vec{r_1}, t_2)$ when $t_2 = t_1 + 2\pi n/\omega$ for n integer. The angular frequency ω can be related to the linear frequency ν as $\omega = 2\pi\nu$. Analogously, the wave vector $\vec{k}$ determines the spatial repetition of the wave, that is, the wavelength λ. The value of $\psi(\vec{r_1}, t_1)$ at position $\vec{r_1}$ and time t_1 is identical to $\psi(\vec{r_2}, t_1)$ when $\vec{k}\cdot\vec{r_2} = \vec{k}\cdot\vec{r_1} + 2\pi n$ with n integer. Unlike a trajectory, a wave is defined at all possible positions and times. Waves can be a scalar or a vectorial function and take real or complex values. For example, Eq. (1.2) is a scalar complex wave of unit amplitude. The waves' dynamical evolution is determined by a set of differential equations. In our *broadcasting* example, Maxwell equations define the electromagnetic field of the emitted radio wave that is given by two vectorial functions, one for the electric field and one for the magnetic field.

Whenever the differential equations that govern the fields are linear, one can apply the superposition principle to explain what happens when two or more fields (waves) traverse simultaneously the same region. The modulus of the total field at each position is related to the amplitudes of the individual waves. In some cases, the modulus of the sum of the amplitudes is much smaller than the sum of the modulus of the amplitudes; this is called destructive interference. In other cases, it is roughly equal to the sum of the modulus of the amplitudes; this is called constructive interference.

1.1.2 Origins of the Quantum Theory

At the end of the nineteenth century, Sir Joseph John Thomson discovered the electron, and in 1911, Ernest Rutherford, a New

Zealander student working in Thomson's laboratory, provided experimental evidence that inside atoms, electrons orbited around a nucleus in a similar manner as planets do around the sun. Rutherford's model of the atom was clearly in contradiction with well-established theories, since classical electromagnetism predicted that orbiting electrons should radiate, gradually lose energy, and spiral inward. Something was missing in the previous explanations, since it seemed that the electron behavior inside an atom could not be explained in terms of classical trajectories. Therefore, alternative ideas needed to be explored to understand atom stability.

In addition, at that time, classical electromagnetism was unable to explain the radiated spectrum of a black body, which is an idealized object that emits a temperature-dependent spectrum of light (like a big fire with different *colors*, depending on the flame temperature). The predicted continuous intensity spectrum of this radiation became unlimitedly large in the limit of large frequencies, resulting in an unrealistic emission of infinite power, which was called the ultraviolet catastrophe. However, the measured radiation of a black body did not behave in this way, indicating that a wave description of the electromagnetic field was also incomplete.

In summary, at the beginning of the twentieth century, it was clear that natural phenomena such as atom stability or black-body radiation, were not well explained in terms of a particle or a wave description alone. It seemed necessary to merge both concepts.

In 1900, Max Planck suggested [4] that black bodies emit and absorb electromagnetic radiation in discrete energies $h\nu$, where ν is the frequency of the emitted radiation and h is the (now-called) Planck constant. Five years later, Einstein used this discovery in his explanation of the photoelectric effect [5], suggesting that light itself was composed of *light quanta* or *photons*[a] of energy $h\nu$. Even though this theory solved the black-body radiation problem, the fact that the absorption and emission of light by atoms are discontinuous was still in conflict with the classical description of the light-matter interaction.

---

[a]In fact, the word *photon* was not coined until 1926, by Gilbert Lewis [6].

In 1913, Niels Bohr [7–9] wrote a revolutionary paper on the hydrogen atom, where he solved the (erroneously predicted in classical terms) instability by postulating that electrons can only orbit around atoms in some particular *nonradiating* orbits. Thus, atom radiation occurs only when electrons *jump* from one orbit to another of lower energy. His *imaginative* postulates were in full agreement with the experiments on spectral lines. Later, in 1924, de Broglie proposed in his PhD dissertation that all particles (such as electrons) exhibit wave-like phenomena like interference or diffraction [1]. In particular, one way to arrive at Bohr's hypothesis is to think that the electron orbiting around the proton is a stationary wave. Since we know that the probability of finding the electron far from the proton is zero, we can impose such spatial boundary conditions on the shape of such a stationary wave. We will obtain that only very particular shapes of the waves (associated to very particular energies) are allowed. Physics at the atomic scale started to be understandable by mixing the concepts of particles and waves. All these advances were later known as the *old quantum theory*. The word *quantum* referred to the minimum unit of any physical entity (e.g., the energy) involved in the interactions at such atomistic scales.

1.1.3 "Wave or Particle?" vs. "Wave and Particle"

In the mid-1920s, theoreticians found themselves in a difficult situation when attempting to advance Bohr's ideas. A group of atomic theoreticians centered on Bohr, Max Born, Wolfgang Pauli, and Werner Heisenberg suspected that the problem went back to trying to understand electron trajectories within atoms. In under two years, a series of unexpected discoveries brought about a scientific revolution [10].

Heisenberg wrote his first paper on quantum mechanics in 1925 [11] and two years later stated his uncertainty principle [12]. It was him, with the help of Born and Pascual Jordan, who developed the first version of quantum mechanics based on a matrix formulation [11, 13–15].

In 1926, Schrödinger published *An Undulatory Theory of the Mechanics of Atoms and Molecules* [16], where, inspired by de Broglie's work [1, 17, 18], he described material points (such as electrons or protons) in terms of a wave solution of the following (wave) equation:

$$i\hbar\frac{\partial\psi(\vec{r},t)}{\partial t} = -\frac{\hbar^2}{2m}\nabla^2\psi(\vec{r},t) + V(\vec{r},t)\psi(\vec{r},t) \tag{1.3}$$

where $V(\vec{r},t)$ is the potential energy *felt* by the electron, and the wave (field) $\psi(\vec{r},t)$ was called the wave function. Schrödinger, at first, interpreted his wave function as a description of the electron charge density $q\,|\psi(\vec{r},t)|^2$ with q the electron charge. Later, Born refined the interpretation of Schrödinger and defined $|\psi(\vec{r},t)|^2$ as the probability density of *finding* the electron in a particular position $\vec{r}$ at time t [10].

Schrödinger's wave version of quantum mechanics and Heisenberg's matrix mechanics were apparently incompatible, but they were eventually shown to be equivalent by Wolfgang Ernst Pauli and Carl Eckart, independently [10, 19].

In order to explain the physics behind quantum systems, the concepts of waves and particles should be merged in some way. Two different routes appeared:

(1) **Wave or particle?**: The concept of a trajectory was, consciously or unconsciously, abandoned by most of the young scientists (Heisenberg, Pauli, Dirac, Jordan, ...). They started a new route, the *wave or particle?* route, where depending on the experimental situation, one has to choose between a wave or a particle behavior. Electrons are associated basically to probability (amplitude) waves. The particle nature of the electron appears when we measure the position of the electron. In Bohr's words, an object cannot be both a wave and a particle at the same time; it must be either one or the other, depending upon the situation. This approach, mainly supported by Bohr, is one of the pillars of the Copenhagen, or orthodox, interpretation of quantum mechanics.

(2) **Wave and particle**: Louis de Broglie, on the other hand, presented an explanation of quantum phenomena where the wave and particle concepts merge at the atomic scale, by assuming that a pilot-wave solution of Eq. (1.3) guides the electron trajectory. This is what we call the Bohmian route. One object cannot be a wave and a particle at the same time, but two can.

The differences between the two routes can be easily seen in the interpretation of the double-slit experiment. A beam of electrons with low intensity (so that electrons are injected one by one) impinges upon an opaque surface with two slits removed on it. A detector screen, on the other side of the surface, detects the position of electrons. Even though the detector screen responds to particles, the pattern of detected particles shows the interference fringes characteristic of waves. The system exhibits, thus, the behavior of both waves (interference patterns) and particles (dots on the screen).

According to the *wave or particle?* route, first the electron presents a wavelike nature alone when the wave function (whose squared modulus gives the probability density of *finding* a particle when a position measurement is done) *travels* through both slits. Suddenly, the wave function *collapses* into a delta function at a (random) particular position on the screen. The particle-like nature of the electron appears, while its wavelike nature disappears. Since the screen positions where *collapses* occur follow the probability distribution dictated by the squared modulus of the wave function, a wave interference pattern appears on the detector screen.

According to the *wave and particle* route, the wave function (whose squared modulus means the particle probability density of *being* at a certain position, regardless of the measurement process) travels through both slits. At the same time, a well-defined trajectory is associated with the electron. Such a trajectory passes through only one of the slits. The final position of the particle on the detector screen and the slit through which the particle passes is determined by the initial position of the particle. Such an initial position is not controllable by the experimentalist, so there is an appearance of randomness in the pattern of detection. The wave function guides the particles in such a way that they avoid those regions in which the interference is destructive and are attracted to the regions in which the interference is constructive, giving rise to the interference pattern on the detector screen. Let us quote the enlightening summary of Bell [20]:

> Is it not clear from the smallness of the scintillation on the screen that we have to do with a particle? And is it not clear, from

> the diffraction and interference patterns, that the motion of the particle is directed by a wave? De Broglie showed in detail how the motion of a particle, passing through just one of two holes in screen, could be influenced by waves propagating through both holes. And so influenced that the particle does not go where the waves cancel out, but is attracted to where they cooperate. This idea seems to me so natural and simple, to resolve the wave-particle dilemma in such a clear and ordinary way, that it is a great mystery to me that it was so generally ignored.

Now, with almost a century of perspective and the knowledge that both routes give exactly the same experimental predictions, it seems that such great scientists took the *strangest* route. Let us imagine that a student asks his or her professor, "What is an electron?" The answer of a (Copenhagen) professor could be, "The electron is not a wave nor a particle. But, do not worry! You do not have to know what an electron is to (compute observable results) pass the exam."[a] If the student insists, the professor might reply, "Shut up and calculate."[b]

Another example of the vagueness of the orthodox formulation can be illustrated by the question that Einstein posed to Abraham Pais: "Do you really think the moon is not there if you are not looking at it?" The answer of a Copenhagen professor, such as Bohr, would be, "I do not need to answer such a question, because you cannot ask me such question experimentally." This answer is technically correct because, from an orthodox point of view, the property of the position of an object is undefined unless we measure it. But, knowing now that an explanation of quantum phenomena can be formulated with well defined positions of particles independently of being measured or not, the previous answer seems a bit impertinent.

---

[a]For example, in the book *Quantum Theory*, [21] written by Bohm before he formulated Bohmian mechanics in 1952, he wrote, when talking about the wave-particle duality: "We find a strong analogy here to the fable of the seven blind men who ran into an elephant: One man felt the trunk and said that 'an elephant is a rope'; another felt the leg and said that 'an elephant is obviously a tree,' and so on."

[b]This quote is sometimes attributed to Dirac, Richard Feynman, or David Mermin [22, 23]. It recognizes that the important content of the orthodox formulation of quantum theory is the ability to apply mathematical models to real experiments.

On the other hand, an alternative (Bohmian) professor would answer, "Electrons are particles whose trajectories are guided by a pilot field which is the wave function solution of the Schrödinger equation. There is some uncertainty in the initial conditions of the trajectories, so that experiments have also some uncertainties." With such a simple explanation, the student would understand perfectly the role of the wave and the particle in the description of quantum phenomena. Furthermore, in the Bohmian interpretation, the position of an electron (or the moon) while we are not looking at it, is always defined, even though it is a hidden variable for experimentalists.

One of the reasons that led the proponents of quantum mechanics to choose the *wave or particle?* route is that the predictions about the positions of electrons are uncertain because the wave function is *spread out* over a volume. This effect is known as the *uncertainty principle*: it is not possible to measure, simultaneously, the exact position and velocity (momentum) of a particle. Therefore, scientists preferred to look for an explanation of quantum effects without the concept of a trajectory that seemed unmeasurable. They constructed a theory to explain the quantum world where the concept of trajectory was not present in the ontology. However, their argumentation to neglect the use of trajectories is, somehow, unfair and unjustified, since it relies on the "principle" that the ontology of a physical theory should not contain entities that cannot be observed.[a]

In addition, everyone with experience on Fourier transforms of conjugate variables recognizes the quantum uncertainty principle as a trivial effect present in any wave theory where the momentum of a particle depends on the slope of its associated wave function. Then, a very localized particle would have a very sharp wave function. In this case such a wave function would have a great slope that implies a large range of possible momenta. On the contrary, if the wave

[a]From a philosophical point of view, this is known as "positivism" or "empiricism" discussed in the introduction and it can be understood as a nonphysical limitation on the possible kinds of theories that we could choose to explain quantum phenomena. For example, the wave function cannot be measured directly in a single experiment but only from an ensemble of experiments. However, there is no doubt that the (complex) wave function comes to be a very useful concept to understand quantum phenomena. Identically, in the de Broglie and Bohm interpretation, the trajectories cannot be directly measured, but they can also be a very interesting tool for understanding quantum phenomena.

function is built from a quite small range of momenta, then it will have a large spatial dispersion.

1.1.4 Louis de Broglie and the Fifth Solvay Conference

Perhaps the most relevant event for the development of the quantum theory was the fifth Solvay Conference, which took place from October 24–29, 1927, in Brussels [24]. As on previous occasions, the participants stayed at *Hôtel Britannique*, invited by Ernest Solvay, a Belgian chemist and industrialist with philanthropic purposes due to the exploitation of his numerous patents. There, de Broglie presented his recently developed pilot wave theory and how it could account for quantum interference phenomena with electrons [24]. He did not receive an enthusiastic reaction from the illustrious audience gathered for the occasion. In the following months, it seems that he had some difficulties in interpreting quantum measurement with his theory and decided to abandon his new pilot wave theory. In fact, one (nonscientific) reason that perhaps forced de Broglie to give up on his theory was that he worked isolated, having little contact with the main research centers in Berlin, Copenhagen, Cambridge, or Munich. By contrast, most of the Copenhagen contributors worked with fluid and constant collaborations among them.

Finally, let us mention that the elements of the pilot wave theory (electrons guided by waves) were already in place in de Broglie's thesis in 1924 [1], before either matrix or wave mechanics existed. In fact, Schrödinger used the de Broglie phases to develop his famous equation (see Eq. (1.3)). In addition, it is important to remark that de Broglie himself developed a single-particle and a many-particle description of his pilot waves, visualizing also the *nonlocality* of the latter [24]. Perhaps, his remarkable contribution and influence have not been fairly recognized by scientists and historians because he abandoned his own ideas rapidly without properly defending them [24, 25].

1.1.5 Albert Einstein and Locality

Not even Einstein gave explicit support to the pilot wave theory [10]. It remains almost unknown that in 1927, the same year that de Broglie published his pilot wave theory [18], Einstein worked out an alternative version of the pilot wave with trajectories determined by many-particle wave functions. However, before the paper appeared in print, Einstein phoned the editor to withdraw it. The paper remains unpublished, but its contents are known from a manuscript [26, 27].

It seems that Einstein, who was unsatisfied with the Copenhagen approach, did not like the pilot wave approach either because both interpretations have this notion of action at a distance: particles that are far away from each other can profoundly and *instantaneously* affect each other. As the father of the theory of relativity, he believed that action at a distance cannot travel faster than the speed of light. Let Bohm explain the difficulties of Einstein with both the Bohmian and the orthodox interpretations [28]:

> In the fifties, I sent [my Quantum Theory book] around to various quantum physicists - including Niels Bohr, Albert Einstein, and Wolfgang Pauli. Bohr didn't answer, but Pauli liked it. Albert Einstein sent me a message that he'd like to talk with me. When we met he said the book had done about as well as you could do with quantum mechanics. But he was still not convinced it was a satisfactory theory. Einstein's objection was not merely that it was statistical. He felt it was a kind of abstraction; quantum mechanics got correct results but left out much that would have made it intelligible. I came up with the causal interpretation (that the electron is a particle, but it also has a field around it. The particle is never separated from that field, and the field affects the movement of the particle in certain ways). Einstein didn't like it, though, because the interpretation had this notion of action at a distance: Things that are far away from each other profoundly affect each other. He believed only in local action.

Einstein, together with Boris Podolsky and Nathan Rosen, presented objections to the orthodox quantum theory in the famous EPR article in 1935, entitled "Can Quantum-Mechanical Description of Physical Reality Be Considered Complete?" [29]. There, they argued that on the basis of the absence of action at a distance, quantum theory must be incomplete. In other words, quantum theory is either nonlocal or incomplete. Einstein believed that locality was a fundamental principle of physics, so he adhered to the view that quantum theory was incomplete. Einstein died in 1955, convinced that a *correct* reformulation of quantum theory would preserve local causality. We will see later that he was wrong in this particular point.

1.1.6 David Bohm and Why the "Impossibility Proofs" were Wrong?

Perhaps the first utility of Bohm's work was the demonstration that the mentioned von Neumann theorem about the "impossible proofs" had limited validity. In 1932, von Neumann put quantum theory on a firm theoretical basis [30]. Some of the earlier works lacked mathematical rigor, and he put the entire theory into the setting of operator algebra. In particular, von Neumann studied the following question: "If the present mathematical formulation of the quantum theory and its usual probability interpretation are assumed to lead to absolutely correct results for every experiment that can ever be done, can quantum-mechanical probabilities be understood in terms of any conceivable distribution over hidden parameters?" von Neumann answered this question negatively. His conclusions, however, relied on the fact that he implicitly restricted his proof to an excessively narrow class of hidden variables, excluding Bohm's hidden variables model. In other words, Bohmian mechanics is a counterexample that disproves von Neumann's conclusions, in the sense that it is possible to obtain the very same predictions of orthodox quantum mechanics with a hidden variables theory [20, 31].

Bohm's formulation of quantum mechanics[a] appeared after the orthodox formalism was fully established. Bohm was, perhaps, the first person to genuinely understand the significance and fundamental implications of the description of quantum phenomena

---

[a]Apart from these works, the history of science has recognized many other relevant contributions by Bohm [32]. As a postgraduate at Berkeley, he developed a theory of plasmas, discovering the electron phenomenon now known as Bohm diffusion. In 1959, with his student Yakir Aharonov, he discovered the Aharonov–Bohm effect, showing how a magnetic field could affect a region of space in which the field had been shielded, although its vector potential did not vanish there. This showed for the first time that the magnetic vector potential, hitherto a mathematical convenience, could have real physical (quantum) effects.

with trajectories guided by waves. Ironically, in 1951, Bohm wrote a book, *Quantum Theory* [21], where he provided "proof that quantum theory is inconsistent with hidden variables" (see page 622 in [21]). In fact, he wrote in a footnote in that section, "We do not wish to imply here that anyone has ever produced a concrete and successful example of such a [hidden variables] theory, but only state that such theory is, as far as we know, conceivable." Furthermore, the book does also contain an unusually long chapter devoted to the *quantum theory of the process of measurement*, where Bohm discusses how the measurement itself can be described from the time evolution of a many-particle wave function, rather than invoking the *wave function collapse*. It seems that Bohm became dissatisfied with the orthodox approach that he had written in his book and began to develop his own causal formulation of quantum theory, which he published in 1952 [2, 3].

The original papers of Bohm in 1952 [2, 3] provide a formal justification of the guidance equation developed 25 years before by de Broglie. Instead of reproducing his exact mathematical development in terms of the quantum Hamilton–Jacobi equation, here, we discuss a very simple explanation of the guidance equation. By a simple mathematical manipulation of the (wave) equation, Eq. (1.3), we can find the local (particle) continuity equation, Eq. (1.1), discussed at the beginning of this section.[a] From the standard definition of the current density $\vec{j}(\vec{r}, t)$, as a product of the particle density $\rho(\vec{r}, t) = |\psi(\vec{r}, t)|^2$ and the velocity $\vec{v}(\vec{r}, t)$, we can exactly obtain the guidance equation that was predicted by de Broglie and Bohm for the particle velocity:

$$\frac{d\vec{r}(t)}{dt} = \vec{v}(\vec{r}, t) = \frac{\vec{j}(\vec{r}, t)}{\rho(\vec{r}, t)} \tag{1.4}$$

There was another important point explained by Bohm in 1952. If one considers an ensemble of trajectories whose initial positions at time $t = 0$ are distributed according to $\rho(\vec{r}, 0) = |\psi(\vec{r}, 0)|^2$, such an ensemble of trajectories will reproduce $\rho(\vec{r}, t) = |\psi(\vec{r}, t)|^2$ at any other time if the trajectories follow the guidance equation, Eq. (1.4). Therefore, we are able to exactly reproduce the time evolution of the wave function solution of Eq. (1.3) from an ensemble of trajectories

[a] See the formal demonstration in Section 1.2.3.2 or in Ref. [33].

guided by waves. Let Bohm himself explain this revolutionary point in the first page of his original reference [2]:

> The usual [Copenhagen] interpretation of the quantum theory is based on an assumption [...] that the physical state of an individual system is completely specified by a wave function that determines only the probabilities of actual results that can be obtained in a statistical ensemble of similar experiments. [...] In contrast, this alternative interpretation permits us to conceive of each individual system as being in a precisely definable state, whose changes with time are determined by definite laws, analogous to (but not identical with) the classical equations of motion. Quantum mechanical probabilities are regarded (like their counterpart in classical mechanics) as only a practical necessity and not as a manifestation of an inherent lack of complete determination in the properties of matter at the quantum level.

Bohm's original papers do also provide a different path to find the trajectories by introducing the polar form of the wave function $\psi(\vec{r}, t) = R(\vec{r}, t)e^{iS(\vec{r},t)/\hbar}$ into the (nonrelativistic) Schrödinger equation, Eq. (1.3). Let us emphasize that $R(\vec{r}, t)$ and $S(\vec{r}, t)$ are real (not complex) functions. Then, after a quite simple manipulation, one obtains from the real part a quantum Hamilton–Jacobi equation:

$$\frac{\partial S(\vec{r}, t)}{\partial t} + \frac{1}{2m}\left(\vec{\nabla} S(\vec{r}, t)\right)^2 + V(\vec{r}, t) + Q(\vec{r}, t) = 0 \tag{1.5}$$

where we have defined the quantum potential $Q(\vec{r}, t)$ as:

$$Q(\vec{r}, t) = -\frac{\hbar^2}{2m}\frac{\vec{\nabla}^2 R(\vec{r}, t)}{R(\vec{r}, t)} \tag{1.6}$$

which is the only difference with respect to the classical Hamilton–Jacobi equation if the velocity is defined as $\vec{v} = \vec{\nabla} S/m$.[a] Such an alternative explanation provides an additional justification for the guidance equation, Eq. (1.4) (see problem 6). It allows a second-order interpretation of Bohmian trajectories in terms of acceleration, forces, and energies. In particular, in this second-order point of view, the new quantum potential $Q(\vec{r}, t)$ is responsible for the deviations of the Bohmian trajectories from the classical

[a]See Section 1.2.3.4 for the detailed formal demonstration.

behavior that can be expected from the classical potential $V(\vec{r}, t)$. As we will discuss later, the quantum potential of a system of N particles will be the responsible for the (nonclassical) nonlocal causality of Bohmian mechanics.

Bohm completed the work of de Broglie in two fundamental aspects. First, as explained before, he demonstrated that Bohmian mechanics leads to exactly the same predictions as the ones obtained by orthodox quantum mechanics. Second, he provided a theory of measurement. He developed an explanation of the measurement problem without invoking the *wave function collapse*. The theory of Bohmian measurement will be discussed in Section 1.4. Some authors argue that if we could change history, allowing Bohm to help de Broglie defend his pilot wave theory in the Solvay Conference, Bohmian mechanics would now be certainly taught at universities [34]. The enlightening work of Bohm, however, appeared 25 years too late, once the Copenhagen interpretation of quantum phenomena was already too well established.

After the Solvay Conference, Bohr, Heisenberg, and their colleagues spread the new interpretation around the world and convinced the vast majority of the physics community that the Copenhagen theory worked with extraordinary precision. A lot of young physicists were attracted to European institutes to study with the "fathers" of this new theory, and during the second quarter of the twentieth century, as good disciples, they spread the Copenhagen interpretation over the entire globe. In his 1976 Nobel lecture, Murray Gell-Mann referred to this question: "Niels Bohr brainwashed an entire generation of physicists into believing that the problem [of the interpretation of quantum mechanics] had been solved fifty years ago" [35].

Finally, we summarize the importance of Bohm's work with another magistral quote from Bell that appears in a 1982 paper entitled "On the Impossible Pilot Wave" and collected in his famous book *Speakable and Unspeakable in Quantum Mechanics* [20]:

> But in 1952 I saw the impossible done. It was in papers by David Bohm. Bohm showed explicitly how parameters could indeed be introduced, into nonrelativistic wave mechanics, with the help of which the indeterministic description could be transformed

into a deterministic one. More importantly, in my opinion, the subjectivity of the "orthodox" version, the necessary reference to the "observer," could be eliminated. Moreover, the essential idea was on that had been advanced already by de Broglie in 1927, in his "pilot wave" picture. But why then had Born not told me of this "pilot wave"? If only to point out what was wrong with it? Why did von Neumann not consider it? More extraordinarily, why did people go on producing "impossibility" proofs after 1952, and as recently as 1978? When even Pauli, Rosenfeld, and Heisenberg, could produce no more devastating criticism of Bohm's version than to brand it as "metaphysical" and "ideological"? Why is the pilot wave picture ignored in textbooks? Should it not be taught, not as the only way, but as an antidote to the prevailing complacency? To show that vagueness, subjectivity, and indeterminism, are not forced on us by experimental facts, but by deliberate theoretical choice?

1.1.7 John Bell and Nonlocality

Along this introduction, we have already mentioned Bell's positive opinion on Bohmian mechanics. Furthermore, his opinion about the Copenhagen interpretation was that the orthodox theory is "unprofessionally vague and ambiguous" [20, 36–38] in so far as its fundamental dynamics is expressed in terms of "words which, however, legitime and necessary in application, have no place in a formulation with any pretension to physical precision" [38].

Bell spent most of his professional career at the European Organization for Nuclear Research (CERN), working almost exclusively on theoretical particle physics and on accelerator design, but found time to pursue a major avocation investigating the foundations of quantum theory.[a] As seen in many of his quotes used in this introduction, his didactic ability to defend Bohmian mechanics against many unjustified attacks has been of extraordinary importance for maintaining the work of de Broglie and Bohm alive among the scientific community. Fortunately, Bell himself had his own reward from this unbreakable support of Bohmian mechanics. His outstanding work on locality and causality was directly inspired by

[a] J. Bell defined himself as "I am a quantum engineer, but on sundays I have principles." Underground colloquium, March 1983.

his deep knowledge of Bohmian mechanics. Bell's theorem has been called "the most profound discovery of science" [39].

Bell's most relevant contribution to physics is probably the demonstration that quantum mechanics is nonlocal, contrarily to what Einstein expected. In 1964, inspired by the EPR paper [29] and Bohm's work on *nonlocal* hidden variables, Bell elaborated a theorem establishing clear mathematical inequalities, now known as Bell inequalities, for experimental results that would be fulfilled by *local* theories but would be violated by *nonlocal* ones [36]. Over the past 30 years, a great number of Bell test experiments have been conducted. These experiments have confirmed that Bell's inequalities are violated (see, for example, Ref. [40]). Therefore, we have to conclude that quantum experimental results cannot be explained with local hidden variable theories. According to Bell, we must accept the real existence, in nature, of faster-than-light causation.

The experimental violation of Bell's inequalities gave direct support, not only to the Copenhagen interpretation, but also to Bohm's formulation of quantum theory, since both are nonlocal theories.

In the Bohmian case, as we have discussed, the quantum potential of N *entangled* particles is defined in a $3N$ configuration space so that an action on the first particle can have an *instantaneous* (i.e., faster-than-light) causal effect on the last particle. In Bohmian mechanics, one can understand that the quantum potential is responsible for the *instantaneous* nonlocal changes on the trajectories of quantum particles. Let Bell explain this point in his own words [20]:

> That the guiding wave, in the general case, propagates not in ordinary three-dimensional (3D) space but in a multidimensional configuration space is the origin of the notorious "nonlocality" of quantum mechanics. It is a merit of the de Broglie–Bohm version [of quantum mechanics] to bring this out so explicitly that it cannot be ignored.

Unfortunately, as already happened with the von Neumann theorem [30], there is an historical misunderstanding about the consequences of Bell's theorems on Bohmian mechanics. Let us mention just one example that appeared in a prestigious journal in 2000 [41]:

> In the mid-1960s John S. Bell showed that if hidden variables existed, experimentally observed probabilities would have to fall below certain limits, dubbed Bell's inequalities. Experiments were carried out by a number of groups, which found that the inequalities were violated. Their collective data came down decisively against the possibility of hidden variables.

The author of this sentence has omitted the adjective "local" when he mentions hidden variables. Therefore, a confident reader, who has no time to read Bell's and Bohm's works, will understand that Bohmian mechanics is refuted by Bell's theorem. However, it is exactly the contrary. Bell's inequalities give direct support to Bohmian mechanics. Unfortunately, this misunderstanding appeared, and continues to appear, in many scientific articles, propagating into textbooks, websites, etc., provoking further comments and replies in the scientific literature. This discussion can give the impression that there still exists some controversy about the validity of Bohmian mechanics for all nonrelativistic quantum phenomena or that there is "something unclear" about it,[a] which is clearly not the case.

1.1.8 Quantum Hydrodynamics

It is sometimes claimed that ideas similar to those developed by de Broglie were put forward by Madelung in 1926 [42]. What Madelung proposed, however, was to regard an electron with mass m and wave function $\psi(\vec{r}, t)$ not as a particle with a determined trajectory but as a continuous fluid with mass density $m|\psi(\vec{r}, t)|^2$ [43, 44]. In Madelung's hydrodynamic interpretation of Eq. (1.3), the fluid velocity coincides mathematically with de Broglie's guiding Eq. (1.4), but the ontological interpretation is quite different.

[a]See, for example, the experience of J. T. Cushing [45] or some recent works, "demonstrating" that Bohmian mechanics was wrong [46], and the comment to the work [47].

Let us return again to the double-slit experiment to understand Madelung's point of view. Orthodox quantum mechanics does not predict what happens to a single electron crossing a double slit, but it predicts what is the statistical probability of detecting electrons when we consider an infinite ensemble of such experiments. In contrast, the proposal from de Broglie and Bohm intends to predict what happens, in principle, to a single electron. At the end of the day, however, because it is not possible to determine the initial position of such a single electron with better uncertainty than that obtained from the initial wave packet spatial dispersion, Bohmian mechanics also provides statistical results. In this regard, Madelung was not interested in dealing with a single-electron trajectory but with the ensemble. When we deal with a very large (infinite) number of particles (trajectories), the particle and current densities that appear in Eq. (1.1) are no longer discrete functions (sum of deltas) anymore but they have to be interpreted as continuous functions. Aerodynamics and hydrodynamics are examples of such continuous material systems. The concept of an individual trajectory becomes irrelevant in these disciplines but Eq. (1.1) is still present. In summary, from an ontological point of view, Madelung's proposal is completely orthodox. However, from a mathematical point of view, its formalism (and computational abilities) is very similar to de Broglie's proposal. Therefore, a reader who does not feel comfortable with the ideas of Bohm and de Broglie can assume Madelung's point of view and use many of the concepts explained in this book as a mathematical computational tool.

1.1.9 Is Bohmian Mechanics a Useful Theory?

Our historical introduction to Bohmian mechanics ends here. The reader has certainly noticed that we have some preference for the Bohmian interpretation over the Copenhagen, orthodox, explanation of quantum phenomena. Obviously, we have a profound and sincere respect for the Copenhagen interpretation of quantum phenomena and its extraordinary computing capabilities. As discussed in the section "What Is a Quantum Theory" in the introduction, a quantum theory connects the ontological and empirical planes. The scientific method is based on experiments. From an empirical point of view, what we have to expect from a physical theory that describes natural phenomena is quite pragmatic. As Bohr said (see page 228 in [48]):

> In my opinion, there could be no other way to deem a logically consistent mathematical formalism than by demonstrating the departure of its consequence from experience or by proving that its predictions did not exhaust the possibilities of observations.

Curiously, similar empirical arguments were used by Bohm to defend his causal theory against its many detractors (see page 18 in [49]):

> In conclusion, the author would like to state that we would admit only two valid reasons for discarding a theory that explains a wide range of phenomena. One is that the theory is not internally consistent, and the second is that it disagrees with experiments.

In summary, from all the discussions done up to here, the only statement that we will use in the rest of the book is that one cannot discard an interpretation of quantum phenomena in terms of wave functions and trajectories because it is *internally consistent* and it *agrees with experiments*. Once such a statement is accepted, other practical questions about Bohmian mechanics appear naturally, for example, is Bohmian mechanics a useful *computational tool* for predicting the results of quantum experiments? Although it provides identical predictions as the orthodox formalism, are there advantages in *understanding/visualizing/explaining* quantum experiments in terms of quantum trajectories? Or, even, is it a recommended formalism for *thinking* about limits and extensions of (nonrelativistic) quantum theory? This book, through several examples in the following chapters, will try to convince the reader that the answer to these three questions is affirmative.

1.2 Bohmian Mechanics for a Single Particle

After the previous historical introduction, we start with the formal presentation of Bohmian mechanics for a single particle. Usually, new scientific knowledge is built from small variations of old ideas. This explains why the initial development of quantum theory (see Section 1.1.3) was so traumatic. Quantum explanations did not evolve from a small variation of classical mechanics but from radically new ideas. The languages of both theories were completely different. Classical theory provides an explanation of a physical experiment on particles in terms of a well-defined trajectory, while the orthodox quantum theory needs a wave function. Here, we will see that both a wave description of an ensemble of classical particles and a description of quantum systems with trajectories are possible. Starting by setting a common language for both classical and quantum theories will certainly improve our understanding on the similarities and differences between them.[a]

1.2.1 Preliminary Discussions

For a reader without any previous knowledge on quantum mechanics, the following sections would be not only a presentation on Bohmian mechanics, but also on quantum mechanics in general. The *Copenhagen* and the *Bohmian* wave functions have exactly the same time evolution (when the same configuration space is used in both cases). In addition, the *Bohmian* wave function is complemented by a *Bohmian* trajectory which appears in a natural way when developing a trajectory-based explanation of quantum mechanics as we will discuss in Section 1.2.3.

The initial formulation of quantum mechanics with Bohmian trajectories developed by de Broglie and Bohm was performed for dynamical systems with associated velocities much slower than the speed of light. Although it is, in principle, possible to extend Bohmian mechanics to relativistic systems (see, for example, Chapter 8 and Refs. [50–52]), we will only deal in this chapter with nonrelativistic systems.

In order to simplify as much as possible our mathematical notation, we will first study a single spinless particle *living* in a one-dimensional (1D) space. The spatial degree of freedom of the particle will be represented by x. The generalization of all the arguments mentioned in this section into a single particle in a 3D

[a]It is very rare to find such descriptions in the standard literature but, in our opinion, it is very important to be able to compare classical and quantum mechanics on an equal footing.

space is quite simple. However, the practical solution of a single-particle Schrödinger equation in a 3D space with x, y and z degrees of freedom has considerable computational difficulties, as we will discuss in Section 1.3 and appendix A.1. In Section 1.3.4 we will consider the role of the spin.

1.2.2 Creating a Wave Equation for Classical Mechanics

In this section we will derive a wave equation similar, although not identical, to the Schrödinger equation for an ensemble of classical trajectories. Certainly, this approach allows us to compare the quantum and classical theories by using a very similar language.

1.2.2.1 Newton's second law

Our starting point to derive such a classical wave equation will be Newton's second law [53]. Let us consider a particle with mass m in a classical system that moves under the action of a potential $V(x)$, where x is the position coordinate. Here, we assume the potential to be time independent to simplify the mathematical treatment. We define the particle trajectory as $x[t]$ and its velocity as $v[t] = dx[t]/dt \equiv \dot{x}[t]$. Since we are considering a classical system, the trajectory of the particle will be obtained from its acceleration, $a[t] = d^2x[t]/dt^2 \equiv \ddot{x}[t]$, from Newton's second law:

$$m\ddot{x}[t] = \left[-\frac{\partial V(x)}{\partial x}\right]_{x=x[t]} \tag{1.7}$$

Since Eq. (1.7) is a second-order differential equation, we need to specify both the initial position $x[t_0] = x_0$ and the initial velocity $v[t_0] = v_0$ of the particle.

1.2.2.2 Hamilton's principle

Apart from Eq. (1.7), there are other alternative ways to describe a classical system. For example, according to Hamilton's principle [53], the trajectory $x_{\text{p}}[t]$[a] solution of Eq. (1.7) between two different times, t_0 and t_f, provides a stationary value for the action function,

---

[a]The suffix "p" means *physical* in order to distinguish from nonphysical trajectories, "np," but it will be omitted, when unnecessary.

$S(x[t]; x_0, t_0; x_f, t_f)$, where $x_0 = x[t_0]$ and $x_f = x[t_f]$. Hereafter, whenever possible, we will omit the dependence of the action function on x_0, t_0, x_f, and t_f:

$$\left[\frac{\delta S(x[t])}{\delta x[t]}\right]_{x[t]=x_p[t]} = 0 \tag{1.8}$$

where the action function is defined as:

$$S(x[t]) = \int_{t_0}^{t_f} L(x[t], \dot{x}[t])dt \tag{1.9}$$

The function $L(x, v)$ is the Lagrangian function:

$$L(x, v) = K(v) - V(x) \tag{1.10}$$

with $K(v) = mv^2/2$ the kinetic energy of the particle. The Lagrangian equation can also be defined as $L(x, v) = pv - H(x, v)$, where $H(x, p) = K(v) + V(x)$ is the Hamiltonian function[a] and $p = mv$ is the (linear) momentum.

Let us discuss the meaning of a stationary (or extremal) value of the integral in Eq. (1.9). We denote the physical trajectory solution of Eq. (1.7) as $x_p[t]$ and choose a slightly different trajectory $x_{np}[t] = x_p[t] + \delta x[t]$ with the same initial and final conditions, that is, the same x_0, t_0, x_f, and t_f. Therefore, we have $\delta x[t_0] = 0$ and $\delta x[t_f] = 0$. See Fig. 1.1 for a schematic representation of both trajectories. Equation (1.8) means that the value of the action function $S_{np} = S(x[t] + \delta x[t]; x_0, t_0; x_f, t_f)$ will always be greater[b] than S_p; that is, $S_{np} > S_p$. Thus, the trajectory that provides a stationary value for the action function is the physical trajectory, $x_p[t]$, while $x_{np}[t] = x_p[t] + \delta x[t]$ is a nonphysical solution incompatible with Eq. (1.7).

The numerical evaluation of Eq. (1.9) to find S_p requires the prior knowledge of the trajectory $x_p[t]$. If we do not know the trajectory, we would have to evaluate $S(x[t]; x_0, t_0; x_f, t_f)$ for all possible trajectories (starting from x_0 and t_0 and ending at x_f and t_f) and choose the one that provides a stationary value. Certainly, it seems that Eq. (1.8) has little practical utility. However, it provides an interesting starting point to develop our classical wave equation.

---

[a] These definitions of Lagrangian and Hamiltonian functions are valid for the simple system described here. In any case, a different definition does not change the main results developed here.

[b] Strictly speaking, the physical trajectory can also correspond to a maximum of the action function. Such a maximum value is also an (stationary) extremal value of Eq. (1.8).

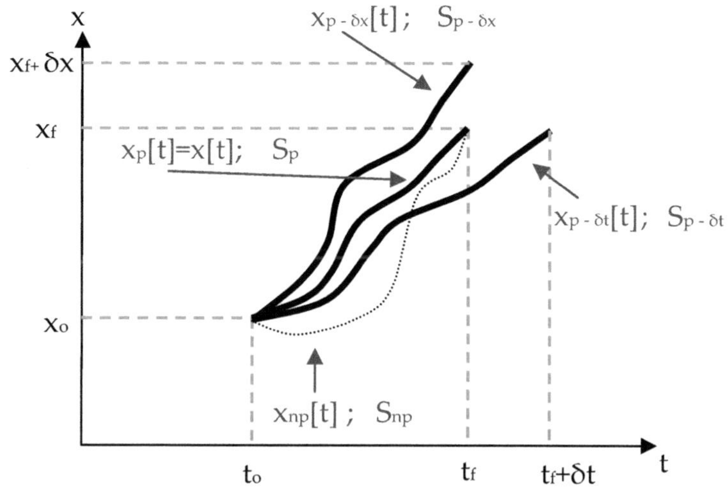

Figure 1.1 Schematic representation of physical (solid lines) and nonphysical (dotted lines) trajectories in the (x, t) plane. The trajectory $x_{\text{p}}[t]$ has the initial time t_0, the final time t_f, the initial position $x_{\text{p}}[t_0] = x_0$, and the final position $x_{\text{p}}[t_f] = x_f$. The trajectory $x_{\text{p}-\delta x}[t]$ is a physical trajectory with identical initial and final conditions as $x_{\text{p}}[t]$, except for a different final position $x_f + \delta x$. The trajectory $x_{\text{p}-\delta t}[t]$ is a physical trajectory with identical initial and final conditions as $x_{\text{p}}[t]$ but reaching the final point at a larger time $t + \delta t$.

1.2.2.3 Lagrange's equation

From Hamilton's principle it is possible to derive the Lagrange's equation [53] that gives us the differential equation that the physical trajectory satisfies.

Let us define a trajectory $x_{\text{np}}[t] = x_{\text{p}}[t] + \delta x[t]$ by adding a small arbitrary displacement $\delta x[t]$ to the physical trajectory. In particular we fix $\delta x[t_0] = 0$ and $\delta x[t_f] = 0$, so both trajectories have the same initial and final conditions. A Taylor expansion of the Lagrange function of Eq. (1.10) around the physical trajectory $x_{\text{p}}[t]$ reads:

$$L(x_{\text{np}}[t], \dot{x}_{\text{np}}[t]) = L(x_{\text{p}}[t], \dot{x}_{\text{p}}[t]) + \left[\frac{\partial L}{\partial x}\delta x[t] + \frac{\partial L}{\partial \dot{x}}\delta \dot{x}[t]\right]_{\substack{x = x_{\text{p}}[t] \\ \dot{x} = \dot{x}_{\text{p}}[t]}} \tag{1.11}$$

The action function of Eq. (1.9) with the Lagrangian of Eq. (1.11) gives two contributions. The first term of the right-hand side (r.h.s.)

of Eq. (1.11) is S_{p}. After performing an integration by parts on the second part and using $\delta x[t_0] = 0$ and $\delta x[t_f] = 0$, we obtain:

$$S(x_{\mathrm{np}}[t]; x_0, t_0; x_f, t_f) = S_{\mathrm{p}} + \int_{t_0}^{t_f} \delta x[t] \times \left(\left[\frac{\partial L}{\partial x}\right]_{\substack{x=x_{\mathrm{p}}[t] \\ \dot{x}=\dot{x}_{\mathrm{p}}[t]}} - \frac{d}{dt}\left[\frac{\partial L}{\partial \dot{x}}\right]_{\substack{x=x_{\mathrm{p}}[t] \\ \dot{x}=\dot{x}_{\mathrm{p}}[t]}} \right) dt \tag{1.12}$$

Hamilton's principle tells us that the action function must take a stationary value for the physical trajectory. This is equivalent to impose that small variations around the physical trajectory do not modify the value of the action, that is, for any (small) variation $\delta x[t]$, $S_{\mathrm{np}} = S_{\mathrm{p}}$. Thus:

$$\left[\frac{\partial L(x,\dot{x})}{\partial x}\right]_{\substack{x=x_{\mathrm{p}}[t] \\ \dot{x}=\dot{x}_{\mathrm{p}}[t]}} - \frac{d}{dt}\left[\frac{\partial L(x,\dot{x})}{\partial \dot{x}}\right]_{\substack{x=x_{\mathrm{p}}[t] \\ \dot{x}=\dot{x}_{\mathrm{p}}[t]}} = 0 \tag{1.13}$$

Equation (1.13) is the Lagrange equation and gives, for a classical system, a differential equation that classical trajectories must fulfill. In order to enlighten the meaning of the Lagrange equation, Eq. (1.13), we can see that by substituting $\partial L/\partial x = -\partial V(x)/\partial x$ and $\partial L/\partial \dot{x} = mv$ in Eq. (1.10), we recover:

$$\left[\frac{\partial V(x,t)}{\partial x}\right]_{\substack{x=x_{\mathrm{p}}[t] \\ \dot{x}=\dot{x}_{\mathrm{p}}[t]}} - \frac{d}{dt}\left[m\frac{dx[t]}{dt}\right]_{\substack{x=x_{\mathrm{p}}[t] \\ \dot{x}=\dot{x}_{\mathrm{p}}[t]}} = 0 \tag{1.14}$$

which is the original Newton's second law of Eq. (1.7). In fact, what we have done is to check that Newton's second law is included in the fundamental Hamilton's principle.

1.2.2.4 Equation for an (infinite) ensemble of trajectories

The formalism based on the action function allows us to find not only a single physical trajectory but also the equation for an (infinite) ensemble of physical trajectories with slightly different initial or final conditions. For example, let us define:

$$S_{\mathrm{p}-\delta t} = S(x_{\mathrm{p}-\delta t}[t]) = \int_{t_0}^{t_f+\delta t} L(x_{\mathrm{p}-\delta t}[t], \dot{x}_{\mathrm{p}-\delta t}[t])dt \tag{1.15}$$

$x_{p-\delta t}[t]$ being a trajectory with identical initial and final positions as $x_p[t]$, but taking a longer time $t_f + \delta t$ (see trajectory $x_{p-\delta t}[t]$ in Fig. 1.1). The new physical trajectory $x_{p-\delta t}[t]$ can be written as $x_{p-\delta t}[t] = x_p[t] + \delta x[t]$ for $t \in [t_0, t_f]$. For larger times, $x_{p-\delta t}[t_f + \delta t] = x_p[t_f + \delta t] + \delta x[t_f + \delta t]$ with $x_p[t_f + \delta t] = x_p[t_f] + \dot{x}_p[t_f]\delta t$, where we have done a first-order Taylor expansion of $x_p[t]$ around t_f for a small δt. Since $x_{p-\delta t}[t_f + \delta t] = x_p[t_f]$, we obtain:

$$\delta x[t_f + \delta t] = -\dot{x}_p[t_f]\delta t \tag{1.16}$$

By following the same steps done after the Taylor expansion in Eq. (1.11), we obtain:

$$S_{p-\delta t} = \int_{t_0}^{t_f+\delta t} L dt + \left[\frac{\partial L}{\partial \dot{x}}\delta x[t]\right]_{t_0}^{t_f+\delta t} + \int_{t_0}^{t_f+\delta t} \delta x[t] \left(\left[\frac{\partial L}{\partial x}\right]_{x=x[t]}^{\dot{x}=\dot{x}[t]} + \frac{d}{dt}\left[\frac{\partial L}{\partial \dot{x}}\right]_{x=x[t]}^{\dot{x}=\dot{x}[t]} \right) dt \tag{1.17}$$

For the first term of the r.h.s. of Eq. (1.17), we obtain:

$$\int_{t_0}^{t_f+\delta t} L dt = \int_{t_0}^{t_f} L dt + \int_{t_f}^{t_f+\delta t} L dt = S_p + L\delta t \tag{1.18}$$

For the second term of the r.h.s. of Eq. (1.17):

$$\left[\frac{\partial L}{\partial \dot{x}}\delta x[t]\right]_{t_0}^{t_f+\delta t} = -\left[\frac{\partial L}{\partial \dot{x}}\right]_{x=x[t_f+\delta t]}^{\dot{x}=\dot{x}[t_f+\delta t]} \dot{x}(t_f)\delta t = -p[t_f+\delta t]\dot{x}[t_f]\delta t \tag{1.19}$$

where we use Eq. (1.16), $\delta x[t_0] = 0$ and $\partial L/\partial \dot{x} = mv = p$, $p = mv$ being the particle momentum.

Finally, since $x[t]$ is also a physical trajectory that fulfills Lagrange equation, from t_0 to $t_f + \delta t$, we obtain for the third term:

$$\int_{t_0}^{t_f+\delta t} \delta x[t] \left[\left[\frac{\partial L}{\partial x}\right]_{x=x[t]}^{\dot{x}=\dot{x}[t]} + \frac{d}{dt}\left[\frac{\partial L}{\partial \dot{x}}\right]_{x=x[t]}^{\dot{x}=\dot{x}[t]} \right] dt = 0 \tag{1.20}$$

Putting together Eqs. (1.18) (1.19), and (1.20), we obtain:

$$S_{p-\delta t} = S_p + L\delta t - p\dot{x}\delta t = S_p - H\delta t \tag{1.21}$$

So, with $\partial S/\partial t_f = \lim_{\delta t \to 0}(S_{p-\delta t} - S_p)/\delta t$ we can conclude that:

$$\frac{\partial S(x_p[t])}{\partial t_f} = -H(x_p[t], \dot{x}_p[t]) \tag{1.22}$$

In summary, if $S_{\rm p}$ is the stationary value of the action function for $x_{\rm p}[t]$, then the variation of the new stationary value of another physical trajectory $x_{{\rm p}-\delta t}[t]$, which has identical initial and final conditions but a slightly modified final time, is equal to the Hamiltonian (with a negative sign) evaluated at the final time of the trajectory $x_{\rm p}[t]$. See problem 1 to discuss a particular example.

Next, we will see that the evaluation of the value $S_{{\rm p}-\delta x}$ when we modify the final position $x_f + \delta x$ without modifying the initial and final times leads also to an interesting result (see Fig. 1.1). Notice that the final position of the new physical trajectory $x_{{\rm p}-\delta x}[t] = x[t] + \delta x[t]$ means:

$$\delta x[t_0] = 0; \quad \delta x[t_f] = \delta x_f \tag{1.23}$$

By following the same steps done after the Taylor expansion in Eq. (1.11), we obtain now:

$$S_{{\rm p}-\delta x} = \int_{t_0}^{t_f} L dt + \left[\frac{\partial L}{\partial \dot{x}} \delta x[t]\right]_{t_0}^{t_f} + \int_{t_0}^{t_f} \delta x[t] \left(\left[\frac{\partial L}{\partial x}\right]_{x=x[t]}^{\dot{x}=\dot{x}[t]} + \frac{d}{dt}\left[\frac{\partial L}{\partial \dot{x}}\right]_{x=x[t]}^{\dot{x}=\dot{x}[t]} \right) dt \tag{1.24}$$

The first term of the r.h.s. of Eq. (1.24) is again $S_{\rm p}$. The second term is

$$\left[\frac{\partial L}{\partial \dot{x}} \delta x[t]\right]_{t_0}^{t_f} = \left[\frac{\partial L}{\partial \dot{x}}\right]_{x=x[t_f]}^{\dot{x}=\dot{x}[t_f]} \delta x_f = p[t_f]\delta x_f \tag{1.25}$$

where we have used Eq. (1.23). The third term is zero. We can conclude that $\partial S/\partial x = \lim_{\delta x \to 0}(S_{{\rm p}-\delta x} - S_{\rm p})/\delta x$ is equal to:

$$\frac{\partial S(x_{\rm p}[t]; x_0, t_0; x, t)}{\partial x} = m\dot{x}_{\rm p}[t] = p_{\rm p}[t] \tag{1.26}$$

In summary, the variation of the stationary value of the action function when we slightly modify the final position of a physical trajectory is equal to the momentum of the trajectory at the final time.[a] To check Eq. (1.26) for a particular case, see problem 2.

---

[a]We have recovered the subindex "p" to emphasize that Eq. (1.26) is only valid for physical trajectories.

1.2.2.5 Classical Hamilton–Jacobi equation

Now, using Eq. (1.22) and substituting the velocity (momentum) into the Hamiltonian by Eq. (1.26), we obtain the well-known Hamilton–Jacobi equation:

$$H\left(x, \frac{\partial S(x,t)}{\partial x}, t\right) + \frac{\partial S(x,t)}{\partial t} = 0 \quad (1.27)$$

We want to use the Hamilton–Jacobi equation, Eq. (1.27), to find all the physical trajectories. For this reason, we have eliminated the dependence of S on the trajectory $x[t]$. In addition, we have also eliminated t_0 and x_0 because we consider that x and t are not parameters but variables. Notice that Eq. (1.27) is valid for physical trajectories so that once we know $S(x, t)$, we are able to directly compute all physical trajectories for all possible initial conditions. Therefore, Eq. (1.27) defines an (infinite) ensemble of trajectories rather than just a single trajectory. However, in most practical cases, the direct solution of the Hamilton–Jacobi equation is much more difficult than using the Newton or Lagrange formulation of classical mechanics. It has limited practical interest. However, it provides a direct theoretical connection with a classical wave equation similar to the Schrödinger equation. See problem 3 for a particular solution of Eq. (1.27).

1.2.2.6 Local continuity equation for an (infinite) ensemble of classical particles

Equation (1.27) can be interesting when dealing with an ensemble of trajectories. For example, it can be used when we have a classical (single-particle) experiment with some practical difficulty in specifying the initial position and velocity of the particle, such that different experimental realizations can have slightly different initial conditions. The ensemble of trajectories solution of Eq. (1.27) will provide a statistical (probabilistic) description of the classical (single-particle) experiment that accounts for the variability in the initial conditions.

For such an experiment, we can reasonably assume that the variability of the initial position will be limited to a particular spatial region. At the initial time $t_0 = 0$, we can define some distribution of the initial position of the particles $R^2(x, 0) \geq 0$.[a] Such an ensemble of trajectories will evolve in time according to Eq. (1.27) so that we will obtain a function $R^2(x, t)$ that describes the particular distribution of particles at any time. The function $R^2(x, t)$ is constructed by "counting" the number of trajectories present inside the interval $(x, x + dx)$ at time t. From Eq. (1.26), we define the particle velocity as $\dot{x}[t] = (1/m)\partial S(x, t)/\partial x$. Finally, we know that all these classical particles will move, in a *continuous* way, from one unit of volume to another. Therefore, as we have extensively discussed in Section 1.1.1, we can ensure that the ensemble of trajectories accomplishes the following *local* conservation law:

$$\frac{\partial R^2(x, t)}{\partial t} + \frac{\partial}{\partial x}\left(\frac{1}{m}\frac{\partial S(x, t)}{\partial x}R^2(x, t)\right) = 0 \qquad (1.28)$$

In summary, Eq. (1.28) just certifies that if a classical particle goes from one point to another, it has to go through all the trajectory positions between these two places.

1.2.2.7 Classical wave equation

Now, we have all the ingredients to develop a wave equation for classical mechanics. In previous paragraphs we have been dealing with two (real) functions, $S(x, t)$ and $R(x, t)$. The first, $S(x, t)$, is the action function and appears in the Hamilton–Jacobi equation, Eq. (1.27). In particular, its spatial derivative determines the velocity of the particles (see Eq. (1.26)). The second, $R(x, t)$, tells us how an ensemble of trajectories is distributed at each time t. It evolves according to the conservation law, Eq. (1.28). In this sense, we can assume that they are some kind of "field" or "wave" that guide classical particles.

We can construct the following classical[b] (complex) wave function $\psi_{cl}(x, t) = R(x, t)\exp(iS(x, t)/\hbar)$. We divide the action function inside the exponential by $\hbar$ (the reduced Planck's constant)

[a]Such a distribution is positive (or zero), but it is not necessary to impose that $R^2(x, 0)$ is normalized to unity.

[b]For simplicity, we assume that $R(x, t)$ and $S(x, t)$ are single valued. This condition is equivalent to assigning, at each time, a unique velocity to each position of physical space. If this is not the case, then we will need different wave functions, that is, a mixed state, to describe the different velocity fields of the classical system.

in order to provide a dimensionless argument. Then, it can be shown that the two previous (real) equations, Eqs. (1.27) and (1.28), for $S(x, t)$ and $R(x, t)$ are equivalent to the following (complex) classical wave equation for $\psi_{cl}(x, t)$:

$$i\hbar\frac{\partial\psi_{cl}(x,t)}{\partial t} = -\frac{\hbar^2}{2m}\frac{\partial^2\psi_{cl}(x,t)}{\partial x^2} + V(x,t)\psi_{cl}(x,t) + \frac{\hbar^2}{2m}\frac{\partial^2|\psi_{cl}(x,t)|/\partial x^2}{|\psi_{cl}(x,t)|}\psi_{cl}(x,t) \quad (1.29)$$

The demonstration of this expression is left as an exercise to the reader (see problem 4). Additionally, see problem 5 for a solution of the classical wave equation, Eq. (1.29), for a free particle.

In conclusion, an ensemble of classical trajectories can be described with a wave function solution of a wave equation. Thus, a common language for classical and quantum mechanics has been obtained by using an (infinite) ensemble of classical trajectories with different initial positions and velocities instead of just one single classical trajectory.

1.2.3 Trajectories for Quantum Systems

Before comparing the classical and quantum wave equations, let us first discuss in this section whether trajectories can be also used to describe quantum systems. We will introduce such trajectories in two different ways: first, as a direct consequence of the local conservation of particles extracted from the Schrödinger equation and, second, directly following the work presented by Bohm in his original paper [2].

1.2.3.1 Schrödinger equation

As discussed in Section 1.1.3, the single-particle Schrödinger equation in a 1D quantum system subjected to a scalar time-dependent potential, $V(x, t)$, is:

$$i\hbar\frac{\partial\psi(x,t)}{\partial t} = -\frac{\hbar^2}{2m}\frac{\partial^2\psi(x,t)}{\partial x^2} + V(x,t)\psi(x,t) \quad (1.30)$$

It is important to emphasize that in the orthodox interpretation of $\psi(x, t)$, Eq. (1.30) does not describe a single experiment but an ensemble of identical (single-particle) experiments. The orthodox meaning of the square modulus of the wave function $|\psi(x,t)|^2$ is the probability density of finding a particle at position x at time t when a measurement is performed. Such probabilities assume an infinite number of identical experiments.

1.2.3.2 Local conservation law for an (infinite) ensemble of quantum trajectories

It is known that there is a *local* continuity equation associated with Eq. (1.30). Let us first mathematically derive it to later discuss its physical implications.

In order to find a local continuity equation, let us work with $\psi(x,t)$ and its complex conjugate $\psi^*(x,t)$. In particular, we can rewrite Eq. (1.30) as:

$$\psi^*(x,t) i\hbar \frac{\partial \psi(x,t)}{\partial t} = -\psi^*(x,t)\frac{\hbar^2}{2m}\frac{\partial^2 \psi(x,t)}{\partial x^2} + \psi^*(x,t)V(x,t)\psi(x,t) \tag{1.31}$$

and the complex conjugate of Eq. (1.31) as:

$$-\psi(x,t) i\hbar \frac{\partial \psi^*(x,t)}{\partial t} = -\psi(x,t)\frac{\hbar^2}{2m}\frac{\partial^2 \psi^*(x,t)}{\partial x^2} + \psi(x,t)V(x,t)\psi^*(x,t) \tag{1.32}$$

From Eqs. (1.31) and (1.32), we obtain:

$$\frac{\partial |\psi(x,t)|^2}{\partial t} = i\frac{\hbar}{2m}\frac{\partial}{\partial x}\left(\psi^*(x,t)\frac{\partial \psi(x,t)}{\partial x} - \psi(x,t)\frac{\partial \psi^*(x,t)}{\partial x}\right) \tag{1.33}$$

We can easily identify Eq. (1.33) as the local conservation of particles discussed in Eq. (1.1) where $\rho(x,t) = |\psi(x,t)|^2$ and the current density, $J(x,t)$, is defined as:

$$J(x,t) = i\frac{\hbar}{2m}\left(\psi(x,t)\frac{\partial \psi^*(x,t)}{\partial x} - \psi^*(x,t)\frac{\partial \psi(x,t)}{\partial x}\right) \tag{1.34}$$

Unlike other wave equations, the Schrödinger equation is compatible with a local conservation of particles due to the fact that $V(x,t)$ is a real function. We have noticed above that we can interpret $\rho(x,t) = |\psi(x,t)|^2$ as a spatial distribution of an ensemble of trajectories. Each trajectory correspond to a different experiment of the single particle system. Then, in spite of dealing with a single particle system, from a statistical point of view, such (very large) ensemble of trajectories can be interpreted as a (very large) ensemble of particles describing simultaneously all possible experiments. The presence of such local conservation of particles is very relevant for us because it justifies our aim to look for an ensemble of continuous trajectories describing $\rho(x, t) = |\psi(x, t)|^2$.

1.2.3.3 Velocity of Bohmian particles

In Bohmian mechanics, the particle velocity is defined as:

$$v(x, t) = \frac{J(x, t)}{|\psi(x, t)|^2} \tag{1.35}$$

where $J(x, t)$ is given by Eq. (1.34). Taking into account that $|\psi(x, t)|^2$ is the distribution of the ensemble of particles, it is easy to show that this velocity definition is compatible with the local continuity equation, Eq. (1.33), and that an ensemble of well-defined trajectories whose initial positions are all selected according to the distribution $|\psi(x, t_0)|^2$ will reproduce $|\psi(x, t)|^2$ at all times (see problem 7).

Notice that one could add a divergence-free term to this velocity, and it would still fulfill Eq. (1.33). Even though this would change the path followed by each individual trajectory, the trajectory distribution and, therefore, the measurement outcomes would remain unchanged. However, if one develops the nonrelativistic limit of a relativistic treatment of Bohmian mechanics, one finds that the divergence-free term does not appear for spin-0-particles and that the velocity is unambiguously defined by Eq. (1.35). In addition, it has also been demonstrated that possible Bohmian paths are naively observable from a large enough ensemble of weak values [54–56] giving full support to the definition in Eq. (1.35). For other values of spin, the situation is a bit more complex [57–59] and far from the scope of this book.

1.2.3.4 Quantum Hamilton–Jacobi equation

Following the path described by Bohm in his original paper [2], we will now start from the quantum wave equation, that is, the

Schrödinger equation, to arrive at a "quantum" Hamilton–Jacobi equation that describes the motion of quantum trajectories.

The first step is to write the quantum (complex) wave function, $\psi(x,t) = \psi_r(x,t) + i\psi_i(x,t)$, in polar form:

$$R^2(x,t) = \psi_r^2(x,t) + \psi_i^2(x,t) \tag{1.36}$$

$$S(x,t) = \hbar \arctan\left(\frac{\psi_i(x,t)}{\psi_r(x,t)}\right) \tag{1.37}$$

In principle, $S(x,t)$, the so-called quantum action, is not well defined when $\psi_r(x,t) = \psi_i(x,t) = 0$, that is at those points where $R(x,t) = 0$, meaning that no particles will reach them.[a]

The quantum Hamilton–Jacobi equation can be found by introducing $\psi(x,t) = R(x,t)\exp(iS(x,t)/\hbar)$ into Eq. (1.30). On the one hand, the imaginary part of the resulting equation gives the *local* conservation law identical to the one shown in Eq. (1.28), which we rewrite here for convenience:

$$\frac{\partial R^2(x,t)}{\partial t} + \frac{\partial}{\partial x}\left(\frac{1}{m}\frac{\partial S(x,t)}{\partial x}R^2(x,t)\right) = 0 \tag{1.38}$$

On the other hand, the real part gives a quantum Hamilton–Jacobi equation:

$$\frac{\partial S(x,t)}{\partial t} + \frac{1}{2m}\left(\frac{\partial S(x,t)}{\partial x}\right)^2 + V(x,t) + Q(x,t) = 0 \tag{1.39}$$

Since the last term in the r.h.s. of the classical wave equation, Eq. (1.29), is not present in the Schrödinger equation, Eq. (1.30), an additional term appears in the quantum Hamilton–Jacobi equation, the so-called quantum potential, defined as:

$$Q(x,t) = -\frac{\hbar^2}{2m}\frac{\partial^2 R(x,t)/\partial x^2}{R(x,t)} \tag{1.40}$$

In conclusion, identical to the classical description of a system obtained from Eq. (1.39), we obtain an interpretation of the wave function solution of the Schrödinger equation as an ensemble of

---

[a] We assume that the wave function is single valued so that $R(x,t)$ is also single valued. However, the definition of $S(x,t)$ has some practical difficulties. In principle, $S(x,t)$ is a multivalued function because the function $\arctan(x)$ itself is a multivalued function. If we want to use Eqs. (1.36) and (1.37) to reconstruct the wave function, then the multivalued problem can be eliminated by imposing an additional restriction on the definition of S [31, 60–63].

quantum trajectories with different initial positions and velocities. The velocity of each trajectory $x[t]$ is defined as:

$$v[t] = \left[\frac{1}{m}\frac{\partial S(x,t)}{\partial x}\right]_{x=x[t]} \tag{1.41}$$

Interestingly, it can be easily shown that this new expression for the quantum velocity is identical to that mentioned in Eq. (1.35). See problem 6 to show that both definitions of the velocity of Bohmian trajectories are identical:

$$v(x,t) = \frac{1}{m}\frac{\partial S(x,t)}{\partial x} = \frac{J(x,t)}{|\psi(x,t)|^2} \tag{1.42}$$

where $J(x,t)$ is defined by Eq. (1.34).

1.2.3.5 A quantum Newton-like equation

If we compute the time derivative of the Bohmian velocity defined in Eq. (1.41), we find a quantum Newton-like equation:

$$\begin{aligned} m\frac{d}{dt}v(x[t],t) &= m\frac{d}{dt}\left[\frac{1}{m}\frac{\partial S}{\partial x}\right]_{x=x[t]} \\ &= \left[\frac{\partial^2 S}{\partial x^2}\right]_{x=x[t]}\dot{x}[t] + \left[\frac{\partial}{\partial x}\frac{\partial S}{\partial t}\right]_{x=x[t]} \end{aligned} \tag{1.43}$$

We can rewrite Eq. (1.43) as:

$$m\frac{d}{dt}v(x[t],t) = \left[\frac{\partial}{\partial x}\left(\frac{1}{2m}\left(\frac{\partial S}{\partial x}\right)^2 + \frac{\partial S}{\partial t}\right)\right]_{x=x[t]} \tag{1.44}$$

Finally, using Eq. (1.39) we get:

$$m\frac{d}{dt}v(x[t],t) = \left[-\frac{\partial}{\partial x}\left(V(x,t) + Q(x,t)\right)\right]_{x=x[t]} \tag{1.45}$$

We conclude here the second route for finding a common language for classical and quantum theories. The quantum (complex) single-particle wave function can be interpreted as an ensemble of trajectories that are all solutions of the same single-particle experiment but with different initial conditions. The quantum trajectories are not solutions of the classical Newton second law with a classical potential but solutions of the quantum Newton second law, Eq. (1.45), where a quantum potential (that accounts for all nonclassical effects) is added to the classical potential.

1.2.4 Similarities and Differences between Classical and Quantum Mechanics

In previous sections, we have provided a common language for classical and quantum theories, in terms of either wave functions or trajectories, to fairly compare differences and similarities between both theories. Here we emphasize that an attempt to establish similarities and differences between classical and quantum mechanics have to be done by comparing either classical and quantum wave functions or by comparing classical and quantum ensembles of trajectories (not by comparing a single classical trajectory with a quantum wave function). The main difference between the mathematical description of the two theories is that the term $Q(x, t)$ that appears in the quantum Hamilton–Jacobi equation, Eq. (1.39), is exactly the same term that appears in the classical wave equation, Eq. (1.29), but with a change of sign. The term $Q(x, t)$ explains the *exotic* properties of quantum systems that are missing in their classical counterparts:

(1) **Differences**

(a) **Quantum superposition**

One can apply the superposition principle to find solutions of the Schrödinger equation, Eq. (1.30), but it cannot be applied to the classical wave equation, Eq. (1.29), since its last term is nonlinear. In fact, if $\psi_1(x, t)$ and $\psi_2(x, t)$ are solutions of the Schrödinger equation, then $a\psi_1(x, t) + b\psi_2(x, t)$ with arbitrary complex a and b is also a solution. Quantum mechanics will keep the (complex) amplitudes a and b constant at any time because of its linearity, which is not the case for classical mechanics.

This is a fundamental difference between classical and quantum systems, which has deep consequences. Quantum mechanics can be developed in a linear vector space, while classical mechanics cannot.

(b) **Quantum wholeness**

The presence of $Q(x, t)$ in the quantum Hamilton–Jacobi equation, Eq. (1.39), implies that Bohmian trajectories depend not only on the classical potential $V(x, t)$ but also on the quantum potential $Q(x, t)$, which is a function of the type of distribution of trajectories associated to different repetitions of the single-particle experiment, $R(x, t)$. In fact, it is the *shape* and not the absolute value of $R(x, t)$ that acts on each individual quantum trajectory. On the contrary, each classical trajectory can be computed independently of the shape of the ensemble. The fact that the dynamics of one quantum trajectory in one particular experiment depends on the ensemble of other trajectories build from other identical experiments is highly counter-intuitive for our classical mind.

This surprising result for quantum mechanics can be illustrated with the double-slit experiment [64]. We assume that the initial wave functions for the classical and quantum ensembles are identical at $t_0 = 0$. Their difference appears in the time evolution of the trajectories. For the quantum trajectories of the ensemble, the shape of the ensemble (whether particles are stopped or not by the double-slit screen) determines the shape of $R(x, t)$, which will affect the dynamics of each trajectory, even those trajectories that are far from the slit (see Fig. 1.2a). On the contrary, the classical Hamilton–Jacobi equation, Eq. (1.27), is totally independent of $R(x, t)$, so a single trajectory is completely independent from the rest of trajectories of other experiments as seen in Fig. 1.2b.

(2) **Similarities**

(a) **Uncertainty**

As far as we deal with an ensemble of (classical or quantum) trajectories, there is an uncertainty in the exact value of any magnitude that can be measured from the ensemble. For example, one can compute the mean value and the standard deviation of the position of the classical ensemble. The classical and quantum uncertainties can have different origins, but both ensembles have uncertainties. The classical uncertainty is due to the technical difficulties in exactly repeating the initial conditions of a particular experiments; the quantum uncertainty has a more intrinsic

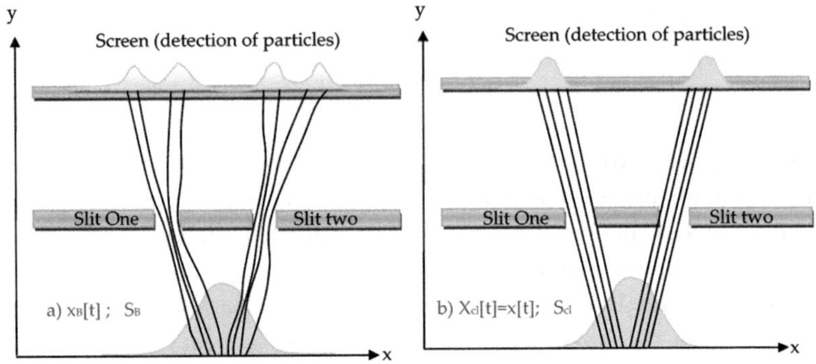

Figure 1.2 A particle is sent toward a thin plate with two slits cut in it. This single-particle experiment is repeated many times with some uncertainty on the initial conditions of the particle positions. The distribution of detected particles on the screen as a function of the position is different if we deal with a quantum (a) or a classical (b) system. In both cases, we use an identical description of the initial ensemble of particles, that is, identical $R(x, t_0)$ and $S(x, t_0)$. However, the classical evolution of the trajectories $x_{cl}[t]$ is described by Eq. (1.27), which is independent of $R(x, t)$, while quantum (Bohmian) trajectories $x_B[t]$ are determined by Eq. (1.39), which depends explicitly on the other particles of the ensemble through $Q(x, t)$ that depends on $R(x, t)$. These differences can also be understood by observing that the evolution of a classical wave function is described by a nonlinear equation, Eq. (1.29), while a quantum wave function evolves with a linear one, Eq. (1.30).

origin [65, 66]. See Ref. [67] for a discussion on how Bohmian mechanics, although being a deterministic theory at the ontological level, provides a quite simple explanation on the origin of the quantum uncertainty at the empirical level.

Even assuming their different physical origin, what we want to emphasize here is that it is not licit to compare the uncertainty of a quantum wave function with the uncertainty of a single classical trajectory, saying that the classical trajectory has no uncertainty and the quantum wave function has. As we have previously discussed, we have to compare the uncertainty of classical and quantum wave functions (or classical and quantum ensembles of trajectories).

(b) **Initial conditions for the (classical or quantum) Hamilton–Jacobi equation**
The classical Newton equation is a second-order differential equation, where both the initial position and the velocity have to be fixed. In the classical or quantum Hamilton–Jacobi equation, it seems that only an initial position is needed because the initial velocity is directly determined by the spatial derivative of the action. However, one can argue that in the Hamilton–Jacobi equation, two initial conditions are fixed, the initial position and the initial wave function (or initial action) that fixes the initial velocity. In other words, for one particular Hamiltonian, even if we fix one particular initial position, then it is possible to obtain different initial velocities if we select different initial wave functions. In conclusion, the discussion that classical trajectories are solved from second-order equations, while Bohmian trajectories from first-order equations, is somehow artificial if one recalls the quantum equilibrium hypothesis.[a]

(c) **Single-valued wave function and multivalued action function**
In principle, the action function solution of the (classical or quantum) Hamilton–Jacobi equation can be multivalued yielding different velocities in a particular x and t point. However, the classical wave or the Schrödinger equations deal with single-valued wave functions. If we want to model (classical or quantum) scenarios with multivalued velocities, then we have to consider several wave functions, one wave function for each possible velocity at x and t.[b] On the contrary, if we only work with a single-valued (classical or quantum) wave function, then, the velocity itself is single-valued everywhere. Therefore, two (classical

[a]The quantum equilibrium hypothesis assumes that the initial positions and velocities of Bohmian trajectories are defined distributed according to the initial wave function [65, 66, 68, 69]. This topic will be further discussed at the end of Section 1.3.5.

[b]In the quantum language, this means working with a density matrix, that is, with mixed states rather than with pure states.

or quantum) trajectories that coincide in one configuration point will have identical velocities. This means that they will follow identical trajectories for any future time. This has the important consequence that all Bohmian trajectories (or classical trajectories) associated with a single-valued wave function cannot cross in the configuration space.

In conclusion, the differences between quantum and classical ensembles of trajectories is not a difference between waves and particles, because both waves and particles can be used to study classical or quantum systems. On the contrary, the difference resides between a linear wave equation (for quantum mechanics), Eq. (1.30), and a nonlinear wave equation (for classical mechanics), Eq. (1.29). One of the most important consequences of such difference is that quantum (Bohmian) trajectories depend on the shape of the ensemble (i.e., quantum wholeness), as seen in Fig. 1.2a, while classical trajectories are independent of the shape of the ensemble, as seen in Fig. 1.2b. This difference between classical and quantum systems has important consequences at a computational level. One can compute a unique classical trajectory. However, because of the quantum wholeness, one needs, in principle, to deal somehow with the whole ensemble of quantum trajectories to know the dynamics of a unique quantum trajectory.

1.2.5 Feynman Paths

In the introduction of this chapter we mentioned that quantum mechanics can be described either in the matrix formulation proposed by Heisenberg and coworkers or in the wave equation formalism developed by Schrödinger. There are alternative representations of quantum phenomena. For example, the Feynman path (see, for instance, Ref. [70]). In this context, the time evolution of a wave function can be written using the Green function (or propagator or transition amplitude) as:

$$\psi(x,t) = \int_{-\infty}^{\infty} G(x_0, t_0; x, t)\psi(x_0, t_0)dx_0 \tag{1.46}$$

Feynman provided an original technique for computing $G(x_0, t_0; x, t)$ from the classical Lagrangian of Eq. (1.10). One considers all

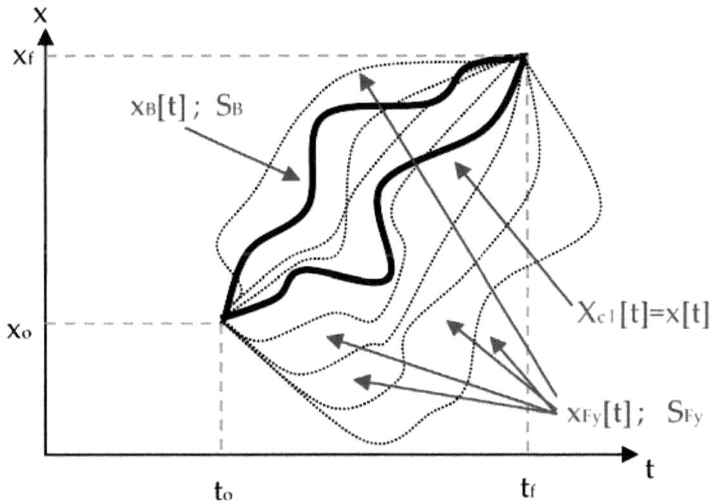

Figure 1.3 Schematic representation of physical (solid lines) and nonphysical (dashed lines) trajectories in the configuration space (x, t) starting at (x_0, t_0) and finishing at (x_f, t_f). In general, Feynman paths $x_{Fy}[t]$ do not provide stationary values of the action function, while the quantum (Bohmian) trajectory $x_B[t]$ provides a stationary value of the quantum action function, S_B in Eq. (1.49), and the classical trajectory $x_{cl}[t]$ a stationary value of the classical action function, S_{cl} in Eq. (1.48).

(physical and nonphysical) paths $x_{Fy}[t]$ that may link the two points (x, t) and (x_0, t_0). See dashed curves in Fig. 1.3. Feynman associated to each path (physical or nonphysical) a complex amplitude $\exp(iS(x[t]; x_0, t_0; x, t)/\hbar)$, where $S(x[t]; x_0, t_0; x, t)$ is defined by Eq. (1.9). We emphasize that we are using not only the trajectory that provides an stationary value of $S(x[t]; x_0, t_0; x, t)$ but all trajectories. Then, Feynman defines the Green function as:

$$G(x_0, t_0; x, t) = C \sum_{\text{all paths}} e^{iS(x[t];x_0,t_0;x,t)/\hbar} \tag{1.47}$$

where C is a normalization constant. It can be demonstrated that the wave function constructed from Eq. (1.46) with Eq. (1.47) reproduces the Schrödinger equation [70]. Certainly, Feynman paths and Bohmian trajectories are completely different. In particular, there is only one Bohmian trajectory $x_B[t]$ solution of Eq. (1.39) that goes from point (x_0, t_0) to point (x_f, t_f), as seen in Fig. 1.3, while there are infinite Feynman paths $x_{Fy}[t]$ that connect both points. In particular, one of the Feynman paths is certainly the Bohmian trajectory $x_B[t]$ and another is the classical trajectory $x_{cl}[t]$. In general, $x_B[t]$ is different from $x_{cl}[t]$. The later is the trajectory that minimizes the classical action function:

$$S(x_{cl}[t]; x_0, t_0; x_f, t_f) = \int_{t_0}^{t_f} \left[\frac{1}{2m}(\dot{x}[t])^2 - V(x[t], t) \right] dt \quad (1.48)$$

while the former is the trajectory that minimizes the quantum action function:

$$S(x_B[t]; x_0, t_0; x_f, t_f) = \int_{t_0}^{t_f} \left[\frac{1}{2m}(\dot{x}[t])^2 - V(x[t], t) - Q(x[t], t) \right] dt \quad (1.49)$$

The derivation of the last expression can be straightforwardly obtained by repeating the steps done from the classical action function to the classical Hamilton–Jacobi equation but using now the quantum Hamilton–Jacobi equation. We conclude that Bohmian trajectories are quite different from Feynman paths. Bohmian trajectories are part of the basic ontology of the theory, while Feynman paths are basically a mathematical tool to compute the probabilities.

On the other hand, one could be interested in discussing whether the Feynman approach could be a more useful computational tool than Bohmian mechanics for solving quantum systems. In principle, it seems that Feynman paths would be less practical because we would have to compute an infinite number of trajectories from one particular initial position and then repeat the procedure for all initial positions. On the contrary, in Bohmian mechanics we only need to compute one quantum trajectory for each initial position. See problems 15, 16, and 17.

1.2.6 Basic Postulates for a Single-Particle

After introducing the reader to the language of trajectories for quantum mechanics (and wave functions for classical mechanics), we can state the basic postulates of the Bohmian theory for a single particle. They summarize in a few sentences all the discussions held so far. In general, the postulates of any physical theory can be presented in different compatible ways. For example, classical mechanics can be postulated from Newton's laws or Hamilton's principle. We follow here the standard presentation of Bohmian mechanics that involves a smaller number of ingredients [31, 61–63, 68, 71].

FIRST POSTULATE: *The dynamics of a single particle in a single experiment in a quantum system is defined by a trajectory $x[t]$ that moves continuously under the guidance of a wave function $\psi(x, t)$. The wave function $\psi(x, t)$ is a solution of the Schrödinger equation:*

$$i\hbar\frac{\partial\psi(x,t)}{\partial t} = -\frac{\hbar^2}{2m}\frac{\partial^2\psi(x,t)}{\partial x^2} + V(x,t)\,\psi(x,t)$$

The trajectory $x[t]$ is obtained by time-integrating the particle velocity $v[t]$ defined as:

$$v(x,t) = \frac{J(x,t)}{|\psi(x,t)|^2}$$

where $J(x, t)$ is the (ensemble) current density given by:

$$J(x,t) = i\frac{\hbar}{2m}\left(\psi(x,t)\frac{\partial\psi^*(x,t)}{\partial x} - \psi^*(x,t)\frac{\partial\psi(x,t)}{\partial x}\right)$$

The initial position $x[t_0]$ and velocity $v[t_0]$ must be specified to completely determine the trajectory $x[t]$.

SECOND POSTULATE (quantum equilibrium hypothesis): *The initial position and velocity of a particular trajectory cannot be known with certainty. When the experiment is repeated many times ($j = 1, \ldots, M$), the initial positions $\{x^j[t_0]\}$ of an ensemble of trajectories $\{x^j[t]\}$, associated to the same $\psi(x, t)$, have to satisfy $R^2(x, t_0) = |\psi(x, t_0)|^2$, that is the number of trajectories of the ensemble between x and $x + dx$ at the initial time t_0 is proportional to $R^2(x, t_0) = |\psi(x, t_0)|^2$. The initial velocity of each trajectory is determined by $v^j[t_0] = J(x^j[t_0], t_0)/|\psi(x^j[t_0], t_0)|^2$.*

The condition on the distribution of the initial position in different experiments can be written mathematically as:

$$R^2(x,t_0) = \lim_{M\to\infty}\frac{1}{M}\sum_{j=1}^{M}\delta(x - x^j[t_0]) \qquad (1.50)$$

where $j = 1, \ldots, M$ is the index of the different trajectories belonging to different experiments. If the set of M positions of different experiments follows the distribution in Eq. (1.50) at the initial time $t = 0$, it is easy to demonstrate that the distribution of positions satisfies $R(x, t)^2 = |\psi(x, t)|^2$ at any time other time t:

$$R^2(x, t) = \lim_{M \to \infty} \frac{1}{M} \sum_{j=1}^{M} \delta(x - x^j[t]) \tag{1.51}$$

provided that the many-particle wave function evolves according to the Schrödinger equation and that the particles move according to the corresponding Bohmian velocity, see problem 7. This property is known as equivariance [63, 66] and it is key for the empirical equivalence between Bohmian mechanics and other quantum theories. Equation (1.51) says that Born's law is always satisfied by counting particles [31, 61–63]. Equation (1.51) also explains why quantum results are unpredictable at the empirical arena [67].

The second postulate needs some additional remarks. It is argued by some authors that the second postulate, the quantum equilibrium hypothesis, is not a necessary postulate of the Bohmian theory. We will briefly explain the reasons in Section 1.3.5 when dealing with the postulates of a many particle system.

These postulates represent a minimalist explanation of the Bohmian interpretation of quantum mechanics without mentioning either the Hamilton–Jacobi equation, Eq. (1.39) or the quantum potential, Eq. (1.40). Certainly, we can formulate Bohmian mechanics without the use of the Hamilton–Jacobi formalism; however, in the authors' opinion, the quantum Hamilton–Jacobi and the quantum potential allows us to improve our understanding of Bohmian mechanics and provides clear arguments for discussing the similarities and differences between classical and quantum theories. Those who dislike the use of the quantum potential (and the quantum Hamilton–Jacobi equation) argue that it *naturally* appears in the Schrödinger equation, but it is not present in other wave equations such as Dirac's one. On the contrary, they state that the concept of Bohmian velocity, as defined in the first postulate, is much more general because it is always present in any wave equation, as far as a local continuity equation for the probability density can be established. This is true. However, in spite of this important limitation, we will show several times in this book that the quantum potential is a quite useful tool.

Finally, let us emphasize that no postulate about measurement is formulated here because, in Bohmian mechanics, no postulate about measurement is needed. For a more detailed explanation, please refer to Refs. [31, 61–63, 68, 71] and Section 1.4. In next section, we will explain how these postulates for single-particle systems are generalized to many-particle systems.

1.3 Bohmian Mechanics for Many-Particle Systems

1.3.1 Preliminary Discussions: The Many Body Problem

Up to now we have only studied a simple system composed of just one particle. However, a single-particle quantum system is some kind of crude idealization of any macroscopic or microscopic system found in nature, which is usually formed by a very large number of interacting particles. Strictly speaking, a single-particle system cannot be measured (one needs to consider the particles of the measuring apparatus that interact with the single-particle system). In this section we will study nonrelativistic many-particle quantum systems with Bohmian trajectories. First of all, let us clarify the meaning of the term "many-particle" in Bohmian mechanics, as it can easily be misinterpreted. As we have explained in the previous section, we need an infinite ensemble of trajectories, $M \to \infty$, to describe the statistics of a single-particle quantum system. These different trajectories should not be confused with different physical particles of the system, since they all refer to different realizations (experiments) of the same single-particle quantum system.

Let us consider now a quantum system with N degrees of freedom, that is, an N body quantum system. We will use a particular variable x_k for each degree of freedom $k = 1, \ldots, N$. The wave function will thus be a function of all $x_1, \ldots, x_N$ variables. Now, a many-particle Bohmian trajectory will involve N interacting physical particles $x_1[t], \ldots, x_N[t]$. The relevant point that allows us to use the adjective "many-particle" is that the N particles interact between them, that is, the interacting potential depends on all possible particle positions, $V(x_1, \ldots, x_N)$, in a nontrivial way. Along this section, in order to simplify our notation, we will use $\vec{x}[t] = (x_1[t], \ldots, x_N[t])$ or $\vec{x} = (x_1, \ldots, x_N)$.

Now, an ensemble of identical experiments will be composed of $M \to \infty$ Bohmian trajectories in the configuration space $\vec{x}^j[t]$, with $j = 1, \ldots, M$. The superscript j refers to the statistical index (counting experiments), while the subscript index will refer to each of the N interacting particles present in each experiment, that is, $\vec{x}^j[t] = (x_1^j[t], \ldots, x_N^j[t])$. A trajectory in the N-dimensional configuration sapce is defined by s set of N trajectories in the physical space. In the rest of chapter, we will use M to refer to the number of experiments and N to the number of physical particles of the quantum system in each experiment.

The first step to obtain the many-particle quantum trajectories is solving the following many-particle Schrödinger equation:

$$i\hbar\frac{\partial\psi(\vec{x},t)}{\partial t} = \left(\sum_{k=1}^{N} -\frac{\hbar^2}{2m}\frac{\partial^2}{\partial x_k^2} + V(\vec{x},t)\right)\psi(\vec{x},t) \tag{1.52}$$

The solution $\psi(\vec{x}, t)$ of this equation is the so-called *many-particle wave function*, that is defined in a N-dimensional space (plus time). The problem of N particles in a 1D space is formally equivalent to the one of a single particle in an N-dimensional space. Here, we will use the term "many-particle" in a wide sense to include the single-particle $N \geq 2$ dimensional case.[a]

Equation (1.52) is analytically unsolvable most of the times, and its numerical integration is out of today's present computer capabilities, even for systems with $N \gtrsim 5$, since we need to compute the wave function $\psi(\vec{x}, t)$ in the N-dimensional configuration space. Let us roughly estimate the hard disk space that we would require to store $\psi(\vec{x}, t)$. Considering, for instance, a system with $N = 10$ particles confined in a 1D region of 10 nm, which we discretize with

[a]The variable N can be defined as the number of particles in a 1D space, or it can be related to the number of particles in a 3D space. In simple words, $\psi(x_1, x_2, x_3, t)$ can be interpreted as three particles in a 1D space or just one particle in a 3D space. From a physical point of view, one particle in a 3D space is a "single-particle" system. However, from the computational point of view, it is equivalent to a three-particle system in 1D.

a spatial step of $\Delta x = 0.1$ nm, we have a grid of 100 points for each dimension x_k. Then, the total number of points in the configuration space for the 10 particles is $100^{10} = 10^{20}$. Using 4 bytes (32 bits) to store the complex value of the wave function at each grid point, the information contained in a 10-particle wave function would require more than 3×10^8 Terabytes (TB) (and more than 3×10^{28} TB for 20 particles). This practical limitation is the main reason why our knowledge of many-particle quantum systems is so poor. This difficulty is the so-called many body problem. In 1929, Dirac wrote the following [72]:

> The general theory of quantum mechanics is now almost complete. The underlying physical laws necessary for the mathematical theory of a large part of physics and the entire chemistry are thus completely known, and the difficulty is only that the exact application of these laws leads to equations much too complicated to be soluble.

Three decades later, Born rephrased the issue [73]:

> It would indeed be remarkable if Nature fortified herself against further advances in knowledge behind the analytical difficulties of the many body problem.

The adjective "many-particle" can certainly be used with classical mechanics for N particles interacting through $V(\vec{x}, t)$. Now, it would be interesting to address the previous problem of storing the state of a 10-particle system but now from the classical mechanics perspective. Solving the problem means finding the many-particle trajectory $\vec{x}^{\text{cl}}[t]$. As discussed at the end of subsection 1.2.4, a classical many-particle trajectory can be computed alone, without knowing the rest of the trajectories of the ensemble (with different initial positions). At a computational level, this feature of classical mechanics implies a dramatic simplification. Basically, at each time step of the simulation, if the analytical expression for the potential is known, we just need to save the position and velocity of each particle. This means 20 real values for the 10-particle systems mentioned above. The classical trajectory of each particle, $x_k^{\text{cl}}[t]$, can be obtained by solving a coupled system of Newton's second laws:

$$m_k \frac{d^2 x_k^{\rm cl}[t]}{dt^2} = \left[-\frac{\partial V(\vec{x}, t)}{\partial x_k} \right]_{\vec{x}=\vec{x}^{\rm cl}[t]} \tag{1.53}$$

Fortunately, to compute a single classical trajectory $\vec{x}^{\rm cl}[t]$, we only need to evaluate the derivative of the many-particle potential $V(\vec{x}, t)$ along the trajectory. In the quantum case, even finding a single trajectory means solving for the entire wave function $\psi(\vec{x}, t)$ that *encapsulates* the information of the entire ensemble of trajectories of all other experiments (the quantum wholeness).

In the scientific literature, there are many attempts to provide reasonable approximations to the many-body quantum problem. Density functional theory [74, 75] and the Hartree–Fock approximation [76–78] are some of the most popular techniques among the scientific community for dealing with the many body problem. In Section 1.3.1 we will discuss whether Bohmian trajectories can help in solving the many body problem. In what follows, we present the basic theory of the many-particle Bohmian trajectories.

1.3.2 Many-Particle Quantum Trajectories

The efforts done in the single-particle section to compare classical and quantum mechanics are valid here. In particular, we could write a classical (many-particle) wave equation or develop (many-particle) trajectories for quantum systems. In this section, we will only explicitly develop the latter. We start by considering (nonrelativistic) spinless particles. In Section 1.3.4 we will introduce spin.

1.3.2.1 Many-particle continuity equation

Following a similar development as we did for the single-particle case, see Section 1.2.3, we can derive the following *local* continuity equation associated to the many-particle Schrödinger equation, Eq. (1.52):

$$\frac{\partial |\psi(\vec{x}, t)|^2}{\partial t} + \sum_{k=1}^{N} \frac{\partial}{\partial x_k} J_k(\vec{x}, t) = 0 \tag{1.54}$$

where we have defined

$$J_k(\vec{x}, t) = i\frac{\hbar}{2m}\left(\psi(\vec{x}, t)\frac{\partial \psi^*(\vec{x}, t)}{\partial x_k} - \psi^*(\vec{x}, t)\frac{\partial \psi(\vec{x}, t)}{\partial x_k}\right) \tag{1.55}$$

as the k-th component of the current density, see problem 7.

Since the many-particle Schrödinger equation is also compatible with a local conservation of particles, we can interpret $|\psi(\vec{x}, t)|^2$ as the spatial distribution of an ensemble of many-particle trajectories assigned to an ensemble of different experiments. The Bohmian velocity of the k-th trajectory is:

$$v_k(\vec{x}, t) = \frac{J_k(\vec{x}, t)}{|\psi(\vec{x}, t)|^2} \tag{1.56}$$

In fact, the strategy followed here to develop Bohmian mechanics can be extended to any quantum equation of motion where a continuity equations holds: first, look for a continuity equation for the probability density and, then, define a velocity for the Bohmian trajectories as the current density divided by the probability density. A particular example will be developed for particles with spin in Section 1.3.4 and Chapter 8.

1.3.2.2 Many-particle quantum Hamilton–Jacobi equation

Alternatively, we can obtain Bohmian mechanics from a quantum Hamilton–Jacobi equation. We start by introducing the polar form of the many-particle wave function $\psi(\vec{x}, t) = R(\vec{x}, t)e^{iS(\vec{x},t)/\hbar}$ into the many-particle Schrödinger equation, Eq. (1.52). Then, after a quite simple manipulation, one obtains from the imaginary part:

$$\frac{\partial R^2(\vec{x}, t)}{\partial t} + \sum_{k=1}^{N} \frac{\partial}{\partial x_k}\left(\frac{1}{m}\frac{\partial S(\vec{x}, t)}{\partial x_k} R^2(\vec{x}, t)\right) = 0 \tag{1.57}$$

where we recognize the velocity of the k-th particle as:

$$v_k(\vec{x}, t) = \frac{1}{m}\frac{\partial S(\vec{x}, t)}{\partial x_k} \tag{1.58}$$

Equations (1.56) and (1.58) are identical. This is shown in problem 6 for a 1D system, but it can straightforwardly be generalized to N dimensions. The real part of the Schrödinger equation leads to a many-particle version of the quantum Hamilton–Jacobi equation:

$$\frac{\partial S(\vec{x}, t)}{\partial t} + \sum_{k=1}^{N} \frac{1}{2m}\left(\frac{\partial S(\vec{x}, t)}{\partial x_k}\right)^2 + V(\vec{x}, t) + Q(\vec{x}, t) = 0 \tag{1.59}$$

where we have defined the quantum potential as:

$$Q(\vec{x}, t) = \sum_{k=1}^{N} Q_k(\vec{x}, t) \tag{1.60}$$

with:

$$Q_k(\vec{x}, t) = -\frac{\hbar^2}{2m} \frac{\partial^2 R(\vec{x}, t)/\partial x_k^2}{R(\vec{x}, t)} \tag{1.61}$$

Again, we can obtain a system of coupled Newton-like equations (one for each component of $\vec{x}$) for the many-particle Bohmian trajectories by computing the time-derivative of the Bohmian velocity of Eq. (1.56):

$$m\frac{d^2 x_k[t]}{dt^2} = \left[-\frac{\partial}{\partial x_k}\left(V(\vec{x}, t) + Q(\vec{x}, t)\right)\right]_{\vec{x}=\vec{x}[t]} \tag{1.62}$$

Both the potential $V(\vec{x}, t)$ and the quantum potential $Q(\vec{x}, t)$ introduce correlations between particles. All physical interactions are essentially expressed as correlations (relations) between the degrees of freedom. In the following, we will discuss some important differences between classical and quantum correlations.

1.3.3 Factorizability, Entanglement, and Correlations

There are important differences between the correlations introduced by the classical potential $V(\vec{x}, t)$ and the quantum potential $Q(\vec{x}, t)$:

(1) In general, the term $V(\vec{x}, t)$ decreases with the distance between particles. A simple example is the Coulomb interaction. However, $Q(\vec{x}, t)$ depends only on the shape of the wave function, not on its value (see Eq. (1.61)). Thus, the quantum potential that we would obtain from Eq. (1.61) using either $R(\vec{x}, t)$ or $aR(\vec{x}, t)$ would be exactly the same, even when $a \to 0$. Thus, the quantum potential can produce a significative interaction between two particles, even if they are very far apart.[a]

[a]These highly nonclassical features of the quantum potential is what led Bohm to argue that the quantum potential interchanges information between systems [32, 60].

(2) In general, the term $V(\vec{x}, t)$ produces classical correlations between different particles. The particular dependence of $V(\vec{x}, t)$ in all variables $x_1, \ldots, x_N$ imposes a restriction on the speed of the interaction. The variations of x_i can only affect x_j after a time larger than $|x_i - x_j|/c$, c being the speed of light. For example, the relation between the particles positions in the electromagnetic interactions ensures that there is no superluminal influence between particles. However, such a restriction is not present in the quantum potential. Thus, very far particles have an *instantaneous* (nonlocal) interaction between them.

The quantum potential is at the origin of all quantum correlations, that is, entanglement, that can imply (nonlocal) faster-than-light interactions when two distant particles are involved. As mentioned in Section 1.1.5, this "spooky action at a distance" is what bothered Einstein about Bohmian mechanics (and quantum mechanics, in general). In 1964, Bell elaborated his famous theorem that established clear experimentally testable mathematical inequalities that would be fulfilled by *local* theories but would be violated by *nonlocal* ones [36]. All experimental results obtained so far confirm that Bell's inequalities are violated. Therefore, contrarily to Einstein's belief, we have to accept the real existence, in nature, of faster-than-light causation.[a] Entanglement is an intrinsic correlation in quantum mechanics (whose complexity and potentialities eventually come from the fact that a N-particle wave function lives in a N-dimensional configuration space) and is at the core of quantum information science, which makes teleportation, quantum communication, quantum cryptography, and quantum computing possible.

To improve our understanding of correlations, let us discuss under which conditions we cannot expect correlations between N particles. Let us focus our attention on a wave function $\psi(\vec{x}, t)$ that can be written as a product of single-particle wave functions

[a]We insist that the experimental violation of Bell's inequalities gives direct support not only to the Copenhagen interpretation but also to the Bohmian one, since the latter is also a nonlocal theory because of the presence of $Q(\vec{x}, t)$.

associated with each of the particles:

$$\psi(\vec{x}, t) = \prod_{k=1}^{N} \psi_k(x_k, t) \tag{1.63}$$

We call such a wave function "factorizable" or "separable". Equation (1.63) expresses the physical independence of the N particles (even though the wave functions $\psi_k(x_k, t)$ may overlap). According to Born's statistical interpretation of wave functions, the squared modulus of a wave function is the probability density of the quantum particle; thus the quantum wave function, Eq. (1.63), corresponds to a system without (classical or quantum) correlations between particles. It is well known that such a solution occurs when the potential in Eq. (1.52) can be written as $V(\vec{x}) = \sum_k V_k(x_k)$. Next, let us reformulate this result with Bohmian trajectories.

We realize that the phase and modulus of Eq. (1.63) are given by:

$$S(\vec{x}, t) = \sum_{k=1}^{N} S_k(x_k, t) \tag{1.64}$$

and

$$R(\vec{x}, t) = \prod_{k=1}^{N} R_k(x_k, t) \tag{1.65}$$

by defining

$$\psi_k(x_k, t) = R_k(x_k, t) e^{i S_k(x_k, t)/\hbar} \tag{1.66}$$

In this case, the many-particle quantum potential can be written as:

$$Q_k(\vec{x}, t) = -\frac{\hbar^2}{2m} \frac{\partial^2 R_k(x_k, t)/\partial x_k^2}{R_k(x_k, t)} \tag{1.67}$$

Then, we can easily deduce from Eq. (1.59) that each $\psi_k(x_k, t)$ is a solution of the following single-particle quantum Hamilton–Jacobi equation:

$$\frac{\partial S_k(x_k, t)}{\partial t} + \frac{1}{2m}\left(\frac{\partial S_k(x_k, t)}{\partial x_k}\right)^2 + V(x_k, t) + Q_k(x_k, t) = 0 \tag{1.68}$$

In addition, each $\psi_k(x_k, t)$ satisfies a conservation law:

$$\frac{\partial R_k^2(x_k, t)}{\partial t} + \frac{\partial}{\partial x_k}\left(\frac{1}{m}\frac{\partial S_k(x_k, t)}{\partial x_k} R_k^2(x_k, t)\right) = 0 \tag{1.69}$$

From these equations we can easily deduce an independent guiding equation for each k-particle:

$$m\frac{dx_k[t]}{dt} = \left(\frac{\partial S_k(x_k, t)}{\partial x_k}\right)_{x_k=x_k[t]} \tag{1.70}$$

showing the absence of correlations between (the different components of the many-particle) Bohmian trajectory.

1.3.4 Spin and Identical Particles

Elementary particles, such as electrons or quarks, have spin, an internal (discrete) degree of freedom that can influence their quantum dynamics in a nontrivial manner. In this section we will briefly explain how to extend (nonrelativistic) Bohmian mechanics to include spin.

1.3.4.1 Single-particle with $s = 1/2$

In the orthodox formulation of quantum mechanics, the state of a single particle with spin s is described by a $2s + 1$ component vector of wave functions with:

$$\vec{\Psi}(\vec{r}, t) = \begin{pmatrix} \Psi_1(\vec{r}, t) \\ \vdots \\ \Psi_{2s+1}(\vec{r}, t) \end{pmatrix} \tag{1.71}$$

where $\vec{r} = (x_1, x_2, x_3)$ represents the 3D position of the particle. The time evolution of this state is no longer governed by the Schrödinger equation but by more involved wave equations such as the Pauli equation for $s = 1/2$ [50, 79, 80].

The strategy to find Bohmian trajectories for spin particles will be the following: we will look for a continuity equation of the probability density and define the Bohmian velocity as the current density divided by the probability density. The idea is quite simple, but the mathematical development can be much more complicated. Therefore, we will focus on a particular example for a spin $s = 1/2$ charged particle, whose vectorial wave function,

$$\vec{\Psi}(\vec{r}, t) = \begin{pmatrix} \Psi_\uparrow(\vec{r}, t) \\ \Psi_\downarrow(\vec{r}, t) \end{pmatrix} \tag{1.72}$$

1.3.4.2 Many-particle system with $s = 1/2$ particles

For a nonrelativistic system of many particles, where the orbital and spin contributions of the Hamiltonian are initially separable,[a] the many-particle wave function will be also separable at all times, that is:

$$\Psi(\vec{r}_1, \ldots \vec{r}_N, t) = \psi(\vec{r}_1, \ldots \vec{r}_N, t)\vec{\chi}(t) \tag{1.84}$$

where $\psi(\vec{r}_1, \ldots \vec{r}_N, t)$ is the time evolution of the orbital state, the solution of the many-particle Schrödinger equation, and $\vec{\chi}(t)$ is the time evolution of the spin part of the system state:

$$\vec{\chi}(t) = \begin{pmatrix} \alpha_{\uparrow_1\uparrow_2\ldots\uparrow_N}(t) \\ \alpha_{\uparrow_1\uparrow_2\ldots\downarrow_N}(t) \\ \vdots \\ \alpha_{\downarrow_1\downarrow_2\ldots\downarrow_N}(t) \end{pmatrix} = \begin{pmatrix} \alpha_1(t) \\ \alpha_2(t) \\ \vdots \\ \alpha_W(t) \end{pmatrix} \tag{1.85}$$

$W \equiv (2s + 1)^N$ is the number of possible combinations of spin projections in one direction for all the particles in the system.[a]

In many practical situations, we are only interested in the evolution of the orbital wave function through the many-particle Schrödinger equation. However, even in this case, there is a peculiar dependence of the dynamics of the particles on the total spin. As explained in many textbooks, there is a pure quantum interaction between identical particles, named "exchange interaction" [79, 81]. This interaction is not classical, and we cannot find a term in the potential energies of the many-particle Hamiltonian of the Schrödinger equation that accounts for it. Alternatively, this new interaction is introduced in the "shape" of the global wave function, through the requirement of a particular symmetry. We say that a many-particle wave function is anti-symmetric when the interchange of the position and spin degrees of freedom associated to two identical fermions (e.g., electrons) results only in a change of the sign of the global wave function, $\vec{\Psi}(\vec{r}_1, \vec{r}_2, \ldots, \vec{r}_N, t)$. Analogously, we say that the many-particle wave function is symmetric when it remains unchanged after the interchange of the degrees of freedom of two identical bosons.

---

[a] For example, in the absence of a magnetic field such that the initial wave function can be written as a product of an orbital part, $\psi(\vec{r}_1, \vec{r}_2, \ldots \vec{r}_N, 0)$, and a spin part, $\vec{\chi}(0)$.

[a] Additionally, one can also look for the projection of the total spin of the system. Both procedures are connected by the Clebsch–Gordan coefficients [79].

The crucial point why we cannot forget about spin when dealing with many-particle Hamiltonians is that for separable wave functions, for example, Eq. (1.84), the symmetry of the orbital part depends on the symmetry of the spin part. For example, since the total wave function must be antisymmetric, two electrons can have a symmetric orbital part if the spin part is antisymmetric, and vice versa. In general, note that the wave function is not separable, and it makes no sense to talk about the symmetry of the orbital and spin parts alone but only of the symmetry of the total wave function. We see that the exchange interaction induces correlations between particles by imposing symmetries (or shapes) to their many-particle wave function.

A standard claim in many quantum mechanics textbooks is that identical particles, for example, two electrons with an antisymmetric wave function, are indistinguishable. It is affirmed that if the particles would have trajectories, they would automatically be distinguishable. In Bohmian mechanics, even with the symmetrization postulate, the adjective "indistinguishable" is inappropriate because one can label one particle's trajectory $\vec{r}_1[t]$ and the other $\vec{r}_2[t]$ and thus distinguish them perfectly at the ontological staff.

Bohmian trajectories can actually help us improve our understanding of the symmetrization postulate. Let us assume a two-electron system with an antisymmetric orbital wave function $\psi(\vec{r}_1, \vec{r}_2, t)$. We assume that an electron labeled 1 with the initial position $\vec{r}_1[0]$ evolves into $\vec{r}_1[t]$ and an another electron labeled 2 evolves from $\vec{r}_2[0]$ to $\vec{r}_2[t]$. Then, it can be easily understood that $\vec{r}_1{}'[0] = \vec{r}_2[0]$ evolves into $\vec{r}_1{}'[t] = \vec{r}_2[t]$ and $\vec{r}_2{}'[0] = \vec{r}_1[0]$ evolves into $\vec{r}_2{}'[t] = \vec{r}_1[t]$. We use primes to notice that $\vec{r}_1[t]$ and $\vec{r}_1{}'[t]$ correspond to trajectories of the same particle with different initial positions ($\vec{r}_1[0]$ or $\vec{r}_2[0]$). This result follows from the symmetry of the velocity, that is, the symmetry of the current density and modulus, when positions of the two electrons are interchanged. In summary, Bohmian trajectories of identical particles are clearly distinguishable in our computation, but the observable (ensemble) results obtained from them are indistinguishable when we interchange their initial positions. Hence, in Bohmian mechanics a set of particles with exchange symmetry are perfectly distinguishable at the ontological plane, while they become indistinguishable at the empirical plane.

1.3.5 Basic Postulates for Many-Particle Systems

The basic postulates of Bohmian mechanics for many-particle systems are exactly the same as those we have developed for a single particle but adding the symmetrization postulate. As done in Section 1.2.6, we also present a minimalist formulation of the postulates.

In order to simplify our description, we assume the simpler many-particle wave function written as:

$$\vec{\Psi}(\vec{r}_1, \ldots, \vec{r}_N, t) = \psi(\vec{r}_1, \ldots, \vec{r}_N, t) \begin{pmatrix} \uparrow_1 \uparrow_2 \cdots \uparrow_N (t) \\ 0 \\ \vdots \\ 0 \end{pmatrix} \tag{1.86}$$

The generalization of the present postulates to include the more general wave functions of Eq. (1.84) follows straightforwardly.

FIRST POSTULATE: *The dynamics of a many-particle quantum system comprises a wave function, Eq. (1.86), whose orbital part* $\psi(\vec{r}_1, \vec{r}_2, \ldots, \vec{r}_N, t)$ *is defined in the configuration space* $\{\vec{r}_1, \vec{r}_2, \ldots, \vec{r}_N\}$ *plus time, and a many-particle trajectory* $\{\vec{r}_1[t], \vec{r}_2[t], \ldots, \vec{r}_N[t]\}$ *that moves continuously under the guidance of the wave function.*

The orbital part of the wave function $\psi(\vec{r}_1, \vec{r}_2, \ldots, \vec{r}_N, t)$ *is a solution of the many-particle Schrödinger equation:*

$$i\hbar \frac{\partial \psi(\vec{r}_1, \ldots, \vec{r}_N, t)}{\partial t} = \left(\sum_{k=1}^{N} -\frac{\hbar^2}{2m} \nabla^2_{\vec{r}_k} + V(\vec{r}_1, \ldots, \vec{r}_N, t) \right) \times \psi(\vec{r}_1, \ldots, \vec{r}_N, t)$$

Each component $\vec{r}_k[t]$ *of the many-particle trajectory* $\{\vec{r}_1[t], \ldots, \vec{r}_N[t]\}$ *is obtained by time-integrating the particle velocity* $\vec{v}_k[t] = \vec{v}_k(\vec{r}_1, \ldots, \vec{r}_N, t)$ *defined from:*

$$\vec{v}_k(\vec{r}_1, \ldots, \vec{r}_N, t) = \frac{\vec{J}_k(\vec{r}_1, \ldots, \vec{r}_N, t)}{|\psi(\vec{r}_1, \ldots, \vec{r}_N, t)|^2}$$

where $|\psi(\vec{r}_1, \ldots, \vec{r}_N, t)|^2 = \psi(\vec{r}_1, \ldots, \vec{r}_N, t)\psi^*(\vec{r}_1, \ldots, \vec{r}_N, t)$ *is the square modulus of the wave function and* $\vec{J}_k = \vec{J}_k(\vec{r}_1, \ldots, \vec{r}_N, t)$ *is the* *k-particle current density*:

$$\vec{J}_k = i\frac{\hbar}{2m}\Big(\psi(\vec{r}_1,\ldots,\vec{r}_N,t)\vec{\nabla}_{\vec{r}_k}\psi^*(\vec{r}_1,\ldots,\vec{r}_N,t) - \psi^*(\vec{r}_1,\ldots,\vec{r}_N,t)\vec{\nabla}_{\vec{r}_k}\psi(\vec{r}_1,\ldots,\vec{r}_N,t)\Big)$$

The initial positions $\{\vec{r}_{10},\vec{r}_{20},\ldots,\vec{r}_{N0}\}$ *and velocities* $\{\vec{v}_{10},\ \vec{v}_{20},\ldots,\ \vec{v}_{N0}\}$ *have to be specified in order to completely determine the many-particle trajectory.*

SECOND POSTULATE (quantum equilibrium hypothesis): *The initial positions* $\{\vec{r}_1^{\,j}[t_0],\vec{r}_2^{\,j}[t_0],\ldots,\vec{r}_N^{\,j}[t_0]\}$ *and velocities* $\{\vec{v}_1^{\,j}[t_0],\vec{v}_2^{\,j}[t_0],\ldots,\ \vec{r}_N^{\,j}[t_0]\}$ *of a particular many-particle j trajectory cannot be known with certainty. When the experiment is repeated many times, these initial positions of an ensemble of trajectories associated with the same* $\psi(\vec{r}_1,\ldots,\vec{r}_N,t)$ *satisfy that the number of trajectories of the ensemble between* $(\vec{r}_1,\ldots,\vec{r}_N)$ *and* $(\vec{r}_1+d\vec{r}_1,\ldots,\vec{r}_N+d\vec{r}_N)$ *at the initial time* t_0 *is proportional to* $R^2(\vec{r}_1^{\,j}[t_0],\ldots,\vec{r}_N^{\,j}[t_0]) = |\psi(\vec{r}_1^{\,j}[t_0],\ldots,\vec{r}_N^{\,j}[t_0],t_0)|^2$*. The initial velocity is determined by* $\vec{v}_k^{\,j}[t_0] = \vec{J}_k(\vec{r}_1^{\,j}[t_0],\ldots,\vec{r}_N^{\,j}[t_0],t_0)/|\psi(\vec{r}_1^{\,j}[t_0],\ldots,\vec{r}_N^{\,j}[t_0],t_0)|^2$.

The condition on the initial position can be written mathematically as:

$$R^2(\vec{r}_1,\ldots,\vec{r}_N,t_0) = \lim_{M\to\infty}\frac{1}{M}\sum_{j=1}^{M}\prod_{k=1}^{N}\delta(\vec{r}-\vec{r}_k^{\,j}[t_0]) \quad \text{for } t=t_0 \tag{1.87}$$

Notice the presence of two indices, the $j = 1,\ldots,M$ for the infinite ensemble of identical experiments and the $k = 1,\ldots,N$ for the N degrees of freedom.

THIRD POSTULATE (symmetrization postulate of quantum mechanics): *If the variables* $\vec{r}_i,\uparrow_i$ *and* $\vec{r}_j,\uparrow_j$ *refer to two identical particles of the ensemble, then the wave function, Eq. (1.86), is either symmetric:*

$$\psi(.,\vec{r}_i,.,\vec{r}_j,.,t)\begin{pmatrix}\uparrow_i\ \ldots\ \uparrow_j\ \ldots\\ 0\\ \vdots\\ 0\end{pmatrix} = \psi(.,\vec{r}_j,.,\vec{r}_i,.,t)\begin{pmatrix}\uparrow_j\ \ldots\ \uparrow_i\ \ldots\\ 0\\ \vdots\\ 0\end{pmatrix} \tag{1.88}$$

if the particles are bosons (every particle with an integer spin 0, 1, 2, ... is a boson), or antisymmetric:

$$\psi(.,\vec{r}_i,.,\vec{r}_j,.,t)\begin{pmatrix} \uparrow_i \cdots \uparrow_j \cdots \\ 0 \\ \vdots \\ 0 \end{pmatrix} = -\psi(.,\vec{r}_j,.,\vec{r}_i,.,t)\begin{pmatrix} \uparrow_j \cdots \uparrow_i \cdots \\ 0 \\ \vdots \\ 0 \end{pmatrix} \tag{1.89}$$

if the particles are fermions (every particle with a half-odd spin 1/2, 3/2, ... is a fermion). In Eq. (1.88) and Eq. (1.89) it is understood that all other degrees of freedom of the other particles remain unchanged.[a]

As we already did in the development of the postulates for a single-particle system, we remind that the second postulate needs some additional remarks since it has been argued by some authors that this postulate, the quantum equilibrium hypothesis, is not a necessary postulate of the Bohmian theory.

First, it has been demonstrated that if one assumes that the many-particle wave function of the whole Universe, the so-called Universal wave function that includes all degrees of freedom of the Universe, satisfies some (typical) conditions then, it follows that each (conditional) wave function of each quantum (sub) system will satisfy its own Eq. (1.87). Thus, the justification about the quantum equilibrium postulate has to be done only for the Universal wave function, not for each individual (sub)system.

Probably, the most accepted view against taking the quantum equilibrium as a postulate comes from the seminal work by Dürr, Goldstein, and Zanghì [63, 66], where the equivariance in any system is discussed from the initial configurations of (Bohmian) particles in the Universe. Using Bohmian mechanics to describe the wave function of the whole Universe, then the wave function associated to any (sub)system is an effective (conditional) wave function of the universal one. Using typicality arguments, they showed that the overwhelming majority of possible selections of initial positions

---

[a]This simple spin vector wave function is clearly symmetric so that the orbital wave function has to be either symmetric or antisymmetric. For general wave functions, such as the one in Eq. (1.84), this postulate implies much more complicated restrictions on the possible orbital and spin wave functions.

of particles in the Universe will satisfy Eq. (1.87) in a subsystem (or Eq. (1.51) for a subsystem with one degree of freedom) [63, 66]. Other authors [82] have attempted to dismiss Eq. (1.87) as a postulate by showing that any initial configuration of Bohmian particles will relax, after some time, to a distribution very close to Eq. (1.51) for a subsystem. Such discussions about quantum equilibrium are far from the scope of this book. Therefore, from a practical point of view, one can postulate Eq. (1.87) (at some initial time) in the Bohmian theory in the same way that Born's law is a postulate in the orthodox theory.

Again, no postulate about measurement is needed, since in Bohmian mechanics measurement is treated as a particular case of interaction between particles (see Section 1.4).

1.3.6 The Conditional Wave Function: Many-Particle Bohmian Trajectories without the Many-Particle Wave Function

In Section 1.2.4, when discussing the similarities and differences between classical and quantum mechanics, we mentioned quantum wholeness (the dependence of each individual Bohmian trajectory on the rest of trajectories through the wave function) as a fundamental difference. This concept acquires an even more dramatic meaning when dealing with many-body systems. A particular Bohmian trajectory of the k-particle $\vec{r}_k^{\,j}[t]$ depends on the rest of infinite $j = 1, \ldots, M$ possible trajectories (with different initial position associated to different experiments) of all $k = 1, \ldots, N$ particles (corresponding to the N degrees of freedom of the quantum system). This property is mathematically manifested by the fact that the many-particle wave function is defined in an N-dimensional configuration space rather than in the usual 3D real space. In a 3D real space one can easily define whether or not two particles are far away. In general, the potential profile that determines the interaction between these two particles decreases with the distance. However, two distant particles that share a common region of the N-dimensional configuration space where the wave function is different from zero have an interaction (entanglement) independently of their distance.

As discussed in Section 1.3.1, the need of computing the many-particle wave function in the N-dimensional configuration space is the origin of the quantum many-body problem. Among the large list of approximations present in the literature to tackle this problem, Bohmian mechanics provides a natural, original and mainly unexplored solution through the use of the conditional wave function [83]. Due to the Bohmian dual description of a quantum system as particles and waves, one can reduce the complexity of a many-particle wave function through the substitution of some of the N degrees of freedom by its corresponding Bohmian trajectories. The new wave function with a reduced number of degrees of freedom is named conditional wave function. For a discussion about the fundamental implications of working with conditional wave functions instead of full many-body wave functions, see Refs. [83, 84]. We will present now some work on the computational abilities of such conditional wave functions.

Hereafter, for simplicity, we return to a definition of the degrees of freedom of the many-particle quantum system in terms of $\{x_1, x_2, \ldots, x_N\}$ instead of $\{\vec{r}_1, \vec{r}_2, \ldots, \vec{r}_N\}$. The direct solution of the many-particle Schrödinger equation, Eq. (1.52), is intractable numerically. On the contrary, differential classical equations of motion deal with solutions in a much smaller configuration space, $\{x_1[t], \ldots, x_{i-1}[t], x_a, x_{i1}[t], \ldots, x_N[t]\}$, where all other trajectories are known parameters, except x_a. In other words, the Newton solution of $x_a[t]$ just needs the spatial dependence of $V(x_1[t], \ldots, x_a, \ldots, x_N[t])$ on the variable x_a. In this section, we summarize the formalism developed by Oriols [85] to compute many-particle Bohmian trajectories without knowing the many-particle wave function. It is a clear example of how Bohmian mechanics can be a powerful computational tool.

The main idea behind the work developed in [85] is that any Bohmian trajectory $x_a[t]$ that is computed from the many-particle wave function $\Psi(\vec{x}, t)$ solution of Eq. (1.52) can be alternatively computed from a much simpler single-particle conditional wave function $\phi_a(x_a, t) = \Psi(x_a, \vec{x}_b[t], t)$. Here, we use the notation $\vec{x} = \{x_a, \vec{x}_b\}$ with $\vec{x}_b = \{x_1, \ldots, x_{a-1}, x_{a+1}, \ldots, x_N\}$ for particle positions and $\vec{x}_b[t] = \{x_1[t], \ldots, x_{a-1}[t], x_{a+1}[t], \ldots, x_N[t]\}$ for Bohmian trajectories.

It is quite trivial to demonstrate the ability of $\phi_a(x_a, t)$ in reproducing $x_a[t]$. By construction, when we use a polar form $\phi_a(x_a, t) = r_a(x_a, t)\ e^{i\ s_a(x_a,t)/\hbar}$, the angle $s_a(x_a, t)$ is identical to the angle $S(x_a, \vec{x}_b, t)$ of $\Psi(x_a, \vec{x}_b, t)$ evaluated at $\{x_a, \vec{x}_b[t]\}$. Therefore, since the velocity of the trajectory $x_a[t]$ is computed from the spatial dependence of $S(x_a, \vec{x}_b, t)$ on x_a when all other positions are fixed at $\vec{x}_b = \vec{x}_b[t]$, the same velocity will be obtained from the spatial dependence of $s_a(x_a, t)$. Interestingly, $\phi_a(x_a, t)$ is solution of a single-particle (pseudo) Schrödinger equation because it depends only on time t and position x_a. Next, our effort will be focused on determining such single-particle equations.

1.3.6.1 Single-particle pseudo-Schrödinger equation for many-particle systems

First of all, we show that any (totally arbitrary) single-valued complex function $\phi_a(x_a, t)$, which has a well-defined second-order spatial derivative and first-order temporal derivative, can be obtained from a Schrödinger-like equation when the following potential $W(x_a, t)$ is used:

$$W(x_a, t) = \frac{i\hbar \frac{\partial \phi_a(x_a,t)}{\partial t} + \frac{\hbar^2}{2m}\frac{\partial^2 \phi_a(x_a,t)}{\partial^2_{x_a}}}{\phi_a(x_a, t)} \tag{1.90}$$

For an arbitrary (complex) function, the potential energy $W(x_a, t)$ can be complex, too. In fact, we are interested in rewriting $W(x_a, t)$ in terms of the polar form of the wave function $\phi_a(x_a, t) = r_a(x_a, t)\ e^{i\ s_a(x_a,t)/\hbar}$. We obtain for the real part:

$$Real[W(x_a, t)] = -\left(\frac{1}{2\ m}\left(\frac{\partial s_a(x_a, t)}{\partial x_a}\right)^2 - \frac{\hbar^2}{2\ m\ r_a(x_a, t)} \times \frac{\partial^2 r_a(x_a, t)}{\partial^2 x_a} + \frac{\partial s_a(x_a, t)}{\partial t}\right) \tag{1.91}$$

From Eq. (1.90), we do also obtain for the imaginary part:

$$Imag[W(x_a, t)] = \frac{\hbar}{2r_a^2(x_a, t)} \times \left(\frac{\partial r_a^2(x_a, t)}{\partial t} + \frac{\partial}{\partial x_a}\left(\frac{r_a^2(x_a, t)}{m}\frac{\partial s_a(x_a, t)}{\partial x_a}\right)\right) \tag{1.92}$$

It can be easily verified that $Imag[W(x_a, t)] = 0$ when the single-particle wave function preserves the norm.

Finally, we are interested in using the above expressions when the single-particle wave function is the one mentioned in the introduction of this subsection, $\phi_a(x_a, t) = \Psi(x_a, \vec{x}_b[t], t)$. In particular, we use $r_a(x_a, t) = R(x_a, \vec{x}_b[t], t)$ and $s_a(x_a, t) = S(x_a, \vec{x}_b[t], t)$. Then, we realize that Eq. (1.90) transforms into:

$$i\hbar\frac{\partial\phi_a(x_a, t)}{\partial t} = \left(-\frac{\hbar^2}{2m}\frac{\partial^2}{\partial^2 x_a} + U_a(x_a, \vec{x}_b[t], t) + G_a(x_a, \vec{x}_b[t], t) + iJ_a(x_a, \vec{x}_b[t], t)\right)\phi_a(x_a, t) \tag{1.93}$$

where we have defined:

$$G_a(\vec{x}, t) = U_b(\vec{x}_b, t) + \sum_{k=1;k\neq a}^{N}\left(\frac{1}{2\,m}\left(\frac{\partial S(\vec{x}, t)}{\partial x_k}\right)^2 + Q_k(\vec{x}, t) - \frac{\partial S(\vec{x}, t)}{\partial x_k}v_k(\vec{x}[t], t)\right) \tag{1.94}$$

with $Q_k(\vec{x}, t)$ as the quantum potential energy defined in Eq. (1.61). The terms $U_a(x_a, \vec{x}_b[t], t)$ and $U_b(\vec{x}_b, t)$ are defined from the many-particle potential that appears in the original many-particle potential energy in Eq. (1.52) as:

$$V(\vec{x}, t) = U_a(x_a, \vec{x}_b, t) + U_b(\vec{x}_b, t) \tag{1.95}$$

In addition, we have defined:

$$J_a(\vec{x}, t) = \sum_{k=1;k\neq a}^{N}\frac{\hbar}{2R^2(\vec{x}, t)}\left(\frac{\partial R^2(\vec{x}, t)}{\partial x_k}v_k(\vec{x}[t], t) - \frac{\partial}{\partial x_k}\left(\frac{R^2(\vec{x}, t)}{m}\frac{\partial S(\vec{x}, t)}{\partial x_k}\right)\right) \tag{1.96}$$

In order to obtain these expressions, one has to carefully evaluate:

$$\frac{\partial S(x_a, \vec{x}_b[t], t)}{\partial t} = \left(\frac{\partial S(x_a, \vec{x}_b, t)}{\partial t}\right)_{\vec{x}_b=\vec{x}_b[t]} + \sum_{k=1;k\neq a}^{N}\frac{\partial S(x_a, \vec{x}_b[t], t)}{\partial x_k}\,v_k(\vec{x}[t], t) \tag{1.97}$$

and use Eq. (1.59) evaluated at $\{x_a, \vec{x}_b[t]\}$. Identically for $\partial R^2(x_a, \vec{x}_b[t], t)/\partial t$ with Eq. (1.57).

The single-particle pseudo-Schrödinger equation, Eq. (1.93), is the main result of this algorithm. Let us discuss the role of each potential term in Eq. (1.93):

(1) The term $U_a(x_a, \vec{x}_b[t], t)$ is a real-valued potential whose explicit dependence on the positions is known. It has to be evaluated from the particular Bohmian position of all particles except $x_a[t]$.
(2) The term $G_a(x_a, \vec{x}_b[t], t)$ is a real-valued potential whose explicit dependence on the positions is unknown (unless we know the many-particle wave function) and needs some educated guess. It takes into account, for example, the exchange interaction between particles.
(3) The term $iJ_a(x_a, \vec{x}_b[t], t)$ is an imaginary-valued potential whose explicit dependence on the positions is also unknown and needs some educated guess. It takes into account that the norm of $\phi_a(x_a, t)$ is not directly conserved (the norm of the many-particle wave function $\Psi(x_a, \vec{x}_b, t)$ is conserved in the entire configuration space, but this is not true for $\Psi(x_a, \vec{x}_b[t], t)$ in the x_a space).

By using Eq. (1.93) for each particle, $x_a[t]$ for $a = 1, \ldots, N$, we obtain a system of N coupled single-particle pseudo-Schrödinger equations that is able to compute many-particle Bohmian trajectories without knowing the many-particle wave function. The great merit of Eq. (1.93) is to demonstrate that such a single-particle solution of a many-particle problem exists, although we do not know exactly the values of the terms $G_a(x_a, \vec{x}_b[t], t)$ and $J_a(x_a, \vec{x}_b[t], t)$. Our algorithms have similarities with the original work on density functional theory [74, 75]. The formidable computational simplification comes at the price that some terms of the potential of the corresponding single-particle Schrödinger equations are unknown, that is, the exchange-correlation functional in density functional theory [74, 75] and, here, the terms in Eqs. (1.94) and (1.96).

1.3.6.2 Example: Application in factorizable many-particle systems

Let's start by discussing which will be the solution $\phi_a(x_a, t)$ of Eq. (1.93) when the many-particle wave function $\Psi(x_a, \vec{x}_b, t)$ is factorizable (i.e., when it can we written as Eq. (1.63)). Then, Eq. (1.68) is valid for each summand of $G_a(x_a, \vec{x}_b[t], t)$ in Eq. (1.93), so it can be written in a compact way as:

$$G_a(x_a, \vec{x}_b[t], t) = -\sum_{k=1;k\neq a}^{N} \frac{dS_k(x_k[t], t)}{dt} \tag{1.98}$$

Then, a (real) time-dependent term (without any spatial dependence) appears in the potential of Eq. (1.93). It can be easily demonstrated that such a term introduces the following time-dependent contribution $\beta_a(t)$ into the phase of $\phi_a(x_a, t)$:

$$\beta_a(t) = -\sum_{k=1;k\neq a}^{N} \int_{t_0}^{t} \frac{dS_k(x_k[t'], t')}{dt'} dt' = -\sum_{k=1;k\neq a}^{N} S_k(x_k[t], t) \tag{1.99}$$

Identically, Eq. (1.69) is valid for each term of $J_a(x_a, \vec{x}_b[t], t)$, so it can be written as:

$$J_a(x_a, \vec{x}_b[t], t) = -\sum_{k=1;k\neq a}^{N} \frac{\hbar}{2}\frac{d}{dt} \ln\left(R_k^2(x_k[t])\right) \tag{1.100}$$

Using $\ln(a\, b) = \ln(a) + \ln(b)$, we obtain a contribution $\alpha_a(t)$ into the phase of $\phi_a(x_a, t)$:

$$\begin{aligned}\alpha_a(t) &= -\hbar \int_{t_0}^{t} \frac{d}{dt'} \ln\left(\prod_{k=1;k\neq a}^{N} R_k(x_k[t'], t')\right) dt' \\ &= -\hbar \ln\left(\prod_{k=1;k\neq a}^{N} R_k(x_k[t], t)\right)\end{aligned} \tag{1.101}$$

Finally, we use that a pure (real or imaginary) time-dependent potential (without spatial dependence) in a Schrödinger-like equation does only introduce a pure (imaginary or real) time-dependent global phase. Thus, we obtain:

$$\begin{aligned}\phi_a(x_a, t) &= \exp\left(\frac{i\, \beta_a(t) - \alpha_a(t)}{\hbar}\right) \psi_a(x_a, t) \\ &= \psi_1(x_1[t], t) \ldots \psi_a(x_a, t) \ldots \psi_N(x_N[t], t)\end{aligned} \tag{1.102}$$

which is, certainly, the expected result. Each term $\psi_k(x_k, t)$ is a single-particle wave function, whose evolution is found from Eq. (1.30) after appropriately defining the initial wave packet at time $t = 0$. Notice the difference between $\psi_a(x_a, t)$ and $\phi_a(x_a, t)$. The former is a single particle wave packet, while the latter has an additional time-dependent function multiplying $\psi_a(x_a, t)$.

1.3.6.3 Example: Application in interacting many-particle systems without exchange interaction

Up to here, we have demonstrated that a Bohmian solution to the many-particle problem exists in terms of a system of coupled single-particle Schrödinger equations (see Eq. (1.93)). The significant computational simplification comes at the price that the terms $G_a(x_a, \vec{x}_b[t], t)$ and $J_a(x_a, \vec{x}_b[t], t)$ of the corresponding single-particle Schrödinger equations are unknown. It is in this sense that we mentioned that these algorithms have similarities with density functional theory [74, 75]. In this subsection we provide a simple approximation for a system of N electrons with Coulomb interaction but without exchange interaction. Later, we will include the exchange interaction.

As mentioned, the solution of Eq. (1.93) needs educated guesses for the terms $G_a(x_a, \vec{x}_b[t], t)$, Eq. (1.94), and $J_a(x_a, \vec{x}_b[t], t)$, Eq. (1.96). Since no exchange interaction is considered, we assume that the origin of the correlations between the a electron and the rest is mainly contained in the term $U_a(x_a, \vec{x}_b[t], t)$. We develop a Taylor expansion of the other two terms, Eqs. (1.94) and (1.96), in the variable x_a around the point $x_a[t]$:

$$G_a(x_a, t) = G_a(x_a[t], t) + \left[\frac{\partial G_a(x_a, t)}{\partial x_a}\right]_{x_a = x_a[t]} (x_a - x_a[t]) + \ldots \tag{1.103}$$

and

$$J_a(x_a, t) = J_a(x_a[t], t) + \left[\frac{\partial J_a(x_a, t)}{\partial x_a}\right]_{x_a = x_a[t]} (x_a - x_a[t]) + \ldots \tag{1.104}$$

The simplest approximation is just a zero-order Taylor term $G_a(x_a, t) \approx G_a(x_a[t], t)$ and $J_a(x_a, t) \approx J_a(x_a[t], t)$.

The conditional wave function $\phi_a(x_a, t)$ solution of Eq. (1.93) can be constructed in two steps. First, by solving Eq. (1.93) without considering the purely time-dependent potential terms, $G_a(x_a[t], t)$ and $J_a(x_a[t], t)$, to find $\tilde{\psi}_a(x_a, t)$:

$$i\hbar\frac{\partial\tilde{\psi}_a(x_a, t)}{\partial t} = \left(-\frac{\hbar^2}{2m}\frac{\partial^2}{\partial x_a^2} + U_a(x_a, \vec{x}_b[t], t)\right)\tilde{\psi}_a(x_a, t) \quad (1.105)$$

where the term $U_a(x_a, \vec{x}_b[t], t)$ is defined in Eq. (1.95). Second, by multiplying the wave function $\tilde{\psi}_a(x_a, t)$ by time-dependent (real or imaginary) values (without any spatial dependence) for the final solution:

$$\phi_a(x_a, t) \approx \tilde{\psi}_a(x_a, t)\exp(z_a(t)) \quad (1.106)$$

with $z_a(t) = i\beta_a(t)/\hbar - \alpha_a(t)\hbar$ defined according to Eq. (1.102). Again, we have used the well-known result that a pure (real or imaginary) time-dependent potential term (without spatial dependence) added into the Hamiltonian of the Schrödinger-like equation does only introduce a pure (imaginary or real) time-dependent global phase, that is, $\exp(z_a(t))$. Since the velocity of Bohmian trajectories does not depend on these pure time-dependent terms $\exp(z_a(t))$, we do not have to compute $\beta_a(t)$ and $\alpha_a(t)$ explicitly.

For example, for a system of $N = 4$ interacting electrons, as depicted in the scheme of Fig. 1.4, we need to solve $N = 4$ Schrödinger-like equations defined by Eq. (1.105) to find $N = 4$ wave functions $\tilde{\psi}_a(x_a, t)$. Many-particle Bohmian trajectories are computed from the Bohmian velocity Eq. (1.35) using $\tilde{\psi}_a(x_a, t)$ to compute the current and the square modulus. The term $U_a(x_a, \vec{x}_b[t], t)$ can be solved from a Poisson equation (if we deal with Coulomb interaction) and it introduces correlations between particles. The initial wave function $\tilde{\psi}_a(x_a, t = 0)$ has to be specified. For example, if we assume that the initial wave function is defined in a region of the space where the many-particle wave function is factorizable, we can use $\tilde{\psi}_a(x_a, 0) = \psi_a(x_a, 0)$.

We will show now an example of the goodness of this simple approximation. We control the strength of the Coulomb interaction between two electrons by changing the lateral area of a 3D electron device, as described in [85]. Figure 1.5 shows the excellent

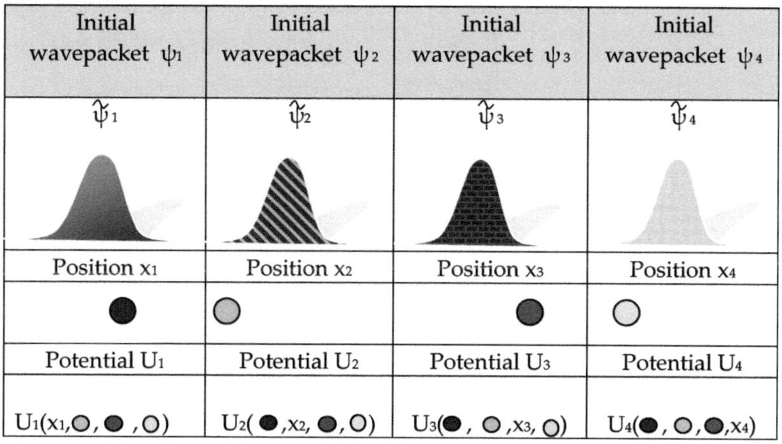

Figure 1.4 For 4 interacting particles without exchange interaction, the present algorithm needs $N = 4$ single-particle wave functions, $\psi_a(x_a, t)$. The subindex a of the wave function is associated with the potential $U_a(x_a, \vec{x}_b[t], t)$ and the initial wave packet. Each wave function $\psi_a(x_a, t)$ determines the a-Bohmian trajectories $x_a[t]$.

agreement between exact Bohmian trajectories and those computed with our algorithm. In Fig. 1.5a, the lateral area is so large that it makes the Coulomb interaction quite negligible. The first electron is transmitted, while the second is reflected. However, as shown in Fig. 1.5c, the smaller lateral area provides strong Coulomb interaction between the electrons, and the second is finally transmitted because of the presence of the first one in the barrier region.

Certainly, we have used the simplest approximation in Eqs. (1.103) and (1.104). Any possible improvement of this approximation will imply an even better agreement between this algorithm and the exact computation.

1.3.6.4 Example: Application in interacting many-particle systems with exchange interaction

Now, we generalize the previous result to an arbitrary system with Coulomb and exchange interactions. For simplicity, we will consider only the wave function defined in Eq. (1.86), that is, only

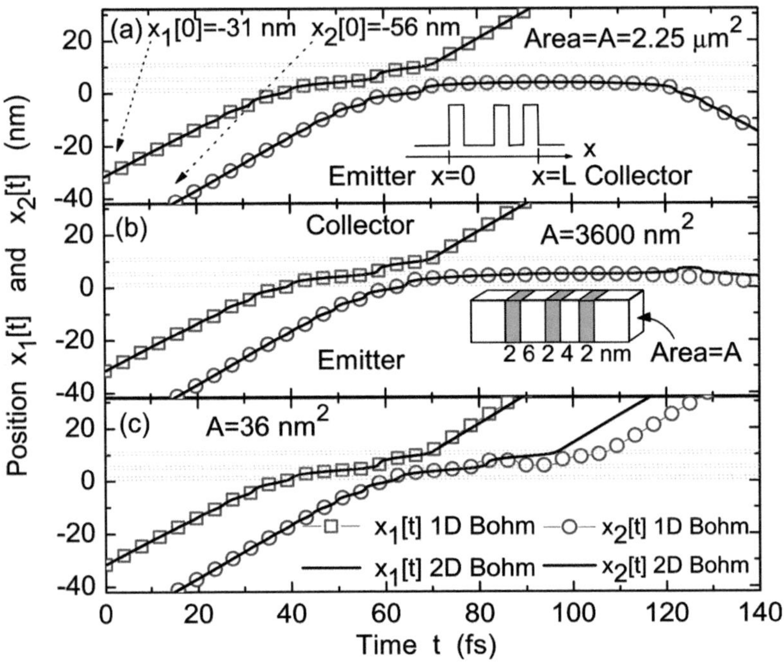

Figure 1.5 Two-particle interacting Bohmian trajectories in a triple barrier tunneling scenario computed from our 1D approach (symbols) and from exact 2D results (solid line) for three different lateral areas that modify the Coulomb interaction. Reprinted with permission from [85]. Copyright 2007 American Physical Society.

the symmetry of the orbital wave function is considered. A simple Taylor approximation for the terms $G_a(x_a, t)$ and $J_a(x_a, t)$, as we assume in Eq. (1.103) and Eq. (1.104), is not valid here. The reason can be easily understood, for example, for a system of (identical) electrons. Due to the Pauli exclusion principle, the modulus of the wave function tends to zero, $R(x_a, \vec{x}_b[t], t) \to 0$, at every position where $x_a \to x_k[t]$. Thus, the term $G_a(x_a, \vec{x}_b[t], t)$ has asymptotes at $x_a \to x_k[t]$ that repel other electrons, that is, $G_a(x_a, \vec{x}_b[t], t) \to \pm\infty$ at the same particular positions x_a, invalidating the approximation Eq. (1.103). For the same reason, $J_a(x_a, \vec{x}_b[t], t) \to \pm\infty$.

Here, the strategy is assuming that an antisymmetric wave function, $\Psi(\vec{x}, t)$, can be constructed from permutations of a many-particle wave function without any symmetry $\Psi_{\text{no-sym}}(\vec{x}, t)$:

$$\Psi(\vec{x}, t) = C \sum_{l=1}^{N!} \Psi_{\text{no-sym}}(x_{p(l)_1}, x_{p(l)_2}, \ldots, x_{p(l)_N}, t)\, s\left(\vec{p}(l)\right) \quad (1.107)$$

The subindex "no-sym" reminds that such wave functions have no spatial symmetric or antisymmetric property. The constant C is a normalization constant that will become irrelevant for the computation of the Bohmian velocity. The sum is over all $N!$ permutations $\vec{p}(l) = \left\{p(l)_1, p(l)_2, \ldots, p(l)_N\right\}$, and $s\left(\vec{p}(l)\right) = \pm 1$ is the sign of the permutations. Then, each wave function $\Psi_{\text{no-sym}}(x_{p(l)_1}, x_{p(l)_2}, \ldots, x_{p(l)_N}, t)$ evaluated at $\vec{x} = \{x_a, \vec{x}_b[t]\}$, can be computed following the previous (no-exchange) algorithm, that is Eqs. (1.105) and (1.106). In particular:

$$\Psi_{\text{no-sym}}(x_{p(l)_1}[t], \ldots, x_{p(l)_j}, \ldots, x_{p(l)_N}[t], t) = \tilde{\psi}_{a,\, p(l)_j}(x_a, t) \exp\left(z_{a,\, \vec{p}(l)}(t)\right) \quad (1.108)$$

where the permutation $\vec{p}(l)$ gives $x_{p(l)_j} = x_a$, that is, $p(l)_j = a$. Now, we have to use two labels in the subindex of $\tilde{\psi}_{a,h}(x_a, t)$ to specify the solution of Eq. (1.105). The first label a accounts for the degree of freedom, that is, the particular trajectory, that we are computing and it also fixes the potential energy $U_a(x_a, \vec{x}_b[t], t)$ in Eq. (1.105). The second label h fixes the initial wave function that we will consider. If the initial many-particle wave function can be defined, in a region without interactions, as:

$$\Psi_{\text{no-sym}}(x_1, \ldots, x_N, 0) = \psi_1(x_1, 0) \ldots \psi_N(x_N, 0) \quad (1.109)$$

then $\tilde{\psi}_{a,\, p(l)_j}(x_a, 0) = \psi_{p(l)_j}(x_a, 0)$. In other words, identical initial wave functions $\Psi_{i,j}(x_i, t_0) = \Psi_{k,j}(x_k, t_0)$ can evolve differently when using $U_i(x_i, \vec{x}_i[t], t)$ or $U_k(x_k, \vec{x}_k[t], t)$.

Finally, using Eq. (1.106), the many-particle wave function $\phi(x_a, t) = \Psi(x_a, \vec{x}_b[t], t)$ can be written as:

$$\phi(x_a, t) = \Psi(x_a \vec{x}_b[t], t) = C \sum_{l=1}^{N!} \tilde{\psi}_{a,\, p(l)_j}(x_a, t) \exp\left(z_{a,\, \vec{p}(l)}(t)\right) s\left(\vec{p}(l)\right) \quad (1.110)$$

Here, the angles $z_{a,\, \vec{p}(l)}(t)$ are relevant and cannot be ignored (only a global phase can be ignored). We emphasize that j is selected according to the condition $p(l)_j = a$, which depends on the index l of the permutations.

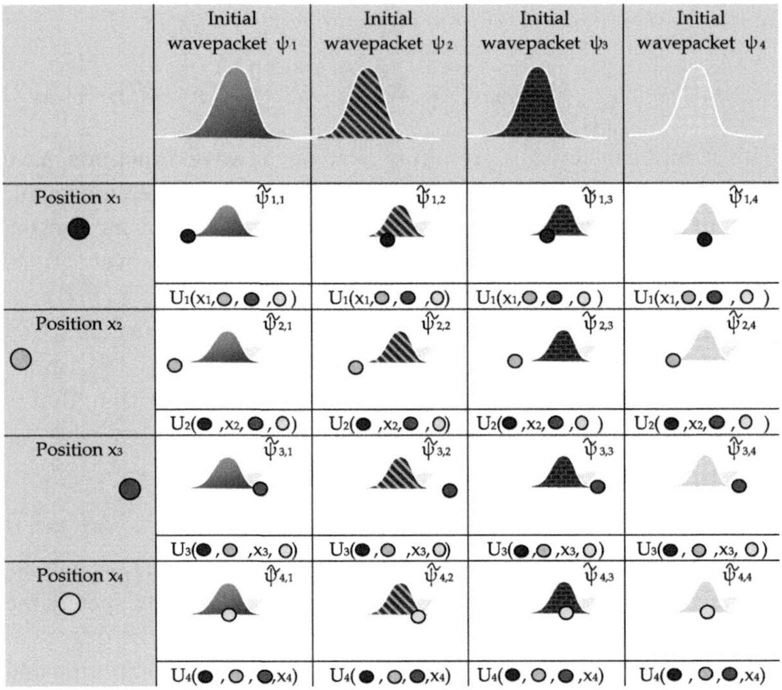

Figure 1.6 When Coulomb plus exchange interactions are considered, a 4·4 matrix of the single-particle wave functions, $\psi_{a,h}(x_a, t)$ (with $a = 1, \ldots, 4$ for the potential energies and $j = 1, \ldots, 4$ for the initial wave packets) is needed to compute (through a Slater determinant) the wave function associated with each Bohmian trajectory.

In fact, we can guarantee the expected symmetry of the quantum wave function by defining appropriately the phases. It is quite simple to realize that the following definition of the angles:

$$\exp\left(z_{a,\vec{p}(l)}(t)\right) = \prod_{k=1,k\neq a}^{N} \tilde{\Psi}_{a,\,p(l)_k}(x_k[t], t) \tag{1.111}$$

accomplishes the symmetry requirements. Finally, putting Eq. (1.111) into Eq. (1.110), we obtain the final wave function:

$$\phi(x_a, t) = C \sum_{l=1}^{N!} \tilde{\Psi}_{1,\,p(l)_1}(x_1[t], t) \ldots \tilde{\Psi}_{a,\,p(l)_a}(x_a, t) \ldots \times \tilde{\Psi}_{N,\,p(l)_N}(x_N[t], t) s(\vec{p}(l)) \tag{1.112}$$

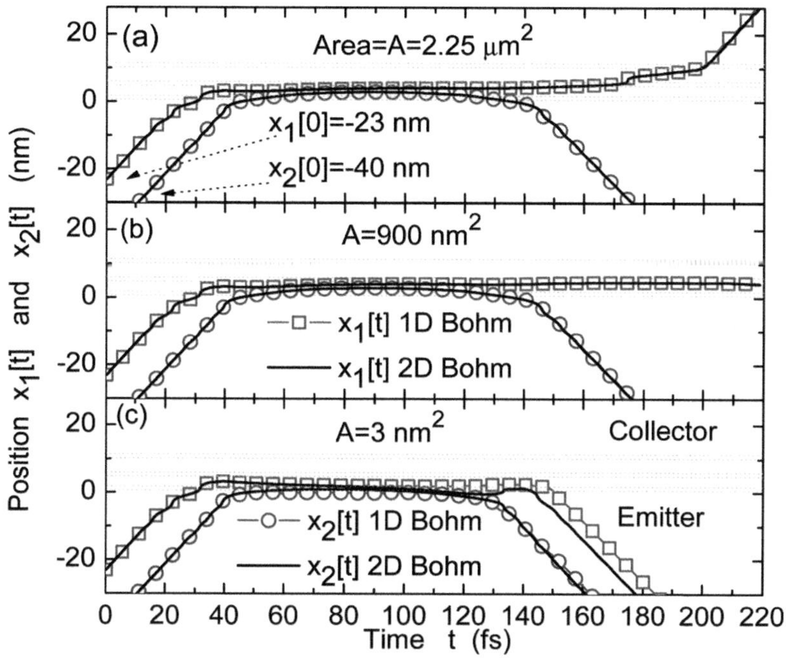

Figure 1.7 Two-particle Bohmian trajectories with Coulomb and exchange interactions computed from our 1D approach (symbols) and from 2D exact results (solid line) for three lateral areas. Reprinted with permission from [85]. Copyright 2007 American Physical Society.

This Eq. (1.112) can be simply computed from a (Slater) determinant of the $N \cdot N$ matrix, $\tilde{\psi}_{a,h}(x_a, t)$ with $a = 1, \ldots, 4$ and $j = 1, \ldots, 4$ for the initial wave packets, which are defined in Fig. 1.6.

The reader can wonder why we need N^2 different wave functions and not $N!$. The reason is because the term $U_a(x_a, \vec{x}_b[t], t)$ in Eq. (1.105) does not change when we interchange the positions of two of the trajectories in $\vec{x}_b[t]$ (only the interchange between $x_a[t]$ and a trajectory of $\vec{x}_b[t]$ becomes relevant). Figure 1.7 shows the excellent agreement between the exact Bohmian trajectories and those computed from this second algorithm. See details in [85].

In order to clarify the meaning of *indistinguishable particles* in the Bohmian language, in Fig. 1.8, we simulate Bohmian trajectories

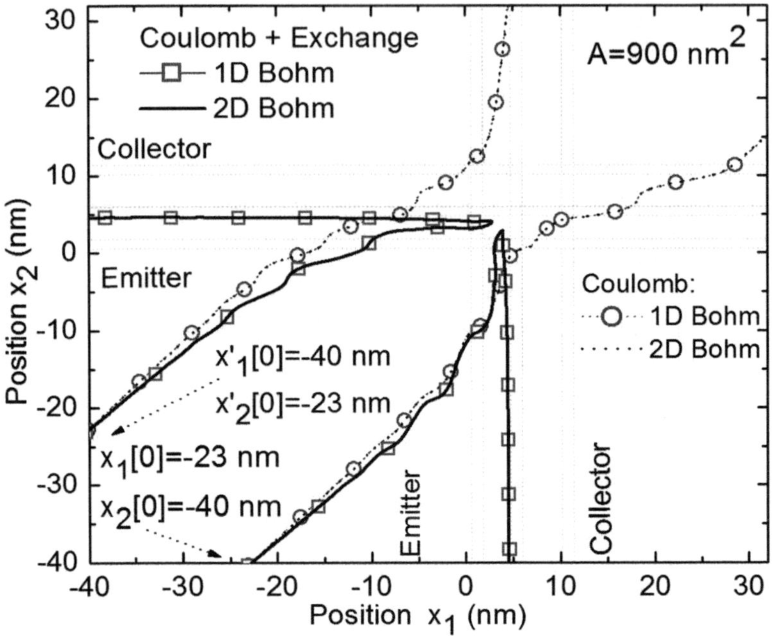

Figure 1.8 Two-particle Bohmian trajectories with Coulomb and exchange interactions computed from our 1D approach (symbols) and from 2D exact results (solid line) when initial positions are interchanged. Reprinted with permission from [85]. Copyright 2007 American Physical Society.

with the initial positions and with interchanged initial positions. By our own construction of the wave function of Eq. (1.112), the velocity has symmetry when we interchange particles, as discussed in Section 1.3.4. Therefore, although we can distinguish each electron by its own trajectories, one cannot discern between them in the final observable results. For example, the number of events where the first trajectory is transmitted and the other reflected is exactly identical to the events where the first is reflected, while the second is transmitted. On the contrary, as seen in Fig. 1.8, this *symmetry* is broken without the exchange interaction algorithm.

1.4 Bohmian Explanation of the Measurement Process

The Bohmian explanation of a quantum measurement process is, perhaps, the most attractive (and also ignored) feature of the Bohmian explanation of the quantum nature [38, 61–63, 86–88]. Let us start by noting that Bell disliked the word "measurement" [38]. He preferred the word "experiment" because "when it is said that something is measured it is difficult not to think of the result as referring to some pre-existing property of the object in question." On the contrary, in an experiment, it is natural to think that everything can change (time-evolve), because of the interactions. Once we admit that the system itself can be modified during the measurement, we can easily understand that the output of a measurement depends on the duration and strength of the measurement process, on the previous measurements, etc.[a]

1.4.1 The Measurement Problem

We have previously talked about the so-called *measurement problem* [89–91] but, what is in fact this measurement problem? It is a problem related to the application of the superposition law in quantum systems. Quantum states of a particle are represented by vectors in a Hilbert space such that linear combinations of them, for example a superposition of macroscopically distinguishable states, also correspond to valid states of the Hilbert space. However, such superposition of states is not always compatible with measurements [91, 92]. The measurement problem can be formulated as the impossibility for a physical quantum theory (in empirical agreement with experiments) to simultaneously satisfy the following three assumptions [90].

(1) The wave function always evolves deterministically according to the linear and unitary Schrödinger equation.

---

[a]Technically, it is stated that any quantum theory compatible with experiments has to be contextual. By construction, Bohmian mechanics is contextual since the measurement is just another type of interaction.

(2) A measurement always finds the physical system in a localized state, not in a superposition of macroscopically distinguishable states.
(3) The wave function is a complete description of a quantum system.

A theory that includes all three assumptions is not empirically compatible with the experimental results. Different physical theories are developed depending on which assumption is ignored [93]. The *measurement problem* appears because none of the proposed solutions fully satisfies the whole scientific community.

The first type of solutions argues that the unitary and linear evolution of the Schrödinger equation is not always valid (such solutions ignore the (1) assumption). For instance, in the instantaneous collapse theories [94] (like the GRW interpretation [95]), a new stochastic equation is used that breaks the superposition principle at the macroscopic level, while still keeping it at the microscopic one [94]. Another possibility is substituting the linear Schrödinger equation by a nonlinear collapse law only when a measurement is performed [96]. This is the well-known orthodox or Copenhagen solution. The dissatisfaction with the Copenhagen solution is that the theory does not clearly specify when, in which circumstances, the linear or nonlinear equation has to be used.

A second type of solution ignores the (2) assumption that a measurement always finds the physical system in a localized state. One then concludes that there are different worlds where different states of the superposition are found. This is the many worlds solution [97, 98] in which the famous Schrödinger's cat is found alive in one world and dead in another.

There is a final kind of solutions that assumes that the wave function alone does not provide a complete description of the quantum state, that is, it ignores the (3) assumption and includes additional elements (hidden variables) in the theory. The most spread of these approaches is Bohmian mechanics explained in this book. In a spatial superposition of two disjoint states for a single-particle system, only the one whose support contains the position of the particle becomes relevant for the dynamics. Notice that, in Bohmian mechanics, it is not mandatory to define which interactions are considered a measurement and which are not. All interactions (implying a measurement or not) are treated identically.

Let us discuss in more detail the differences between an orthodox and a Bohmian measurement. As we have already explained, in Bohmian mechanics, the evolution of the wave function is determined by the Schrödinger equation (when the appropriate Hamiltonian is used). This is not true for the orthodox formulation of the measurement because one linear equation (the Schrödinger equation) is used for nonmeasuring interactions, while a different nonlinear equation (the collapse law) is needed to tackle measuring interactions.

1.4.1.1 The orthodox measurement process

In any attempt to study a quantum system, one usually separates the quantum system itself and the rest. The separation is arbitrary, but mandatory in any practical computation since it is impossible to deal with all degrees of freedom of the system and the rest. This separation is also typical in the modeling of a quantum measurement. Let us notice that, in principle, the origin of the separation between the quantum system and the measuring apparatus is technical; it is not a direct consequence of the quantum theory that we select to describe the quantum world. However, it is true that, because of the ontology of the Copenhagen theory, the measurement process is explained by a proper separation between the system and the measuring apparatus (see Fig. 1.9b), while the Bohmian theory tends to consider system and apparatus altogether (see Fig. 1.9a).

The typical orthodox prediction of some experimental property of the quantum system is described through the use of a proper operator $\hat{G}$ whose *eigenvalues* give the possible outcomes of the measurement. When we measure a particular eigenvalue, the initial wave function is transformed into an *eigenfunction* of the operator. This is the so-called von Neumann (or projective) measurement. Thus, the time evolution of the wave function of a quantum system is governed by two (quite) different laws:

(1) The first dynamical evolution is given by the Schrödinger equation. This dynamical law is deterministic in the sense that

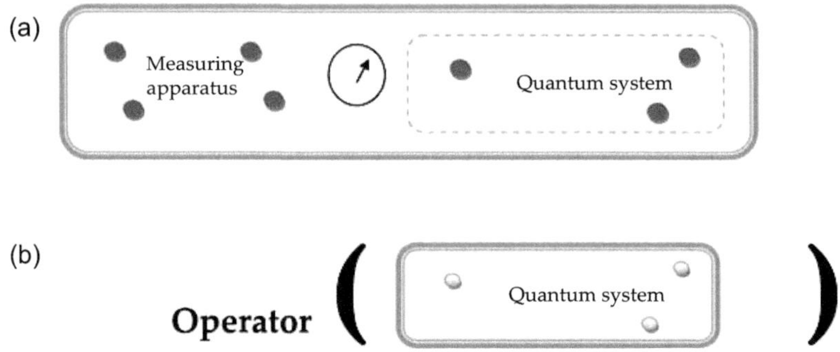

Figure 1.9 (a) The Bohmian measurement is better explained by assuming that the quantum system and the measuring apparatus are two interacting systems. (b) The typical orthodox measurement is better explained by assuming that only the quantum system is explicitly simulated, while the measuring apparatus is substituted by a proper operator acting on the wave function of the system.

the final wave function of the quantum system is perfectly determined when we know the initial wave function and the Hamiltonian of the quantum system.

(2) The second dynamical law is called the *collapse* of the wave function. The collapse is a process that occurs when the wave function interacts with a measuring apparatus. The initial wave function before the measurement is substituted by one of the *eigenstates* of the particular operator $\hat{G}$. Contrarily to the dynamical law given by the Schrödinger equation, the collapse is not deterministic, since the final wave function is randomly selected among the operator's eigenstates.

The duality in the equation of motions (linear or nonlinear) of a quantum system in the orthodox interpretation is certainly a persistent controversial issue. There are many scientist unsatisfied with this solution of the *measurement problem* [38]. As repeatedly stressed by Bell, the orthodox theory is unprofessional because it does not explain with total accuracy which parts of the entire quantum system form the measuring apparatus and which the quantum system itself. It is even not clear if the measuring apparatus needs some kind of human activity (*with a PhD?* [38]) to be accepted as a proper measuring apparatus. The problematical way in which measurement is treated in the orthodox formulation has been stressed by Bell [20]:

> The concept of "measurement" becomes so fuzzy on reflection that it is quite surprising to have it appearing in physical theory at the most fundamental level. Less surprising perhaps is that mathematicians, who need only simple axioms about otherwise undefined objects, have been able to write extensive works on quantum measurement theory-which experimental physicists do not find it necessary to read. [...] Does not any analysis of measurement require concepts more fundamental than measurement? And should not the fundamental theory be about these more fundamental concepts?

In short, Bell argued that the separation between the quantum system and the measuring apparatus in Fig. 1.9b is arbitrary. For practical computations, the *encapsulation* of the rest of the world (except the quantum system) into a mathematical entity called an operator, $\hat{G}$, is a very clever *trick* that allows for straightforward calculations of the results of quantum measurements [99, 100] without considering the rest of the world. However, conceptually, the role played by the operator $\hat{G}$ as a fundamental part of the orthodox theory is at the origin of the criticism against the orthodox solution of the measurement problem.

The reader can argue that, since we have said that the separation between the system and the rest is always arbitrary, one can always enlarge what we consider as a system (see Fig. 1.9a) also in the orthodox formulation of the measurement. This would not solve the measurement problem in the orthodox formulation. Certainly, we can enlarge the system in the orthodox quantum formulation of the measurement process by including as a system part of what was defined previously as an apparatus. The degrees of freedom of the new system will interact with the degrees of freedom of the old system through a linear equation of motion (Schrödinger equation). Therefore, we will always need *something* external to the enlarged system (some new operator $\hat{G}$) to provide the *collapse* of the wave function of the enlarged system.

1.4.1.2 The Bohmian measurement process

In Bohmian theory, the measurement process is treated just as any other quantum process of interacting particles and the previous measurement difficulties of the orthodox interpretation simply disappear. At least, from a conceptual point of view. There is no need to introduce operators [61–63, 99, 100]. Here, the entire quantum system is described by a many-particle trajectory plus a many-particle wave function (rather than a wave function alone). The wave function and the trajectory are both associated with the entire system, that is, the quantum system plus the measuring apparatus. Then, there is one dynamical law for the evolution of the wave function and another for the evolution of the trajectory:

(1) The Schrödinger equation (with the appropriate Hamiltonian of the quantum system plus the measuring apparatus) determines the time evolution of the wave function independently of whether a measurement process takes place or not.
(2) The time evolution of the particle is determined by the time integration of the Bohmian velocity independently of whether a measurement process takes place or not.

For example, imagine that some kind of pointer indicates the measured quantity; then, the particles (degrees of freedom) that form the pointer must be present in the Hamiltonian.[a] In this sense, in Bohmian mechanics, a physical quantum system must be described by a many-particle Hamiltonian. Once the Bohmian trajectories associated with the positions of the pointer are known, the value of the measurement is already predicted. We just need knowledge of the positions of the pointer particles. Again, Bell provided us with one of his didactic sentences [20]:

> In physics the only observations we must consider are position observations, if only the positions of instrument pointers. It is a great merit of the de Broglie–Bohm picture to force us to

[a]In modern electronic measuring devices, the pointer could be represented by a seven-segment array of light-emitting diode (LED) displays, each one with two possible states, ON and OFF. When electrons are present inside the PN interface of one of the LEDs, a radiative transition of the electrons from the conduction to the valence band produces light corresponding to the ON state. The absence of electrons is associated with an OFF state.

> consider this fact. If you make axioms, rather than definitions and theorems, about the "measurement" of anything else, then you commit redundancy and risk inconsistency.

Therefore, a proper modeling of a Bohmian measurement just needs the explicit consideration of the degrees of freedom of the pointer in the many-particle wave function and many-particle Bohmian trajectories that define the entire system. See Fig. 1.10. In particular, we have to introduce into the Hamiltonian the interaction of the particles of the pointer with the rest of the particles of the system. The back reaction of the measurement process on the wave function is trivially considered. Certainly, a Hamiltonian with or without the measuring apparatus will provide a different evolution of the quantum system wave function. For this reason, Bell preferred the word "experiment" instead of "measurement."

The Bohmian and the orthodox explanations of a measurement produce the same probabilistic predictions. However, the mathematical implementation of the equations of motion in each case is quite different. The orthodox quantum theory requires an operator to describe the effect of the measuring apparatus, but this operator is not needed in the Bohmian explanation.

To be fair, the simplicity of the Bohmian measurement is true at the conceptual level explained above. However, any practical attempt to directly implement the Bohmian protocol will have to face with the *many body problem* discussed in Section 1.3.1. A pointer in a measuring apparatus is a macroscopic object with a number of particles on the order of Avogadro's number ($\approx 10^{23}$ particles). Can we simulate such large number of particles and look at the "position of the instrument pointers" to get the measured valued as suggested by Bell? From a practical point of view, the answer is no because of the computational burden. Approximations reducing the simulated degrees of freedom, implying the artificial separation of the whole Bohmian system (as depicted in Fig. 1.9b) into systems plus apparatus (as depicted in Fig. 1.9a), will be mandatory.

1.4.2 Theory of the Bohmian Measurement Process

Now, we provide a mathematical demonstration of the equivalence between orthodox and Bohmian measurements for a typical ensemble over identical experiments. In the standard interpretation

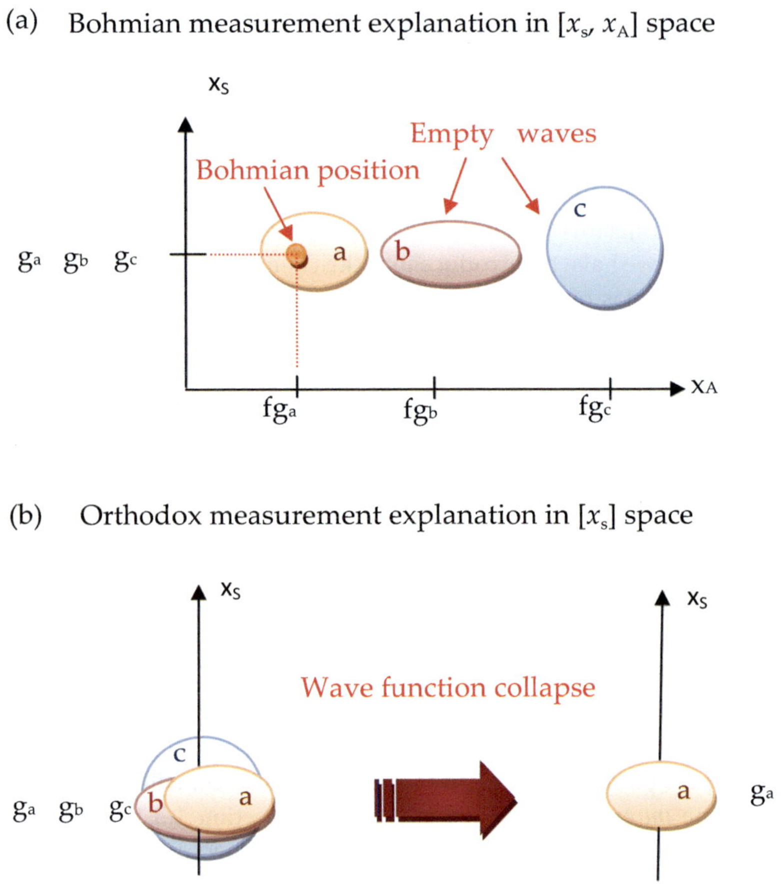

Figure 1.10 (a) Bohmian measurement in the $[x_S, x_A]$ space: From the nonoverlapping many-particle (system + apparatus) wave function, only the g_a part of the wave function where the trajectory is present is needed to compute the evolution of the Bohmian system. (b) Orthodox measurement in the $[x_S]$ space: The (system) wave function collapses into the g_a part when the measurement takes place.

of quantum theory, the von Neumann (projective) measurement process is defined in a particular quantum region, which is defined through the degrees of freedom $\vec{x}_S$. See Fig. 1.10b. The state of the quantum system in this particular region is determined through the wave function $\psi_S(\vec{x}_S, t)$. The process of measuring a particular magnitude is mathematically defined through an operator, for example $\hat{G}$, acting on the wave function. The possible outcomes of the measurement process correspond to one of the possible eigenvalues g of this operator that satisfy the equation $\hat{G}\psi_g(\vec{x}_S) = g\psi_g(\vec{x}_S)$ with $\psi_g(\vec{x}_S)$, an eigenvector of this operator. The set $\psi_g(\vec{x}_S)$ forms an orthonormal basis of the Hilbert space of the quantum system so that the arbitrary initial wave function can be written as:

$$\psi_S(\vec{x}_S, t) = \sum_g c_g(t)\psi_g(\vec{x}_S) \tag{1.113}$$

with $c_g(t) = \int \psi_S(\vec{x}_S, t)\psi_g^*(\vec{x}_S)$ a complex value with the only restriction $\sum_g |c_g(t)|^2 = 1$ in order to ensure that $\psi_S(\vec{x}_S, t)$ is well normalized. When measuring the eigenvalue g_a the total wave function $\psi_S(\vec{x}_S, t)$ collapses into $\psi_{g_a}(\vec{x}_S)$. Then, the probability of obtaining g_a in the measuring apparatus is just $P_{g_a} = |c_{g_a}(t)|^2$. To avoid unnecessary complications, hereafter, we have assumed that the basis $\psi_g(\vec{x}_S)$ has no degeneracy.

In order to understand how this (von Neumann) measurement can be described by means of Bohmian mechanics, we use the Schrödinger equation plus a trajectory, defined both in the enlarged configuration space that includes the quantum system and the measuring apparatus. There is no privileged status for the *observer*, that is, the measuring apparatus, over the quantum system. This is the reason why some authors describe Bohmian theory under the title of "quantum mechanics without observers" [87].

Surprisingly, this idea was, somehow, anticipated in the book *Quantum Theory*, published in 1951 [21], by Bohm. Although in this book Bohm followed an orthodox presentation of quantum mechanics, he also provided a detailed and conceptually intricate treatment of the measurement process in Chapter 22. First, Bohm demanded in unequivocal terms a quantum theoretical treatment of the entire process of measurement within the Schrödinger equation. In particular, he argued that a good measuring apparatus will force the wave function to decouple into several nonoverlapping wave packets. Each one of these wave packets can be associated to one particular value of the measurement process. Second, he supplemented his measurement process by adding some obscure discussion about decoherence where only one of the several wave packets survives (as we mentioned at the end of Section 1.4.1.1, even after enlarging the system, one has to invoke somehow the *collapse* of the wave function). After 1951, when Bohm presented his "hidden variables" theory [2, 3], he kept the first part of his previous explanation of the measurement process and then provided a simple explanation about the final selection of the available wave packets: the selected wave packet is the one that contains the trajectory,[a] see Fig. 1.10a.

Our mathematical demonstration on how Bohmian mechanics explains the (von Neumann) measurement follows [31, 61, 63, 86, 99]. In order to be able to define the measurement process in the Bohmian formalism, apart from the degrees of freedom $\vec{x}_S$, we need the degrees of freedom of the positions $\vec{x}_A$ of the pointer belonging to a measuring apparatus.[b] Thus, we define a total wave function $\Phi_T(\vec{x}_S, \vec{x}_A, t)$ in a larger configuration space that includes the quantum region plus the measuring region, $\{\vec{x}_S, \vec{x}_A\}$. According to the Bohmian postulates, we do also select a trajectory $\{\vec{x}_S[t], \vec{x}_A[t]\}$ in this larger configuration space. The (Bohmian) time evolution of the total wave function and the total trajectory is what we need to explain the (von Neumann) measurement process described earlier.

There are some necessary conditions that the time-evolution of the entire Bohmian system has to satisfy to say that a measuring apparatus is able to correctly determine the eigenvalues g. First, the pointer positions $\vec{x}_A[t]$ of such an apparatus have to be restricted to a particular region, $\vec{x}_A[t] \in S_{g_1}$, every time that the quantum system is in the eigenstate $\psi_{g_1}(\vec{x}_S)$. Let us define $\Phi_{g_1}(\vec{x}_S, \vec{x}_A, t)$ as the total wave function that fits with the property that any experiment whose

[a] Very roughly speaking, Bohmian mechanics avoids the need to select randomly the final wave packet because the initial position of the trajectory has already selected the final wave packet at the beginning.

[b] In fact, we would have to include all others degrees of freedom of the "environment," $\vec{x}_E$, needed to be able to assume that the considered entire system is a closed one described by a pure (not mixed) state. In addition, as mentioned in Section 1.4.1.2, we emphasize that a pointer in a measuring apparatus is a macroscopic object with a number of particles on the order of Avogadro's number, not the single degree of freedom $\vec{x}_A[t]$ used here. A single degree of freedom cannot be considered a pointer because it is not a macroscopic object (we do not *see* $\vec{x}_A[t]$). In any case and without loss of generality, we perform here a conceptual demonstration with a single degree of freedom $\vec{x}_A[t]$ for the pointer.

quantum system is described by $\psi_{g_1}(\vec{x}_S)$ implies that the pointer points in the particular region, $\vec{x}_A[t] \in S_{g_1}$. A second necessary condition for a good measuring apparatus of the eigenvalue g is that $S_{g1} \cap S_{g2} = 0$ when we measure. We have defined the restricted region allowed by the pointer positions associated with a second eigenstate, $\psi_{g_2}(\vec{x}_S)$, as $\vec{x}_A[t] \in S_{g_2}$. This implies that the states $\Phi_{g_1}(\vec{x}_S, \vec{x}_A, t)$ and $\Phi_{g_2}(\vec{x}_S, \vec{x}_A, t)$ do not overlap in the larger configuration space during the measurement.

Since the eigenstates $\psi_g(\vec{x}_S)$ form a complete basis, we can decomposed any function, and in particular each $\Phi_g(\vec{x}_S, \vec{x}_A, t)$, into the following sum:

$$\Phi_g(\vec{x}_S, \vec{x}_A, t) = \sum_{g'} f_{g'}(\vec{x}_A, t)\ \psi_{g'}(\vec{x}_S), \tag{1.114}$$

with $f_g(\vec{x}_A, t) = \int \Phi_g(\vec{x}_S, \vec{x}_A, t)\psi_g^*(\vec{x}_S)d\vec{x}_S$. However, from our previous discussion about the properties of a "good" von Neumann measurement apparatus, $\Phi_g(\vec{x}_S, \vec{x}_A, t)$ cannot be a sum over different eigenstates $\psi_g(\vec{x}_S)$ in Eq. (1.114) because then the measuring apparatus would provide the same pointer position $\vec{x}_A[t] \in S_g$ for different eigenstates. Thus, during the time of the measurement, the only good decomposition for $\Phi_g(\vec{x}_S, \vec{x}_A, t)$ is:

$$\Phi_g(\vec{x}_S, \vec{x}_A, t) = f_g(\vec{x}_A, t)\ \psi_g(\vec{x}_S) \tag{1.115}$$

We emphasize that $f_g(\vec{x}_A, t)$ is a normalized function because $\Phi_g(\vec{x}_S, \vec{x}_A, t)$ and $\psi_g(\vec{x}_S)$ are also normalized functions in their respective configurations spaces. By its own construction, $f_{g_1}(\vec{x}_A, t) \cap f_{g_2}(\vec{x}_A, t) = 0$ during the measuring time. Thus, even if $\psi_{g_1}(\vec{x}_S)$ and $\psi_{g_2}(\vec{x}_S)$ overlap, the states $\Phi_{g_1}(\vec{x}_S, \vec{x}_A, t)$ and $\Phi_{g_1}(\vec{x}_S, \vec{x}_A, t)$ do not overlap in the larger configuration space. See Fig. 1.10a. Thus, we can ensure that a general function in the quantum system, Eq. (1.113), can be rewritten in the whole, that is, quantum plus measuring, configuration space associated to a good measuring apparatus as:

$$\Phi_T(\vec{x}_S, \vec{x}_A, t) = \sum_{g} c_g(t)\ \ f_g(\vec{x}_A, t)\ \psi_g(\vec{x}_S) \tag{1.116}$$

In summary, during the time of measurement, the only total wave functions that can *live* in the entire quantum system that includes a good (in this case a projective or Von-Neumann type) measuring apparatus of the eigenvalues g are the ones written in Eq. (1.116).

An example of such wave functions is depicted in Fig. 1.10a. It is important to notice that Eq. (1.116) implies no restriction on the wave function $\psi_S(\vec{x}_S, t)$ but only on the total wave function $\Phi_T(\vec{x}_S, \vec{x}_A, t)$. If these restrictions are not respected, we can find other types of total wave functions in the configuration space $\{\vec{x}_S, \vec{x}_A\}$, but they would be incompatible with stating that we have an apparatus that is able to measure the eigenvalue g with certainty at time t.

After this discussion, we can show quite trivially how the von Neumann measurement is exactly reproduced with Bohmian mechanics. As we have mentioned, apart from the wave function in Eq. (1.116), we have to select an initial trajectory $\{\vec{x}_S[0], \vec{x}_A[0]\}$. Such a trajectory will evolve with the total wave function, and during the measurement, the particle trajectory $\{\vec{x}_S[t], \vec{x}_A[t]\}$ will be situated in only one of the nonoverlapping wave packets of Eq. (1.116), for example $f_{g_a}(\vec{x}_A, t)\psi_{g_a}(\vec{x}_S)$ as depicted in Fig. 1.10a. Thus, the pointer positions will be situated in $\vec{x}_A[t] \in S_{g_a}$ and we will conclude with certainty that the eigenvalue of the quantum system is g_a. In addition, the subsequent evolution of this trajectory can be computed from $f_{g_a}(\vec{x}_A, t)\psi_{g_a}(\vec{x}_S)$ alone. In other words, we do not need the entire wave function Eq. (1.116) because the particle velocity can be computed from $f_{g_a}(\vec{x}_A, t)\psi_{g_a}(\vec{x}_S)$. The rest of circles of Fig. 1.10a are empty waves that do not overlap with $f_{g_a}(\vec{x}_A, t)\psi_{g_a}(\vec{x}_S)$ so that they have no effect on the velocity of the Bohmian particle.[a] This is the simple explanation of how the complicated orthodox collapse is interpreted within Bohmian mechanics. This discussion is depicted in Fig. 1.10.

We have mentioned before that the von Neumann measurement predicts that the probability of finding g_a is $P_{g_a} = |c_{g_a}(t)|^2$. We can easily recover this result from our Bohmian picture. This probability is obtained after the repetition of the same experiment, with the same $\Phi_T(\vec{x}_S, \vec{x}_A, t)$, many times. The second (quantum equilibrium) postulate of Bohmian mechanics (see Section 1.3.5) ensures that the modulus of the wave function will be reproduced by the

[a]Let us remind that, in a realistic measuring apparatus, the number of particles of the pointer is around the Avogadro's number. Therefore, the wave functions $f_g(\vec{x}_A, t)\psi_g(\vec{x}_S)$ *live* in a $N = 10^{23}$ configuration space where the possibility of a coincidence of two empty waves is almost zero.

trajectories when we repeat the same experiment many times. Then, the probability of finding the eigenvalue g_a, that is, the eigenstate $\psi_{g_a}(\vec{x}_S)$, is just:

$$P_{g_a} = \frac{\iint |c_{g_a}(t) f_{g_a}(\vec{x}_A, t)\psi_{g_a}(\vec{x}_S)|^2 d\vec{x}_S d\vec{x}_A}{\iint \sum_g |c_g(t) f_g(\vec{x}_A, t)\psi_g(\vec{x}_S)|^2 d\vec{x}_S d\vec{x}_A}$$
$$= \frac{|c_{g_a}(t)|^2}{\sum_g |c_g(t)|^2} = |c_{g_a}(t)|^2 \qquad (1.117)$$

where we have explicitly used the condition that the each $\Phi_{g_1}(\vec{x}_S, \vec{x}_A, t)$ does not overlap with the others for the computation of the modulus of the total wave function, $|\Phi_T(\vec{x}_S, \vec{x}_A, t)|^2$, in the denominator. Additionally, we use $\int |\psi_g(\vec{x}_S)|^2 d\vec{x}_S = 1$, $\int |f_g(\vec{x}_A, t)|^2 d\vec{x}_A = 1$, and $\sum_g |c_g(t)|^2 = 1$. Thus, the same probabilistic predictions are obtained from Bohmian and orthodox quantum formulations when a von Neumann measurement is performed.

We want to enlarge the explanation of the role played by the empty waves belonging to the wave packet of Eq. (1.116), which do not contain the particle. In principle, one can argue that such empty waves can evolve and, in later times, overlap with the original wave that contains the particle. If we are interested in doing subsequent (i.e., two times) quantum measurements, a good measuring apparatus has to avoid these "spurious" overlaps. This can be understood as an additional condition for qualifying our measuring apparatus as a good apparatus. In the examples of the subsequent subsections 1.4.2.1 and 1.4.2.2, we will realize that the assumption of nonoverlapping between empty and full waves can be very reasonable in many practical implementations of the measuring apparatus. In fact, in a real system with a very large number of particles associated to the apparatus (or the environment), the probability that such overlap occurs in the configuration space of all the particles is almost null. See a discussion on page 79 in Ref. [86]. If the empty wave packets later coincide (by putting some reflected mirrors for example), then, the apparatus described above would not be a good apparatus to measure eigenvalues. On the contrary, this apparatus (with the mirrors) would be a very interesting/enlightening design to test quantum interference effects with waves and particles. The double-slit experiment is the most famous example of interference effects between empty and full wave packets.

Finally, let us mention that we have only considered von Neumann (i.e., projective) measurements. Other types of measurements are also possible, which do not collapse into an eigenstate (for example, the generalized or weak measurements). Obviously, such measurements can also be explained within Bohmian mechanics with an extension of the simple ideas for projective measurements discussed here [61, 86, 99].

1.4.2.1 Example: Bohmian measurement of the momentum

Let us explain the momentum measurement of the stationary state of an electron initially in an energy eigenstate of a square well of size L. For example, $\psi_n(x) = C\sin(n\pi x/L)$ within the well and zero elsewhere, C being the normalization constant and n an integer (denoting the vibrational state). Since the wave function is real, we obtain $p = \partial S(x)/\partial x = 0$. Thus, the Bohmian particle is at rest, meaning that its velocity and kinetic energy are zero. However, we know that for a high enough value of n, the previous wave function $\psi(x)$ can be roughly approximated by a sum of two momentum eigenstates with eigenvalues $\pm n\hbar/L$. Therefore, when we do a (Copenhagen) momentum measurement, we will obtain an outcome $\pm n\hbar/L$ and the previous wave function will *collapse* into one of the two momentum eigenstates.

We have shown that the Bohmian measurement has to provide the same eigenvalues and eigenstates. How? The answer is that according to the Bohmian measurement, we have to specify in detail how the measurement process develops; which is the Hamiltonian that provokes the measurement.

A measurement of the momentum, for example, can be undertaken by removing the walls that form the well and detecting the electron somewhere in a screen far from the initial walls. The time interval between removing the walls and detecting the particles will allow us a computation of the "electron velocity." The time-dependent process of removing the walls will imply that the stationary wave function will evolve into two time-dependent wave packets moving on opposite directions, which will become completely separated in space. The particle will end in one wave packet or the other with a momentum very close to $\pm n\hbar/L$, the sign depending on which wave packet the initial position of the particle enters [3]. Finally, the electron will be detected far from the walls after the appropriate interval of time. For this particular experiment, we are sure that the empty wave will not affect the wave that carries the particle. As mentioned earlier, the relevant point is that we have been able to measure the same eigenstates and eigenvalues predicted by the orthodox interpretation without invoking the *collapse* of the wave function as a fundamental law.

In order to understand the Bohmian explanation of this experiment, it is important to discuss the roles played by the kinetic and quantum potentials. The initial Bohmian kinetic energy is zero, while the initial quantum potential is $Q(x, t) = (n\hbar/L)^2/(2m)$ for the initial state $\psi(x) = C\ \sin(n\pi x/L)$. On the contrary, the final energy of the system appears in the form of kinetic energy of electrons $(n\hbar/L)^2/(2m)$, while the final quantum potential from Eq. (1.6) is negligible $Q(x, t) \sim 0$. See problem 10.

The reader will realize that we have not follow all the indications of the previous section. In principle, for a formal Bohmian discussion of measurement, we would have to discuss the wave function and particles of the whole quantum plus measuring apparatus system. However, in this section we have avoided the wave function and trajectory related to the measuring apparatus. In this example, the screen is the measuring apparatus of the position of the system's particle. The part of the measuring process involving the screen has not been explicitly described in our example (we could obviously explain why a proper coupling of the degrees of freedom of the system and the screen will imply that after impinging with the screen, the system wave function becomes a wave packet highly localized around the position of the measurement, but we have not done it because it is quite obvious).

This example clearly shows that what we get from a measurement depends on the measurement itself. We have explained an experiment to measure the velocity of the particle. In fact, the Bohmian particle was at rest (with zero velocity) before we start the measurement. However, because of the type of measurement that we have designed, the Bohmian particle acquires a nonzero velocity. What we get from the measurement is exactly this nonzero value of the velocity, but this is not the pre-existing velocity of the particle before the measurement. Contrarily to a classical measurement, in the quantum world, in general, we cannot get information of a quantum system from a measurement without providing some type of perturbation on it. This property of quantum phenomena is named contextuality.

1.4.2.2 Example: Sequential Bohmian measurement of the transmitted and reflected particles

We provide another didactic example of how Bohmian mechanics explains sequential measurements. The wave packets in Fig. 1.11 represent the solution of the (unitary) Schrödinger equation for a wave packet incident upon a tunneling barrier, at three different times. The initial wave packet (with norm equal to one) is divided into a transmitted plus a reflected wave packet. According to the Copenhagen interpretation, when the system is measured at time t_1, a nonunitary evolution appears in the wave function. Thus, randomly, for example the reflected wave packet disappears. Only the transmitted wave packet describes the electron at time t_1. Then, when the system is measured again at t_2, the electron is only represented by the transmitted wave packet.

Alternatively, the same evolution can be explained with Bohmian mechanics. The initial position of the Bohmian trajectory is selected according to the quantum equilibrium hypothesis at the initial time t_0. Let us assume that we select an initial position that corresponds to a trajectory that is finally transmitted. Then, at times t_1 and t_2, the evolution of the trajectory is only determined by the transmitted wave packet because the reflected wave packet is an "empty wave" that has no effect on the evolution of the trajectory. Here, we have implicitly assumed that the eigenstates of the measuring apparatus are transmitted wave packets for positive currents, and reflected ones for negative currents.

As expected, the probability of measuring, first, the particle as being transmitted at time t_1 and measuring, after, the same particle as being reflected at time t_2 is zero either with orthodox or Bohmian mechanics.

Again, the reader can argue that we have not enlarged the system including the measuring apparatus. Next, we provide a more

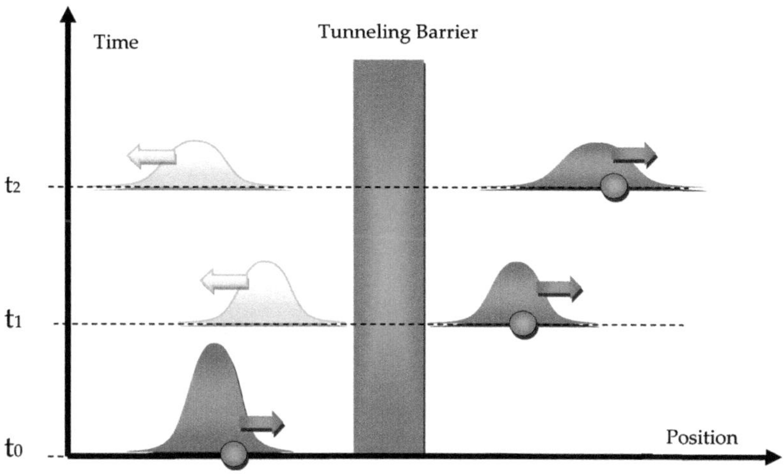

Figure 1.11 Schematic explanation of the ability of Bohmian mechanics to discuss the unitary and nonunitary evolution of a wave packet incident upon a tunneling barrier for three different times $t_0 < t_1 < t_2$.

formal description of the Bohmian measurement of the charge for a tunneling electron including the quantum system plus the measuring apparatus. The quantum system is defined by an electron labeled as $x_1[t]$ impinging on a tunneling barrier. Behind the barrier there is a measuring device named the "transmitted charge detector" and modeled as a single degree of freedom $x_2[t]$, which can detect the successful transmission of an electron. The dynamic of the system plus apparatus includes, first, an interaction of the electron with the potential barrier and, subsequently, an interaction with the transmitted charge detector. It is important to stress that both interactions are regarded at the very same level within Bohmian mechanics. The measurement interaction introduces a channelization of the wave function in the configuration space such that the desired property of the quantum system (here, whether the electron is reflected or transmitted) can be read off from the final position $x_2[t]$ of a particle, which plays the role of the pointer of the apparatus. The interaction between the electron and the pointer can be modeled as:

$$H_{int} = \lambda Q(x_1) P_{x_2} = -i\hbar\lambda Q(x_1)\frac{\partial}{\partial x_2}, \tag{1.118}$$

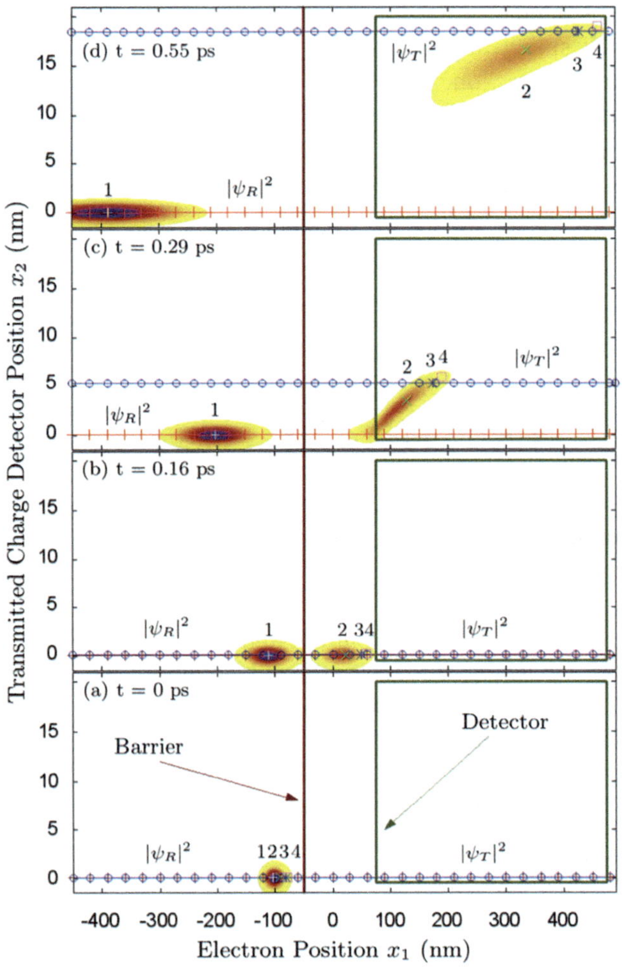

Figure 1.12 Time evolution of the squared modulus of $\Psi(x_1, x_2, t)$ at four different times. The configuration space region where the *transmitted charge detector* is present is indicated by a rectangle and the barrier by a solid line. The + line indicates the squared modulus of the conditional wave function $|\psi_R|^2 = |\Psi(x_1, x_2^{\alpha=1}[t], t)|^2$, while the $\odot$ line corresponds to $|\psi_T|^2 = |\Psi(x_1, x_2^{\alpha=3}[t], t)|^2$. Four trajectories $\{x_1^\alpha[t], x_2^\alpha[t]\}$ with different initial positions are presented with $\square$, $*$, $\times$ and $+$. The actual detector position associated with the reflected trajectory (+) with $\alpha = 1$ does not move because there is no interaction between this trajectory and the detector. Reprinted with permission from [101]. Copyright 2013 Springer Nature.

where $P_{x_2} = -i\hbar\partial/\partial x_2$ is the momentum operator of the detector and $\lambda = 50$ nm/ps is the interaction constant. $Q(x_1)$ is a function that is equal to zero when the electron is outside the detector, ($x_1 <$ 75 nm in Fig. 1.12), and is equal to one when the particle is inside the detector ($x_1 > 75$ nm).[a] In Fig. 1.12 the region in the configuration space in which $Q(x_1)$ is different from zero is represented by a rectangle in the right of the figure. The many-particle Schrödinger equation reads

$$i\hbar\frac{\partial\Psi(x_1, x_2, t)}{\partial t} = \left(-\frac{\hbar^2}{2m}\frac{\partial^2}{\partial x_1^2} - \frac{\hbar^2}{2M}\frac{\partial^2}{\partial x_2^2} + U(x_1) - i\hbar\lambda Q(x_1)\frac{\partial}{\partial x_2}\right) \times \Psi(x_1, x_2, t), \tag{1.119}$$

where m is the effective mass of the electron, M is the mass of the apparatus pointer and $U(x_1)$ is the external potential energy barrier seen as a vertical line in the middle of Fig. 1.12.

The main feature of a transmitted charge detector is that the center of mass of the wave function in the x_2 direction has to move if the electron is transmitted and it has to be at rest if the electron is reflected. We solve Eq. (1.119) numerically considering as initial wave function the products of two gaussian wave packets, i.e. $\Psi(x_1, x_2, 0) = \psi(x_1, 0)\phi(x_2, 0)$. All details of this simulation can be found in Ref. [101, 102]. In particular we are considering $M = 75000\ m$. In Fig. 1.12 the numerical solution of the squared modulus of $\Psi(x_1, x_2, t)$ is plotted at four different times. At the initial time $t = 0$, Fig. 1.12a, the entire wave function is at the left of the barrier. At a later time t_0 the wave function has split up into reflected and transmitted parts due to the barrier, see Fig. 1.12b. Then, because the electron has not yet arrived at the transmitted charge detector, the wave function has the following form:

$$\Psi(x_1, x_2, t_0) = [\psi_T(x_1, t_0) + \psi_R(x_1, t_0)]\,\phi(x_2, t_0). \tag{1.120}$$

After that, Figs. 1.12c and d, the interaction of the detector with the transmitted part of the wave function appears. For time $t > t_0$ the transmitted part of the wave function is shifted up in the x_2 direction while the reflected part does not move. The interaction with the apparatus thus produces two channels in the configuration

[a]The transition of $Q(x_1)$, from zero to one, is softly done in order to minimize the perturbation of the quantum system as explained in Ref. [101, 102].

space, one corresponding to the electron being transmitted and the other corresponding to the electron being reflected, getting an entangled superposition among the electron and the apparatus.

In Fig. 1.12 we also plot the actual positions of the system and detector $\{x_1[t], x_2[t]\}$ for four different possible initial positions $\{x_1[0], x_2[0]\}$, corresponding to four distinct runs of the experiment (labelled by $\alpha = 1, \ldots, 4$). Of the four possible evolutions shown, three show the electron being transmitted ($\alpha = 2, 3, 4$) and one being reflected ($\alpha = 1$). While the pointer position $x_2[t]$ does not move for the reflected particle, its evolution for the transmitted ones clearly shows a movement. In conclusion, looking at the *detector* position we can perfectly certify if the particle has been reflected ($x_1[t] < -50$ nm and $x_2[t] = 0$ nm) or transmitted ($x_1[t] > -50$ nm and $x_2[t] \approx 15$ nm). We hope the reader will realize how trivially we have been able to explain the measurement, using only a *channelized* (unitary) time-evolution of 2D wave function plus two Bohmian trajectories in the physical space, one trajectory for the system and another for the measuring apparatus.

Once we have solved the complete problem of the measurement in the (2D) configuration space, we can describe the same measurement in (1D) physical space with the help of the *conditional wave function*. The key point illustrated here is that the collapse of the one-particle wave function for the electron arises naturally and automatically in Bohmian mechanics. It is simply a consequence of slicing the unitary-evolving (2D) wave function Ψ along the (moving) line $x_2 = x_2[t]$, resulting $\psi_1(x_1, t) = \Psi(x_1, x_2[t], t)$. In Fig. 1.12 we have plotted two solid horizontal lines corresponding to a slice of the wave function at two different values of $x_2[t]$. In Fig. 1.13 we report the evolution of these (time-dependent) slices of the many-particle wave function, the *conditional wave function* for the electron, for the trajectories $\alpha = 1$ and $\alpha = 3$ from Fig. 1.12. We clearly see that if the particle is reflected, as it is the case for $\alpha = 1$, the position of the pointer does not change with time and, after the interaction with the detector has been performed, the electron's *conditional wave function* includes only a reflected part. See Figs. 1.13c and d. On the other hand, when the particle is transmitted (e.g., $\alpha = 3$), it is the reflected part of the *conditional wave function* which collapses away, leaving only the transmitted packet. See Figs. 1.13g

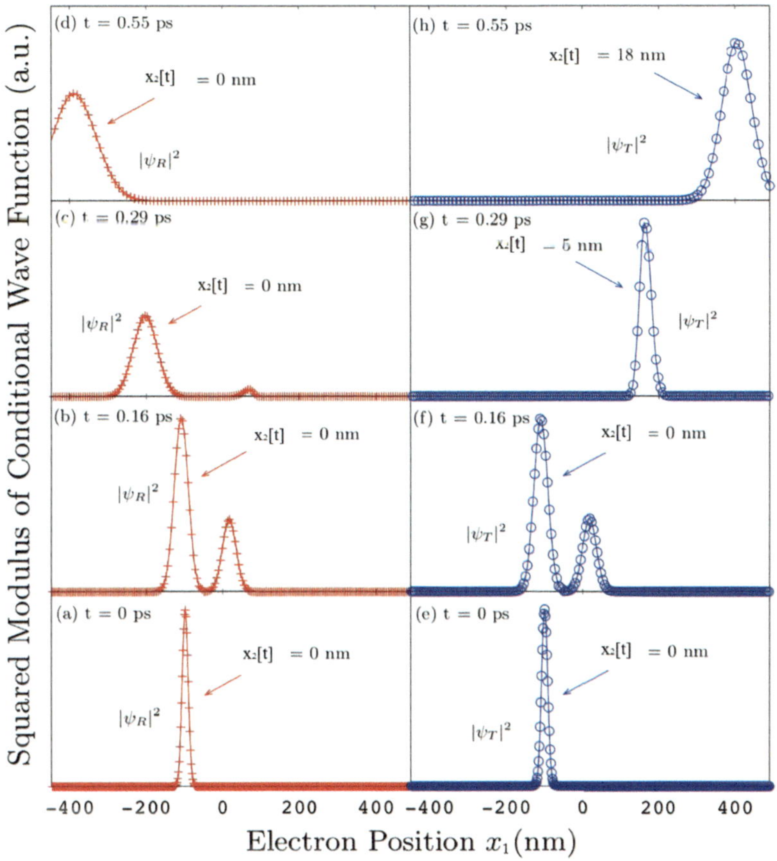

Figure 1.13 The + line in (a), (b), (c) and (d) is the time evolution of the squared modulus of the conditional wave function associated to the trajectory $\alpha = 1$ in Fig. 1.12, i.e., $\psi_R = |\Psi(x_1, x_2^{\alpha=1}[t], t)|$. The $\odot$ line in (e), (f), (g) and (h) is the squared modulus of the conditional wave function associated to the trajectory $\alpha = 3$ in Fig. 1.12. i.e. $\psi_T = |\Psi(x_1, x_2^{\alpha=3}[t], t)|$. The actual detector position $x_2[t]$ is plotted at each time in order to compare these results with those in Fig. 1.12. Reprinted with permission from [101]. Copyright 2013 Springer Nature.

and h. Note that the evolution of $\psi_1(x_1, t)$ (the electron's *conditional wave function*) is not unitary, even though the evolution of Ψ is.

While a wave function formulation of quantum mechanics provides only statistical information about the experimental results, with the help of the Bohmian trajectories, we have been able to recover the individual result of each experiment. In fact for each experiment the pointer of the detector is either moving (corresponding to a transmitted electron) or not (reflected electron), while an ensemble of repeated experiments (where the initial positions of the particles, both the electron $x_1[0]$ and the detector $x_2[0]$, are selected according to the squared modulus of the wave function at the initial time $|\Psi(x_1, x_2, 0)|^2$) reproduce the same statistical results.

Finally, by looking at figures 1.12 and 1.13, we conclude that the Bohmian theory provides a quite simple explanation on when, why and how the (conditional) wave function collapses. The Bohmian explanation of the measurement, with waves and particles, demystifies the orthodox postulate of the collapse law by explaining it as a trivial result of making a transition from a large configuration space (quantum system plus apparatus) to a smaller one (quantum system alone). To be fair, in spite of its extraordinary simplicity to explain the collapse law, the model explained here for measuring the transmitted charge has still an important approximation. We have assumed that the pointer is just one degree of freedom $x_2[t]$. As we have already mentioned, a realistic model for the pointer has to involve a very large number of particles (on the order of Avogadro's number $\approx 10^{23}$). Then, we will also be able to conclude that the macroscopic pointer has a classical behavior. However, such type of simulation involves facing again with the many body problem mentioned in Section 1.3.1. Some preliminary discussions on the classical-quantum transition for macroscopic objects can be found in Refs. [55, 103–106].

1.4.3 The Evaluation of a Mean Value in Terms of Hermitian Operators

1.4.3.1 Why Hermitian operators in Bohmian mechanics?

In the previous section, we have emphasized the ability of Bohmian mechanics to address quantum phenomena just from the Schrödinger equation and the equation of motion of the trajectories. The operators, which are an indispensable tool in the orthodox formulation of quantum mechanics to deal with the measurement process, become unnecessary in the Bohmian formulation. Once one realizes that operators are unnecessary, the conceptual difficulties associated with the wave function collapse of the orthodox formulation disappear (as we have seen in passing from Fig. 1.12 to Fig. 1.13). However, we have mentioned that the Bohmian measurement process explained above (with an idealized single-particle pointer) has limited practical utility. Including the measuring apparatus into the entire Hamiltonian is not an easy task, because it implies increasing the number of degrees of freedom. The exact Bohmian simulation of the measurement process is, most of the time, a quite difficult computational task.

In these circumstances, the use of a Hermitian operator acting only on the wave function of the quantum system with the ability of providing the outcomes of the measurement process without the explicit simulation of the measuring apparatus is very welcomed. Let us emphasize that we are talking only at the computational level. In simple words, operators are not needed in Bohmian mechanics (the operators are part of the Copenhagen ontology, but they are not part of the Bohmian ontology), but they are a very helpful mathematical trick in practical computational issues. These ideas are emphasized by Goldstein, Dürr, Teufel, and coworkers when they refer to the "naive realism about operators" [61, 99, 100]. Somehow, we are arguing here that the Copenhagen theory can be transformed into an effective model of the Bohmian theory where trajectories are deliberatively ignored for computational/practical proposals. Then, although we have a computational ignorance about the position of such electron, the idea of an (unmeasured) electron traversing a tunneling barrier would be fully supported by such effective quantum theory.

1.4.3.2 Mean value from the list of outcomes and their probabilities

We now want to reproduce the mean value, obtained from orthodox Hermitian operators, within Bohmian mechanics. Someone who dislikes what we have explained up to here about the Bohmian measurement can just focus on how Bohmian trajectories exactly reproduce the mean value of orthodox Hermitian operators. For simplicity, we will consider single-particle systems. The generalization to many-particle systems can be done straightforwardly.

First, according to the orthodox measurement, we assume that a particular experiment is described by the quantum system wave function $\psi(x, t)$ alone. After a large number, M, of repetitions of the same experiment, we can elaborate a large list $i = 1 \ldots, I$ of the possible outcomes $\{a_i\}$ with their occurrence probabilities $\{P_i\}$. From this data we can compute the mean value as:

$$\langle \hat{A} \rangle_\psi = \sum_{i=1}^{I} a_i P_i \tag{1.121}$$

This is the standard (classical or quantum) definition of the mean value.

1.4.3.3 Mean value from the wave function and the operators

According to the orthodox quantum theory, an observable is associated with a Hermitian operator $\hat{A}$ that describes the measurement process. Such an operator fulfills $\hat{A}|u_i\rangle = a_i|u_i\rangle$, $\{|u_i\rangle\}$ being its eigenstates. Therefore, $P_i = \langle\psi|u_i\rangle\langle u_i|\psi\rangle$ so that we can write:

$$\langle \hat{A} \rangle_\psi = \langle\psi| \left(\sum_{i=1}^{I} a_i |u_i\rangle\langle u_i| \right) |\psi\rangle = \langle\psi| \hat{A} |\psi\rangle \tag{1.122}$$

where we have used $a_i|u_i\rangle = \hat{A}|u_i\rangle$ and we have identified the expression $\sum_{i=1}^{I} |u_i\rangle\langle u_i|$ as the identity because of the orthonormality of the eigenstates. Therefore, we can compute the mean value of an ensemble of experiments from Eq. (1.122) by only knowing the wave function and the operator (without knowing either the eigenstates or the outcome values and their probabilities). Thus, we have been able to compute mean values within the orthodox quantum interpretation without directly mentioning the wave function collapse.

1.4.3.4 Mean value from Bohmian mechanics in the position representation

We can always write the Hermitian operator $\hat{A}$ and the mean value $\langle \hat{A} \rangle_\psi$ in the position representation. Then, the mean value of this operator over the wave function $\psi(x,t)$ is given by:

$$\langle \hat{A} \rangle_\psi = \int_{-\infty}^{\infty} \psi^*(x,t) \hat{A}\left(x, -i\hbar\frac{\partial}{\partial x}\right) \psi(x,t) dx \tag{1.123}$$

Alternatively, the same mean value can be computed from Bohmian mechanics by defining a spatial average of a "local" magnitude $A_B(x)$ weighted by $R^2(x,t)$:

$$\langle \hat{A} \rangle_\psi = \int_{-\infty}^{\infty} R^2(x,t) A_B(x) dx \tag{1.124}$$

In order to obtain the same value with Eqs. (1.123) and (1.124), we can easily identify the local mean value $A_B(x)$ as:

$$A_B(x) = Real\left(\left[\frac{\psi^*(x,t)\hat{A}\left(x, -i\hbar\frac{\partial}{\partial x}\right)\psi(x,t)}{\psi^*(x,t)\psi(x,t)}\right]_{\psi(x,t)=R(x,t)e^{i\frac{S(x,t)}{\hbar}}}\right) \tag{1.125}$$

We take only the real part, $Real()$, because we know that the mean value must be real, and Eq. (1.125) without $Real()$ could take complex values.

1.4.3.5 Mean value from Bohmian trajectories

For practical purposes, we will compute the mean value using Eq. (1.124) with a large $j = 1, \ldots, M$ number of Bohmian trajectories with different initial positions. We will select the initial position $x^j[t_0]$ of the Bohmian trajectories according to the second postulate. Therefore, we can use Eq. (1.50) to write $R^2(x,t)$ in Eq. (1.124). Finally, we obtain:

$$\langle \hat{A} \rangle_\psi = \lim_{M\to\infty} \frac{1}{M} \sum_{j=1}^{M} A_B(x^j[t]). \tag{1.126}$$

By construction, in the limit $M \to \infty$, the value of Eq. (1.126) is identical to the value of Eq. (1.124).

Now, we provide several examples of how some common mean values are computed from the orthodox quantum formalism and from Bohmian trajectories. First, we compute the mean value of the position:

$$\langle x \rangle_\psi = \int_{-\infty}^{\infty} \psi^*(x,t) x \psi(x,t) dx \tag{1.127}$$

with $x_B(x) = x$ so that:

$$\langle x \rangle_\psi = \int_{-\infty}^{\infty} R^2(x,t) x dx \tag{1.128}$$

Identically, the mean value of the momentum:

$$\langle p \rangle_\psi = \int_{-\infty}^{\infty} \psi^*(x,t) \left(-i\hbar \frac{\partial}{\partial x}\right) \psi(x,t) dx \tag{1.129}$$

with $p_B(x) = \partial S(x,t)/\partial x$:

$$\langle p \rangle_\psi = \int_{-\infty}^{\infty} R^2(x,t) \frac{\partial S(x,t)}{\partial x} dx \tag{1.130}$$

For the classical potential, we have:

$$\langle V \rangle_\psi = \int_{-\infty}^{\infty} \psi^*(x,t) V(x,t) \psi(x,t) dx \tag{1.131}$$

with $V_B(x,t) = V(x,t)$ so that:

$$\langle V \rangle_\psi = \int_{-\infty}^{\infty} R^2(x,t) V(x,t) dx \tag{1.132}$$

Now, we compute the mean value of the kinetic energy:

$$\langle K \rangle_\psi = \int_{-\infty}^{\infty} \psi^*(x,t) \left(-\frac{\hbar^2}{2m}\frac{\partial^2}{\partial x^2}\right) \psi(x,t) dx \tag{1.133}$$

It is important to notice that the local mean value of the kinetic energy takes into account the Bohmian kinetic energy plus the quantum potential. In particular, $K_B(x)$ can be obtained from the expression:

$$K_B(x) = Real\left(-\frac{R(x,t)e^{-iS(x,t)/\hbar} \frac{\hbar^2}{2m}\left(\frac{\partial}{\partial x}\right)^2 R(x,t)e^{iS(x,t)/\hbar}}{R^2(x,t)}\right) \tag{1.134}$$

The real part[a] of K_B is:

$$K_B = \frac{1}{2m}\left(\frac{\partial S(x,t)}{\partial x}\right)^2 + Q(x,t) \tag{1.135}$$

[a] It can be demonstrated quite easily that the imaginary part of Eq. (1.134) is equal to the spatial derivative of the current that becomes zero when integrated over all space. We use $J(x = \pm\infty, t) = 0$, which is always valid for wave functions that are normalized to unity, but it is not true for other types of wave functions such as plane waves.

so that, finally, we obtain the Bohmian expression of the mean kinetic energy of the ensemble of trajectories:

$$\langle K \rangle_{\psi} = \int_{-\infty}^{\infty} R^2(x,t) \left(\frac{1}{2m} \left(\frac{\partial S(x,t)}{\partial x} \right)^2 + Q(x,t) \right) dx \quad (1.136)$$

See problem 10 for a detailed calculation and the explanation of the meaning of the two terms.

Finally, we compute the mean value of the current density operator. First, let us remark that the probability density operator can be written as $|x\rangle\langle x|$ and its expected mean value is $\langle \psi|x\rangle\langle x|\psi\rangle = |\psi(x,t)|^2$ or, in the Bohmian language, $\langle \psi|x\rangle\langle x|\psi\rangle = R^2(x,t)$. The (Hermitian) current operator can be written as $\hat{J} = 1/(2m)(|x\rangle\langle x|\hat{p} + \hat{p}|x\rangle\langle x|)$. It can be easily demonstrated that:

$$\langle J \rangle_{\psi} = J(x,t) = v(x,t) R^2(x,t) = \lim_{M\to\infty} \frac{1}{M} \sum_{j=1}^{M} v(x^j[t]) \delta(x - x^j[t]). \quad (1.137)$$

The average value of the current density depends on the position, and it is equal to the average Bohmian velocity multiplied by the squared modulus of $R(x,t)$. At a particular position "x," this current is just the sum of all particles that reside around this position $x = x^j[t]$ at time t, multiplied by the Bohmian velocity at such position and time.

1.4.3.6 On the meaning of local Bohmian operators $A_B(x)$

It is important to emphasize that the local Bohmian operators $A_B(x)$ are not the eigenvalues of the operator $\hat{A}$. In general, the eigenvalues are not position dependent, while $A_B(x)$ are. The expression $A_B(x)$ is what is needed to compute the mean values of $\hat{A}$ with Eq. (1.124). This is its exact meaning. We have encapsulated all the effects of the measuring apparatus on the system into $A_B(x)$. Somehow, the use of $A_B(x)$ seems to emphasize that we are measuring a pre-existent property of the system, while we have learnt during this section devoted to the measurement that an interaction between system and apparatus during the time taken for doing the measurement (see Fig. 1.12) is mandatory.

There is a well-known example, mentioned in the previous Section 1.4.2.1, that emphasizes the differences between eigenvalues and local Bohmian operators. We consider the particle between two walls, separated by a distance L, whose wave function is $\psi(x) = C\ \sin(n\pi x/L)$ within the walls and zero elsewhere. Since the wave function is real, we obtain $p = \partial S(x)/\partial x = 0$. Thus, the local Bohmian momentum is $p_B(x) = \partial S(x,t)/\partial x = 0$, and the mean value in Eq. (1.130) is also zero. On the contrary, the wave function can be written as:

$$\psi(x) \approx C' \left(e^{-i\pi x/L} - e^{i\pi x/L}\right) \tag{1.138}$$

We know that $e^{i\ n\pi x/L}$ and $e^{-i\ n\pi x/L}$ are eigenstates of the momentum operator with eigenvalues $a_1 = n\hbar\pi/L$ and $a_2 = -n\hbar\pi/L$, respectively. Thus, the probabilities for each of these eigenvalues are identical, and the mean value of the momentum from Eq. (1.121) is again zero.

In conclusion, in general, $A_B(x)$ cannot be identified with a_i. However, by construction, the mean values computed from a_i and $A_B(x)$ are identical. A similar discussion about the evaluation of the mean value of the quantum power can be found in Ref. [107].

In this Section 1.4.3, we have only discussed how the mean values (i.e., the first moment) obtained from orthodox quantum mechanics can be exactly reproduced within Bohmian mechanics. More complicated higher moments (e.g., the second moment called the variance, whose positive square root is the standard deviation) computed from orthodox quantum mechanics can be equivalently reproduced within Bohmian mechanics, as we anticipated in Section 1.4.2.2 [61, 99, 100]. With the appropriate care on the contextuality of quantum phenomena, two-time correlations functions can also be computed in the same way. See also Ref. [108].

1.5 Concluding Remarks

In this chapter we have provided a fully comprehensive and didactic overview of the Bohmian formulation of quantum mechanics that can be useful for understanding the following chapters and also for any newcomers interested in using Bohmian mechanics in their daily research activity.

To be fair, most of the research on quantum phenomena is developed with the orthodox machinery, and few significative contributions have been performed with Bohmian mechanics. Many researchers ignore Bohmian mechanics, or they decide not to use it because they believe that someone demonstrated that there is something strange about Bohmian trajectories. In the introduction of this book, we illustrated this situation using Hans Christian Andersen's tale "The Emperor's New Clothes." The tale explains how a naked emperor strolling in procession on the streets believes he is well dressed, until a small child cries out, "The emperor is naked!" Children tales end happily with a moral that pretends to improve their behavior. Almost a century has passed, and we are still somewhere in the middle of the Bohmian tale. Bohm and Bell were the first to show to the scientific community that "Bohmian mechanics is a correct formulation of quantum phenomena." As in the tale and despite the whispering among the townsfolk, it seems that the idea that the orthodox formulation is the only licit tool to analyze quantum phenomena has gone on and the "orthodox emperor" continues to walk more proudly than ever.

It is still not clear what ending awaits the Bohmian tale and what moral it will teach us. In our opinion, Bohmian mechanics will only be able to have a happy ending once it can be explicitly demonstrated that a particular meaningful quantum problem is better calculated (in terms of computational abilities) or formulated within Bohmian mechanics than within its orthodox counterpart. Perhaps, some new experiments will allow us to elucidate which one better captures the reality underneath. As we have mentioned in the section "What Is a Quantum Theory" in the introduction, even if we could demonstrate in the future that either the Copenhagen or the Bohmian theories is wrong (or both), the practical utility of these theories in their range of validity would not dismiss.

We hope that a happy ending for the Bohmian theory is not very far and that this chapter (and the entire book) can contribute to shorten the time needed to arrive to it. If so, we might learn a moral similar to the one deduced from Andersen's tale: "It is useful to look for answers outside of the main stream." Sometimes the scientific revolutions are hidden there.

Acknowledgments

This work has been partially supported by FEDER and the "Ministerio de Ciencia e Innovación" through the Spanish Project TEC2015-67462-C2-1-R, the Generalitat de Catalunya (2014 SGR-384), the European Union's Horizon 2020 Grant agreement no: 604391 of the Flagship initiative "Graphene-Based Revolutions in ICT and Beyond" and European Union's Horizon 2020 research and innovation programme under the Marie Skłodowska-Curie grant agreement No 765426 (TeraApps). We want to thank Fabio L. Traversa and Guillermo Albareda for their help with the discussion of Bohmian computational algorithms. We also want to thank Alfonso Alarcón and Albert Benseny for their help in editing and reviewing much of this chapter and Enrique Colomés for the design and numerical simulations of the cover. Finally, we want to sincerely thank Ramón Corbalán, Sheldon Goldstein, Basile Hiley, Hrvoje Nikolić, Ward Struyve, Travis Norsen, Nino Zanghì, Roderick Tumulka, Damiano Marian, Adriano Orefice, and David Peat for very fruitful discussions.

1.6 Problems and Solutions

Problem 1 *Check that Eq. (1.22) is fulfilled for a free particle with* $V(x, t) = 0$. *We define the initial time as* $t_0 = 0$ *and the initial position as* $x_0 = 0$.

Solution. For a free particle, the Lagrangian is just the kinetic energy with a constant velocity, $v[t] = v_0$. We have $S_{\rm p} = K(v_0)t_f$ so that $S_{\rm p} = mx_f{}^2/(2t_f)$ and $S_{{\rm p}-\delta t} = mx_f{}^2/(2(t_f + \delta t))$. Notice that modifying the final time, without modifying the final position, means changing the initial velocity. If we use the Taylor expansion $1/(1+x) \simeq 1 - x$, then we find $S_{{\rm p}-\delta t} = mx_f{}^2\ (1 - \delta t/t_f)/(2t_f)$. Finally, $S_{{\rm p}-\delta t} - S_{\rm p} = -mx_f{}^2\ \delta t/(2t_f{}^2)$. This is the kinetic energy with a negative sign that, in our simple system, is also the Hamiltonian with a negative sign. Therefore, Eq. (1.22) is exactly recovered.

Problem 2 *Following the example of problem 1, test Eq. (1.26).*

Solution. The new value will be $S_{\mathrm{p}-\delta x} = m(x_f + \delta x)^2/(2t_f)$. If we neglect $(\delta x)^2$, we obtain $S_{\mathrm{p}-\delta x} = m{x_f}^2/(2t_f) + mx_f\delta x/t_f$. So that $S_{\mathrm{p}-\delta x} - S_{\mathrm{p}} = m\, v\delta\, x$. Finally, Eq. (1.26) is recovered.

Problem 3 *Solve Eq. (1.27) for the free particle case developed in problems 1 and 2 to find $S(x_0, t_0; x, t)$.*

Solution. Equation Eq. (1.27) can be rewritten as $\partial S(x, t)/\partial t = -H(x, t) = -E(v_0)$. The system is conservative, and the value of the Hamiltonian at any time is equal to the initial kinetic energy $E(v_0) = mv_0^2/2$. Therefore, we can transform Eq. (1.27) into $(\partial S(x, t)/\partial x)^2/(2m) = E(v_0)$ so that $\partial S/\partial x = \pm\sqrt{2mE(v_0)} = mv_0$. If we assume that the initial velocity (energy) is independent of the initial (final) position, then we easily integrate the partial derivatives of S to obtain finally, $S(x, t) = \pm mv_0 x - E(v_0)t$.

As expected, this action function corresponds to trajectories with constant velocity $\pm v_0$, that is, $x[t] = x_0 \pm v_0\ t$. The important point of our calculation here is that we have found all trajectories with an arbitrary initial position x_0 but a fixed initial velocity $\partial S/\partial x = \pm mv_0$. Let us emphasize that, although the classical function S is multivalued, if the initial velocity was $+v_0$, then we can use this knowledge and keep always the positive velocity for such trajectories.

Problem 4 *Show that the two real Eqs. (1.27) and (1.28) are equivalent to the following complex classical wave wave equation, also written in Eq. (1.29):*

$$i\hbar\frac{\partial\psi_{cl}(x, t)}{\partial t} = -\frac{\hbar^2}{2m}\frac{\partial^2\psi_{cl}(x, t)}{\partial x^2} + V(x, t)\psi_{cl}(x, t) + \frac{\hbar^2}{2m}\frac{\partial^2|\psi_{cl}(x, t)|/\partial x^2}{|\psi_{cl}(x, t)|}\psi_{cl}(x, t) \qquad (1.139)$$

Solution. We start the formal demonstration by introducing the polar form of the wave function, $\psi_{cl}(x, t) = R(x, t)\exp(iS(x, t)/\hbar)$, into Eq. (1.29). We have explicitly included the time in the scalar potential of Eq. (1.29) in order to compare it with the Schrödinger equation. Then, we evaluate the temporal and spatial derivatives:

$$\frac{\partial}{\partial t}\left(Re^{iS/\hbar}\right) = \left(\frac{\partial R}{\partial t} + \frac{i}{\hbar}R\frac{\partial S}{\partial t}\right)e^{iS/\hbar} \tag{1.140}$$

$$\frac{\partial^2}{\partial x^2}\left(Re^{iS/\hbar}\right) = \left(\frac{\partial^2 R}{\partial x^2} + 2\frac{i}{\hbar}\frac{\partial R}{\partial x}\frac{\partial S}{\partial x} + \frac{i}{\hbar}R\frac{\partial^2 S}{\partial x^2} + \frac{-1}{\hbar^2}R\left(\frac{\partial S}{\partial x}\right)^2\right)e^{iS/\hbar} \tag{1.141}$$

Now, if we introduce Eqs. (1.140) and (1.141) into Eq. (1.29) and eliminate the factor $e^{iS/\hbar}$, which is present in all terms, we obtain:

$$i\hbar\frac{\partial R}{\partial t} - R\frac{\partial S}{\partial t} = -\frac{\hbar^2}{2m}\frac{\partial^2 R}{\partial x^2} - i\frac{\hbar}{m}\frac{\partial R}{\partial x}\frac{\partial S}{\partial x} - i\frac{\hbar}{2m}R\frac{\partial^2 S}{\partial x^2} + \frac{1}{2m}R\left(\frac{\partial S}{\partial x}\right)^2 + V\ R + \frac{\hbar^2}{2m}\frac{\partial^2 R}{\partial x^2} \tag{1.142}$$

The real part of Eq. (1.142) is:

$$-R\frac{\partial S}{\partial t} = \frac{1}{2m}R\left(\frac{\partial S}{\partial x}\right)^2 + V\ R \tag{1.143}$$

which is exactly the Hamilton–Jacobi Eq. (1.27).

On the other hand, the imaginary part of equation Eq. (1.142) gives:

$$\frac{\partial R}{\partial t} = -\frac{1}{m}\frac{\partial R}{\partial x}\frac{\partial S}{\partial x} - \frac{1}{2m}R\frac{\partial^2 S}{\partial x^2} \tag{1.144}$$

Now, if we multiply both terms of Eq. (1.144) by $2R$, we directly recover the conservation law of Eq. (1.28). Notice how the constant $\hbar$ does not appear either in Eq. (1.143) or in Eq. (1.144).

Problem 5 *Show that $\psi_{cl}(x,t) = \exp(i(\pm px - Et)/\hbar)$ is a solution of the classical wave equation of Eq. (1.29) for the free-particle system with $V(x,t) = 0$.*

Solution. We compute the first two terms of the classical wave Eq. (1.29).

$$i\hbar\frac{\partial}{\partial t}\left(e^{iS/\hbar}\right) = i\hbar\frac{i}{\hbar}Ee^{iS/\hbar} \tag{1.145}$$

$$-\frac{\hbar^2}{2m}\frac{\partial^2}{\partial x^2}e^{iS/\hbar} = -\frac{\hbar^2}{2m}\left(\frac{i}{\hbar}\right)^2 p^2 e^{iS/\hbar} \tag{1.146}$$

Since the other two terms are zero, we obtain $E = p^2/(2\ m)$. Thus, $\psi_{cl}(x, t) = \exp(i(\pm px - Et)/\hbar)$ is a solution of the classical wave Eq. (1.29).

Surprisingly, a plane wave, apart from a free-particle quantum solution of the Schrödinger equation, is also a licit solution of a wave equation for classical mechanics. The fact that both classical and quantum solutions are identical is due to the fact that for the particular solution discussed here, $R - 1$, the last term of Eq. (1.29), that is, the quantum potential, is zero. Whether we want to model a free particle as a plane wave or as an ensemble of trajectories is just a matter of convenience.

Problem 6 *Show the following identity for the computation of the Bohmian velocity.*

$$v(x, t) = \frac{1}{m}\frac{\partial S(x, t)}{\partial x} = \frac{J(x, t)}{|\psi(x, t)|^2}$$

where $J(x, t)$ is defined by Eq. (1.34).

Solution. We start by performing the time derivative of $S(x, t)$:

$$\frac{\partial S(x, t)}{\partial x} = \hbar\frac{\partial}{\partial x}\tan^{-1}\left(\frac{\psi_i(x, t)}{\psi_r(x, t)}\right)$$

$$\frac{\partial S(x, t)}{\partial x} = \hbar\frac{\frac{\partial}{\partial x}\frac{\psi_i(x,t)}{\psi_r(x,t)}}{1 + \left(\frac{\psi_i(x,t)}{\psi_r(x,t)}\right)^2}$$

$$\frac{\partial S(x, t)}{\partial x} = \hbar\frac{\psi_r(x, t)\frac{\partial\psi_i(x,t)}{\partial x} - \psi_i(x, t)\frac{\partial\psi_r(x,t)}{\partial x}}{\psi_r^2(x, t) + \psi_i^2(x, t)}$$

Finally, using the following identity:

$$\psi(x, t)\frac{\partial\psi^*(x, t)}{\partial x} - \psi^*(x, t)\frac{\partial\psi(x, t)}{\partial x} = \frac{2}{i}\left(\psi_r(x, t)\frac{\partial\psi_i(x, t)}{\partial x} - \psi_i(x, t)\frac{\partial\psi_r(x, t)}{\partial x}\right)$$

we obtain that both definitions of the velocity of quantum trajectories are identical:

$$v(x, t) = \frac{1}{m}\frac{\partial S(x, t)}{\partial x} = \frac{J(x, t)}{|\psi(x, t)|^2}$$

where $J(x, t)$ is defined by Eq. (1.34).

Problem 7 *Show that the Bohmian velocity definition*

$$v(x,t) = \frac{J(x,t)}{|\psi(x,t)|^2}$$

given in Eq. (1.35) fulfills the continuity equation, Eq. (1.33):

$$\frac{\partial|\psi(x,t)|^2}{\partial t} + \frac{\partial J(x,t)}{\partial x} = 0$$

$J(x,t)$ being the current defined in Eq. (1.34):

$$J(x,t) = i\frac{\hbar}{2m}\left(\psi(x,t)\frac{\partial\psi^*(x,t)}{\partial x} - \psi^*(x,t)\frac{\partial\psi(x,t)}{\partial x}\right)$$

As a conclusion, we learn that a proper (i.e., quantum equilibrium) ensemble of Bohmian trajectories will reproduce $R(x,t)^2 = |\psi(x,t)|^2$ at any time.

Solution. We rewrite the continuity equation as:

$$\frac{\partial}{\partial t}\int_x^{x+\Delta x} R^2(x',t)dx' + J(x+\Delta x,t) - J(x,t) = 0$$

By using the Bohmian definition of the velocity, we rewrite the above equation as:

$$\frac{\partial}{\partial t}\int_x^{x+\Delta x} R^2(x',t)dx' = -R^2(x+\Delta x,t)v(x+\Delta x,t) + R^2(x,t)v(x,t)$$

Now, we make use of the ensemble of Bohmian trajectories by using Eq. (1.50) to write $R^2(x+\Delta x,t)$ and $R^2(x,t)$ as a sum of delta functions at $t = t_0$. Finally, we obtain:

$$\begin{aligned}\frac{\partial}{\partial t}\int_x^{x+\Delta x} R^2(x',t)dx' &= +\lim_{M\to\infty}\frac{1}{M}\sum_{i=1}^{M}\delta(x - x_i[t])v_i(t)\\ &\quad - \lim_{M\to\infty}\frac{1}{M}\sum_{i=1}^{M}\delta(x+\Delta x - x_i[t])v_i(t)\end{aligned} \tag{1.147}$$

We can rewrite δ as time derivatives of the Heaviside (or unit step) functions Θ:

$$\begin{aligned}\frac{\partial}{\partial t}\int_x^{x+\Delta x} R^2(x',t)dx' &= +\lim_{M\to\infty}\frac{1}{M}\sum_{i=1}^{M}\frac{\partial}{\partial t}\\ &\quad \times\left(\Theta(x_i[t]-x)\,\Theta(x+\Delta x - x_i[t])\right)\end{aligned} \tag{1.148}$$

Thus, counting the number of Bohmian i particles inside the region $[x, x + \Delta x]$ does exactly reproduce the square modulus of the wave function and its time-dependent variations. In conclusion, if we know that the ensemble of Bohmian trajectories reproduces $R(x, t_0)$ at the particular time t_0, then it will reproduce $R(x, t)$ at any other time:

$$R^2(x, t) = \lim_{M \to \infty} \frac{1}{M} \sum_{i=1}^{M} \delta(x - x_i[t])$$

This is precisely the reason why it is claimed that Bohmian mechanics exactly reproduces the position measurement of orthodox quantum mechanics. Since an ensemble of Bohmian trajectories reproduces $\psi(x, t)$ at any time, the ensemble of trajectories will also reproduce the mean value of other observables when they are written in the position representation.

Problem 8 *Demonstrate using Bohmian trajectories that the (average) dwell time $\langle \tau \rangle$ that a particle described by the wave function $\psi(\vec{r}, t)$ spends in a particular region $\vec{r} \in \Omega$ can be written in an orthodox language as:*

$$\langle \tau \rangle = \int_{-\infty}^{\infty} dt \int_{\Omega} dv |\psi(\vec{r}, t)|^2 \tag{1.149}$$

Solution. For a particular Bohmian trajectory $\vec{r}_k[t]$, the time spent in a particular region τ_k for this k-particle can be unambiguously written as:

$$\tau_k = \int_{t_k^i}^{t_k^o} dt = \int_{-\infty}^{\infty} dt \Theta(t - t_k^i) \Theta(t_k^o - t)$$

where t_k^i and t_k^o are the times when the electron enters and leaves the volume Ω, respectively. The function $\Theta(t)$ is a Heaviside function (or unit step function). We can relate the Heaviside function to the delta function δ as:

$$\Theta(t - t_k^i) \Theta(t_k^o - t) = \int_{\Omega} dv \delta(\vec{r} - \vec{r}_k[t])$$

In order to obtain the average value of the dwell time, $\langle \tau \rangle$, we have to make a sum over the ensemble $k = 1, \ldots, M$ of Bohmian trajectories:

$$\langle \tau \rangle = \lim_{M\to\infty} \frac{1}{M} \sum_{k=1}^{M} \tau_k = \lim_{M\to\infty} \frac{1}{M} \sum_{k=1}^{M} \int_{-\infty}^{\infty} dt \int_{\Omega} dv \delta(\vec{r} - \vec{r}_k[t])$$

Finally, by noting that the ensemble of Bohmian trajectories reproduces the modulus of the wave function:

$$|\psi(\vec{r}, t)|^2 = \lim_{M\to\infty} \frac{1}{M} \sum_{k=1}^{M} \delta(\vec{r} - \vec{r}_k[t])$$

we obtain the expected results, Eq. (1.149). See more information on tunneling times in [111, 112].

Problem 9 *Let us consider a single Bohmian trajectory* $x_k[t]$ *accounting for the quantum dynamics of a 1D particle in a potential* $V(x)$. *Show that the left probability density:*

$$P_L \equiv \int_{-\infty}^{x_k[t]} |\psi(x, t)|^2 \, dx$$

is constant, which, in turn, means that 1D Bohmian trajectories do not cross each other.

Solution. We are interested in showing that:

$$\frac{dP_L}{dt} = \frac{d}{dt} \int_{-\infty}^{x_k[t]} dx |\psi(x, t)|^2 = 0$$

We know that:

$$\frac{d}{dt} \int_{-\infty}^{x_k[t]} dx |\psi(x, t)|^2 = |\psi(x, t)|^2 \frac{dx_k[t]}{dt} + \int_{-\infty}^{x_k[t]} dx \frac{\partial |\psi(x, t)|^2}{\partial t} = 0$$

By the continuity equation, Eq. (1.33), with the current density in Eq. (1.34), we obtain:

$$\frac{d}{dt} P_L = |\psi(x, t)|^2 \frac{dx_k[t]}{dt} - \int_{-\infty}^{x_k[t]} dx \frac{\partial J(x, t)}{\partial x} = 0$$

We finally obtain $\frac{d}{dt} P_L = 0$, that is, P_L is a constant when the Bohmian trajectory moves according to:

$$\frac{dx_k[t]}{dt} = \frac{J(x_k[t], t)}{|\psi(x_k[t], t)|^2}$$

This is exactly the Bohmian velocity of Eq. (1.42).

Problem 10 *Show that the expected value of the kinetic energy, where the kinetic operator is given by* $\hat{T} = -(\hbar^2/2m)\nabla^2$*, yields:*

$$\langle \hat{T} \rangle = \langle Q \rangle + \frac{\langle p^2 \rangle}{2m} \tag{1.150}$$

where Q is the quantum potential. Equation (1.150) indicates that the quantum potential Q could be also interpreted as an additional quantum kinetic energy.

Solution. By writing the wave function in polar form, $\psi = Re^{iS/\hbar}$, we calculate the expression of the expected value of the kinetic energy by averaging over all the possible positions of the particle, that is, in a (infinite) volume V:

$$\begin{aligned}
\langle \hat{T} \rangle &= \int \psi^* \hat{T} \psi dV = \int \psi^* \left[-\frac{\hbar^2}{2m} \nabla^2 \psi \right] dV \\
&= \int Re^{-iS/\hbar} \left[-\frac{\hbar^2}{2m} \nabla^2 (Re^{iS/\hbar}) \right] dV \\
&= \int Re^{-iS/\hbar} \left[-\frac{\hbar^2}{2m} \vec{\nabla} \cdot \left(\vec{\nabla} Re^{iS/\hbar} + \frac{i}{\hbar} R \vec{\nabla} S e^{iS/\hbar} \right) \right] dV \\
&= \int R \left[-\frac{\hbar^2}{2m} \left(\nabla^2 R + \frac{2i}{\hbar} \vec{\nabla} R \cdot \vec{\nabla} S + \frac{i}{\hbar} R \nabla^2 S - \frac{1}{\hbar^2} R (\nabla S)^2 \right) \right] dV \\
&= \int R^2 \left(-\frac{\hbar^2}{2m} \frac{\nabla^2 R}{R} \right) dV - \frac{i\hbar}{2m} \int \vec{\nabla} \cdot \left(R^2 \vec{\nabla} S \right) dV \\
&\quad + \int R^2 \left(\frac{1}{2m} (\vec{\nabla} S)^2 \right) dV
\end{aligned} \tag{1.151}$$

The first and third terms in the r.h.s. of Eq. (1.151) are, respectively, $\langle Q \rangle$ and $\frac{\langle p^2 \rangle}{2m}$. The application of the divergence theorem on the integral in the second term, leads us to:

$$\int \vec{\nabla} \cdot \left(R^2 \vec{\nabla} S \right) dV = \oint \left(R^2 \vec{\nabla} S \right) \cdot \vec{n} dS = 0, \tag{1.152}$$

S being the surface boundary enclosing the volume of integration and $\vec{n}$ the outward pointing unit vector perpendicular to it. Equation (1.152) is zero since the population must be conserved, and there can be no flux of probability outside the (infinite) volume V.

Thus, we see that the expected value of the quantum kinetic energy operator corresponds to the sum of the expected values of the classical kinetic energy $\frac{\langle p^2 \rangle}{2m}$ and the quantum potential. This discussion is related to what we have explained in Section 1.4.2.1.

Problem 11 *Show that for a stationary state, $\psi(x,t) = \varphi(x)e^{-iEt/\hbar}$, that is, for an eigenstate of the Hamiltonian, it follows that:*

$$P^2(x) = 2m\,[E - V(x) - Q(x)] \tag{1.153}$$

where $P(x) = \partial S/\partial x$ is the Bohmian momentum and $E = -\partial S/\partial t$ the energy eigenvalue of the corresponding energy eigenstate. Note that Eq. (1.153) suggests that the semiclassical (i.e., WKB) approximation for the particle momentum given by $p(x) = \pm\sqrt{2m[E - V(x)]}$ is accurate at those positions where the quantum potential is relatively small.

Solution. For an eigenstate of the Hamiltonian, it follows that:

$$\frac{\partial R}{\partial t} = 0$$

$$\frac{\partial S}{\partial t} = -E$$

Then, by using the quantum Hamilton–Jacobi Eq. (1.39):

$$\frac{\partial S}{\partial t} = -V - \frac{1}{2m}\left(\frac{\partial S}{\partial x}\right)^2 - Q$$

with

$$Q = -\frac{\hbar^2}{2m}\frac{1}{R}\frac{\partial^2 R}{\partial x^2}$$

one obtains

$$P^2(x) = 2m\,[E - V(x) - Q(x)]$$

where we have used the Bohmian definition of the momentum $P(x) = \partial S/\partial x$.

Problem 12 *Recently, there have been several investigations where a quantum trajectory extension of Bohmian mechanics has been performed taking the quantum action S to be complex [124, 126]. Assuming the following expression for the wave function:*

$$\psi(x,t) = e^{i\frac{S(x,t)}{\hbar}}$$

derive the complex quantum Hamilton–Jacobi equation associated with it.

Then, assuming that $v(x, t) \equiv (\vec{\nabla} S(x, t))/m$ and taking the time derivate of this Hamilton–Jacobi equation, obtain the Newtonian-like equation of motion for the velocity:

$$\frac{dv[x_0, t]}{dt} = -\frac{\vec{\nabla} V}{m} + \frac{i\hbar}{2m}\nabla^2 v \qquad (1.154)$$

where the quantum features, such as nonlocality, are manifested in the second term of the r.h.s. of Eq. (1.154).

Solution. Casting $\psi(x, t) = e^{i\frac{S(x,t)}{\hbar}}$ into the time-dependent Schrödinger equation yields:

$$\frac{\partial S}{\partial t} + \frac{1}{2m}(\vec{\nabla} S)^2 + V = \frac{i\hbar}{2m}\nabla^2 S$$

where

$$Q_C \equiv -\frac{i\hbar}{2m}\nabla^2 S$$

is the "new" quantum potential. The first step to obtain the "new" Newtonian-like equation of motion is to take the spatial derivative of the complex quantum Hamilton–Jacobi equation:

$$m\left(\frac{\partial v}{\partial t} + v\vec{\nabla} v\right) - \frac{i\hbar}{2}\vec{\nabla}^2 v = -\vec{\nabla} V$$

Then, by identifying the total derivative of the velocity:

$$\frac{dv}{dt} = \frac{\partial v}{\partial t} + v\vec{\nabla} v$$

one finally arrives at

$$\frac{dv[x_0, t]}{dt} = -\frac{\vec{\nabla} V}{m} + \frac{i\hbar}{2m}\nabla^2 v$$

The terms at the r.h.s. can be identified as the classical and the quantum force acting on the particle, respectively.

Problem 13 *The Gross–Pitaevskii equation governs the dynamics of a Bose–Einstein condensate (BEC) in the so-called mean field approximation. In the zero temperature limit, it reads:*

$$i\hbar\frac{\partial \psi(\vec{r}, t)}{\partial t} = \left[-\frac{\hbar^2}{2m}\nabla^2 + V(\vec{r}, t) + g\,|\psi(\vec{r}, t)|^2\right]\psi(\vec{r}, t) \qquad (1.155)$$

where g is the nonlinear interaction between the atoms of the BEC, $V(\vec{r}, t)$ is the trapping potential, and the BEC wave function is *normalized to the number of atoms as* $\int d^3\vec{r}\,|\psi(\vec{r},t)|^2 = N$. *Note that for* $g = 0$, *one formally recovers the Schrödinger equation. Apply Madelung's formulation to the BEC by writing down its wave function as* $\psi(\vec{r},t) = \sqrt{n(\vec{r},t)}e^{i\theta(\vec{r},t)}$ *and obtain the continuity and Hamilton–Jacobi equations associated with Eq. (1.155).*

Solution. First of all, notice that following the usual nomenclature of the literature on this Eq. (1.155), $\psi(\vec{r},t) = \sqrt{n(\vec{r},t)}e^{i\theta(\vec{r},t)}$, here we use different functions for the role of $R(\vec{r},t)$ and $S(\vec{r},t)$ that are related, but not identical, to $= n(\vec{r},t)$ and $\theta(\vec{r},t)$, respectively. Following the same steps as in Section 1.2.3.4, it is straightforward to obtain:

$$m\frac{\partial \vec{v}}{\partial t} = -\vec{\nabla}\left(V + ng + \frac{1}{2}mv^2 - \frac{\hbar}{2m}\frac{1}{\sqrt{n}}\nabla^2\sqrt{n}\right) \quad (1.156)$$

$$\frac{\partial n}{\partial t} = -\vec{\nabla}(n\vec{v}) \quad (1.157)$$

where we have defined $\vec{v}(\vec{r},t) \equiv \frac{\hbar}{m}\vec{\nabla}\theta(\vec{r},t)$. Equation (1.156) is the classical Euler equation for an inviscid compressible fluid with an additional pressure term whose origin is purely quantum (the last term in this expression) that is very often called *quantum pressure*. These equations will be used later in Chapter 3.

Problem 14 *Use the many-particle Schrödinger Eq. (1.52) to deduce a many-particle local continuity equation.*

Solution. In order to find a local continuity equation, let us work with $\psi(\vec{x},t)$ and its complex conjugate $\psi^*(\vec{x},t)$. In particular, we can rewrite Eq. (1.30) as:

$$\psi^*(\vec{x},t)i\hbar\frac{\partial \psi(\vec{x},t)}{\partial t} = \psi^*(\vec{x},t)\left(\sum_{k=1}^{N} -\frac{\hbar^2}{2m}\frac{\partial^2}{\partial x_k^2}\psi(\vec{x},t)\right) + \psi^*(\vec{x},t)V(x,t)\psi(\vec{x},t)$$

and the complex conjugate as:

$$-\psi(\vec{x},t)i\hbar\frac{\partial \psi^*(\vec{x},t)}{\partial t} = \psi(\vec{x},t)\left(\sum_{k=1}^{N} -\frac{\hbar^2}{2m}\frac{\partial^2}{\partial x_k^2}\psi^*(\vec{x},t)\right) + \psi(\vec{x},t)V(x,t)\psi^*(\vec{x},t)$$

When the first equation is subtracted from the second, we obtain:

$$\frac{\partial|\psi(\vec{x},t)|^2}{\partial t}+\sum_{k=1}^{N} J_k(\vec{x},t)=0$$

where we have used:

$$-\psi^*(\vec{x},t)\frac{\hbar^2}{2m}\frac{\partial^2}{\partial x_k^2}\psi(\vec{x},t)+\psi(\vec{x},t)\frac{\hbar^2}{2m}\frac{\partial^2}{\partial x_k^2}\psi^*(\vec{x},t)$$
$$=\frac{\hbar^2}{2m}\frac{\partial}{\partial x_k}\left(\psi(\vec{x},t)\frac{\partial}{\partial x_k}\psi^*(\vec{x},t)-\psi^*(\vec{x},t)\frac{\partial}{\partial x_k}\psi(\vec{x},t)\right)$$

and defined:

$$J_k(\vec{x},t)=i\frac{\hbar}{2m}\left(\psi(\vec{x},t)\frac{\partial\psi^*(\vec{x},t)}{\partial x_k}-\psi^*(\vec{x},t)\frac{\partial\psi(\vec{x},t)}{\partial x_k}\right)$$

as the k-th component of the current density.

Problem 15 *Use the continuity equation to show the following relation for the modulus of a wave function $\psi(x,t)$ evaluated at the position $x^\alpha[t_1]$ at time t_1 and at the position $x^\alpha[t_2]$ at time t_2:*

$$|\psi(x^\alpha[t_2],t_2)|=\exp\left(-\int_{t_1}^{t_2}\left[\frac{1}{2}\frac{\partial v(x,t)}{\partial x}\right]_{x^\alpha[t]}dt\right)|\psi(x^\alpha[t_1],t_1)|$$

Notice that we are discussing the evolution of $|\psi(x^\alpha[t],t)|$ along a single trajectory $x^\alpha[t]$.

Solution. From the continuity equation, we get:

$$\frac{\partial|\psi(x,t)|^2}{\partial t}+\frac{\partial}{\partial x}\left(|\psi(x,t)|^2 v(x,t)\right)=0$$

which can be rewritten as:

$$2|\psi(x,t)|\frac{\partial|\psi(x,t)|}{\partial t}+|\psi(x,t)|^2\frac{\partial v(x,t)}{\partial x}+2v(x,t)|\psi(x,t)|\frac{\partial|\psi(x,t)|}{\partial x}=0$$

For one particular trajectory $x^\alpha[t]$ then $v(x^\alpha[t],t)=\frac{dx^\alpha[t]}{dt}$. Then, we get:

$$\frac{d|\psi(x^\alpha[t],t)|}{dt}=\frac{\partial|\psi(x^\alpha[t],t)|}{\partial t}+v(x^\alpha[t],t)\left[\frac{\partial|\psi(x,t)|}{\partial x}\right]_{x^\alpha[t]}$$
$$=-|\psi(x^\alpha(t),t)|\frac{1}{2}\left[\frac{\partial v_B(x,t)}{\partial x}\right]_{x^\alpha[t]}$$

Then, finally:

$$\frac{d}{dt}ln\left(|\psi(x^{\alpha}[t],t)|\right)=-\frac{1}{2}\left[\frac{\partial v(x,t)}{\partial x}\right]_{x^{\alpha}[t]}$$

which means that:

$$|\psi(x^{\alpha}[t_2],t_2)|=\exp\left(-\int_{t_1}^{t_2}\left[\frac{1}{2}\frac{\partial v(x,t)}{\partial x}\right]_{x^{\alpha}[t]}dt\right)|\psi(x^{\alpha}[t_1],t_1)|$$

Problem 16 *Use the Hamilton–Jacobi equation to show the following relation for the quantum action $S(x,t)$ evaluated at the position $x^{\alpha}[t_1]$ at time t_1 and at the position $x^{\alpha}[t_2]$ at time t_2:*

$$S(x^{\alpha}[t_2],t_2)=S(x^{\alpha}[t_1],t_1)+\int_{t_1}^{t_2}\left(\frac{1}{2}m\cdot v(x^{\alpha}[t],t)^2-V(x^{\alpha}[t],t)\right.$$
$$\left.-Q(x^{\alpha}[t],t)\right)dt$$

Solution. We get:

$$\frac{dS(x^{\alpha}[t],t)}{dt}=\left[\frac{\partial S(x,t)}{\partial x}\right]_{x^{\alpha}[t]}v(x^{\alpha}[t],t)+\left[\frac{\partial S(x,t)}{\partial t}\right]_{x^{\alpha}[t]}$$

where $\frac{\partial S(x,t)}{\partial x}=m\cdot v(x,t)$ and $\frac{\partial S(x,t)}{\partial t}=-m\cdot v(x,t)^2/2-V(x,t)-Q(x,t)$. Then, we get:

$$\frac{dS(x^{\alpha}[t],t)}{dt}=m\cdot v(x^{\alpha}[t],t)^2-m\cdot v(x^{\alpha}[t],t)^2/2-V(x^{\alpha}[t],t)$$
$$-Q(x^{\alpha}[t],t)$$

and finally:

$$S(x^{\alpha}[t_2],t_2)=S(x^{\alpha}[t_1],t_1)+\int_{t_1}^{t_2}\left(\frac{1}{2}m\cdot v(x^{\alpha}[t],t)^2\right.$$
$$\left.-V(x^{\alpha}[t],t)-Q(x^{\alpha}[t],t)\right)dt$$

Problem 17 *Use the continuity equation and the Hamilton–Jacobi equation to show the following relation for the wave function $\psi(x,t)$ evaluated at the position $x^{\alpha}[t_1]$ at time t_1 and at the position $x^{\alpha}[t_2]$ at time t_2:*

$$\frac{\psi(x^{\alpha}[t_2],t_2)}{\psi(x^{\alpha}[t_1],t_1)}=\exp\left(\int_{t_1}^{t_2}\left[\frac{i}{\hbar}\left(\frac{1}{2}m\cdot v(x,t)^2-V(x,t)-Q(x,t)\right)\right.\right.$$
$$\left.\left.-\frac{1}{2}\frac{\partial v(x,t)}{\partial x}\right]_{x^{\alpha}[t]}dt\right)$$

Notice that we are discussing the evolution of $\psi(x^{\alpha}[t], t)$ along a single trajectory $x^{\alpha}[t]$.

Solution. We write $\psi(x^{\alpha}[t_2], t_2) = |\psi(x^{\alpha}[t_2], t_2)| \exp(\frac{i}{\hbar} S(x^{\alpha}[t_2], t_2))$ and use the results of problem 15 for the modulus and the results of problem 16 for the quantum action to write:

$$\frac{\psi(x^{\alpha}[t_2], t_2)}{\psi(x^{\alpha}[t_1], t_1)} = exp\left(\int_{t_1}^{t_2} \left[\frac{i}{\hbar} \left(\frac{1}{2} m \cdot v(x, t)^2 - V(x, t) - Q(x, t) \right) - \frac{1}{2} \frac{\partial v(x, t)}{\partial x} \right]_{x^{\alpha}[t]} dt \right)$$

Once we have finished the demonstration, it is interesting to realize the differences between the expression above and the typical Feynman path integral developed in Section 1.2.5 evaluated at $x^{\alpha}[t_2]$:

$$\psi(x^{\alpha}[t_2], t_2) = \int_{-\infty}^{\infty} G(x_0, t_0; x^{\alpha}[t_2], t_2) \psi(x_0, t_0) dx_0$$

The evaluation of $\psi(x^{\alpha}[t_2], t_2)$ does only require the value of $\psi(x^{\alpha}[t_1], t_1)$ and the knowledge of $v(x, t)$, $V(x, t)$, $Q(x, t)$ and $\frac{1}{2} \frac{\partial v(x,t)}{\partial x}$ along the path of the single trajectory $x^{\alpha}[t]$. No information about other trajectories departing from different positions $x \neq x^{\alpha}[t_1]$ at time t_1 are required.

A.1 Appendix: Numerical Algorithms for the Computation of Bohmian Mechanics

In this section, we will describe the basic numerical algorithms available in the literature for the computation of Bohmian mechanics. An excellent source of information on this issue can be found in Ref. [43]. There are many possible algorithms that can be classified following different schemes. One can distinguish between analytical or synthetic algorithms.[a] An analytical algorithm, first, solves the Schrödinger equation and, then, evaluates the Bohmian velocity directly from the numerical wave function. In

[a]The names "analytical" and "synthetic" are related to philosophical concepts. The original name of analytic propositions referred to those propositions that were true simply by virtue of their meaning, while synthetic propositions were not.

Table A.1 1D and 2D Bohmian computational algorithms (TD = time dependent; TI = time independent; SE = Schrödinger equation; QHJE = quantum Hamilton–Jacobi equation; BT = Bohmian trajectries)

		Analytical		Synthetic	
		no-trajectory	**trajectory**	**no-trajectory**	**trajectory**
Spatial 1D	Time-dependent	—	$TDSE_{1D} + BT$	$TDQHJE_{1D}$	$TDQHJE_{1D} + BT$
	Time-independent	—	$TISE_{1D} + BT$	$TIQHJE_{1D}$	$TIQHJE_{1D} + BT$
Spatial 2D	Time-dependent	—	$TDSE_{2D} + BT$	$TDQHJE_{2D}$	$TDQHJE_{2D} + BT$
	Time-independent	—	$TISE_{2D} + BT$	$TIQHJE_{2D}$	$TIQHJE_{2D} + BT$

such algorithms, the computational difficulties are directly related to the numerical solution of the Schrödinger equation. Alternatively, one can choose computational algorithms where the Schrödinger equation is substituted by the quantum Hamilton–Jacobi equation (synthetic approach). Then, the difficulties of the algorithms are intrinsic to the numerical resolution of the (nonlinear) Hamilton–Jacobi scheme. In addition, the quantum Hamilton–Jacobi equation can be solved with or without trajectories. Another classification divides algorithms between time-dependent and time-independent ones. From a computational point of view, the former is basically related to an eigenvalue problem, while the latter is related to an initial value problem. Finally, a classification between 1D spatial grids and 2D (or even 3D) is also possible.[a] In Table A.1, we have summarized the characteristics of the numerical algorithms that we discuss.

Much of our attention in this section will be devoted to time-dependent scenarios. In such scenarios, either the Schrödinger equation or the quantum Hamilton–Jacobi equation can be written as a first-order differential equation in time of the form:

$$\frac{dW(t)}{dt} = f(W(t), t) \tag{A.1}$$

[a]Certainly, the direct solution of more than three degrees of freedom is very computationally demanding. To the best of our knowledge, the only Bohmian algorithm that takes into account correlations for more than three degrees of freedom is the one mentioned in Section 1.3.6. This many-particle Bohmian algorithm does not fit with the previous classification because it is a mixed analytical and synthetic algorithm.

where $W(t)$ can take the role of $\psi(x,t)$, $R(x,t)$ and $S(x,t)$). $f(W(t),t)$ is a generic function that depends on $W(t)$ and time t. We define a mesh for the time variable $t_j = j\Delta t$ for $j = 1,\ldots,M$ and assume that the initial value of $W(0)$ is known. Then, one can use an explicit finite difference of the time derivative:

$$\left.\frac{dW(t)}{dt}\right|_{t=t_j} \approx \frac{W(t_{j+1}) - W(t_j)}{\Delta t} \tag{A.2}$$

giving a simple and explicit numerical solution of the time-dependent Eq. (A.1) as:

$$W(t_{j+1}) = W(t_j) + f(W(t_j),t_j)\Delta t \tag{A.3}$$

where the solution $W(t_{j+1})$ at t_{j+1} can be found directly from the knowledge of $W(t_j)$ and $f(W(t_j),t_j)$ at the previous time t_j. These kinds of solutions are named explicit solutions (also known as forward Euler method) and are, in general, unstable because the accumulated error during successive time integrations grows exponentially. Only for a very dense time mesh, that is, a very small Δt, the solution becomes conditionally stable.

On the contrary, we can look for an *implicit* approximation of the derivative:

$$\left.\frac{dW(t)}{dt}\right|_{t=t_j} \approx \frac{W(t_j) - W(t_{j-1})}{\Delta t} \tag{A.4}$$

that provides an implicit solution (also known as backward Euler method) of the time-dependent Eq. (A.1) as:

$$W(t_j) = W(t_{j-1}) + f(W(t_j),t_j)\Delta t \tag{A.5}$$

The practical numerical solution of Eq. (A.5) depends on the type of equation. For instance, for the Schrödinger equation, the function $f(W(t_j),t_j)$ is linear with respect to $W(t_j)$, and then, the relationships between $W(t_j)$ and $W(t_{j-1})$ can be written as a set of coupled linear equations, one for each grid point. Then, the entire system of linear equations can be written as the product of a (known) matrix and an (unknown) vector. The (unknown) vector is formed by the unknown value of $W(t_j)$ at each grid point. The solution follows from inverting the matrix. On the contrary, for the quantum Hamilton–Jacobi equation, where $f(W(t_j),t_j)$ is nonlinear, one has to look for the numerical solution of $W(t_j) - W(t_{j-1}) -$

Table A.2 Bohmian time-dependent computational algorithms

	Type of solution	SE→ linear $f(W(t), t)$	QHJE→ nonlinear $f(W(t), t)$
Implicit $\frac{dW(t)}{dt}$	Unconditional stable	Matrix inversion	Newton-like algorithm
Explicit $\frac{dW(t)}{dt}$	Conditional stable	Explicit solution	Explicit solution
	Unstable	No solution	No solution

$f(W(t_j), t_j)\Delta t = 0$ through a Newton-like procedure. Apart from the important increase of the computational burden, both types of implicit solutions provide a nongrowing error independently of the value of Δt. For this reason, they are called unconditionally stable algorithms. See a summary of the classification in Table A.2.

Hereafter, we will explain in detail some of these numerical algorithms, discussing their main advantages and drawbacks. We will mention some very simple algorithms that can be useful for those newcomers who are just interested in computing Bohmian trajectories.

A.1.1 Analytical Computation of Bohmian Trajectories

According to the first postulate of Bohmian mechanics, the computation of trajectories is quite simple once the wave function is known. As we mentioned in the introduction, here, the utility of using Bohmian trajectories is that allows for an intuitive and different explanation of many quantum phenomena. In particular, in this section, we will discuss three simple numerical methods for solving the 1D Schrödinger equation:

$$i\hbar\frac{\partial\psi(x,t)}{\partial t} = -\frac{\hbar^2}{2m}\frac{\partial^2\psi(x,t)}{\partial x^2} + V(x,t)\psi(x,t) \tag{A.6}$$

A.1.1.1 Time-dependent Schrödinger equation for a 1D space (TDSE$_{1D}$-BT) with an explicit method

In this subsection, we will explain a very simple explicit algorithm to compute the time-dependent Schrödinger equation. Apart from a mesh in the time variable, an additional mesh is imposed on the 1D spatial degree of freedom with a constant step, Δx, that is, $x_k = k\,\Delta x$ for $k = 1, \ldots, N$. Then, we define $\psi(x,t)|_{x=x_k;t=t_j} = \psi_j(x_k)$, and the temporal and spatial derivatives present in the time-dependent Schrödinger equation can be approximated by:

$$\left.\frac{\partial \psi(x,t)}{\partial t}\right|_{x=x_k;t=t_j} \approx \frac{\psi_{j+1}(x_k) - \psi_{j-1}(x_k)}{2\Delta t} \tag{A.7}$$

$$\left.\frac{\partial^2 \psi(x,t)}{\partial x^2}\right|_{x=x_k;t=t_j} \approx \frac{\psi_j(x_{k+1}) - 2\psi_j(x_k) + \psi_j(x_{k-1})}{\Delta x^2} \tag{A.8}$$

Inserting Eqs. (A.7) and (A.8) into Eq. (A.6), we obtain the following simple recursive expression:

$$\begin{aligned}\psi_{j+1}(x_k) = \psi_{j-1}(x_k) &+ i\frac{\hbar \Delta t}{\Delta x^2 m}\left(\psi_j(x_{k+1}) - 2\psi_j(x_k) + \psi_j(x_{k-1})\right) \\ &- i\frac{2\Delta t}{\hbar} V_j(x_k)\psi_j(x_k)\end{aligned} \tag{A.9}$$

Once we know the wave function at the particular times t_j and t_{j-1} for all spatial positions in the mesh, we can compute the wave function at next time t_{j+1}, using Eq. (A.9). This algorithm solves an initial-value problem. The recursive application of Eq. (A.9) provides the entire time evolution of the wave packet. Equation (A.9) is not valid for the first, x_1, and last, x_N, points. To avoid discussions on the boundary conditions, we can use a very large spatial simulating box so that the entire wave packet is contained in it at any time. Then the wave function at the borders is negligible.

As indicated in Tables A.1 and A.2, this explicit solution can be unstable, and its error grows in each recursive application of Eq. (A.9). To provide a (conditional) stable solution we have to deal with small values of Δt and Δx. For example, to study electron transport in nanoscale structures as depicted in Figs. A.1 and A.2, this recursive procedure provides accurate results (the norm of the wave packets is conserved with high precision) when Δx is on the order of 1 or 2 Å and the temporal step, Δt, is around 10^{-16}s.

In order to define the initial value of the wave function, we can consider that the wave packet at times $t = \{t_1, t_1 + \Delta t\}$ evolves in a flat potential region contained in a much larger simulation box. Then, for example, we can define the initial wave function as a time-

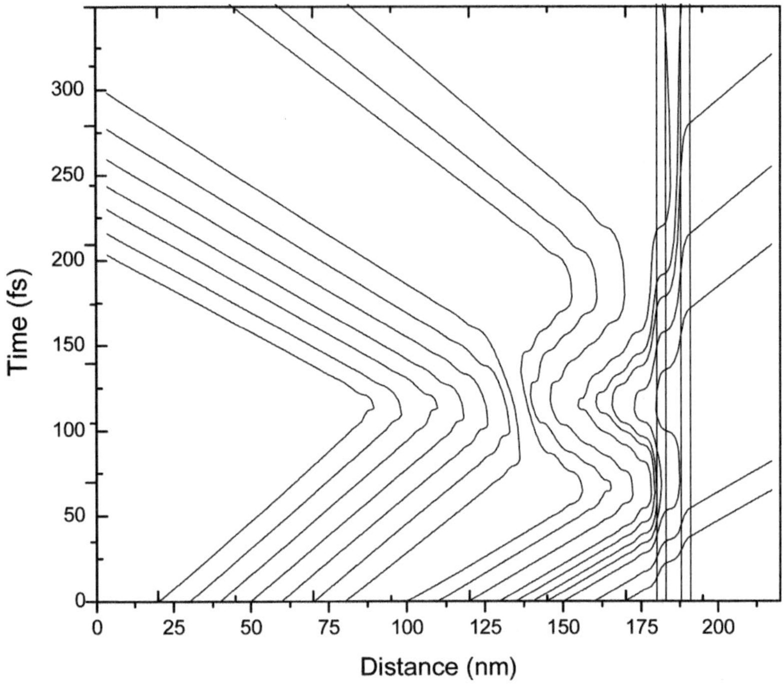

Figure A.1 Representative Bohmian trajectories associated with double-packet scattering in a double-barrier potential. The position of the barriers is indicated by vertical lines. Reprinted with permission from [112]. Copyright 1996 American Physical Society.

dependent Gaussian wave packet [33]:

$$\psi(x,t) = \left(\frac{2a^2}{\pi}\right)^{1/4} \frac{e^{i\phi} e^{i(k_c(x-x_0))}}{\left(a^4 + \frac{4\hbar^2(t-t_1)^2}{m^2}\right)^{1/4}} \times \exp\left(-\frac{\left[x - x_0 - \frac{\hbar k_c}{m}(t-t_1)\right]^2}{a^2 + \frac{2i\hbar(t-t_1)}{m}}\right) \tag{A.10}$$

where a is the spatial dispersion of the wave packet, m the particle effective mass, x_0 the central position of the wave packet at the initial time t_1, $k_c = \sqrt{\frac{2mE}{\hbar^2}}$ the central wave vector in the x direction related with the central energy E, and $\phi = -\theta - \hbar k_c{}^2 t/(2m)$ with $\tan(2\theta) = 2\hbar t/(ma^2)$ (see [33]). In particular, at the initial time

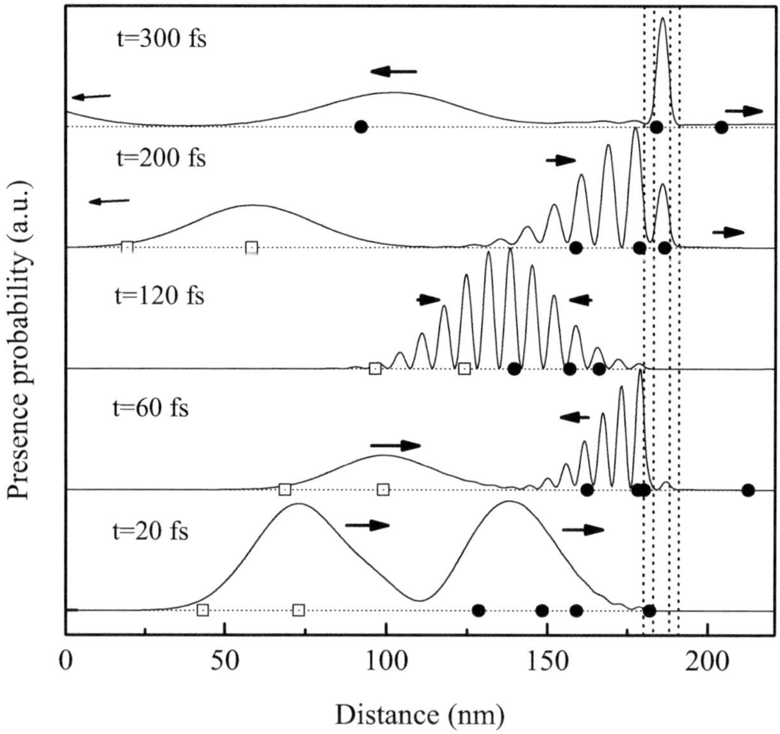

Figure A.2 Picture motion of the same double-packet wave function considered in Fig. A.1, calculated by numerical integration of the time-dependent Schrödinger equation. Five representative "snapshots" obtained at different times are shown with the vertical scale arbitrarily changed in each case for clarity. Although the norm of the wave function is always unity, it does not seem so because of the changes in the scale. The marks are visual aids that indicate the position of some related Bohmian trajectories shown in Fig. A.1. The double-barrier position is indicated by the vertical dashed lines, and the arrows indicate the sense of motion of the two packets. Reprinted with permission from [112]. Copyright 1996 American Physical Society.

$t = t_1$, we obtain the simplified expression:

$$\psi(x, t = t_1) = \left(\frac{2}{\pi a^2}\right)^{1/4} e^{i(k_c(x-x_0))} \exp\left(-\frac{(x-x_0)^2}{a^2}\right) \qquad \text{(A.11)}$$

We define the wave packet spatial dispersion $\sigma_x = a/\sqrt{2}$ and the wave packet width in the reciprocal space as $\sigma_k = 1/\sigma_x$.

To finish this practical computation of Bohmian trajectories from the direct solution of the time-dependent wave function, we can discuss some features of Bohmian trajectories related to the fact that they do not cross in the configuration space. We consider a quite "exotic" initial wave packet. We use a sum of two Gaussian wave packets defined by Eq. (A.10) with different central positions and central (momenta) wave vectors [112]. The wave packet is certainly quite exotic because it describes just one particle. In Figs. A.1 and A.2 we see that Bohmian trajectories can be reflected for two different reasons: first, because of their interaction with the classical potential (the particles collide with the barrier) and, second, because of the collision with other trajectories traveling in the opposite direction (Bohmian trajectories do not cross). The second process is responsible for the reflection of those particles of the first packet, which never reach the barrier, and for the reflection of the entire second packet. These collisions between Bohmian particles are related to the quantum potential in regions where the classical potential is zero, but for them to occur, there should be particles coming from right to left. In this regard, if the initial wave packet is prepared as a superposition of eigenstates incident from left to right (as is always assumed in scattering thought experiments), and the classical potential is zero on the left-hand side (l.h.s.) of the barrier, then finding particles coming from the r.h.s. in this region will be at least very uncommon.

A.1.1.2 Time-independent Schrödinger equation for a 1D space (TISE$_{1D}$) with an implicit (matrix inversion) method

Next, we consider another very simple computational algorithm similar to the one discussed in the previous section (1D, analytic and explicit) but for time-independent systems. First, let us discuss under which circumstances we can expect that a time-independent Schrödinger equation can provide a reasonable physical description. Let us assume that the wave function $\psi(x, t)$ can be written as a product of a spatial function $\varphi(x)$ and a temporal function $\zeta(t)$, that is, $\psi(x, t) = \varphi(x)\zeta(t)$. Then, the Schrödinger equation, Eq. (A.6), with a time-independent potential $V(x)$ can be written as:

$$\frac{i\hbar\frac{\partial\zeta(t)}{\partial t}}{\zeta(t)} = \frac{-\frac{\hbar^2}{2m}\frac{\partial^2\varphi(x)}{\partial x^2} + V(x)\varphi(x)}{\varphi(x)} \tag{A.12}$$

Since the l.h.s. of Eq. (A.12) only depends on t and the r.h.s. of Eq. (A.12) only depends on x, both sides are equal to a constant that we refer as E. Therefore, we obtain two separate equations:

$$i\hbar\frac{\partial\zeta(t)}{\partial t} = E\zeta(t) \tag{A.13}$$

$$-\frac{\hbar^2}{2m}\frac{\partial^2\varphi(x)}{\partial x^2} + V(x)\varphi(x) = E\varphi(x) \tag{A.14}$$

The first of these equations has a trivial analytical solution $\zeta(t) = e^{i\frac{Et}{\hbar}}$. However, the time-independent Schrödinger equation needs a numerical method to be solved for arbitrary shapes of the potential $V(x)$.

The second-order derivative can be approximated as:

$$\left.\frac{\partial^2\varphi(x)}{\partial x^2}\right|_{x=x_k} \approx \frac{\varphi(x_{k+1}) - 2\varphi(x_k) + \varphi(x_{k-1})}{\Delta x^2} \tag{A.15}$$

Then, Eq. (A.14) can be written in a matrix representation as:

$$\left(-\frac{\hbar^2}{2m}\hat{\nabla}^2 + (V(x_k) - E)\,\hat{I}\right)\vec{\varphi} = 0 \tag{A.16}$$

where $\vec{\varphi} = (\varphi(x_2), \ldots, \varphi(x_{N-1}))^T$ is a $N-2$ vector, $\hat{I}$ is the $(N-2)\times(N-2)$ identity matrix and $\hat{\nabla}^2$ is the $(N-2)\times(N-2)$ matrix constructed from Eq. (A.15) as:

$$\hat{\nabla}^2 = \frac{1}{\Delta x^2}\begin{pmatrix} -2 & 1 & 0 & 0 & \ddots \\ 1 & -2 & 1 & 0 & \ddots \\ \ddots & \ddots & \ddots & \ddots & \ddots \\ \ddots & 0 & 1 & -2 & 1 \\ \ddots & 0 & 0 & 1 & -2 \end{pmatrix} \tag{A.17}$$

The solution of $\vec{\varphi}$ can be found by inverting the matrix equation, Eq. (A.16). The (nonzero) solution of Eq. (A.16) requires that the determinant of the matrix be equal to zero. The determinant requirement is called the characteristic equation (less often, the secular equation). In our case, the determinant can only be zero for some particular values of the energy E (i.e., quantized energies). Thus, this algorithm finds the eigenvalues E and eigenvectors $\vec{\varphi}$, typical of a boundary value problem.[a]

In the previous vector definition, $\vec{\varphi}$, only $N-2$ unknowns are considered. We have assumed that $\varphi(x_1) = 0$ and $\varphi(x_N) = 0$. This implies that $\vec{\varphi}$ is a real vector. Then, if we select an eigenstate, we realize that the current density is zero, $J(x, t) = 0$, because $\varphi(x_k) = \varphi(x_k)^*$. All Bohmian velocities are zero, $\partial S/\partial x = 0$, meaning that Bohmian trajectories do not "move." This result is obviously consistent with the fact that the amplitude $R(x, t)$ of the global wave function $\psi(x, t) = \varphi(x)e^{-iEt/\hbar}$ is time independent. One is tempted to affirm that there is no kinetic energy for these Bohmian trajectories. However, as we see in Eq. (1.136), the evaluation of the mean value of the kinetic energy takes into account not only the Bohmian velocity but also the quantum potential. We can explain this point, differently, from the quantum Hamilton–Jacobi equation, Eq. (1.39). In our particular time-independent case, we have an energy conservation law:

$$V(x) - \frac{\hbar^2}{2m}\frac{\partial^2 R(x)/\partial x^2}{R(x)} = E$$

where we have used $\partial S/\partial t = -E$ in Eq. (1.39). Then, we can obtain the following (time-independent Schrödinger-like) equation for the real wave function $R(x)$:

$$-\frac{\hbar^2}{2m}\frac{\partial^2 R(x)}{\partial x^2} + V(x)R(x) = E\ R(x) \tag{A.18}$$

Comparing this expression with the previous Eq. (A.14), we realize that in this particular problem (for which all Bohmian trajectories have zero velocity), the quantum potential takes the role of a "kinetic" energy.

[a] The practical implementation of this algorithm can be done, for example, quite easily with the software MATLAB. The instruction **[V,D] = eig(A)** produces matrices of eigenvalues (D) and eigenvectors (V) of matrix A, so that $A * V = V * D$. Equation (A.16) can be rewritten as $A * V = V * D$.

A.1.1.3 Time-independent Schrödinger equation for a 1D space (TISE$_{1D}$) with an explicit method

We can also look for a very simple explicit algorithm to find the solution $\vec{\varphi}$ of the time-independent Schrödinger equation. Equation (A.14) is a homogeneous second-order differential equation that can be written as:

$$\frac{d^2}{dx^2}\varphi(x) = f(x)\psi(x) \tag{A.19}$$

where $f(x)$ is the constant of proportionality between the solution and the second derivative at each point:

$$f(x) = \frac{2m}{\hbar^2}(V(x) - E) \tag{A.20}$$

From the Taylor expansion for $\varphi(x_{k+1})$ around x_k, we get:

$$\begin{aligned}\varphi(x_{k+1}) = \varphi(x_k) &+ \Delta x \left.\frac{d\varphi(x)}{dx}\right|_{x=x_k} + \frac{\Delta x^2}{2!}\left.\frac{d^2\varphi(x)}{dx^2}\right|_{x=x_k} \\ &+ \frac{\Delta x^3}{3!}\left.\frac{d^3\varphi(x)}{dx^3}\right|_{x=x_k} + \frac{\Delta x^4}{4!}\left.\frac{d^4\varphi(x)}{dx^4}\right|_{x=x_k} \\ &+ \frac{\Delta x^5}{5!}\left.\frac{d^5\varphi(x)}{dx^5}\right|_{x=x_k} + O(h^6)\end{aligned} \tag{A.21}$$

Identically, for $\varphi(x_{k-1})$ around x_k, we get:

$$\begin{aligned}\varphi(x_{k-1}) = \varphi(x_k) &- \Delta x \left.\frac{d\varphi(x)}{dx}\right|_{x=x_k} + \frac{\Delta x^2}{2!}\left.\frac{d^2\varphi(x)}{dx^2}\right|_{x=x_k} \\ &- \frac{\Delta x^3}{3!}\left.\frac{d^3\varphi(x)}{dx^3}\right|_{x=x_k} + \frac{\Delta x^4}{4!}\left.\frac{d^4\varphi(x)}{dx^4}\right|_{x=x_k} \\ &- \frac{\Delta x^5}{5!}\left.\frac{d^5\varphi(x)}{dx^5}\right|_{x=x_k} + O(\Delta x^6)\end{aligned} \tag{A.22}$$

The sum of these two equations gives:

$$\begin{aligned}\varphi(x_{k+1}) + \varphi(x_{k-1}) = 2\varphi(x_k) &+ \Delta x^2 \left.\frac{d^2\varphi(x)}{dx^2}\right|_{x=x_k} \\ &+ \frac{\Delta x^4}{12}\left.\frac{d^4\varphi(x)}{dx^4}\right|_{x=x_k} + O(\Delta x^6)\end{aligned} \tag{A.23}$$

We use Eq. (A.19) to define, directly, the second-order derivative. We compute the fourth-order derivative from a second-order derivative written as a difference quotient. Finally, we find a recursive expression for the wave function $\varphi(x_{k-1})$:

$$\varphi(x_{k-1}) = \frac{\left[2 + \frac{10}{12}\Delta x^2 f(x_k)\right]\varphi(x_k) - \left[1 - \frac{\Delta x^2}{12} f(x_{k+1})\right]\varphi(x_{k+1})}{1 - \frac{\Delta x^2}{12} f(x_{k-1})} \tag{A.24}$$

We have explicitly written $\varphi(x_{k-1})$ in Eq. (A.24) because later, in a numerical example, we will be interested in specifying the boundary conditions of $\varphi(x)$ at $x \rightarrow \pm\infty$. The Numerov method [109, 110] provides a numerical algorithm to solve ordinary differential equations of the second order in which the first-order term does not appear. It is a fourth-order linear multistep method. The above expression, Eq. (A.24), allows a simple solution of $\varphi(x_{k-1})$ once we know $\varphi(x_k)$, $\varphi(x_{k+1})$, $f(x_k)$, $f(x_{k+1})$, and $f(x_{k-1})$. Finally, repeating the process, we can determine the entire solution.

We will apply this last explicit computational algorithm for tunneling scenarios. For the numerical simulations, we will use $\Delta x = 1$ Å. Again, the amplitude of the global wave function $\psi(x,t) = \varphi(x)e^{-iEt/\hbar}$ is time independent. Thus, the current density $J(x,t)$, from Eq. (1.34), is also time independent. Therefore, from the continuity equation, Eq. (1.28), we deduce that $J(x,t)$ is also uniform (position-independent). Then, the velocity $v(x) =$ constant$/R^2(x)$ does not depend on time but depends on position. We have $S(x,t) \neq 0$. The velocity can be computed from:

$$\frac{\partial S(x,t)}{\partial x} = \sqrt{2m(E - V(x) - Q(x))} \tag{A.25}$$

See problem 11 for a detailed calculation. This quantum expression, Eq. (A.25), is different from the similar expression extracted from the classical Hamilton–Jacobi equation $\partial S(x,t)/\partial x = \sqrt{2m(E - V(x))}$. When $E < V(x)$, we will not be able to find a (real) velocity in a classical system (classical particles are unable to pass through spatial regions where the potential barrier is higher than its energy). Interestingly, when $E < V(x)$ in Eq. (A.25), we will still find (real) velocities because of the term $Q(x)$ in Eq. (A.25) (Bohmian particles are able to pass through spatial regions where the potential barrier is higher than its energy). This is the well-known tunneling effect.

As a numerical example, let us concentrate on (time-independent) scattering states traveling from left to right in a double-barrier structure. For a simulation box $0 < x < 40$ nm, these states can be written as:

$$\varphi(x) = \frac{1}{\sqrt{2\pi}}\left(e^{ikx} + r(k)e^{-ikx}\right) \quad ; \quad x < 0 \text{ nm} \tag{A.26}$$

$$\varphi(x) = \frac{1}{\sqrt{2\pi}}\left(t(k)e^{ikx}\right) \quad ; \quad x > 40 \text{ nm} \tag{A.27}$$

where $r(k)$ and $t(k)$ are (unknown) complex coefficients accounting for reflection and transmission, respectively. However, up to an irrelevant (complex) constant that can be fixed by "normalization," we know the results $\varphi(x)$ at $x \rightarrow \pm\infty$. In the central region, we can use the Numerov numerical method using the fact that the wave function and its derivative have to be continuous in the matching points between the regions where we have analytical or numerical knowledge of the wave functions.

In Fig. A.3, we have represented one Bohmian trajectory for a state incident from left to right upon a double-barrier potential profile. Its energy $E = 0.05$ eV corresponds to the first resonant level of the double barrier [113]. In this particular case, where the transmission coefficient is very close to unity, the results obtained within the Bohmian approach are quite compatible with our intuitive understanding of the tunneling phenomenon: the particle is transmitted, and its velocity decreases in the well. We do also plot the total potential-that is, the sum $V(x) + Q(x)$-which is lower than the electron energy $E = 0.05$ eV.

However, the situation is far from intuitive for nonresonant states. In Fig. A.3b, we have represented one of these nonresonant trajectories for the same potential profile. Now, although the transmission coefficient is much smaller than unity for the nonresonant states, they present the same features as the resonant ones: all Bohmian trajectories are transmitted through the barrier. This fact can be easily understood from a mathematical point of view. As we mentioned, the current and particle densities are time independent and positive everywhere. So, the Bohmian velocity is always positive. In this regard, although Bohmian trajectories associated with scattering states perfectly reproduce the presence probability and the current density, they do not reproduce our particle-intuitive

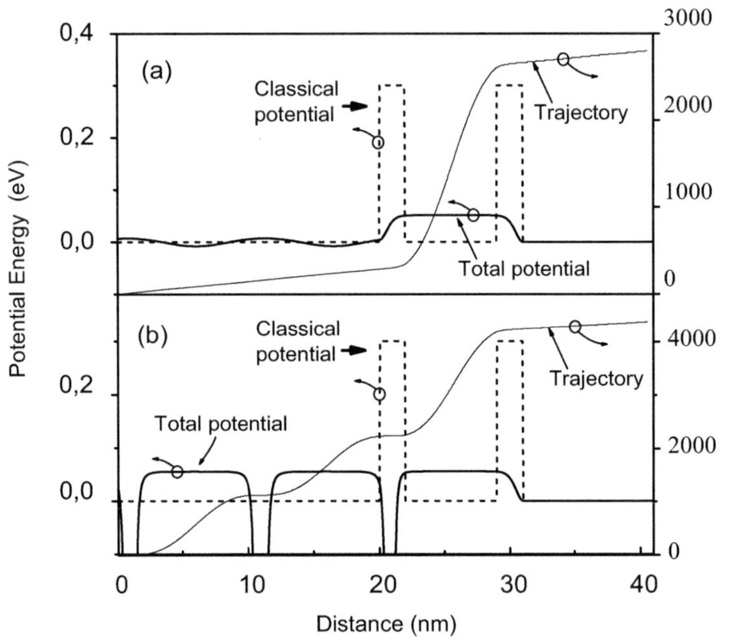

Figure A.3 Bohmian trajectories associated with stationary scattering states impinging on a typical GaAs/AlGaAs RTD with 2 nm barriers of 0.3 eV separated by a 7 nm well. The classical potential and the total potential (the sum of classical and quantum terms) are also depicted in dashed and solid lines, respectively. (a) Resonant eigenstate, $E_k = 0.05$ eV; (b) nonresonant eigenstate, $E_k = 0.06$ eV. Reprinted with permission from [113]. Copyright 1996 Elsevier. *Abbreviation*: RTD, resonant tunneling diode.

understanding of the tunneling phenomenon, since in principle, we would expect reflected as well as transmitted trajectories. The origin of this lack of intuitive behavior of Bohmian trajectories is only due to the initial wave function that we select. The velocity of Bohmian particles is neither related to the incident electron nor related to the reflected one but to the average of both because the wave function itself is defined as a weighted sum of waves with positive and negative momentum. In particular, if the barrier is infinitely high, then the reflection coefficient is equal to one and Bohmian's velocity is zero (one would expect positive and negative electron velocities).

Transport is better formulated as a time-dependent process. A time dependent picture of transport involves a wave packet with a well defined initial velocity and, after the interaction, a well-defined reflected (or transmitted) velocity, but not a unique velocity with an average of incident and reflected velocities.

All the 1D numerical algorithms previously discussed can be easily generalized to 2D and 3D systems (see, for example, the solution of the Schrödinger equation for computing Bohmian trajectories in atomic scenarios in Refs. [114, 115]). However, the computational burden of such techniques with finite-difference meshes is enormous, and the solution using adaptative meshes seems interesting. For example, the triangular mesh generated using an algorithm inspired by Ref. [116], with Delaunay refinement algorithms, yields a final mesh that is surprisingly well shaped, resulting in an excellent support to solve partial differential equations, as it was remarked in Ref. [117].

A.1.2 Synthetic Computation of Bohmian Trajectories

Up to here, we have only discussed analytical algorithms to find Bohmian trajectories. However, it is possible to compute Bohmian trajectories without knowing the wave function. The key point is using the polar form $\psi(\vec{r}, t) = R(\vec{r}, t)\exp(iS(\vec{r}, t)/\hbar)$ to obtain the quantum Hamilton–Jacobi equations. As we anticipated, these new equations are nonlinear in $R(\vec{r}, t)$ and $S(\vec{r}, t)$. We will now discuss the advantages and disadvantages of numerically solving the quantum Hamilton–Jacobi equation, Eq. (1.39). Hereafter, we will analyze two different algorithms to solve numerically Eq. (1.39). First, directly solving these equations (the Eulerian formalism) or some equivalent variable change. Second rewriting the Hamilton–Jacobi equations within the Langrangian formalism and solving them.

The first point that we want to clarify is that, although the function $S(\vec{r}, t)$ is in principle multivalued, the numerical solution of $S(\vec{r}, t)$ has a unique value at each point $\vec{r}$ and t. The reason is because of an important theorem on the existence and uniqueness of solutions to first-order equations with given initial-value problems. See page 70 in [118].

A.1.2.1 Time-dependent quantum Hamilton–Jacobi equations ($TDQHJE_{1D}$) with an implicit (Newton-like fixed Eulerian mesh) method

In the Eulerian formalism, we use a fixed grid. We are interested in providing a discussion valid for 1D, 2D, and 3D spatial scenarios. Therefore, we define a grid in real space through a set of discrete vectors $\vec{p}_i$ for $i = 1, \ldots, N$-in particular, $\vec{p}_i = (x_i, y_i, z_i)$ for a 3D space, $\vec{p}_i = (x_i, y_i)$ for a 2D space, and $\vec{p}_i = (x_i)$ for 1D. In order to provide a compact notation, we define:

$$\vec{R}_j = \left(R_j(\vec{p}_1), \ldots, R_j(\vec{p}_N)\right)^T \tag{A.28}$$

$$\vec{S}_j = \left(S_j(\vec{p}_1), \ldots, S_j(\vec{p}_N)\right)^T \tag{A.29}$$

$$\vec{V}_j = \left(V_j(\vec{p}_1), \ldots, V_j(\vec{p}_N)\right)^T \tag{A.30}$$

where $R_j(\vec{p}_i) = R(\vec{r}, t)$ at $t = t_j$ and $\vec{r} = \vec{p}_i$ (identically for $S_j(\vec{p}_i) = S(\vec{r}, t)$ and $V_j(\vec{p}_i) = V(\vec{r}, t)$). We have also used a mesh in time $t_j = j\Delta t$ for $j = 1, \ldots, M$

In the definition of the temporal derivatives of the quantum Hamilton–Jacobi, we can use an *implicit* expression, Eq. (A.5), or an *explicit* one, Eq. (A.3). The explicit expression will lead to an algorithm similar to the one described in Section A.1.1.1 with the same advantages (simplicity) and disadvantages (i.e., conditional stable). In this section, we will explore the *implicit* route. Then, the discrete version of the quantum Hamilton–Jacobi equations, Eq. (1.39), reads:

$$\begin{gathered}\frac{\vec{R}_j^2 - \vec{R}_{j-1}^2}{\Delta t} + \hat{\nabla}\left(\frac{\vec{R}_j^2 \hat{\nabla} \vec{S}_j}{m}\right) = 0 \\ \frac{\vec{S}_j - \vec{S}_{j-1}}{\Delta t} + \frac{(\hat{\nabla}\vec{S}_j)^2}{2m} - \frac{\hbar^2}{2m}\frac{\hat{\nabla}^2 \vec{R}_j}{\vec{R}_j} + \vec{V}_j = 0\end{gathered} \tag{A.31}$$

The operations such as power, multiplication, and division between the vectors $\vec{X}_j$ and $\vec{Y}_j$ are understood over each point in the spatial mesh, that is:

$$\vec{X}_j \vec{Y}_j = [X_j(\vec{p}_1)Y_j(\vec{p}_1), \ldots, X_j(\vec{p}_N)Y_j(\vec{p}_N)]^T \tag{A.32}$$

Here the symbol $\hat{\nabla}$ is the matrix representation of the differential operator $\vec{\nabla}$ in the 1D, 2D, or 3D spatial mesh. The representation of $\hat{\nabla}$ depends in general on the choice of the spatial mesh and on how the spatial derivatives are approximated. For example, in a 1D system with a uniform mesh step size Δx, the discrete version of the spatial derivative in one dimension is:

$$\frac{\partial R(x,t)}{\partial x} = \frac{R_j(\vec{p}_{i+1}) - R_j(\vec{p}_{i-1})}{2\Delta x}, \tag{A.33}$$

with $\vec{p}_i = x_0 + i\,\Delta x$. Then, the nabla operator in one dimension $\hat{\nabla}_{1\mathrm{D}}$ is represented by the following matrix:

$$\hat{\nabla}_{1\mathrm{D}} = \frac{1}{2\Delta x}\begin{pmatrix} 2 & -2 & 0 & 0 & \ddots \\ 1 & 0 & -1 & 0 & \ddots \\ \ddots & \ddots & \ddots & \ddots & \ddots \\ \ddots & 0 & 1 & 0 & -1 \\ \ddots & 0 & 0 & -2 & 2 \end{pmatrix} \tag{A.34}$$

Notice that we have changed the definition of the spatial derivative from Eq. (A.33) in x_1 and x_N. In two or three dimensions the matrices still have many zeros with the entries 1 and -1 properly located.

Hereafter, as we mentioned, we only discuss the computational problems of the *implicit* solution of the Hamilton–Jacobi equations. Such an equation is nonlinear with respect to $\vec{R}_j$ and $\vec{S}_j$, and the matrix inversion explained in Section A.1.1.2 cannot be done. Then, in order to find the numerical solution of the discrete version of Eq. (A.1.2.1), we have to employ some Newton-like method. If we rewrite in compact form the Eq. (A.1.2.1) as:

$$F(\vec{R}_j, \vec{S}_j) = 0 \tag{A.35}$$

then we have to find the roots (zeros) of the nonlinear expression in Eq. (A.35). It is proved that the sequence:

$$\begin{pmatrix} \vec{R}_j^{(h+1)} \\ \vec{S}_j^{(h+1)} \end{pmatrix} = -J_F^{-1}(\vec{R}_j^{(h)}, \vec{S}_j^{(h)}) \begin{pmatrix} \vec{R}_j^{(h)} \\ \vec{S}_j^{(h)} \end{pmatrix} \tag{A.36}$$

where $J_F(\vec{R}_j^{(h)}, \vec{S}_j^{(h)})$ is the Jacobian of $F(\vec{R}_j^{(h)}, \vec{S}_j^{(h)})$, for suitable initial vectors $\vec{R}_j^{(0)}$ and $\vec{S}_j^{(0)}$, converges quadratically to the solution of Eq. (A.35) for $h \to \infty$. Thus the key of this approach is to solve the linear system in Eq. (A.36). However evaluating explicitly the Jacobian:

$$J_F(\vec{R}_j, \vec{S}_j) =$$
$$\begin{pmatrix} \frac{2}{\Delta t}\hat{diag}(\vec{R}_j) + \hat{\nabla}\left(\hat{diag}\left(\frac{2\vec{R}_j\hat{\nabla}\vec{S}_j}{m}\right)\right) & \hat{\nabla}\hat{diag}\left(\frac{\vec{R}_j^2}{m}\right)\hat{\nabla} \\ -\frac{\hbar^2}{2m}\hat{diag}\left(\frac{1}{\vec{R}_j}\right)\hat{\nabla}^2 - \hat{diag}\left(\frac{\hat{\nabla}^2\vec{R}_j}{\vec{R}_j^2}\right) & \frac{1}{\Delta t}\hat{diag}(\vec{S}_j) + \hat{diag}\left(\frac{\hat{\nabla}\vec{S}_j}{m}\right)\hat{\nabla} \end{pmatrix} \tag{A.37}$$

where:

$$\hat{diag}(\vec{X}_j) = \begin{pmatrix} X_j(\vec{p}_1) & 0 & 0 & 0 & \ddots \\ 0 & X_j(\vec{p}_2) & 0 & 0 & \ddots \\ \ddots & \ddots & \ddots & \ddots & \ddots \\ \ddots & 0 & 0 & X_j(\vec{p}_{N-1}) & 0 \\ \ddots & 0 & 0 & 0 & X_j(\vec{p}_N) \end{pmatrix} \tag{A.38}$$

we can readily recognize that for small components of $\vec{R}_j$, the Jacobian is ill conditioned and the system of Eq. (A.36) is impossible to be solved numerically. Unfortunately this is the case of localized particles where $\vec{R}_j$ become exponentially small far from the central position of the particles. This problem does not appear with the *explicit* algorithm for solving the quantum Hamilton–Jacobi equations. The problem can also be avoided by choosing the variable change $\psi(\vec{r}, t) = \exp(C(\vec{r}, t) + iS(\vec{r}, t)/\hbar)$ allowing to the equations:

$$\frac{\partial}{\partial t}C(\vec{r}, t) = -\frac{1}{2m}\left(\nabla^2 S(\vec{r}, t) + 2\vec{\nabla}S(\vec{r}, t)\cdot\vec{\nabla}C(\vec{r}, t)\right) \tag{A.39}$$

$$\frac{\partial}{\partial t}S(\vec{r}, t) = -\frac{(\vec{\nabla}S(\vec{r}, t))^2}{2m} + \frac{\hbar^2}{2m}\left(\nabla^2 C(\vec{r}, t) + (\vec{\nabla}C(\vec{r}, t))^2\right) - V(\vec{r}, t)$$

However, the advantage of this method carries out also its disadvantage. The limitation of this approach is that the functions $S(\vec{r}, t)$ and $C(\vec{r}, t)$ strongly vary in the entire simulation space. Thus when we account for a realistic situation for which a large simulation space is used, the computational problem can become very hard because the number of space mesh points $\vec{p}_k$ will be quite large to obtain an accurate solution.

A.1.2.2 Time-dependent quantum Hamilton–Jacobi equations ($TDQHJE_{1D}$) with an explicit (Lagrangian mesh) method

A different approach to solve the quantum Hamilton–Jacobi equation is based on the Lagrangian picture. The basic idea is to use as a grid the Bohmian trajectories themselves. In this picture, the point $\vec{p}_k$ is no longer a fixed spatial point but a time-dependent point $\vec{p}_k = \vec{p}_k(t)$ with $k = 1, \ldots, N$ moving with instantaneous velocity $\vec{v}_k(t)$. For a detailed analysis see Refs. [119–122].

To pass from the Eulerian to the Lagrangian hydrodynamic picture, we must be careful with the time derivative of the functions $R(\vec{p}_k(t), t)$, $S(\vec{p}_k(t), t)$ and $V(\vec{p}_k(t), t)$. We have to use the chain rule:

$$\frac{d}{dt} = \frac{\partial}{\partial t} + \vec{v}(\vec{r}, t) \cdot \vec{\nabla} \tag{A.40}$$

which includes the *convective* term $\vec{v}(\vec{r}, t) \cdot \vec{\nabla}$ with the velocity defined:

$$\vec{v}(\vec{r}, t) = \frac{1}{m}\vec{\nabla}S(\vec{r}, t) \tag{A.41}$$

The new equations derived from the quantum Hamilton–Jacobi equations (or equivalently from Eq. (A.1.2.1)) can be written as:

$$\frac{d}{dt}\vec{p}_k(t) = \vec{v}(\vec{p}_k(t), t) \tag{A.42}$$

$$\frac{d}{dt}S(\vec{p}_k(t), t) = \frac{(\vec{\nabla}S(\vec{p}_k(t), t))^2}{2m} + \frac{\hbar^2}{2m}\frac{\nabla^2 R(\vec{p}_k(t), t)}{R(\vec{p}_k(t), t)} - V(\vec{p}_k(t), t) \tag{A.43}$$

$$\vec{v}(\vec{p}_k(t), t) = \frac{1}{m}\vec{\nabla}S(\vec{p}_k(t), t) \tag{A.44}$$

$$\frac{d}{dt}R(\vec{p}_k(t), t) = -\frac{1}{2}R(\vec{p}_k(t), t)\vec{\nabla} \cdot \vec{v}(\vec{p}_k(t)) \tag{A.45}$$

The differential system in Eqs. (A.42–A.45) is written for the trajectory $\vec{p}_k(t)$ with a particular initial position $\vec{p}_k(0)$. However to (numerically) reproduce the wave function $\psi(\vec{r}, t) = R(\vec{r}, t)\exp(iS(\vec{r}, t)/\hbar)$ defined in the entire space, we have to deal with $k = 1, \ldots, N$ trajectories with $N \to \infty$. The distribution of their initial positions $\vec{p}_k(0)$ has to be obtained according to the squared modulus of the initial wave function $|\psi(\vec{p}_k(0), 0)|^2$, that is, the quantum equilibrium hypothesis described in Section 1.2.6. See also Ref. [31]. Thus, by construction we are using a mesh that is continuously adapting to the wave function time evolution because the Bohmian trajectories themselves follow the wave function.

In this picture the strategy to solve Eqs. (A.42–A.45) can be very complicated [119–122]. However, some simple consideration can be done in order to understand the limits of this approach. The method widely used [119–122] to integrate Eqs. (A.42–A.45) is to solve them in cascade with the following approximations: at time t_j the new trajectories positions are computed through:

$$\vec{p}_{j,k} = \vec{p}_{j-1,k} + \Delta t \vec{v}_{j-1,k} \tag{A.46}$$

with $\vec{p}_{j,k} = \vec{p}_k(t_j)$. In the updated grid defined by $\vec{p}_{j,k}$, the variables $S_j(\vec{p}_{j,k})$, $\vec{v}_j(\vec{p}_{j,k})$, and $\vec{R}_j(\vec{p}_{j,k})$ are evaluated by:

$$\begin{aligned}
S_j(\vec{p}_{j,k}) &= S_{j-1}(\vec{p}_{j,k}) + \Delta t \left[\frac{(\hat{\nabla}_{j-1} S_{j-1}(\vec{p}_{j,k}))^2}{2m} \right. \\
&\quad \left. + \frac{\hbar^2}{2m} \frac{\hat{\nabla}^2_{j-1} R_{j-1}(\vec{p}_{j,k})}{R_{j-1}(\vec{p}_{j,k})} - \vec{V}_{j-1}(\vec{p}_{j,k}) \right] \\
R_j(\vec{p}_{j,k}) &= R_{j-1}(\vec{p}_{j,k}) - \Delta t \left[\frac{1}{2m} R_{j-1}(\vec{p}_{j,k}) \hat{\nabla}^2_{j-1} \vec{S}_{j-1}(\vec{p}_{j,k}) \right] \\
v_j(\vec{p}_{j,k}) &= \frac{1}{m} \hat{\nabla}_j \vec{S}_j(\vec{p}_{j,k})
\end{aligned}$$

where $\hat{\nabla}_j$ is the gradient operator evaluated in $\vec{p}_{j,k}$ through efficient interpolation methods [119–122]. It can be realized that we have used an (unstable) explicit (forward Euler) time integration method [123] to avoid the computational burden associated with implicit methods with nonlinear equations discussed before. This limitation (small temporal step to ensure stable solutions) is made worse since the time step must be further reduced when some trajectory approaches points for which the potential V and/or the quantum potential Q strongly varies provoking large variations of some $\vec{v}_j$.

The variables S and $\vec{v}$ vary in the entire simulation space; thus when we account for a realistic situation such that a large simulation space is used, the computational problem can become very hard due to the fact that the number of space mesh points $\vec{p}_k$ can diverge to obtain an accurate solution. On the contrary, this method has some fundamental advantage over methods using fix grid points. In the latter, the extension of 1D algorithms to 2D and 3D algorithms means increasing exponentially the number of grip points. However, in the former methods (i.e., when the grid points are Bohmian trajectories), the number of points can be chosen by the user. Thus, for example, one can simulate a 1D system with 1,000 Bohmian trajectories (grid points) and a 2D system with 3,000 Bohmian trajectories (grid points).

A.1.3 More Elaborated Algorithms

We have presented very simple and easily implementable codes for computing Bohmian trajectories in simple scenarios. These algorithms can be used for newcomers to be introduced to the computation of quantum dynamics with Bohmian mechanics. Certainly, there are many more elaborated and "professional" computational algorithms in the literature. An excellent source of information on this issue can be found in the book written by R. E. Wyatt entitled *Quantum Dynamics with Trajectories: Introduction to Quantum Hydrodynamics* [43].

To end this section, we want to briefly mention some original and powerful techniques, inspired by Bohmian mechanics, available in the literature, whose explicit description is far from the scope of this introductory appendix. Recently, B. Poirier showed how Bohmian trajectories can be computed for 1D eigenstates without the wave function [125]. The work of N. Makri shows the possibility of using imaginary time algorithms to estimate eigenstates and eigenvalues [127]. With a similar goal, I. P. Christov uses quantum trajectories to develop time-dependent quantum Monte Carlo algorithms [128, 129]. We also mention the work of E. L. Bittner [130] developed in Chapter 5 in this book. The work presented in Section 1.3.6 can also be interpreted as an approximating technique for computing many-particle systems [85]. This last work has been used by the group of Oriols to develop a powerful quantum electron transport simulator. (http://europe.uab.es/bitlles). See Refs. [132, 133] or Chapter 6 for practical results on electron correlation in nanoelectronics. Another line of work is also devoted to the approximation of the quantum potential. See, for example, the work of S. Garashchuk and V. A. Rassolov [131]. The correlation between classical and quantum systems has also been studied extensively. See, for example, Ref. [134–136]. We do also highlight the work by D. J. Tannor and coworkers on the use of imaginary quantum trajectories (complex action) for quantum dynamics [124, 126]. We also want to notice the important amount of work done by many people in the field of ab-initio molecular dynamics, specially for tackling the coupled electron-nuclear motion in nonadiabatic electronic transitions [137–143]. Finally, we mention the more recent work by L. B. Madsen and co-workers in the field of bosonic systems [144].

All the works mentioned have been elaborated during the past 18 years after the pioneering work of Wyatt [119, 145]. Much work is still needed to evaluate the real capability of synthetic applications of Bohmian mechanics as a computational tool.

References

1. L. de Broglie, *Recherches sur la théorie des quantas*, PhD thesis, University of Paris (1924); see also [17].
2. D. Bohm, *A Suggested Interpretation of the Quantum Theory in Terms of "Hidden" Variables I*, Physical Review **85**, 166, (1952).
3. D. Bohm, *A Suggested Interpretation of the Quantum Theory in Terms of "Hidden" Variables II*, Physical Review **85**, 180, (1952).
4. M. Planck, *On the Law of Distribution of Energy in the Normal Spectrum*, Annalen der Physik **4**, 553 (1901).
5. A. Einstein, *Über einen die Erzeugung und Verwandlung des Lichtes betreffenden heuristischen Gesichtspunkt*, Annalen der Physik **17**, 132 (1905).
6. G. N. Lewis, *The Conservation of Photons*, Nature **118**(2), 874 (1926); See also http://www.nobeliefs.com/photon.htm
7. N. Bohr, *On the Constitution of Atoms and Molecules, Part I*, Philosophical Magazine **26**, 1 (1913).
8. N. Bohr, *On the Constitution of Atoms and Molecules, Part II Systems Containing Only a Single Nucleus*, Philosophical Magazine **26**, 476 (1913).
9. N. Bohr, *On the Constitution of Atoms and Molecules, Part III Systems Containing Several Nuclei*, Philosophical Magazine **26**, 857 (1913).

10. B. L. van der Waerden, *Sources of Quantum Mechanics* (Dover Publications, 1968).
11. W. Heisenberg, *Über quantentheoretishe Umdeutung kinematisher und mechanischer Beziehungen*, Zeitschrift für Physik **33**, 879 (1925); English translation in Ref. [10].
12. W. Heisenberg, *Über den anschaulichen Inhalt der quantentheoretischen Kinematik und Mechanik*, Zeitschrift für Physik **43**(3–4), 172 (1927).
13. M. Born, *Zur Quantenmechanik der Stovorgänge*, Zeitschrift für Physik **37**, 863 (1926).
14. M. Born and P. Jordan, *Zur Quantenmechanik*, Zeitschrift für Physik **34**, 858 (1925); English translation in Ref. [10].
15. M. Born, W. Heisenberg, and P. Jordan, *Zur Quantenmechanik II*, Zeitschrift für Physik **35**, 557 (1925); English translation in Ref. [10].
16. E. Schrödinger, *An Undulatory Theory of the Mechanics of Atoms and Molecules*, Physical Review **28**, 1049 (1926).
17. L. de Broglie, *Recherches sur la théorie des quantas*, Annales de Physique **3**, 22 (1925).
18. L. de Broglie *La mécanique ondulatorie et la structure atomique de la matière et du rayonnement*, Journal de Physique et du Radium **8**, 225 (1927).
19. B. L. van der Waerden, *From Matrix Mechanics and Wave Mechanics to Unified Quantum Mechanics*, Notices of the AMS **44**, 323 (1997).
20. J. S. Bell, 1987, *Speakable and Unspeakable in Quantum Mechanics* (Cambridge University Press, Cambridge, 1987).
21. D. Bohm, *Quantum Theory* (Dover Publications, New York, 1951).
22. N. D. Mermin, *What's Wrong with This Pillow?*, Physics Today **42**(4), 9 (1989).
23. N. D. Mermin, *Could Feynman Have Said This?*, Physics Today **57**(5), 10 (2004).
24. G. Bacciagaluppi and A. Valentini, *Quantum Theory at the Cross-roads: Reconsidering the 1927 Solvay Conference* (Cambridge University Press, Cambridge, 2009).
25. L. de Broglie, *Tentative d'interpretation causale et non-lineairie de la mecanique ondulatorie* (Gauthier-Villars, Paris, 1956).
26. http://www.alberteinstein.info/db/ViewDetails.do?DocumentID = 34338; explained also in [27].

27. P. R, Holland, *What's Wrong with Einstein's 1927 Hidden-Variable Interpretation of Quantum Mechanics?*, Foundations of Physics, **35**(2) 177 (2005).
28. D. Bohm, on an interview in 1987; see http://www.fdavidpeat.com/interviews/bohm.htm
29. A. Einstein, B. Podolsky, and N. Rosen, *Can Quantum Mechanical Description of the Physical Reality Be Considered Complete?* Physics Review **47**, 777 (1935).
30. J. von Neumann, *Mathematische Grundlagen der Quantenmechanik* (Springer Verlag, Berlin 1932); English translation by: R. T. Beyer, *Mathematical Foundations of Quantum Mechanics* (Princeton University Press, Princeton, 1955).
31. P. R. Holland, *The Quantum Theory of Motion: An account of the de Broglie-Bohm Causal Interpretation of Quantum mechanics* (Cambridge University Press, Cambridge, 1993).
32. D. Peat, *Infinite Potential: the life and times of David Bohm* (Helix Books, Addison-Wesley, 1997).
33. C. Cohen-Tanoudji, B. Diu, and F. Laloë, *Quantum Mechanics* (Wiley, New York, 1977).
34. H. Nikolic, *Would Bohr Be Born if Bohm Were Born before Born?*, American Journal of Physics **76**, 143 (2008).
35. M. Gell-Mann, *The Nature of the Physical Universe: The 1976 Nobel Conference* (Wiley, New York, 1979); M. Gell-Mann presented his Nobel lecture but did not submit the manuscript for inclusion in the official volume. He explains: *I did have a written version of the lecture that I gave in Sweden, but I was not satisfied with it and did not submit it ... I tried to write a better one, including an adequate discussion of quarks, and agonized over it for months, but in the end, I did not finish it in time for it to be included in the volume.*
36. J. S. Bell, *On the Einstein Podolsky Rosen Paradox*, Physics **1**, 195 (1964).
37. J. S. Bell, *On the Impossible Pilot Wave*, Foundations of Physics **12**, 989 (1982). Reprinted in [20].
38. J. S. Bell, *Against Measurement*, Physics World **3**, 33 (1990).
39. H. P. Stapp, *Are Superluminal Connections Necessary?*, Il Nuovo Cimento **40B**, 191 (1977).
40. A. Aspect, P. Grangier, and G. Roger, *Experimental Realization of Einstein-Podolsky-Rosen-Bohm Gedankenexperiment: A New Violation of Bell's Inequalities*, Physical Review Letters **49**(2), 91 (1982).

41. D. Kleppner and R. Jackiw, *One Hundred Years of Quantum Physics*, Science **289**(5481), 893 (2000).
42. E. Madelung, *Quantentheorie in hydrodynamischer Form*, Zeitschrift fur Physik **40**, 322 (1926).
43. R. E. Wyatt *Quantum Dynamics with Trajectories: Introduction to Quantum Hydrodynamics* (Springer, 2005).
44. T. Takabayasi, *On the Formulation of Quantum Mechanics Associated with Classical Pictures*, Progress of Theoretical Physics **8**, 143 (1952).
45. It is written in Ref. [71] by Professor J. T. Cushing, himself, that he submitted a research proposal to the US National Science Foundation *for theoretical work to be done, within the framework of Bohm's version of quantum theory, on some foundational questions in quantum mechamics* and received the following evaluation *The subject under consideration, the rival Copenhagen and causal (Bohm) interpretations of the quantum theory, has been discussed for many years and in the opinion of several members of the Physics Division of the NSF, the situation has been settled. The causal interpretation is inconsistent with experiments which test Bell's inequalities. Consequently ... funding ... a research program in this area would be unwise.*
46. G. Brida, E. Cagliero, G. Falzetta, M. Genovese, M. Gramegna, and E. Predazzi, *Biphoton Double-Slit Experiment*, Physical Review A **68**, 033803 (2003).
47. X. Oriols, *Comment on "Biphoton Double-Slit Experiment"*, Physical Review A **71**, 017801 (2005).
48. P. A. Schilpp, Discussion with Einstein on epistemological problems in atomic physics in *The Library of Living Philosophers, Volume 7. Albert Einstein: Philosopher-Scientist*, pp. 199–241 (Open Court, La Salle-Illinois).
49. D. Bohm, *Proof that Probability Density Approaches* $|\psi|^2$ *in Causal Interpretation of Quantum Theory*, Physical Review **89**, 458 (1953).
50. W. Struyve, *The de Broglie–Bohm Pilot-Wave Interpretation of Quantum Theory* PhD thesis, Universiteit Gent (2004) and also in arXiv:quant-ph/0506243.7
51. D. Dürr, S. Goldstein, R. Tumulka, and N. Zanghì, *Bohmian Mechanics and Quantum Field Theory*, Physical Review Letters **93**(9), 090402 (2004).

52. D. Dürr, S. Goldstein, R. Tumulka, and N. Zanghì, *Trajectories and Particle Creation and Annihilation in Quantum Field Theory*, Journal of Physics A - Matemathical and General **36**(14), 4143 (2003).
53. R. P. Feynman, R. B. Leighton, and M. Sands, *The Feynman Lectures on Physics, I, II and III* (Addison-Wesley, New York, 1963).
54. H. M. Wiseman, *Grounding Bohmian mechanics in weak values and bayesianism* New Journal of Physics **9**, 165 (2007).
55. D. Marian, N. Zanghì, X. Oriols, *Weak Values from Displacement Currents in Multiterminal Electron Devices*, Physical Review Letters **116**, 110404 (2016) and also in arXiv:1508.00248.
56. E. Colomés, Z.Zhan, X. Oriols *Comparing Wigner, Husimi and Bohmian distributions: Which one is a true probability distribution in phase space?*, Journal of Computational Electronics **14**, 894 (2015) and also in arXiv1507.07828.
57. W. Struyve, *Pilot-wave theory and quantum fields*, Reports on Progress in Physics **73** 106001 (2010).
58. P. R. Holland, *Uniqueness of Paths in Quantum Mechanics*, Physical Review A **60**, 4326 (1999).
59. P. R. Holland and C. Philippidis, *Implications of Lorentz Covariance for the Guidance Equation in Two-Slit Quantum Interference*, Physical Review A **67**, 062105 (2003).
60. D. Bohm and B. J. Hiley. *The Uundivided Universe: An Ontological Intepretation of Quantum Theory* (Routledge & Kegan Paul, London, 1993).
61. D. Dürr and S. Teufel, *Bohmian Mechanics: The Physics and Mathematics of Quantum Theory* (Spinger, 2009).
62. A. Benseny, G. Albareda, A. S. Sanz, J. Mompart, and X. Oriols, *Applied Bohmian mechanics*, European Physical Journal D **68**, 286 (2014).
63. D. Dürr, S. Goldstein, and N. Zanghì, *Quantum Mechanics without Quantum Philosophy* (Springer-Verlag, Berlin, 2013).
64. C. Philippidis, C. Dewdney, and B. J. Hiley, *Quantum Interference and the Quantum Potential*, Il Nuovo Cimento **52B**, 15, (1979).
65. S. Goldstein and W. Struyve, *On the Uniqueness of Quantum Equilibrium in Bohmian Mechanics*, Journal of Statistical Physics **128**5, 1197 (2007).
66. D. Dürr, S. Goldstein, and N. Zanghì, *Quantum Equilibrium and the Origin of Absolute Uncertainty*, Journal of Statistical Physics **67**(5–6), 843 (1992).

67. A. Solé, X. Oriols, D. Marian, and N. Zanghì, *How Does Quantum Uncertainty Emerge from Deterministic Bohmian Mechanics?*, Fluctuation and Noise Letters **15**, 1640010 (2016).
68. A. Valentini, *Pilot-Wave Theory: An Alternative Approach to Modern Physics* (Cambridge University Press, Cambridge, 2006).
69. S. Colin and W. Struyve, *Quantum Non-Equilibrium and Relaxation to Equilibrium for a Class of the de Broglie–Bohm-Type Theories*, New Journal of Physics **12**, 043008 (2010).
70. R. P. Feynman and A. R. Hibbs,*Quantum Mechanics and Path Integrals* (McGraw-Hill, New York, 1965).
71. J. T. Cushing, A. I. Fine, and S. Goldstein, *Bohmian Mechanics and Quantum Theory: An Appraisal* (Kluwer, Dordrecht, 1996).
72. P. A. M. Dirac, *Quantum Mechanics of Many-Electron Systems*, Proceedings of the Royal Society of London **A 123**, 714 (1929).
73. M. Born, *The Classical Mechanics of the Atom* (Ungar, New York, 1960).
74. P. Hohenberg and W. Kohn, *Inhomogeneous Electron Gas*, Physical Review **136**(3B), B864 (1964).
75. W. Kohn and L. J. Sham, *Self-Consistent Equations Including Exchange and Correlation Effects*, Physical Review **140**(4A), A1133 (1965).
76. D. R. Hartree, *The Wave Mechanics of an Atom with a Non-Coulomb Central Field. Part I. Theory and Methods* Mathematical Proceedings of the Cambridge Philosophical Society **24**(01), 89 (1928).
77. D. R. Hartree, *The Wave Mechanics of an Atom with a Non-Coulomb Central Field. Part II. Some Results and Discussion* Mathematical Proceedings of the Cambridge Philosophical Society **24**(01), 111 (1928).
78. V. Fock, *Näherungsmethode zur Lösung des quanten mechanichanischen Mehrkörperproblems* , Zeitschrift für Physik **61**, 126 (1930).
79. L. D. Landau and E. M. Lifschitz, *Quantum Mechanics Non-Relativistic Theory* (Pergamon Press, London, 1958).
80. C. Colijn and E. R. Vrscay, *Spin-Dependent Bohm Trajectories for Hydrogen Eigenstates*, Physical Letters A **300**, 334 (2002).
81. J. J. Sakurai, *Modern Quantum Mechanics* (Addison-Wesley, 1994).
82. M. D. Towler, N. J. Russell, and A. Valentini, *Time scales for dynamical relaxation to the Born rule*, Proceedings of the Royal Society of London Series A **468** 990 (2011).
83. T. Norsen, *The Theory of (Exclusively) Local Beables*, Foundations of Physics **40**, 1858 (2010).

84. T. Norsen, D. Marian, X. Oriols, *Can the wave function in configuration space be replaced by single-particle wave functions in physical space?*, Synthese **192 (10)**, 3125 (2015).
85. X. Oriols, *Quantum-Trajectory Approach to Time-Dependent Transport in Mesoscopic Systems with Electron-Electron Interactions*, Physical Review Letters **98**, 066803 (2007).
86. D. Bohm and B. J. Hiley, *The Undivided Universe* (Routledge, New York, 1993).
87. S. Goldstein, *Quantum mechanics without observers-part one*, Physics Today **51**(3), 42 (1998).
88. J. S. Bell, *On the Problem of Hidden Variables in Quantum Theory*, Reviews of Modern Physics **38**, 447 (1966); reprinted in [20].
89. H. D. Zeh, *On the Interpretation of Measurement in Quantum Theory*, Foundations of Physics **1**, 69 (1970).
90. T. Maudlin, *Three measurement problems*, Topoi **14**, 7 (1995).
91. D. Bohm and J. Bub, *A Proposed Solution of the Measurement Problem in Quantum Mechanics by a Hidden Variable Theory*, Reviews of Modern Physics **38**, 453 (1966).
92. H. Nikolić, *Classical mechanics without determinism*, Foundation of Physics Letters **19**, 553 (2006).
93. N. Herbert, *Quantum reality* (Anchor Books, New York 1984).
94. A. Bassi, K. Lochan, S. Satin, T. P. Singh and H. Ulbricht, *Models of wave-function collapse, underlying theories, and experimental tests*, Reviews of Modern Physics **85**, 471 (2013).
95. G. C. Ghirardi, A. Rimini, and T. Weber, *Unified Dynamics for Microscopic and Macroscopic System*, Physical Review D **34**, 470 (1986).
96. N. Bohr, *Uber die Serienspektra der Element*, Zeitschrift fur Physik **2**, 423 (1920).
97. H. Everett, *Relative State Formulation of Quantum Mechanics*, Reviews of Modern Physics **29**, 454 (1957).
98. D. Wallance, *Chaos, Decoherence, and Branching*, in *The Emergent Multiverse: Quantum Theory according to the Everett Interpretation*, Oxford University Press, Oxford (2012).
99. M. Daumer, D. Dürr, S. Goldstein, and N. Zanghì, *Naive Realism about Operators*, Probability, Dynamics and Causality **45**(2–3), 379 (1997).
100. D. Dürr, S. Goldstein and N. Zanghì, *Quantum Equilibrium and the Role of Operators as Observables in Quantum Theory*, Journal of Statistical Physics **116**, 9595 (2004).

101. G. Albareda, D. Marian, A. Benali, S. Yaro, N. Zanghì, X. Oriols *Time-resolved Electron Transport with Quantum Trajectories*, Journal of Computational Electronics **12**, 405 (2013).

102. D. Marian, E. Colomés, Z. Zhan, X. Oriols, *Quantum Noise from a Bohmian perspective: fundamental understanding and practical computation in Electronic Devices*, Journal of Computational Electronics **14**, 114 (2015) and also in arXiv:1410.0530.

103. Xavier Oriols and Albert Benseny, *Conditions for the classicality of the center of mass of many-particle quantum states*, New Journal of Physics **19** 063031 (2017) and also in arXiv:1602.03988.

104. A. Benseny, D. Tena, X. Oriols, *On the Classical Schrodinger Equation*, Fluctuation and Noise Letters **15 (03)** 1640011 (2016) and also in arXiv1607.00168.

105. X. Oriols, *Can decoherence make quantum theories unfalsifiable? Understanding the quantum-to-classical transition without it*, Journal of Physics: Conference Series **701**, 12004 (2016).

106. X Oriols, *Special Issue on Quantum and Classical Frontiers of Noise*, Fluctuations and Noise Letters **15**, 1602001 (2016).

107. D. H. Kobe, *Quantum Power in de Broglie–Bohm Theory*, Journal of Physics A: Mathematical Theoretical **40**, 5155 (2007).

108. J. G. Muga, J. P. Palao, and R. Sala, *Average local values and local variances in quantum mechanics*, Physics Letters A **238**, 90, (1998).

109. B. V. Numerov, *A Method of Extrapolation of Perturbations* Monthly Notices of the Royal Astronomical Society **84**, 592 (1924).

110. B. V. Numerov, *Note on the Numerical Integration of* $d^2x/dt^2 = f(x, t)$, Astronomische Nachrichten **230**, 359 (1927).

111. C. R. Leavens, *Time of Arrival in Quantum and Bohmian Mechanics*, Physical Review A **58**2, 840 (1998).

112. X. Oriols, F. Martín, and J. Suñé, *Implications of the Noncrossing Property of Bohm Trajectories in One-Dimensional Tunneling Configurations*, Physical Review A **54**, 2594 (1996).

113. X. Oriols, F. Martín, and J. Suñé, *Oscilatory Bohm Trajectories in Ressonant Tunneling Structures*, Solid State Communication **99**, 123 (1996)

114. A. Benseny, J. Bagudà, S. Fernández-Vidal, R. Corblán, A. Picón, L. Roso, G. Birkl, and J. Mompart, *Atomtronics with Holes: Coherent Transport of an Empty Site in a Triple-Well Potential*, Physical Review A **82**, 013604 (2010).

115. A. Picón, A. Benseny, J. Mompart, J. R. Vázquez Aldana, L. Plaja, G. F. Calvo, and L. Roso, *Transferring Spin and Orbital Angular Momenta of Light to Atoms*, New Journal of Physics **12**, 083053 (2010).
116. P. O. Persson and G. Strang, *A Simple Mesh Generator InMatlab*, SIAM Review **46**(2), 329 (2004).
117. M. R. Pinto, *Comprehensive Semiconductor Device Simulation for Silicon ULSI*, PhD thesis, Stanford University (1990).
118. L. Perko, *Differential Equations and Dynamical Systems* (Springer Verlag, New York, 1991).
119. C. L. Lopreore and R. E. Wyatt, *Quantum Wave Packet Dynamics with Trajectories*, Physical Review Letters **82**, 5190 (1999).
120. R. E. Wyatt and E. R. Bittner, *Quantum Wave Packet Dynamics with Trajectories: Implementation with Adaptive Lagrangian Grids*, Journal of Chemical Physics **113**, 8898 (2000).
121. R. E. Wyatt, D. J. Kouri, and D. K. Hoffman, *Quantum Wave Packet Dynamics with Trajectories: Implementation with Distributed Approximating Functionals*, Journal of Chemical Physics **112**, 10730 (2000).
122. D. Nerukh and J. H. Frederick, *Multidimensional Quantum Dynamics with Trajectories: A Novel Numerical Implementation of Bohmian Mechanics*, Chemical Physics Letters **332**(1–2), 145 (2000).
123. J. C. Striwerda, *Finite Difference Scheme and Partial Differential Equation* (Wadsworth & Brooks/Cole, 1989).
124. Y. Goldfarb, I. Degani, and D. J. Tannor, *Bohmian Mechanics with Complex Action: A New Trajectory-Based Formulation of Quantum Mechanics*, Journal of Chemical Physics **125**, 231103 (2006).
125. B. Poirier, *Bohmian mechanics without pilot waves*, Chemical Physics **3760**, 4 (2010).
126. W. Koch, F. Grossmann, and D. J. Tannor, *Trajectory Based Non-Markovian Dissipative Tunneling*, Physical Review Letters **105**, 230405 (2010).
127. J. Liu and N. Makri, *Bohm's Formulation in Imaginary Time: Estimation of Energy Eigenvalues*, Molecular Physics **103**, 1083 (2005).
128. I. P. Christov, *Time Dependent Quantum Monte Carlo: Preparation of the Ground State*, New Journal of Physics **9**, 70 (2007).
129. I. P. Christov, *Dynamic Correlations with Time-Dependent Quantum Monte Carlo*, Journal of Chemical Physics **128**, 244106 (2008).

130. E. R. Bittner, *Quantum Tunneling Dynamics Using Hydrodynamic Trajectories*, Journal of Chemical Physics **112** 22, 9703 (2000).
131. S. Garashchuk and V. A. Rassolov, *Quantum Dynamics with Bohmian Trajectories: Energy Conservation Approxiamtion to the Quantum Potential* Chemical Physics Letters **376**, 358 (2003).
132. G. Albareda, H. López, X. Cartoixà, J. Suñé, and X. Oriols, *Time-Dependent Boundary Conditions with Lead-Sample Coulomb Correlations: Application to Classical and Quantum Nanoscale Electron Device Simulators*, Physical Review B **82**, 085301 (2010).
133. G. Albareda, J. Suñé, and X. Oriols, *Many-Particle Hamiltonian for Open Systems with Full Coulomb Interaction: Application to Classical and Quantum Time-Dependent Simulations of Nanoscale Electron Devices*, Physical Review B **79**, 075315 (2009).
134. E. Gindensperger, C. Meier, and J. A. Beswick, *Mixing Quantum and Classical Dynamics Using Bohmian Trajectories*, Journal of Chemical Physics **113** 21, 9369 (2000).
135. D. Marian, E. Colomés, X. Oriols, *Time-dependent exchange and tunneling: detection at the same place of two electrons emitted simultaneously from different sources*, Journal of Physics: Condensed Matter **27**, 245302 (2015) and also in arXiv:1408.1990.
136. E. Colomés, D. Marian, X. Oriols, *Quantum noise with exchange and tunneling: Predictions for a two-particle scattering experiment with time-dependent oscillatory potentials*, Journal of Statistical Mechanics: Theory and experiment **5**, 054009 (2016) and also in arXiv:1511.00608.
137. B. E. F. Curchod, I. Tavernelli, and U. Rothlisberger, *Trajectory-based solution of the nonadiabatic quantum dynamics equations: an on-the-fly approach for molecular dynamics simulations* Physical Chemistry Chemical Physics **13**, 3231 (2011).
138. G. Albareda, et al., *Conditional Born-Oppenheimer Dynamics: Quantum Dynamics Simulations for the Model Porphine*, The Journal of Physical Chemistry Letters **6**, 1529 (2015).
139. G. Albareda, et al., *Correlated electron-nuclear dynamics with conditional wave functions* Physical Review letters **113**, 083003 (2014).
140. G. Albareda, A. Abedi, I. Tavernelli, A. Rubio, *Universal steps in quantum dynamics with time-dependent potential-energy surfaces: Beyond the Born-Oppenheimer picture*, Physical Review A **94**, 062511 (2016).
141. B. Gu, and I. Franco, *Partial hydrodynamic representation of quantum molecular dynamics*, The Journal of Chemical Physics **146**, 194104 (2017).

142. E. Gindensperger, C. Meier, J. A. Beswick, *Mixing Quantum and Classical Dynamics Using Bohmian Trajectories*, Journal of Chemical Physics **113**, 9369 (2000).
143. O. V. Prezhdo and C. Brooksby, *Quantum Backreaction through the Bohmian Particle* Physical Review Letters **86**, 3215 (2001).
144. T. A. Elsayed, K. Mølmer, L. B. Madsen, *Solving the Quantum Many-Body Problem with Bohmian Trajectories* in arXiv:1706.00818.
145. R. E. Wyatt, *Wave Packet Dynamics on Adaptive Moving Grids*, Journal of Chemical Physics **117** 21, 9569 (2002).

Chapter 2

Hydrogen Photoionization with Strong Lasers

Albert Benseny,[a,b] Antonio Picón,[c,d] Jordi Mompart,[b] Luis Plaja,[c] and Luis Roso[e]

[a] *Department of Physics and Astronomy, Aarhus University, Ny Munkegade 120, DK-8000, Aarhus C, Denmark*
[b] *Departament de Física, Universitat Autònoma de Barcelona (UAB), E-08193 Bellaterra, Spain*
[c] *Departamento de Física Aplicada, University of Salamanca (USAL), E-37008 Salamanca, Spain*
[d] *Institut de Ciències Fotòniques (ICFO), The Barcelona Institute of Science and Technology, E-08860 Castelldefels (Barcelona), Spain*
[e] *Centro de Láseres Pulsados Ultracortos Ultraintensos (CLPU), E-37008 Salamanca, Spain*
albert.benseny-cases@oist.jp; jordi.mompart@uab.cat

The invention of lasers in the early sixties triggered the development of the rich field of nonlinear optics. Similarly, the development of lasers with ultrahigh intensities, i.e., when light's electric field is larger than the attraction between electrons and nuclei, has boosted extreme nonlinear optics to a relevant and promising discipline in physics. Besides its technological potentialities, strong field physics is also challenging in a fundamental level, leading to the reconsideration of the basic processes of light-matter interactions. Also, the complexity of the exact solutions of the nonperturbative

Applied Bohmian Mechanics: From Nanoscale Systems to Cosmology (Second Edition)
Edited by Xavier Oriols and Jordi Mompart

ISBN 978-981-4800-10-5 (Hardcover), 978-0-429-29474-7 (eBook)
www.jennystanford.com

equations has encouraged the development of simple approaches based in semiclassical concepts, that offer invaluable insight into the basic processes involved. In this sense, Bohmian trajectories offer an interesting extension of the semiclassical treatment, since their description is purely quantum and, yet, they retain the intuitiveness of a classical description. In this chapter we study the interaction of a hydrogen atom with short and intense laser pulses. By first using a simplified one-dimensional model, we show how Bohmian trajectories can be used in the calculation of the above-threshold ionization (ATI) and harmonic generation spectra. We also present a full three-dimensional (3D) model to study the interaction of the atom with beams bearing orbital angular momentum (OAM) and use the trajectories to gain physical insight into the absorption of angular momentum.

2.1 Introduction

2.1.1 A Brief Overview of Photoionization

There is little question about the fundamental role of optics in the historical development of the physical sciences. Light is an obvious tool for visualization, but it is also a peculiar physical system by itself: it propagates in vacuum with a fixed velocity, its wave nature can be made evident with simple experiments, while other phenomena are more conveniently explained with a particle description. Thus, the study of light propagation and its interaction with matter lies in the foundations of electromagnetism, special relativity and quantum physics.

The discovery of the photoelectric effect by Philipp Lénárd, and its explanation by Albert Einstein in terms of light quanta [1] (it was not until 1926 that the name *photon* was proposed by Gilbert Lewis) was essential in the genesis of quantum physics. About 20 years later, the photoelectric problem was still taken as a model to develop time-dependent perturbation theory by Paul Dirac [2]. The possibility of inducing two-photon transitions was predicted theoretically in 1931 in Maria Göppert-Mayer's doctoral dissertation [3], laying the foundations of multiphoton processes.

The absorption of multiple photons by a system is often followed by the emission of the accumulated energy in a single photon in the nonlinear process known as optical harmonic generation. To induce these nonlinearities, the optical sources have to be intense enough, and thus, the observation of these processes had to wait for the advent of lasers in 1960. The first experimental detection of the second optical harmonic generation was reported in 1961 by Franken et al. [4].

The first classical calculations of multiphoton ionization rates were presented by H. Barry Bebb and Albert Gold in a seminal paper published in 1966 [5]. The following years witnessed the subsequent refinement of these calculations, which were specially complicated due to the sums over the large set of intermediate states. However, other authors studied approaches different from perturbation theory, taking into account ultrastrong laser intensities [6–9], paving the way for the theoretical study of the ultrastrong lasers field [10].

Laser techniques, such as Q-switching, mode-locking, and chirped pulse amplification [11], increased the intensity of laser pulses to the nonperturbative regime, and the first experiments applying them to photoionization [12, 13] showed the emergence of high-order multiphoton transitions with energies well above the ionization threshold. Also, the first reports on high-order harmonic generation showed unexpected features [14, 15], such as the emergence of a plateau of several harmonics with similar intensities, followed by an abrupt cut-off.

The nonperturbative features of the harmonic spectrum stimulated the theoretical work in very different directions. Semiclassical or classical models [16–18] were proposed to explain the most salient features of the spectrum. In these models, harmonics are generated by the ionized electrons that subsequently rescatter with the ion when the field force changes its sign. The case of harmonic generation is a paradigmatic example of an extraordinary simplification of a physical process in the limit of strong nonperturbative interactions, where a naïve expectation would predict an increasing complexity. Semiclassical trajectories are extracted from the classical equations of motion and a time-dependent phase according with their instantaneous action is assigned to them. When applicable, the classical or semiclassical interpretations of quantum physics are extraordinarily powerful in terms of intuition but, also, in terms of computing. In this context, Bohmian trajectories offer an interesting extension of the semiclassical treatment, as they include the fully quantum dynamical effects given by the quantum potential. As a result, their dynamics reproduce faithfully the probability distribution of the quantum wave function at any time.

2.1.2 The Computational Problem of Photoionization

Photoionization is one of the main consequences of the interaction of an atom (or molecule) with an electromagnetic pulse. In the nonrelativistic regime, the exact quantum description of this interaction is given by the Schrödinger equation, that for a quantum system consisting of N particles, reads:

$$i\hbar\frac{\partial\psi(\vec{r}_1,\ldots,\vec{r}_N,t)}{\partial t} = H\psi(\vec{r}_1,\ldots,\vec{r}_N,t), \tag{2.1}$$

where $\psi(\vec{r}_1,\ldots,\vec{r}_N,t)$ is the wave function that describes the quantum state of the system, whose dynamics is governed by the Hamiltonian H. One can easily realize that to obtain the complete dynamics of a 3D system of N particles, Eq. 2.1 must be integrated over $3N$ dimensions. The integration time scales exponentially with the number of dimensions, which heavily limits the size of the systems that can be simulated. See Ref. [19] for a recent review on the methods used. In principle, even the full dynamics of a "simple" hydrogen atom cannot be simulated, since its wave function lives in a six-dimensional space: three dimensions for the electron plus three dimension for the proton. Furthermore, in the case of strong fields, perturbative approaches are useless to solve Eq. 2.1, since we are considering a regime where the electromagnetic term and the Coulomb potential are of the same order of magnitude.

With the aim of reducing the computational costs, a series of approximations are usually performed. One of the most commonly used in the study of atomic dynamics is the fixed-nuclei approximation, which consists in taking the atom nuclei as point particles at fixed positions, and thus, considering only the dynamics of the electrons. For a hydrogen atom interacting with an ultraintense and ultrashort light pulse, the assumption of a fixed nucleus is quite natural due to the mass of the proton being almost 2,000 times larger than that of the electron. Thus, the proton being so much heavier than the electron, the motion of the nucleus will be much slower than that of the electron. In addition, if the dynamics is rewritten in terms of the center of mass and relative motions, the coordinate of the center of mass will practically coincide with the position of the nucleus, while the relative coordinate with the electron's location. Then, in the approximation of a local homogeneous electromagnetic field, the center of mass dynamics will not couple to the electromagnetic field.

Thus, typically, the computation of atomic photoionization consists in the study of electron dynamics around a fixed charged nucleus interacting with a strong light field. For a hydrogen atom, the most general description of this system, will consist of a 3D integration of the Schrödinger equation. The study of the exact dynamics of more complex atoms is very computationally demanding and only helium has been studied so far [20]. Lithium still remains beyond present computational capabilities. Thus, a simplified atomic model, restricting the electron dynamics to one or two dimensions is frequently used in the literature to overcome this limit. For instance, a one-dimensional approach allows to compute exactly the three-particle problem (lithium) with a desktop computer [21]. Even though this is a drastic approximation, it retains the main physics of the photoionization dynamics and allows for a simple study of the light-atom interaction.

2.1.3 Photoionization with Bohmian Trajectories

The hydrogen atom has been extensively studied in the literature with Bohmian trajectories as a prototype system to investigate photoionization [22]. Simple one-dimensional models have been used for instance to perform straightforward calculations of the ATI and harmonic-generation emission spectra [23, 24], as well as for an insight into the role of the quantum potential in photoionization [25, 26], or even to study the chaotic behavior of classical and Bohmian trajectories [27]. Moreover, a self-consistent method based on the quantum Hamilton–Jacobi formalism has been proposed to study hydrogen photoionization [28]. Higher-dimensional models have also been implemented, in particular, to study the sub-cycle photoionization dynamics, which gives rise to multiple ionization bursts each optical cycle. The trajectories pinpoint the origin of these electron ionization structures [29]. Other works have studied the transfer of angular momentum to a single hydrogen atom with helical beams [30]. The Bohmian formalism has also been used to address ionization concepts such as dwell time [31] and exit point [32], as they have very natural descriptions inside this formulation. Moreover, a method for calculating ionization yields has been proposed, which is based on simple criteria from the Bohmian trajectories' energy, and gives good agreement with the usual methods [33].

Perhaps the most studied feature of photoionization with Bohmian trajectories are the dynamics associated to high-order harmonic generation. It has been shown that trajectories from different parts of the electron wave packet contribute to different parts of the harmonic spectrum: trajectories which ionize and oscillate with the field provide a good representation of the low-frequency part of the spectrum, while the inner ones are responsible for the plateau and cut-off for the high-frequency harmonics [34]. This has been confirmed by more in-depth studies with long-range and short-range potentials which have also assessed that the inner trajectories affect nonlocally the outer ones [35, 36]. In this context, an efficient method for obtaining the harmonic-generation spectrum from as little as 20 trajectories has been proposed [37]. Trajectories were also used to illustrate the effect of the multiple recollisions of the electron with the nucleus in some features, such as subpeaks, of the harmonic generation spectrum [38, 39]. More complex processes, such as the appearance of multiple plateaus and the extension of the cut-off during the harmonic generation process have been studied in two-color laser fields [40]. From a more fundamental point of view, the destructive interference of the even harmonics has also been investigated, as their associated phases are random due to the trajectory positions, making them disappear from the spectrum [41].

An approximate method based on Bohmian trajectories to study multielectronic atoms in ultrastrong fields has been developed by using the conditional wave function formalism [42]. This time-dependent quantum Monte Carlo method can be applied to study both dynamical [43, 44] and ground state problems [45]. This technique has been applied to the study of the dynamics of three-dimensional helium [46], as well as to investigate correlations and nonlocality in multielectron states [47, 48].

Molecule photoionization has also been studied with a Bohmian formalism. The interaction of a H_2^+ ionic molecule with an intense infrared laser pulse presents multiple bursts ionization within a half-cycle of a laser field [49], which can be understood from the fast electron dynamics inside the molecule, as can be seen from the Bohmian trajectories of the electron [50]. A similar study in the ionization of a one-dimensional H_2 molecule [51], easily allows to distinguish between two different classes of ionization, correlated and uncorrelated, and elucidates from where does the ejected electron originate. The high-order harmonic generation in H_2^* has also been studied with Bohmian trajectories, finding that the minimum in the spectrum is due to the interference of the trajectories around the two atoms of the molecule [52].

This chapter is separated in two main parts. In the first one, Section 2.2 we will focus on a simple one-dimensional model for the hydrogen atom, and we will review, following the lines of Refs. [53] and [54] the use of Bohmian trajectories for the calculation of harmonic-generation and ATI spectra. For the latter, we will present a novel method in which the wave function is reconstructed from the trajectories.

In the second part, Section 2.3, we will investigate in detail the interaction of a hydrogen atom with intense Gaussian and Laguerre-Gaussian (LG) light pulses, by means of 3D numerical simulations. We will study the dependence of the angular momentum exchange on the light polarization and OAM. The Bohmian equations for an electron in an electromagnetic field will be derived to calculate the Bohmian trajectories of the evolution in order to gain physical insight into the absorption of angular momentum by the hydrogen atom. In particular, we will illustrate how the electron absorbs the angular momentum due to the light polarization, i.e., spin, and that due to its transverse profile, i.e., its OAM. These results were published [55] and constituted the first study of photoionization in three dimensions addressed with Bohmian trajectories.

Throughout this chapter we will make use of atomic units for the numerical values of physical quantities. However, we will keep the fundamental constants in all the formulae to be consistent with the rest of the book. In atomic units, the electron mass m_e, the electron charge e, Planck's reduced constant $\hbar$ (that has units of angular momentum), and Coulomb's force constant $K = \frac{1}{4\pi\epsilon_0}$ are chosen to be 1. Some notable derived units are:

- length: $a_0 = \dfrac{\hbar^2}{Km_e e^2}$ (Bohr radius),
- energy: $E_h = \dfrac{K^2 m_e e^4}{\hbar^2}$ (Hartree energy),
- time: $\dfrac{\hbar}{E_h}$,
- electric field: $\dfrac{E_h}{ea_0}$.

In these units, the speed of light is equivalent to $c = Ke^2/\hbar\alpha_{\mathrm{f}} = 1/\alpha_{\mathrm{f}} \simeq 137.036$, being α_{f} the fine structure constant.

2.2 One-Dimensional Photoionization of Hydrogen

2.2.1 The Physical Model

In this section we will first present the 3D Schrödinger equation for a hydrogen atom interacting with light to later restrict its dynamics to one dimension, parallel to the light polarization. In any case, we will consider the atom nucleus position to be fixed and consider only the electron dynamics.

In the length gauge,[a] the Schrödinger equation takes the form:

$$i\hbar\frac{\partial\psi(\vec{r},t)}{\partial t} = \left[\frac{\vec{p}^2}{2m} + q\vec{r}\cdot\vec{E}(\vec{r},t) + V(\vec{r})\right]\psi(\vec{r},t), \tag{2.2}$$

[a]One must take into account that the length gauge is only valid within the dipolar approximation, that we will apply in this section. Alternatively, one could use the velocity gauge as it will be done in Section 2.3.

where m and q are, respectively, the mass and the (negative, $q = -e$) charge of the electron, $\psi(\vec{r}, t)$ is the electron wave function, $\vec{p} = -i\hbar\vec{\nabla}$ the linear momentum operator, $\vec{E}(\vec{r}, t)$ the light electric field, and $V(\vec{r})$ the Coulomb potential due to the interaction with the hydrogen nucleus:

$$V(\vec{r}) = -\frac{1}{4\pi\epsilon_0}\frac{e^2}{|\vec{r}|}. \tag{2.3}$$

In this section, we will apply both the transverse spatial and the electric dipole approximations to the light field. On the one hand, the transverse spatial approximation consists in taking the light beam waist much larger than the electron wave packet. For a light beam with a Gaussian transverse profile, this is equivalent to consider the electric field constant in the transverse plane, i.e., $\vec{E}(\vec{r}, t) = \vec{E}(z, t)$, z being the beam-propagating direction. On the other hand, in the electric dipole approximation, the light wavelength is assumed to be much larger than the electron wave packet (note that the wavelength for visible light is around 500 nm, while $a_0 \simeq 0.05$ nm), and then, the electric field is taken constant along the propagation axis, i.e., $\vec{E}(\vec{r}, t) = \vec{E}(x, y, t)$. One must be careful with the simultaneous application of both approximations, since then, the entire spatial dependence of $\vec{E}$, and therefore of the magnetic vector potential $\vec{A} = -\int \vec{E}dt$, would be ignored, which is equivalent to neglecting all magnetic field effects, since $\vec{H} = \vec{\nabla} \times \vec{A}(t) = 0$.

Taking the light polarization and electron dynamics to be along the x axis, the one-dimensional Schrödinger equation, Eq. 2.2, reads:

$$i\hbar\frac{\partial\psi(x, t)}{\partial t} = -\frac{\hbar^2}{2m}\frac{\partial^2\psi(x, t)}{\partial x^2} + q\,x\,E(t)\psi(x, t) + V(x)\psi(x, t). \tag{2.4}$$

In this case, it is convenient to redefine the Coulomb potential of Eq. 2.3, since a direct reduction to one dimension would have an inevitable singularity at $x = 0$. Thus, it is usually substituted by a soft core potential[a] of the form:

$$V(x) = -\frac{1}{4\pi\epsilon_0}\frac{e^2}{\sqrt{a^2 + x^2}}, \tag{2.5}$$

---

[a] This potential is also named Rochester potential in the literature after Eberly and coworkers from the University of Rochester first introduced it [62].

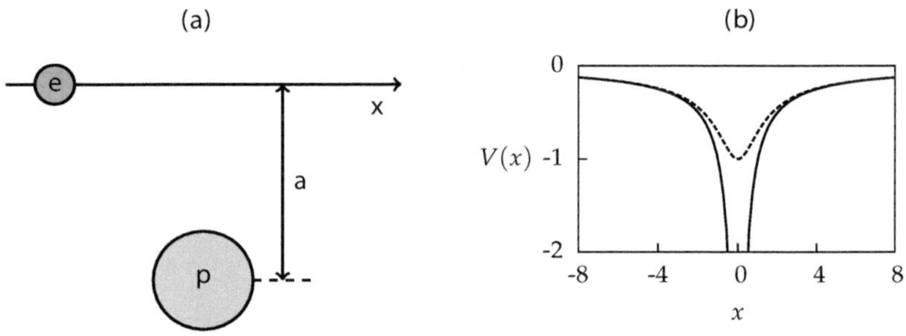

Figure 2.1 (a) Sketch of the one-dimensional model of the hydrogen atom. The electron (e) dynamics is restricted to a line separated a distance a from the nucleus (p). (b) Comparison between the full Coulomb potential in one dimension, i.e., Eq. 2.3 with $|\vec{r}| = |x|$, (solid curve), and the soft core potential of Eq. 2.5 for $a = 1$ (dashed curve).

where a is a parameter of the order of the Bohr radius that is normally selected by fitting the energy of the one-dimensional model ground state to the hydrogen ground state energy. This potential, see Fig. 2.1, is equivalent to restricting the electron dynamics to a line that is at a distance a of the nucleus. Asymptotically, it behaves like the Coulomb potential, but it is smoothed around the atom nucleus position.

For the electric field of the light, we will assume a pulse profile given by:

$$E(t) = E_0 U_{\text{env}}(t) \cos(\omega_0 t), \tag{2.6}$$

E_0 being the electric field amplitude, ω_0 the light frequency, and $U_{\text{env}}(t)$ the temporal envelope of the pulse. We define $\tau = 2\pi/\omega_0$ as the light period and $T = N_{\text{cyc}}\tau$ as the total pulse duration, with N_{cyc} the number of cycles. For short pulses, such as $N_{\text{cyc}} = 3$, the time envelope will be taken as a quadratic sine function:

$$U_{\text{env}}^{\text{short}}(t) = \sin^2\left(\frac{\pi t}{T}\right). \tag{2.7}$$

For longer pulses, $N_{\text{cyc}} > 6$, we will consider the following envelope: The pulse ramps up in three cycles, and stays at maximum amplitude until the last three cycles when it ramps down, i.e.,

$$U_{\text{env}}^{\text{long}}(t) = \begin{cases} \sin^2\left(\frac{\pi t}{6\tau}\right) & t \leq 3\tau \\ 1 & (N_{\text{cyc}} - 3)\tau > t > 3\tau \\ \sin^2\left(\frac{\pi(N_{\text{cyc}}\tau - t)}{6\tau}\right) & t > (N_{\text{cyc}} - 3)\tau \end{cases}. \tag{2.8}$$

We integrate numerically Eq. 2.4 by means of the Crank–Nicolson algorithm in a grid with a fixed spatial step of $\Delta x = 0.3$ au and find the evolution of the electron wave function $\psi(x, t)$ under the light pulse described in Eq. 2.6. We then obtain the evolution of a set of $N_{\rm traj} = 8192$ Bohmian trajectories, $\{x_k[t]\}$, whose initial distribution is taken uniform in the region of the x axis where $|\psi(x, 0)|^2$ is relevant, namely in $x \in (-10, 10)$ au. Note that the probability of finding the electron outside this region is less than 10^{-7}. We assign to each trajectory a weight $\gamma_k = \gamma_0 \, |\psi(x_k[t = 0])|^2$, with γ_0 chosen so $\sum_{k=1}^{N_{\rm traj}} \gamma_k = 1$. We define the expected value of an operator of the form $\mathcal{A} = \mathcal{A}(x)$ as:

$$\langle \mathcal{A} \rangle_\psi = \int_{-\infty}^{\infty} \mathcal{A}(x) \, |\psi(x, t)|^2 \, dx, \tag{2.9}$$

$$\langle \mathcal{A} \rangle_{\rm t} = \sum_{k=1}^{N_{\rm traj}} \mathcal{A}(x_k(t)) \gamma_k. \tag{2.10}$$

For $N_{\rm traj} \rightarrow \infty$, $\langle \mathcal{A} \rangle_\psi = \langle \mathcal{A} \rangle_{\rm t}$.

2.2.2 Harmonic Generation

From the point of view of energy conservation, harmonic generation can be seen as the absorption of n photons with frequency ω_0, followed by the emission of a photon with higher frequency $n\omega_0$ (see Fig. 2.2). However, this simplified picture gives only little information about the emitted spectrum (the allowed emission frequencies), and no information about the dynamics. For systems with inversion symmetry, the matter response is an odd function of the field amplitude and, therefore, only odd harmonics can be emitted.

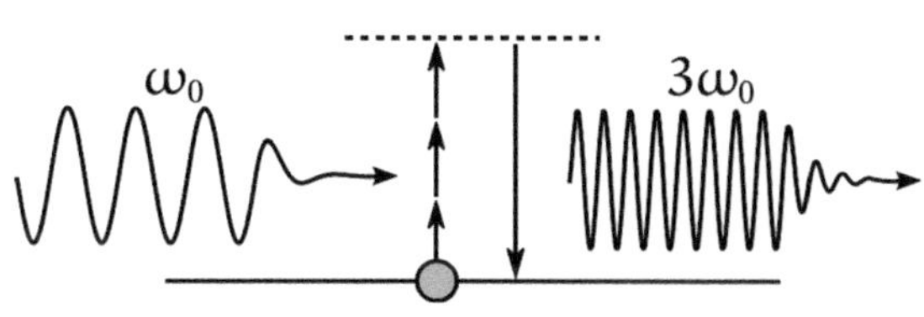

Figure 2.2 Schematic view of the harmonic generation process.

In a classical picture, the high-order harmonic generation process can be explained as an acceleration of the detached electron by the electromagnetic field and its subsequent recollision with the nucleus, giving rise to the emission of higher order harmonic radiation. When the light interacts with a large number of atoms, the emitted radiation is dominated by the coherent part, that can be described using classical electromagnetism. Larmor's formula gives the radiated electromagnetic power per solid angle by an accelerated charge [63]:

$$\frac{dP(t)}{d\Omega} = \frac{e^2}{4\pi c^3} |\vec{a}(t)|^2 \sin^2\theta, \tag{2.11}$$

where θ is the angle between the electron acceleration and the incident light propagation axis.

Therefore, and according to the correspondence principle, one needs to evaluate the expected value of the electron acceleration to determine the coherent part of the harmonic spectrum from the quantum formulation. This can be done in two ways: (1) either by calculating the second derivative of the expected value of the position or (2) by calculating the time derivative of the expected value of the electron momentum. By recalling Ehrenfest theorem, the latter can be rewritten as:

$$\begin{aligned}\langle\vec{a}\rangle &= \frac{1}{m}\frac{d}{dt}\langle\vec{p}\rangle = \frac{1}{i\hbar m}\int \psi^*(\vec{r},t)\,[\vec{p},H]\,\psi(\vec{r},t)d\vec{r} \\ &= -\frac{1}{m}\int \psi^*(\vec{r},t)\left(\vec{\nabla}V(\vec{r}) + q\,\vec{E}(t)\right)\psi(\vec{r},t)d\vec{r} \\ &= -\frac{1}{m}\left\langle \vec{\nabla}V(\vec{r}) + q\,\vec{E}(t)\right\rangle.\end{aligned} \tag{2.12}$$

where the commutator $[\vec{p}, H]$ has been calculated as[a]:

$$\begin{aligned}[\vec{p},H] &= \left[\vec{p}, \frac{p^2}{2m} + V(\vec{r}) + q\,\vec{r}\cdot\vec{E}(t)\right] \\ &= \left[\vec{p}, \frac{p^2}{2m}\right] + \left[\vec{p}, V(\vec{r}) + q\,\vec{r}\cdot\vec{E}(t)\right] \\ &= -i\hbar\left(\vec{\nabla}V(\vec{r}) + q\,\vec{E}(t)\right).\end{aligned} \tag{2.13}$$

Thus, Eq. 2.12 tells us that we can evaluate the electron acceleration from the total force (the gradient of the total potential

[a]By using the commutator property $[A, f(B)] = [A, B]\,\frac{df(B)}{dB}$ with $A = \vec{p}$ and $B = \vec{r}$.

acting on the electron) divided by the electron mass. The power spectrum emitted by the electron will be given by the Fourier transform of the expected values of $\frac{d^2x}{dt^2}$ and $\vec{\nabla}V + qE$:

$$P_{a,\alpha}(\omega) \propto \left| \mathrm{FT}\left[\frac{d^2}{dt^2}\langle x \rangle \right] \right|^2 \propto \left| \omega^2\, \mathrm{FT}\left[\langle x \rangle_\alpha \right] \right|^2, \qquad (2.14)$$

$$P_{F,\alpha}(\omega) \propto \left| \mathrm{FT}\left[\left\langle \vec{\nabla}V(\vec{r}) + qE(t) \right\rangle_\alpha \right] \right|^2 \qquad (2.15)$$

$\mathrm{FT}[f]$ is the Fourier transform of $f(t)$ that we will numerically calculate via a fast Fourier transform algorithm [64]. The subindex α indicates whether the expected values are calculated from the trajectories (t) or from the wave function (ψ), cf. Eqs. 2.9 and 2.10.

Thus, we integrate Eq. 2.2 with the light pulse given in Eq. 2.6 to obtain the evolution of the wave function and from it we calculate the evolution of the trajectories $\{x_k[t]\}$ by integrating the Bohmian velocity:

$$v_k[t] = \frac{dx_k[t]}{dt} = \frac{\hbar}{m}\mathrm{Im} \left. \frac{\vec{\nabla}\psi(\vec{r},t)}{\psi(\vec{r},t)} \right|_{x_k[t]}. \qquad (2.16)$$

Figure 2.3a shows the time evolution of the electron wave function and Fig. 2.3b of a subset of the trajectories ensemble. Notice that some of the trajectories escape from the nucleus, and should be associated with ionization. Even though it appears that a considerable number of trajectories are ionized, they correspond to the wave function wings and have smaller weights γ_k, accounting for a final 12% of ionization probability for the parameters used in the simulation. In Fig. 2.3c there is an expanded view of the trajectories closer to the nucleus, the most relevant ones for the harmonic generation. We also plot the time evolution of the mean value of the trajectories position, $\langle x \rangle_{\mathrm{t}}$, showing an anharmonic oscillation.

Figure 2.4a presents the power spectrum obtained from Eq. 2.14 using the trajectories, i.e., from $\langle x \rangle_{\mathrm{t}}$. As expected, we see that there is emission at frequencies $n\omega_0$ with $n = 1$, 3 and 5. For longer pulses, higher harmonic orders would appear. In order to compare these results with the standard calculations, we plot in Fig. 2.4b-d the difference between the results of Fig. 2.4a and the power spectra calculated from Fig. 2.4b $\langle x \rangle_\psi$, Fig. 2.4c $\left\langle \vec{\nabla}V + qE \right\rangle_{\mathrm{t}}$

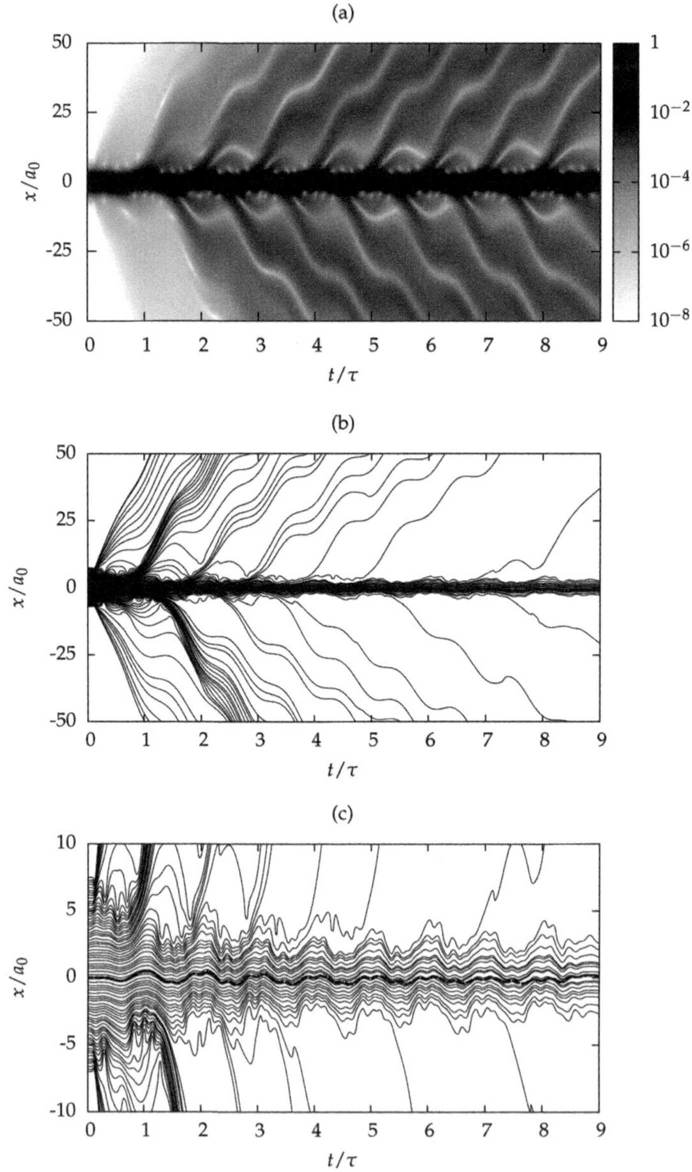

Figure 2.3 Simulation of the photoionization of a one-dimensional hydrogen atom with a pulse with $\omega_0 = 0.148$ au, $E_0 = 0.06$ au and $N_{\text{cyc}} = 9$. (a) Time evolution of $|\psi(x, t)|^2$. (b) Bohmian trajectories associated to $|\psi(x, t)|^2$. For clarity, we only plot a 76-trajectories subset of the ensemble. (c) Expanded view from (b) with the mean value $\langle x \rangle_t$ (thick dashed curve).

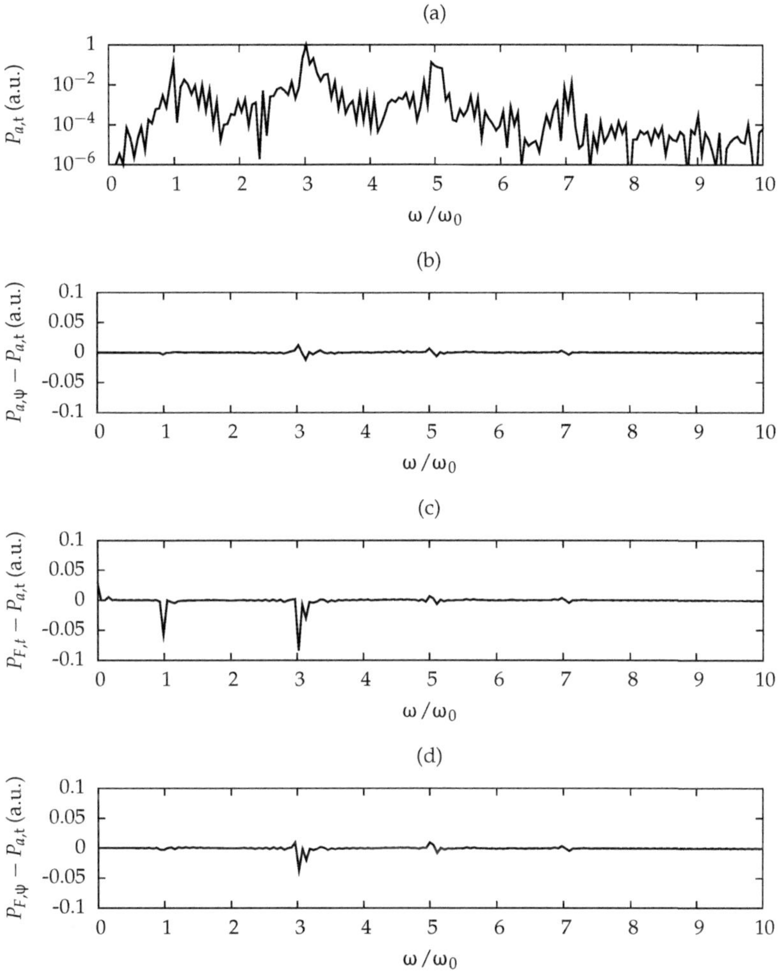

Figure 2.4 Harmonic generation power spectra for the simulation shown in Fig. 2.3. (a) Power spectrum $P_{a,t}(\omega)$. (b-d) Absolute differences between the results from (a) and (b) $P_{a,\psi}(\omega)$, (c) $P_{F,t}(\omega)$ and (d) $P_{F,\psi}(\omega)$. See Eqs. 2.14 and 2.15 for the definition of these spectra.

and Fig. 2.4d $\left\langle \vec{\nabla} V + qE \right\rangle_{\psi}$, by means of Eqs. 2.9 and 2.10. It is clearly seen that the differences between the different methods are very small.

We have, thus, presented a simple method by which we can calculate the harmonic emission spectra of a hydrogen electron accelerated by a light field. For a large number of trajectories, the results converge to the ones obtained from the wave function, for both the calculations from the electron position and the force acting on the electron. However, trajectories have allowed us to visualize the dynamics of both photoionization and harmonic generation processes. The combination of the method here discussed with the quantum trajectory algorithms in Section 1.3.6 could be straightforwardly applied to more complex atoms to reduce the required computational resources.

2.2.3 Above Threshold Ionization

In Einstein's photoelectric effect [1], see Fig. 2.5a, ionization takes place by the absorption of a photon with an energy $\hbar\omega_0$ larger than the ionization potential, I_P, i.e., the minimum energy that the atomic electron has to acquire to reach the continuum. The remaining energy $\hbar\omega_0 - I_P$, is converted into kinetic energy for the ejected

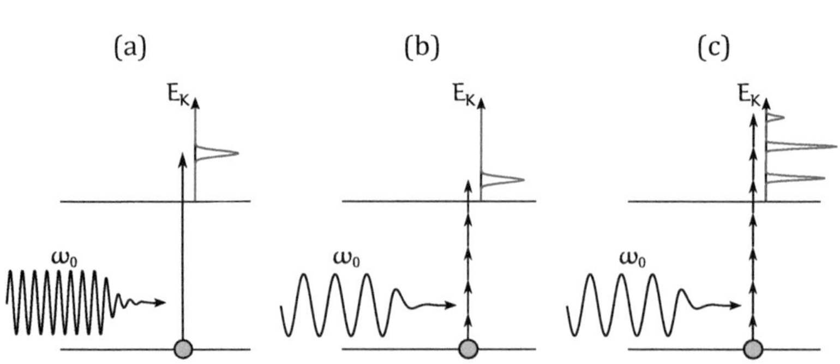

Figure 2.5 Schematic view of (a) the photoelectric effect, (b) multiphoton ionization, and (c) ATI processes. In each case, the ejected electron distribution of kinetic energies E_K is sketched. Notice that in the ATI case, the allowed kinetic energies are separated by $\hbar\omega_0$

electron. From this point of view, the threshold photon frequency required to produce photoionization is $\omega_{th} = I_P/\hbar$.

Maria Göppert-Mayer predicted as early as 1931 [3] the possibility of achieving photoionization by the simultaneous absorption of multiple photons of a frequency ω_0 smaller than ω_{th} (see Fig. 2.5b). In this case, the required number of absorbed photons is the minimum n_{min} that fulfills $n_{min}\hbar\omega_0 > I_P$. Thus, the ejected electron has kinetic energy $n_{min}\hbar\omega_0 - I_P$.

ATI is the process by which an electron ionizes in a multiphoton process, absorbing more photons than this needed minimum n_{min} (see Fig. 2.5c). The possible kinetic energies that the ejected electron has are $n\hbar\omega_0 - I_P$, with $n > n_{min}$.

One way to compute the ATI spectrum is from the velocity distribution of the ionized electron. In Ref. [54], Lai and coworkers, calculate the ATI spectrum by making a histogram of the velocities of the Bohmian trajectories of the system. After letting them evolve in the laser field, the energy of those trajectories far away from the nucleus was calculated as

$$E = E_p + \frac{1}{2} m \langle v \rangle^2 , \tag{2.17}$$

where E_p is the ponderomotive energy of the electron inside the field

$$E_p = \frac{e^2 E_0}{4m\omega_0^2}. \tag{2.18}$$

However, as it is shown in Ref. [65], in general, the momentum distribution of the trajectories does not correspond with the distribution of the Bohmian trajectories velocities. Therefore, the approach taken by Lai et al. might not be fully accurate.

The usual method for obtaining the ATI spectrum consists in evaluating the momentum distribution from the Fourier transform of the electron wave function after it has interacted with the light field and is ionized far from the nucleus. In order to consider only the ionized part of the electron, the wave function around the nucleus is usually removed before taking the Fourier transform. In our case, we will remove the wave function in $|x| < 20$ au. Thus, the probability of having an electron with kinetic energy $E_K = \hbar^2/2m$ is:

$$P(E_K) = \left|\psi(|\vec{k}|, T)\right|^2. \tag{2.19}$$

Note that this includes all $\vec{k}$ components with $|\vec{k}| = \sqrt{2mE_K}/\hbar$.

We propose here an alternative method to calculate the ATI spectrum by using Bohmian trajectories. The main idea is to reconstruct the entire (complex) electron wave function from the ensemble of trajectories and obtain the momentum distribution by taking its Fourier transform. As in the previous section, we find the evolution of $\psi(x, t)$ by integrating Eq. 2.2, and, from it, the evolution of the trajectories. At the end of the pulse we reconstruct the wave function from the trajectories positions $x_k[T]$ and velocities $v_k[T]$. To reconstruct the wave function we need to find both the modulus $R(x)$ and the phase $S(x)$. $R(x)$ will be given by the positions of the trajectories, and $S(x)$ by their velocities.

The modulus over the points of the grid, $R(x_i)$, is found by adding a Gaussian function for each of the trajectories (see Fig. 2.6), taking into account their weights γ_k:

$$R(x_i) = \sum_{k=1}^{N_{\text{traj}}} \frac{\gamma_k}{\sqrt{2\pi\sigma_t^2}} e^{-\frac{(x_i - x_k)^2}{2\sigma_t^2}}, \tag{2.20}$$

where σ_t is the width of the Gaussians taken to be, 0.5 au, i.e., of the order of the Δx space grid increment.

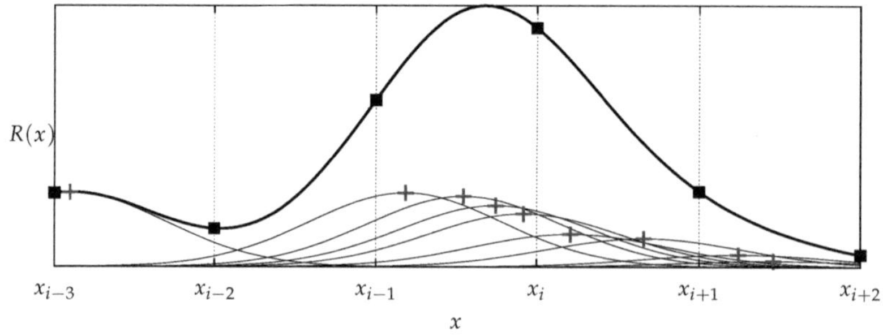

Figure 2.6 Sketch of the reconstruction of $R(x)$. Adding the Gaussian functions (thin curves), defined for each pair $\{x_k[T], \gamma_k\}$ (crosses), we reconstruct the modulus of the wave function (thick curve) that we evaluate on the grid (squares).

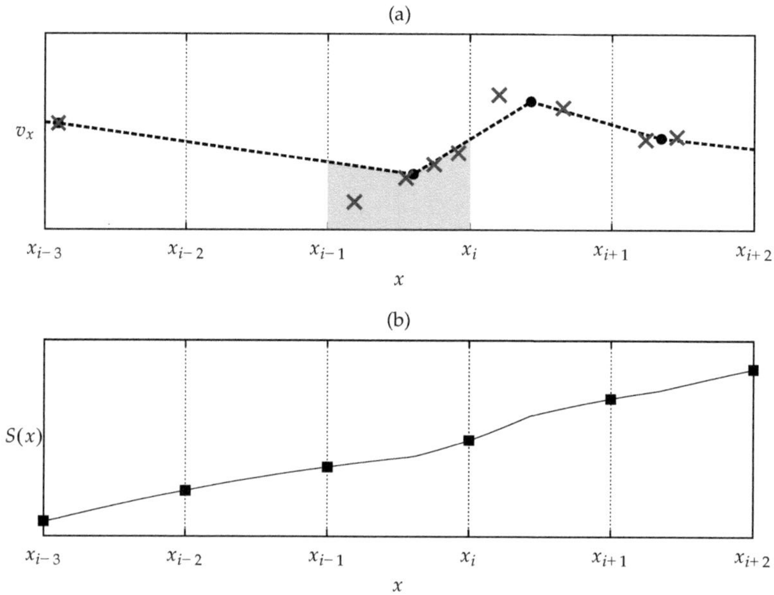

Figure 2.7 Sketch of the reconstruction of $S(x)$. (a) Crosses indicate final pairs $\{x_k[T], v_k[T]\}$, circles indicate the averages of these pairs in each space grid interval that yield the interpolating function $v_{\text{int}}(x)$ (dashed curve). The gray area is the integral of the interpolating function between x_{i-1} and x_i that gives $S(x_i) - S(x_{i-1})$. (b) Integral of $v_{\text{int}}(x)$ (thin line) to reconstruct $S(x)$, evaluated on the grid (squares).

The phase is found by integrating the velocity field, as shown in Fig. 2.7a. From each interval of the space grid, we average the velocity and position of all the trajectories within it (crosses). Then, by joining the points found in this way (circles), we create an interpolating function, $v_{\text{int}}(x)$ (black dashed line). Finally, we integrate over $v_{\text{int}}(x)$ between every two grid points (gray area) in order to obtain the phase on the grid (see Fig. 2.7b):

$$S(x_i) = S(x_{i-1}) + \int_{x_i}^{x_{i+1}} v_{\text{int}}(x)dx. \tag{2.21}$$

Since it is an arbitrary choice, we take $S(x_0) = 0$.

The results for the ATI spectrum calculated from Eq. 2.19 for a pulse with $N_{\text{cyc}} = 3$, $\omega_0 = 0.148$ au, $E_0 = 0.1$ au, and temporal envelope defined by Eq. 2.7 are plotted in Fig. 2.8. Figure 2.9 shows

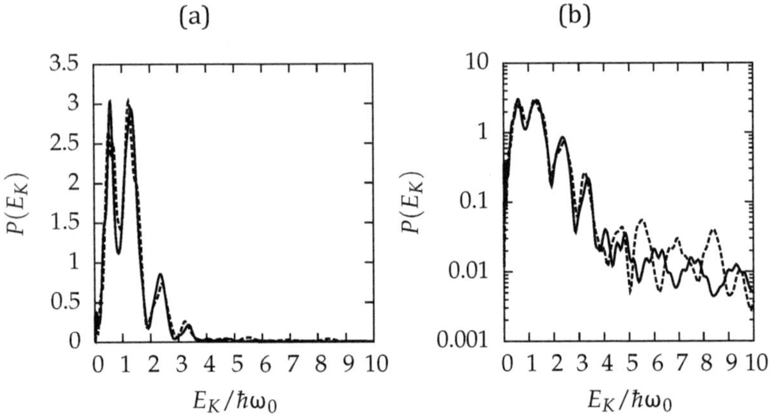

Figure 2.8 ATI spectra obtained from the original wave function (solid curve) and the wave function reconstructed from the trajectories (dashed curve), for a pulse of 3 cycles. Results are shown in (a) linear and (b) logarithmic scales.

the results for a pulse with $N_{\text{cyc}} = 9$, $\omega_0 = 0.148$ au, $E_0 = 0.1$ au, and temporal envelope defined by Eq. 2.8.

As expected, the obtained ATI spectra consist of different peaks separated by an energy $\hbar\omega_0$, the second peak being as strong as the first. We see a good agreement between the two methods, being the first three peaks reproduced in the reconstructed wave function.

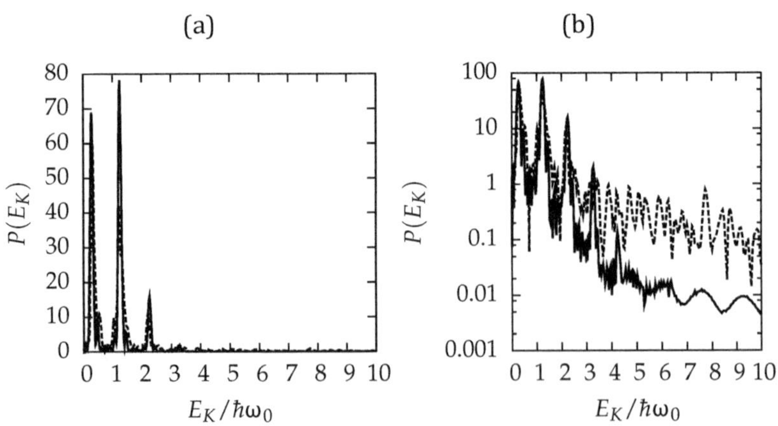

Figure 2.9 The same as in Fig. 2.8, but for a pulse 9 cycles long.

Nonetheless, as the logarithmic plot reveals, the reconstructed wave function is too noisy to reproduce higher-order peaks.

As a conclusion, we have presented a very simple method to reconstruct the modulus and the phase of the wave function from the trajectories that applied to the ATI problem, gives accurate results. As in the harmonic generation scenario, the combination of this method with previously discussed trajectory algorithms could be used to extend it beyond the hydrogen case.

2.3 Hydrogen Photoionization with Beams Carrying Orbital Angular Momentum

2.3.1 Physical System

In the past few years a great interest has been focused on helical light beams that are able to transport spin and orbital angular momentum (OAM) in their propagation direction [66, 67] such as Laguerre–Gaussian (LG) beams. In fact, the transfer of OAM to matter has been already reported experimentally for a Bose–Einstein condensate [68] and for an ensemble of cold atoms [69], and has been investigated for molecular systems [70]. OAM has also attracted attention from the standpoint of applications ranging from optical tweezers [71] to novel quantum information protocols [72].

At variance with previous works, where the OAM was coupled to the center of mass of an atomic ensemble, here we focus on the interaction of an OAM light pulse with a single atom allowing for electronic transitions to both bound states and the continuum. In this context, we have derived novel selection rules for light-matter interactions [73] where more than one $\hbar$ unit of angular momentum per photon is exchanged.

Following our previous study [55], we will investigate the system sketched in Fig. 2.10: a hydrogen atom interacting with an ultrashort and ultraintense light pulse carrying OAM whose spatial profile will be expressed as an LG mode [66, 67]. We will simulate the dynamics of this interaction by means of 3D numerical simulations, and investigate the angular momentum exchange in different photoionization scenarios as well as its manipulation through the

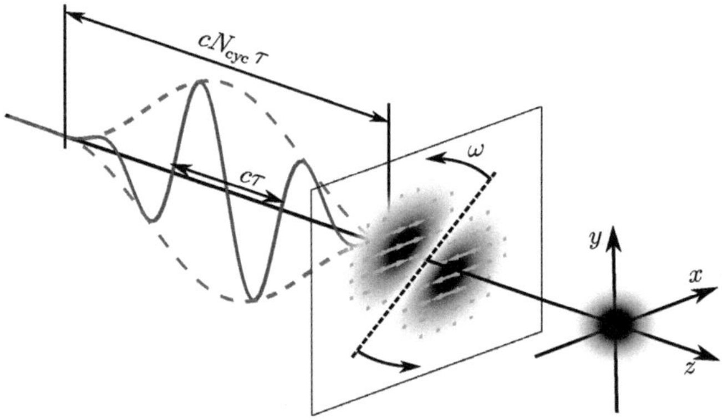

Figure 2.10 Sketch of the physical system consisting of a light pulse interacting with a hydrogen atom at the origin, where the time pulse dependence and its associated initial parameters are introduced. As an example, we plot the transverse profile of the vector potential in Eq. 2.24 for a linearly polarized LG mode with topological charge $\ell = 1$. Reprinted with permission from Ref. [55]. Copyright 2010 IOP Publishing Ltd and Deutsche Physikalische Gesellschaft.

light polarization (linear or circular). Bohmian trajectories obtained from the evolution of the quantum state will give physical insight into the dynamics and the absorption of angular momentum by the atom.

At variance with the previous section, we will now work in the velocity gauge, including the effects of both electric and magnetic fields. Thus, the Schrödinger equation reads:

$$i\hbar\frac{\partial\psi(\vec{r},t)}{\partial t} = \frac{1}{2m}\left[\vec{p} - q\vec{A}(\vec{r},t)\right]^2 \psi(\vec{r},t) + V(\vec{r})\psi(\vec{r},t), \quad (2.22)$$

where $\vec{A}(\vec{r},t) = -\int_0^t \vec{E}(\vec{r},t')dt'$ is the vector potential describing the electromagnetic field of light.

By expanding Eq. 2.22 and, taking into account that $\vec{A}(\vec{r},t)$ is a transversal field, i.e., $\vec{\nabla}\cdot\vec{A}(\vec{r},t) = 0$, we obtain:

$$\begin{aligned} i\hbar\frac{\partial\psi(\vec{r},t)}{\partial t} &= -\frac{\hbar^2}{2m}\nabla^2\psi(\vec{r},t) + \frac{i\hbar q}{m}\vec{A}(\vec{r},t)\cdot\vec{\nabla}\psi(\vec{r},t) \\ &\quad + V(\vec{r})\psi(\vec{r},t) + \frac{q^2}{2m}\vec{A}(\vec{r},t)^2\psi(\vec{r},t). \quad (2.23) \end{aligned}$$

The terms in the right-hand side (rhs) of Eq. 2.23 are, respectively, the field-free kinetic term, the interaction of the electron with the laser field (responsible for the emission/absorption processes), the Coulomb interaction with the nucleus and the ponderomotive potential.

The ponderomotive energy is the cycle averaged kinetic energy of a free electron in the electromagnetic field. Its gradient gives rise to the ponderomotive force, that tends to drive the electrons out of the zones where the laser is more intense. This term is sometimes wrongly neglected, claiming that it is small for transversally smooth pulses. In some cases, however, it can become very important, and be responsible for effects such as the trapping of the atom inside of the light pulse [55]. In any case, when the vector potential is position independent, $\vec{A}(\vec{r}, t) = \vec{A}(t)$, the ponderomotive term has no effect on the dynamics, since it can be removed from Eq. 2.23 by performing a unitary transformation on the wave function $\psi(\vec{r}, t) \to e^{-i\frac{q^2}{2m\hbar}\int_0^t \vec{A}^2(t')dt'}\psi(\vec{r}, t)$, i.e., taking a time dependent energy origin. This is not the case for LG beams, and we will show that it leads to effects such as the appearance of new selection rules for photon absorption.

The light pulse is considered to be propagating along the z direction, and its temporal envelope, see Fig. 2.10, is parametrized by a quadratic sine function, with frequency $\omega_e = \pi/N_{cyc}\tau$, where N_{cyc} and τ are the cycle number and the period of the carrier wave, respectively. Hence, the vector potential can be cast in the form:

$$\begin{aligned}\vec{A}(\vec{r}, t) = \vec{A}_0\, w_0 \sin^2\left(\frac{\omega_e}{c}(z + a_0) - \omega_e t\right) \\ \times \left[e^{i\frac{\omega_0}{c}(z-ct)+i\chi}\, LG_{\ell,p}(\rho, \phi, z) + \text{c.c.}\right] \\ \times \left[\theta_H(z - ct + \pi c/\omega_e + a_0) - \theta_H(z - ct + a_0)\right], \end{aligned} \quad (2.24)$$

$\theta_H(x)$ being the Heaviside step function, χ the "global" carrier-envelope phase, $\omega_0 = 2\pi/\tau$ the frequency of the carrier wave, w_0 is the beam waist at $z = 0$, and $\vec{A}_0$ the vector amplitude of the wave. $LG_{\ell,p}(\rho, \phi, z)$ are the LG modes, that read:

$$\begin{aligned}LG_{\ell,p}(\rho, \phi, z) = \sqrt{\frac{2p!}{\pi(|\ell| + p)!}}\frac{1}{w(z)} L_p^{|\ell|}\left(\frac{2\rho^2}{w^2(z)}\right)\left(\frac{\sqrt{2}\rho}{w(z)}\right)^{|\ell|} \\ \times \exp\left(-\frac{\rho^2}{w^2(z)}\right)\exp\left[i\ell\phi + i\frac{k_0\rho^2}{2R(z)} + i\Phi_G(z)\right],\end{aligned} \quad (2.25)$$

where $\rho = \sqrt{x^2+y^2}$, $\phi = \arctan(y/x)$, $k_0 = \omega_0/c$ is the carrier wave number, $w(z) = w_0\sqrt{1+(z/z_0)^2}$ is the beam width, $R(z) = z\left[1+(z_0/z)^2\right]$ is the phase-front radius, $z_0 = k_0 w_0^2/2$ is the Rayleigh range, $\Phi_G(z) = -(2p+|\ell|+1)\arctan(z/z_0)$ is the Gouy phase, and $L_p^{|\ell|}(\xi)$ are the associated Laguerre polynomials:

$$L_p^{|\ell|}(\xi) = \sum_{m=0}^{p}(-1)^m \frac{(|\ell|+p)!}{(p-m)!\,(|\ell|+m)!\,m!}\,\xi^m. \tag{2.26}$$

The indices $\ell = 0, \pm 1, \pm 2, \ldots$ and $p = 0, 1, 2, \ldots$ correspond to the topological charge and the number of nonaxial radial nodes of the mode. Note that for $\ell = p = 0$, we recover the standard Gaussian beam. As shown in Refs. [66, 67], in the paraxial regime LG modes carry a discrete OAM of $\ell\hbar$ units per photon along their propagation direction. We would also like to emphasize the fact that, at variance with the plane wave case, the LG wave front (equal phase surface) forms in space part of a helicoidal surface given by $\ell\phi+kz =$ const. As a consequence, the "local" carrier-envelope phase and the electric field amplitude that interacts with each part of the electron wave function depend on the transverse position. In the Coulomb gauge, the vector potential can have a polarization component in the propagation direction [66], which is usually neglected. A realistic solution should include this longitudinal polarization component, which becomes particularly relevant near the vortex axis. However, to first order, both components will give independent dynamics, allowing us to explicitly study only the transverse part.

The light polarization in the xy plane is included in $\vec{A}_0 \equiv A_0(c_x\,\vec{e}_x + c_y\,\vec{e}_y)$, where c_x and c_y are complex numbers satisfying $|c_x|^2 + |c_y|^2 = 1$. For example, for a circular polarization, $c_x = 1/\sqrt{2}$ and $c_y = is/\sqrt{2}$, with $s = \pm 1$. Note that here we make use of the notation of Ref. [74], where a left (right) circularly polarized beam carries angular momentum $s\hbar$ with $s = +1$ ($s = -1$) along the propagation direction. A linear polarization forming an angle ϕ with the x axis corresponds to $c_x = \cos\phi$ and $c_y = \sin\phi$, that is an equal superposition of the two circular polarizations.

2.3.2 Bohmian Equations in an Electromagnetic Field

Here we will derive the Bohmian equations of motion for the trajectories in the velocity gauge. Our starting point is Eq. 2.23:

$$i\hbar\frac{\partial\psi}{\partial t} = -\frac{\hbar^2}{2m}\nabla^2\psi + \frac{i\hbar q}{m}\vec{A}\cdot\vec{\nabla}\psi + \frac{i\hbar q}{2m}(\vec{\nabla}\cdot\vec{A})\psi + \frac{q^2}{2m}\vec{A}^2\psi + q\,V\psi. \tag{2.27}$$

We have dropped the dependences of $\vec{A}$, ψ, and V to simplify the notation. In order to obtain the Bohmian equations in an arbitrary electromagnetic field, we have not removed the term with $\frac{i\hbar q}{2m}\vec{\nabla}\cdot\vec{A}$.

As in the standard formulation of Bohmian mechanics, we write the (complex) wave function in polar form

$$\psi = Re^{iS/\hbar}, \tag{2.28}$$

$R \equiv R(\vec{r}, t)$ and $S \equiv S(\vec{r}, t)$, respectively, being the modulus and phase of the wave function. By inserting Eq. 2.28 into the Schrödinger equation, Eq. 2.27, (and cancelling $e^{iS/\hbar}$ on both sides of the equation) we get:

$$\begin{aligned} i\hbar\frac{\partial R}{\partial t} - R\frac{\partial S}{\partial t} &= -\frac{\hbar^2}{2m}\nabla^2 R - \frac{i\hbar}{2m}R\nabla^2 S - \frac{i\hbar}{m}\vec{\nabla}R\cdot\vec{\nabla}S \\ &\quad + \frac{1}{2m}R\left(\vec{\nabla}S\right)^2 + \frac{i\hbar q}{m}\vec{A}\cdot\vec{\nabla}R - \frac{q}{m}R\vec{A}\cdot\vec{\nabla}S \\ &\quad + \frac{i\hbar q}{2m}R\vec{\nabla}\cdot\vec{A} + \frac{q^2}{2m}\vec{A}^2 R + qVR. \end{aligned} \tag{2.29}$$

Taking the real part of Eq. 2.29 we obtain:

$$\frac{\partial S}{\partial t} = \frac{\hbar^2}{2m}\frac{\nabla^2 R}{R} - \frac{1}{2m}\left(\vec{\nabla}S - q\vec{A}\right)^2 - qV, \tag{2.30}$$

and its imaginary part (multiplied by $2R$ on both sides) reads:

$$\frac{\partial R^2}{\partial t} = -\frac{1}{m}\vec{\nabla}\left[R^2\left(\vec{\nabla}S - q\vec{A}\right)\right]. \tag{2.31}$$

Equations 2.30 and 2.31 lead us to a new definition of the Bohmian velocity:

$$\vec{v} = \frac{1}{m}\left[\vec{\nabla}S - q\vec{A}\right]. \tag{2.32}$$

It is straightforward to see that in absence of field ($\vec{A} = 0$), we recover the standard definition for the Bohmian velocity.

2.3.3 Selection Rules

The Hamiltonian from Eq. 2.22 can be split as $\mathcal{H} = \mathcal{H}_0 + \mathcal{H}_I + \mathcal{H}_{II}$, where $\mathcal{H}_0$ is the free hydrogen Hamiltonian (kinetic and Coulomb terms), $\mathcal{H}_I \equiv -q(\vec{p}\cdot\vec{A}+\vec{A}\cdot\vec{p})/2m$ and $\mathcal{H}_{II} \equiv q^2\vec{A}^2/2m$. Decomposing the electron state in spherical harmonics with the quantization axis in the light propagation direction, the transition probability between any two spherical harmonics, $Y_{L_i}^{M_i}$ and $Y_{L_f}^{M_f}$, given by the first interaction Hamiltonian term can be written as $\int Y_{L_f}^{M_f *}\mathcal{H}_I Y_{L_i}^{M_i}\, d\Omega$. For a light beam carrying $\ell\hbar$ units of OAM, see expression Eq. 2.24, and assuming the dipolar approximation ($\lambda \gg a_0$) and the transverse spatial approximation ($w_0 \gg a_0$), the selection rules associated to the $\mathcal{H}_I$ interaction [73] are:

$$|\Delta L| \leq |\ell| + 1, \tag{2.33}$$

$$\Delta L + |\ell| \text{ is odd}, \tag{2.34}$$

$$\Delta M = \ell + s. \tag{2.35}$$

It comes clear from Eqs. 2.33–2.35, that for $\ell = 0$ we recover the usual selection rules for plane waves. Furthermore, for $\ell \neq 0$ we can excite transitions with an angular momentum exchange of more than one unit of $\hbar$. In photon terms, these selection rules can be thought as the absorption of a photon carrying a total angular momentum $\ell + s$ in the propagation direction, where s accounts for the spin angular momentum.

Proceeding in a similar manner with $\mathcal{H}_{II}$, one obtains [73]:

$$|\Delta L| \leq 2|\ell|, \tag{2.36}$$

$$\Delta L \text{ is even}, \tag{2.37}$$

$$\Delta M = 2\ell. \tag{2.38}$$

In the case of plane waves, the contribution of this interaction Hamiltonian is just a constant term, yielding a ponderomotive force, but peculiarly, for $\ell \neq 0$ it can produce transitions between spherical harmonics. Since selection rules in Eqs. 2.36–2.38 appear from a term with $\vec{A}^2$, we expect them to be associated with two-photon processes, and, therefore, will be less significant than those associated with selection rules in Eqs. 2.33–2.35, since the intensity near the vortex is relatively small.

In the following section, where the complete dynamics of the system are numerically solved, selection rules given by Eqs. 2.33–2.35 and 2.36–2.38 will be essential to understand the interaction process. We would like to remark that these selection rules apply to other scenarios beyond photoionization, such as transitions between bound states.

2.3.4 Numerical Simulations

Initially, before the interaction with the light pulse, we assume the hydrogen electron to be in the system's ground state, i.e., $\psi_0(\vec{r}) = \psi(\vec{r}, t = 0) = e^{-r/a_0}/\sqrt{\pi a_0^3}$. We will consider the atom to be centered around the light propagation axis, i.e., at $\rho = 0$, corresponding to the maximum amplitude of the Gaussian profile or, alternatively, the LG vortex position, and choose carrier-envelope phase to be $\chi = 0$. In all the following cases we consider the pulse, see Eq. 2.24, to have $N_{cyc} = 3$, a carrier frequency of $\omega_0 = 1$ au ($2\pi \times 6.57 \times 10^{15}$ s^{-1}, ultraviolet) and period $\tau = 2\pi$ au (152 as). We assume a beam waist that satisfies the paraxial regime, $w_0 = 9 \times 10^4$ au (4.79 μm), which is much larger than the typical size of the atom ($w_0 \gg a_0$). For simplicity, since the OAM of the light is given only by the topological charge ℓ, we will restrict ourselves to LG modes with $p = 0$. In this section we will make use of neither the electric dipole nor the transverse spatial approximation.

We express the electron quantum state at each time as $|\psi(t)\rangle = c_0 |\psi_0\rangle + |\delta\psi(t)\rangle$, where $|\delta\psi(t)\rangle$ is the excited part of the state and $c_0 = \langle\psi_0|\psi(t)\rangle$ is the ground state probability amplitude. Since the field has a frequency $\omega_0 = 1$ au, larger than the bound hydrogen energy (0.5 au), we expect that, at the end of the pulse, the excited part will account for ionization. Thus, we define the ionization probability as $P_I = |\langle\delta\psi(t)|\delta\psi(t)\rangle|^2 = 1 - |c_0|^2$. To clearly observe the effects of the light on the atom, we will only plot the probability distribution of the excited part of the electron state.

By means of the Crank–Nicolson algorithm, we integrate Schrödinger Eq. 2.22 in a 3D grid of $N_x \times N_y \times N_z = 400 \times 400 \times 250$ points, covering a space of (–20, 20) au × (–20, 20) au in the polarization plane and of (–12.5, 12.5) au in the propagation direction. We take absorbing boundary conditions in order to avoid unphysical reflections when the ionized wave function reaches the borders of the integration grid. The losses due to these absorbing boundaries will be, in most cases, relatively small and should be associated with ionization.

The initial positions of the trajectories $\{\vec{x}_k[0]\}$ are distributed randomly following the probability density function of the hydrogen ground state: $R^2(\vec{r}, 0) = \frac{1}{\pi a_0^3} e^{-2r/a_0}$. During the evolution we will calculate average values for different observables such as the average position of the electron $\langle \vec{r} \rangle$ or its angular momentum $\langle \vec{L} \rangle = \langle \vec{r} \times (\vec{p} - q\,\vec{A}) \rangle$. Moreover, we will project numerically the quantum state onto the spherical harmonics in order to check the agreement with the previously presented selection rules.

2.3.4.1 Gaussian pulses

In this section, we consider a Gaussian pulse, i.e., $\ell = 0$, $p = 0$, and we take the electric field amplitude to be $A_0\,\omega_0 = 3 \times 10^4$ au for both cases of linear and circular polarization.

The results of the simulation with a pulse linearly polarized in the x direction, plotted in Fig. 2.11, show that the electron begins to be ionized in the first cycle and, at the end of the pulse, 30% of the quantum electron state is ionized (see Fig. 2.11b). We also see in Fig. 2.11c that there is no exchange of angular momentum in the z direction since the light pulse is linearly polarized and does not carry OAM. Figure 2.11d shows that the most excited spherical harmonics are Y_1^1 and Y_1^{-1}, in complete agreement with selection rules discussed in Section 2.3.3, since the linear polarization is a superposition of left and right circular polarizations.

The motion of the electron wave packet is unraveled in the trajectories plotted in Fig. 2.11e. Trajectories describe oscillations at the field frequency along the polarization axis following (with a $\frac{\pi}{2}$ phase delay) the oscillation of the electric field. Some of the trajectories, i.e., the ones that start far enough from the nucleus, get ionized near the electric field maxima. The mean value of the position oscillates in the x axis, while it remains at the origin of the yz plane, except for a small shift (of the order of 10^{-2} au) on the z direction due to the magnetic field, in agreement with previous

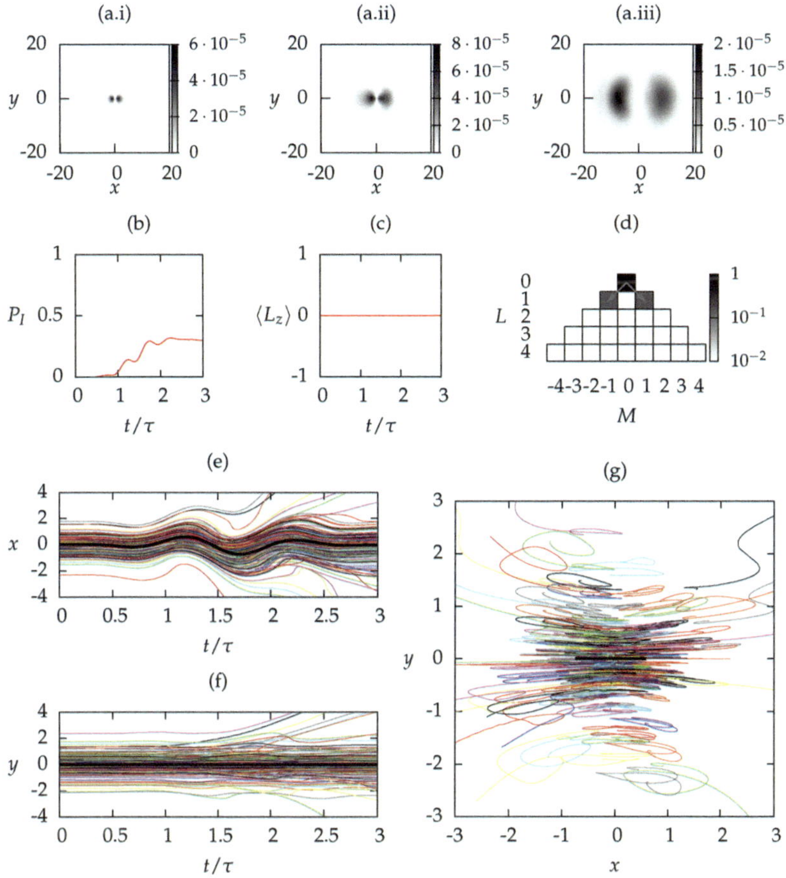

Figure 2.11 Results of a simulation with a Gaussian pulse ($\ell = 0$, $p = 0$) linearly polarized in the x direction with electric field amplitude $A_0\,\omega_0 = 3 \times 10^4$ au. For the rest of the parameters, see text. (a) Projection of the excited state $|\delta\psi(t)\rangle$ onto the plane xy for $t = \tau$, $t = 2\tau$, and $t = 3\tau$. Time evolution of (b) the ionization probability P_I and (c) the expected value of the angular momentum along the z axis, $\langle L_z \rangle$. (d) Population of spherical harmonics Y_L^M at the end of the pulse, arrows correspond to allowed transitions (see Section 2.3.3). Bohmian trajectories projection onto the (e) x and (f) y axis over time and in the (g) xy plane. The thick black curves correspond to the electron position mean value. Reprinted with permission from Ref. [55]. Copyright 2010 IOP and Deutsche Physikalische Gesellschaft.

results [75]. A shift around this value will also appear in all the following scenarios.

The Coulomb potential acting over the trajectories that are closer to the nucleus retains them in the nucleus vicinity. The ones that distance from it make the trajectory cloud expand. This can also be seen as the spreading of the quantum state function far from the nucleus where the Coulomb potential vanishes.

In Fig. 2.12 we present the results of a simulation with a right circular polarized pulse. We have checked that a left circular polarization yields the expected symmetric results.

As in the case of linear polarization, the ionization probability reaches about 30% at the end of the pulse. At variance with the linearly polarized Gaussian pulse, which does not transfer angular momentum in the z direction to the electron, in Fig. 2.12c we see that, after the interaction, the electron has $\langle L_z \rangle = -0.3$ au ($-0.3\,\hbar$). As the selection rules predict, we expect that a right (left) circularly polarized light mostly excites the Y_1^{-1} (Y_1^1) spherical harmonic, which is in accordance with the numerical results in Fig. 2.12d. Note that the electron is excited to $L = 1$ states through one-photon processes and to $L = 2$ states through two-photon processes.

The trajectories depicted in Fig. 2.12e describe oscillations along the two polarization axis. It is noteworthy that for each of the trajectories, the main dynamics in these two axis are independent. The mean value of the electron position shows that the electric field displaces the entire trajectory cloud from the propagation axis and makes it rotate clockwise (since it is right circularly polarized) around its initial position. This is due to the fact that the electric field is almost uniform in the transverse plane, therefore acting in the same way over all the trajectories.

2.3.4.2 Laguerre–Gaussian pulses

From now on we consider pulses bearing OAM, e.g., with $\ell = 1$ and $p = 0$, both the cases with linear and circular polarization.

We consider the atom to be centered at the light vortex, where the electric field amplitude is zero and increases linearly in the vicinity of the singularity. Thus, we require very intense lasers to affect the atom. We choose a pulse with $A_0\,\omega_0 = 1.4 \times 10^4$ au, corresponding

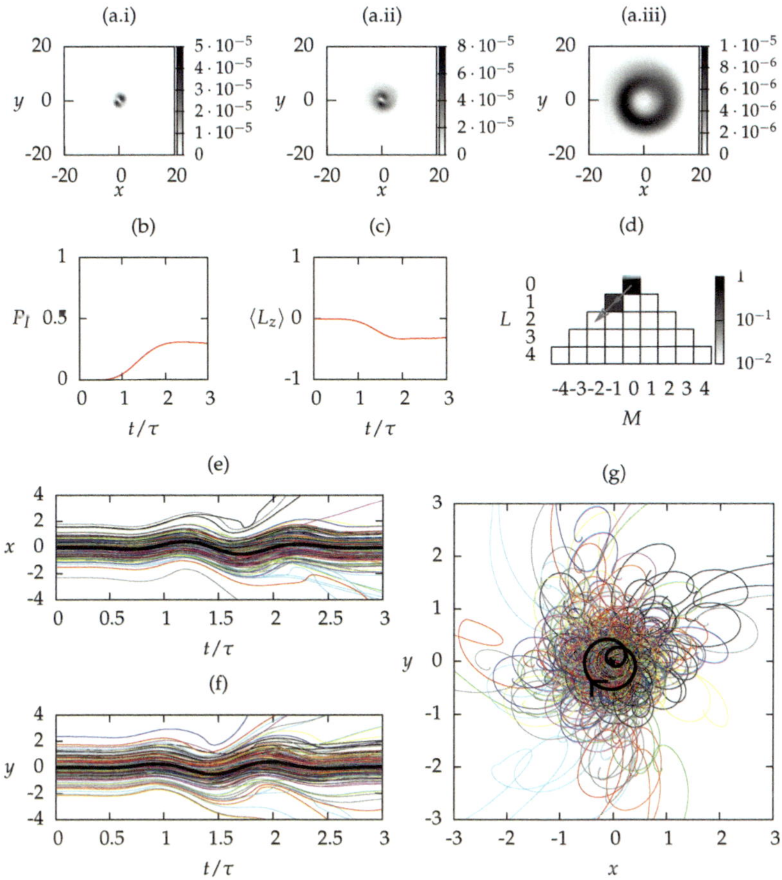

Figure 2.12 As in Fig. 2.11 for a Gaussian pulse ($\ell = 0$) right circularly polarized in the xy plane. The arrow in (g) represents the rotation direction of the mean position. Reprinted with permission from Ref. [55]. Copyright 2010 IOP Publishing Ltd and Deutsche Physikalische Gesellschaft.

to a peak intensity (at the beam waist) of 6.7×10^{24} W/cm^2, although the electric field at a distance of 1 au from the vortex is only 0.16 au. Albeit the considered light intensity is unreachable nowadays, the reader must take into account that very strong lasers are under construction [58, 76], planning to reach up to 10^{26} W/cm^2 at 800 nm wavelength.

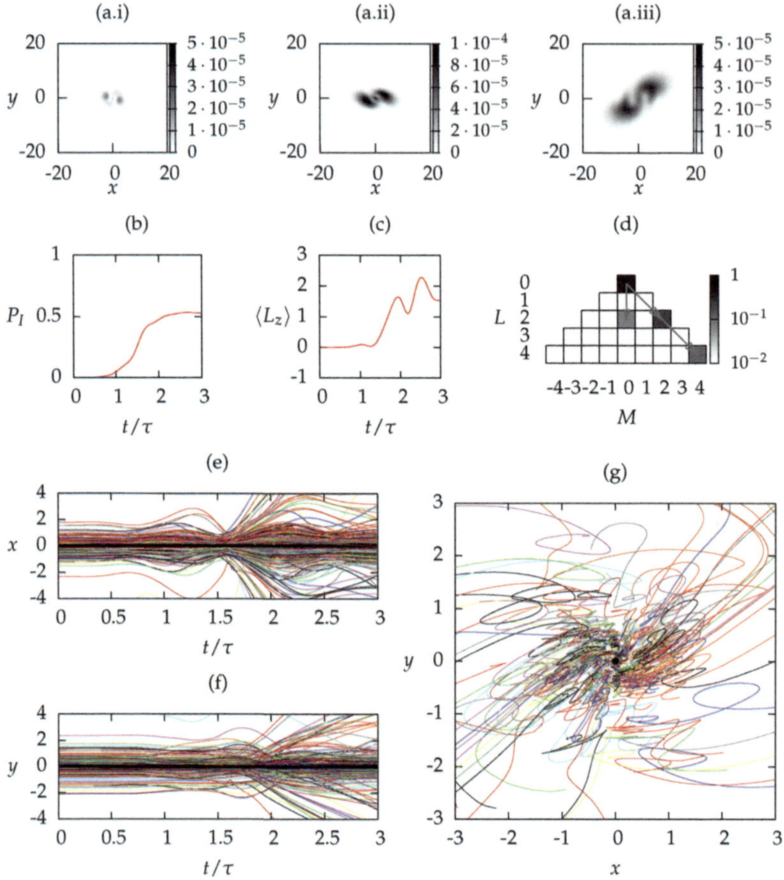

Figure 2.13 The same as for Fig. 2.11 but with an LG mode with $\ell = 1$ linearly polarized in the x direction with electric field amplitude $A_0\,\omega_0 = 1.4 \times 10^4$ au. Reprinted with permission from Ref. [55]. Copyright 2010 IOP and Deutsche Physikalische Gesellschaft.

In Fig. 2.13 we show the results of a simulation with an LG pulse with $\ell = 1$ polarized in the x direction. We have checked that due to the symmetry of the problem, the interaction with an $\ell = -1$ pulse leads to the expected symmetric results.

As shown in Fig. 2.13a, during the interaction with the pulse the electron is ionized, but it remains trapped around the light vortex, due to the ponderomotive potential of the pulse. Once the electric field intensity decreases, the electron state begins to expand, as it can be seen in Fig. 2.13a-iii or in the trajectories depicted in Fig. 2.13e. Note that even though the pulse is linearly polarized in the x direction, the OAM breaks the symmetry on the y axis. At the end of the pulse the ionization probability is 53%.

It is interesting to point out the total angular momentum transferred to the electron (see Fig. 2.13c). The electron starts in the ground state, with zero angular momentum and, as the pulse interacts with the electron, it oscillates, reaching, at the end of the pulse, a finite amount of angular momentum of 1.53 au ($1.53\,\hbar$), at variance with the linearly polarized Gaussian case where no exchange of angular momentum took place. Thus, we expect that the populated electron excited states at the end of the pulse bear angular momentum. The spherical harmonics populations plotted in Fig. 2.13d show perfect agreement with the selection rules from Section 2.3.3, since the linear polarization is a superposition of the two circular polarizations, i.e., $\Delta M = \ell \pm 1$. Note that the electron is excited to $L = 2$ states through one-photon processes and to $L = 4$ states through two-photon processes.

The mean motion of the electron state is negligible. Nevertheless, the dynamics is far from simple, due to the electric field having a spatial profile where it points in opposite directions in opposite points in the polarization plane. A peculiar effect is that, even though the polarization is linear in the x direction, the trajectories are forced to acquire a velocity in the y direction in order to rotate counterclockwise due to the light $\ell = +1$ OAM, giving rise to the absorption of angular momentum by the electron. Notice that in the circularly polarized Gaussian case, the rotation of the trajectories around the propagation axis took place in sequence with each other, whereas here each of the trajectories rotates independently around the axis.

We now consider the same pulse, but circularly polarized. At variance with the Gaussian circularly polarized beam, the dynamics for the left and right circular polarization cases are different due to the relative sign between the light OAM and polarization components of the angular momentum along the z direction. This becomes clear from the results shown in Fig. 2.14a and Fig. 2.15a: whereas the right circular polarization preserves the circular

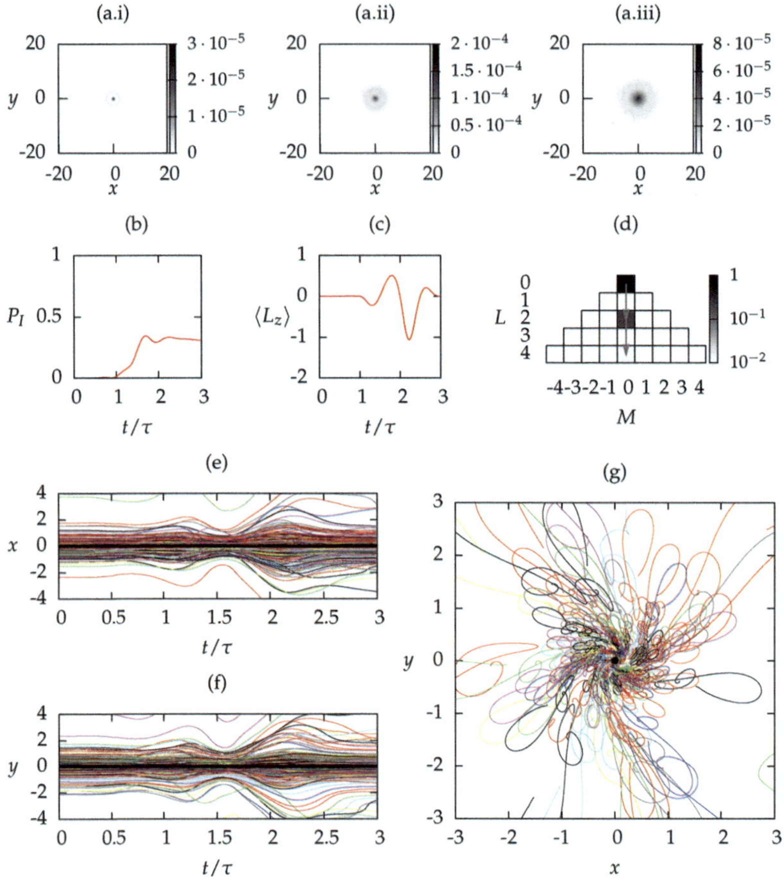

Figure 2.14 The same as for Fig. 2.13 but with right circular polarization. Reprinted with permission from Ref. [55]. Copyright 2010 IOP and Deutsche Physikalische Gesellschaft.

symmetry of the excited quantum state, the left circular polarization gives rise to a more complex behavior due to the exchange of angular momentum. Note that in the case of right circular polarization ($s = -1$), the spin and OAM angular components compensate, while in the left circular polarization case ($s = 1$), they add up.

Figures 2.14b and 2.15b show that, even though the electric field intensity is the same for both cases, in the left circular polarization case, the ionization is approximately twice that in the case of right

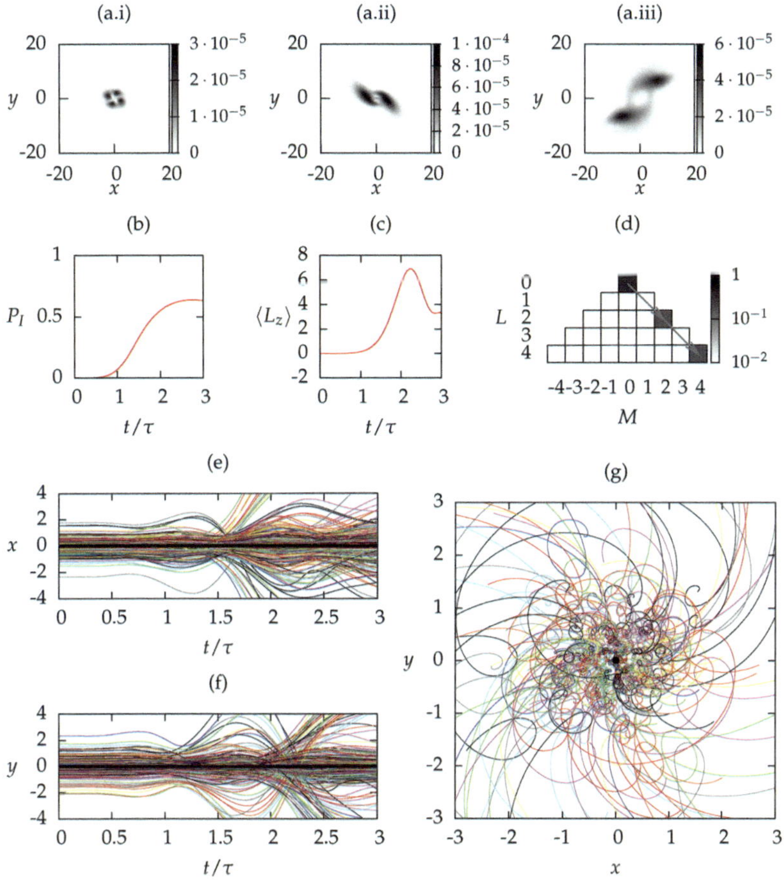

Figure 2.15 The same as for Fig. 2.13 but with left circular polarization. Reprinted with permission from Ref. [55]. Copyright 2010 IOP and Deutsche Physikalische Gesellschaft.

circular polarization (63% and 31%, respectively). We attribute this enhancement of the ionization to the fact that, in this particular case, an absorption of an ($\ell = +1, s = +1$) photon is more probable than the absorption of an ($\ell = +1, s = -1$) photon. At the end of the pulse, the expected value for L_z in the left circular polarization case is 3.35 au, while becomes null for the right circular polarization, as can be seen in Figs. 2.14c and 2.15c.

The transfer of angular momentum to the electron is clearly seen in Figs. 2.14d and 2.15d, in perfect agreement with the previously discussed selection rules. We can check that a right circular polarization only excites states with $M = 0$, as $\Delta M = \ell + s = 0$, therefore not transferring any OAM in the z direction to the atom. On the other hand, a left circular polarization excites transitions obeying $\Delta M = \Delta L = \ell + s = 2$.

As in the linear polarization case, the main dynamics are concentrated on the polarization plane and the mean motion of the electron state is again negligible. The absorption of angular momentum by the electron is clearly depicted in the trajectories, while in the left circular polarization, the electric field of the light exerts a torque on the trajectory cloud forcing it to rotate around the propagation axis, the combination of OAM and right circular polarization partially inhibits this rotation.

2.4 Conclusions

In this chapter we have investigated the interaction of a hydrogen atom with an intense laser field. In the first part of the chapter, we have focused on a simple one-dimensional model for the hydrogen atom interacting with a light field. Bohmian trajectories have been used to perform simple calculations of the harmonic emission spectra, that have yielded results equivalent to the ones obtained from the standard methods. For the ATI problem, we have presented a novel method to reconstruct the modulus and the phase of the wave function from the trajectories that gives the expected energy spectrum for the ejected electrons. Furthermore, trajectories have allowed us to visualize the dynamics of both the photoionization and the harmonic generation processes.

In the second part we have investigated in detail the transfer of angular momentum to an atom interacting with an intense light pulses carrying both spin and orbital angular momenta, by means of 3D numerical simulations. Notably, we have shown that it is possible to exchange more than one unit of $\hbar$ of angular momentum per photon along the propagation direction of the pulse fulfilling $\Delta M = \ell + s$, where ℓ is the topological charge of the pulse and $s = \pm 1$ its polarization contribution. To get physical insight into the dynamics of the electron wave function and the absorption of angular momentum, we have computed the Bohmian trajectories, displaying in each scenario the dynamics of the electron in the polarization plane. In particular, we have illustrated how the electron absorbs both the angular momentum due to the light polarization, i.e., spin, and that due to its transverse profile, i.e., OAM.

We have seen that Bohmian trajectories can be useful for both making calculations in the dynamics of quantum systems and obtaining physical insight into such dynamics. All the calculations presented in this chapter could be combined with the quantum trajectory algorithms discussed in the first chapter of this book to extend them beyond the hydrogen case and reduce the required computational resources. Thus, Bohmian trajectories are a very promising tool for the study of multielectron atoms in electromagnetic fields. Furthermore, we have also seen that Bohmian trajectories allow for a simple visualization the complex dynamics of this processes, which is sometimes concealed behind the wave function.

Acknowledgments

We thank Alfonso Alarcón, Andreas Becker, Ramón Corbalán, Gabriel Fernández Calvo, Xavier Oriols, and Javier Rodríguez Vázquez de Aldana for fruitful discussions. We acknowledge support by the Spanish Ministry of Science and Innovation (MICINN) under contracts FIS2008-02425, FIS2009-09522, FIS2007-29091-E, CSD2007-00013 (Consolider projects SAUUL), CSD2006-00019 (Consolider projects QOIT), and FIS2014-57460-P, the Catalan Government under contracts SGR2009-00347 and SGR2014-1639 and Junta de Castilla y León under contract SA146A08. Albert Benseny acknowledges financial support from the Villum Foundation.

References

1. A. Einstein, *Über einen die Erzeugung und Verwandlung des Lichtes betreffenden heuristischen Gesichtspunkt*, Annalen der Physik, **17**, 132 (1905).
2. P. A. M. Dirac, *The Fundamental Equations of Quantum Mechanics*, Proceedings of the Royal Society, **A114**, 243 (1927).
3. M. Göppert-Mayer, *Elementary File with Two Quantum Fissures*, Annals of Physics, **9**, 273 (1931).
4. P. A. Franken, A. E. Hill, C. W. Peters, and G. Weinreich, *Generation of Optical Harmonics*, Physical Review Letters, **7**, 118 (1961).
5. H. B. Bebb and A. Gold, *Multiphoton Ionization of Hydrogen and Rare-Gas Atoms*, Physical Review, **143**, 1 (1966).
6. L. V. Keldysh, *Ionization in Field of a Strong Electromagnetic wave*, Soviet Physics JETP, **20**, 1307 (1965).
7. F. H. M. Faisal, *Multiple Absorption of Laser Photons by Atoms*, Journal of Physics B, **6**, L89 (1973).
8. H. R. Reiss, *Effect of an Intense Electromagnetic Field on a Weakly Bound System*, Physical Review A, **22**, 1786 (1980).
9. A. Becker and F. H. M. Faisal, *Intense-Field Many-Body S-Matrix Theory*, Journal of Physics B, **38**, R1 (2005).
10. P. Agostini and L. F. DiMauro, *The Physics of Attosecond Light Pulses*, Reports on the Progress of Physics, **67**, 813 (2004); Th. Brabec and F. Krausz, *Intense Few-Cycle Laser Fields: Frontiers of Nonlinear Optics*, Review of Modern Physics, **72**, 545 (2000); Th. Brabec (ed.), *Strong Field Laser Physics*, (Springer, 2008); F. Grossmann, *Theoretical Femtosecond Physics: Atoms and Molecules in Strong Laser Fields*, (Springer, 2008); Various authors, *The Fifth International Conference on Inertial Fusion Sciences and Applications (IFSA2007)*, Journal of Physics: Conference Series, **112** (2008); G. A. Mourou, T. Tajima, and S. V. Bulanov, *Optics in the Relativistic Regime*, Review of Modern Physics, **78**, 309 (2006); M. Wegener, *Extreme Nonlinear Optics: An Introduction*, (Springer, 2005).
11. D. Strickland and G. Mourou, *Compression of Amplified Chirped Optical Pulses*, Optics Communications, **56**, 219 (1985).
12. A. Weingartshofer, E. M. Clarke, J. K. Holmes, and C. Jung, *Experiments on Multi-photon Free-Free Transitions*, Physical Review A, **19**, 2371 (1979).
13. P. Agostini, F. Fabre, G. Mainfray, G. Petite, and N. K. Rahman, *Free-Free Transitions Following Six-Photon Ionization of Xenon Atoms*, Physical Review Letters, **42**, 1127 (1979).

14. A. McPherson, G. Gibson, H. Jara, J. Johann, I. A. McIntyre, K. Boyer, and C. K. Rhodes, *Studies of Multiphoton Production of Vacuum-Ultraviolet Radiation in the Rare Gases*, Journal of the Optical Society of America, **4**, 595 (1987).

15. M. Ferray, A. L'Huillier, X. F. Li, L. A. Lompre, G. Mainfray, and C. Manus, *Multiple-Harmonic Conversion of 1064-nm Radiation in Rare-Gases*, Journal of Physics B, **21**, L31 (1988).

16. K. J. Schafer, B. Yang, L. F. DiMauro, and K. C. Kulander, *Above Threshold Ionization Beyond the High Harmonic Cutoff*, Physical Review Letters, **70**, 1599 (1993).

17. P. B. Corkum, *Plasma Perspective on Strong Field Multiphoton Ionization*, Physical Review Letters, **71**, 1994 (1993).

18. M. Lewenstein, Ph. Balcou, M. Yu. Ivanov, Anne L'Huillier, and P. B. Corkum, *Theory of High-Harmonic Generation by Low-Frequency Laser Fields*, Physical Review A, **49**, 2117 (1994).

19. K. L. Ishikawa and T. Sato, *A Review on ab initio Approaches for Multielectron Dynamics*, IEEE Journal of Selected Topics in Quantum Electronics, **21**, 1 (2015).

20. J. S. Parker, K. J. Meharg, G. A. McKenna, and K. T. Taylor, *Single-Ionization of Helium at Ti: Sapphire Wavelengths: Rates and Scaling Laws*, Journal of Physics B, **40**, 1729 (2007); J. S. Parker, G. S. J. Armstrong, M. Boca, and K. T. Taylor, *From the UV to the Static-Field Limit: Rates and Scaling Laws of Intense-Field Ionization of Helium*, Journal of Physics B, **42**, 134011 (2009).

21. C. Ruiz, L. Plaja, and L. Roso, *Lithium Ionization by a Strong Laser Field*, Physical Review Letters, **94**, 063002 (2005).

22. A. Benseny, G. Albareda, A. S. Sanz, J. Mompart, and X. Oriols, *Applied Bohmian Mechanics*, European Physics Journal D, **68**, 286 (2014).

23. X. Y. Lai, Q. Y. Cai and M. S. Zhan, *Above-Threshold Ionization Photoelectron Spectrum from Quantum Trajectory*, European Physics Journal D, **53**, 393 (2009).

24. X. Y. Lai, Q. Y. Cai, and M. S. Zhan, *Bohmian Mechanics to High-Order Harmonic Generation*, Chinese Physics B, **19**, 020302 (2010).

25. X. Y. Lai, Q. Y. Cai, and M. S. Zhan, *From a Quantum to a Classical Description of Intense Laser-Atom Physics with Bohmian Trajectories*, New Journal of Physics, **11**, 113035 (2009).

26. S. S. Wei, S. Y. Li, F. M. Guo, Y. J. Yang, and B. B. Wang, *Dynamic Stabilization of Ionization for an Atom Irradiated by High-Frequency Laser Pulses Studied with the Bohmian-Trajectory Scheme*, Physical Review A, **87**, 063418 (2013).

27. F. H. M. Faisal and U. Schwengelbeck, *De Broglie-Bohm Formulation of Quantum Mechanics, Quantum Chaos and Breaking of Time-Reversal Invariance*, Pramāna **51**, 585 (1998).
28. P. Botheron and B. Pons, *Self-consistent Bohmian Description of Strong Field-Driven Electron Dynamics*, Physical Review A, **82**, 021404(R) (2010).
29. H. Z. Jooya, D. A. Telnov, P. C. Li, and S. I. Chu, *Exploration of the Subcycle multiphoton Ionization Dynamics and Transient Electron Density Structures with Bohmian Trajectories*, Physical Review A, **91**, 063412 (2015).
30. A. Picón, A. Benseny, J. Mompart, J. R. Vázquez de Aldana, L. Plaja, G. F. Calvo, and L. Roso, *Transferring Orbital and Spin Angular Momenta of Light to Atoms*, New Journal of Physics, **12**, 083053 (2010).
31. F. M. Guo, J. G. Chen, S. Y. Li, and Y. J. Yang, *Calculation of the Duration of the Atomic Tunneling Ionization in a Strong Electrostatic Field by Using Bohmian Trajectories Approach*, Chinese Physics Letters, **32**, 073201 (2015).
32. I. A. Ivanov, C. H. Nam, and K. T. Kim, *Exit Point in the Strong Field Ionization Process*, Scientific Reports, **7**, 39919 (2017).
33. Y. Song, Y. Yang, F. Guo, and S. Li, *Revisiting the Time-Dependent Ionization Process through the Bohmian-Mechanics Method*, Journal of Physics B, **50**, 095003 (2017).
34. Y. Song, F. M. Guo, S. Y. Li, J. G. Chen, S. L. Zeng, and Y. J. Yang, *Investigation of the Generation of High-Order Harmonics through Bohmian Trajectories*, Physical Review A, **86**, 033424 (2012).
35. J. Wu, B. B. Augstein, and C. Figueira de Morisson Faria, *Local Dynamics in High-Order-Harmonic Generation Using Bohmian Trajectories*, Physical Review A, **88**, 023415 (2013).
36. J. Wu, B. B. Augstein, and C. Figueira de Morisson Faria, *Bohmian-Trajectory Analysis of High-Order-Harmonic Generation: Ensemble Averages, Nonlocality, and Quantitative Aspects*, Physical Review A, **88**, 063416 (2013).
37. Y. Song, S. Y. Li, X. S. Liu, F. M. Guo, and Y. J. Yang, *Investigation of Atomic Radiative Recombination Processes by the Bohmian-Mechanics method*, Physical Review A, **88**, 053419 (2013).
38. P. C. Li, Y. L. Sheu, H. Z. Jooya, X. X. Zhou, and S. I. Chu, *Exploration of Laser-Driven Electron-Multirescattering Dynamics in High-Order Harmonic Generation*, Scientific Reports, **6**, 32763 (2016).

39. H. Z. Jooya, D. A. Telnov, and S. I. Chu, *Exploration of the Electron Multiple Recollision Dynamics in Intense Laser Fields with Bohmian Trajectories*, Physical Review A, **93**, 063405 (2016).

40. H. Z. Jooya, D. A. Telnov, P. C. Li, and S. I. Chu, *Investigation of the Characteristic Properties of High-Order Harmonic Spectrum in Atoms Using Bohmian Trajectories*, Journal of Physics B, **48**, 195401 (2015).

41. J. Peatross and J. Johansen, *Can Even-Order Laser Harmonics Exhibited by Bohmian Trajectories in Symmetric Potentials be Observed?*, Optics Express, **22**, 576 (2014).

42. I. P. Christov, *Polynomial-Time-Scaling Quantum Dynamics with Time-Dependent Quantum Monte Carlo*, Journal of Physical Chemistry A, **113**, 6016 (2009).

43. I. P. Christov, *Correlated Non-perturbative Electron Dynamics with Quantum Trajectories*, Optics Express, **14**, 6906 (2006).

44. I. P. Christov, *Time-Dependent Quantum Monte Carlo and the Stochastic Quantization*, Journal of Chemical Physics, **127**, 134110 (2007).

45. I. P. Christov, *Time-Dependent Quantum Monte Carlo: Preparation of the Ground State*, New Journal of Physics, **9**, 70 (2007).

46. I. P. Christov, *Correlated Electron Dynamics with Time-Dependent Quantum Monte Carlo: Three-Dimensional Helium*, Journal of Chemical Physics, **135**, 044120 (2011).

47. I. P. Christov, *Dynamic Correlations with Time-Dependent Quantum Monte Carlo*, Journal of Chemical Physics, **128**, 244106 (2008).

48. I. P. Christov, *Exploring Quantum Non-locality with de Broglie-Bohm Trajectories*, Journal of Chemical Physics, **136**, 034116 (2012).

49. N. Takemoto and A. Becker, *Multiple Ionization Bursts in Laser-Driven Hydrogen Molecular Ion*, Physical Review Letters, **105**, 203004 (2010).

50. N. Takemoto and A. Becker, *Visualization and Interpretation of Attosecond Electron Dynamics in Laser-Driven Hydrogen Molecular Ion Using Bohmian Trajectories*, Journal of Chemical Physics, **134**, 074309 (2011).

51. R. Sawada, T. Sato, and K. L. Ishikawa, *Analysis of Strong-Field Enhanced Ionization of Molecules Using Bohmian Trajectories*, Physical Review A, **90**, 023404 (2014).

52. J. Wang, B. B. Wang, F. M. Guo, S. Y. Li, D. J. Ding, J. G. Chen, S. L. Zeng, and Y. J. Yang, *Probing Dynamic Interference in High-Order Harmonic Generation from Long-Range Molecular Ion: Bohmian Trajectory Investigation*, Chinese Physics B, **23**, 053201 (2014).

53. X. Y. Lai, Q. Y. Cai, and M. S. Zhan, *Bohmian Mechanics to High-Order Harmonic Generation*, Chinese Physics B **19**, 020302 (2010).
54. X. Y. Lai, Q. Y. Cai, and M. S. Zhan, *Above-Threshold Ionization Photoelectron Spectrum from Quantum Trajectory*, European Physics Journal D, **53**, 393 (2009).
55. A. Picón, A. Benseny, J. Mompart, J. R. Vázquez de Aldana, L. Plaja, G. F. Calvo, and L. Roso, *Transferring Orbital and Spin Angular Momenta of Light to Atoms*, New Journal of Physics, **12**, 083053 (2010).
56. I. P. Christov, *Correlated Non-perturbative Electron Dynamics with Quantum Trajectories*, Optics Express, **14**, 6906 (2006).
57. I. P. Christov, *Time-Dependent Quantum Monte Carlo: Preparation of the Ground State*, New Journal of Physics, **9**, 70 (2007).
58. T. Tajima, Scientific Advisory Committee: Report on the ELI Science (2009).
59. X. Y. Lai, Q. Y. Cai, and M. S. Zhan, *From a Quantum to a Classical Description of Intense Laser-Atom Physics with Bohmian Trajectories*, New Journal of Physics, **11**, 113035 (2009).
60. P. Botheron and B. Pons, *Self-consistent Bohmian Description of Strong Field-Driven Electron Dynamics*, Physical Review A, **82**, 021404(R) (2010).
61. N. Takemoto and A. Becker, *Visualization and Interpretation of Attosecond Electron Dynamics in Laser-Driven Hydrogen Molecular Ion Using Bohmian Trajectories*, Journal of Chemical Physics, **134**, 074309 (2011).
62. J. Javanainen, J. H. Eberly, and Q. Su, *Numerical Simulations of Multiphoton Ionization and Above-Threshold Electron Spectra*, Physical Review A, **37**, 3430 (1988); Q. Su and J. H. Eberly, *Model Atom for Multiphoton Physics*, Physical Review A, **44**, 5997 (1991).
63. J. D. Jackson, *Classical Electrodynamics* (Wiley, New York, 1974).
64. W. H. Press, S. A. Teukolsky, W. T. Vetterling, and B. P. Flannery, *Numerical Recipes. The Art of Scientific Computing, 3rd Ed.*, (Cambridge University Press, 2007).
65. M. Davidović, D. Arsenović, M. Božic, A. S. Sanz, and S. Miret-Artés, *Should Particle Trajectories Comply with the Transverse Momentum Distribution?*, European Physical Journal Special Topics, **160**, 95 (2008).
66. L. Allen, M. W. Beijersbergen, R. J. C. Spreeuw, and J. P. Woerdman, *Orbital Angular Momentum of Light and the Transformation of Laguerre-Gaussian Laser Modes*, Physical Review A, **45**, 8185 (1992).

67. G. F. Calvo, A. Picón, and E. Bagan, *Quantum Field Theory of Photons with Orbital Angular Momentum*, Physical Review A, **73**, 013805 (2006).
68. M. F. Andersen, C. Ryu, P. Cladé, V. Natarajan, A. Vaziri, K. Helmerson, and W. D. Phillips, *Quantized Rotation of Atoms from Photons with Orbital Angular Momentum*, Physical Review Letters, **97**, 170406 (2006).
69. R. Inoue, N. Kanai, T. Yonehara, Y. Miyamoto, M. Koashi, and M. Kozuma, *Entanglement of Orbital Angular Momentum States between an Ensemble of Cold Atoms and a Photon*, Physical Review A, **74**, 053809 (2006).
70. M. van Veenendaal and I. McNulty, *Prediction of Strong Dichroism Induced by X Rays Carrying Orbital Momentum*, Physical Review Letters, **98**, 157401 (2007).
71. S. Franke-Arnold, L. Allen, and M. Padgett, *Advances in Optical Angular Momentum*, Laser & Photonics Reviews, **2**, 299 (2008).
72. D. L. Andrews, *Structured Light and Its Applications* (Academic Press, London, 2008).
73. A. Picón, J. Mompart, J. R. Vázquez de Aldana, L. Plaja, G. F. Calvo, and L. Roso, *Photoionization with Orbital Angular Momentum Beams*, Optics Express, **18**, 3660 (2010).
74. C. Cohen-Tannoudji and A. Kostler, *Optical Pumping*, in Progress in Optics **25**, edited by E. Wolf (North-Holland, Amsterdam), 33 (1966).
75. J. R. Vázquez de Aldana and L. Roso, *Magnetic Field Effects in Strong Field Ionization of Single-Electron Atoms: Three-Dimensional Numerical Simulations*, Laser and Particle Beams, **20**, 185–193 (2002).
76. ELI Delivery Consortium: https://eli-laser.eu/

Chapter 3

Atomtronics: Coherent Control of Atomic Flow via Adiabatic Passage

Albert Benseny,[a,b] Joan Bagudà,[b] Xavier Oriols,[c] Gerhard Birkl,[d] and Jordi Mompart[b]

[a]*Department of Physics and Astronomy, Aarhus University, Ny Munkegade 120, DK-8000, Aarhus C, Denmark*
[b]*Departament de Física, Universitat Autònoma de Barcelona (UAB), E-08193 Bellaterra, Spain*
[c]*Departament d'Enginyeria Electrònica, Universitat Autònoma de Barcelona (UAB), E-08193 Bellaterra, Spain*
[d]*Institut für Angewandte Physik, Technische Universität Darmstadt (TUD), D-64289 Darmstadt, Germany*
abc@phys.au.dk; jordi.mompart@uab.cat

Atomtronics is one of the foremost bottom-top approaches to quantum information science, where neutral atoms in optical, magnetic, and electric potentials form the building blocks of coherent diode and transistor-like devices. In this context, we discuss in this chapter an efficient and robust technique to coherently transport a single neutral atom, a single hole, or even a Bose–Einstein condensate (BEC) between the two extreme traps of a triple-well optical potential. Solving the corresponding quantum dynamics in terms of Bohmian trajectories, we investigate this adiabatic transport process and present a very counterintuitive effect: by slowing down

Applied Bohmian Mechanics: From Nanoscale Systems to Cosmology (Second Edition)
Edited by Xavier Oriols and Jordi Mompart

ISBN 978-981-4800-10-5 (Hardcover), 978-0-429-29474-7 (eBook)
www.jennystanford.com

the total time duration of the transport process, it is possible to achieve sudden particle accelerations yielding ultrahigh atomic velocities.[a]

3.1 Introduction

3.1.1 Atomtronics

Electronics is the branch of science and technology that deals with the controlled flow of electrons. Most of today's electronic devices use semiconductor diodes and transistors to manipulate the electronic current. Notably, the miniaturization of these devices is getting closer and closer to the single atom scale, reaching the limit of Moore's law [1].[b] Nowadays, electronics top-bottom approach is crossing the frontier between the classical and the quantum realm, where quantum effects such as quantum tunneling and quantum correlations, i.e., entanglement, are unavoidable. In particular, the transition from the classical to the quantum world constitutes one of the physical scenarios where the Bohmian formulation can give more insight into the underlying physics. See Chapter 1 for a detailed discussion of these aspects and Chapter 6 for a particular application of Bohmian trajectories to quantum electronic devices.

Alternatively, the beginning of the twenty-first century has brought the development of techniques to isolate and manipulate individual neutral atoms [2], allowing to follow a bottom-top approach where quantum systems acquire classical features as their size and their coupling with the environment increase. In fact, in the past few years a lot of attention has been devoted to the field of *atomtronics* [3], where atomic matter waves in optical [4–7], electric [8], and magnetic [9] potentials play an analogous role to electrons in electronic devices. In this particular context, several proposals on coherent diode- and transistor-like behaviors for ultracold atoms [10, 11] and BECs [12] both in double- and triple-well optical potentials as well as in optical lattices have been

---

[a]This chapter is dedicated to the memory of Rafa Martínez.

[b]Presently, the scientific community is searching for completely different alternatives to standard semiconductor devices, in what is called the "more than Moore" route.

deeply investigated with applications ranging from atomic cooling to quantum information processing [13]. Thus, neutral atoms offer a wide range of possible applications such as, for instance, interfacing classical electronic devices to quantum computers.

The time for coherent atomtronics is already here [14]. This has mainly become possible due to two different reasons: (i) techniques for cooling and trapping atoms are by now very well established [15], and (ii) neutral atoms are comparatively less sensitive to decoherence than charged particles since their interaction with the "classical" environment can be almost completely inhibited. Neutral atoms can be stored and manipulated in optical lattices, standard dipole traps, and microtraps. In particular, magnetic and optical microtraps offer an interesting perspective for storing and manipulating arrays of atoms with the eventual possibility to scale, parallelize, and miniaturize the atomtronic devices. Moreover, optical microtraps can take advantage of the fact that most of the current techniques used in atom optics and laser cooling are based on the optical manipulation of atoms. In fact, the possibility to store and to selectively address single optical microtraps, as well as initializing and reading out the quantum states in each of the sites, has been experimentally reported [4–7].

3.1.2 Three-Level Atom Optics

In this regard, there is a need for the development of novel techniques to control the coherent flow of matter waves in optical and magnetic traps based on tunneling devices. To that aim, we introduced a set of coherent techniques [16] to efficiently transport matter waves between the two extreme traps of a triple-well potential by adiabatically following an energy eigenstate of the system, the so-called spatial dark state that, ideally, only involves the vibrational ground states of the two extreme wells. We named these techniques three-level atom optics techniques, since the interference of (single) atom matter waves is at their core and the adiabatic transport process is the matter wave analogue of the three-level quantum optical technique known as simulated Raman adiabatic passage (STIRAP) [17]. The main goal of the matter wave STIRAP technique consists in achieving an efficient and robust

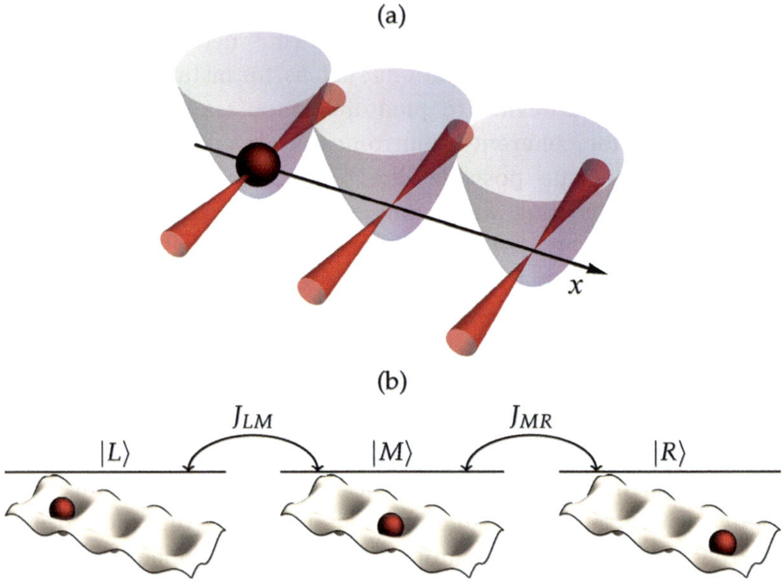

Figure 3.1 (a) Three red-detuned lasers are used to create a triple-well potential for a single atom initially located in the left trap. We will assume control over the temporal variation of the trap distances, i.e., the position of the laser beam foci, and, therefore, over the tunneling of the atom between the traps. (b) Sketch of the three-level atom optics system consisting of the localized states of the atom in the left, middle and right traps, i.e., $|L\rangle$, $|M\rangle$, and $|R\rangle$, and the tunneling rates between them J_{LM} and J_{MR}.

transfer of a single atom from the left to the right trap via the tunneling interaction, i.e., atom transport between the extreme traps should be performed with a fidelity close to 100%, while being almost nonsensitive to the fluctuations of the control parameter values.

To illustrate the basic ideas behind the matter wave STIRAP technique, let us start with a very simple three-level model for a single atom in a triple-well potential (see Fig. 3.1a). By varying the distance between the traps or modifying the height of the barriers between the potential wells, the atom can tunnel from one trap to another. We assume three identical traps with tunneling only occurring between neighboring wells, and that the dynamics of the three-level system are restricted to the vibrational ground states of the three traps, i.e., $|L\rangle$, $|M\rangle$, and $|R\rangle$ (see Fig. 3.1b). Then the system will be described by the Hamiltonian:

$$H = \frac{\hbar}{2}\begin{pmatrix} 0 & J_{LM} & 0 \\ J_{LM} & 0 & J_{MR} \\ 0 & J_{MR} & 0 \end{pmatrix}, \tag{3.1}$$

where J_{LM} (J_{MR}) is the tunneling rate of the atom between the left and middle (middle and right) traps.

The atom dynamics can be controlled by modifying the time dependences of the distances $d_{LM}(t)$ (between the left and middle traps), and $d_{MR}(t)$ (between the middle and right traps) which, in turn, affect the tunneling rates J_{LM} and J_{MR}. In particular, the tunneling rate between the ground vibrational states of two traps can be straightforwardly calculated as the energy splitting between the lowest symmetric and antisymmetric energy eigenstates of the corresponding double-well. For two piece-wise truncated harmonic traps, this yields [16]

$$J(\alpha d) = \omega_x \frac{-1 + e^{(\alpha d)^2/4}\left(1 + \alpha d\sqrt{\pi}\,\mathrm{erfc}(\alpha d/2)/2\right)}{\sqrt{\pi}\left(e^{(\alpha d)^2/2} - 1\right)/\alpha d}, \tag{3.2}$$

d being the distance between the centers of the two traps and α^{-1} the width of the vibrational ground state of an isolated trap (see Eq. 3.8 later). The coherent coupling between two orthogonal states of a quantum system gives rise to oscillations of their probability amplitudes, such as the Rabi oscillations of a two-level atom interacting with a laser field. In the matter wave case, when a trap with an atom and an empty trap are approached, the atomic population oscillates between the two traps at the tunneling frequency J. For $\alpha d \gtrsim 9$ and the time scales considered throughout the chapter, the tunneling rate between adjacent traps is almost negligible and they can be considered as isolated.

Adiabatic transport via the matter wave STIRAP technique is based on the fact that one of the three eigenstates of the Hamiltonian in Eq. 3.1 involves only the ground states of the two extreme traps:

$$|D(\Theta)\rangle = \cos\Theta|L\rangle - \sin\Theta|R\rangle, \tag{3.3}$$

where the mixing angle Θ is defined as $\tan\Theta = J_{LM}/J_{MR}$ such that for $\Theta = 0$, one obtains $|D(0)\rangle = |L\rangle$, and for $\Theta = \pi/2$,

$|D(\pi/2)\rangle = -|R\rangle$. The state $|D(\Theta)\rangle$ is known as the spatial dark state, where dark accounts for the fact that in the optical STIRAP, this state is decoupled from the light beams.

Thus, starting with the atom in the left trap, i.e., in state $|L\rangle$, and following Eq. 3.3, it is possible to transfer it to the right trap, $|R\rangle$, by adiabatically varying the mixing Θ angle from 0 to $\pi/2$. This corresponds to approaching and separating first the right trap to the middle trap and, with a delay, the left trap to the middle one (see Fig. 3.2a) such that there is a time period when J_{MR} decreases while J_{LM} increases (see Fig. 3.2b). This counterintuitive sequence moves the atom directly from $|L\rangle$ to $|R\rangle$ with an ideally null probability to be in $|M\rangle$, i.e., the ground state of the middle trap (see Fig. 3.2c). As its optical analog, the matter wave STIRAP technique does not require an accurate control on the system parameters as long as the process is adiabatic (slow) enough [16, 17].

An experimental realization of matter wave STIRAP with ultracold atoms has been recently reported in a Lieb optical lattice [18]. Moreover, many extensions of these three-level atom optics techniques to atoms in vortex states [19], to systems composed of interacting atoms [11, 40], to atomic wave packets propagating in dipole waveguides [21, 22], to BECs [23–26], to electronic transport in quantum dot systems [27, 28], and to superconductors [29] have been investigated. Also, the performance of the matter wave STIRAP for a single atom or a BEC has been studied by means of optimal control methods [30]. Moreover, by exploiting the wave analogies between classical and quantum systems, light transfer in an engineered triple-well optical waveguide by means of the classical analog of the matter wave STIRAP has been experimentally reported [31, 32]. For a review of spatial adiabatic passage techniques, we refer the reader to [20].

3.1.3 Adiabatic Transport with Trajectories

While the simplified system described in the previous section is useful to illustrate the basic ideas of adiabatic transport based on matter wave STIRAP, a realistic treatment accounting for deviations

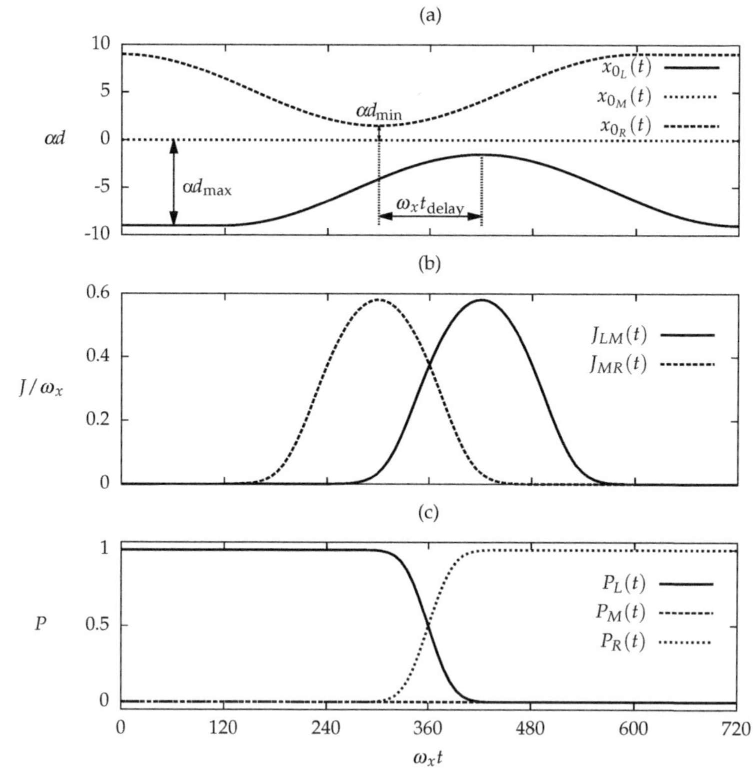

Figure 3.2 Time dependence of (a) trap positions, (b) tunneling rates, and (c) trap populations $P_i = |\langle\psi(t)|i\rangle|^2$ for the matter wave STIRAP technique. The middle trap has been taken to be static at $x_{0_M} = 0$, while displacing only the two extreme traps. $d_{LM} = |x_{0_M} - x_{0_L}|$ and $d_{MR} = |x_{0_R} - x_{0_M}|$ denote the distance between left and middle and between middle and right traps, respectively. Both distances are modeled with a cosine function truncated at the minimum separation. The tunneling rates shown in (b) have been calculated by means of Eq. 3.2. The success of the atom transport process from the left to the right trap is plotted in (c), where the main signature of the STIRAP technique can be seen: the transport process takes place without exciting the population of the middle trap. Time is given in units of ω_x^{-1}, ω_x being the corresponding harmonic well-trapping frequency, and distances are given in units of $\alpha^{-1} = \sqrt{\hbar/(m\omega_x)}$, m being the mass of the atom.

from perfect adiabaticity as well as coupling to excited vibrational states and direct coupling from the left to the right trap requires the integration of the Schrödinger equation in position space. Furthermore, the matrix formulation we used does not allow a direct computation of the Bohmian trajectories, since they require knowledge of the spatial dependence of the wave function. This is the approach to the transport problem that we will follow from now on. As we will see in Section 3.3, there is one peculiarity in the previously discussed adiabatic three-level atom optics techniques, namely, the existence at all times of a node in the spatial dark state at the position of the middle trap that seems to have unexpected implications. The real dynamics will follow closely (but not perfectly) the adiabatic evolution, with the dynamical state remaining very close (but not identical) to the spatial dark state, which does not overlap with the middle trap ground state, i.e., $\langle D(\Theta)|M\rangle = 0$. Therefore, at all times, there will be a spatial region located at the middle trap where the population is minimum and arbitrarily close to zero during the evolution.

From this observation, one might conclude that it is possible to transport a single atom or even a BEC *directly* from the left trap to the right one without transiting the center region. In fact, M. Rab et al. reached what is, from our point of view, an incorrect conclusion [24]:

> Classically it is impossible to have transport without transit, i.e., if the points 1, 2, and 3 lie sequentially along a path then an object moving from 1 to 3 must, at some time, be located at 2. For a quantum particle in a three-well system it is possible to transport the particle between wells 1 and 3 such that the probability of finding it at *any time* in the classical accessible state in well 2 is negligible.

Clearly, quantum *transport without transit* is in contradiction with the continuity equation that derives from the Schrödinger or the Gross–Pitaevskii equations. However, their unraveling in terms of quantum trajectories, namely, the Bohm [33] and Madelung's hydrodynamic [34, 35] formulations, while being equivalent to standard quantum mechanics when averaged over the complete set of initial conditions, provides a very clear physical picture of continuity. Therefore, since Bohmian trajectories give a correct description of the behavior of quantum systems, we will use them here to investigate the transport of a single atom, a single hole, and a BEC between the two extreme traps of a triple-well potential to show that matter wave transport always implies transit.

There are different approaches to the study of transport in quantum systems, but most of them present unsurmountable difficulties when trying to associate a velocity to the transport process. This is not the case of Bohmian mechanics, since it provides an unambiguous way to calculate velocities and, in contrast to other approaches such as the Feynman path-integral or Wigner formulations, always yields a positive-definite answer when calculating transport times [36]. Note that Bohm's velocity is equivalent to the standard definition of velocity $\vec{v} = \vec{j}/\rho$, where $\vec{j}$ is the particle current and $\rho = \psi^*\psi$. This velocity is, in general, associated with a combination of a reflected and a transmitted beams [37], but in our adiabatic transport process we only have a transmitted "beam".

The system studied in this chapter presents an extremely counterintuitive effect, by slowing down the total matter wave STIRAP sequence, it is possible to achieve sudden particle accelerations yielding ultrahigh velocities. In fact, it has been shown that Bohm's velocity calculated from the Dirac equation is always bounded by the speed of light [38, 39]. Thus, in the limit of quasi perfect adiabaticity, relativistic corrections to the Schrödinger equation might be needed to properly address the transport process, while avoiding superluminal matter wave propagation.

The rest of the chapter is structured as follows. In Section 3.2 we will present the physical system under investigation, where atoms are trapped in optical potentials created by arrays of microlenses. Section 3.3 deals with the study of the adiabatic transport of a single atom in a triple-well potential by means of Bohmian trajectories. We then proceed to investigate, in Section 3.4, the adiabatic transport of two interacting atoms in a triple-well potential where the atomic interaction and the atoms spin are used as control parameters to engineer coherent atomic diodes and transistors. By means of the Madelung quantum hydrodynamic equations of motion, we will investigate in Section 3.5 the adiabatic transport of a BEC. Finally, we will present the conclusions in Section 3.6.

The main ideas behind the coherent control of matter wave tunneling based on adiabatic passage that we will discuss in this chapter are already present in the literature [11, 16, 20, 22], while their Bohmian formulation has been published in Refs. [11, 41].

3.2 Physical System: Neutral Atoms in Optical Microtraps

In this section we will briefly describe the main features of the optical microtrap array system [4, 5] used to isolate and manipulate individual neutral atoms that we will consider during most of this chapter (see Fig. 3.3). These microtraps are created by illuminating a set of microlenses with a red-detuned laser beam such that in each of the foci of the individual lenses, neutral atoms can be stored by means of the dipole force. By illuminating the microlenses with two independent laser beams, it is possible to generate two sets of traps which can be approached or separated by changing the relative angle between the two lasers, allowing the atoms to tunnel between different microtraps.

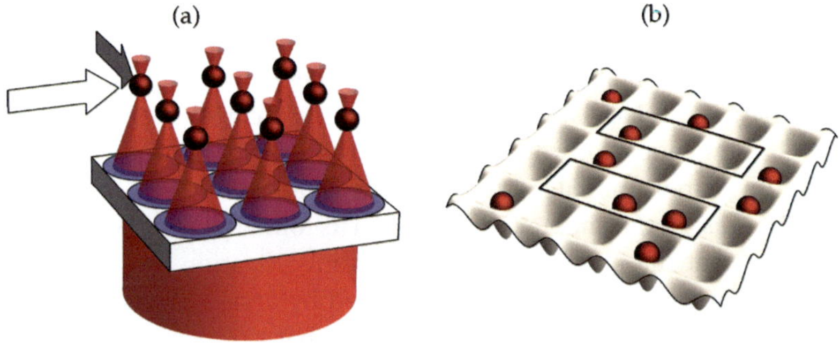

Figure 3.3 Sketch of (a) the optical microlenses array used to trap neutral atoms in their foci and (b) resulting two-dimensional array of optical traps partially loaded with neutral atoms. The traps are separated enough so it is possible to manipulate individual traps with additional lasers. In (b), matter wave STIRAP techniques can be applied to move either atoms in the array or, alternatively empty sites, i.e., holes. We will mainly focus on one-dimensional arrays of three traps with either one atom or one hole initially in one of its extremes.

To be specific, we consider the two-dimensional (2D) array of optical microtraps discussed by Dumke et al. [4], where one atom per trap [42], single site addressing, as well as the ability to approach columns (or, alternatively, rows) of traps yielding coherent atomic transport have been demonstrated [5]. Thus, we assume that tunneling between sites occurs only in the column movement direction (namely, x) and therefore the main dynamics will be restricted to one dimension. Typically, the engineered optical potentials have a Gaussian intensity profile, although for simplicity and in order to reduce the time needed for the numerical simulations we will consider here (truncated) harmonic traps.[a] For the preparation of the initial state we will assume the ability to place a single neutral atom in the ground state of a specific trap. Although we will require only three traps, the use of trap arrays has the advantage of allowing several experiments to be performed in parallel.

3.2.1 One-Dimensional Hamiltonian

For n_A identical atoms, the dynamics of the system is governed by the Schrödinger equation:

$$i\hbar\frac{d}{dt}\psi(x_1, \ldots, x_{n_A}, t) = H\psi(x_1, \ldots, x_{n_A}, t) \tag{3.4}$$

with the Hamiltonian:

$$H = \sum_{k=1}^{n_A} -\frac{\hbar^2}{2m}\frac{\partial^2}{\partial x_k^2} + \sum_{k=1}^{n_A} V(x_k, t) + \sum_{\substack{k,l=1 \\ k>l}}^{n_A} U(x_k, x_l), \tag{3.5}$$

m being the mass of the atoms, V the trapping potential and U the interaction between every pair of atoms. We assume, for simplicity, that the trapping potential consists of an array of n truncated harmonic wells centered at positions $\{x_{0_i}(t)\}$, $i = 1, 2, \ldots, n$:

$$V(x, t) = \frac{1}{2}m\omega_x^2 \min_i \left\{(x - x_{0_i}(t))^2\right\}. \tag{3.6}$$

The distances between the traps centers will be given by $d_i = |x_{0_{i+1}} - x_{0_i}|$. When dealing with more than one atom, we will consider

---

[a] This assumption is well justified in the limit of large trap separations or deep enough traps.

cold collisional interactions such that the interaction between every pair of neutral atoms can be modeled by a contact potential of the form [43]:

$$U(x_k, x_l) = 2\hbar a_s \omega_p \delta(x_k - x_l), \tag{3.7}$$

with a_s the s-wave scattering length, ω_p the transverse trapping frequency, and $\delta(x)$ the Dirac delta function that we numerically implement as a very narrow Gaussian with unit area.

Neutral atoms will be assumed to be initially in the vibrational ground state of each trap, whose wave functions take the form:

$$\phi_i(x) = \langle x|i\rangle = \sqrt[4]{\frac{m\omega_x}{\pi\hbar}} e^{-m\omega_x(x-x_{0_i})^2/2\hbar} = \frac{\sqrt{\alpha}}{\sqrt[4]{\pi}} e^{-\alpha^2(x-x_{0_i})^2/2}, \tag{3.8}$$

where we have defined $\alpha \equiv \sqrt{m\omega_x/\hbar}$ as the inverse of the width of the ground state.

3.3 Adiabatic Transport of a Single Atom

In this section, we will use Bohmian trajectories to investigate the coherent and adiabatic transport of a single atom between the two extreme traps of a triple-well potential. We will assume that the atom is initially located in the left trap and that the matter wave STIRAP sequence for the trap distances is applied in such a way that the mixing angle Θ (see Eq. 3.3), is smoothly varied from 0 to $\pi/2$.

3.3.1 The Matter Wave STIRAP Paradox with Bohmian Trajectories

With the temporal variation of the traps shown in Fig. 3.2a, Fig. 3.4 shows the evolution of (a) the trapping potentials and (b) the atomic probability density $|\psi(x, t)|^2$ by integrating the corresponding 1D Schrödinger equation. It is clearly seen in Fig. 3.4b that the atom is transferred from the left to the right trap with an almost null population of the central region around $\alpha x \sim 0$. This ideally direct transport from the left to the right trap is the main signature of the matter wave STIRAP technique and seems to imply the failure of the quantum continuity equation. In fact, the adiabaticity condition

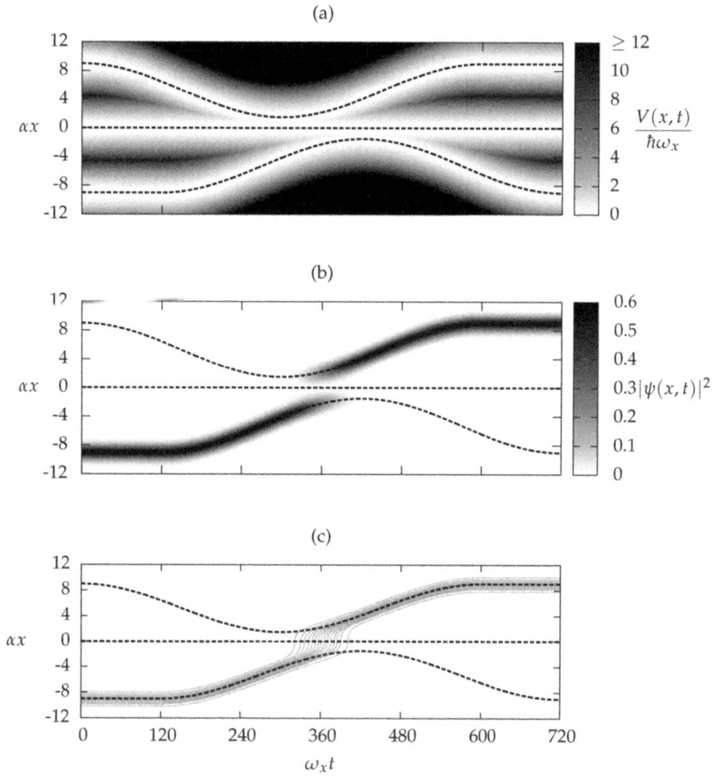

Figure 3.4 Evolution of (a) the potential $V(x, t)$ during the matter wave transport, (b) the density probability $|\psi(x, t)|^2$, and (c) Bohmian trajectories obtained from the evolution of $\psi(x, t)$. Dashed lines represent the temporal variation of the traps given in Fig. 3.2.

means that the atomic state will be very close to the ideal spatial dark state given in Eq. 3.3 and, therefore, that the population in the middle trap can be made arbitrarily small by increasing the total time of the atom transport. If so, the whole wave function will transit through a quasi node.

To elucidate this apparent breakdown of continuity in the matter wave STIRAP technique, we have computed in Fig. 3.4c the corresponding Bohmian trajectories. There is one peculiarity in this transport process: the fact that all trajectories must cross through a quasi node implies that they must do it at a high velocity, since the probability density in the central region is always very small, illustrated by the slope of the trajectories around the position of the middle trap center. Note that the region where Bohmian trajectories move faster, around $\alpha x \sim 0$, corresponds to the trapping potential minima of the central well. Note also that the velocity of an individual Bohmian trajectory is not a quantum observable, since one should average over all the trajectories. In any case, Bohmian trajectories clearly show that the continuity equation is perfectly fulfilled in such a way that the more adiabatic is the transfer process, the faster Bohmian trajectories cross through the central region.

3.3.2 Velocities and Accelerations of Bohmian Trajectories

Figure 3.5 shows an enlarged view of the previously discussed Bohmian trajectories around the central region, together with their velocities and accelerations. Each trajectory corresponds to a different initial position of the atom in the left trap. It is clearly seen from Fig. 3.5a,b that trajectories try to minimize the time spent in the central region (corresponding to the middle trap) by speeding up. The velocity that the trajectories reach in the middle region is on the order of 10 ω_x/α, two orders of magnitude faster than the mean wave packet velocity (dashed curve in Fig. 3.5b). Note that all trajectories achieve similar maximum velocities, although at different times, depending on their initial position inside the wave function. A measurement of the mean velocity of these trajectories at each point x_p in the middle region can be calculated as:

$$\langle v(x_p) \rangle = \frac{1}{N_t} \sum_k v_k[x_p], \tag{3.9}$$

where N_t is the number of trajectories in our simulation and $v_k[x_p]$ is the velocity of the k-th trajectory as it passes through x_p. These velocities are uniquely defined because the trajectories move from left to right, crossing each point only once. The distribution $v(x_p)$ is plotted in Fig. 3.6a. The numerical data has been fitted by a Lorentzian curve of maximum value $\bar{v}_{\max}$ and width Δx, which are shown in Figs. 3.6b,c.

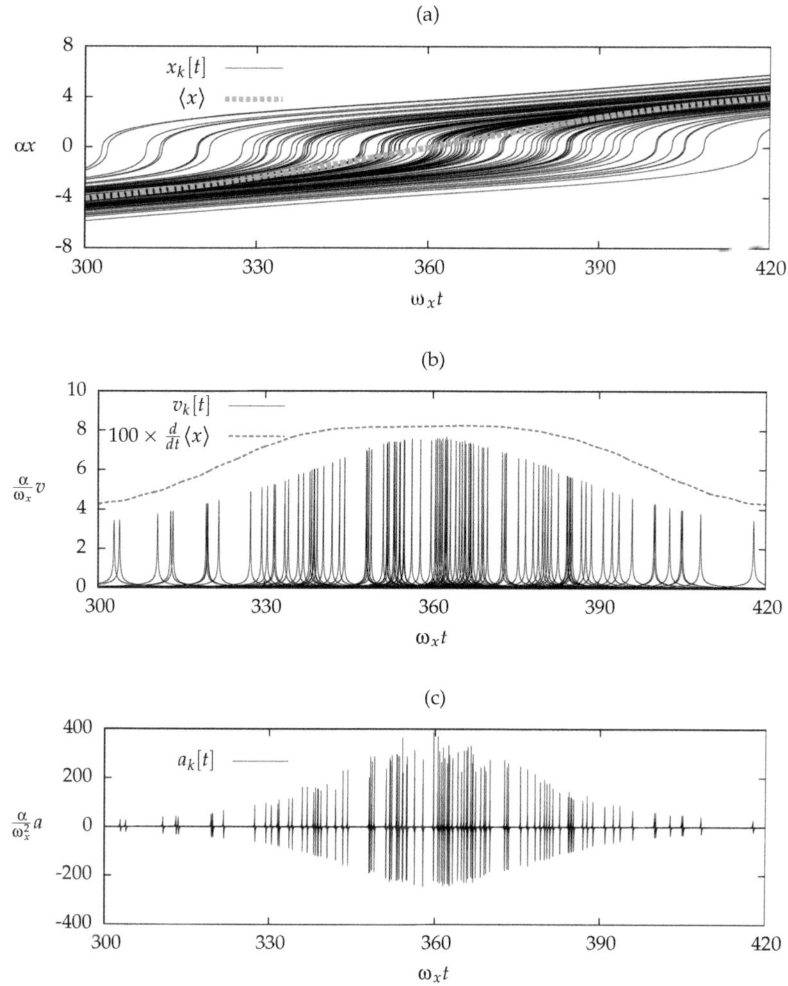

Figure 3.5 (a) Positions, (b) velocities, and (c) accelerations (in units of $\alpha^{-1}\omega_x^2$) of the Bohmian trajectories for the matter wave STIRAP sequence. We plot with dashed curves in (a) the mean value of the position of the atom wave packet and in (b) its mean velocity (multiplied by a factor of 100 for clarity).

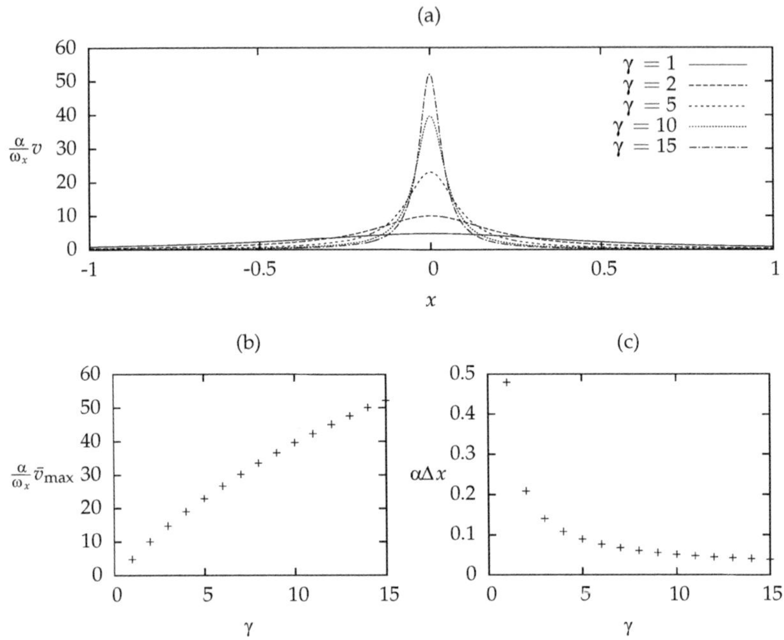

Figure 3.6 Effect of the change of the total time of the matter wave STIRAP sequence for different cases done γ times slower. $\gamma = 1$ for the simulations of Figs. 3.4 and 3.5. (a) Velocity distributions according to Eq. 3.9 for different γ. (b) Maximum and (c) width of the velocity distributions versus γ.

Slowing the total matter wave STIRAP sequence, the transport process will become more adiabatic, and the instantaneous state of the atom will remain closer to the ideal dark state. Thus, the middle trap will be less populated, resulting in an increase of the velocity of the trajectories in the central region. Figure 3.6b shows that by slowing down the transport process, the maximum velocity of the trajectories, $\bar{v}_{\max}$, reached in the central region certainly grows. As we see from Fig. 3.6b a very counterintuitive process appears: by increasing the time for the transport of the entire wave packet, the velocity that each of the trajectories reaches increases. Furthermore, as the maxima of the velocity distributions increase, their widths decrease (see Fig. 3.6c), such that their area, i.e., the product $\bar{v}_{\max}\Delta x$, remains constant.

Nothing indicates that this increase of the peak velocity might have any limit and, since we are using here the nonrelativistic Schrödinger equation, could even surpass the speed of light. However, as discussed by Leavens and Mayato [38] in their investigations of the tunneling effect, superluminal tunneling times are *an artifact of using the nonrelativistic Schrödinger equation*. With a correct relativistic description, e.g., by using the Dirac equation, Bohmian velocities cannot surpass the speed of light. Therefore, the appearance of superluminal trajectories in our problem would mean that our system would no longer be correctly described by the Schrödinger equation and the adiabatic transport might break down. It is remarkable that in the limit where the transport is done very slow one should take relativistic corrections into account to properly describe the problem.

In our opinion, a possible method to detect these ultrahigh velocities could be the use of a Doppler radar, in a similar fashion as the police track the speed of a vehicle. According to Section 1.4 about Bohmian measurement, we would have to introduce the interaction between the measuring apparatus (the radar) and the atom in the traps. This formal discussion is out of the scope of this chapter, but let us point out the intriguing results that could be obtained. Let us assume that the atom to be transferred between the two extreme traps has internal electronic structure, e.g., a two-level system with ω_A being its transition frequency. From a classical perspective, the Doppler effect consists in the following: for a laser beam of frequency ω_0 counterpropagating with respect to the atom velocity, the laser frequency that the atom perceives is given by $\omega_0' = \omega_0(1+v/c)$, where v is the atom velocity and c the speed of light. The scattering and even the absorption of photons is a very nonlinear process with respect to the laser detuning $\Delta = \omega_0' - \omega_A$ being maximum at $\Delta = 0$. We expect that by taking $\omega_0 \ll \omega_A$, a sudden increase of the atom velocity around $\alpha x \sim 0$ should result in an increase of the number of scattered photons and/or the fluorescence due to light's absorption. The back action of the field on the atoms could even inhibit the adiabatic transport process.

Another interesting effect is that the increase in the velocities is very abrupt, leading to very high accelerations (and decelerations) on the order of $\sim 200\omega_x^2/\alpha$ (see Figure 3.5c). It is then also an open problem to investigate the matter wave STIRAP technique for a charged particle, i.e., could the adiabatic transport of an ion lead to the emission of radiation?

3.4 Adiabatic Transport of a Single Hole

3.4.1 Hole Transfer as an Array-Cleaning Technique

Quantum gases trapped in optical potentials [4–7], e.g., microtrap arrays or optical lattices, have attracted considerable attention since they fulfill all the basic requirements for quantum information processing [44]. Quantum registers with single-site addressing of about a hundred of qubits [4] and cluster-entangled states of thousands of atoms [6] have been reported, respectively, in 2D optical microtrap arrays and 3D optical lattices. Furthermore, the loading of quantum gases into three-dimensional (3D) optical lattices, achieving the Mott insulator regime with ideally one atom per site for both bosons [45] and fermions [46], has been experimentally demonstrated, reaching one of the main goals for quantum information processing with neutral atoms. In most quantum computation proposals with trapped neutral atoms, a defect-free quantum system (where all sites of the lattice are occupied by exactly one atom) is needed to start the information processing. Therefore, it is necessary to remove empty sites from the physical area of computation.

With this purpose, we investigate in this section single-occupancy dipole trap arrays loaded with a gas of either fermions or bosons presenting some isolated defects consisting in empty sites, i.e., holes (see Fig. 3.7). We will engineer a transport scheme, based on the matter wave STIRAP technique, that will allow us to transport holes among traps. This technique can be used to clean faulty loaded arrays of traps by moving some defects (empty sites) away from the area of interest, or for the construction of a coherent single hole diode and a coherent single hole transistor [11].

By means of the integration of the Schrödinger equation, we will investigate the adiabatic transport of a hole between the two extreme traps of a triple-well potential. Furthermore, the Bohmian

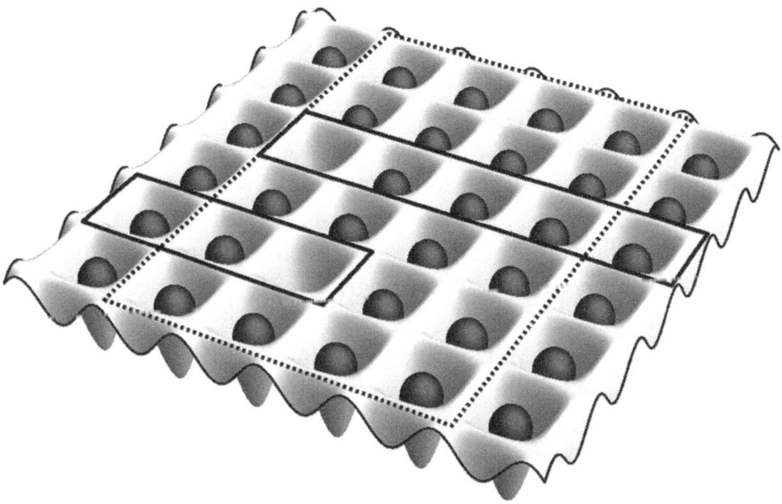

Figure 3.7 Sketch of a 2D array of optical traps loaded with one neutral atom per site but exhibiting two defects with empty sites. matter wave STIRAP techniques will be applied to move empty sites, i.e., holes, away from the physical area of computation that is defined by the atomic qubits inside the dotted rectangular area. We will focus on 1D arrays of n traps with one hole in one of its extremes. In the picture, these domains are tagged with solid-outlined rectangles of $n = 3$ and $n = 5$.

formulation of quantum mechanics will be used to gain detailed physical insight into the transport process. Since the simulation of systems with more than two atoms is very computationally demanding, we will only present here results for the three-trap case, even though this transport scheme has been shown to be extendable to an array of arbitrary length by means of the Hubbard model [11].

3.4.2 Adiabatic Transport of a Hole in an Array of Three Traps

3.4.2.1 Three-level approximation description

As in Section 3.3, we will start by introducing a three-level model that will allow us to transport a hole between the two extreme traps of a triple-well potential. We will take a subdomain of the optical trap array shown in Fig. 3.7, consisting in three in-line traps with one empty site and two identical (fermionic or bosonic) atoms. As in the previous section, we will consider a triple-well potential but now with two atoms in it, leaving an empty trap. Thus, in the following lines, we will make use of a hole description similar to the previously introduced by Mompart et al. [47]. For this description to be valid, the following conditions must be fulfilled: (i) each trap contains, at most, one atom; (ii) all atoms are cooled down to the vibrational ground state of each trap; and (iii) tunneling is adiabatically controlled to strongly suppress the probability of double occupancy.

Starting from the Mott-insulator regime [45, 46], conditions (i) and (ii) are directly fulfilled. To satisfy condition (iii) for identical fermions, we assume that all atoms have parallel spins such that the Pauli exclusion principle forbids the double occupancy of any vibrational state. For bosons, we consider that the s-wave-scattering length that governs the interaction between atoms is large enough to, in the adiabatic limit, inhibit double occupancy. From now on, throughout this section we will consider a spin-polarized collection of neutral atoms and refer to the case where the spatial wave function is antisymmetric (symmetric) as the fermionic (bosonic) case. In fact, for the 1D case, spin-polarized fermions and hardcore bosons will present equivalent dynamics [48].

For fermionic atoms, the previously mentioned conditions restrict the system dynamics to the Hilbert space generated by the lowest three energy eigenstates (see Fig. 3.8) whose spatial wave functions read:

$$|\widetilde{L}^F\rangle \equiv \frac{1}{\sqrt{2}}\left[|M\rangle_1|R\rangle_2 - |R\rangle_1|M\rangle_2\right], \tag{3.10}$$

$$|\widetilde{M}^F\rangle \equiv \frac{1}{\sqrt{2}}\left[|R\rangle_1|L\rangle_2 - |L\rangle_1|R\rangle_2\right], \tag{3.11}$$

$$|\widetilde{R}^F\rangle \equiv \frac{1}{\sqrt{2}}\left[|L\rangle_1|M\rangle_2 - |M\rangle_1|L\rangle_2\right], \tag{3.12}$$

where $|j\rangle_k$ is the (time-dependent) state of the k-th atom localized in the j-th trap, with its associated wave function $\phi_j(x_k) = \langle x_k|j\rangle$. States $|\widetilde{i}^F\rangle$ with $i = L, M, R$ account for the fermionic hole being in the left, middle, and right trap, respectively. Their associated wave functions $\widetilde{\phi}_i^F(x_1, x_2) = \langle x_1, x_2|\widetilde{i}^F\rangle$ live in a 2D space. Note that Eqs. 3.10–3.12 are antisymmetric since we have assumed that

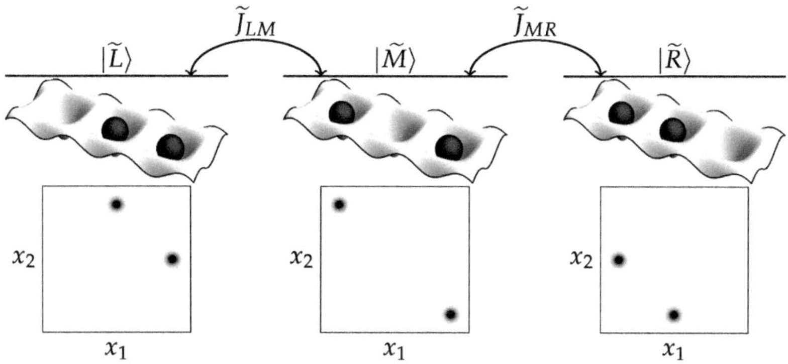

Figure 3.8 Three-level simplified description of the system for a hole. See the text for the definition of states and coupling strengths. The bottom row represents the joint probability distribution in the configuration space, for the states $|\langle x_1, x_2|\widetilde{i}^F\rangle|^2, i = L,\ M,\ R$. Note that there are two spots due to the antisymmetrization of the wave function. Reprinted with permission from Benseny et al. [11]. Copyright 2010 American Physical Society.

the spin state of the atoms is symmetric. In the bosonic case, the localized states for the hole, $|\widetilde{i}^B\rangle$, are given by the symmetrized versions of Eqs. 3.10–3.12, namely

$$|\widetilde{L}^B\rangle \equiv \frac{1}{\sqrt{2}}\left[|M\rangle_1|R\rangle_2 + |R\rangle_1|M\rangle_2\right], \tag{3.13}$$

$$|\widetilde{M}^B\rangle \equiv \frac{1}{\sqrt{2}}\left[|R\rangle_1|L\rangle_2 + |L\rangle_1|R\rangle_2\right], \tag{3.14}$$

$$|\widetilde{R}^B\rangle \equiv \frac{1}{\sqrt{2}}\left[|L\rangle_1|M\rangle_2 + |M\rangle_1|L\rangle_2\right]. \tag{3.15}$$

In this basis, the dynamics of the system will be governed by the Hamiltonian:

$$H_{\text{hole}} = \frac{\hbar}{2}\begin{pmatrix} 0 & \widetilde{J}_{\text{LM}}(t) & 0 \\ \widetilde{J}_{\text{LM}}(t) & 0 & \widetilde{J}_{\text{MR}}(t) \\ 0 & \widetilde{J}_{\text{MR}}(t) & 0 \end{pmatrix}, \tag{3.16}$$

$\widetilde{J}_i$ being the hole-tunneling rate between two adjacent traps (see Fig. 3.8). Since atom and hole transport processes are complementary, the tunneling rate for the hole is equal to the tunneling rate for the atom (see Eq. 3.2), i.e., $\widetilde{J}_i = J_i$.

Hamiltonian 3.16 is formally equivalent to Hamiltonian 3.1 and therefore has a dark state that only involves the two states where the hole is in the extreme traps, $|\widetilde{L}^F\rangle$ and $|\widetilde{R}^F\rangle$. Thus, one can transport the hole between these two states by adiabatically following the dark state by approaching the traps as shown in Fig. 3.2a.

3.4.2.2 Numerical simulations

The previously discussed three-level approach has been introduced to illustrate the main ideas behind the hole transport scheme. Nevertheless, and in order to be accurate, in what follows we will numerically solve the Schrödinger equation in real space.

We have performed an exact simulation, i.e., a numerical integration of the Schrödinger equation of the adiabatic transport process of a single hole in a triple-well potential with two identical fermions. The initial state is $\widetilde{\phi}(x_1, x_2, t = 0) = \widetilde{\phi}_L^F(x_1, x_2) = \langle x_1, x_2|\widetilde{L}^F\rangle$ with $\alpha d_1 = \alpha d_2 = 9$, while the expected final state is, up to a global phase, $\widetilde{\phi}(x_1, x_2, t = T) = \widetilde{\phi}_R^F(x_1, x_2) = \langle x_1, x_2|\widetilde{R}^F\rangle$ with $\alpha d_1 = \alpha d_2 = 9$. As in the previous section, we have taken the middle trap to be static at $x = 0$ while displacing only the two extreme traps, since the hole transport sequence is the same as for the atom transport (see Fig. 3.2). Thus, the protocol starts by first approaching the two occupied traps to later approach the empty trap to the middle one.

Figure 3.9 shows different snapshots for the temporal evolution of the two-fermion joint probability distribution $|\widetilde{\phi}(x_1, x_2, t)|^2$. During this evolution, the diagonal $x_1 = x_2$ is not populated since it is forbidden due to the Pauli exclusion principle. The probability density is mirrored at both sides of this diagonal due to the antisymmetrization of the wave function. Notice how the transport succeeds, transferring the atoms from the state corresponding to the hole being on the left, $\widetilde{\phi}_L^F(x_1, x_2)$, to the state corresponding to the hole being on the right $\widetilde{\phi}_R^F(x_1, x_2)$.

3.4.3 Hole Transport Fidelity

To characterize the transport between the different traps we define here their associated fidelities. These definitions will be also very useful for the characterization of the atomtronic devices introduced

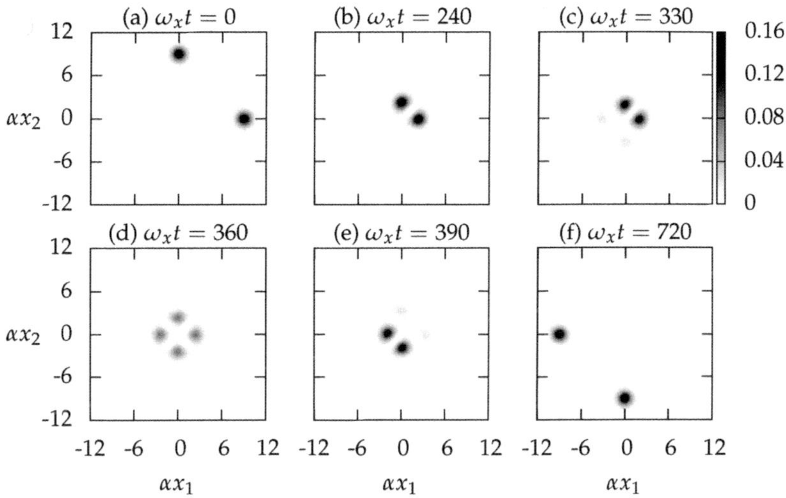

Figure 3.9 Coherent transport of a fermionic hole in a triple-well potential. Snapshots of the two-fermion joint probability distribution $|\widetilde{\phi}(x_1, x_2, t)|^2$ at times $\omega_x t = 0, 240, 330, 360, 390$, and 720. The initial and final states correspond, respectively, to $\widetilde{\phi}(x_1, x_2, t=0) = \widetilde{\phi}_L^F(x_1, x_2)$ and $\widetilde{\phi}(x_1, x_2, t=T) = \widetilde{\phi}_R^F(x_1, x_2)$. The trap approaching scheme is shown in Fig. 3.2a. Reprinted with permission from Benseny et al. [11]. Copyright 2010 American Physical Society.

in Section 3.4.5. Thus, by starting with the hole in trap $i = L, M, R$ and performing the temporal evolution, we check the population of each trap $j = L, M, R$. We denote, for either a fermionic (F) or a bosonic (B) hole case, the state of the system at the end of the process as $\widetilde{\phi}_i^{F/B}(x_1, x_2, T)$, T being the total duration of the STIRAP sequence. The final population of the state with the hole in trap j, $\widetilde{\phi}_j^{F/B}(x_1, x_2)$ will be given by its product with the evolved wave function, namely

$$F_{i\to j}^{F/B} = \left| \iint \widetilde{\phi}_j^{F/B*}(x_1, x_2)\widetilde{\phi}_i^{F/B}(x_1, x_2, T)dx_1dx_2 \right|^2 . \qquad (3.17)$$

Thus, $F_{i\to j}^{F/B}$ is the fidelity of the fermionic (F) or bosonic (B) transport process of the hole between traps i and j after applying the hole matter wave STIRAP sequence (see Fig. 3.2a). Note that for the matter wave adiabatic transport process, we want to maximize $F_{L\to R}^{F/B}$.

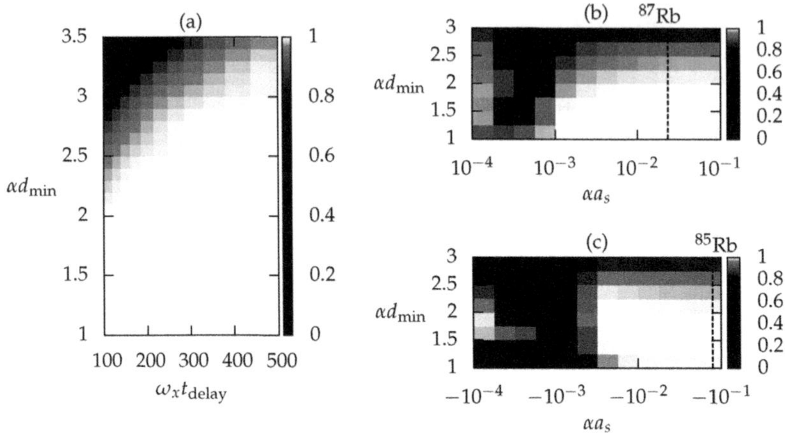

Figure 3.10 (a) $F^F_{L\to R}$ in the parameter plane $\{t_{\rm delay}, d_{\rm min}\}$. (b) $F^B_{L\to R}$ for the bosonic case in the parameter plane $\{a_s, d_{\rm min}\}$ for a_s positive and (c) for a_s negative. Scattering lengths of ^{85}Rb ($\alpha a_s = -7.98 \times 10^{-2}$) and ^{87}Rb ($\alpha a_s = 2.32 \times 10^{-2}$) are indicated by dashed black lines. The temporal variation of the traps and the system parameters are given in Fig. 3.2. For the bosonic case, the transverse trapping frequency is taken to be $\omega_p = 24\omega_x$ [5]. Reprinted with permission from Benseny et al. [11]. Copyright 2010 American Physical Society.

Figure 3.10a depicts $F^F_{L\to R}$ in the parameter plane $\{t_{\rm delay}, d_{\rm min}\}$ (see Fig. 3.2 for the definition of these two parameters), showing that the hole transport is robust and efficient provided that the adiabaticity condition is fulfilled.

We have also simulated the hole transfer process for the case of bosonic atoms by integrating the corresponding Schrödinger equation. Figure 3.10b,c shows $F^B_{L\to R}$ in the parameter plane $\{a_s, d_{\rm min}\}$. As shown in the figure, the adiabatic transfer process succeeds for $\alpha a_s = -7.98 \times 10^{-2}$ and $\alpha a_s = 2.32 \times 10^{-2}$ corresponding, respectively, to the s-wave-scattering length, see Eq. 3.7, of ^{85}Rb and ^{87}Rb [5], while it breaks down for weaker interactions since then double occupancy starts to play a dominant role. As expected, for large absolute values of the s-wave-scattering length, bosons become hardcore and then their dynamics are equivalent to that of the fermionic case.

3.4.4 Bohmian Trajectories for the Hole Transport

As we have shown, due to the adiabatic following of the dark state, the hole is transferred from the left to the right trap with an ideally negligible probability to be in the middle one. Therefore, the signature that the hole has been transferred through the matter wave STIRAP technique is that the counterdiagonal $x_1 = -x_2$ is practically not populated (see Fig. 3.9c–e). However, resorting to the continuity equation associated with the two-atom matter wave, the corresponding wave function must cross at some point this *forbidden* counterdiagonal. To gain physical insight into this particular feature of the adiabatic transport process, we will discuss now the previous simulations by means of Bohmian trajectories.

Figure 3.11 shows a set of quantum trajectories calculated from the time evolution of Fig. 3.9. Their initial positions (see Fig. 3.11a), were randomly distributed according to $|\widetilde{\phi}_L^F(x_1, x_2)|^2$. As expected, Fig. 3.11b reveals that the time evolution of the quantum trajectories follows the evolution of the wave function, ending up distributed according to $|\widetilde{\phi}_R^F(x_1, x_2)|^2$ (see Fig. 3.11c). In Fig. 3.11b we have also plotted the atomic probability distribution for the intermediate time $\omega_x t = 360$ corresponding to the fourth snapshot in Fig. 3.9d. Note also that in order to allow for the easy visualization of the transport process, we only show in Fig. 3.11b a reduced number of quantum trajectories.

As we discussed in the previous section, when the trajectories reach a forbidden region, they increase their velocity in order to reduce the density of trajectories per unit time. Clearly, when crossing the counterdiagonal, each quantum trajectory suddenly accelerates (see Fig. 3.11d). In addition, we can observe the trajectories making a detour from the central region of the counterdiagonal (see Fig. 3.11b). This is because it is favorable for the trajectories to cross the counterdiagonal away from the center, as the probability density is significantly smaller there.

3.4.5 Atomtronics with Holes

Making use of the fact that the hole transfer process here presented is spatially nonsymmetric, we will now discuss both a coherent

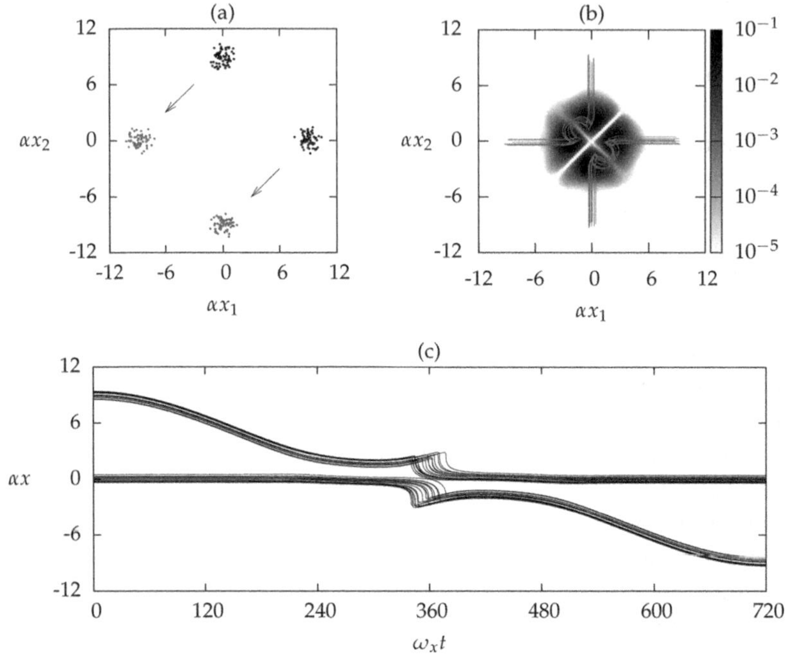

Figure 3.11 Bohmian trajectories corresponding to the temporal evolution of the system shown in Fig. 3.9. (a) Initial (black) and final (gray) distributions of quantum trajectories (cf. Fig. 3.9a and Fig. 3.9f). (b) Evolution of the trajectories in configuration space showing the joint probability distribution $|\widetilde{\phi}^F(x_1, x_2, T/2)|^2$. (c) Evolution of the trajectories as a function of time. Reprinted with permission from Benseny et al. [11]. Copyright 2010 American Physical Society.

single-hole diode and a coherent single-hole transistor in a triple-well potential. The hole transfer from left to right and vice versa strongly depends on both the two-atom collisional interaction and the exchange interaction, and therefore, both interactions will be used here to control the diode and transistor operation regimes.

3.4.5.1 Single-hole diode

In this section we will design a single-hole diode by using the collisional interaction between two bosons as a control parameter to allow the hole transport from left to right, and inhibit the transport

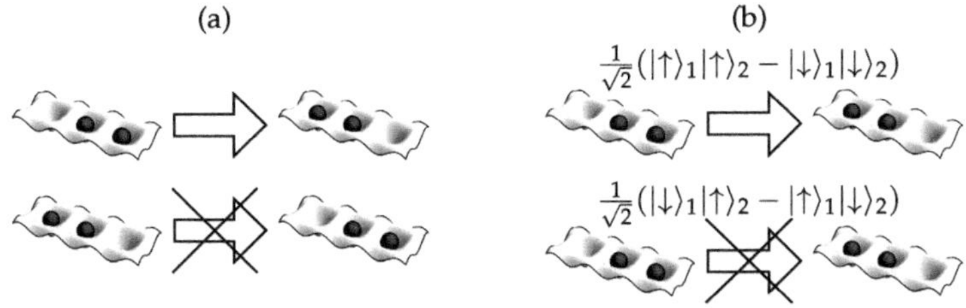

Figure 3.12 Sketch of the atomtronics devices we will discuss. (a) Single-hole diode: for identical trap-approaching schemes but with appropriately tuned interactions between two bosonic atoms, the hole will be transported from the left to the right trap, but the inverse process will not succeed. (b) Single-hole transistor for two noninteracting fermionic atoms: by performing identical trap-approaching schemes, the hole transport from left to right will succeed or will be inhibited depending on whether the spin state of the two atoms is symmetric or antisymmetric. Reprinted with permission from Benseny et al. [11]. Copyright 2010 American Physical Society.

from right to left (see Fig. 3.12a). Thus, Fig. 3.13a shows the fidelity of the bosonic hole transport processes $F^B_{L\to R}$, $F^B_{R\to L}$, and $F^B_{R\to M}$ against the strength of the s-wave-scattering length, that can be tuned by adjusting a Feshbach resonance [49]. The parameter values for the temporal variation of the traps are taken as in Fig. 3.2 such that the fidelity of the hole transport process from left to right, $F^B_{L\to R}$ (circles in Fig. 3.13a), is larger than 0.99 above a certain threshold value for the scattering length, indicated by point A in Fig. 3.13a, i.e., when the interaction is strong enough and the bosons become hard-core.

By performing the same trap-approaching scheme but with the hole starting on the right trap, the process corresponding to the hole being transferred from the right to the left trap (triangles in Fig. 3.13a) is inhibited ($F^B_{R\to L} \sim 0$ at point B) or succeeds ($F^B_{R\to L} \sim 1$ at point C), depending on the value of αa_s.

We thus define the fidelity of the diode as

$$F_D = F^B_{L\to R}\left(1 - F^B_{R\to L}\right). \qquad (3.18)$$

since this fidelity is maximal when the bosonic hole is transported from left to right and, simultaneously, the opposite process consisting in the hole transport from right to left is inhibited, i.e., $F_D = 1$ when $F^B_{L\to R} = 1$ and $F^B_{R\to L} = 0$.

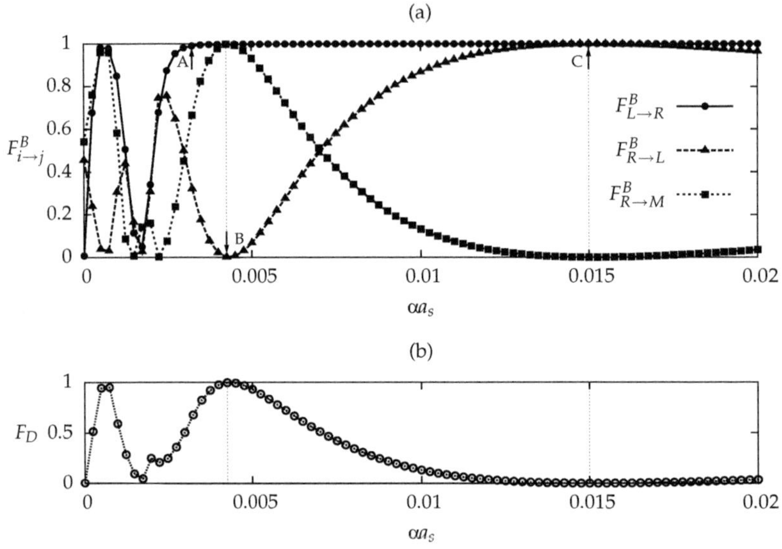

Figure 3.13 Diode for a bosonic hole. As a function of the s-wave-scattering length: (a) fidelities of hole transport processes $F^B_{L\to R}$ (circles), $F^B_{R\to L}$ (triangles), and $F^B_{R\to M}$ (squares) and (b) diode fidelity F_D (see Eq. 3.18) in the text. See text for the definitions of points A, B, and C. The rest of the parameter values are as in Fig. 3.2. Reprinted with permission from Benseny et al. [11]. Copyright 2010 American Physical Society.

As can be seen in Fig. 3.13b, that shows F_D as a function of the scattering length, by tuning the product of the inverse of the size of the wave function and the s-wave-scattering length to $\alpha a_s \sim 4.25 \times 10^{-3}$, where $F_D \sim 1$ (corresponding to point B), we obtain a scheme that transports the hole from the left to the right trap but transfers a hole from the right trap to the middle one (see squares in Fig. 3.13a). Note that an ideal diodic behavior where the hole ends at the right trap, no matter if initially it was at the left or the right trap, would violate the unitarity of the quantum evolution.

Figure 3.14 shows the trajectories for a simulation where the hole transfer from right to left is inhibited. The trajectories show Rabi-like oscillations between the different traps, ending with the hole at the middle trap (one atom in the left and one in the right trap). The frequency of these oscillations depends on the interaction

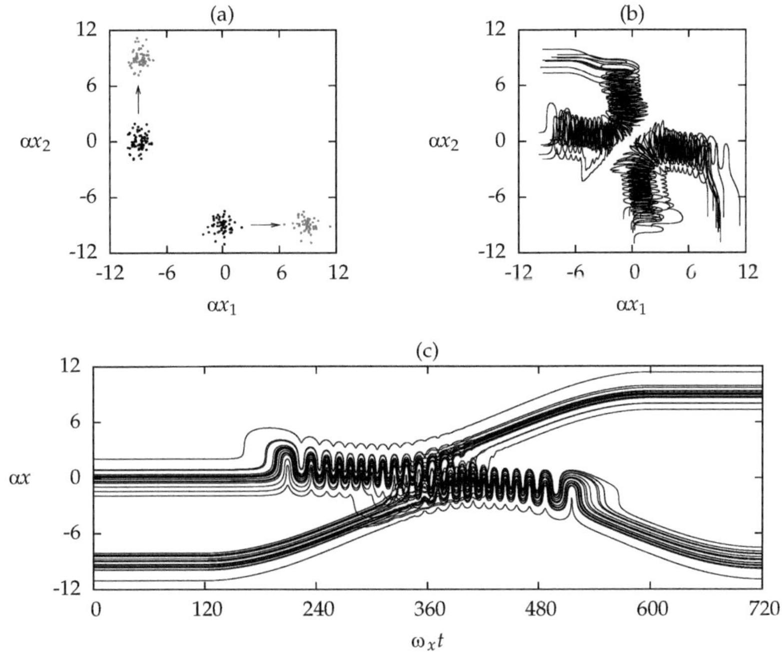

Figure 3.14 Bohmian trajectories from a simulation for a bosonic hole starting in the right trap. The s-wave-scattering length is tuned to $\alpha a_s = 4.25 \times 10^{-3}$ to inhibit the transfer of the hole to the left trap. (a) Initial (black) and final (gray) distributions of quantum trajectories. (b) Evolution of the trajectories in configuration space. (c) Evolution of the trajectories as a function of time.

strength between the atoms, and this allows us to control the diodic process.

3.4.5.2 Single-hole transistor

Figure 3.10a shows that in the fermionic case, the hole transport from left to right achieves high fidelities. On the other hand, Fig. 3.10b,c show that in the weakly interacting bosonic case, i.e., for $a_s \to 0$, the hole transport does not perfectly succeed. In fact, for $a_s = 0$, the fidelity $F^B_{L\to R}$ vanishes with the hole ending in a superposition between being in the left and middle traps. From the previous observations, it is possible to engineer a single-hole transistor where the spin state of the atoms is used to control the hole current from the left to the right trap.

For instance, it is straightforward to check that for two fermions in the middle and right traps (hole in the left), the state with symmetric spin state $|S\rangle = (|\uparrow\rangle_1|\uparrow\rangle_2 - |\downarrow\rangle_1|\downarrow\rangle_2)/\sqrt{2}$ and an antisymmetric spatial state ($|\widetilde{L}^F\rangle$, cf. Eq. 3.10),

$$|S\rangle|\widetilde{L}^F\rangle = \frac{1}{2}[|\uparrow M\rangle_1|\uparrow R\rangle_2 - |\uparrow R\rangle_1|\uparrow M\rangle_2 - |\downarrow M\rangle_1|\downarrow R\rangle_2 + |\downarrow R\rangle_1|\downarrow M\rangle_2], \quad (3.19)$$

and the state with antisymmetric spin state $|A\rangle = (|\downarrow\rangle_1|\uparrow\rangle_2 - |\uparrow\rangle_1|\downarrow\rangle_2)/\sqrt{2}$ and symmetric spatial state ($|\widetilde{L}^B\rangle$),

$$|A\rangle|\widetilde{L}^B\rangle = \frac{1}{2}[|\downarrow M\rangle_1|\uparrow R\rangle_2 + |\downarrow R\rangle_1|\uparrow M\rangle_2 - |\uparrow M\rangle_1|\downarrow R\rangle_2 - |\uparrow R\rangle_1|\downarrow M\rangle_2], \quad (3.20)$$

are coupled via a spin flip on the atom in the middle trap, i.e., $|\uparrow M\rangle_k \leftrightarrow |\downarrow M\rangle_k$. A similar argument for bosonic atoms can be done between the states $|S\rangle|\widetilde{L}^B\rangle$ and $|A\rangle|\widetilde{L}^F\rangle$.

This control over the system behavior between the bosonic and fermionic cases allows us to create a coherent hole transistor scheme where the matter wave STIRAP sequence from the left to the right trap succeeds or is inhibited depending on the spin state of the atoms. The case for two noninteracting fermions is depicted in Fig. 3.12b. Thus, the figure of merit corresponds to the maximization of the transistor fidelity defined as:

$$F_T = F^F_{L\to R}\left(1 - F^B_{L\to R}\right). \quad (3.21)$$

since it will be maximal when $F^F_{L\to R} \sim 1$ and $F^B_{L\to R} \sim 0$. As we have discussed, for parameters of Fig. 3.2a, $F_T > 0.99$ for noninteracting atoms.

To further test the fidelity of the atomtronic transistor, we have calculated F_T, computing $F^{F/B}_{L\to R}$, adding a jitter in the trap positions (to simulate experimental imperfections, see Fig. 3.15a) such as:

$$d_i^s(t) = d_i^0(t) + A_s \cos\omega_s t, \quad (3.22)$$

$d_i^0(t)$ being the distance between the i-th and $i+1$-th trap shown in Fig. 3.2a. A_s and ω_s represent, respectively, the amplitude and the frequency of the jitter. The results of F_T for different values of

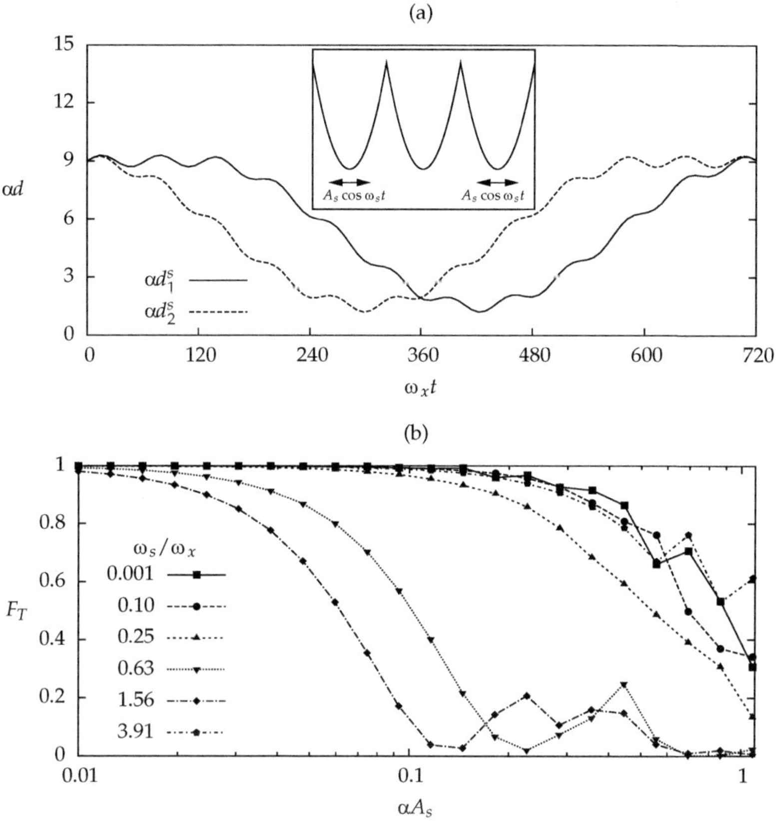

Figure 3.15 Single-hole transistor. (a) Temporal variation of the trap distances shown in Fig. 3.2a with an added jittering of amplitude $\alpha A_s = 0.3$ and frequency $\omega_s = 0.1\omega_x$. The inset shows a sketch of the three traps and its simulated jittering. (b) Fidelity F_T of the transistor (see Eq. 3.21) as a function of the amplitude of position jitter for different jitter frequencies. $a_s = 0$, and the rest of the parameter values are as in Fig. 3.2. Reprinted with permission from Benseny et al. [11]. Copyright 2010 American Physical Society.

A_s and ω_s are plotted in Fig. 3.15b and indicate that for small jitter amplitudes the transistor still works with high fidelity ($F_T > 0.99$) for a wide range of frequencies, except for those close to ω_x (trapping frequency) that, as expected, excite the atoms to higher vibrational states.

3.5 Adiabatic Transport of a Bose–Einstein Condensate

The transport scheme introduced in Section 3.3 can be extended to the transport of a BEC. In that context, Rab et al. [24] analyzed the spatiotemporal dynamics of a BEC confined in a harmonic trap which has been split into three regions, or wells, via the addition of two parallel narrow Gaussian barriers placed near the center of the trap and equidistant to it (see Fig. 3.16). These two barriers can be created, for instance, with two blue-detuned lasers and their height controlled with the intensity of such lasers. For the purpose of a qualitative analysis, the system is first reduced to a three-level model, the three levels representing the condensate on the ground state of each of the three wells, while the tunneling rate between two neighbor wells provides the coupling between them and is controlled through the height of the barriers. By an appropriate time evolution of the barriers, this three-level system is able to undergo a STIRAP-like process [16], consisting on the adiabatic evolution of the quantum state from, e.g., the left trap level to the right one with a very small population at any time on the middle trap level. This is numerically verified by Rab et al. [24]. In this case, the middle trap ground state does not have the same energy as the ground states of the left and right traps, but the matter wave STIRAP technique still works because the two left and right traps ground states do have the same energy (analogously to the two-photon resonance of the quantum optical STIRAP [17]). The authors then perform a more realistic analysis via the spatiotemporal integration of the Gross–Pitaevskii equation, obtaining essential agreement with the previous simplified model, i.e., the population at any time on the middle spatial region can be made arbitrarily small by increasing the duration—and thus the adiabaticity—of the process.

As we pointed out in the introductory discussion of this chapter, Rab et al. [24] conclude that the BEC can be transported between the left and right traps without transiting the middle trap. They call this effect "transport without transit," suggesting that the BEC does not cross the middle region. If that were to be true, it could be considered a teleportation mechanism.

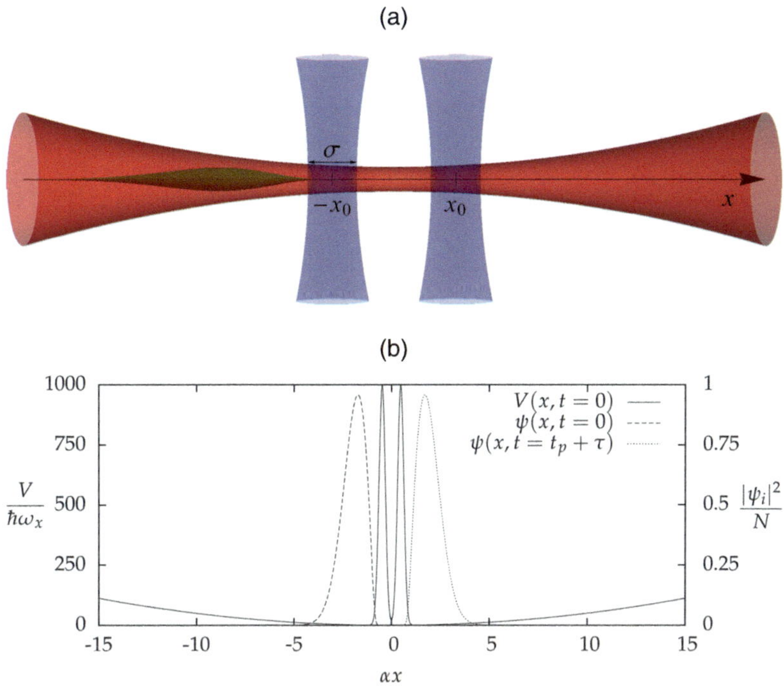

Figure 3.16 (a) Sketch of the physical implementation of the triple-well potential. A red-detuned (attractive) laser is used to generate a harmonic trapping, and two blue-detuned (repulsive) lasers create the barriers that define the three trapping regions for the BEC. We will assume control over the temporal variation of the intensity of the blue-detuned lasers (that will affect the heights of the trap barriers). In the matter wave analogue of the STIRAP technique, the main goal will consist in an efficient and robust transfer of the BEC initially in the left trap, as in the figure, to the right trap via the tunneling interaction. (b) Potential felt by the BEC consisting of a harmonic potential and two Gaussian barriers centered at $\pm x_0$, see Eq. 3.28 with the probability distributions of the initial and final states of the BEC for the matter wave STIRAP protocol.

In this section we will investigate the matter wave transport scheme introduced by Rab et al. [24], by means of Madelung's hydrodynamic model of quantum mechanics [34], formally equivalent to Bohmian mechanics, since its fluid analogy seems to be the most adequate in the context of a BEC.

3.5.1 Madelung Hydrodynamic Formulation

In the limit of zero temperature and assuming the mean field approximation, the BEC dynamics can be properly described by a wave function ψ that obeys the Gross–Pitaevskii equation [50, 51]:

$$i\hbar\frac{\partial\psi(\vec{r},t)}{\partial t} = \left[-\frac{\hbar^2}{2m}\nabla^2 + V(\vec{r},t) + g\left|\psi(\vec{r},t)\right|^2\right]\psi(\vec{r},t), \quad (3.23)$$

where g is the nonlinear interaction between the atoms [51] and $V(\vec{r},t)$ is the trapping potential. Note that in this formulation we take the BEC wave function to be normalized to the number of atoms, i.e., $\int d^3\vec{r}\,|\psi(\vec{r},t)|^2 = N$.

Formally, Madelung's formulation is obtained from the following transformation for the condensate wave function:

$$\psi(\vec{r},t) = \sqrt{n(\vec{r},t)}e^{i\theta(\vec{r},t)}. \quad (3.24)$$

Substituting Eq. 3.24 into the Gross–Pitaevskii equation and separating real and imaginary parts, we are led to the following two equations:

$$m\frac{\partial\vec{v}}{\partial t} = -\vec{\nabla}\left(V + ng + \frac{1}{2}mv^2 - \frac{\hbar}{2m}\frac{1}{\sqrt{n}}\nabla^2\sqrt{n}\right), \quad (3.25)$$

$$\frac{\partial n}{\partial t} = -\vec{\nabla}\cdot(n\vec{v}), \quad (3.26)$$

where we have defined

$$\vec{v}(\vec{r},t) \equiv \frac{\hbar}{m}\vec{\nabla}\theta(\vec{r},t). \quad (3.27)$$

Equation 3.25 is the classical Euler equation for an inviscid compressible fluid with an additional pressure term whose origin is purely quantum and which is sometimes called *quantum pressure*. Equation 3.26 is simply the continuity equation for the condensate density and ensures that the fluid transport is always achieved in a continuous fashion.

3.5.2 Numerical Simulations

To study the 1D dynamics of the BEC undergoing matter wave STIRAP, we model the BEC trapping potential as [24]

$$V(x,t) = \frac{m}{2}\omega_x^2 x^2 + V_{12}(t)e^{-\frac{(x+x_0)^2}{2\sigma^2}} + V_{23}(t)e^{-\frac{(x-x_0)^2}{2\sigma^2}}, \quad (3.28)$$

$V_{12}(t)$ and $V_{23}(t)$ being the height of the potential barriers, located at $-x_0$ and x_0, respectively (see Fig. 3.16b). The initial barrier heights separating the traps will be considered large enough to inhibit tunneling, and lowering them will allow us to make the BEC tunnel among the traps. The BEC will be initially in the ground state of the left trap, and the matter wave STIRAP technique will be used to transport it to the right trap.

The temporal profile of the potential barriers height is taken as

$$V_{23}(t) = \begin{cases} V_{\max} & t \leq 0 \\ 16(V_{\max} - V_{\min})\left(\frac{t}{t_p} - \frac{1}{2}\right)^4 + V_{\min} & 0 < t < t_p \\ V_{\max} & t \geq t_p \end{cases}, \quad (3.29)$$

$$V_{12}(t) = V_{23}(t - \tau), \quad (3.30)$$

with τ being the time delay between the pulsing of $V_{12}(t)$ and $V_{23}(t)$, always taken to be $\tau = t_p/10$. Expressions 3.29, 3.30 show that, again, the matter wave STIRAP transport is achieved by favoring first the tunneling between the middle and right traps and then the tunneling between the left and middle traps, with a total adiabatic transport time $T \equiv t_p + \tau$.

For the numerical simulations, we take parameter values similar to those given by Rab et al. [24]: $N = 1000$, $V_{\min} = 5\hbar\omega_x$, $V_{\max} = 10^3\hbar\omega_x$, $\sigma = 0.16\alpha^{-1}$, $x_0 = 0.48\alpha^{-1}$, $g_{1D} = g/(2\pi\alpha_r^{-2}) = 0.31$, α_r^{-1} being the size of the BEC in the radial direction. t_p will be typically taken as $\omega_x t_p = 1000$, but we will make simulations for different values to study the effects of slowing down the process.

We integrate the Gross–Pitaevskii Eq. 3.23 with the potential barrier pulses defined in Eqs. 3.29, 3.30, with the initial state set on the left trap obtained by imaginary time evolution (see Fig. 3.16b). The results for the condensate density over time are plotted in Fig. 3.17. The BEC wave function is transferred from the left to the right traps, presenting, as in the case for the transport of a single atom, a quasi node in the region of the central trap. We plot the current lines associated with Eq. 3.27 for this evolution in Fig. 3.18, observing basically the same dynamical behavior as in the single-atom case.

What makes it possible to have an arbitrarily small population at any time on the middle region, while keeping the transport

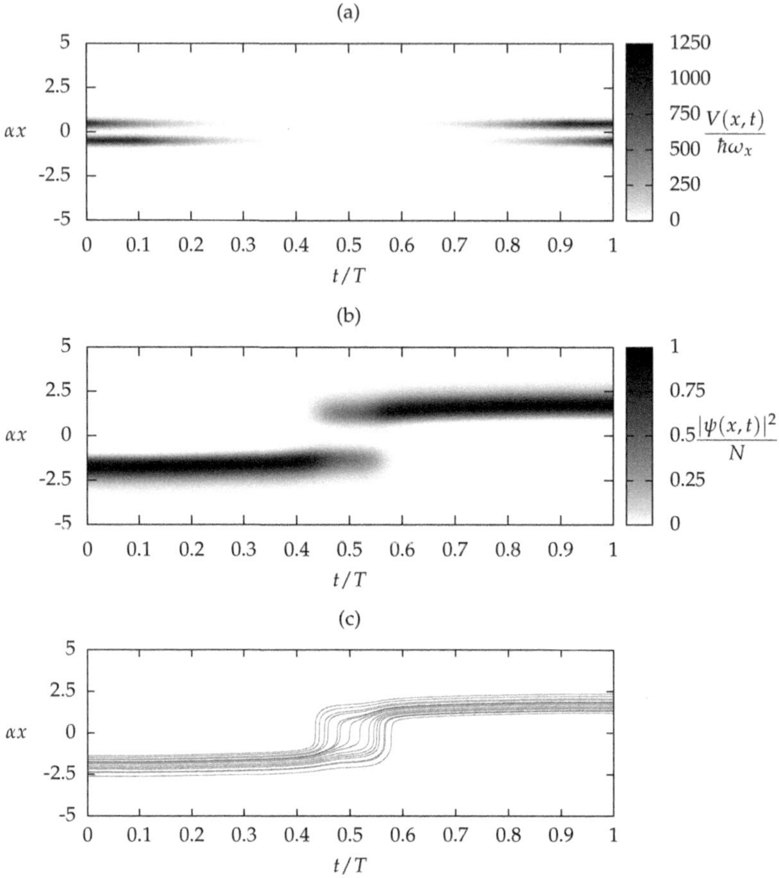

Figure 3.17 Adiabatic transport of a BEC in a triple-well potential. Time evolution of (a) the potential profile $V(x, t)$, (b) the condensate density $|\psi(x, t)|^2$, (c) the current lines calculated with Eq. 3.27.

mechanism working, is clearly visualized in terms of the fluid analogy of the wave function: as the evolution is made more adiabatic and the middle population decreases, the fluid velocity at the central point increases, just as a classical fluid increases its velocity when it passes through a narrow conduct. In order to obtain a quantitative argument, we have performed various simulations for different times of t_p, and for each one we have calculated the

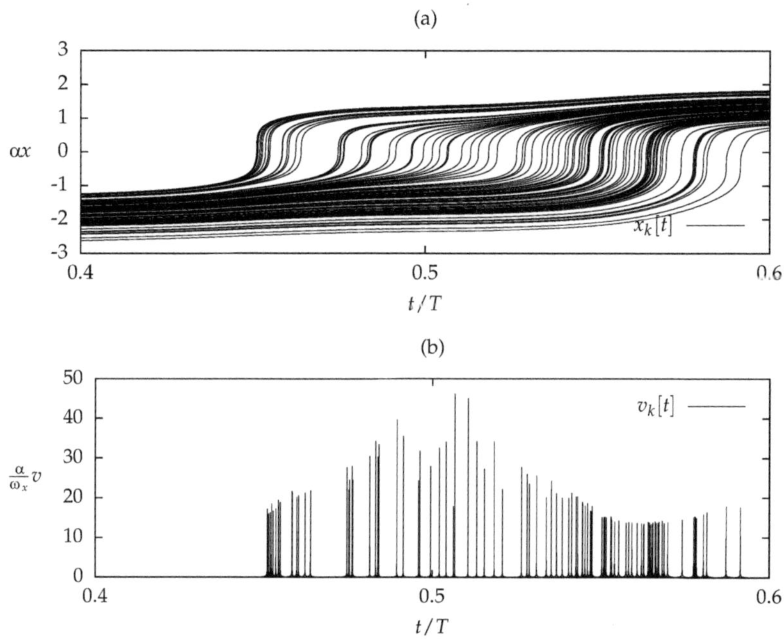

Figure 3.18 (a) Enlarged view of Fig. 3.17c with a larger number of current lines. (b) Fluid velocities associated to the current lines. The position is given in units of α^{-1} and the velocities in units of $\alpha^{-1}\omega_x$.

maximum velocity[a] that the fluid attains at the central point, the integrated flux through that point, i.e.,

$$F_0 \equiv \int_0^T v(x = 0, t)\,|\psi(x = 0, t)|^2\,dt, \tag{3.31}$$

and the maximum population on the central trap:

$$N_2(t) = \int_{-x_0}^{x_0} |\psi(x, t)|^2\,dx, \tag{3.32}$$

during the evolution. The results are plotted in Fig. 3.19. We see that while the central population always remains below 5% of the total number of atoms, the maximum velocity grows, ensuring a constant value for the integrated flux, which is close to the total number of atoms. A final value for this quantity equal to N means that all the condensate that has been transported from left trap to the right, has transited through the middle point.

[a]In these cases, the velocity, v, is the reduction of Eq. 3.27 to one dimension.

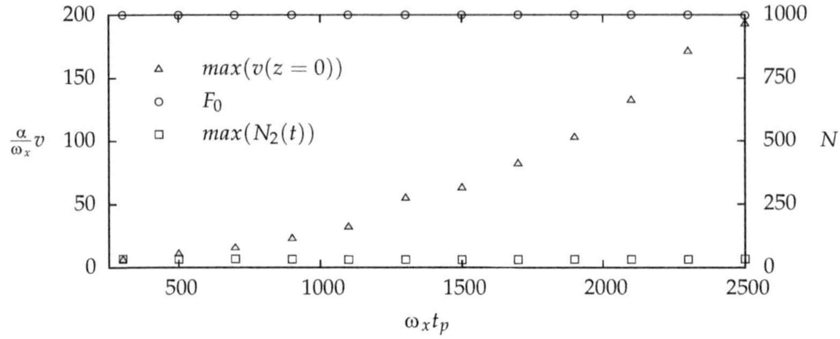

Figure 3.19 Results for different simulations with different times t_p. Triangles: Maximum velocity of the fluid in units of ω_x/α. Circles: Total flux through the central point F_0. Squares: Maximum population in the middle region max($N_2(t)$). Velocities are given in units of $\alpha^{-1}\omega_x$.

3.6 Conclusions

Using the matter wave analogue of the quantum optical STIRAP technique [16, 17], we have discussed an efficient and robust method to coherently transport either atoms or empty sites, i.e., holes, in triple-well potentials. The coherent transport process consists in adiabatically following a spatially delocalized dark state by an appropriate temporal control of the tunneling rates. These techniques could be used to prepare defect-free trap domains to, eventually, perform quantum computations or to design atomtronic devices. We have first introduced the transport process in a simplified three-state model to, later on, simulate it with exact numerical integrations of the Schrödinger equation for time dependent potentials.

The matter wave STIRAP process involves the adiabatic following of one of the time-dependent eigenstates of the Hamiltonian that happens to have a node on the central region, and for this reason the population at any time on this middle region can be made very small. However, exact adiabaticity is never strictly reached, and while for certain problems this fact could possibly be neglected, here it plays a crucial role. The existence of nodes causes, in principle, no problems to the integration of the Schrödinger equation, but they are critical in those alternative integration methods based on quantum trajectories that derive an expression for a velocity from the phase of the wave function. This is the case of the Bohmian or the Madelung's hydrodynamic formulations [34, 35]. This allows to partially describe the dynamics of the system in terms of a velocity field that follows classical-like equations, thus providing a causal point of view which in some cases may help to bring out certain aspects of a specific problem that are hidden under the usual interpretation.

As the main result we have shown that the velocity of the trajectories is unbounded at the quasi node, i.e., it grows without any apparent limit as the adiabaticity of the STIRAP process is increased. What makes this fact singular in the present case is that all the wave function transits through this point, as is ensured by the continuity equation, and indeed does it with this very high velocity. We have demonstrated that it is implicit in the Schrödinger and Gross–Pitaevskii equations that *matter wave transport without transit* is forbidden. Furthermore, since the matter wave STIRAP is based on the nonrelativistic Schödinger equation, deviations from relativistic predictions are expected to occur for certain parameters. It is remarkable that relativistic corrections may be needed in an adiabatic context that, from an "orthodox" point of view, does not seem to imply high energies or velocities.

Acknowledgments

We thank Verònica Ahufinger, Alfonso Alarcón, Albert Bramón, Ramón Corbalán, Gabriele De Chiara, Kai Eckert, Sònia Fernández-Vidal, Maciej Lewenstein, Ricard Menchón, Antonio Picón, Luis Roso, and Daniel Viscor for fruitful discussions. The authors gratefully acknowledge financial support from the Spanish Ministry of Economy and Competitiveness, MINECO, (FIS2014-57460-P, FIS2017-86530-P) and from the Generalitat de Catalunya (SGR2017-1646). Xavier Oriols acknowledges support from Fondo Europeo de Desarrollo Regional (FEDER), the "Ministerio de Ciencia e Innovación" through the Spanish Project TEC2015-67462-C2-1-R and the European Union's Horizon 2020 research and innovation program under grant agreement No Graphene Core2 785219 and under the

Marie Sklodowska-Curie grant agreement No 765426 (TeraApps). Albert Benseny acknowledges financial support from the Villum Foundation.

References

1. G. E. Moore, *Cramming More Components onto Integrated Circuits*, Electronics, **38**, 8 (1965).
2. D. Frese, B. Ueberholz, S. Kuhr, W. Alt, D. Schrader, V. Gomer, and D. Meschede, *Single Atoms in an Optical Dipole Trap: Towards a Deterministic Source of Cold Atoms*, Physical Review Letters, **85**, 3777 (2000); J. Beugnon, C. Tuchendler, H. Marion, A. Gaëtan, Y. Miroshnychenko, Y. R. P. Sortais, A. M. Lance, M. P. A. Jones, G. Messin, A. Browaeys, and P. Grangier, *Two-Dimensional Transport and Transfer of a Single Atomic Qubit in Optical Tweezers*, Nature Physics, **3**, 696 (2007).
3. B. T. Seaman, M. Krämer, D. Z. Anderson, and M. J. Holland, *Atomtronics: Ultracold-Atom Analogs of Electronic Devices*, and references therein, Physical Review A, **75**, 023615 (2007).
4. R. Dumke, M. Volk, T. Müther, F. B. J. Buchkremer, G. Birkl, and W. Ertmer, *Micro-Optical Realization of Arrays of Selectively Addressable Dipole Traps: A Scalable Configuration for Quantum Computation with Atomic Qubits*, Physical Review Letters, **89**, 097903 (2002); G. Birkl, F. B. J. Buchkremer, R. Dumke, and W. Ertmer, *Atom Optics with Microfabricated Optical Elements*, Optics Communications, **191**, 67 (2001); F. B. J. Buchkremer, R. Dumke, M. Volk, T. Müther, G. Birkl, and W. Ertmer, *Quantum Information Processing with Microfabricated Optical Elements*, Laser Physics, **12**, 736 (2002); A. Lengwenus, J. Kruse, M. Volk, W. Ertmer, and G. Birkl, *Coherent Manipulation of Atomic Qubits in Optical Micropotentials*, Applied Physics B **86**, 377 (2007).
5. A. Lengwenus, J. Kruse, M. Schlosser, S. Tichelmann, and G. Birkl, *Coherent Transport of Atomic Quantum States in a Scalable Shift Register*, Physical Review Letters, **105**, 170502 (2010); J. Kruse, C. Gierl, M. Schlosser, and G. Birkl, *Reconfigurable Site-Selective Manipulation of Atomic Quantum Systems in Two-Dimensional Arrays of Dipole Traps*, Physical Review A, **81**, 060308(R) (2010).
6. O. Mandel, M. Greiner, A. Widera, T. Rom, T. W. Hänsch, and I. Bloch, *Controlled Collisions for Multi-Particle Entanglement of Optically Trapped Atoms*, Nature (London), **425**, 937 (2003).

7. N. Schlosser, G. Reymond, I. Protsenko, and P. Grangier, *Sub-Poissonian Loading of Single Atoms in a Microscopic Dipole Trap*, Nature (London), **411**, 1024 (2001); S. Kuhr, W. Alt, D. Schrader, M. Müller, V. Gomer, and D. Meschede, *Deterministic Delivery of a Single Atom*, Science, **293**, 278 (2001); M. Karski, L. Förster, J. M. Choi, W. Alt, A. Widera, and D. Meschede, *Nearest-Neighbor Detection of Atoms in a 1D Optical Lattice by Fluorescence Imaging*, Physical Review Letters, **102**, 053001 (2009); M. Anderlini, P. J. Lee, B. L. Brown, J. Sebby-Strabley, W. D. Phillips, and J. V. Porto, *Controlled Exchange Interaction between Pairs of Neutral Atoms in an Optical Lattice*, Nature (London), **448**, 452 (2007); D. D. Yavuz, P. B. Kulatunga, E. Urban, T. A. Johnson, N. Proite, T. Henage, T. G. Walker, and M. Saffman, *Fast Ground State Manipulation of Neutral Atoms in Microscopic Optical Traps*, Physical Review Letters, **96** 063001 (2006); K. D. Nelson, X. Li, and D. S. Weiss, *Imaging Single Atoms in a Three-Dimensional Array*, Nature Physics, **3**, 556 (2007).
8. P. Krüger, X. Luo, M. W. Klein, K. Brugger, A. Haase, S. Wildermuth, S. Groth, I. Bar-Joseph, R. Folman, and J. Schmiedmayer, *Trapping and Manipulating Neutral Atoms with Electrostatic Fields*, Physical Review Letters, **91**, 233201 (2003); T. Kishimoto, H. Hachisu, J. Fujiki, K. Nagato, M. Yasuda, and H. Katori, *Electrodynamic Trapping of Spinless Neutral Atoms with an Atom Chip*, Physical Review Letters, **96**, 123001 (2006).
9. M. Trinker, S. Groth, S. Haslinger, S. Manz, T. Betz, I. Bar-Joseph, T. Schumm, and J. Schmiedmayer, *Multilayer Atom Chips for Versatile Atom Micromanipulation*, Applied Physics Letters, **92**, 254102 (2008), L. Della Pietra, S. Aigner, Ch. vom Hagen, S. Groth, I. Bar-Joseph, H. Lezec, and J. Schmiedmayer, *Designing Potentials by Sculpturing Wires*, Physical Review A, **75**, 063604 (2007).
10. A. Ruschhaupt and J. G. Muga, *Atom Diode: A Laser Device for a Unidirectional Transmission of Ground-State Atoms*, Physical Review A, **70**, 061604(R) (2004); A. Ruschhaupt and J. G. Muga, *Three-Dimensional Effects in Atom Diodes: Atom-Optical Devices for One-Way Motion*, Physical Review A, **76**, 013619 (2007); J. J. Thorn, E. A. Schoene, T. Li, and D. A. Steck, *Experimental Realization of an Optical One-Way Barrier for Neutral Atoms*, Physical Review Letters, **100**, 240407 (2008); G. N. Price, S. T. Bannerman, K. Viering, E. Narevicius, and M. G. Raizen, *Single-Photon Atomic Cooling*, Physical Review Letters, **100**, 093004 (2008); R. A. Pepino, J. Cooper, D. Z. Anderson, and M. J. Holland, *Atomtronic Circuits of Diodes and Transistors*, Physical Review Letters, **103**, 140405 (2009).

11. A. Benseny, S. Fernández-Vidal, J. Bagudà, R. Corbalán, A. Picón, L. Roso, G. Birkl, and J. Mompart, *Atomtronics with Holes: Coherent Transport of an Empty Site in a Triple-Well Potential*, Physical Review A, **82**, 013604 (2010).
12. A. Micheli, A. J. Daley, D. Jaksch, and P. Zoller, *Single Atom Transistor in a 1D Optical Lattice*, Physical Review Letters, **93**, 140408 (2004); J. A. Stickney, D. Z. Anderson, and A. A. Zozulya, *Transistorlike Behavior of a Bose–Einstein Condensate in a Triple-Well Potential*, Physical Review A, **75**, 013608 (2007); J. Y. Vaishnav, J. Ruseckas, C. W. Clark, and G. Juzelunas, *Spin Field Effect Transistors with Ultracold Atoms*, Physical Review Letters, **101**, 265302 (2008).
13. D. Jaksch, H. J. Briegel, J. I. Cirac, C. W. Gardiner, and P. Zoller, *Entanglement of Atoms via Cold Controlled Collisions*, Physical Review Letters, **82**, 1975 (1999); D. Jaksch, J. I. Cirac, P. Zoller, S. L. Rolston, R. Cote, and M. D. Lukin, *Fast Quantum Gates for Neutral Atoms*, Physical Review Letters, **85**, 2208 (2000).
14. R. Dumke, Z. Lu, J. Close, N. Robins, A. Weis, M. Mukherjee, G. Birkl, C. Hufnagel, L. Amico, and M. G. Boshier, *Roadmap on Quantum Optical Systems*, J. Opt. **18**, 093001 (2016); L. Amico, G. Birkl, M. Boshier, and L.-C. Kwek, *Focus on Atomtronics-Enabled Quantum Technologies*, New Journal of Physics, **19**, 020201 (2017).
15. H. J. Metcalf and P. van der Straten, *Laser Cooling and Trapping (Graduate Texts in Contemporary Physics)*, (Springer, 1999).
16. K. Eckert, M. Lewenstein, R. Corbalán, G. Birkl, W. Ertmer, and J. Mompart, *Three-Level Atom Optics via the Tunneling Interaction*, Physical Review A, **70**, 023606 (2004).
17. K. Bergmann, H. Theuer, and B. Shore, *Coherent Population Transfer Among Quantum States of Atoms and Molecules*, Review of Modern Physics, **70**, 1003 (1998); N. V. Vitanov, A. A. Rangelov, B. W. Shore, K. Bergmann, *Stimulated Raman Adiabatic Passage in Physics, Chemistry and Beyond*, Review of Modern Physics, **89**, 015006 (2017).
18. S. Taie, T. Ichinose, H. Ozawa, and Y. Takahashi, *Spatial Adiabatic Passage of Massive Quantum Particles*, arXiv:1708.01100 [cond-mat.quant-gas].
19. S. McEndoo, S. Croke, J. Brophy, and Th. Busch, *Phase Evolution in Spatial Dark States*, Physical Review A, **81**, 043640 (2010); R. Menchon-Enrich, S. McEndoo, J. Mompart, V. Ahufinger, and Th. Busch, *Tunneling-Induced Angular Momentum for Single Cold Atoms*, Physical Review A, **89**, 013626 (2014); J. Polo, A. Benseny, Th. Busch, V. Ahufinger and J. Mompart, *Transport of Ultracold Atoms between Concentric Traps*

via Spatial Adiabatic Passage, New Journal of Physics, **18**, 015010 (2016).

20. R. Menchon-Enrich, A. Benseny, V. Ahufinger, A. D. Greentree, Th. Busch, and J. Mompart, *Spatial Adiabatic Passage: A Review of Recent Progress*, Reports on Progress in Physics, **79**, 074401 (2016).
21. Y. Loiko, V. Ahufinger, R. Menchon-Enrich, G. Birkl, and J. Mompart, *Coherent Injecting, Extracting, and Velocity Filtering of Neutral Atoms in a Ring Trap via Spatial Adiabatic Passage*, European Physical Journal D, **68**, 147 (2014).
22. K. Eckert, J. Mompart, R. Corbalán, M. Lewenstein, and G. Birkl, *Three Level Atom Optics in Dipole Traps and Waveguides*, Optics Communications, **264**, 264 (2006).
23. E. M. Graefe, H. J. Korsch, and D. Witthaut, *Mean-Field Dynamics of a Bose–Einstein Condensate in a Time-Dependent Triple-Well Trap: Nonlinear Eigenstates, Landau-Zener Models, and Stimulated Raman Adiabatic Passage*, Physical Review A, **73**, 013617 (2006).
24. M. Rab, J. H. Cole, N. G. Parker, A. D. Greentree, L. C. L. Hollenberg, and A. M. Martin, *Spatial Coherent Transport of Interacting Dilute Bose Gases*, Physical Review A, **77**, 061602R (2008).
25. V. O. Nesterenko, A. N. Nikonov, F. F. de Souza Cruz, and E. L. Lapolli, *STIRAP Transport of Bose–Einstein Condensate in Triple-Well Trap*, Laser Physics, **19**, 616 (2009).
26. C. Ottaviani, V. Ahufinger, R. Corbalán, and J. Mompart, *Adiabatic Splitting, Transport, and Self-Trapping of a Bose–Einstein Condensate in a Double-Well Potential*, Physical Review A, **81**, 043621 (2010).
27. A. D. Greentree, J. H. Cole, A. R. Hamilton, L. C. L. Hollenberg, *Coherent Electronic Transfer in Quantum Dot Systems Using Adiabatic Passage*, Physical Review B, **70**, 235317 (2004).
28. J. H. Cole, A. D. Greentree, L. C. L. Hollenberg, and S. Das Sarma, *Spatial Adiabatic Passage in a Realistic Triple-Well Structure*, Physical Review B, **77**, 235418 (2008).
29. J. Siewert, T. Brandes, and G. Falci, *Adiabatic Passage with Superconducting Nanocircuits*, Optics Communications, **264**, 435 (2006).
30. A. Negretti, A. Benseny, J. Mompart, and T. Calarco, *Speeding Up the Spatial Adiabatic Passage of Matter Waves in Optical Microtraps by Optimal Control*, Quantum Information Processing, **12**, 1439 (2013).
31. S. Longhi, G. Della Valle, M. Ornigotti, and P. Laporta, *Coherent Tunneling by Adiabatic Passage in an Optical Waveguide System*, Physical Review B, **76**, 201101 (2007).

32. R. Menchon-Enrich, A. Llobera, V. J. Cadarso, J. Mompart, and V. Ahufinger, *Adiabatic Passage of Light in CMOS-Compatible Silicon Oxide Integrated Rib Waveguides*, IEEE Photonics Technology Letters, **24**, 536 (2011); R. Menchon-Enrich, A. Llobera, J. Vila-Planas, V. J. Cadarso, J. Mompart, and V. Ahufinger, *Light Spectral Filtering Based on Spatial Adiabatic Passage*, Light: Science & Applications, **2**, e90 (2013).
33. D. Bohm, *A Suggested Interpretation of the Quantum Theory in Terms of "Hidden" Variables. I*, Physical Review, **85**, 166 (1952); D. Bohm, *A Suggested Interpretation of the Quantum Theory in Terms of "Hidden" Variables. II*, Physical Review, **85**, 180 (1952).
34. E. Madelung, *Quantentheorie in Hydrodynamischer Form*, Zeitschrift für Physik, **40** 322 (1926).
35. T. Takabayasi, *On the Formulation of Quantum Mechanics Associated with Classical Pictures*, Progress in the Theory of Physics, **8**, 143 (1952); T. Takabayasi, *Remarks on the Formulation of Quantum Mechanics with Classical Pictures and on Relations between Linear Scalar Fields and Hydrodynamical Fields*, Progress in the Theory of Physics, **9**, 187 (1952).
36. R. Landauer and Th. Martin, *Barrier Interaction Time in Tunneling*, Review of Modern Physics, **66** 217 (1994).
37. A. M. Jayannavar, *A Note on Traversal Time for Tunneling*, Pramāna **29**, 341 (1987).
38. C. R. Leavens and R. Sala Mayato, *Are Predicted Superluminal Tunneling Times an Artifact of Using the Nonrelativistic Schrödinger Equation?*, Annals of Physics, **7** 662 (1998).
39. D. V. Tausk and R. Tumulka, *Can We Make a Bohmian Electron Reach the Speed of Light, at Least for One Instant?*, Journal of Mathematical Physics, **51**, 122306 (2010).
40. C. J. Bradly, M. Rab, A. D. Greentree, and A. M. Martin, *Coherent Tunneling via Adiabatic Passage in a Three-Well Bose-Hubbard System*, Physical Review A, **85**, 053609 (2012); A. Benseny, J. Gillet, and Th. Busch, *Spatial Adiabatic Passage via Interaction-Induced Band Separation*, Physical Review A, **93**, 033629 (2016); I. Reshodko, A. Benseny, and Th. Busch, *Robust Boson Dispenser: Quantum State Preparation in Interacting Many-Particle Systems*, Physical Review A, **96**, 023606 (2017); A. Benseny, I. Reshodko, and Th. Busch, *Entanglement in Spatial Adiabatic Processes for Interacting Atoms*, Few-Body Systems, **59**, 48 (2018).
41. A. Benseny, J. Bagudà, X. Oriols and J. Mompart, *Need for Relativistic Corrections in the Analysis of Spatial Adiabatic Passage of Matter Waves*, Physical Review A, **85**, 053619 (2012).

42. M. Schlosser, S. Tichelmann, J. Kruse, and G. Birkl, *Scalable Architecture for Quantum Information Processing with Atoms in Optical Micro-Structures*, Quantum Information Processing, **10**, 907 (2011); M. Schlosser, J. Kruse, C. Gierl, S. Teichmann, S. Tichelmann, and G. Birkl, *Fast Transport, Atom Sample Splitting and Single-Atom Qubit Supply in Two-Dimensional Arrays of Optical Microtraps*, New Journal of Physics, **14**, 123034 (2012).
43. T. Calarco, E. A. Hinds, D. Jaksch, J. Schmiedmayer, J. I. Cirac, and P. Zoller, *Quantum Gates with Neutral Atoms: Controlling Collisional Interactions in Time-Dependent Traps*, Physical Review A, **61**, 022304 (2000).
44. D. P. DiVicenzo, *The Physical Implementation of Quantum Computation*, Fortschritte der Physik, **48**, 771 (2000); I. Walmsley and P. Knight, *Quantum Information Science*, Optics & Photonics News, **13**, 42 (2002).
45. M. Greiner, O. Mandel, T. Esslinger, T. W. Hänsch, and I. Bloch, *Quantum Phase Transition from a Superfluid to a Mott Insulator in a Gas of Ultracold Atoms*, Nature (London), **415**, 39 (2002).
46. R. Jördens, N. Strohmaier, K. Günter, H. Moritz, and T. Esslinger, *A Mott Insulator of Fermionic Atoms in an Optical Lattice*, Nature (London), **455**, 204 (2008).
47. J. Mompart, L. Roso, and R. Corbalán, *Coherent Population Trapping in Two-Electron Three-Level Systems with Aligned Spins*, Physical Review Letters, **88**, 023603 (2002).
48. S. Sachdev, *Quantum Phase Transitions* (Cambridge University Press, United Kingdom, 1999); M. Lewenstein, A. Sanpera, V. Ahufinger, B. Damski, A, Sen, and U. Sen, *Ultracold Atomic Gases in Optical Lattices: Mimicking Condensed Matter Physics and Beyond*, Advances in Physics, **56**, 243 (2007).
49. C. Chin, R. Grimm, P. Julienne, and E. Tiesinga, *Feshbach Resonances in Ultracold Gases*, Review of Modern Physics, **82**, 1225 (2010).
50. E. P. Gross, *Structure of a Quantized Vortex in Boson Systems*, Nuovo Cimento **20**, 454 (1961); L. P. Pitaevskii, *Vortex Lines in an Imperfect Bose Gas*, Soviet Physics JETP **13**, 451 (1961); E. P. Gross, *Hydrodynamics of a Superfluid Condensate*, Journal of Mathematical Physics, **4**, 195 (1963).
51. L. Pitaevskii and S. Stringari, *Bose–Einstein Condensation (The International Series of Monographs on Physics)*, (Oxford University Press, USA, 2003).

Chapter 4

Bohmian Pathways into Chemistry: A Brief Overview

Ángel S. Sanz
Department of Optics, Faculty of Physical Sciences,
Universidad Complutense de Madrid, Pza. Ciencias 1,
Ciudad Universitaria, 28040 Madrid, Spain
a.s.sanz@fis.ucm.es

Perhaps because of the popularity that trajectory-based methodologies have always had in chemistry and the important role they have played, Bohmian mechanics has been increasingly accepted within this community, particularly in those areas of the theoretical chemistry based on quantum mechanics, e.g., quantum chemistry, chemical physics, or physical chemistry. From a historical perspective, this evolution is remarkably interesting, particularly when the scarce applications of Madelung's former hydrodynamical formulation, dating back to the late 1960s and the 1970s, are compared with the many different applications available at present. As also happens with classical methodologies, Bohmian trajectories are essentially used to described and analyze the evolution of chemical systems, to design and implement new computational propagation techniques, or a combination of both. In the first case, Bohmian trajectories have the advantage that they avoid

Applied Bohmian Mechanics: From Nanoscale Systems to Cosmology (Second Edition)
Edited by Xavier Oriols and Jordi Mompart

ISBN 978-981-4800-10-5 (Hardcover), 978-0-429-29474-7 (eBook)
www.jennystanford.com

invoking typical quantum-classical correspondence to interpret the corresponding phenomenon or process, while in the second case quantum-mechanical effects appear by themselves, without the necessity to include artificially quantization conditions. Rather than providing an exhaustive revision and analysis of all these applications (excellent monographs on the issue are available in the literature for the interested reader, which can be consulted in the bibliography here supplied), this chapter has been prepared in a way that it may serve the reader to acquire a general view (or impression) on how Bohmian mechanics has permeated the different traditional levels or pathways to approach molecular systems in chemistry: electronic structure, molecular dynamics and statistical mechanics. This is done with the aid of some illustrative examples—theoretical developments in some cases and numerical simulations in other cases.

4.1 Introduction

Quantum mechanics has acquired a prominent role over the years in different areas of the theoretical chemistry, including the quantum chemistry, the physical chemistry or the chemical physics, as well as other areas at the borderline between chemistry and physics. Hence, it is not surprising that the idea of analyzing and understanding quantum-mechanically chemical systems by means of a hydrodynamic language, which dates back to the late 1960s and early 1970s [1–6], has also permeated all these areas. Bohmian mechanics [7–17], the widespread denomination that is now used for the hydrodynamic picture or reformulation of the Schrödinger equation [18]—formerly suggested by Madelung in 1926 [19]—has thus made its own pathways throughout chemistry [15, 17, 20, 21]. This is not a minor issue taking into account that chemistry is at present a sort of crossroad for a number of overlapping disciplines, such as atomic and molecular physics, optics and quantum optics, solid state physics, condensed matter physics, chemical physics, nanotechnology, biochemistry or molecular chemistry, and more recently also different branches associated with the so-called quantum technologies [22], such as quantum information and computation [23, 24], chemicals and materials quantum design [25–29] or quantum biology [30–33].

A question that naturally arises is why Bohmian mechanics now receives more attention than about half a century ago, when it was hardly considered a worthless hidden-variable theory almost relegated to oblivion—a source of much controversy within the quantum foundations community (something that unfortunately has survived to date without a truly deep physical justification). Perhaps a suitable answer can be given taking on a pragmatic view on what Bohmian mechanics is and what Bohmian mechanics is not in simple terms, that is, at a very basic (that is, pragmatic) level, without entering a lengthy discussion on the issue. In this sense, on the one hand, we find a hydrodynamic model that describes the evolution (actually, *diffusion*) of a quantum system in time throughout the corresponding configuration space—just in the same way classical hydrodynamic systems behave—in compliance with Madelung's view [19]. The particularity of the quantum system versus a classical hydrodynamic one is that, a priori, if no further specifications are made, quantum pure states evolve as ideal, inviscid fluids [34]. This does not introduce any new interpretation for the quantum mechanics based on the presence or not of external observers, since the mathematics of the Schrödinger equation are just the mathematics of the diffusion equation with a pure imaginary diffusion constant, as noticed by Fürth [35] or Comisar [36]. Actually, the idea of devising quantum systems as quantum fluids has also been a viewpoint shared by others apart from Madelung, as Landau [37] or London [38], and is of common use in solid state physics and condensed matter physics (actually, in many cases, Madelung is not even credited for it).

On the other hand, we have de Broglie's pilot wave theory [39–41], which after the 5th Solvay Conference and further (rather tough) discussions has reached us as the (oversimplified) concept of de Broglie's hypothesis: electrons (and any kind of material system, in general, from tiny electrons or neutrons to large, complex organic molecular compounds) display both particle-like and wave-like properties in the same way that light is also characterized by both properties (after Einstein's 1905 explanation for the photoelectric effect). However, the idea formerly supported by de Broglie in 1924 (and taken as a basis by Schrödinger in the derivation of his famous equation) is that quantum systems consist of two subjects: a carrier (pilot) wave field and a carried singularity (point-like) representing the electron. Unlike Madelung's proposal, de Broglie's one already implies a certain way to understand or "interpret" quantum systems, and therefore it found too much opposition at the Solvay Conference, eventually being rejected in its former version. However, from the hydrodynamic perspective mentioned above, there is nothing wrong with it, since de Broglie's approach is, in the end, only a way to describe with a quantum language the evolution of, saying it in simple terms, a leaf released on the surface of a stream: the leaf conveys information about the evolution of the stream, but tells nothing about the evolution of the elementary compounds of such a stream, which is still in compliance with the Copenhagen (Bohr's) view. These streamlines thus allow a visualization of the evolution of the quantum fluid just as the path pursued by a (classical) tracer particle [42, 43] (e.g., charcoal dust) is used to follow the flow of a classical fluid [44, 45]. Since 2005, experiments with bouncing droplets in vibrating fluids, independently performed by the groups of Yves Couder and Emmanuel Fort in Paris [46–52] and John Bush in Boston [53–55], have become nice illustrations of a classical analog de Broglie's proposal.

Madelung and de Broglie devised their approaches during the early days of quantum mechanics, when the theory was still being developed. About a quarter of a century afterwards, Bohm proposed another analogous model to challenge the formal and conceptual grounds of quantum mechanics, which was already a widely accepted theory. It is interesting to note, though, that Bohm's approach, formerly suggested as a counterproof to von Neumann's theorem on the impossibility of (local) hidden-variable theories compatible with quantum mechanics [7], gathers the flavors of both Madelung's hydrodynamic formulation and the insight of de Broglie's pilot-wave theory. The most remarkable aspect of Bohm's suggestion is the fact that it emphasizes very clearly that quantum mechanics is intrinsically nonlocal, thus introducing a cutting-edge view of this theory that transcends the understanding of physical systems according to the principle of locality. This was the main reason for its rejection by the mid-1950s, until Bell decided to analyze such implications and formalize them about a decade later, producing his brilliant proof on the nonlocal nature of quantum mechanics in terms of an inequality. Some applications of Bohm's started appearing by the late 1970s and during the 1980s and 1990s, where different physical problems typical of the quantum mechanics were translated into Bohm's trajectory-based model. Nonetheless, the model was strongly attached to interpretational (metaphysical-like) connotations related to the measurement problem in quantum mechanics [56], and therefore it was often neglected or not taken too seriously by the community. The landscape started changing by the late 1990s, when it underwent a remarkable boost through its applications and potential interests in chemistry, not only to describe and explain the dynamics exhibited by quantum systems, but also as a source of new computational tools, where the key element is precisely the Bohmian trajectory.

Also the negative metaphysical ("surrealistic" [57]) connotations associated with Bohm's trajectories started changing, since, in the end, they are properly defined from a formal viewpoint. For those familiarized with the treatment and solution of partial differential equations, such trajectories are just the curves arising from the method of characteristics, well-known in mathematical physics to solve finite-order partial differential equations [58]. In quantum mechanics, the method of quantum characteristics is applied, by analogy to its classical counterpart, to the flows in the configuration space defined by the Weyl-Wigner transform [59, 60], which eventually gives rise to the appearance of an infinite-order, with respect to the classical case, through the Moyal product [61] (in the classical limit, the bracket that defines this product becomes the Poisson bracket, which implies an order reduction due to the disappearance of nonlocal correlations). Alternatively, by virtue of a suitable decoupling between the phase and the amplitude of the wave function, Schrödinger's equation can be recast in the form of a differential equation susceptible to also be solved by the method of characteristics, either in terms of real-valued trajectories [15, 62–64] or complex-valued ones [65–72].

From a historical perspective, it is remarkable the fact that Bohmian mechanics entered chemistry from Madelung's approach instead of Bohm's one, more closely connected, perhaps, to physics due to its deep metaphysical (interpretational) connotations discussed above, which are absent in Madelung's one. Madelung's approach, moreover, is in compliance with the tradition in chemistry of appealing to pictorial representations of the systems under study (the pictorial view enabled by Bohmian mechanics was also a positive feature remarked by Bell [73]). Setting as our "coordinate origin" the two categories pointed out by Wyatt [15], namely *analytic* and *synthetic*, this acceptance and future development is even more understandable. Within the analytic approach [16, 17], first the wave function is propagated and then the Bohmian trajectories are obtained "on the fly" from its evolution. The goal of this approach is essentially interpretational, in a sense that it is used to understand what is going on with the system along its evolution in time. This kind of information complements the information obtained by means of other standard quantum techniques, thus providing a wider understanding of the underlying phenomenology. The synthetic approach [15], on the other hand, is aimed at computing quantum-mechanical quantities "on the fly" by directly synthesizing the Bohmian trajectories, that is, acting the other way around, without any need for predetermining the wave function. In this case, the equations of motion for the trajectories are integrated alongside the quantum Hamilton–Jacobi equation (which rules the trajectory dynamics) and the continuity equation (which rules the "ensemble" dynamics).

The applications revisited in this chapter constitute a brief account on both approaches, analytic and synthetic, within chemistry. Due to the many problems where it has been applied and their different nature, the discussion will be limited to some illustrative examples that cover a series of aspects that are discussed with more detail in monographs on the issue [15, 17, 20, 21] as well as in the bibliography here provided. With this in mind, the chapter has been organized as follows. An overview on the different levels at which molecular systems are traditionally studied is presented in Section 4.2, starting from the Born–Oppenheimer approximation [74] and then revisiting (in general terms) the methodologies available to deal with both electronic configuration [75–81] and dynamics, distinguishing in this latter case between molecular dynamics for a few degree-of-freedom systems [82, 83] and statistical mechanics (many-body problems) [84–89]. Analogously, a discussion on some fundamental aspects of Bohmian mechanics is introduced in Section 4.3. Some specific illustrations of the application of Bohmian mechanics to problems typical of the levels mentioned before are described in Section 4.4. Finally, a general, summarizing reflection on the matter in Section 4.5 concludes this chapter.

4.2 Approaching Molecular Systems at Different Levels

Molecular systems typically include a rather large number of degrees of freedom, which makes unaffordable their direct study. Different approximations and/or approaches have thus been considered to cope with this inconvenience. Usually, given the different time-scales involved in the dynamics of the different components, namely nuclei and electrons, the first important approximation is the so-called *Born–Oppenheimer approximation* [74], which allows us to simplify the study of any molecular system (this notion is taken here in a broad sense, thus including simple molecules, solid surfaces, polymeric chains, clusters, crystalline structures, etc.) by splitting it into its nuclear and electronic parts. This division has led to three levels or types of methodologies in chemistry to study and explain molecular systems: electronic structure [75–81], dynamics [82, 83] and statistics [84–89].

Electronic structure methods (e.g., valence bond theory, ab initio methods, density functional theory, etc.) are aimed at obtaining properties associated with the electronic configuration, such as the chemical bonding and intermolecular forces [81]. Basically all these methods are based on quantum mechanics and are included under what is known as quantum chemistry. Molecular dynamics and statistical mechanics methods are based either in classical mechanics or in quantum mechanics, and essentially constitute the scope of the chemical physics and physical chemistry. Molecular dynamics methods, e.g., wave packet propagation methods, are used to describe the properties associated with the nuclear motion (dynamics), necessary to understand chemical reaction processes, for example. In this case, the role of electrons manifests through effective potential functions, namely potential energy surfaces, generated from their bonding to the nuclei. At this point, it is worth highlighting that, for practical purposes, a division has been considered between a few degree-of-freedom treatments, which are henceforth denoted as molecular dynamics methods, and those to deal with many-body problems, which will be referred to as statistical mechanics methods. This convenient classification allows us to distinguish between methods and problems where the full system dynamics is relevant, from those where we are more interested in statistical properties or we only need to focus our attention on a part of the total system (neglecting the remaining "environmental" degrees of freedom). Bearing this in mind, when the number of nuclei involved is relatively large, we shall talk of statistical methods (e.g., molecular dynamics, Monte Carlo, path integrals, etc.), which include methodologies and theories developed to tackle open quantum systems, complex systems, systems far from equilibrium, etc., all of them approximate.

Of course, sometimes the Born–Oppenheimer approximation cannot be used and the above division can no longer be considered. In such cases, one needs to consider a full treatment of the molecular system (electronic plus nuclear parts) using on-the-fly methods that take into account the interaction between electrons and nuclei at the same time in a sort of two-step feedback process: electrons accommodate to a certain nuclei configuration, which is then used to evolve the nuclei and, from the latter new rearrangement, a new electron configuration is determined, and so on, until the full dynamical-configurational simulation is completed. Dealing with this kind of systems is often complicated and highly computationally demanding, although different methods have been developed to deal with them [90–92], but this is a topic that goes beyond the scope of this chapter.

4.2.1 The Born–Oppenheimer Approximation

Molecular systems are usually characterized by a large number of degrees of freedom, which makes complicated and even computationally prohibitive their study. To circumvent this inconvenience we can take advantage of the remarkable mass difference between nuclei and electrons ($m_n \gg m_e$), which physically translates into very different time scales ruling their respective dynamics (electrons are faster than nuclei) as well as very different kinetic energies ($\langle p_n^2 \rangle / 2m_n \ll \langle p_e^2 \rangle / 2m_e$). Accordingly, at a first level of approximation, it is reasonable to assume that nuclei are almost at rest, while electrons move under the action of the effective potential field generated by point-like particles (the nuclei) that occupy definite positions. This is the basic idea behind the Born–Oppenheimer approximation, which we can also find in other contexts where there is an important asymmetry between time scales and that serves us to simplify the dynamical study of different sets of degrees of freedom. For example, in molecular dynamics it can be used to describe some degrees of freedom according to Newton's equation (classical trajectories) under the action of some effective generalized potential functions, while other degrees of freedom evolve quantum-mechanically (Bohmian trajectories). A simple illustrative example of this splitting will be seen with more detail in Section 4.4.5. The same idea can also be found in the context of the theory of open quantum systems [86–89], where large systems can be split up into two subsystems, each ruled by its own dynamical time scales (see Section 4.2.4). Depending on the difference between these scales and the coupling strength between the two subsystems, more or less complex treatment can be devised to tackle the corresponding problem.

More specifically, the Born–Oppenheimer approximation arises in chemistry as a need for solving one of its major problems: finding solutions of the nonrelativistic, time-independent Schrödinger equation that describes a molecular system:

$$\begin{aligned} &\hat{H}\,\Phi_\ell(\vec{R}_1, \vec{R}_2, \ldots, \vec{R}_M, q_1, q_2, \ldots, q_N) \\ &= E_\ell \Phi_\ell(\vec{R}_1, \vec{R}_2, \ldots, \vec{R}_M, q_1, q_2, \ldots, q_N). \end{aligned} \tag{4.1}$$

The energy eigenvalues that we extract from this equation serve us to determine the electronic configuration of the system and, therefore, chemical and physical properties, such as bonding, electric, magnetic, phase transitions and states of matter, geometry, etc. In Eq. 4.1, Φ_ℓ stands for the wave function of the ℓth state of the molecular system, which depends on the $3N$ space coordinates

$\{\vec{r}_i\}_{i=1}^N$ and the N spin coordinates $\{s_i\}_{i=1}^N$ of its electrons, collectively denoted as $\{q_i\}_{i=1}^N$, and the $3M$ space coordinates of its nuclei $\{\vec{R}_A\}_{A=1}^M$; E_ℓ is the eigenenergy associated with the state Φ_ℓ; and $\hat{H}$ is the Hamiltonian describing the M nuclei and N electrons as well as their interactions. In the absence of external fields (e.g., gravitational, electromagnetic, etc.), $\hat{H}$ reads (in atomic units) as:

$$\hat{H} = -\frac{1}{2}\sum_{i=1}^{N}\nabla_{\vec{r}_i}^2 + \sum_{i=1}^{N}\sum_{j>i}^{N}\frac{1}{r_{ij}} - \frac{1}{2}\sum_{A=1}^{M}\frac{\nabla_{\vec{R}_A}^2}{M_A} + \sum_{A=1}^{M}\sum_{B>A}^{M}\frac{Z_A Z_B}{R_{AB}} - \sum_{i=1}^{N}\sum_{A=1}^{M}\frac{Z_A}{r_{iA}}, \tag{4.2}$$

where $r_{ij} = |\vec{r}_i - \vec{r}_j|$ is the distance between the ith and $j$electrons (the same holds for r_{iA} and R_{AB} for the electron-nucleus and nucleus-nucleus distances, respectively). The first two terms in Eq. 4.2 are the total electron kinetic energy and the electron-electron repulsive potential interaction, respectively, and represent the (full) electronic contribution to the Hamiltonian, which we generically denote as $\hat{h}_e$. Similarly, the third and fourth terms describe the total nuclear kinetic energy and the nucleus-nucleus repulsive potential interaction, respectively, which we gather under the label $\hat{H}_n$. Finally, the fifth term gives the attractive electron-nucleus electrostatic interaction, which will be labeled as $\hat{V}_{en}$, since it couples the electronic and nuclear motions. Accordingly, Eq. 4.2 can be expressed in a more compact form as:

$$\hat{H} = \hat{H}_n + \hat{h}_e + \hat{V}_{en}, \tag{4.3}$$

which has the same form of the generic Hamiltonians potentials in the theory of open quantum systems [86–89]:

$$\hat{H} = \hat{H}_S + \hat{H}_E + \hat{V}_{\text{int}}, \tag{4.4}$$

where the full system is partitioned into two subsystems: the system of interest (S) and a surrounding environment or bath (E), with their interaction being described by the term $\hat{V}_{\text{int}}$ ($= \hat{V}_{en}$).

At this level of approximation (it is here where the Born–Oppenheimer approximation comes into play), the nuclei are assumed to have zero motion. Therefore, for a certain fixed position of all the involved nuclei, i.e., what we call a nuclear configuration, only the second and third terms in (4.3) are relevant, i.e.:

$$\hat{H}_e = \hat{h}_e + \hat{V}_{en}. \tag{4.5}$$

The description of the molecular system thus only depends on the distribution of the electrons (*electronic cloud*) around the nuclei, which play the role of a fixed environment that acts upon the electrons through $\hat{V}_{en}$, but that do not feel any back-action from the electrons. The total wave function of the system can then be *factorized* and represented as the product of two separate wave functions, one accounting for the electronic cloud and the other for the ensemble of nuclei. That is, we have:

$$\Phi_\ell(\vec{R}_1, \vec{R}_2, \ldots, \vec{R}_M, q_1, q_2, \ldots, q_N) \\ \approx \varphi^{(\ell)}_{\mathrm{n,M}}(\vec{R}_1, \vec{R}_2, \ldots, \vec{R}_M) \otimes \phi^{(\ell)}_{\mathrm{e,N}}(q_1, q_2, \ldots, q_N), \tag{4.6}$$

where $\varphi^{(\ell)}_{\mathrm{n,M}}$ and $\phi^{(\ell)}_{\mathrm{e,N}}$ denote M- and N-particle wave functions, respectively. Notice that each electronic coordinate depends parametrically on all the nuclear ones, i.e., $q_i = q_i(\vec{R}_1, \vec{R}_2, \ldots, \vec{R}_M)$, for $i = 1, 2, \ldots, N$, but not vice versa. After this factorization, the total energy of the system, obtained from the partitioned Hamiltonian (4.2), reads as:

$$E_{\mathrm{tot}} = E_n + E_e, \tag{4.7}$$

which, as can be seen, consists of the sum of two separate terms. The first term:

$$E_n = \sum_{A=1}^{M} \sum_{B>A}^{M} \frac{Z_A Z_B}{R_{AB}}, \tag{4.8}$$

is a constant associated with the nucleus-nucleus Coulomb-type repulsive interaction, which can be determined either classically or quantum-mechanically, depending on the degree of sophistication of the method considered (and the accuracy requested). It is a constant, because it only depends on the position of the nuclei, which is assumed to be, by hypothesis, fixed. The second term:

$$E^{(\ell)}_e = \langle \phi^{(\ell)}_e | \hat{H}_e | \phi^{(\ell)}_e \rangle, \tag{4.9}$$

corresponds to the expectation value of the Hamiltonian (4.5) in the electronic state $\phi^{(\ell)}_e$. Unlike the nuclear term, the electronic one changes each time the nuclear configuration changes; all values generated in this way provide us with the potential energy surface associated with the molecular system under study, from which we eventually determine dynamical properties (e.g., ionization and photoionization, isomerization, reaction and scattering rates, electron transfer properties, diffraction, etc.).

4.2.2 Electronic Configuration

As seen above, once the nuclear and electronic parts of the full molecular Hamiltonian are specified and split up by virtue of the Born–Oppenheimer approximation, the next step consists in solving the Schrödinger equation for the electronic Hamiltonian (4.5). The main goal of electronic structure methods is to obtain "exact" quantum energy states by further factorizing the problem, recasting the total polyelectronic wave function $\phi_{e,N}^{(\ell)}$ as a product of uncorrelated, single-electron wave functions, $\phi_{e,1}^{(\ell)}$:

$$\phi_{\mathrm{e,N}}^{(\ell)}(q_1, q_2, \ldots, q_N) \quad \longrightarrow \quad \Pi_{j=1}^{N}\phi_{\mathrm{e,1,j}}^{(\ell)}(q_j), \tag{4.10}$$

where the products combine in the form of Slater determinants in order to preserve the corresponding electron antisymmetry properties. These wave functions are plugged into their corresponding *reduced* Schrödinger equations, characterized by a single-particle Hamiltonian, $\hat{h}_{e,j}$, which is a function of an effective potential where the interaction with the other N-1 electrons is accounted for as a sort of average or mean effective field—in this sense, the N electrons are commonly regarded as noninteracting. Then, after a variational process in which the energy, assumed to be a functional of the single-particle probability density, is required to be an extremum, both the "correct" wave function (4.10) and energy (4.7) are found.

After proceeding in that way, the electronic structure of the molecular system is determined, which is the same to say how electrons distribute throughout the system. With this information we can obtain the system chemical properties, such as chemical bonding, intermolecular interactions, electronic charge distributions, dipole and multipole moments or vibrational/rotational frequencies, for instance. Now, only in the case of the hydrogen atom, the corresponding Schrödinger equation can be exactly solved; for any other atomic or molecular system, involving three or more particles, different approximate approaches have been proposed in the literature [76–80] to tackle the problem of determining their electronic configurations. These approaches or methodologies arise basically from two models [93]: the *wave model* and the *density matrix model*.

In the *wave model* each atom is assumed to be a point-like, positively charged nucleus surrounded by an electronic cloud, which is associated with a certain orbital and represents the electron probability distribution in configuration space. The two types of theories arisen from this model are the *valence bond theories* and the *molecular orbital theories*. Valence bond theories [94] are based on the assumption that orbitals primarily localize on each nuclei (atomic orbitals), thus they essentially focus on pairwise interactions, and therefore establish a direct link with the classical chemical models based on the drawings of bonds typical of Lewis structures. Though not very common nowadays, the concepts such as *resonance* [95] (chemical aromaticity) and orbital *hybridization* [96] arise from this theory.

Molecular orbital theories [76–78], on the contrary, consider delocalized orbitals covering the whole molecule (molecular orbitals). Thus, the molecular orbital wave function is expressed as a linear combination of atomic orbitals (Slater determinants). These atomic orbitals can be hydrogen-like or Gaussian functions. The basic method within the molecular orbital approach is the Hartree–Fock (HF) method, an ab initio approach based on assuming that $\phi_{\mathrm{e,N}}^{(\ell)}$ is given by a Slater determinant, which leads to a set of N coupled monoelectronic Schrödinger equations. In these equations, the electron-electron repulsion interactions involved in $\hat{h}_e$ are accounted for by a *mean field* interaction (i.e., the averaged action of the N-1 remaining electrons), thus neglecting *electron correlation*. In those cases where large molecules are considered and the HF method becomes inoperative, one can apply the so-called semiempirical molecular quantum-mechanical methods (e.g., the Hückel and extended Hückel methods or the Pariser–Parr–Pople method), where a simple Hamiltonian plus a set of fitting parameters are used to adjust the experimental data. On the other hand, within the ab initio stream, there are different routes to tackle the problem of the electron correlation, and therefore to improve the HF results, such as the post-Hartree–Fock methods (e.g., configuration interaction, coupled cluster or Møller–Plesset), the multiconfigurational self-consistent field or the multireference configuration interaction.

Regarding the *density matrix model* [79, 97], the interest in developing this type of methodology arises as a need from the very high computational demand involved in the study of large molecular systems by means of ab initio methodologies. Thus, although the latter are more accurate physically, this accuracy is sacrificed in order to get computational advantages. The first attempt in this direction is the Thomas–Fermi model [98, 99], where the electron kinetic energy of an atom is expressed as a functional of the atom electronic distribution. This model is the precursor of the modern *density functional theory* [79] (DFT), where the fundamental physical information about the molecular system is obtained from a single-particle density in a three-dimensional space, derived variationally either within a time-independent framework [79] (ground state) or a time-dependent one [100, 101] (excited states). Nowadays, DFT is a very popular calculation method in quantum chemistry, as well as in solid state physics, due to its computational simplicity combined with its (numerical) accuracy. Also within the density matrix model, it is worth stressing the role of those approaches which are aimed at including the electronic correlation, utilizing for such a purpose a *reduced density matrix* formalism [102–116].

By changing the positions of the nuclei, one obtains different values of the electronic energy, thus describing a multidimensional energy surface parameterized in terms of the nuclei positions, $\{\vec{R}_A\}_{A=1}^{M}$. This is the so-called *potential energy surface* or *Born–Oppenheimer surface*, which reflects the configuration of the molecular system and determines its dynamical behavior (i.e., its physical/chemical properties). An alternative approach to obtain the potential energy surface of a molecular system is by using *molecular mechanics* or *force field methods*. Unlike the previously described methods, molecular mechanics is not based on quantum mechanics but on a classical treatment of systems. In this case, the system is modeled by a collection of atoms (by "atom" it is assumed a single unity rather than a true physical atom) held together by spring-like force functions (force fields) and a corresponding set of fitting parameters. Each atom is characterized by a radius (usually, the van der Waals one), a charge and polarizability, and the potential energy surface is obtained by minimizing some energy functional at different nuclei positions using some optimization method (e.g., the Metropolis algorithm or the Monte Carlo method). Strictly speaking, these methods cannot be considered electronic structure ones, since electrons are not explicitly taken into account. However, they are assumed to be in an optimal distribution around the fixed nuclei positions and, therefore, are implicitly responsible for the nuclear structural arrangement. Furthermore, in order to improve the performance of these methods, hybrid quantum mechanical molecular mechanics approaches are considered [117], which combine on-the-fly molecular mechanics with forces evaluated quantum-mechanically from either ab initio or DFT calculations.

Finally, it is also worth mentioning the availability of schemes based on topology, devised to analyze the structures (potential energy surfaces) obtained and determine, for example, eventual *reaction paths* or *minimum energy paths*, i.e., the paths connecting reactants with products following steepest descent curves [118] (curves with the fastest descent from a potential hill to a valley). These paths are used to understand and interpret chemical reactions as well as to determine kinetic constants and, therefore, reaction rates without involving any dynamical simulation. This is part of what is known as mathematical chemistry.

4.2.3 Dynamics of "Small" Molecular Systems

Once the electronic structure—or, equivalently, the effective interaction among nuclei due to the combination of the electromagnetic interaction and the quantum nature of the electrons—is determined and characterized by a potential energy surface, the next step consists in obtaining properties associated with the motion of the nuclei, which occurs at slower timescales—actually, many times it happens that the motion refers to some associated effective degrees of freedom, e.g., vibrational or rotational modes, rather than the nuclei themselves, which allows to further simplify the study of the full system. This procedure not only provides dynamical information about the system, but also allows us to test the accuracy of the potential energy surfaces, previously obtained by means of the electronic structure methods described in previous section, by comparing the corresponding simulations with the experimental data recorded (e.g., the potential energy surface characterizing a certain metal surface can be compared with the real surface by computing the associated Bragg peaks and then comparing this result with the experimental diffraction pattern). And, through this procedure, we also test the accuracy and suitability of the electronic configurations computed.

To tackle the problem of the quantum dynamics, there are basically two approaches [119, 120]: time-independent and time-dependent. Due to the complexity involved in solving the time-dependent Schrödinger equation, historically the time-independent approach was first developed. By the 1960s, different time-dependent classical and quantum-mechanical numerical schemes started to develop, this also being the starting point of computational chemistry. This is the reason why former calculations and numerical schemes in electronic structure as well as in dynamics, with the obtention of spectral lines or collision cross sections, were mostly developed within the first 25 years of quantum mechanics. However, in the case of the dynamics, it is worth stressing that part of the work was already done: the time-independent Schrödinger equation is analogous to the Helmholtz equation that appears in classical wave theory and electromagnetism, so many solutions were already known since the nineteenth century [121]. In the case of hard-wall-like potential models, one deals with standard boundary condition problems, either in bound systems to find spectral lines, or in the continuum to determine scattering cross sections.

For space-dependent potentials, such as those describing bound systems or tunneling problems [122, 123], electromagnetism also provided a well-known methodology: the Jeffrey–Wentzel–Kramers–Brillouin (JWKB) method [124–128]. More recently, different numerical methods have been developed in the literature to solve directly the time-independent Schrödinger equation and determine the bound levels associated with (bound) potentials [129, 130], which are based either in the so-called "shooting and matching" strategies or in variational techniques. In either case one can discretize the space or consider some basic functions (e.g., Gaussians, sinc-functions) and express the solution as a linear combination of these functions. Also, as an analytical tool, classical trajectories have been employed, taking into account the correspondence between eigenstates (bound states) and classical periodic orbits [131, 132].

In the case of processes and phenomena in the continuum, soon after the establishment of quantum mechanics, a quantum scattering theory was developed [133, 134]. Essentially, in this theory an incoming particle is represented by a coherent plane wave, $e^{i\vec{k}\cdot\vec{r}}$, at some initial time at $t \to -\infty$. At $t = 0$, the particle collides with a target and gets diffracted, which translates into an outgoing wave function consisting of a superposition of scattering states (also coherent plane waves) at $t \to +\infty$. Apart from the analytical partial waves approach, there is also a number of numerical methods which have been developed in the literature [119, 120], such as the close-coupling method, which basically consists of splitting the time-independent Schrödinger equation into a set of coupled one-dimensional Schrödinger equations accounting for each of the eventual diffraction channels. Also, as in the case of bound systems, classical trajectories have been used to determine, for example, diffraction probabilities [135, 136]; although diffraction is a quantum concept, in principle it is possible to obtain good estimates in scattering [137–139] and preionization [140–142].

Regarding the time-dependent domain, the strategies developed [83, 120, 143, 144] are aimed at solving the time-dependent Schrödinger equation directly, which allow to monitor the system evolution along time from the initial to the final asymptotic state, instead of passing directly from the former to the latter. Usually, wave packet propagation methods are the most commonly used to to carry out these studies, although other approximate techniques have also been developed in order to find a balance between accuracy and computational demand. These approaches result interesting, for example, to monitor the passage from reactants to products in a chemical reaction (e.g., in proton transfer reactions, isomerization reactions or coherent-control schemes), and are usually accompanied and complemented by classical, trajectory-based studies (quantum-classical correspondence schemes).

4.2.4 Statistical Approach to Large (Complex) Molecular Systems

Typically, the dynamical studies mentioned above are related to systems which are considered as isolated from a surrounding medium or *environment*. However, real physical and chemical systems interact, exchanging energy with their surroundings, and deviations from their isolated behavior are observed. A very remarkable effect, in this regard, is the broadening undergone by atomic and molecular spectral lines, for example, which is a signature of the life-time associated with the corresponding quasi-bound (stationary) states. Nevertheless, we also find such situations when measures involving bulk properties (e.g., the viscosity of liquid water or its crystallization) are involved. A connection between the microscopic description and the macroscopic observations has then to be established. Obviously, a full quantum mechanical calculation involving the system of interest plus its environment results unaffordable computationally. Thus, different approximated statistical approaches have been developed in the literature, based on the theory of open quantum systems, where the main analysis tool being the correlation function, which allows to extract the most relevant properties associated with the system. The correct way to tackle a statistical problem in quantum mechanics is through the Liouville–von Neumann equation, which is the equivalent to Schrödinger's equation, although for the *density matrix operator* or, simply, *density matrix*, $\hat{\rho}(t) = \sum_\alpha |\Psi_\alpha(t)\rangle\langle\Psi_\alpha(t)|$, instead of the wave function, $|\Psi(t)\rangle$. However, full many degree-of-freedom, quantum mechanical dynamical problems result prohibitive computationally, and different strategies have been developed in the literature to deal with them. Within the different methodologies commonly employed, there are basically three schemes, which correspond to the three levels at which a quantum statistical problem can be solved: quantum mechanical, mixed/hybrid quantum classical, and fully classical. Nevertheless, independently of the level of accuracy of our statistical description, there is always a very clear distinction between system and environment, this being the starting point in any of these approaches. In this regard, it is worth mentioning that this is also the partitioning philosophy used in some recent ab initio approaches: instead of studying in detail a full macromolecular structure, the total system is partitioned into the reactive (or more active) part and the surrounding, which is described in terms of an effective, *embedding* potential.

It is also worth mentioning that, depending on the properties of the environment, we can refer to it in different ways [87]. For example, usually, when it consists of a few degrees of freedom, it is commonly referred as environment, while the term *reservoir* is used when there is a large number of degrees of freedom. If it is in thermal equilibrium, i.e., its properties remain stationary with time (except for small fluctuations)—in other words, this is the limit where the environment "forgets" information about its initial state and stays in equilibrium at a certain temperature—then we employ the term *thermal bath*. For example, a harmonic oscillator coupled to another harmonic oscillator acts as an environment with respect to the latter because there is an energy and coherence transfer from the latter to the former. If instead of one harmonic oscillator, we couple an infinite number of oscillators, the environment becomes a reservoir and, if such collection of oscillators is at thermal equilibrium, as in the so-called *Caldeira–Leggett model* [145–147], we will have a thermal bath acting on the system oscillator.

Within the first level, the quantum mechanical one, one starts from the Liouville–von Neumann equation. Proceeding algebraically according to some assumptions on the timescales associated with system and environment, one can reformulate the problem in terms of a master equation describing only the system evolution and where the effects of the environment appear in terms of some dissipative operators or dissipators after assuming the Markovian approximation (i.e., the detailed dynamics of the environment becomes irrelevant, and only its effects are important regarding the system dynamics). This is achieved by partially tracing over the environment states, which leads to the so-called *reduced density matrix*, $\hat{\rho}_S$, describing the system (or subsystem) of interest or *reduced system*. Note that this splitting between system and environment as a function of their corresponding timescales (i.e., $\hat{\rho} = \hat{\rho}_S \otimes \hat{\rho}_E$) is not other than the Born–Oppenheimer approximation considered in ab initio methods to separate the electronic motion from the nuclear one (see Section 4.2.1). This procedure thus gives rise to very well-known master equations for the system reduced density matrix, such as the *Lindblad equation* or its version in the Markovian weak-coupling limit, namely the so-called *Redfield equation* [148–151]. But also one can obtain master equations to describe the evolution of system expectation values starting from the Heisenberg representation, such as the *quantum Langevin equation* [84, 152, 153], which is the Heisenberg equation of motion for the coordinate of a Brownian particle coupled to a thermal bath. At this stage, apart from analytical derivations, there are basically two methods to solve these equations quantum-mechanically: the *path integral method* [154–157] and the *quantum state diffusion theory* [89]. In the first case, one directly attacks the problem of the density matrix, while in the latter one solves a stochastic Schödinger equation, obtained from the transformation $\hat{\rho}_S = (1/N)\sum_{k=1}^{N} |\Psi_k\rangle\langle\Psi_k|$, where the Ψ_k are (stochastic) solutions of such an equation [158].

Whenever the environment dynamics are relevant, the Markov approximation cannot be applied. Then, we are in the next level, that of the mixed or hybrid Quantum classical methods, where usually the system dynamics are accounted for quantum-mechanically, while the environment dynamics are described by means of classical mechanics, this splitting also being based in the Born–Oppenheimer approximation, as mentioned above. Different methods and techniques have been developed in the literature to tackle this problem, such as mean field theories [159], surface-hopping methods [160], or semiclassical initial value representations [161–171], which allow to describe both the effects of the environment over the system and vice versa in spite of their different (classical or quantum) mechanical evolution. In this regard, it is worth stressing that, depending on whether interatomic interactions can be described by a potential energy surface or not, we may have adiabatic or nonadiabatic dynamics. In the latter case, the interaction between different coupled potential energy surfaces (corresponding to different electronic molecular states) has to be included into the corresponding dynamical description [172], thus not satisfying the Born–Oppenheimer approximation, which implies finding good descriptors of the coupling between the electronic and nuclear parts of the Hamiltonian. Now, for certain cases, one can further proceed and represent the system also classically, thus leading to the third level, which is constituted by purely classical methodologies (i.e., based on Newton's equations), such as simple molecular dynamics simulations, or refinements of them, such as the classical Wigner method or the linearized initial value representation, where a particular choice of the initial conditions is considered (though the propagation is fully classical). Sometimes, molecular dynamics calculations are combined with DFT in order to obtain, on the fly, the potential energy surfaces governing the nuclear motion, as it happens in the *Car–Parrinello method* [173, 174].

4.3 Bohmian Mechanics

4.3.1 Fundamentals

For simplicity, but without loss of generality, we are going to start with the Schrödinger equation associated with a nonrelativistic particle (system) of mass *m*. The general solutions to this equation are given in the form of time-dependent complex-valued fields, which we generally call wave functions and represent the state of the system (quantum states) at a given time throughout the full accessible configuration space associated with the physical system studied. From these solutions we extract all relevant physical information about the quantum system, that is, intensities and coherence-based properties (interference effects). Without entering here a discussion on the well-known debate about the completeness of the wave function (actually, not necessary at all), but taking into account that such information is encoded in the probability density (i.e., the square of the amplitude of the wave function) and the phase of the wave function, we can recast the latter in terms of two real-valued fields. This is actually a closer approach to Schrödinger's derivation of his equation from Hamilton's formulation of classical mechanics and de Broglie's novel idea of associating a wave to nonzero mass systems (the so-called *Hamiltonian analogy* [118]).

From a formal viewpoint, in principle, if we consider time-independent, eigenvalue problems, Schrödinger's equation acquires the form of a general wave equation (Helmholtz equation), which has exactly the same solutions as the homologous equations we find in electromagnetism or classical wave mechanics [58, 121]. However, in the time-dependent case this analogy is not direct. If the Helmholtz equation is rewritten in paraxial form, it is true that it resembles Schrödinger's equation, although with one of the coordinates playing the role of the evolution parameter (time). However, in general, what we really observe is a closer resemblance to the heat equation, since it is also a parabolic partial differential equation, as diffusion equations are, although with the particularity that the diffusion coefficient is a pure imaginary constant. This fact was already stressed by Fürth [35] shortly after the equation appeared and later on by Comisar [36]. Nelson also introduced [175] a Markovian approach from which Schrödinger's equation arose as a result of diffusion process in a subquantum medium.

Taking into account the previous discussion, let us consider the nonlinear transformation:

$$\Psi(\vec{r}, t) = R(\vec{r}, t)\, e^{iS(\vec{r},t)/\hbar}, \tag{4.11}$$

which allows us to recast the complex-valued field Ψ in terms of two real-valued fields, one accounting for the amplitude of such a complex field, and the other one for the phase that any complex quantity introduces. The same transformation is also considered for the conjugate complex (we go from two field variables to another two field variables), with the only difference being that the phase acquires a minus sign. The equations of motion for the amplitude (R) and phase (S) fields arise after substitution of the wave fields (Ψ and Ψ^*) into the time-dependent Schrödinger equation and its complex form, respectively. After a brief rearrangement of terms, adding and subtracting these equations, we obtain their imaginary and real parts, which read, respectively, as

$$\frac{\partial \rho}{\partial t} + \vec{\nabla} \cdot \left(\rho \, \frac{\vec{\nabla} S}{m} \right) = 0, \tag{4.12}$$

$$\frac{\partial S}{\partial t} + \frac{(\vec{\nabla} S)^2}{2m} + V_{\rm eff} = 0. \tag{4.13}$$

In Eq. 4.13, the third term on the left-hand side:

$$V_{\text{eff}} = V + Q = V - \frac{\hbar^2}{2m}\frac{\nabla^2 R}{R}$$
$$= V - \frac{\hbar^2}{4m}\left[\frac{\nabla^2 \rho}{\rho} - \frac{1}{2}\left(\frac{\vec{\nabla}\rho}{\rho}\right)^2\right], \qquad (4.14)$$

is an *effective potential* resulting from the sum of the external potential function V (particle-particle interactions in electronic structure problems and potential energy surfaces in dynamics ones; other external contributions, such as the application of external laser fields or gravity, are also included), and the *quantum potential*, Q. The latter is responsible for transferring the *nonlocality* property of quantum mechanics [176, 177] (which manifests through the phenomenon of *quantum entanglement* [178, 179]) to the particle's motion due to its dependence on the quantum state (through the probability density ρ).

As it can be noticed, Eq. 4.12 is the well-known probability conservation equation, which states that for a pure state the probability must preserve along time, but also that if we specify a certain region of the configuration space, the probability contained in it must also remain constant along time if its boundaries vary in time in a prescribed way [180]. In classical mechanics [83], this variation takes place according to the evolution in phase space of (Newtonian) trajectories with initial conditions selected on the boundary; the evolution of the trajectories generates tubes in phase space where the probability is confined all the way through. Quantum-mechanically the same also holds [180] if we choose as the confining trajectories the ones that arise from the equation of motion:

$$\vec{v} = \frac{\vec{J}}{\rho} = \frac{\vec{\nabla}S}{m}, \qquad (4.15)$$

although the preservation of the probability takes places in the system configuration space instead of a generalized phase space. This transport equation arises naturally in the standard formulations of quantum mechanics if we divide the quantum flux or probability current [181, 182]:

$$\vec{J} = \frac{\hbar}{2mi}\left(\Psi^*\nabla\Psi - \Psi\nabla\Psi^*\right), \qquad (4.16)$$

by the probability density. In other words, physically it means that the transport or diffusion of probability through the system configuration space can be described as a velocity vector field multiplying the probability density, in compliance with usual diffusion (transport) equations. Nonetheless, the way how Bohm introduced that quantity in 1952 was postulating it once the quantum or generalized Hamilton–Jacobi equation (4.13) was settled [7]. That is, he established a direct analogy with classical mechanics, where an analogous equation arises within the Hamilton–Jacobi formulation, namely the Jacobi momentum law. In the classical case, S denotes the classical (mechanical) action, which is a hypersurface in configuration space associated with the evolution of the system directly linked to the geometric phase in the *eikonal approximation* [118]; in quantum mechanics, the phase field S is a direct analog of the classical action, although it is specified in the configuration space of the system. It is by virtue of this constraint (i.e., to restrict the motion to the configuration space instead of a phase space with coordinates and momenta independent of each other) that nonlocality arises (or manifests) in quantum mechanics.

Notice that the trajectories that one obtains by integration (in time) of Eq. 4.15 describe the flow of the probability throughout the corresponding configuration space. That is, such trajectories are just probability streamlines, analogous to those that we find in classical hydrodynamics (this is precisely the point emphasized by Madelung in 1926). However, one can also be tempted to identify such trajectories with the actual motion of real particles in order to establish a (alternative) quantum theory without observers. Although this has been a major issue of debate over years and, actually, one of the main reasons why Bohmian mechanics has been generally rejected, there is nothing in the theory, not even current experiments, supporting such a link. The only accessible information to date in an experiment is the quantum flux, as it has been recently shown [183], which is in compliance with the usual theory [184].

Equation 4.15 together with the time-dependent Schrödinger equation are the two basic equations within the analytic approach discussed above [15], for they allow us to obtain the evolution of the wave function Ψ as well as the associated Bohmian trajectories, necessary to interpret the flow of the probability or, equivalently, the motion of a swarm of *noninteracting* quantum particles [16, 17, 144, 185]. On the other hand, from a computational viewpoint, schemes aimed at integrating partial equations are always affected by numerical instabilities, as happens with Eqs. 4.12 and 4.13. Hydrodynamically speaking [186, 187], these equations are expressed within an *Eulerian framework*, i.e., from a fixed space point we will observe the quantum fluid passing by. So, an easier way to proceed consists in taking advantage of Eq. 4.15 to define the Lagrange time derivative:

$$\frac{d}{dt} = \frac{\partial}{\partial t} + \vec{v} \cdot \vec{\nabla}. \tag{4.17}$$

Recasting Eqs. 4.12, 4.13, and 4.15 in terms of this operator, which renders the set of coupled equations of motion:

$$\frac{d\rho}{dt} = -\rho \vec{\nabla} \cdot \vec{v}, \tag{4.18}$$

$$\frac{d\vec{v}}{dt} = -\frac{\vec{\nabla} V_{\text{eff}}}{m}, \tag{4.19}$$

$$\frac{dS}{dt} = \frac{1}{2} mv^2 - V_{\text{eff}}, \tag{4.20}$$

or, equivalently:

$$\frac{d\rho}{dt} = -\rho \vec{\nabla} \cdot \vec{v}, \tag{4.21}$$

$$\frac{dS}{dt} = \frac{1}{2} mv^2 - V_{\text{eff}}, \tag{4.22}$$

$$\frac{d\vec{r}}{dt} = \vec{v} = \frac{\vec{\nabla} S}{m}. \tag{4.23}$$

This set of equations constitutes the core of the so-called (Bohmian) *quantum trajectory method* [15], an algorithm based on Bohmian mechanics to obtain quantum-mechanical results. It was firstly proposed [62–64] following the scheme of Eqs. 4.18, 4.19 and 4.20. However, evaluating the gradient of the force is an important source of numerical error and the second scheme was considered. Alternatively, to avoid these drawbacks, one can rederive this set of equations in complex form [65–72], by considering a complex phase field, which already include information about the amplitude of the wave function [188].

4.3.2 Nonlocality and Entanglement

Dealing with many-body problems, as it is the case when molecular systems are involved, unavoidably leads to the question of the appearance of *quantum correlations* among the different degrees of freedom necessary to specify the (quantum) state of the system. Typically the series of approximations that are considered to tackle these problems (starting from the Born–Oppenheimer one, whenever it is required) implies a breach of such correlations. Although this is an efficient route from a problem-solving viewpoint, physically it means there is a certain loss of quantum coherence in the system, which is not going to be shared anymore by the different separated parts. In other words, the degree of *entanglement* [178, 179] between such parts is totally lost–whatever happens to one of the parts of the full system will only have classical-like consequences on the others, and vice versa.

To better understand the implications of entanglement and the possibility to nonlocally share information among different parts of a quantum system is precisely through Bohmian mechanics. Consider the N-body wave function in the position space:

$$\Psi(\vec{r}_1, \vec{r}_2, \ldots, \vec{r}_N, t) = R(\vec{r}_1, \vec{r}_2, \ldots, \vec{r}_N, t)\, e^{iS(\vec{r}_1, \vec{r}_2, \ldots, \vec{r}_N, t)/\hbar}. \quad (4.24)$$

The *reduced* single-particle probability density for the ith particle (or, in general, degree of freedom) is defined as:

$$\tilde{\rho}_i(\vec{r}_i, t) = \int \cdots \int \rho(\vec{r}_1, \vec{r}_2, \ldots, \vec{r}_N, t) d\vec{r}_1 d\vec{r}_2 \cdots d\vec{r}_{i-1} d\vec{r}_{i+1} \cdots d\vec{r}_N, \quad (4.25)$$

where $\rho = R^2$. This reduced probability density describes the statistics of the ith particle ignoring the statistics of the remaining particles, although due to the correlation among them any change in the configuration of the system will have consequences on this particle. To this particle we can also assign an equation of motion:

$$\vec{v}_i = \frac{\vec{\nabla}_{\vec{r}_i} S}{m}, \quad (4.26)$$

which takes into account the full wave function and therefore the motion of the other particles. This procedure gives rise to a set of N equations of motion coupled through the "wholistic" phase S, where the evolution of a particle will be strongly nonlocally influenced by the others (apart from other classical like interactions through V). This entanglement becomes more apparent through the quantum potential:

$$Q = -\frac{\hbar^2}{2m}\sum_{i=1}^{N}\frac{\nabla^2_{\vec{r}_i} R}{R}, \tag{4.27}$$

where $Q = Q(\vec{r}_1, \vec{r}_2, \ldots, \vec{r}_N, t)$, which is nonseparable and therefore strongly nonlocal.

As a simple illustration of these ideas, consider the case of two particles which interact, at some time, through a potential V. The two-particle wave function describing this system is:

$$\Psi(x, y, t) = R(x, y, t)\, e^{iS(x,y,t)/\hbar}, \tag{4.28}$$

with $\rho(x, y, t) = R^2(x, y, t)$ being the total probability density, and $\tilde{\rho}_1(x, t)$ and $\tilde{\rho}_2(y, t)$ the reduced probability densities associated with each particle. The trajectories for these particles are obtained from:

$$\vec{v}_1 = \frac{1}{m}\frac{\partial S(x, y, t)}{\partial x}, \qquad \vec{v}_2 = \frac{1}{m}\frac{\partial S(x, y, t)}{\partial y}, \tag{4.29}$$

which are implicitly influenced by the quantum potential:

$$Q(x, y, t) = -\frac{\hbar^2}{2m}\frac{1}{R(x, y, t)}\left[\frac{\partial^2 R(x, y, t)}{\partial x^2} + \frac{\partial^2 R(x, y, t)}{\partial y^2}\right]. \tag{4.30}$$

According to Schrödinger [178, 179], after the interaction, even if the two particles are very far apart one from another, the wave function (4.28) becomes nonfactorizable, i.e.:

$$\Psi(x, y, t) \neq \psi(x, t) \otimes \psi(y, t), \tag{4.31}$$

and the motion of both particles will remain entangled. Different works in the literature analyze the trajectory correlation among entangled particles [189–195], the most recent one within the many-body context of transport phenomena in mesoscopic systems [196].

This does not happen, however, when the wave function is factorizable in terms of single-particle partial wave functions:

$$\Psi(x, y, t) = \psi_1(x, t)\psi_2(y, t) = R_1(x, t)R_2(y, t)\, e^{i[S_1(x,t)+S_2(y,t)]/\hbar}, \tag{4.32}$$

since $\rho(x, y, t) = \tilde{\rho}_1(x, t)\tilde{\rho}_2(y, t)$ is a direct product of (partial) densities, while the particles' equations of motion:

$$\vec{v}_1 = \frac{1}{m}\frac{\partial S_1(x, t)}{\partial x}, \qquad \vec{v}_2 = \frac{1}{m}\frac{\partial S_2(y, t)}{\partial y}, \tag{4.33}$$

are uncoupled. This can also be readily inferred from the quantum potential, which is fully separable:

$$\begin{aligned} Q(x, y, t) &= -\frac{\hbar^2}{2m}\frac{1}{R_1(x, t)}\frac{\partial^2 R_1(x, t)}{\partial x^2} - \frac{\hbar^2}{2m}\frac{1}{R_2(y, t)}\frac{\partial^2 R_2(y, t)}{\partial y^2} \\ &= Q_1(x, t) + Q_2(y, t). \end{aligned} \tag{4.34}$$

In general, for an N-particle wave function:

$$\Psi(\vec{r}_1, \vec{r}_2, \ldots, \vec{r}_N, t) = \Pi_{i=1}^N \psi_i(\vec{r}_i, t) = \Pi_{i=1}^N R_i(\vec{r}_i, t)\, e^{iS_i(\vec{r}_i, t)/\hbar}, \tag{4.35}$$

the ith-particle equation of motion is:

$$\vec{v}_i = \frac{\vec{\nabla}_{\vec{r}_i} S_i}{m}, \tag{4.36}$$

and the total quantum potential will be given by the direct addition of the quantum potentials associated with each separate particle:

$$Q = -\frac{\hbar^2}{2m}\sum_{i=1}^{N}\frac{\nabla_{\vec{r}_i}^2 R_i}{R_i} = \sum_{i=1}^{N} Q_i. \tag{4.37}$$

After using the Born–Oppenheimer approximation, splitting electronic from nuclear degrees of freedom, the essence of electronic structure methodologies is precisely finding the electronic configuration of molecular systems by appealing to the idea of separability, from ab initio methods (wave function approach) to the density functional theory (probability density approach). Of course, once an optimal basis set is found, correlations will appear when appealing to the appropriate symmetrization conditions. For example, in the case of ab initio methodologies, Slater determinants are considered to provide the correct symmetry to the linear combinations of orbitals formed to describe the configuration of the corresponding molecular system (see Section 4.2.2).

There is another interesting issue concerning correlation and the Born–Oppenheimer approximation. Notice that although Eq. 4.6 is separable, due to the dependence of the electronic coordinates, $\vec{r}_i$, on the nuclear ones, $\vec{R}_A$, the eventual trajectories for the electrons will depend on the nuclei positions (in agreement with the fact that the potential energy surface varies throughout space), but the same will not happen for the nuclear trajectories with respect to the electrons. Again here we find the problem of the back-reaction, although this time it is within the Bohmian context.

4.3.3 Weak Values and Equations of Change

In the recent years, the concept of *weak value* [197] has been a subject of much interest, getting some popularity within the quantum metrology and quantum foundations communities. For example, by means of measurements of the weak value it is possible to perform tomography of the quantum state associated with a single photon [198] and also to determine the streamlines describing the (averaged) flow of probability in Young's two slit experiment [183]. However, this idea, now in fashion, is directly related to the well-known quantum flux [181, 182] and to the averaging of two-state transition probabilities [199, 200]. However, the essence behind this idea can be traced back to 1970s, to a work published by Hirschfelder [201], where we also find a certain connection to the later application of quantum hydrodynamics to time-dependent DFT (see Section 4.4.1). According to this author, the framework presented in this work facilitates the study of problems involving external electric and magnetic fields (as seen in previous section) by grounding it on the concept of *equation of change* for arbitrary quantum mechanical properties in configuration space. These equations can be used to study the energy flow from one part of a molecule to another, the nature of molecular collisions or the magnetic properties of molecules. The first two cases are somehow summarized by the applications which will be discussed in next section regarding the molecular dynamics. The third type of application, on the other hand, has been exploited in the literature by different authors [202–212] to understand the magnetic properties of molecules within a framework that encompasses electronic structure and topology.

Although in a brief manner, let us here consider how the ideas underlying quantum hydrodynamics emerge in a very elegant fashion within the framework proposed by Hirschfelder [201]. To start with, let us define an arbitrary property $\mathcal{S} \equiv \Psi_1^* \hat{\mathcal{O}} \Psi_2$ (do not confuse this calligraphic $\mathcal{S}$ denoting a certain property with the normal S referring to the phase of a wave function), where both Ψ_1 and Ψ_2 satisfy the same time-dependent Schrödinger equation (i.e., both are governed by the same Hamiltonian) and $\hat{\mathcal{O}}$ is an (arbitrary) operator associated with a quantum observable. If $\Psi_2 = \Psi_1$, the integral of $\mathcal{S}$ over configuration space, denoted by $\bar{\mathcal{S}}$, describes the *expectation value* associated with $\mathcal{O}$, while for $\Psi_2 \neq \Psi_1$, $\bar{\mathcal{S}}$ gives a *transition probability* or *rate* (from the state Ψ_1 to a state Ψ_2) governed by $\mathcal{O}$. It can be easily shown that $\mathcal{S}$ satisfies the equation of change:

$$\frac{\partial}{\partial t}\left(\Psi_1^* \hat{\mathcal{O}} \Psi_2\right) = \Psi_1^* \left\{ \frac{\partial \hat{\mathcal{O}}}{\partial t} - \frac{i}{\hbar}\left[\hat{\mathcal{O}}, \hat{H}\right] \right\} \Psi_2 - \frac{i}{\hbar}\left[\Psi_1^* \hat{H} \hat{\mathcal{O}} \Psi_2 - (\hat{H}\Psi_1)^\dagger \hat{\mathcal{O}} \Psi_2\right]. \tag{4.38}$$

Specifically, for a Hamiltonian such as:

$$\hat{H} = \frac{\hat{p}^2}{2m} + \hat{V}(\hat{\vec{r}}), \tag{4.39}$$

it can be shown [201] that Eq. 4.38 takes the functional form:

$$\frac{\partial}{\partial t}\left(\Psi_1^* \hat{\mathcal{O}} \Psi_2\right) + \vec{\nabla} \cdot \vec{F} = \mathcal{K}, \tag{4.40}$$

where

$$\vec{F} = \frac{1}{2}\left[\Psi_1^* \dot{\hat{\vec{r}}} \hat{\mathcal{O}} \Psi_2 - (\dot{\hat{\vec{r}}} \Psi_1)^\dagger \hat{\mathcal{O}} \Psi_2\right] \tag{4.41}$$

and

$$\mathcal{K} = \Psi_1^* \left\{ \frac{\partial \hat{\mathcal{O}}}{\partial t} - \frac{i}{\hbar}\left[\hat{\mathcal{O}}, \hat{H}\right] \right\} \Psi_2 \tag{4.42}$$

are, respectively, the flux of $\mathcal{S}$ and its rate of production, and

$$\dot{\hat{\vec{r}}} = -\frac{i}{\hbar}\left[\hat{\vec{r}}, \hat{H}\right] = \frac{\hat{\vec{p}}}{m}. \tag{4.43}$$

Now, if $\hat{\mathcal{O}}$ is a linear Hermitian operator, physically the most significant form of $\mathcal{S}$ will be:

$$\mathcal{S} = \mathrm{Re}\left[\Psi_1^* \hat{\mathcal{O}} \Psi_2\right] = \frac{1}{2}\left[\Psi_1^* \hat{\mathcal{O}} \Psi_2 + (\hat{\mathcal{O}} \Psi_1)^\dagger \Psi_2\right]. \tag{4.44}$$

This is because $\mathcal{S}$, as given by (4.44), can be considered itself an observable associated with a certain linear Hermitian operator, $\hat{\mathcal{O}}^H$, in such a way that the integral over configuration space of $\Psi_1^*\hat{\mathcal{O}}^H\Psi_2$ is precisely (4.44). In particular, this associated operator reads as:

$$\hat{\mathcal{O}}^H(\vec{r},\vec{r}') = \frac{1}{2}\left[\hat{\mathcal{O}}(\vec{r}')\delta(\vec{r}'-\vec{r}) + \delta(\vec{r}'-\vec{r})\hat{\mathcal{O}}(\vec{r}')\right], \tag{4.45}$$

and, therefore, one readily notices that:

$$\int \Psi_1^*(\vec{r}')\hat{\mathcal{O}}^H(\vec{r},\vec{r}')\Psi_2(\vec{r}')d\vec{r}' = \frac{1}{2}\,\mathrm{Re}[\Psi_1^*\hat{\mathcal{O}}\Psi_2]. \tag{4.46}$$

Because $\mathcal{S}$ is an observable with respect to a linear Hermitian operator, it is called a *subobservable*, which obeys the equation of change:

$$\frac{\partial}{\partial t}\,\mathrm{Re}[\Psi_1^*\hat{\mathcal{O}}\Psi_2] = \mathrm{Re}\left[\Psi_1^*\,\frac{\partial\hat{\mathcal{O}}}{\partial t}\,\Psi_2\right] + \frac{1}{\hbar}\,\mathrm{Im}\left[\Psi_1^*\hat{\mathcal{O}}\hat{H}\Psi_2 - (\hat{H}\Psi_1)^\dagger\hat{\mathcal{O}}\Psi_2\right]. \tag{4.47}$$

Note that this equation is precisely the expectation value of $\dot{\hat{\mathcal{O}}}^H$.

Let us now consider the expression (4.44) for $\mathcal{S}$ and the particular case $\Psi_2 = \Psi_1 = \Psi$. If $\hat{\mathcal{O}} = \mathbb{I}$, the subobservable corresponds to the probability density $\mathcal{S} = \Psi^*\Psi = \rho$. Substituting this condition into Eq. 4.47, we find the well-known continuity equation:

$$\frac{\partial\rho}{\partial t} + \vec{\nabla}\cdot\vec{J} = 0, \tag{4.48}$$

where

$$\vec{J} \equiv \mathrm{Re}[\Psi^*\dot{\hat{\vec{r}}}\Psi] = \left(\frac{\hbar}{2mi}\right)\left(\Psi^*\vec{\nabla}\Psi - \Psi\vec{\nabla}\Psi^*\right) \tag{4.49}$$

is the quantum probability density current or quantum flux (see Section 4.3). The equation of change for this latter quantity is obtained assuming $\hat{\mathcal{O}} = \hat{\vec{p}}/m$:

$$\frac{\partial\vec{J}}{\partial t} + \vec{\nabla}\cdot\mathbf{F} = -\frac{1}{m}\,\rho\vec{\nabla}V, \tag{4.50}$$

where

$$\mathbf{F} = \frac{1}{4m^2}\left[\Psi^*\vec{p}\vec{p}\Psi + \Psi\vec{p}\vec{p}\Psi^* + (\vec{p}\Psi)^\dagger(\vec{p}\Psi) + (\vec{p}\Psi)(\vec{p}\Psi)^\dagger\right] \tag{4.51}$$

is a tensor quantity accounting for the flux of the quantum probability density current. According to London [38], the velocity $\vec{v}$, as defined in Section 4.3, cannot be a subobservable since there is no linear Hermitian operator $\hat{O}^H$ for which $\vec{v}$ is an expectation value. However, an equation of motion for this quantity can be obtained using both (4.48) and (4.50), which reads as:

$$m\rho \frac{d\vec{v}}{dt} = -\rho\vec{\nabla}V - \vec{\nabla} \cdot \mathbf{P}. \tag{4.52}$$

As can be readily noticed, this equation is (4.19), which arises when its right-hand side is conveniently rearranged in terms of the effective potential. As it is given, (4.52) describes an incompressible fluid, with:

$$\mathbf{P} = -\left(\frac{\hbar^2}{4m}\right)\rho\vec{\nabla}\vec{\nabla}\ln\rho = m\mathbf{F} - \frac{m}{\rho}\vec{J}\vec{J} \tag{4.53}$$

being the *quantum pressure tensor* introduced by Takabayasi in the 1950s [10, 11], whose effects will be discussed in Section 4.4.3.

4.4 Applications

Covering all aspects of the topic under consideration would take more than a single chapter due to the many different applications that have been developed in the literature. Thus, instead of trying to go for such an ambitious programme, the following sections are aimed at providing an illustration on how Bohmian mechanics has been or can be used to tackle different problems related to the aspects described in Section 4.2.

4.4.1 Time-Dependent DFT: The Quantum Hydrodynamic Route

Although it is not well known in the Bohmian literature, one of the former (and earlier) applications of Bohmian mechanics in its hydrodynamic version was the design of time-dependent DFT algorithms [213–233]. The formal grounds of this approach, known as *quantum fluid dynamics DFT* (QFD-DFT), rely on a set of hydrodynamical equations [214–218], having the advantage of being able to deal with dynamical processes evolving in time in terms of single-particle time-dependent equations [227–232]. As happens with time-independent DFT, the density is also determined by solving a single-particle noninteracting time-dependent Schrödinger equation [217], which approximately, but rather accurately, deals with both static and dynamic correlations between electrons. This methodology thus facilitates the computation of ab initio electron densities for N-electron systems, with the advantage of delivering both density and energy with a superior accuracy to HF methods at a comparable computational overload. For a more detailed historical account on this approach and its achievements, the interested reader may consult Ref. [234].

In order to understand how QFD-DFT works, consider the polynuclear and polyelectronic time-dependent Schrödinger equation:

$$\begin{aligned} &\hat{H}\Psi(\vec{R}_1, \vec{R}_2, \ldots, \vec{R}_M, q_1, q_2, \ldots, q_N, t) \\ &= i\hbar\frac{\partial\Psi(\vec{R}_1, \vec{R}_2, \ldots, \vec{R}_M, q_1, q_2, \ldots, q_N, t)}{\partial t}. \end{aligned} \tag{4.54}$$

As in time-independent DFT, the purpose of time-dependent DFT is to provide solutions to the electronic part of Ψ, also assumed to be separable, as (4.6), although now:

$$|\phi_{\mathrm{e,N}}(q_1, q_2, \ldots, q_N, t)|^2 dq_1 dq_2 \cdots dq_N dt, \tag{4.55}$$

represents the probability that electrons 1, 2, ..., N will be found simultaneously within the volume elements $dq_1, dq_2, \ldots, dq_N$, respectively, during the time interval dt. Taking this into account and then further proceeding as in DFT, one finds that the time-dependent polyelectronic wave function can be expressed as a combination of product of single monoelectronic time-dependent states:

$$\phi_{\mathrm{e,N}}(q_1, q_2, \ldots, q_N, t) \quad \longrightarrow \quad \Pi_{i=1}^{N}\phi_{\mathrm{e,1,i}}(q_i, t), \tag{4.56}$$

and the time plays the role of a parameter. In this regard, considering different times could be associated, within the time-independent DFT approach, with assuming different states, each labeled with a different index k. The process to obtain the electronic density is thus similar to that in time-independent DFT, but considering subsequent times, which is done within the quantum hydrodynamics framework, where the quantum probability is understood as a quantum fluid.

As an illustration of the method, we are going to analyze the calculation of a density for an N-electron system under the influence of an external time-dependent periodic force, formerly studied by Bartolotti [214–216]. In the time-independent case, assuming the electrons are noninteracting, we need N orbitals to describe the electrons, this being equivalent to assume that each electron can be described by N different orthonormal orbitals ϕ_i (i.e., $\langle\phi_i|\phi_k\rangle = \delta_{ik}$), and therefore:

$$\sum_{i=1}^{N} |\phi_i(\vec{r})|^2 = \rho(\vec{r}) \tag{4.57}$$

is the "exact" density of the system. As in standard DFT [235], the ϕ_i are obtained after minimization of the kinetic energy of the N-electron system:

$$T_s[\{\phi_i\}] = -\frac{1}{2}\sum_{i=1}^{N}\int \phi_i(\vec{r})\left(\nabla^2\phi_i(\vec{r})\right) d\vec{r}, \tag{4.58}$$

which renders the Euler–Lagrange equation:

$$-\frac{1}{2}\nabla^2\phi_i + v_{\text{eff}}\phi_i = \epsilon_i\phi_i. \tag{4.59}$$

As can be noticed, (4.59) is a single-particle time-independent Schrödinger equation, where v_{eff} is an effective potential including the external (nuclear) interaction as well as the averaged action of the remaining N-1 electrons, and ϵ_i is the associated eigenenergy—ϵ_i can also be interpreted as the Lagrange multiplier that insures (4.57) is satisfied by the ϕ_i. Also note in (4.59) that, if this equation is divided by ϕ_i, it will read as:

$$Q_i + v_{\text{eff}} = \epsilon_i, \tag{4.60}$$

where:

$$Q_i(\vec{r}) = -\frac{1}{2}\frac{\nabla^2\phi_i(\vec{r})}{\phi_i(\vec{r})} \tag{4.61}$$

is an *effective quantum potential* associated with the state ϕ_i. This means that the sum of this quantum potential to the effective one results in the bound-state energy ϵ_i.

Now, in the time-dependent case, consider the system is described by the (time-dependent) orbitals $\phi_i(\vec{r}, t)$, from which we want to obtain the (time-dependent) density $\rho(\vec{r}, t)$. In polar form these orbitals read as:

$$\phi_i(\vec{r}, t) = R_i(\vec{r}, t)\, e^{iS_i(\vec{r}, t)}, \tag{4.62}$$

with:

$$\int_t \int R_i(\vec{r}, t) R_k(\vec{r}, t) d\vec{r} = \delta_{ik}, \tag{4.63}$$

where $\int_t$ denotes a time-averaged integration over one period of time. The kinetic energy associated with this (noninteracting) N-electron system reads [215] as:

$$T_s[\{R_i, S_i\}]_t = -\frac{1}{2}\sum_{i=1}^{N} \int_t \int \Big\{ R_i(\vec{r}, t)\nabla^2 R_i(\vec{r}, t) \\ - R_i^2(\vec{r}, t)\left(\vec{\nabla} S_i(\vec{r}, t)\right)^2 \Big\} d\vec{r}. \tag{4.64}$$

Here, the constraints are:

$$\sum_{i=1}^{N} R_i^2(\vec{r}, t) = \rho(\vec{r}, t), \tag{4.65}$$

as in the time-independent case, and the conservation of the number of particles:

$$\sum_{i=1}^{N} \frac{\partial R_i^2}{\partial t} \left(= \frac{\partial \rho}{\partial t} \right) = -\vec{\nabla} \cdot \vec{J}, \tag{4.66}$$

where $\vec{J}$ is the single-particle quantum probability density current vector. Thus, after minimizing (4.64) with respect to the R_i (subjected to the previous constraints), we find the Euler–Lagrange equation:

$$-\frac{1}{2}\nabla^2 R_i + v_{\text{eff}} R_i = \epsilon_i R_i, \tag{4.67}$$

where $v_{\text{eff}}(\vec{r}, t)$ is the Lagrange multiplier associated with the constraint defined in (4.65), and the $\epsilon_i(\vec{r}, t)$ are the Lagrange multipliers associated with the conservation of the number of particles given by (4.63) and (4.66). The $\epsilon_i(\vec{r}, t)$ can be split out as a sum of two terms:

$$\epsilon_i(\vec{r}, t) = \epsilon_i^{(0)} + \epsilon_i^{(1)}(\vec{r}, t), \tag{4.68}$$

where $\epsilon_i^{(0)}$ arise from the normalization constraint and $\epsilon_i^{(1)}$ are the Lagrange multipliers associated with the charge-current conservation defined by (4.66). Also note that, analogously to (4.59), (4.67) can now be expressed as:

$$Q_i(\vec{r}, t) = -\frac{1}{2}\frac{\nabla^2 R_i(\vec{r}, t)}{R_i(\vec{r}, t)} \tag{4.69}$$

if we divide by R_i, where Q_i is the time-dependent effective quantum potential associated with the state ϕ_i.

On the other hand, minimizing $T_s[\{R_i, S_i\}]_t$ with respect to S_i, subject to the constraint:

$$\frac{\partial S_i}{\partial t} = -\epsilon_i, \tag{4.70}$$

gives rise to the Euler–Lagrange equation:

$$\frac{\partial R_i^2}{\partial t} + \vec{\nabla} \cdot (R_i^2 \vec{\nabla} S_i) = 0. \tag{4.71}$$

The scheme based on the coupled equations (4.67) and (4.71) thus provides a means of determining the exact time-dependent density of the system, noticing that:

$$\vec{J}(\vec{r}, t) = \sum_{i=1}^{N} R_i^2(\vec{r}, t)\vec{\nabla} S_i(\vec{r}, t). \tag{4.72}$$

Moreover, if the time-dependence is turned off, the QFD-DFT approach correctly reduces to the standard time-independent DFT results: since $\vec{\nabla} S_i$ vanishes, (4.66), (4.70) and (4.71) will be identically satisfied and (4.64) will reduce to the time-independent kinetic energy of the N-electron system.

More recently, Elsayed et al. [236] have proposed a Bohmian-based many-body approach to study the breathing dynamics of a many-boson system enclosed in a trap with both long and short range interactions, without appealing to typical mean field approximations [237, 238]. This approach is based on the concept of conditional wave function [196] and its accuracy has been tested by comparing with the standardized, numerically exact multiconfigurational time-dependent Hartree (MCDTH) method [239].

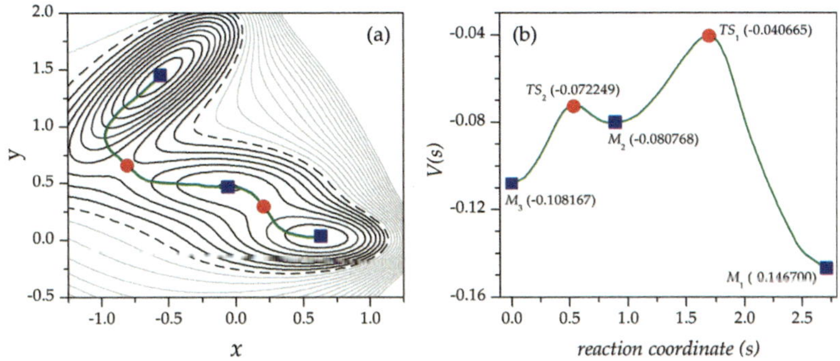

Figure 4.1 (a) Contour plot of the Müller–Brown potential energy surface and reaction path (green thick line) joining the reactants minimum with the products one. In the way how the surface is defined, the black contours represent equipotential lines for negative values of the energy, while the gray ones stand for positive energies; the zero-energy separatrix appears as a dashed line. (b) Energy profile along the reaction path. In both panels, the maxima indicate the position of transition states (TS) and are denoted by red circles, while the minima associated with the different molecular states (M) are indicated with blue squares.

4.4.2 Bound System Dynamics: Chemical Reactivity

As an example of bound dynamics describing a chemical reaction, let us consider a prototype potential model in reaction-path finding, namely the Müller–Brown potential energy surface [240–244]. As seen in Fig. 4.1b, this potential energy surface has three minima (blue squares), M_1, M_2 and M_3, corresponding to the products, intermediate and reactants states, respectively, and two transition states (red circles), TS_1 and TS_2, separating products from preequilibrium and the latter from reactants, respectively. The physical meaning of these singular points is better understood by inspecting the energy profile displayed in Fig. 4.1b. This profile is taken along the reaction path (green line), plotted as a function of the arc-length, which is defined in simple (and standard) terms as:

$$s(x, y) \approx \sum_{i=1}^{N} \sqrt{\Delta x_i^2 + \Delta y_i^2} = \sum_{i=1}^{N} \sqrt{(x_i - x_{i-1})^2 + (y_i - y_{i-1})^2}, \tag{4.73}$$

where the initial point of the curve, (x_0, y_0), is M_3, the final point is M_1, and $(x, y) = (x_N, y_N)$ is some intermediate point along the path. Furthermore, the reaction path in this case corresponds to the steepest descent curve from the relative maxima (TSs) to the adjacent minima (Ms).

Regarding the dynamical analysis, the Müller–Brown potential energy surface is assumed to describe a process governed by a proton transfer, for instance, Schrödinger's equation is applied to a particle with mass $m = 1,836$ (atomic units, where $m_e = 1$ is the electron mass). Quantum-mechanically, the initial state is specified by an initial Gaussian wave packet:

$$\Psi_0(x, y) = \frac{1}{\sqrt{2\pi\sigma_x\sigma_y}} e^{-(x-x_0)^2/4\sigma_x^2-(y-y_0)^2/4\sigma_y^2+ip_{x,0}(x-x_0)/\hbar+ip_{y,0}(y-y_0)/\hbar}, \tag{4.74}$$

with $\sigma_x^2 = \sigma_y^2 = \sigma_0^2 = 0.0125$ and $(p_{x,0}, p_{y,0}) = (-p_0, p_0)$. Physically, from a classical view point, this wave packet is assumed to represent a Gaussian random distribution of initial positions around the minimum of the products state, and a (Gaussian) Boltzmann distribution of initial momenta [which can be readily obtained from the momentum representation of the wave packet (4.74)], with dispersion $\hbar/2\sigma_0$ along both directions and average momenta $(p_{x,0}, p_{y,0})$.

The value p_0 is considered as a control parameter to determine the effectiveness of the transfer process as a function of the initial energy carried by the proton. The initial conditions for the Bohmian trajectories are obtained by randomly sampling $\rho_0 = |\Psi_0|^2\delta(p_x - p_{x,0})\delta(p_y - p_{y,0})$. To compare with, alternative classical statistical simulations are also considered in order to emphasize different quantum-classical aspects. To this end, two samples of classical trajectories are considered. In one of them, trajectories are initially distributed according to the Wigner distribution associated with (4.74):

$$\begin{aligned}\rho_{0,cl}^{(1)} &= \rho_W(x, p_x, y, p_y) \\ &\propto e^{-(x-x_0)^2/2\sigma_x^2-\sigma_x^2(p_x-p_{x,0})^2/\hbar^2-(y-y_0)^2/2\sigma_y^2-\sigma_y^2(p_y-p_{y,0})^2/\hbar^2}.\end{aligned} \tag{4.75}$$

This distribution introduces a dispersion in momenta such that the classical average energy:

$$\bar{E}_{cl} = \int E(x, p_x, y, p_y)\rho_W(x, p_x, y, p_y)dxdydp_xdp_y$$
$$= \frac{p_0^2}{m} + \bar{V} + \bar{\delta} \approx \frac{1}{N}\sum_{i=1}^{N} E_i(x_0^i, y_0^i), \quad (4.76)$$

coincides with the (quantum) expectation value of the energy, $\bar{E}_{\mathrm{q}} = \langle \hat{H} \rangle$. In Eq. 4.76, $\bar{V} = \sum_{i=1}^{N} V_i(x_0^i, y_0^i)/N$ is the average value of the potential energy, $\bar{\delta} = \hbar^2/4m\sigma^2$ is a sort of internal energy related to the *spreading ratio* of a Gaussian wave packet [245], and the last sum runs over all (classical/quantum) particles considered [with (x_0^i, y_0^i) denoting their corresponding initial positions]. In the second ensemble, the classical trajectories are distributed according to $\rho_{0,cl}^{(2)} = \rho_0$, just as in the Bohmian case. By means of a simple calculation, it can be shown that $\bar{\delta}$ vanishes and therefore the average (ensemble) energy becomes smaller.

The first calculation considered is a measure of reaction probabilities, which provides information about the amount of products formed with time, as well as the formation rate or reaction velocity if we measure the slope of such a function. The reaction probability is defined in the form of a restricted norm [244, 246]:

$$\mathcal{P}(t) \equiv \int_{\Sigma} |\Psi(x, y, t)|^2 dxdy, \quad (4.77)$$

where Σ is the space region above the border line separating products from preequilibrium/reactants, here chosen to be:

$$y_{R\rightarrow P}(x) = 0.8024x + 1.2734. \quad (4.78)$$

From a Bohmian viewpoint, Eq. 4.77 has to be interpreted [244, 246] as the fraction of trajectories N_{Σ} that penetrate into Σ at a time t with respect to the total number N initially considered:

$$\mathcal{W}(t) \equiv \frac{N_{\Sigma}(t)}{N}. \quad (4.79)$$

This quantity will approach $\mathcal{P}(t)$ in the limit $N \rightarrow \infty$ whenever the initial conditions of the trajectories will be sampled according to ρ_0. Classically, the analog of Eq. 4.79 is also considered to determine a classical rate of production of products, using a subscript "cl" to distinguish this quantity from its quantum-mechanical counterpart [i.e., $\mathcal{W}_{cl}(t)$ for the classical fraction]. Probability can flow backwards from products to reactants [1–3], mainly in bound potentials, thus leading to multiple crossings of $y_{R\to P}(x)$ by the same quantum/classical trajectory. However, working with individual trajectories brings in an advantageous feature: one can determine uniquely when a single particle is in the products region, and therefore neglect its count in (4.79) at subsequent times. Hence, another interesting quantity is the fraction of trajectories going from reactants to products without recrossing the border line at subsequent times:

$$\bar{\mathcal{W}}(t) \equiv \frac{\bar{N}_\Sigma(t)}{N}. \tag{4.80}$$

Assuming one could extract the products formed during the reaction by some chemical or physical procedure, $\bar{\mathcal{W}}(t)$ would provide the maximum amount of products at each time and, at $t \to \infty$, it would render the maximum amount of products which can be extracted from the reaction given a certain initial state.

The fractions $\mathcal{W}$ (solid line) and $\bar{\mathcal{W}}$ (dashed line) are displayed in Figs. 4.2a and b for $p_0 = 4$ and $p_0 = 10$, respectively. In each panel the three different initial distributions mentioned above are shown: Bohmian trajectories according to ρ_0 (circles) and classical trajectories according to $\rho_{0,cl}^{(1)}$ (triangles) and $\rho_{0,cl}^{(2)}$ (inverted triangles), using a total of 5×10^4 trajectories in all cases. For $p_0 = 4$, $\bar{E} - \bar{\delta}$ is well below TS_2 and TS_1, and therefore dynamics should mainly proceed via tunneling according to standard quantum mechanics. This means that one would expect $\mathcal{W}$ to be larger asymptotically than $\mathcal{W}_{cl}^{(1)}$ and $\mathcal{W}_{cl}^{(2)}$. However, in Fig. 4.2a we observe that $\mathcal{W}$ is between $\mathcal{W}_{cl}^{(1)}$ and $\mathcal{W}_{cl}^{(2)}$, i.e., there is an excess of classical trajectories which can pass the transition states and reach products. The reason for this behavior is that classical distributions can explore many initial conditions, which eventually may imply energies of individual trajectories higher than the energy of the transition states. Eventually this leads to the formation of products even in cases where the average (ensemble) energy is below the energy of the transition states. This effect will be more relevant in

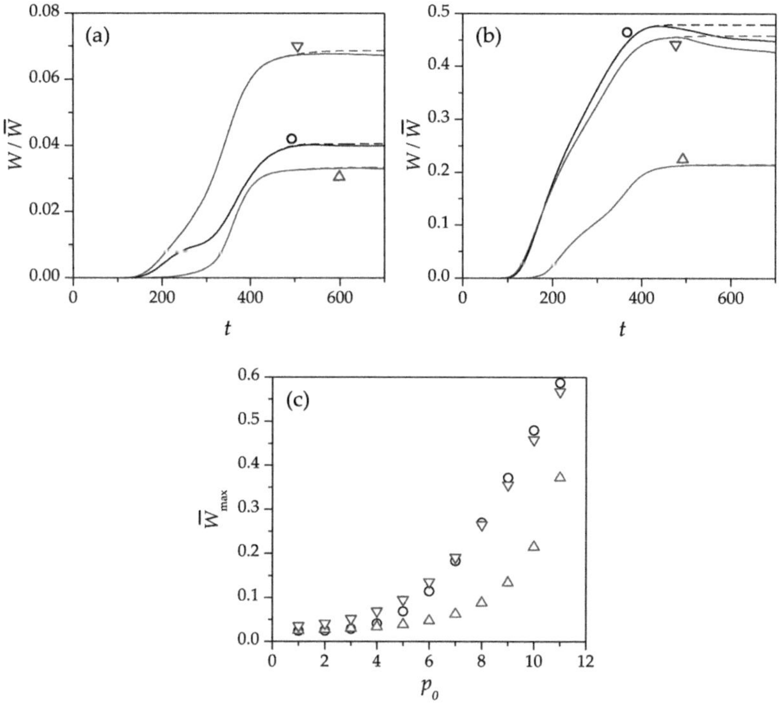

Figure 4.2 Reaction probabilities $\mathcal{W}$ (solid curve) and $\bar{\mathcal{W}}$ (dashed curve) for (a) $p_0 = 4$ and (b) $p_0 = 10$, and three different initial distributions: quantum trajectories distributed according to ρ_0 (circles) and classical trajectories distributed according to ρ_0 (triangles) and ρ_W (inverted triangles). In panel (c), $\bar{\mathcal{W}}$ at $t \approx 700$ as a function of p_0: $\bar{\mathcal{W}}_{\text{Bohm}}$ (circles), $\bar{\mathcal{W}}^{cl}_{\text{Bohm}}$ (triangles) and $\bar{\mathcal{W}}^{cl}_{\text{Wigner}}$ (inverted triangles).

the case of trajectories distributed according to $\rho^{(1)}_{0,cl}$ than in the case of $\rho^{(2)}_{0,cl}$. On the other hand, for Bohmian trajectories the dynamics is very different due to the presence of the quantum potential: for low p_0, the wave packet spreads faster than it propagates [245], thus favoring the formation of ripples by interference which will hinder the passage of Bohmian trajectories to products — remember that in Bohmian mechanics tunneling cannot be understood as in standard quantum mechanics, but as a process mediated by an "effective" time-dependent barrier [247], V_{eff}. Nevertheless, in both cases, quantum and classical, we note that the maximum formation of products, $\bar{\mathcal{W}}$ is almost the same as $\mathcal{W}$.

For $p_0 = 10$, however, $\bar{E} - \bar{\delta}$ is above TS_1, and hence a larger amount of products is expected, as it can be seen in Fig. 4.2b, where we observe that the Bohmian distribution and the classical one $\rho_{0,cl}^{(1)}$ provide similar values (although the latter goes below the former). In this case, although tunneling may still be active, the direct passage is going to control the dynamics in both cases, classical and quantum mechanical. Note that in the Bohmian case the propagation is now faster than the spreading of the wave packet and, therefore, more trajectories can be promoted to products before interference starts to influence the dynamics. Regarding $\bar{\mathcal{W}}$, we find a trend similar to $\mathcal{W}$, but the difference between the asymptotic values of these magnitudes has increased due to the larger energy (on average) carried by the particles, which favors the recrossing. Only for the distribution $\rho_{0,cl}^{(2)}$ such a difference is negligible, since there is not much energy in excess.

In order to compare the maximum amount of products formed in a more general way, in Fig. 4.2c there is a comparison among the three initial distributions for a range of p_0. As it can be noticed, the formation of products is more efficient classically than quantum-mechanically for low values of p_0, while this behavior changes as p_0 becomes larger (the switching appears [244] for $p_0 \sim 8$, just when $\bar{E}$ approximately coincides with the energy of the transition state T_1, which connects the preequilibrium with products).

The dynamics associated with the process here described are illustrated in Fig. 4.3, where a series of snapshots display the time evolution of the probability density associated with a Gaussian wave packet (equally spaced darker contours) starting on the reactants minimum with $p_0 = 9$. As time proceeds, the wave packet evolves essentially along the direction indicated by the reaction path (thicker solid line). However, as can be seen, particularly in the plots at $t = 100$, 200, and 300, the motion of the wave packet is quite similar to that shown by a water stream when flowing along a river bed, trying to burst its banks each time it reaches a meander. Specifically, at $t = 300$, we observe that the main stream deviates remarkably from the reaction path, although later this quantum fluid gets

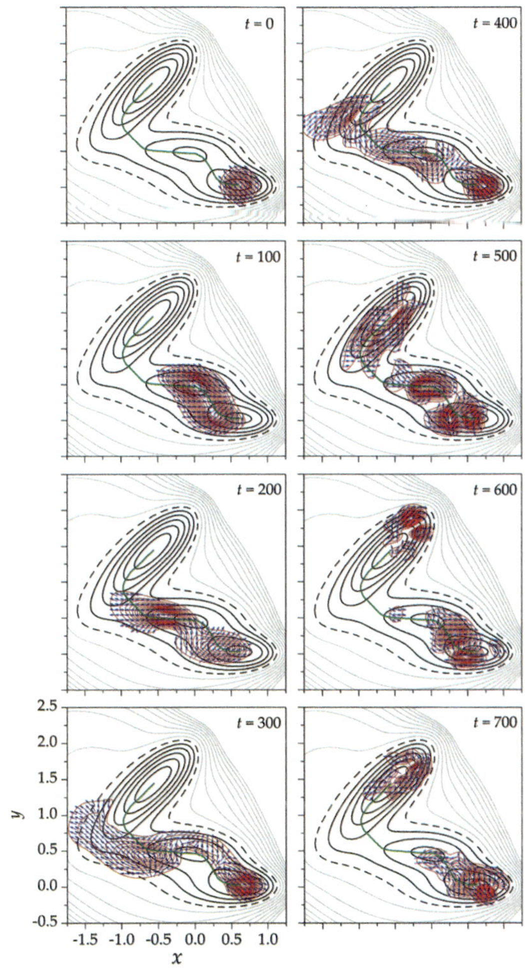

Figure 4.3 Time evolution of the probability density (red contours) associated with an initial Gaussian wave packet starting on the reactants minimum with $p_0 = 9$ (see text for details). The superimposed (blue) arrows show the directionality of the quantum flux, thus providing a clue on the ensemble (Bohmian) trajectory dynamics. Several contours of the Müller–Brown potential energy surface (with the same meaning as in Fig. 4.3) as well as the reaction path have also been included for a better understanding of the time evolution of the probability density and the associated flux.

back and moves towards the products region (see plot at $t = 500$). This deviation beyond the reaction path, which becomes larger with p_0, was called *quantum bobsled effect* by Marcus [248]. This effect, observed by McCullough and Wyatt [1–3] when studying the dynamics of the collinear $H+H_2$ reaction dynamics, arises from a direct analogy to the deviation undergone by a bobsled from the center of the track where it moves through due to its inertia (a *positive centrifugal effect*, according to Marcus [248]). Following with the analogy of the river, we observe that, after some time, when the passage to products has occurred the quantum fluid evolves more slowly and only shape changes are noticeable, just the same case after a region of "brave waters". At this stage, if the dynamics continues, there can be some evolution from products to reactants (and vice versa), but there are no important inertial effects.

The information provided by the probability density results very valuable in order to determine the evolution of the system. However, Bohmian trajectories present some advantages which cannot be noticed studying only the wave packet dynamics. The case analyzed here is a clear example. Instead of displaying trajectories, which would result quite messy after some time, we have superimposed an arrow map on the region occupied by the wave packet at each time. The direction of the arrows indicate at each particular point of space the directionality of the vector field $\vec{v}$ and, therefore, the local direction of the flow (in this regard, Bohmian trajectories will be tangent to these vectors). According to these maps, we observe how from a situation where all the arrows are aligned along the same direction at $t = 0$, we pass to another at $t = 100$, where part of the arrows point downwards and part along the direction indicated by the reaction path. This happens precisely because the wave packet has collided with the first "meander"; something similar can be seen at $t = 200$, when the wave packet collides with the second "meander" and so on. The arrows also allow us to detect the presence of *quantum vorticity* [12, 249, 250], i.e., the vortical motion that appears whenever the wave function displays a node. In this case, the arrows twist around the node, giving rise to a *quantum whirlpool effect* [1–3].

4.4.3 Scattering Dynamics: Young's Two-Slit Experiment

In order to illustrate a practical application of Bohmian mechanics now in the case of a scattering system, we are going to consider a model of Young's experiment with electrons where the interaction between the electrons and the two slits is described by a repulsive potential energy surface (barrier) [119, 185]:

$$V(x, y) = \left(V_0 - \frac{1}{2} m\omega^2 y^2 + \frac{m^2 \omega^4 y^4}{16 V_0} \right) e^{-x^2/\alpha^2}, \tag{4.81}$$

where $\alpha = 25$, $\omega = 600$, $V_0 = 8000$, and m is the electron mass. As the ingoing wave function we consider a quasi-plane or quasi-monochromatic initial wave function, with energy $\langle E \rangle_i \simeq 500$, which is constructed by linearly superimposing a series of identical Gaussian wave packets along the direction parallel to the y-axis. This wave function is then launched perpendicularly to the double-slit from a distance $\langle x \rangle_0 = -400$ (far enough from the potential energy surface, where $V \approx 0$). The interest in this kind of models relies on the fact that physical diffracting systems are constituted by atoms or molecules which interact with the diffracted electrons and, therefore, a description of the experiment including the interaction potentials or potential energy surfaces results closer to the real experiment [251–254]. In Fig. 4.4a, a contour plot of the probability density after the collision and diffraction from the two slits is shown. As can be noted, for this incidence energy, there is a large portion of the wave function which is back-scattered, while an angular distribution of very well-defined diffracted (forward-scattered) peaks appear behind the slits (for $y > 0$). The corresponding angular diffraction pattern is represented in Fig. 4.4b, where the Gaussian-like envelope can be associated with the particular form chosen for the "holes" of the slits [144, 185].

In order to understand the dynamics that is taking place within this diffraction scenario (but that can also be extrapolated to any other scattering scenario), now we consider it under a Bohmian or quantum hydrodynamical, where the probability density describes a swarm of noninteracting particles. Thus, consider the three panels of Fig. 4.5, where only a half of the trajectories (those corresponding to the upper slit) have been represented to make clearer the figures (moreover, their incident parts are not shown either), taking

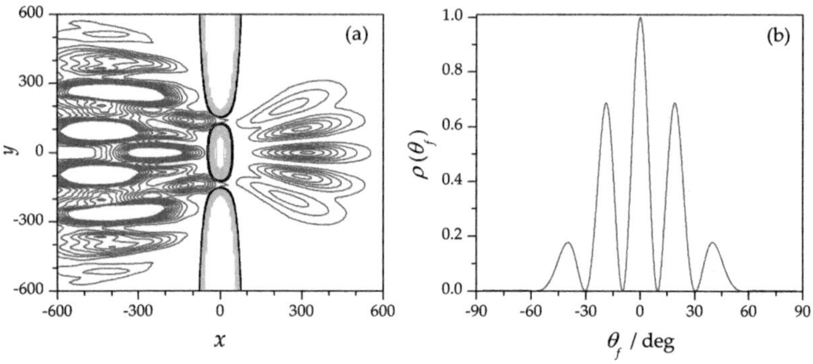

Figure 4.4 (a) Contour plot of the probability density associated with an electron crossing the two slits modeled by the potential energy surface (4.81) at the end of the simulation, i.e., the time when the interference pattern is already formed [119, 185]. Given the high-backscattering component, for a better understanding the contours are truncated to a certain value that allows an optimal visualization of the transferred density. Moreover, the contour plot of the potential energy surface is also displayed, with black contours for equipotentials lesser than or equal to $\langle E \rangle_i \simeq 500$, and gray for higher values. (b) Angular distribution of the intensity computed at the end of the simulation (see text for details).

advantage of the reflection symmetry with respect to $y = 0$ (in Bohmian mechanics, the trajectories coming from each slit cannot intersect because of the *noncrossing flux* property [245]; see below). The ensembles of trajectories in each panel have been chosen to sample three regions of the initial probability density, i.e., varying their initial position y_0, but with the same x_0-position: (a) $x_0 = \langle x \rangle_0 - 100$, (b) $x_0 = \langle x \rangle_0$, and (c) $x_0 = \langle x \rangle_0 + 100$. As can be readily seen, the initial position is fundamental in the quantum dynamics displayed by the trajectories, for it leads to very different behaviors due to the action of Q, something which does not happen in classical mechanics (provided classical trajectories are launched far enough from the interaction region, their behavior would be independent on their x_0-position [246]). Note that those trajectories launched at distances further away from the double-slit barrier will not reach it (see Fig. 4.5a), while only those started at closer distances will be able to surmount it (see Fig. 4.5c). In the case of the trajectories started at intermediate positions (see Fig. 4.5b), they will remain

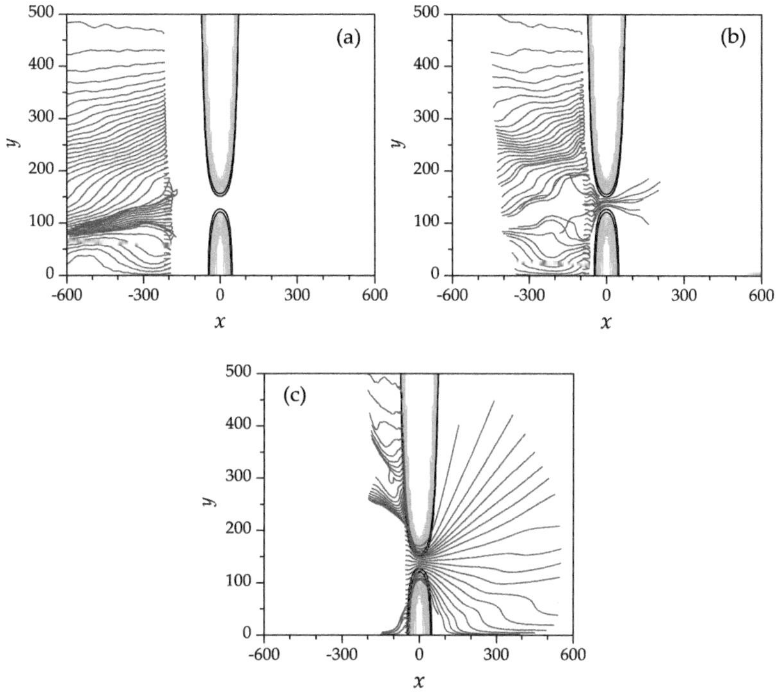

Figure 4.5 Bohmian trajectories illustrating the probability flow in the Young-type experiment with the potential energy surface (4.81). Three different sets of initial conditions have been chosen, which are shown in the three corresponding panels. In these panels, the initial value along the y direction is always the same, with x_0 ranging from 0 to about 500, while the three specific values have been chosen for y_0 with respect to the initial probability density: (a) at the rearmost part, (b) at the center, and (c) in the foremost part (see text for details). The propagation is up to the final time in the simulation, which corresponds to the case shown in Fig. 4.4. For a better visualization in panels (a) and (b), only the back-scattered or diffracted part of the trajectories has been plotted. Moreover, the contour plot of the potential energy surface is also displayed, with black contours for equipotential lines lesser than or equal to $\langle E \rangle_i \simeq 500$, and gray for higher values.

near the barrier until the "pressure" exerted by those coming behind will diminish and either they will be able to move backwards or pass through the slits. This is a general effect which can also be observed in other scattering problems [246, 249], which can be understood in terms of a sort of *quantum pressure* (within a hydrodynamical viewpoint). The presence of a quantum pressure can be better understood when the quantum Euler equation (4.19) is recast in terms of the velocity components. Then, the quantum force (arising from the quantum potential) can be rearranged [10–12] as:

$$F_i = -\frac{1}{\rho}\frac{\partial \sigma_{ik}}{\partial r_k}, \tag{4.82}$$

where i, $k = x, y, z$, with:

$$\sigma_{ik} \equiv -\frac{\hbar^2 \rho}{4m}\frac{\partial^2 \ln \rho}{\partial r_i \partial r_k} \tag{4.83}$$

being a quantum pressure tensor, in analogy to the classical pressure tensor $p\delta_{ik}$. Thus, taking into account this hydrodynamical scenario, it is clear that electrons will tend to move towards those regions with lower values of the quantum pressure.

The number of particles passing through the slits is a function of the energy E_i as well as the particular form of the potential energy surface. However, given a configuration of the potential energy surface, it is clear that for a certain incidence energy, those electrons associated with the rearmost part of the incident probability density (swarm) will be more likely to get diffracted. The fraction of diffracted electrons will then increase as the incidence energy also increases, which means that electrons coming from layers behind will also start to get diffracted. But, more importantly, by means of this Bohmian picture, we are able to determine which electrons are going to contribute to diffraction and, more specifically, to which diffraction peak [255], something that cannot be known within the standard quantum framework. As seen in Fig. 4.5c, the interference behind the slits manifests by grouping or channeling the trajectories along the diffraction angles (along which electrons display an essentially free motion) undergoing a strong "repulsive"-like behavior whenever they approach regions of fast variation of the quantum potential. The formation of these channels within the so-called Fraunhofer region [255] is a direct consequence of the information that the quantum potential transmits to particles about the status of each slit (either open or close). This is also related with the noncrossing flux property [245] of Bohmian mechanics mentioned above, which arises from the fact that two (Bohmian) trajectories cannot coincide on the same space point at the same time due to the single-valuedness of S (except on a nodal point). In this way, for the symmetric (with respect to the $y = 0$) double slit we are considering, all those trajectories started above $y = 0$ will not be able to cross to the region $y < 0$, for this would violate the noncrossing flux property.

4.4.4 Effective Dynamical Treatments: Decoherence and Reduced Bohmian Trajectories

An interesting description of Bohmian mechanics arises from the field of decoherence and the theory of open quantum systems. Here, in order to extract useful information about the system of interest, one usually computes its associated reduced density matrix by tracing the total density matrix, $\hat{\rho}(t)$, over the environment degrees of freedom. In the configuration representation and for an environment constituted by N particles, the system reduced density matrix is obtained after integrating $\hat{\rho}(t) \equiv |\Psi(t)\rangle\langle\Psi(t)|$ over the $3N$ environment degrees of freedom, $\{\vec{r}_i\}_{i=1}^{N}$:

$$\tilde{\rho}(\vec{r}, \vec{r}', t) = \int \langle \vec{r}, \vec{r}_1, \vec{r}_2, \ldots, \vec{r}_N | \Psi(t)\rangle \\ \times \langle \Psi(t) | \vec{r}', \vec{r}_1, \vec{r}_2, \ldots, \vec{r}_N \rangle d\vec{r}_1 d\vec{r}_2 \cdots d\vec{r}_N. \quad (4.84)$$

The system (reduced) quantum density current can be derived from this expression, being:

$$\tilde{\vec{J}}(\vec{r}, t) \equiv \frac{\hbar}{m} \operatorname{Im}[\vec{\nabla}_{\vec{r}} \tilde{\rho}(\vec{r}, \vec{r}', t)] \Big|_{\vec{r}'=\vec{r}}, \quad (4.85)$$

which satisfies the continuity equation:

$$\dot{\tilde{\rho}} + \vec{\nabla} \tilde{\vec{J}} = 0. \quad (4.86)$$

In Eq. 4.86, $\tilde{\rho}$ is the diagonal element (i.e., $\tilde{\rho} \equiv \tilde{\rho}(\vec{r}, \vec{r}, t)$) of the reduced density matrix and gives the measured intensity [256].

Taking into account Eqs. 4.85 and 4.86, now we define the velocity field, $\dot{\vec{r}}$, associated to the (reduced) system dynamics as:

$$\tilde{\vec{J}} = \tilde{\rho} \dot{\vec{r}}, \quad (4.87)$$

which is analogous to the Bohmian velocity field. Now, from Eq. 4.87, we define a new class of quantum trajectories as the solutions to the equation of motion:

$$\dot{\vec{r}} \equiv \frac{\hbar}{m} \frac{\text{Im}[\vec{\nabla}_{\vec{r}}\tilde{\rho}(\vec{r}, \vec{r}', t)]}{\text{Re}[\tilde{\rho}(\vec{r}, \vec{r}', t)]}\Bigg|_{\vec{r}'=\vec{r}} . \quad (4.88)$$

These new trajectories are the so-called *reduced quantum trajectories* [194, 195], which are only related to the system reduced density matrix. As shown in [194, 195], the dynamics described by Eq. 4.88 leads to the correct intensity (with its time evolution being described by Eq. 4.86) when the statistics of a large number of particles is considered. Moreover, it is also straightforward to show that Eq. 4.88 reduces to the well-known expression for the velocity field in Bohmian mechanics when there is no interaction with the environment. An application of this approach to the analysis of dephasing by incoherence in Young's experiment and its influence on the transverse momentum transfer [257] can be found in [258].

4.4.5 Pathways to Complex Molecular Systems: Mixed Bohmian-Classical Mechanics

As mentioned in Section 4.2.4, there are different hybrid approaches to deal with many degree-of-freedom systems, where the system degrees of freedom are described quantum-mechanically, while the bath ones are accounted for classically. Among them, we find methods such as the mean field approximation [159] or the surface hopping trajectories [160]. In these methods, the key point is the implementation of the so-called *back-reaction* [259], i.e., the action of the system over the bath, since the contrary is simple and it is usually done in terms of a time-dependent potential, which is a function of the bath coordinates (following the Born–Oppenheimer approximation scheme). Here, we are going to analyze this problem within the framework of the so-called mixed quantum Bohmian approach [260–263].

Consider a two interacting systems X and Y (or degrees of freedom), each specified by the coordinates x and y, respectively, and with associated masses m_x and m_y. Proceeding as in Section 4.3, with the polar ansatz (4.11) and the corresponding two-dimensional

Schrödinger equation, Eqs. 4.12 and 4.13 read as:

$$\frac{\partial S}{\partial t} + \frac{1}{2m_x}\left(\frac{\partial S}{\partial x}\right)^2 + \frac{1}{2m_y}\left(\frac{\partial S}{\partial y}\right)^2 = -V_{\text{eff}}, \tag{4.89}$$

$$\frac{\partial R^2}{\partial t} + \frac{1}{m_x}\frac{\partial S}{\partial x}\frac{\partial R^2}{\partial x} + \frac{1}{m_y}\frac{\partial S}{\partial y}\frac{\partial R^2}{\partial y} = -R^2\left(\frac{1}{m_x}\frac{\partial^2 S}{\partial x^2} + \frac{1}{m_y}\frac{\partial^2 S}{\partial y^2}\right), \tag{4.90}$$

where the quantum potential has the form:

$$Q(x, y, t) = -\frac{\hbar^2}{2m_x}\frac{1}{R}\frac{\partial^2 R}{\partial x^2} - \frac{\hbar^2}{2m_y}\frac{1}{R}\frac{\partial^2 R}{\partial y^2}, \tag{4.91}$$

which is nonfactorizable (see Section 4.3.2). Let us recast Eq. 4.89 in its Eulerian form by applying the operators $\partial/\partial x$ and $\partial/\partial y$ to both sides of this equation. This gives rise to two coupled equations:

$$\frac{\partial}{\partial t}\left(\frac{\partial S}{\partial x}\right) + v_x\frac{\partial^2 S}{\partial x^2} + v_y\frac{\partial^2 S}{\partial y \partial x} = -\frac{\partial V_{\text{eff}}}{\partial x}, \tag{4.92}$$

$$\frac{\partial}{\partial t}\left(\frac{\partial S}{\partial y}\right) + v_y\frac{\partial^2 S}{\partial y^2} + v_x\frac{\partial^2 S}{\partial x \partial y} = -\frac{\partial V_{\text{eff}}}{\partial y}, \tag{4.93}$$

where $v_x = p_x/m_x = (1/m_x)(\partial S/\partial x)$ and $v_y = p_y/m_y = (1/m_y)(\partial S/\partial y)$. Taking into account the definition of the Lagrange time-derivative (see Section 4.3), these two equations can be recast in a more familiar Newtonian form:

$$m_x\frac{d^2 x}{dt^2} = -\frac{\partial V_{\text{eff}}}{\partial x}, \tag{4.94}$$

$$m_y\frac{d^2 y}{dt^2} = -\frac{\partial V_{\text{eff}}}{\partial y}. \tag{4.95}$$

So far no approximation has been invoked. Let us now consider, for instance, that $m_y \gg m_x$, which physically means that for the timescale ruling the behavior of X, the system Y behaves almost classically. This slower evolution along the y coordinate means that the second space-derivatives of S and R with respect to this coordinate will be negligible. Accordingly, Eqs. 4.92 and 4.93 can be recast as:

$$\frac{\partial}{\partial t}\left(\frac{\partial \tilde{S}}{\partial x}\right) + \frac{1}{m_x}\frac{\partial \tilde{S}}{\partial x}\frac{\partial^2 \tilde{S}}{\partial x^2} + \frac{1}{m_y}\frac{\partial \tilde{S}}{\partial y}\frac{\partial^2 \tilde{S}}{\partial y \partial x} = -\frac{\partial \tilde{V}_{\text{eff}}}{\partial x}, \tag{4.96}$$

$$\frac{\partial}{\partial t}\left(\frac{\partial \tilde{S}}{\partial y}\right) + \frac{1}{m_x}\frac{\partial \tilde{S}}{\partial x}\frac{\partial^2 \tilde{S}}{\partial x \partial y} = -\frac{\partial \tilde{V}_{\text{eff}}}{\partial y}, \tag{4.97}$$

where $\tilde{S}$ and $\tilde{R}$ denote the approximate values of S and R, respectively. In these equations, $\tilde{V}_{\text{eff}}$ is the corresponding approximate effective potential, with:

$$\tilde{Q}(x, t|y) = -\frac{\hbar^2}{2m_x}\frac{1}{\tilde{R}}\frac{\partial^2 \tilde{R}}{\partial x^2}, \tag{4.98}$$

where $(x, t|y)$ means that $\tilde{Q}$ depends on y implicitly through a sort of parametrization—as it also happens in the Born–Oppenheimer approximation (see Section 4.2.1). On the other hand, (4.90) becomes the approximate continuity equation:

$$\frac{\partial \tilde{R}^2}{\partial t} + \frac{\partial}{\partial x}\left(\frac{\tilde{R}^2}{m_x}\frac{\partial \tilde{S}}{\partial x}\right) + \frac{1}{m_y}\frac{\partial \tilde{S}}{\partial y}\frac{\partial \tilde{R}^2}{\partial y} = 0. \tag{4.99}$$

Evaluating (4.96) and (4.99) along the quasi-classical trajectory $y(t)$ allows us to define the pseudo-Lagrangian time-derivative operator:

$$\frac{d}{dt} = \frac{\partial}{\partial t} + v_y\frac{\partial}{\partial y}, \tag{4.100}$$

and, therefore, to recast those equations as:

$$\frac{d}{dt}\left(\frac{\partial \tilde{S}}{\partial x}\right) + \left(\frac{1}{m_x}\frac{\partial \tilde{S}}{\partial x}\right)\left(\frac{\partial^2 \tilde{S}}{\partial x^2}\right) = -\frac{\partial \tilde{V}_{\text{eff}}}{\partial x}, \tag{4.101}$$

$$\frac{d\tilde{R}^2}{dt} + \frac{\partial}{\partial x}\left(\tilde{R}^2\ \frac{1}{m_x}\frac{\partial \tilde{S}}{\partial x}\right) = 0, \tag{4.102}$$

which satisfy the pseudo-Schrödinger equation:

$$i\hbar\ \frac{d\tilde{\Psi}(x, y(t), t)}{dt} = \left[-\frac{\hbar^2}{2m_x}\frac{\partial^2}{\partial x^2} + V(x, y(t))\right]\tilde{\Psi}(x, y(t), t), \tag{4.103}$$

where $\tilde{\Psi} = \tilde{R}e^{i\tilde{S}/\hbar}$. As can be noticed, the dimensionality of the full quantum problem reduces to the (subspace) dimensionality associated with X, since the classical-like coordinate y acts as a time-dependent parameter (the external potential V has become time-dependent in virtue of this parametrization). On the other hand, Y evolves according to a quasi-classical Newtonian equation:

$$m_y\frac{d^2 y}{dt^2} = \frac{\partial}{\partial y}\left(V(x, y, t) + \tilde{Q}(x, t|y)\right), \tag{4.104}$$

which arises from Eq. 4.97 after applying the pseudo-Lagrangian operator (note that it can also be obtained from Eq. 4.95 after the corresponding approximation is considered) and is integrated after getting the solution $\tilde{\Psi}(x, y(t), t)$ from (4.103).

As briefly mentioned above, the method here described constitutes a sort of translation of the Born–Oppenheimer rule to the case of slow and fast degrees of freedom. This quantum-classical approach is commonly used when dealing with quantum open systems, whenever some degrees of freedom display a slower dynamics than other, which are faster and in many cases can be treated even as a surrounding noise function (thermal bath). Within the Bohmian literature, approaches to deal with the many-body problem have been suggested, for instance, by Garashchuk and Rassolov [264–266], Bittner [267], Makri [268–270] and, more recently, Franco [271].

4.5 Concluding Remarks

Looking back, if there is a healthy and increasing interest at present in Bohmian mechanics, it could be said it partly comes from the important numerical work developed by Wyatt and colleagues since the end of the 1990s. Up to that time Bohmian mechanics had been applied to some physical cases of interest, disproving that trajectories were not incompatible at all with quantum mechanics beyond just metaphysical discussions. From a pragmatic viewpoint, leaving aside interpretational issues concerning hidden variables, the reality of the wave function and the quantum realm, such important contributions were probably marginalized because, in the end, they were only another way to describe what was already known. If a formulation renders the same, what is it worth for? Beyond quantum foundations aspects, this might have been a wondering question—a rather reasonable and understandable one, after all—in people's minds (reviewers, many of them) that made them reluctant to accept a trajectory-based description of quantum phenomena. Notice that something similar also happened when the first time-dependent quantum-mechanical simulations started appearing by the end of the 1960s [272], making possible the impossible, namely providing a neat and unambiguous visualization of the full quantum process, since the initial state up to the final one. Quantum mechanics was not a "black box" anymore, as claimed by Bohr and followers over decades. With time it was understood that these pictures detailing the process could be used to extract valuable information about it, which could not be done (or not, at least, in an easy manner) with the other more widespread time-independent techniques. Nowadays this type of simulations constitute a consolidated standard tool to approach quantum-mechanical systems, from simple academic-type problems to very complex ones. Something similar has also happened with Bohmian mechanics as soon as it has started being perceived as a worth exploring and exploiting numerical methodology, particularly in chemistry, as Wyatt and other pioneers in the field have shown along the about last twenty years. In this regard, it is possible to find Bohmian-based algorithms to determine electronic structure as well as to describe the dynamics of many-body systems.

Besides the computational aspect, the interpretational one has also gained in relevance over the last decades, just as soon as relatively complex (more realistic) problems have started being considered, from atomic and molecular scattering to nanodevices, from nonlinear optics to electron microscopy. For years, molecular dynamics and statistics have mainly relied on the computation of classical trajectories (based on both Newtonian and Hamiltonian schemes) and, in general, the tools of classical mechanics. Good for the purpose, but approximate (at different degrees, depending on the method and/or problem considered) because they lack the essence of quantum mechanics, namely coherence and all the related properties, such as interference, tunneling, or entanglement. With the extensive use and application of quantum mechanics in the development of electronic structure methods, first, and then in molecular dynamics through wave packet propagation methods, better and more accurate solutions were obtained, of course. However, interpretations of the corresponding outcomes still had to make use of classical argumentations based on the longstanding quantum-classical correspondence thinking established by Bohr back in the 1920s. Bohmian mechanics has allowed to change the landscape, providing a correct trajectory-based interpretation for quantum phenomena (compared to classical approaches), that is, in consonance with the evolution of the corresponding wave functions and without introducing any extra *ad hoc* artifacts (e.g., quantization rules over classical ensembles). Of course, this does not mean at all that we should now make the mistake of only seeking for Bohmian trajectories to the detriment of trying to still establish bridges between quantum and classical dynamics with the aid of classical trajectories. So far, the quantum world has been and still is disconcerting and challenging to us (and perhaps it will remain so for long), hence any tool to explore it should always be welcome rather than banned.

Getting back to chemistry, it is perhaps pertinent to conclude this chapter quoting the very last paragraph of Walter Kohn's Nobel Lecture [273]:

> Looking into the future I expect that wave function-based and density based theories will, in complementary ways, continue not only to give us quantitatively more accurate results, but also contribute to a better physical/chemical understanding of the electronic structure of matter.

Probably Bohmian mechanics, that is, the hydrodynamic picture of quantum mechanics will play a major role in this regard. This is something yet to come, although some signs are already on the way, principally the change in the perception we have at present of this approach.

Acknowledgments

This work has been financially supported by the Spanish MINECO (grant No. FIS2016-76110-P).

References

1. E. A. McCullough and R. E. Wyatt, *Quantum Dynamics of the Collinear (H, H_2) Reaction*, Journal of Chemical Physics, **51**, 1253 (1969).

2. E. A. McCullough and R. E. Wyatt, *Dynamics of the Collinear $H + H_2$ Reaction. I. Probability Density and Flux*, Journal of Chemical Physics, **54**, 3578 (1971).
3. E. A. McCullough and R. E. Wyatt, *Dynamics of the Collinear $H + H_2$ Reaction. II. Energy Analysis*, Journal of Chemical Physics, **54**, 3592 (1971).
4. J. O. Hirschfelder, A. C. Christoph, and W. E. Palke, *Quantum Mechanical Streamlines. I. Square Potential Barrier*, Journal of Chemical Physics, **61**, 5435 (1974).
5. J. O. Hirschfelder, C. J. Goebel, and L. W. Bruch, *Quantized Vortices Around Wavefunction Nodes. II*, Journal of Chemical Physics, **61**, 5456 (1974).
6. J. O. Hirschfelder and K. T. Tang, *Quantum Mechanical Streamlines. III. Idealized Reactive Atom-Diatomic Molecule Collision*, Journal of Chemical Physics, **64**, 760 (1976).
7. D. Bohm, *A Suggested Interpretation of the Quantum Theory in Terms of "Hidden" Variables. I*, Physical Review, **85**, 166 (1952).
8. D. Bohm, *A suggested Interpretation of the Quantum Theory in Terms of "Hidden" Variables. II*, Physical Review, **85**, 180 (1952).
9. D. Bohm, *Proof That Probability Density Approaches $|\psi|^2$ in Causal Interpretation of the Quantum Theory*, Physical Review, **89**, 458 (1953).
10. T. Takabayasi, *On the Formulation of Quantum Mechanics Associated with Classical Pictures*, Progress in the Theory of Physics, **8**, 143 (1952).
11. T. Takabayasi, *Remarks on the Formulation of Quantum Mechanics with Classical Pictures and on Relations between Linear Scalar Fields and Hydrodynamical Fields*, Progress in the Theory of Physics, **9**, 187 (1953).
12. P. R. Holland, *The Quantum Theory of Motion* (Cambridge University Press, Cambridge, 1993).
13. D. Dürr, *Bohmsche Mechanik als Grundlage der Quantenmechanik* (Springer, Berlin, 2001).
14. D. Dürr and S. Teufel, *Bohmian Mechanics* (Springer, Berlin, 2009).
15. R. E. Wyatt, *Quantum Dynamics with Trajectories* (Springer, New York, 2005).
16. A. S. Sanz and S. Miret-Artés, *A Trajectory Description of Quantum Processes. I. Fundamentals*, Lecture Notes in Physics **850** (Springer, Berlin, 2012).

17. A. S. Sanz and S. Miret-Artés, *A Trajectory Description of Quantum Processes. II. Applications*, Lecture Notes in Physics **831** (Springer, Berlin, 2014).

18. A. S. Sanz, *Bohm's Approach to Quantum Mechanics: Alternative Theory or Practical Picture?*, Frontiers of Physics, **14**, 11301 (2019).

19. E. Madelung, *Quantentheorie in hydrodynamischer Form*, Zeitschrift für Physik, **40**, 332 (1926).

20. P. K. Chattaraj, *Quantum Trajectories* (CRC Taylor and Francis, New York, 2010).

21. K. H. Hughes and G. Parlant, *Quantum Trajectories* (CCP6, Daresbury, UK, 2011).

22. J. P. Dowling and G. J. Milburn, *The Second Quantum Revolution*, Philosophical Transactions of the Royal Society of London, Series A, **361**, 1655 (2003).

23. M. A. Nielsen and I. L. Chuang, *Quantum Computation and Quantum Information* (Cambridge University Press, Cambridge, 2000).

24. J. Olson, Y. Cao, J. Romero, P. Johnson, P.-L. Dallaire-Demers, N. Sawaya, P. Narang, I. Kivlichan, M. Wasielewski, and A. Aspuru-Guzik, *Quantum Information and Computation for Chemistry*, NSF Workshop Report, arXiv:1706.05413 (2017).

25. M. E. Eberhart, *Quantum Mechanics and Molecular Design in the Twenty First Century*, Foundations of Chemistry, **4**, 201 (2002).

26. M. E. Eberhart and D. P. Clougherty, *Looking for Design in Materials Design*, Nature Materials, **3**, 659 (2004).

27. A. Lehmann and C. D. Maranas, *Molecular Design Using Quantum Chemical Calculations for Property Estimation*, Industrial & Engineering Chemistry Research, **43**, 3419 (2004).

28. T. Zhou, D. Huang, and A. Caflisch, *Quantum Mechanical Methods for Drug Design*, Current Topics in Medicinal Chemistry, **10**, 33 (2010).

29. M. Pavone, A. M. Ritzmann, and E. A. Carter, *Quantum-Mechanics-Design Principles for Solid Oxide Fuel Cell Cathode Materials*, Energy & Environmental Science, **4**, 4933 (2011).

30. N. Lambert, Y.-N. Chen, Y.-C. Cheng, C.-M. Li, G.-Y. Chen, and F. Nori, *Quantum Biology*, Nature Physics, **9**, 10 (2012).

31. Th. Ritz, S. Adem, and K. Schulten, *A Model for Photoreceptor-Based Magnetoreception in Birds*, Biophysical Journal, **78**, 707 (2000).

32. G. S. Engel, T. R. Calhoun, E. L. Read, T.-K. Ahn, T. Mančal, Y.-C. Cheng, R. E. Bankenship, and G. R. Fleming, *Evidence for Wavelike Energy Transfer*

through Quantum Coherence in Photosynthetic Systems, Nature, **446**, 782 (2007).

33. M. Sarovar, A. Ishizaki, G. R. Fleming, and K. B. Whaley, *Quantum Entanglement in Photosynthetic Light-Harvesting Complexes*, Nature Physics, **6**, 462 (2010).
34. I. Bialynicki-Birula, M. Cieplak, and J. Kamisnki, *Theory of Quanta* (Oxford University Press, Oxford, 1992).
35. R. Fürth, *Über einige Beziehungen zwischen klassischer Statistik und Quantenmechanik*, Zeitschrift für Physik, **81**, 143 (1933).
36. G. G. Comisar, *Brownian-Motion Model of Nonrelativistic Quantum Mechanics*, Physical Review, **138**, B1332 (1965).
37. L. Landau, *The Theory of Superfluidity of Helium II*, Journal of Physics (USSR), **5**, 71 (1941).
38. F. London, *Planck's Constant and Low Temperature Transfer*, Review of Modern Physics, **17**, 310 (1945).
39. L. de Broglie, *Remarques sur la nouvelle Mecánique ondulatoire*, Comptes Rendes, **183**, 273 (1926).
40. L. de Broglie, *Sur la possibilité de relier les phénomènes d'interférence et de diffraction à la théorie des quanta de lumière*, Comptes Rendes, **183**, 447 (1926).
41. L. de Broglie, *Sur le rôle des ondes continues Ψ en Mécanique ondulatoire*, Comptes Rendes, **185**, 380 (1927).
42. T. H. Solomon, E. R. Weeks, and H. L. Swinney, *Observation of Anomalous Diffusion and Lévy Flights in a Two-Dimensional Rotating Flow*, Physical Review Letters, **71**, 3975 (1993).
43. J. C. Sommerer, H.-C. Ku, and H. E. Gilreath, *Experimental Evidence for Chaotic Scattering in a Fluid Wake*, Physical Review Letters, **77**, 5055 (1996).
44. A. S. Sanz and S. Miret-Artés, *Quantum Phase Analysis with Quantum Trajectories: A Step towards the Creation of a Bohmian Thinking*, American Journal of Physics, **80**, 525 (2012).
45. A. S. Sanz, *Quantumness beyond Quantum Mechanics*, Journal of Physics: Conference Series, **361**, 012016 (2012).
46. Y. Couder, S. Protière, E. Fort, and A. Boudaoud, *Dynamical Phenomena: Walking and Orbiting Droplets*, Nature, **437**, 208 (2005)
47. S. Protière, A. Boudaoud, and Y. Couder, *Particle-Wave Association on a Fluid Interface*, Journal of Fluid Mechanics, **554**, 85 (2006).

48. Y. Couder and E. Fort, *Single-Particle Diffraction and Interference at a Macroscopic Scale*, Physical Review Letters, **97**, 154101 (2006)
49. S. Protière and Y. Couder, *Orbital Motion of Bouncing Drops*, Physics of Fluids, **18**, 091114 (2006).
50. A. Eddi, E. Fort, F. Moisy, and Y. Couder, *Unpredictable Tunneling of a Classical Wave-Particle Association*, Physical Review Letters, **102**, 240401 (2009).
51. E. Fort, A. Eddi, A. Boudaoud, J. Moukhtar, and Y. Couder, *Path-Memory Induced Quantization of Classical Orbits*, Proceedings of the National academy of Sciences of the United States of America, **108**, 17515 (2010).
52. Y. Couder, A. Boudaoud, S. Protière, and E. Fort, *Walking Droplets: A form of Wave-Particle Duality at Macroscopic Scale?*, Europhysics News, **41**, 14 (2010).
53. J. W. M. Bush, *Quantum Mechanics Writ Large*, Proceedings of the National academy of Sciences of the United States of America, **107**, 17455 (2010).
54. D. M. Harris, J. Moukhtar, E. Fort, Y. Couder, and J. W. M. Bush, *Wavelike Statistics from Pilot-Wave Dynamics in a Circular Corral*, Physical Review E, **88**, 011001(R) (2013).
55. J. W. M. Bush, *Pilot-Wave Hydrodynamics*, Annual Review of Fluid Mechanics, **47**, 269 (2015).
56. W. H. Zurek and J. A. Wheeler, *Quantum Theory of Measurement* (Princeton University Press, Princeton, NJ, 1983).
57. B. G. Englert, M. O. Scully, G. Süssmann, and H. Walther, *Surrealistic Bohm Trajectories*, Zeitschrift fur Naturforschung, **47a**, 1175 (1992).
58. R. Courant and D. Hilbert, *Methods of Mathematical Physics* (Wiley-Interscience, New York, 1966).
59. H. Weyl, *Quantenmechanik und Gruppentheorie*, Zeitschrift für Physik, **46**, 1 (1927).
60. E. P. Wigner, *On the Quantum Correction for Thermodynamic Equilibrium*, Physical Review, **40**, 749 (1932).
61. J. E. Moyal, *Quantum Mechanics as a Statistical Theory*, Mathematical Proceedings of the Cambridge Philosophical Society, **45**, 99 (1949).
62. R. E. Wyatt, *Quantum Wavepacket Dynamics with Trajectories: Wavefunction Synthesis Along Quantum Paths*, Chemical Physics Letters, **313**, 189 (1999).

63. C. L. Lopreore and R. E. Wyatt, *Quantum Wave Packet Dynamics with Trajectories*, Physical Review Letters, **82**, 5190 (1999).
64. C. L. Lopreore and R. E. Wyatt, *Quantum Wave Packet Dynamics with Trajectories: Reflections on a Downhill Ramp Potential*, Chemical Physics Letters, **325**, 73 (2000).
65. Y. Goldfarb, I. Degani, and D. J. Tannor, *Bohmian Mechanics with Complex Action: A New Trajectory-Based Formulation of Quantum Mechanics*, Journal of Chemical Physics, **125**, 231103 (2006).
66. Y. Goldfarb and D. J. Tannor, *Interference in Bohmian Mechanics with Complex Action*, Journal of Chemical Physics, **127**, 161101 (2007).
67. Y. Goldfarb, J. Schiff, and D. J. Tannor, *Unified Derivation of Bohmian Methods and the Incorporation of Interference Effects*, Journal of Chemical Physics, **111**, 10416 (2007).
68. Y. Goldfarb, I. Degani, and D. J. Tannor, *Semiclassical Approximation with Zero Velocity Trajectories*, Chemical Physics, **338**, 106 (2007).
69. C.-C. Chou and R. E. Wyatt, *Computational Method for the Quantum Hamilton–Jacobi Equation: Bound States in One Dimension*, Journal of Chemical Physics, **125**, 174103 (2006).
70. C.-C. Chou and R. E. Wyatt, *Quantum Trajectories in Complex Space: One-Dimensional Stationary Scattering Problems*, Journal of Chemical Physics, **128**, 154106 (2008).
71. C.-C. Chou and R. E. Wyatt, *Quantum Streamlines within the Complex Quantum Hamilton–Jacobi Formalism*, Journal of Chemical Physics, **129**, 124113 (2008).
72. C.-C. Chou and R. E. Wyatt, *Complex-Extended Bohmian Mechanics*, Journal of Chemical Physics, **132**, 134102 (2010).
73. J. S. Bell, *Speakable and Unspeakable in Quantum Mechanics* (Cambridge University Press, Cambridge, 1987).
74. M. Born and J. R. Oppenheimer, *Zur Quantentheorie der Molekeln*, Annalen der Physik, **84**, 457 (1927).
75. B. H. Bransden and C. J. Joachain, *Physics of Atoms and Molecules* (Longman Scientific & Technical, Essex, 1983).
76. I. N. Levine, *Quantum Chemistry* (Prentice Hall, Upper Saddle River, NJ, 2000), 5th Ed.
77. A. Szabo and N. S. Ostlund, *Modern Quantum Chemistry* (Dover, Publications, Mineola, NY, 1996).
78. P. Fulde, *Electron Correlations in Molecules and Solids* (Springer, Berlin, 2002), 3rd Ed.

79. W. Koch and M. C. Holthausen, *A Chemist's Guide to Density Functional Theory* (Wiley-VCH, Weinheim, 2001), 2nd Ed.
80. R. M. Martin, *Electronic Structure: Basic Theory and Practical Methods* (Cambridge University Press, Cambridge, 2004).
81. I. G. Kaplan, *Intermolecular Interactions: Physical Picture, Computational Methods and Model Potentials* (John Wiley & Sons, Chichester, 2006).
82. J. Z. H. Zhang, *Theory and Application of Quantum Molecular Dynamics* (World Scientific, Singapore, 1999).
83. D. J. Tannor, *Introduction to Quantum Mechanics* (University Science Books, Sausalito, CA, 2006).
84. W. H. Louisell, *Quantum Statistical Properties of Radiation* (John Wiley & Sons, New York, 1973).
85. D. A. McQuarrie, *Statistical Mechanics* (Harper & Row, New York, 1976).
86. U. Weiss, *Quantum Dissipative Systems* (World Scientific, Singapore, 2008), 3rd Ed.
87. H.-P. Breuer and F. Petruccione, *The Theory of Open Quantum Systems* (Oxford University Press, New York, 2002).
88. H. Carmichael, *An Open Systems Approach to Quantum Optics* (Springer-Berlag, Berlin, 1993).
89. I. Percival, *Quantum State Diffusion* (Cambridge University Press, Cambridge, 1998).
90. H. Nakai, *Simultaneous Determination of Nuclear and Electronic Wave Functions without Born–Oppenheimer Approximation: Ab initio NO+MO/HF Theory*, International Journal of Quantum Chemistry, **86**, 511 (2002).
91. H. Nakai, *Nuclear Orbital Plus Molecular Orbital Theory: Simultaneous Determination of Nuclear and Electronic Wave Functions without Born–Oppenheimer Approximation*, International Journal of Quantum Chemistry, **107**, 2849 (2007).
92. L. Veis, J. Višňák, H. Nishizawa, H. Nakai, and J. Pittner, *Quantum Chemistry beyond Born–Oppenheimer Approximation on a Quantum Computer: A Simulated Phase Estimation Study*, International Journal of Quantum Chemistry, **116**, 1328 (2016).
93. M. Head-Gordon, *Quantum Chemistry and Molecular Processes*, Journal of Physical Chemistry, **100**, 13213 (1996).
94. S. S. Shaik and P. C. Hiberty, *A Chemist's Guide to Valence Bond Theory* (Wiley-Interscience, New Jersey, 2007).

95. L. Pauling, *The Application of the Quantum Mechanics to the Structure of the Hydrogen Molecule and Hydrogen Molecule-Ion and to Related Problems*, Chemical Reviews, **5**, 173 (1928).
96. L. Pauling, *The Nature of the Chemical Bond. Application of Results Obtained from the Quantum Mechanics and from a Theory of Paramagnetic Susceptibility to the Structure of Molecules*, Journal of the American Chemical Society, **53**, 1367 (1931).
97. K. Blum, *Density Matrix Theory and Applications* (Plenum Press, New York, 1981).
98. L. H. Thomas, *The Calculation of Atomic Fields*, Proceedings of the Cambridge Philosophical Society, **23**, 542 (1927).
99. E. Fermi, *Un metodo statistico per la determinazione di alcune priopretà dell'atomo*, Rendus Accademia Nazionale dei Lincei, **6**, 602 (1927).
100. M. A. L. Marques and E. K. U. Gross, *Time-Dependent Density Functional Theory*, Annual Review of Physical Chemistry, **55**, 427 (2004).
101. S. Botti, A. Schindlmayr, R. Del Sole, and L. Reining, *Time-Dependent Density-Functional Theory for Extended Systems*, Reports on the Progress in Physics, **70**, 357 (2007).
102. H. Nakatsuji, *Equation for Direct Determination of Density Matrix*, Physical Review A, **14**, 41 (1976).
103. H. Nakatsuji and K. Yasuda, *Direct Determination of the Quantum-Mechanical Density Matrix Using the Density Equation*, Physical Review Letters, **76**, 1039 (1996).
104. K. Yasuda and H. Nakatsuji, *Direct Determination of the Quantum-Mechanical Density Matrix Using the Density Equation. 2*, Physical Review A, **56**, 2648 (1997).
105. C. Valdemoro, *Approximating the 2nd-Order Reduced Density-Matrix in Terms of the 1st-Order One*, Physical Review A, **45**, 4462 (1992).
106. C. Valdemoro, *Contracting and Calculating Traces over the N-Electron Space: Two Powerful Tools for Obtaining Averages*, International Journal of Quantum Chemistry, **60**, 131 (1996).
107. C. Valdemoro, *Electron Correlation and Reduced Density Matrices*, in *Correlation and Localization*, P. R. Surjan (Ed.) (Springer, Berlin, 1999), pp. 187–200.
108. C. Valdemoro, L. M. Tel, D. R. Alcoba, E. Pérez-Romero, and F. J. Casquero, *Some Basic Properties of the Correlation Matrices*, International Journal of Quantum Chemistry, **90**, 1555 (2002).

109. D. R. Alcoba and C. Valdemoro, *Spin Structure and Properties of the Correlation Matrices Corresponding to Pure Spin States: Controlling the S-Representability of These Matrices*, International Journal of Quantum Chemistry, **102**, 629 (2005).

110. E. R. Davidson, *Reduced Density Matrices in Quantum Chemistry* (Academic Press, New York, 1976).

111. A. J. Coleman and V. I. Yukalov, *Reduced Density Matrices: Coulson's Challenge* (Springer-Verlag, New York, 2000).

112. J. Cioslowski (Ed.), *Many-Electron Densities and Reduced Density Matrices* (Kluwer, Dordrecht, 2000).

113. D. A. Mazziotti, *Contracted Schrödinger Equation: Determining Quantum Energies and Two-Particle Density Matrices without Wave Functions*, Physical Review A, **57**, 4219 (1998).

114. D. A. Mazziotti, *Pursuit of N-Representability for the Contracted Schrödinger Equation through Density-Matrix Reconstruction*, Physical Review A, **60**, 3618 (1999).

115. D. A. Mazziotti, *Anti-Hermitian Contracted Schrödinger Equation: Direct Determination of the Two-Electron Reduced Density Matrices of Many-Electron Molecules*, Physical Review Letters, **97**, 143002 (2006).

116. D. A. Mazziotti (Ed.), *Reduced-Density-Matrix Mechanics with Applications to Many-Electron Atoms and Molecules*, Advanced Chemical Physics, **134** (Wiley, New York, 2007).

117. P. Sherwood, *Hybrid Quantum Mechanics/Molecular Mechanics Approaches*, in *Modern Methods and Algorithms of Quantum Chemistry*, J. Grotendorst (Ed.) (John von Neumann Institute for Computing, Jülich, 2000), 2nd Ed.

118. M. Born and E. Wolf, *Principles of Optics* (Cambridge University Press, Cambridge, 1999), 7th Ed.

119. R. Guantes, A. S. Sanz, J. Margalef-Roig, and S. Miret-Artés, *Atom-Surface Diffraction: A Trajectory Description*, Surface Science Reports, **53**, 199 (2004).

120. A.S. Sanz and S. Miret-Artés, *Selective Adsorption Resonances: Quantum and Stochastic Approaches*, Physics Reports, **451**, 37 (2007).

121. P. M. Morse and H. Feshbach, *Methods of Theoretical Physics* (McGraw Hill, New York, 1953).

122. M. Razavy, *Quantum Theory of Tunneling* (World Scientific, New Jersey, 2003).

123. J. Ankerhold, *Quantum Tunneling in Complex Systems* in *Springer Tracts in Modern Physics* (Springer, New York, 2007), p. 224.
124. H. Jeffreys, *On Certain Approximate Solutions of Linear Differential Equations of the Second Order*, Proceedings of the London Mathematical Society (2), **23**, 428 (1925).
125. G. Wentzel, *Eine Verallgemeinerung der Quantenbedingungen für die Zwecke der Wellenmechanik*, Zeitschrift für Physik, **38**, 518 (1926).
126. H. A. Kramers, *Wellenmechanik und halbzahlige Quantisierung*, Zeitschrift für Physik, **39**, 828 (1926).
127. L. Brillouin, *La mécanique ondulatoire de Schrödinger; une méthode générale de résolution par approximations successives*, Comptes Rendus **183**, 24 (1926).
128. L. Brillouin, *Sur un type gènéral de problèmes, permettant la séparation des variables dans la mécanique ondulatoire de Schrödinger*, Comptes Rendus **183**, 270 (1926).
129. C. Cerjan (Ed.), *Numerical Grid Methods and Their Application to Schrödinger's Equation* (Kluwer Academic, Amsterdam, 1993).
130. N. J. Giordano, *Computational Physics* (Prentice Hall, Upper Saddle River, NJ, 1997).
131. M. C. Gutzwiller, *Chaos in Classical and Quantum Mechanics* (Sprigner-Verlag, New York, 1990).
132. E. J. Heller, *Bound-State Eigenfunctions of Classically Chaotic Hamiltonian Systems: Scars of Periodic Orbits*, Physical Review Letters, **53**, 1515 (1984).
133. R. G. Newton, *Scattering Theory of Waves and Particles* (McGraw-Hill, New York, 1966).
134. M. S. Child, *Molecular Collision Theory* (Dover, New York, 1996).
135. C. J. Ray and J. M. Bowman, *Quasiclassical Trajectory Calculations of He-LiF(001) Diffraction Scattering*, Journal of Chemical Physics, **63**, 5231 (1975).
136. C. J. Ray and J. M. Bowman, *Quasiclassical Studies of Rigid Rotor-Solid Surface Diffraction Scattering*, Journal of Chemical Physics, **66**, 1122 (1976).
137. D. Farías, C. Díaz, P. Rivière, H. F. Busnego, P. Nieto, M. F. Somers, G. J. Kroes, A. Salin, and F. Martín, *In-Plane and out-of-Plane Diffraction of H_2 from Metal Surfaces*, Physical Review Letters, **93**, 246104 (2004).

138. C. Díaz, M. F. Somers, G. J. Kroes, H. F. Busnego, A. Salin, and F. Martín, *Quantum and Classical Dynamics of H_2 Scattering from Pd(111) at Off-Normal Incidence*, Physical Review B, **72**, 035401 (2005).

139. C. Díaz, H. F. Busnego, P. Rivière, D. Farías, P. Nieto, M. F. Somers, G. J. Kroes, A. Salin, and F. Martín, *A Classical Dynamics Method for H_2 Diffraction from Metal Surfaces*, Journal of Chemical Physics, **122**, 154706 (2005).

140. G. Delgado-Barrio, P. Villarreal, P. Mareca, and G. Albelda, *Three-Dimensional Vibrational Predissociation of the van der Waals Complex He· · · $I_2(B)$. A Quasiclassical Study*, Journal of Chemical Physics, **78**, 280 (1983).

141. J. Rubayo-Soneira, A. Garía-Vela, G. Delgado-Barrio, and P. Villarreal, *Vibrational Predissociation of I_2-Ne. A Quasiclassical Dynamical Study*, Chemical Physics Letters, **243**, 236 (1995).

142. A. Garía-Vela, J. Rubayo-Soneira, G. Delgado-Barrio, and P. Villarreal, *Quasiclassical Dynamics of the I_2-Ne_2 Vibrational Predissociation: A Comparison with Experiment*, Journal of Chemical Physics, **104**, 8405 (1996).

143. C. Leforestier, R. H. Bisseling, C. Cerjan, M. D. Feit, R. Friesner, A. Guldberg, A. Hammerich, G. Jolicard, W. Karrlein, H.-D. Meyer, N. Lipkin, O. Roncero, and R. Kosloff, *A Comparison of Different Propagation Schemes for the Time Dependent Schrödinger Equation*, Journal of Computational Physics, **94**, 59 (1991).

144. A. S. Sanz, F. Borondo, and S. Miret-Artés, *Particle Diffraction Studied Using Quantum Trajectories*, Journal of Physics: Condensed Matter, **14**, 6109 (2002).

145. V. B. Magalinskii, *Dynamical Model in the Theory of the Brownian Motion*, Soviet Physics JETP-USSR, **9**, 1381 (1959).

146. A. O. Caldeira and A. J. Leggett, *Influence of Dissipation on Quantum Tunneling in Macroscopic Systems*, Physical Review Letters, **46**, 211 (1981).

147. A. O. Caldeira and A. J. Leggett, *Quantum Tunneling in a Dissipative System*, Annals of Physics (NY), **149**, 374 (1983).

148. R. K. Wangsness and F. Bloch, *The Dynamical Theory of Nuclear Induction*, Physical Review, **89**, 728 (1953).

149. A. G. Redfield, *On the Theory of Relaxation Processes*, IBM Journal of Research and Development, **1**, 19 (1957).

150. A. G. Redfield, *The Theory of Relaxation Processes*, Advanced Magnetic Resonance, **1**, 1 (1965).

151. W. T. Pollard, A. K. Felts, and R. A. Friesner, *The Redfield Equation in Condensed-Phase Quantum Dynamics*, Advanced Chemical Physics, **93**, 77 (1996).
152. G. W. Ford, M. Kac, and P. Mazur, *Statistical Mechanics of Assemblies of Coupled Oscillators*, Journal of Mathematical Physics, **6**, 504 (1965).
153. G. W. Ford and M. Kac, *On the Quantum Langevin Equation*, Journal of Statistical Physics, **46**, 803 (1987).
154. R. P. Feynman, *Space-Time Approach to Non-relativitic Quantum Mechanics*, Review of Modern Physics, **20**, 367 (1948).
155. R. P. Feynman and A. R. Hibbs, *Quantum Mechanics and Path Integrals* (McGraw-Hill, New York, 1965).
156. R. P. Feynman, *Statistical Mechanics* (W. A. Benjamin, Reading, MA, 1972).
157. L. S. Schulman, *Techniques and Applications of Path Integrals* (Wiley, New York, 1981).
158. A. S. Sanz, *Effective Markovian Description of Decoherence in Bound Systems*, Canadian Journal of Chemistry, **92**, 168 (2014).
159. G. D. Billing, *Classical Path Method in Inelastic and Reactive Scattering*, International Reviews in Physical Chemistry, **13**, 309 (1994).
160. J. C. Tully, *Nonadiabatic Molecular-Dynamics*, International Journal of Quantum Chemistry, **25**, 299 (1991).
161. W. H. Miller, *Semiclassical Theory of Atom-Diatom Collisions: Path Integrals and the Classical S Matrix*, Journal of Chemical Physics, **53**, 1949 (1970).
162. W. H. Miller, *Classical S Matrix: Numerical Application to Inelastic Collisions*, Journal of Chemical Physics, **53**, 3578 (1970).
163. W. H. Miller, *Quantum and Semiclassical Theory of Chemical Reaction Rates*, Faraday Discussions, **110**, 1 (1998).
164. W. H. Miller, *The Semiclassical Initial Value Representation: A Potentially Practical Way for Adding Quantum Effects to Classical Molecular Dynamics Simulations*, Journal of Physical Chemistry A, **105**, 2942 (2001).
165. W. H. Miller and B. M. D. D. Jansen op de Haar, *A New Basis Set Method for Quantum Scattering Calculations*, Journal of Chemical Physics, **86**, 6213 (1987).
166. J. Z. H. Zhang, S.-I. Chu, and W. H. Miller, *Quantum Scattering via the S-Matrix Version of the Kohn Variational Principle*, Journal of Chemical Physics, **88**, 6233 (1988).

167. M. A. Sepúlveda and F. Grossmann, *Time-Dependent Semiclassical Mechanics*, Advanced Chemical Physics, **96**, 191 (1996).

168. N. Makri, *Quantum Dissipative Dynamics: A Numerically Exact Methodology*, Journal of Physical Chemistry A, **102**, 4414 (1998).

169. J. Ankerhold, M. Salteer, and E. Pollak, *A study of the Semiclassical Initial Value Representation at Short Times*, Journal of Chemical Physics, **116**, 5925 (2002).

170. E. Pollak and J. Shao, *Systematic Improvement of Initial Value Representations of the Semiclassical Propagator*, Journal of Physical Chemistry A, **107**, 7112 (2003).

171. E. Pollak and S. Miret-Artés, *Thawed Semiclassical IVR Propagators*, Journal of Physics A, **37**, 9669 (2004).

172. S. Nielsen, R. Kapral, and G. Ciccotti, *Non-Adiabatic Dynamics in Mixed Quantum-Classical Systems*, Journal of Statistical Physics, **101**, 225 (2000).

173. R. Car and M. Parrinello, *Unified Approach for Molecular Dynamics and Density-Functional Theory*, Physical Review Letters, **55**, 2471 (1985).

174. T. D. Kuhne, M. Krack, F. R. Mohamed, and M. Parrinello, *Efficient and Accurate Car-Parrinello-like Approach to Born–Oppenheimer Molecular Dynamics*, Physical Review Letters, **98**, 066401 (2007).

175. E. Nelson, *Derivation of the Schrödinger Equation from Newtonian Mechanics*, Physical Review, **150**, 1079 (1966).

176. J. S. Bell, *On the Problem of Hidden Variables in Quantum Mechanics*, Review of Modern Physics, **38**, 447 (1966).

177. A. S. Sanz and S. Miret-Artés, *Aspects of Nonlocality from a Quantum Trajectory Perspective: A WKB Approach to Bohmian Mechanics*, Chemical Physics Letters, **445**, 350 (2007).

178. E. Schrödinger, *Discussion of Probability Relations between Separated Systems*, Proceedings of the Cambridge Philosophical Society, **31**, 555 (1935).

179. E. Schrödinger, *Probability Relations between Separated Systems*, Proceedings of the Cambridge Philosophical Society, **32**, 446 (1936).

180. A. S. Sanz and S. Miret-Artés, *On the Unique Mapping Relationship between Initial and Final Quantum States*, Annals of Physics, **339**, 11 (2013).

181. D. Bohm, *Quantum Theory* (Prentice-Hall, New York, 1951).

182. L. I. Schiff, *Quantum Mechanics* (McGraw-Hill, Singapore, 1968), 3rd Ed.

183. S. Kocsis, B. Braverman, S. Ravets, M. J. Stevens, R. P. Mirin, L. K. Shalm, and A. M. Steinberg, *Observing the Average Trajectories of Single Photons in a Two-Slit Interferometer*, Science, **332**, 1170 (2011).

184. A. S. Sanz, M. Davidović, M. Božić, and S. Miret-Artés, *Understanding Interference Experiments with Polarized Light through Photon Trajectories*, Annals of Physics, **325**, 763 (2010).

185. A. S. Sanz and S. Miret-Artés, *Atom-Surface Diffraction: A Quantum Trajectory Description*, in *Quantum Dynamics of Complex Molecular Systems*, D. A. Micha and I. Burghardt (Eds.) (Springer, Berlin, 2007).

186. P. M. Gerhardt, R. J. Gross, and J. I. Hochstein, *Fundamentals of Fluid Mechanics* (Addison-Wesley, New York, 1992).

187. J. H. Spurk, *Fluid Mechanics* (Springer-Verlag, Berlin, 1997).

188. A. Benseny, G. Albareda, A. S. Sanz, J. Mompart, and X. Oriols, *Applied Bohmian Mechanics*, European Physical Journal D, **68**, 286 (2014).

189. C. Dewdney, *Nonlocally Correlated Trajectories Two-Particle Quantum Mechanics*, Foundations of Physics, **18**, 867 (1988).

190. M. M. Lam and C. Dewdney, *Locality and Nonlocality in Correlated Two-Particle Interferometry*, Physics Letters A, **150**, 127 (1990).

191. E. Guay and L. Marchildon, *Two-Particle Interference in Standard and Bohmian Quantum Mechanics*, Journal of Physics A, **36**, 5617 (2003).

192. K. Na and R. E. Wyatt, *Quantum Hydrodynamic Analysis of Decoherence: Quantum Trajectories and Stress Tensor*, Physics Letters A, **306**, 97 (2002).

193. K. Na and R. E. Wyatt, *Quantum Hydrodynamic Analysis of Decoherence*, Physica Scripta, **67**, 169 (2003).

194. A. S. Sanz and F. Borondo, *A Quantum Trajectory Description of Decoherence*, European Physical Journal D, **44**, 319 (2007).

195. A. S. Sanz and F. Borondo, *Contextuality, Decoherence and Quantum Trajectories*, Chemical Physics Letters, **478**, 301 (2009).

196. X. Oriols, *Quantum-Trajectory Approach to Time-Dependent Transport in Mesoscopic Systems with Electron-Electron Interactions*, Physical Review Letters, **98**, 066803 (2007).

197. Y. Aharonov, D. Z. Albert, and L. Vaidman, *How the Result of a Measurement of a Component of the Spin of a Spin*-$\frac{1}{2}$ *Particle can Turn Out to be 100*, Physical Review Letters, **60**, 1351 (1988).

198. J. S. Lundeen, B. Sutherland, A. Patel, C. Stewart, and Ch. Bamber, *Direct Measurement of the Quantum Wavefunction*, Nature, **474**, 188 (2011).

199. H. M. Wiseman, *Grounding Bohmian Mechanics in Weak Values an Bayesianism*, New Journal of Physics, **9**, 165 (2007).

200. B. J. Hiley, *Weak values: Approach through the Clifford and Moyal algebras*, Journal of Physics: Conference Series, **361**, 012014 (2012).

201. J. O. Hirschfelder, *Quantum Mechanical Equations of Change. I*, Journal of Chemical Physics, **68**, 5151 (1978).

202. R. F. W. Bader, *Quantum Topology of Molecular Charge Distributions. III. The Mechanics of an Atom in a Molecule*, Journal of Chemical Physics, **73**, 2871 (1980).

203. J. A. N. F. Gomes, *Delocalized Magnetic Currents in Benzene*, Journal of Chemical Physics, **78**, 3133 (1983).

204. J. A. N. F. Gomes, *Topological Elements of the Magnetically Induced Orbital Current Densities*, Journal of Chemical Physics, **78**, 4585 (1983).

205. P. Lazzeretti, *Ring Currents*, Progress in Nuclear Magnetic Resonance Spectroscopy, **36**, 1 (2000).

206. S. Pelloni, F. Faglioni, R. Zanasi, and P. Lazzeretti, *Topology of Magnetic-Field-Induced Current-Density Field in Diatropic Monocyclic Molecules*, Physical Review A, **74**, 012506 (2006).

207. S. Pelloni, P. Lazzeretti, and R. Zanasi, *Spatial Ring Current Model of the [2.2]Paracyclophane Molecule*, Journal of Physical Chemistry A, **111**, 3110 (2007).

208. S. Pelloni, P. Lazzeretti, and R. Zanasi, *Topological Models of Magnetic Field Induced Current Density Field in Small Molecules*, Theoretical Chemistry Accounts, **123**, 353 (2009).

209. S. Pelloni and P. Lazzeretti, *Spatial Ring Current Model for the Prismane Molecule*, Journal of Physical Chemistry A, **112**, 5175 (2008).

210. S. Pelloni and P. Lazzeretti, *Topology of Magnetic-Field Induced Electron Current Density in the Cubane Molecule*, Journal of Chemical Physics, **128**, 194305 (2008).

211. S. Pelloni and P. Lazzeretti, *Ring Current Models for Acetylene and Ethylene Molecules*, Chemical Physics, **356**, 153 (2009).

212. I. García Cuesta, A. Sánchez de Merás, S. Pelloni, and P. Lazzeretti, *Understanding the Ring Current Effects on Magnetic Shielding of Hydrogen and Carbon Nuclei in Naphthalene and Anthracene*, Journal of Computational Chemistry, **30**, 551 (2009).

213. F. Bloch, *Bremsvermögen von Atomen mit mehreren Elektronen*, Zeitschrift für Physik, **81**, 363 (1933).

214. L. J. Bartolotti and J. C. Mollmann, *4th Order Time-Dependent Variation Perturbation-Theory Based on the Hydrodynamic Analogy*, Molecular Physics, **38**, 1359 (1979).

215. L. J. Bartolotti, *Time-Dependent Extension of the Hohenberg-Kohn-Levy Energy-Density Functional*, Physical Review A, **24**, 1661 (1981)

216. L. J. Bartolotti, *Time-Dependent Kohn-Sham Density-Functional Theory*, Physical Review A, **26**, 2243 (1982).

217. E. Runge and E. K. U. Gross, *Density-Functional Theory for Time-Dependent Systems*, Physical Review Letters, **52**, 997 (1984).

218. B. M. Deb and S. K. Ghosh, *Schrödinger Fluid-Dynamics of Many-Electron Systems in a Time-Dependent Density-Functional Framework*, Journal of Chemical Physics, **77**, 342 (1982).

219. B. M. Deb and P. K. Chattaraj, *How Can Density Functional Theory be Excited from the Ground State?*, Proceedings of the Indian Academy of Science, **99**, 67 (1987).

220. B. M. Deb and P. K. Chattaraj, *Quantum Fluid Density Functional Theory of Time-Dependent Phenomena – Ion Atom Collisions* Chemical Physics Letters, **148**, 550 (1988).

221. B. M. Deb and P. K. Chattaraj, *Density-Functional and Hydrodynamical Approach to Ion-Atom Collisions through a New Generalized Nonlinear Schrödinger Equation*, Physical Review A, **39**, 1696 (1989).

222. B. M. Deb, P. K. Chattaraj, and S. Mishra, *Time-Dependent Quantum-Fluid Density-Functional Study of High-Energy Proton-Helium Collisions*, Physical Review A, **43**, 1248 (1991).

223. B. Kr. Dey and B. M. Deb, *Time-Dependent Quantum Fluid-Dynamics of the Photoionization of the He Atom Under an Intense Laser Field*, International Journal of Quantum Chemistry, **56**, 707 (1995).

224. B. Kr. Dey and B. M. Deb, *A Theoretical Study of the High-Order Harmonics of a 200 nm Laser from* H^{-2} *and* HeH^{+}, Chemical Physics Letters, **276**, 157 (1997).

225. B. Kr. Dey and B. M. Deb, *Stripped Ion-Helium Atom Collision Dynamics Within a Time-Dependent Quantum Fluid Density Functional Theory*, International Journal of Quantum Chemistry, **67**, 251 (1998).

226. B. Kr. Dey and B. M. Deb, *Direct ab initio Calculation of Ground-State Electronic Energies and Densities for Atoms and Molecules through a Time-Dependent Single Hydrodynamical Equation*, Journal of Chemical Physics, **110**, 6229 (1999).

227. G. P. Lawes and N. H. March, *Approximate Differential-Equation for Calculating the Electron-Density in Closed Shell Atoms and in Molecules*, Physica Scripta, **21**, 402 (1980).

228. B. M. Deb and S. K. Ghosh, *New Method for the Direct Calculation of Electron-Density in Many-Electron Systems. 1. Application to Closed-Shell Atoms*, International Journal of Quantum Chemistry, **23**, 1 (1983).

229. M. Levy, J. P. Pardew, and V. Sahni, *Exact Differential Equation for the Density and Ionization Energy of a Many-Particle System*, Physical Review A, **30**, 2745 (1984).

230. N. H. March, *The Local Potential Determining the Square Root of the Ground-State Electron-Density of Atoms and Molecules from the Schrödinger Equation*, Physics Letters A, **113**, 476 (1986).

231. G. Hunter, *The Exact One-Electron Model of Molecular-Structure*, International Journal of Quantum Chemistry, **29**, 197 (1986).

232. M. Levy and H. Ou-Yang, *Exact Properties of the Pauli Potential for the Square Root of the Electron Density and the Kinetic Energy Functional*, Physical Review A, **38**, 625 (1988).

233. M. McClendon, *Real-Space Diffusion Theory of Multiparticle Quantum Systems*, Physical Review A, **38**, 5851 (1988).

234. A. S. Sanz, X. Giménez, J. M. Bofill, and S. Miret-Artés, *Time-Dependent Density Functional Theory from a Bohmian Perspective*, in *Chemical Reactivity Theory*, P. K. Chattaraj (Ed.) (Taylor & Francis, New York, 2009).

235. W. Kohn and L. J. Sham, *Self-Consistent Equations Including Exchange and Correlation Effects*, Physical Review, **140**, A1133 (1965).

236. T. A. Elsayed, K. Mølmer, and L. B. Madsen, *Entangled Quantum Dynamics of Many-Body Systems using Bohmian Trajectories*, Scientific Reports, **8**, 12704 (2018).

237. O. E. Alon, A. I. Streltsov, and L. S. Cederbaum, *Multiconfigurational Time-Dependent Hartree Method for Bosoms: Many-Body Dynamics of Bosonic Systems*, Physical Review A, **77**, 033613 (2008).

238. V. Bolsinger, S. Krönke and P. Schmelcher, *Beyond Mean-Field Dynamics of Ultra-Cold Bosonic Atoms in Higher Dimensions: Facing the Challenges with a Multiconfigurational Approach*, Journal of Physics B, **50**, 034003 (2017).

239. M. H. Beck, A. Jäckle, G. Worth, and H.-D. Meyer, *The Multiconfiguration Time-Dependent Hartree (MCTDH) Method: A High Efficient Algorithm for Propagatin Wavepackets*, Physics Reports, **324**, 1 (2000).

240. K. Müller and L. D. Brown, *Location of Saddle Points and Minimum Energy Paths by a Constrained Simplex Optimization Procedure*, Theoretica Chimie Acta, **53**, 75 (1979).

241. R. Crehuet and J. M. Bofill, *The Reaction Path Intrinsic Reaction Coordinate Method and the Hamilton–Jacobi Theory*, Journal of Chemical Physics, **122**, 234105 (2005).

242. A. Aguilar-Mogas, X. Giménez, and J. M. Bofill, *Finding Reaction Paths Using the Potential Energy as Reaction Coordinate*, Journal of Chemical Physics, **128**, 104102 (2008).

243. W. Quapp, *Chemical Reaction Paths and Calculus of Variations*, Theoretical Chemistry Accounts, **121**, 227 (2008).

244. A. S. Sanz, X. Giménez, J. M. Bofill, and S. Miret-Artés, *Understanding Chemical Reactions within a Generalized Hamilton–Jacobi Framework*, Chemical Physics Letters, **478**, 89 (2009); Erratum Chemical Physics Letters, **488**, 235 (2010).

245. A. S. Sanz and S. Miret-Artés, *A Trajectory-Based Understanding of Quantum Interference*, Journal of Physics A, **41**, 435303 (2008).

246. A. S. Sanz and S. Miret-Artés, *Quantum Trajectories in Elastic Atom-Surface Scattering: Threshold and Selective Adsorption Resonances*, Journal of Chemical Physics, **122**, 014702 (2005).

247. C. Dewdney and B. J. Hiley, *A Quantum Potential Description of One-Dimensional Time-Dependent Scattering from Square Barriers and Square Wells*, Foundations of Physics, **12**, 27 (1982).

248. R. A. Marcus, *On the Analytical Mechanics of Chemical Reactions. Quantum Mechanics Of Linear Collisions*, Journal of Chemical Physics, **45**, 4493 (1966).

249. A. S. Sanz, F. Borondo, and S. Miret-Artés, *Quantum Trajectories in Atom-Surface Scattering with Single Adsorbates: The Role of Quantum Vortices*, Journal of Chemical Physics, **120**, 8794 (2004).

250. A. S. Sanz, F. Borondo, and S. Miret-Artés, *Role of Quantum Vortices in Atomic Scattering from Single Adsorbates*, Physical Review B, **69**, 115413 (2004).

251. W. Schöllkopf and J. P. Toennies, *Nondestructive Mass Selection of Small van der Waals Clusters*, Science, **266**, 1345 (1994).

252. R. E. Grisenti, W. Schöllkopf, J. P. Toennies, G. C. Hegerfeldt, and T. Köhler, *Determination of Atom-Surface van der Waals Potentials from Transmission-Grating Diffraction Intensities*, Physical Review Letters, **83**, 1755 (1999).

253. R. E. Grisenti, W. Schöllkopf, J. P. Toennies, G. C. Hegerfeldt, T. Köhler, and M. Stoll, *Determination of the Bond Length and Binding Energy of the Helium Dimer by Diffraction from a Transmission Grating*, Physical Review Letters, **85**, 2284 (2000).

254. R. E. Grisenti, W. Schöllkopf, J. P. Toennies, J. R. Manson, T. A. Savas, and H. I. Smith, *He-Atom Diffraction from Nanostructure Transmission Gratings: The Role of Imperfections*, Physical Review A, **61**, 033608 (2000).

255. A. S. Sanz, F. Borondo, and S. Miret-Artés, *Causal Trajectories Description of Atom Diffraction by Surfaces*, Physical Review B, **61**, 7743 (2000).

256. A. S. Sanz, F. Borondo, and M. Bastiaans, *Loss of Coherence in Double-Slit Diffraction Experiments*, Physical Review A, **71**, 42103 (2005).

257. A. S. Sanz, M. Davidović, and M. Božić, *Full Quantum Mechanical Analysis of Atomic Three-Grating Mach-Zehnder Interferometry*, Annals of Physics, **353**, 205 (2015).

258. A. Luis and A. S. Sanz, *What Dynamics Can be Expected for Mixed States in Two-Slit Experiments?*, Annals of Physics, **357**, 95 (2015).

259. O. V. Prezhdo and C. Brooksby, *Relationship between Quantum Decoherence Times and Solvation Dynamics in Condensed Phase Chemical Systems*, Physical Review Letters, **86**, 3215 (2001).

260. E. Gindensperger, C. Meier, and J. A. Beswick, *Mixing Quantum and Classical Dynamics Using Bohmian Trajectories*, Journal of Chemical Physics, **113**, 9369 (2000).

261. E. Gindensperger, C. Meier, and J. A. Beswick, *Quantum-Classical Dynamics Including Continuum States Using Quantum Trajectories*, Journal of Chemical Physics, **116**, 8 (2002).

262. E. Gindensperger, C. Meier, J. A. Beswick, and M.-C. Heitz, *Quantum-Classical Description of Rotational Diffractive Scattering Using Bohmian Trajectories: Comparison with Full Quantum Wave Packet Results*, Journal of Chemical Physics, **116**, 10051 (2002).

263. C. Meier and J. A. Beswick, *Femtosecond Pump-Probe Spectroscopy of I_2 in a Dense Rare Gas Environment: A Mixed Quantum/Classical Study of Vibrational Decoherence*, Journal of Chemical Physics, **121**, 4550 (2004).

264. S. Garashchuk and V. Rassolov, *Semiclassical Dynamics Based on Quantum Trajectories*, Chemical Physics Letters, **364**, 562 (2002).

265. S. Garashchuk and V. Rassolov, *Modified Quantum Trajectory Dynamics Using a Mixed Wave Function Representation*, Journal of Chemical Physics, **121**, 8711 (2004).

266. S. Garashchuk, V. Rassolov, and O. Prezhdo, *Semiclassical Bohmian Dynamics*, Reviews in Computational Chemistry, **27**, 287 (2011).

267. E. R. Bittner, *Quantum Initial Value Representation Using Approximate Bohmian Trajectories*, Journal of Chemical Physics, **119**, 1358 (2003).

268. Y. Zhao and N. Makri, *Bohmian versus Semiclassical Description of Interference Phenomena*, Journal of Chemical Physics, **119**, 60 (2003).

269. J. Liu and N. Makri, *Forward-Backward Quantum Dynamics for Time Correlation Functions*, Journal of Physical Chemistry A, **108**, 806 (2004).

270. J. Liu and N. Makri, *Monte Carlo Bohmian Dynamics from Trajectory Stability Properties*, Journal of Physical Chemistry A, **108**, 5408 (2004).

271. B. Gu and I. Franco, *Partial Hydrodynamic Representation of Quantum Molecular Dynamics*, Journal of Chemical Physics, **146**, 194104 (2017).

272. A. Goldberg, H. M. Schey, and J. L. Schwartz, *Computer-Generated Motion Pictures of One-Dimensional Quantum-Mechanical Transmission and Reflection Phenomena*, American Journal of Physics, **36**, 454 (1968).

273. http://nobelprize.org/nobel_prizes/chemistry/laureates/1998/kohn-lecture.pdf

Chapter 5

Adaptive Quantum Monte Carlo Approach States for High-Dimensional Systems

Eric R. Bittner,[a] Donald J. Kouri,[a] Sean Derrickson,[a,b] and Jeremy B. Maddox[c]

[a]*Department of Chemistry, University of Houston, Houston, TX 77204-5003, USA*
[b]*Oceaneering International, Inc., 11911 FM 529 Rd, Houston, TX 77041, USA*
[c]*Department of Chemistry, Western Kentucky University, 1906 College Heights Blvd, Bowling Green, KY 42101, USA*
bittner@uh.edu

We present an overview of our work in developing hydrodynamic-based methods for studying the structure and quantum dynamics of rare-gas clusters. We use a hydrodynamical approach based on the Bohm description of quantum mechanics (QM) to satisfy an orbital-free density functional-like Euler–Lagrange equation for the ground state of the system. In addition, we use an information theoretical approach to obtain the optimal density function derived from a series of statistical sample points in terms of density approximates. These are then used to calculate an approximation to the quantum force in the hydrodynamic description. We also show how this approach can be extended to finite temperature and use this to

Applied Bohmian Mechanics: From Nanoscale Systems to Cosmology (Second Edition)
Edited by Xavier Oriols and Jordi Mompart

ISBN 978-981-4800-10-5 (Hardcover), 978-0-429-29474-7 (eBook)
www.jennystanford.com

examine the thermodynamic properties of rare-gas clusters with up to 100 atoms.

5.1 Introduction

In many-body systems, quantum mechanical effects become important when the de Broglie wavelength of a given atom becomes comparable to the characteristic interatomic interaction length, σ. If we define a characteristic temperature as $T^* = \epsilon / k_B$ in terms of the potential well depth, ϵ, one arrives at what is termed the de Boer ratio [100],

$$\Lambda = \frac{\lambda(T^*)}{\sigma} = \frac{\hbar}{\sigma\sqrt{m\epsilon}}, \tag{5.1}$$

which is a useful quantity in determining whether or not quantum mechanical effects are important in a given physical system. In the strictly classical limit, $\Lambda = 0$, while systems with $\Lambda > 0.3$ should be considered as strongly quantum mechanical.

Atomic and molecular clusters provide almost ideal laboratory systems for studying quantum versus classical dynamical and structural effects as a function of the size and scale of a given system. For example, the de Boer ratio for Ar clusters $\Lambda_{Ar} \approx 0.03$. As such, quantum effects in Ar clusters and liquids play an insignificant role, and their properties can be accurately modeled using classical molecular dynamics techniques. However, for lighter atoms such as Ne, where $\Lambda_{Ne} \approx 0.1$, quantum mechanical effects are be important, especially at low temperatures. Quantum corrections are important even in equilibrium calculations and finite temperature calculations because the quantum character strongly affects the thermodynamics via changes in the ground state structure due to increasing zero-point energies. For example, quantum corrections have been shown to lower solid-to-liquid transition temperatures by approximately 10%, and the zero-point energy for small clusters can equal up to 35% of the classical binding energy.

In this chapter, we present our development of a Bohmian trajectory-based approach for computing the quantum mechanical structure, energetics, and thermodynamics of multiatom systems.

These systems are important test cases for both theoretical and computational studies since they are at the crossing point between mesoscopic and microscopic length scales. Hence, quantum confinement and delocalization effects may yet play an important role in their structure and properties. From a computational point of view, these are particularly challenging systems since they have too many degrees of freedom to be tackled by more standard basis set approaches. Furthermore, many of their interesting quantum properties are only manifest at very low temperature, making them challenging systems for path-integral Monte Carlo-based approaches due to the potentially infinitely long imaginary time integration required to reach $T = 0$.

We shall first review the salient features of the Bohmian approach, focusing upon how one might use it to develop new computational approaches for many-body systems. We then present a variational approach that finds the quantum ground state for N-atom rare clusters using a statistical modeling approach for determining a best estimate of the quantum potential for a multidimensional system. We then extend this approach to finite temperature and present results for systems as large as Ne_{100}. In the last part of the chapter, we present an idea based upon supersymmetric SUSY) QM that extends our approach to excited states. We conclude with a discussion of the prospects of our methods.

5.2 Mixture Modeling Approach

Perhaps the most fundamental objective in statistical analysis pertains to the development of probabilistic models that can explain and predict the observations of interesting physical processes. The capacity to estimate the effectiveness of a statistical model goes hand in hand with the ability to improve its explanatory and predictive powers. Problems related to this idea are encountered throughout the biological, physical, and social sciences. In some cases, it is possible to construct a model that incorporates prior knowledge and experience in terms of a few (or more often many!) adjustable parameters. The primary goal is then to find a particular set of parameters that best explains the observed data and can predict the likely outcome of new observations. The mathematical formalism which quantifies these notions is provided within the Bayesian construction of statistical analysis [68]. In the Bayesian approach, probabilities are treated subjectively as a degree of belief rather than a frequency of observation. Though this distinction is somewhat controversial, Bayesian statistics are crucially important to probabilistic learning [23], decision-making theory, and statistical inference problems. In the quantum physics literature, Bayesian probabilities have recently been addressed in connection with a diverse range of problems, including many-body potential energy surfaces [70, 71], the control of open quantum systems [104], quantum tomography [27], measurement theory in quantum logic devices [59–61, 99], and quantum Monte Carlo simulations [19, 56].

In this section we develop an approximate methodology for estimating the multidimensional quantum distribution function associated with a statistical ensemble of space-time trajectories. The scheme that we propose is built upon a parameterized Gaussian model for the quantum density. We explore the advantages and limitations of this model and outline an iterative procedure based upon Bayesian probability theory for finding a set of Gaussian parameters that mimics the true density function. This fitted density is then used to compute an approximate quantum force that drives the ensemble of trajectories. We show how this approach can be used to determine the ground state density and energy of a multidimensional quantum mechanical system.

5.2.1 Motivation for a Trajectory-Based Approach

According to Newton's second law, the physical motion of a particle is characterized by a space-time path $\vec{r}(t)$ that satisfies the equations of motion:

$$\dot{\vec{r}}(t) = \vec{p}(t)/m, \tag{5.2}$$

$$\dot{\vec{p}}(t) = -\vec{\nabla} V(\vec{r}(t)), \tag{5.3}$$

where m is the mass of the particle and $-\vec{\nabla} V(\vec{r})$ is the force associated with an externally applied potential energy field. In principle, the position $\vec{r}$ and momentum $\vec{p}$ of a classical particle can be determined with arbitrary precision, and it is well known that exactly specifying the initial conditions $\vec{r}_0 = \vec{r}(0)$ and $\vec{p}_0 = \vec{p}(0)$ will completely determine the particle's trajectory for all time.

In QM, however, the situation is quite different. The precision with which $\vec{r}$ and $\vec{p}$ can be simultaneously known is limited by Heisenberg's uncertainty principle: $\Delta x \Delta p \geq \hbar/2$. The dynamical properties of a particle are embodied in a complex wave function $\psi(\vec{r}, t)$ that satisfies the time-dependent Schrödinger equation:

$$i\hbar\partial_t \psi(\vec{r}, t) = -\frac{\hbar^2}{2m}\vec{\nabla}^2\psi(\vec{r}, t) + V(\vec{r})\psi(\vec{r}, t). \tag{5.4}$$

Traditionally, the wave function is interpreted as a time-dependent amplitude associated with the instantaneous probability of finding the particle in an infinitesimal volume of space $d^3\vec{r}$ about the point $\vec{r}$. From this point of view, individual physical particles are treated as statistical objects and the notion that particles follow definite paths in space-time is apparently a meaningless concept in QM.

One way to rationalize this disparity among the classical and quantum theories is due to the Feynman path integral approach to QM [41]. According to Feynman's analysis, a path $\vec{r}(t)$ connecting two points in space-time is associated with a complex phase factor $\phi = \exp(iS_{cl}/\hbar)$ determined by the classical action:

$$S_{cl}[\vec{r}(t)] = \int_0^t ds \, \frac{1}{2}m\dot{\vec{r}}^2 - V(\vec{r}) \tag{5.5}$$

along the path. The probability amplitude with which a particle makes a transition from an initial point $(\vec{r}_0, 0)$ to some final point $(\vec{r}_t, t)$ is expressed as the sum of phase factors over all possible paths connecting the two points. For a quantum mechanical particle, no particular path is preferred; therefore, we must consider an infinite number of paths in order to compute the transition probability. For macroscopic objects, however, the classical action is much larger than $\hbar$, implying that the net contribution of phase in the transition amplitude is due to the path that minimizes the classical action. Hence, the most probable path for a macroscopic object will be a trajectory that also satisfies the classical equations of motion. Feynman's treatment is particularly enlightening because it allows us to discuss both classical and quantum mechanical phenomena on an equal footing, that is, in terms of an ensemble of all possible paths that effectively reduces to the classical trajectory in the limit that $\hbar$ is small.

5.2.1.1 Bohmian interpretation

Another formulation of QM in terms of an ensemble of paths is due to the work of de Broglie [31–33] and Bohm [15–17, 52]. In the de Broglie–Bohm interpretation of QM one assumes that a quantum mechanical system is physically composed of two parts: a wave *and* a point particle. Mathematically, the wave is represented by a wave function $\psi(\vec{r}, t)$ that satisfies Eq. 5.4 and is associated with the probability density $\rho(\vec{r}, t) = |\psi(\vec{r}, t)|^2$ for finding the particle when its exact position is unknown. Regardless of whether or not it can be observed, the particle always follows a precisely defined trajectory. The wave function plays a direct physical role in this by influencing the particle's trajectory through the introduction of a nonlocal "quantum potential" that gives rise to all nonclassical behavior including zero-point energy, tunneling, and self-interference effects.

While the Bohm interpretation of QM is still a subject of philosophical controversy and stimulating debate, it has also, over the last several years, attracted attention in the chemical physics community as a viable framework for the development of novel trajectory-based computational methodologies.

To see this influence, we begin by writing the wave function in complex polar form:

$$\psi(\vec{r}, t) = \rho(\vec{r}, t)^{1/2} \exp(iS(\vec{r}, t)/\hbar), \tag{5.6}$$

where the density $\rho(\vec{r}, t)$ and phase $S(\vec{r}, t)$ (quantum action) are real functions dependent upon space and time. The particle is assumed to follow a definite trajectory $\vec{r}(t)$ along which its velocity:

$$\begin{aligned}\dot{\vec{r}} &= \vec{v}(\vec{r}(t), t)\\ &= \vec{\nabla} S(\vec{r}(t), t)/m\end{aligned} \tag{5.7}$$

is determined by the phase amplitude of the wave function. For notational convenience we will drop the explicit dependence on $\vec{r}$ and t. Substituting Eqs. 5.6 and 5.7 into Schrödinger's equation and equating the real and imaginary components yields a pair of coupled partial differential equations:

$$\partial_t \rho = -\vec{\nabla} \cdot (\rho \, \vec{v}), \tag{5.8}$$

$$\partial_t S = -(Q + V) - \frac{1}{2} m \vec{v}^2, \tag{5.9}$$

which are easily identified as the continuity equation and a generalized Hamilton–Jacobi equation, respectively. The quantum potential term $Q = Q(\vec{r}, t)$, given by:

$$Q = -\frac{\hbar^2}{4m} \left(\frac{\vec{\nabla}^2 \rho}{\rho} - \frac{1}{2} \left(\frac{\vec{\nabla} \rho}{\rho} \right)^2 \right), \tag{5.10}$$

distinguishes Eq. 5.9 from a purely classical equation and encapsulates the nonlocal influence of ψ on the trajectory of the particle. The quantum potential is often interpreted as an internal energy or "shape" energy associated with the curvature of the quantum density. We can explicitly include Q into the equations of motion for the particle by introducing a material time derivative:

$$d_t f = \partial_t f + \vec{v} \cdot \vec{\nabla} f \tag{5.11}$$

such that the field equations of motion for ρ and S are given by:

$$d_t \rho = -\rho \, \vec{\nabla} \cdot \vec{v}, \tag{5.12}$$

$$d_t S = \frac{1}{2} m \vec{v}^2 - (Q + V), \tag{5.13}$$

where the notation d_t signifies the time rate of change in the inertial reference frame of a particle moving along the trajectory $\vec{r}(t)$ with velocity $\vec{v}(\vec{r}(t), t)$. It is evident from the right-hand side of Eq. 5.13 that $d_t S$ represents a generalized Lagrangian $\mathcal{L}(\vec{r}(t), t)$ for the quantum mechanical system. Taking $\vec{\nabla} \mathcal{L}$, we have the relations:

$$\dot{\vec{v}} = -\vec{\nabla}(Q + V)/m, \tag{5.14}$$

$$\dot{\vec{r}} = \vec{v} = \vec{\nabla} S/m, \tag{5.15}$$

which appear to be a quantum mechanical analogue of Newton's equations that bears a quantum force term $F_Q = -\vec{\nabla} Q$ supplementing the classical force. We make a special note here to emphasize that the fields ρ, S, $\vec{v}$, and all their derivatives in Eqs. 5.12–5.15 are implicitly evaluated along the path $\vec{r} = \vec{r}(t)$, and this path is the trajectory of a physically real point particle, at least within the Bohmian paradigm.

5.2.1.2 Quantum hydrodynamic trajectories

It is fairly obvious from Eqs. 5.12–5.15 that if the wave function for a system were known, one would simply need to choose an initial position vector $\vec{r}(0)$ to completely specify the state of the system for all time. While this feature of the Bohmian picture is indeed a useful interpretive tool, we are still faced with the often formidable task of solving the time-dependent Schrödinger equation.

To remedy this situation, we begin by formally introducing an ensemble of quantum trajectories:

$$\mathcal{R}(t) = \{\vec{r}_1(t), \ldots, \vec{r}_n(t)\}, \tag{5.16}$$

which are distinguished from one another by their initial positions at time $t = 0$. Just as one can construct a conceptual ensemble of fictitious fluid particles to represent a classical fluid, we have established an analogous ensemble of probability fluid elements to sustain the quantum density, phase, and velocity field. The theoretical groundwork for the evolution of these "particles" has already been laid out. One notable exception is that Eqs. 5.12 and 5.13 are now to be defined over a whole ensemble of quantum trajectories that simultaneously satisfy the set of differential equations given by:

$$\dot{\vec{v}}_i = \vec{\nabla}\mathcal{L}(\vec{r}_i(t), t), \tag{5.17}$$

$$\dot{\vec{r}}_i = \vec{v}_i. \tag{5.18}$$

Given an initial wave function $\psi(\vec{r}_i(0), 0)$ defined over the ensemble of points $\mathcal{R}(0)$, we can simultaneously solve Eqs. 5.12 and 5.13 to reconstruct the wave function at a later time t according to the relation:

$$\psi(\vec{r}_i(t), t) = \psi(\vec{r}_i(0), 0) \exp\left(-\frac{1}{2}\int_0^t \vec{\nabla}\cdot\vec{v}(\vec{r}_i(s), s)\, ds\right) \times \exp\left(\frac{i}{\hbar}\int_0^t \mathcal{L}(\vec{r}_i(s), s)\, ds\right), \tag{5.19}$$

where the wave function is represented point-wise along the ensemble of paths $\mathcal{R}(t)$. We emphasize that there is no approximation in Eq. 5.19, and it is a formal solution to the Schrödinger equation evaluated explicitly over a set of hydrodynamic-like quantum trajectories.

A few comments on the behavior of quantum trajectories are in order. First, the single-valuedness of $\psi(\vec{r}, t)$ requires that quantum trajectories must not intersect one another. If two trajectories were to cross, it would imply that the wave function has two distinct values of phase at the same point in space-time. Similarly, trajectories are not allowed to cross through nodal regions of the wave function where the phase is discontinuous and the probability of finding a particle is zero. This restriction is strictly upheld by the influence of the quantum force, which is very intense around nodal surfaces. Moreover, if the ensemble of trajectories are initially distributed according to the probability density $\rho(\vec{r}, 0)$, then the ensemble will be representative of $\rho(\vec{r}, t)$ for all time. This is simply a consequence of the statistical assumption on $\psi(\vec{r}, t)$ and is consistent with the continuity of quantum probability density. Finally, in the limit that $\hbar \to 0$, the ensemble equations of motion decouple, and the trajectories will evolve independently of one another according to Newton's equations.

In essence, the Bohm picture depicts a quantum mechanical system in terms of an ensemble of correlated particle trajectories. Though the true particle follows a unique trajectory, it is inextricably coupled to an ensemble of alternate paths by the influence of the wave function acting through a quantum force.

5.2.1.3 Computational considerations

Numerical solutions of the time-dependent Schrödinger equation are traditionally obtained by calculating the short-time quantum propagator using fast Fourier transforms [69], finite basis sets, or discrete variable representations [73]. Typically, the computational overhead associated with these techniques scales exponentially with the dimensionality of the physical problem. Trajectory-based methodologies, on the other hand, offer tremendous numerical scaling advantages, especially for high-dimensional systems where traditional techniques are not feasible. In particular, the Bohm interpretation of QM has inspired a growing number of theoretical and computational studies involving a wide range of problems such as reactive scattering dynamics [95, 107], tunneling systems [11, 12, 77, 92], mixed quantum/classical simulations [46–48, 91], electronic transitions [20, 108, 109], photodissociation [38, 53, 85,

103], mixed quantum states [21, 22], and quantum dissipation [80, 81, 110].

The quantum trajectory method (QTM), developed by Wyatt and coworkers [77, 107], incorporates the ideas of computational fluid dynamics to solve the hydrodynamic field equations over a discrete ensemble of quantum fluid elements. Using finite element methods, the fluid particles are arranged into small neighborhoods over which a moving weighted least squares (MWLS) fitting procedure [75] is used to locally expand the hydrodynamic fields ρ (more typically $\log\rho$), S, and $\vec{v}$ in a simple polynomial basis. Once the fields and their derivatives are known, the integrals in Eq. 5.19 along with solutions to Eqs. 5.17–5.18 are evaluated over a short time step. While this strategy scales almost linearly with the number of trajectories, its versatility in practice is hindered by the fact that the ensemble of particles generally tends to become extremely disorganized for anharmonic systems making it difficult, if not impossible, to fit the quantum hydrodynamic fields. This becomes especially apparent around the nodes of the wave function, where the fitting errors will oftentimes cause the quantum trajectories to cross one another leading to spurious numerical results. Very recently [54, 101], new methods in adaptive grids have been developed to formulate a reconstruction of the wave function over an ensemble of generalized hydrodynamic trajectories that avoids the problem with quantum nodes and provides a much more stable framework for solving the hydrodynamic equations of motion.

Another application of the Bohmian approach includes the development of semiclassical approximation strategies for including quantum effects into otherwise classical calculations. Garashchuk and Rassolov [42, 43] have recently presented a semiclassical methodology based upon Bohm trajectories that is formally insensitive to trajectory crossings and also avoids explicitly solving the continuity equation. In this approximate methodology, the quantum density is convoluted with a minimum uncertainty Gaussian wave packet and expanded in a linear combination of Gaussian functions:

$$\rho(x) \approx f(x) = \sum_n c_n^2 \exp(-a_n^2(x - X_n)). \tag{5.20}$$

The Gaussian parameters $s = \{c_n, X_n, a_n\}$ in Eq. 5.20 are determined by minimizing the functional:

$$F = \int (\rho(x) - f(x))^2 dx \tag{5.21}$$

using an iterative procedure which explicitly involves solving the set of equations $\partial F/\partial s_k = 0$. The parameterized density leads to an approximate quantum potential (AQP) that is used to propagate an ensemble of trajectories. Garashchuk and Rassolov have presented results for Eckart barrier tunneling that improve upon the Herman–Kluck semiclassical IVR method and are shown to agree quite well with exact quantum mechanical results.

5.2.2 Density Estimation

5.2.2.1 The mixture model

Suppose that $\mathcal{R} = \{\vec{r}_1, \ldots, \vec{r}_N\}$ is an ensemble of Bohm fluid elements that statistically represents a multidimensional quantum probability density. Such a distribution of data points can be generated from a Metropolis sampling procedure or perhaps from the output of a quantum Monte Carlo simulation [28, 50]. In order to propagate these particles in time, we must evaluate ρ and its derivatives for every member in the ensemble. Instead of solving the hydrodynamic field equations explicitly, we intend to extract this information directly from the ensemble of trajectories.

We assume that the quantum density can be represented by a mixture model [45, 86] determined by summing a finite number, M, of Gaussian components or "clusters." The mixture model decomposition is expressed as a sum of joint probabilities:

$$\rho(\vec{r}) = \sum_m^M p(\vec{r}, c_m), \tag{5.22}$$

where $p(\vec{r}, c_m)$ is the probability that a randomly chosen member of $\mathcal{R}$ has the configuration $\vec{r}$ and is a variant of the mth Gaussian cluster designated by c_m. Each Gaussian cluster is parameterized by a weight $p(c_m)$, a mean position vector $\vec{\mu}_m$, and a vector of variances $\vec{\sigma}_m^2$. We can also replace the variance vector with a full covariance matrix $\mathbf{C}_m$, if necessary.

By definition [44], each joint probability in Eq. 5.22 is related to a pair of conditional probabilities according to the relation:

$$p(\vec{r}, c_m) = p(c_m)\,p(\vec{r}|c_m) = \rho(\vec{r})\,p(c_m|\vec{r}), \tag{5.23}$$

where the forward conditional probability $p(\vec{r}|c_m)$ refers to the probability that a randomly chosen variant of c_m has the configuration $\vec{r}$. Conversely, the posterior probability $p(c_m|\vec{r})$ refers to the probability that the configuration point $\vec{r}$ is a variant of the cluster c_m. In probability theory the factors $\rho(\vec{r})$ and $p(c_m)$ are marginal probabilities; however, we shall simply refer to them as the quantum density and weight of the mth Gaussian cluster, respectively. The expansion weights are strictly positive semidefinite and sum to unity. Substituting the first equality of Eq. 5.23 into Eq. 5.22, we have:

$$\rho(\vec{r}) = \sum_{m}^{M} p(c_m)\,p(\vec{r}|c_m), \tag{5.24}$$

where we can specify the form of $p(\vec{r}|c_m)$ to reflect our belief that $\rho(\vec{r})$ is a mixture of Gaussian components. We explore this approximation with two different Gaussian cluster models.

The first model assumes that each cluster is completely separable and takes the form of a product over the N_d-dimensional configuration space:

$$p(\vec{r}|c_m) = \prod_{d}^{N_d} \sqrt{\frac{1}{2\pi\sigma_{m,d}^2}}\, e^{-(\vec{r}_d-\mu_{m,d})^2/(2\sigma_{m,d}^2)}. \tag{5.25}$$

The second model explicitly takes into account nonseparable correlations in configuration space and incorporates the full covariance matrix:

$$p(\vec{r}|c_m) = \sqrt{\frac{||\mathbf{C}_m^{-1}||}{(2\pi)^{N_d}}}\, e^{-(\vec{r}-\vec{\mu}_m)^T.\mathbf{C}_m^{-1}.(\vec{r}-\vec{\mu}_m)/2}. \tag{5.26}$$

In comparison with the separable case, the fully covariant model can represent more complicated density structures with fewer clusters; however, this is at the cost of greater computational expense. For low-dimensional systems it is advantageous to use the fully covariant model, but in high dimensions it is much more efficient to use a larger number of separable clusters. The principle at work here is related to the idea of collective correspondence discussed by Heller [51] regarding the cooperative effort of overlapping Gaussian wave packets to describe position-momentum correlations in phase space. It is also feasible to construct a mixture model that incorporates any combination of covariant and separable degrees of freedom, especially if there is reason to do so based on the symmetry of the physical problem.

5.2.2.2 Expectation maximization

Now that we have established a model to work with, the trick is to determine the Gaussian parameters $p(c_m)$, $\vec{\mu}_m$, and $\mathbf{C}_m$ (or σ_m). The mean position vector and covariance matrix of the clusters are defined by the moments of the forward conditional probabilities:

$$\vec{\mu}_m = \int \vec{r}\ p(\vec{r}|c_m)\, d\vec{r}, \tag{5.27}$$

$$\mathbf{C}_m = \int (\vec{r} - \vec{\mu}_m)^T (\vec{r} - \vec{\mu}_m)\ p(\vec{r}|c_m)\, d\vec{r}. \tag{5.28}$$

For the separable case, the variances are given by the diagonal elements $\sigma^2_{m,i} = (\mathbf{C}_m)_{ii}$. Rearranging Eq. 5.23 and substituting into Eqs. 5.27–5.28, we can write these parameters as:

$$\vec{\mu}_m = \int \vec{r}\ \frac{\rho(\vec{r})\, p(c_m|\vec{r})}{p(c_m)}\, d\vec{r}, \tag{5.29}$$

$$\mathbf{C}_m = \int (\vec{r} - \vec{\mu}_m)^T (\vec{r} - \vec{\mu}_m) \frac{\rho(\vec{r})\, p(c_m|\vec{r})}{p(c_m)}\, d\vec{r}, \tag{5.30}$$

which are easily approximated by a pair of Monte Carlo sums over the ensemble of Bohmian particles:

$$\vec{\mu}_m \approx \frac{1}{Np(c_m)} \sum_n^N \vec{r}_n\, p(c_m|\vec{r}_n), \tag{5.31}$$

$$\mathbf{C}_m \approx \frac{1}{Np(c_m)} \sum_n^N (\vec{r}_n - \vec{\mu}_m)^T (\vec{r}_n - \vec{\mu}_m)\, p(c_m|\vec{r}_n). \tag{5.32}$$

A similar expression for the expansion weights in terms of a sum over $\mathcal{R}$ is given by:

$$p(c_m) \approx \frac{1}{N} \sum_n^N p(c_m|\vec{r}_n). \tag{5.33}$$

The posterior terms $p(c_m|\vec{r}_n)$ for each data point in Eqs. 5.31–5.33 are evaluated directly from the forward probabilities according to Bayes' formula:

$$p(c_m|\vec{r}_n) = \frac{p(c_m)\,p(\vec{r}_n|c_m)}{\sum_m p(c_m)\,p(\vec{r}_n|c_m)}. \tag{5.34}$$

In some sense, the ensemble of particles can be viewed as a data set that catalogs the results of many successive measurements on an ensemble of identically prepared quantum systems. Each member of the ensemble wields an equal amount of information describing the underlying probability distribution. The key to understanding how this information is distributed among the Gaussian clusters is contained within Bayes' formula. From a Bayesian viewpoint the numerator in Eq. 5.34 essentially boils down to a measure of how well the cluster c_m describes the fluid element with configuration $\vec{r}_n$. The sum in the denominator is a measure how well the particle at $\vec{r}_n$ is described by all of the clusters. The ratio of the two quantities then determines the fraction of explanatory information that the particle gives to the mth cluster. Hence, the cluster which best describes $\vec{r}_n$ will have the largest posterior probability for that point.

The circular structure in Eqs. 5.22–5.34 provides the framework for an iterative procedure known as the expectation maximization (EM) algorithm [35, 45, 86] that seeks to find a set of parameters that gives the best estimate for the density of $\mathcal{R}$. Computing the forward and posterior probabilities determines how well an arbitrarily parametrized mixture model is *expected* to represent the ensemble. Evaluating the sums in Eqs. 5.31–5.33 gives rise to a new set of parameters that is said to *maximize* the log-likelihood:

$$L = \log \prod_n \rho(\vec{r}_n) \tag{5.35}$$

of the distribution. A likelihood is a probability measure referring to the outcome of an event that is already known to have occurred. The log-likelihood of the distribution is a measure of how well the overall density model describes the whole collection of data points. The EM algorithm works very much like the variational principle, in that, there is a likelihood equation defined over parameter space:

$$\vec{\nabla}_{c_m} L = 0, \tag{5.36}$$

such that L is a maximum for models that are effective in describing the ensemble's distribution. Furthermore, it can be shown that the update rules in Eqs. 5.31–5.33 move the clusters through parameter space in the direction along $\vec{\nabla}_{c_m} L$, that is, in the direction that improves the density estimate. The cycle of estimating the expected distribution function, and maximizing the log-likelihood is repeated iteratively until a satisfactory estimate of the density is achieved.

It is important to realize that finding the maximum likelihood estimate of a distribution is not always a well-defined problem. In fact, there are generally multiple roots to the likelihood equation, and it is not necessarily guaranteed that there is a global maximum. While this is an important problem, our main concern here is simply to find an *acceptable* set of parameters that approximately represents the quantum density. However, one problem that we will need to address concerns the number, M, of Gaussian clusters used in the density estimate. For a Gaussian wave packet evolving in a parabolic potential field the answer is simple, but in general we will never really know how many clusters to use. When a wave packet bifurcates at a potential barrier, it will often develop complicated oscillations and nodal structures that are impossible to capture with Gaussians. Though there are statistical methods for "guessing" the number of components in a statistical data set, we do not incorporate them here. Instead, we simply try to use a minimum number of Gaussian clusters that gives reasonable results.

The overall scheme of the mixture model approximation and EM algorithm is as follows: First we generate the ensemble of probability fluid elements, usually a Gaussian density packet, via some appropriate sampling technique. The EM algorithm is initialized by choosing a set of parameters for a pre-set number of Gaussian clusters. Typically the initial clusters are given a uniform weight $p(c_m) = 1/M$. The mean position vectors are randomly selected from the domain of the ensemble. The initial variances are chosen to be large enough to encompass the entire ensemble and the cross terms $(\mathbf{C}_m)_{i,j}$ are zero. We cycle through the EM routine until the parameters converge to an acceptable density estimate. Convergence can be evaluated in a number of ways by monitoring the cluster parameters, the conditional probabilities, the log-likelihood, or any combination thereof.

5.2.3 Computational Results

5.2.3.1 Bivariate distribution with multiple nonseparable Gaussian components

To illustrate some of the points in the previous section, we demonstrate the convergence of the EM algorithm using a known probability distribution function. In Fig. 5.1 we have plotted the contours of a bivariate probability distribution function $\rho(x, y)$ consisting of four equally weighted nonseparable Gaussian components. The solid lines reflect the half-width contours of each

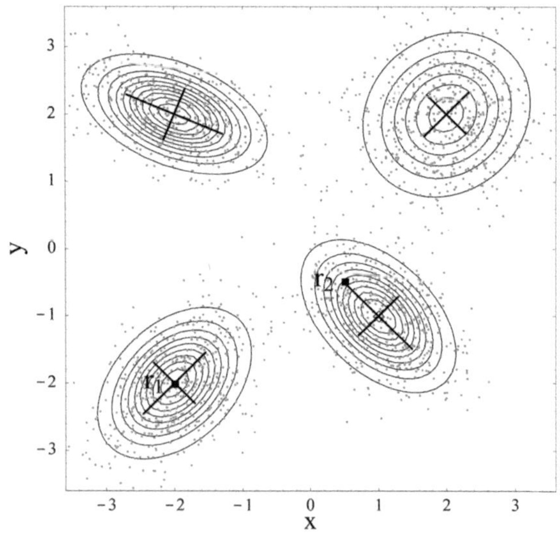

Figure 5.1 The contours reflect a multivariate probability distribution comprised of four Gaussian components with nonzero xy covariances. The solid black lines represent the half-widths of the Gaussian components. The ensemble of gray points are generated from the exact probability distribution function via a Metropolis sampling procedure and are taken as the input data for the EM algorithm. The black squares labeled r_1 and r_2 are tagged data points discussed later in Figs. 5.4 and 5.5. Reprinted with permission from Maddox and Bittner [82]. Copyright 2003 American Institute of Physics.

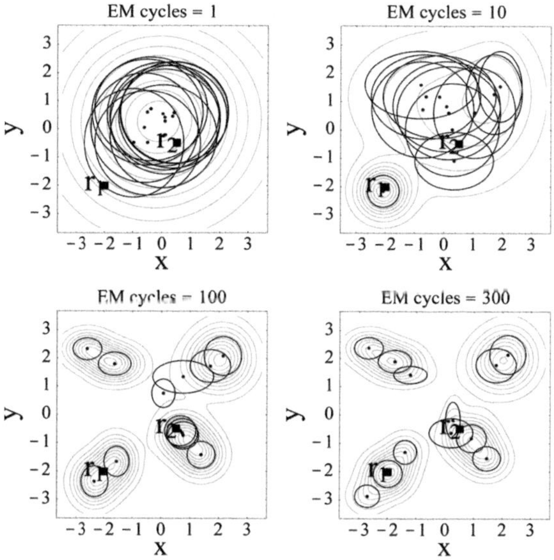

Figure 5.2 This figure illustrates the EM algorithm for the data pictured in Fig. 5.1 using 16 separable Gaussian clusters. The contours reflect the approximated probability density during the course of the EM fitting routine. The black dots correspond to the average position of the Gaussian cluster, and the solid black ellipses represents the Gaussian half-width contours. Reprinted with permission from Maddox and Bittner [82]. Copyright 2003 American Institute of Physics.

component and their orientation with respect to the x and y axes. The gray points correspond to an ensemble $\mathcal{R}$ of 2000 variants of $\rho(x, y)$, which were randomly generated using a Metropolis sampling algorithm. Two of these data points, labeled $\vec{r}_1$ and $\vec{r}_2$, have been tagged for later discussion.

In Fig. 5.2 we show the evolution of 16 separable Gaussian clusters over the course of the EM fitting algorithm. The contour plots indicate the relative intensity of the fitted density at various stages of the EM fit. The black dots and ovals correspond to the μ_m's and σ_m-contours for the individual Gaussian clusters. The initial random guess for the clusters is not illustrated. After 1 EM cycle, the clusters tend to aggregate near the mean of $\mathcal{R}$, and the estimated density does not reflect any details of the exact distribution. After about 10 EM cycles, the clusters begin sorting

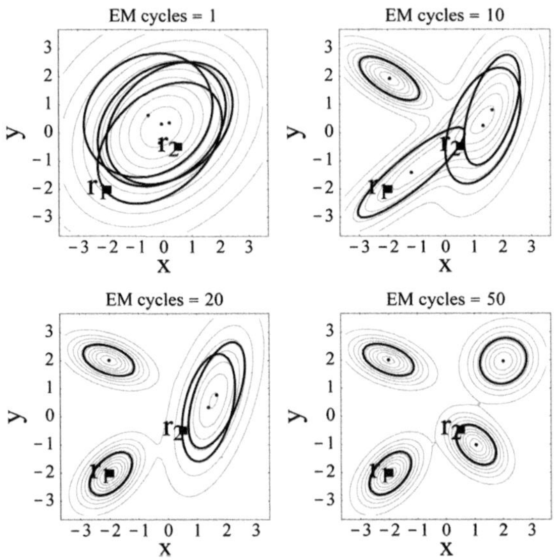

Figure 5.3 This figure illustrates the EM algorithm for the data pictured in Fig. 5.1 using four nonseparable Gaussian clusters. Compared to the separable case, the fully covariant model gives much more accurate results with less clusters and fewer EM cycles. Reprinted with permission from Maddox and Bittner [82]. Copyright 2003 American Institute of Physics.

out the structure of the density. By 100 EM cycles, the clusters have found all four Gaussian components of the distribution and are struggling to recover the proper covariance in each component. The separable clusters are hindered in this because they have no freedom to rotate in the xy plane and must work collectively to capture the xy correlations. At 400 cycles we can see that the separable clusters have performed fairly well in finding the positions and relative orientations of the density components; however, the estimated density is somewhat distorted from the true distribution. The density estimate can be improved by including more sampling points and more clusters, but this also increases the computational demand.

Figure 5.3 illustrates the performance of the fully covariant model using four nonseparable clusters to describe the same data set. As expected, the fully covariant model performs much better than the separable case because the exact probability distribution is rigorously a mixture of four equally weighted nonseparable

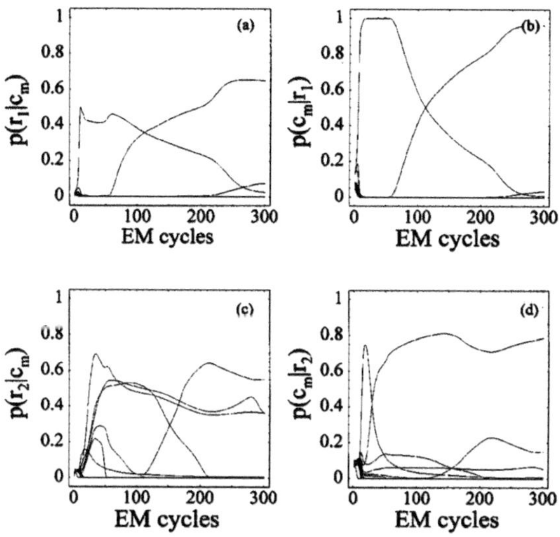

Figure 5.4 Plot (a) shows 16 separable forward probabilities at the tagged data point $\vec{r}_1$ as a function EM cycles. Plot (b) shows the corresponding posterior probabilities. Plots (c) and (d), respectively, depict the separable forward and posterior probabilities at the data point at $\vec{r}_2$. Reprinted with permission from Maddox and Bittner [82]. Copyright 2003 American Institute of Physics.

Gaussian components. After 1 EM cycle, the nonseparable clusters also collect near the mean of $\mathcal{R}$; however, they immediately develop nonzero off-diagonal covariances. Between 10-20 EM cycles, the clusters locate the individual density components. By 50 cycles, the clusters have established a stable configuration, which very closely mimics the true probability distribution.

Essentially, the EM algorithm performs a parallel search over the Gaussian parameter space and looks for regions where the clusters will be most effective in describing the data points. To help quantify the collective effort of the clusters, we examine the forward and posterior probabilities at the tagged data points $\vec{r}_1$ and $\vec{r}_2$. First, notice in Fig. 5.2 that there are never more than one or two Gaussian clusters centered near the point $\vec{r}_1$. Figures 5.4a and 5.4b, respectively, show how the $p(\vec{r}_1|c_m)$'s and $p(c_m|\vec{r}_1)$'s evolve for the separable cluster model. Both plots indicate that for the first 75 EM cycles, there is really only one cluster which dominates the density estimate at $\vec{r}_1$. The posteriors are particularly interesting

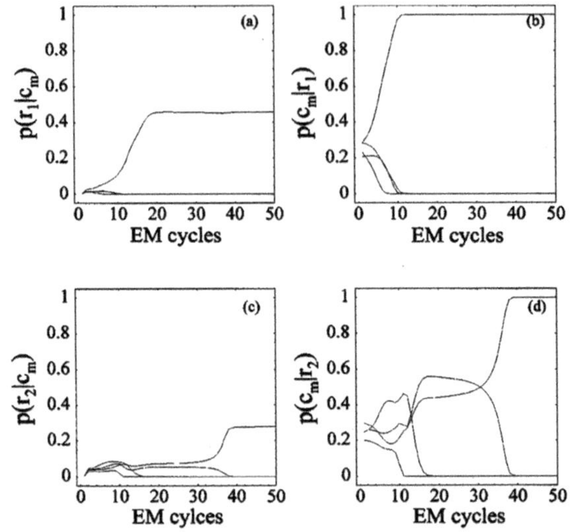

Figure 5.5 Plot (a) shows four nonseparable forward probabilities at the tagged data point $\vec{r}_1$ as a function EM cycles. Plot (b) shows the corresponding posterior probabilities. Plots (c) and (d), respectively, depict the four nonseparable forward and posterior probabilities at the data point at $\vec{r}_2$. Reprinted with permission from Maddox and Bittner [82]. Copyright 2003 American Institute of Physics.

because they reflect the fraction of explanatory power a data point gives to each of the cluster. The curve crossing at roughly 105 EM cycles reflects that the originally dominant cluster is eventually displaced by a different cluster that becomes nearly centered at $\vec{r}_1$. The situation for $\vec{r}_2$ is analogous to that for $\vec{r}_1$ but is complicated by the fact that there is greater overlap between multiple clusters. The forward and posterior probabilities at the point $\vec{r}_2$ are shown in Figs. 5.16c and 5.16d, respectively. Ultimately one cluster dominates the density estimate at $\vec{r}_2$, however, this is true to a lesser extent than at $\vec{r}_1$.

In Fig. 5.5 we plot the forward and posterior probabilities at $\vec{r}_1$ and $\vec{r}_2$ for the fully covariant model. It is clear that the behavior of the nonseparable Gaussian clusters is consistent with the separable ones. The exception to this is that the fully covariant clusters converge to a stable configuration in fewer EM cycles. This is due to the obvious fact there are many more equally good arrangements for

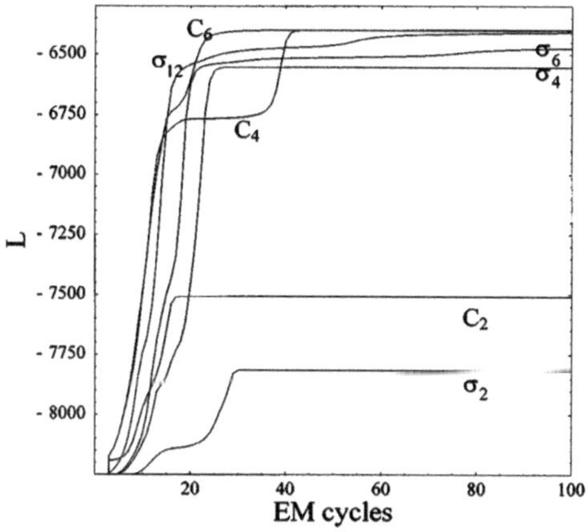

Figure 5.6 Plot of the log likelihood verses number of EM cycles for various density estimates of the data illustrated in Fig. 5.15. The notation σ_m and $\mathbf{C}_m$ refers to a density fit performed with m Gaussian clusters using the separable and fully covariant model, respectively. Reprinted with permission from Maddox and Bittner [82]. Copyright 2003 American Institute of Physics.

16 nonseparable Gaussians than there are for only 4 nonseparable Gaussians. Another way of saying this is that the log-likelihood has many more local maxima for the $M = 16$ separable model compared to the $M = 4$ nonseparable model. This point is highlighted by Fig. 5.6, where we plot the log-likelihood L for several different density fits. The separable cases are designated with a σ_m, where the integer m indicates how many Gaussians were used to perform the fit. Likewise, $\mathbf{C}_m$ refers to a fully covariant cluster fit with m nonseparable Gaussian components. The plateaus in the log-likelihood indicate that the EM algorithm is converging upon a root of the likelihood equation. It is conceivable that the EM algorithm could essentially become stuck at a local maximum or even a saddle point that does not give a particularly good density estimate. For these situations it is necessary to incorporate a small random perturbation in the cluster parameters in order move the fit away from such anomalous regions of parameter space. Another problem is that a cluster might become too focused on a single data point. This is described as a root of the likelihood equation lying on the exterior

of parameter space. When this happens, the variance and weight of the cluster become exceedingly small and in numerical applications will often round to zero, causing some terms in Eqs. 5.25–5.26 and 5.31–5.32 to diverge. This problem can be avoided in practice by adding a small fraction to the diagonal covariances in Eq. 5.32. This imposes an artificial boundary in parameter space that forces the Gaussian clusters away from the exterior roots.

5.2.4 The Ground State of Methyl Iodide

Now that we have highlighted some key features of the mixture model approximation and EM algorithm, we turn our attention to a problem with more physical merits. In order to propagate the quantum ensemble in time we must compute both the classical and quantum forces acting on the ensemble particles. Given a maximum likelihood estimate for ρ in the form of Eq. 5.25 or 5.26, it is a fairly straightforward exercise in book-keeping to compute an approximate quantum force:

$$F_Q = \frac{\hbar}{4m}\left(\frac{\vec{\nabla}^3\rho}{\rho} - \frac{\vec{\nabla}(\vec{\nabla}\rho\cdot\vec{\nabla}\rho)}{2\rho^2} - \left(\frac{\vec{\nabla}^2\rho}{\rho} - \frac{\vec{\nabla}\rho\cdot\vec{\nabla}\rho}{\rho^2}\right)\frac{\vec{\nabla}\rho}{\rho}\right) \tag{5.37}$$

in terms of the Gaussian parameters. The quantum and classical forces are then used to drive the ensemble of trajectories by integrating Eqs. 5.17 and 5.18 over a short time step using a Verlet leapfrog-type method. The EM algorithm is repeated using the previously fit cluster parameters as the starting point. Recycling the old parameters significantly decreases the number of EM cycles required to obtain convergence in the next density estimate. The whole process of alternating between EM cycles and Verlet steps continues until we have integrated the equations of motion to some appropriate final time.

For a Gaussian density packet evolving on a parabolic potential surface, the mixture model approximation requires only one cluster and is exact for all time. For nontrivial problems, however, the quantum density will generally exhibit a very complicated structure in configuration space. Clearly, the mixture model approximation will not be able to capture the exact intricacies of a realistic quantum distribution. Consequently, it is not feasible, using the present formulation of our methodology, to obtain numerically accurate quantum densities for nonstationary systems. Ground state quantum densities, on the other hand, are characteristically much simpler than their excited state and nonstationary counterparts. We believe our approach will be most useful for determining the ground state properties of high-dimensional systems.

For stationary systems the quantum force exactly counterbalances the classical force, and the ensemble of quantum trajectories does not evolve in time. The ground state can then be realized from a nonstationary state by adding a small damping term to Eq. 5.17:

$$\dot{\vec{v}} = F_Q - \vec{\nabla} V/m - \gamma \vec{v}, \tag{5.38}$$

where γ represents a small dissipative coefficient. This fictitious friction, in turn, causes the ensemble particles to lose a small amount of kinetic energy at each time step in the simulation. For a classical ensemble, the distribution collapses to a delta function(s) centered about the minimum energy point(s) of the potential surface. For the quantum mechanical ensemble, however, as the distribution becomes increasingly narrow, the quantum force becomes very strong and requires the ensemble to maintain some minimum finite width. At longer simulation times an equilibrium is reached, and the resulting distribution is representative of the ground state quantum density. The corresponding ground state energy can be resolved to within the statistical error of a Monte Carlo integration over the ensemble elements.

To illustrate this, we demonstrate the convergence of an initial Gaussian ensemble to the ground state distribution for the CH_3-I stretching/bending modes of the lowest electronic state of methyl iodide. For our purposes this provides a nontrivial anharmonic potential surface to test our methodology. The vibrational system is treated as a single-particle ($m = 20000$ amu) evolving on a two-dimensional (2D) model potential energy surface developed by Shapiro and Bersohn [97]. The potential energy curves for this anharmonic surface are depicted in Fig. 5.7 by the gray contour lines. In Fig. 5.7a we illustrate a numerically exact representation of

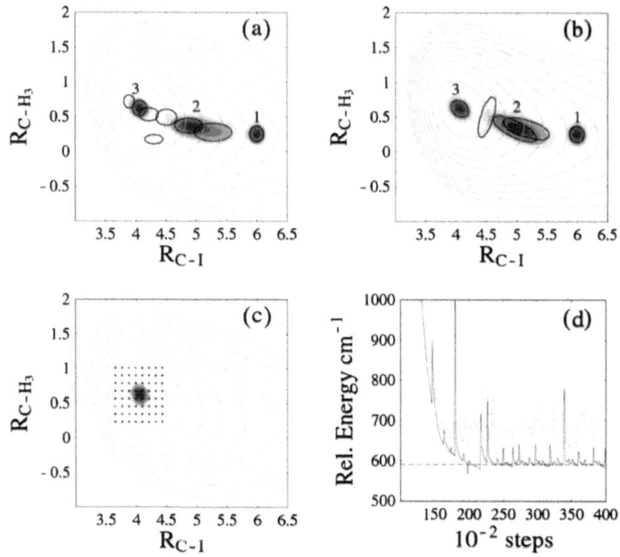

Figure 5.7 Plots (a) and (b) show the relaxation of a Gaussian wave packet in an anharmonic potential well for both the separable and fully covariant models, respectively. The gray contours reflect the potential energy curves for a model of CH_3I. The shaded contours indicate the shape of the approximated density after (1) 0, (2) 10 000, and (3) 40 000 Verlet time steps, respectively. The solid curves represent the half-width contours of the Gaussian clusters. Plot (c) shows the numerically accurate DVR ground state and the associated grid of quadrature points. Plot (d) shows the energy of the estimated density as a function of time steps. The dotted and solid data corresponds to the separable and nonseparable models, respectively, while the dashed horizontal line represents the DVR energy. Reprinted with permission from Maddox and Bittner [82]. Copyright 2003 American Institute of Physics.

the ground state density obtained by diagonalizing the Hamiltonian of the system using a 2D discrete variable representation (DVR). The grid points indicate the minimum number of Chebychev quadrature points required to obtain convergence in the lowest-energy eigenvalue. Obviously, a much larger grid would be necessary to perform a dynamical calculation on this system.

Figures 5.7b and 5.7c illustrate the estimated density for both the separable and fully covariant models, respectively. The black ovals represent the half-width contours of the Gaussian clusters. There are four clusters in the separable case and two for the fully covariant model. The various contour plots labeled (1), (2), and (3) correspond to snapshots of the estimated density at different points in the simulation. For both models, the initial density (1) is Gaussian, and all but one of the clusters are redundant. As the ensemble is propagated the individual clusters behave differently from one another. The contours (2) show the quantum density at an intermediate time after roughly 10 000 Verlet time steps ($\delta t =$ 1 atomic time unit). At longer times an equilibrium is achieved and the contours (3) are representative of the quantum ground state. In Fig. 5.7d we plot the energy of the system relative to the bottom of the potential well as a function of the number of Verlet time steps. The DVR energy at 591cm^{-1} serves as a benchmark and is indicated by the dashed horizontal line. The dotted and solid energy curves are for the separable and nonseparable models, respectively. Dropping the first 20000 time steps we find that the average energy for the separable case is 665.2 ± 33.6 cm^{-1}, which is well above the DVR energy. The average energy for the nonseparable model falls just barely within reach of the DVR energy at 580.0 ± 10.1 cm^{-1}. The sharp energy spikes for the nonseparable calculation are due to anomalous changes in the cluster parameters such as a sudden jump in μ_m or rotation of $\mathbf{C}_m$. These effects do not pose a significant problem since the clusters quickly respond to correct the abnormalities within a few time steps. For the sake of comparison we have also performed the same calculation for a mixture model with four fully covariant clusters. The average energy for the equilibrated system improved only slightly, but the statistical variation doubled 581.6 ± 21.7 cm^{-1}. The final arrangements of the clusters for the $M = 2$ and $M = 4$ covariant models are not discernible, and the additional clusters do not significantly aid or disrupt the global density fit. However, they do contribute to the sporadic deviations in the equilibrium energy.

In the following sections we present some of our work in applying this approach to study the thermodynamic properties of mesoscopic rare-gas clusters and conclude by presenting an extension for computing excitation energies.

5.3 Quantum Effects in Atomic Clusters at Finite Temperature

Rare-gas clusters approximated by the simple Lennard–Jones (LJ) Pair-wise potential are ideal test cases for many-body simulations [10, 79, 94], providing a useful benchmark for new methods. In addition, rare-gas clusters are often used to probe the transition from microscopic to macroscopic properties in atomic systems. The mesoscopic regime has many unique properties wide variety of theoretical and experimental techniques [8, 24, 58, 66].

In this section, we extend the quantum hydrodynamic method developed in the previous section [37] to study the nature of quantum effects for mesoscopic systems at finite temperature through the entropic functional given by Mermin [88, 89]. Our approach assumes that the configurational density $n(\vec{r}_1, \ldots, \vec{r}_N)$ can be represented with a superposition of statistical approximates, $p(\mathbf{r}_1, \ldots, \vec{r}_N, c_m)$. The algorithm then uses a Bayesian analysis to determine the best statistical approximates given a statistical sampling of the density. It then uses a grid-free hydrodynamic adaptive approach to relax sample points that make up a statistical sampling of the quantum density to the ground state equilibrium density.

In what follows, we present a brief overview of the grid-free adaptive hydrodynamic approach for computing the quantum ground state density for a system of N nuclei introduced earlier and then show how it can be extended to finite temperature. We also give a review of the Bayesian analysis used to deduce the best set of m statistical approximates from a statistical sampling of the density. We then show the quantum hydrodynamical scheme used to adapt the sample points towards a minimal energy configuration. We will then present results on clusters of neon of up to 37 atoms ($N = 37$) for temperatures from 0 K to 30 K, which spans the solid to liquid transition for bulk Ne ($T_m = 24.56$ K and $T_b = 27.07$ K). In the present work we will demonstrate that quantum effects can indeed be captured with our hydrodynamic method at finite temperature and that quantum effects lead to some thermodynamic behavior for small, symmetric clusters.

5.4 Quantum Structures at Zero and Finite Temperature

5.4.1 Zero Temperature Theory

The Euler–Lagrange equation for the motion of our particles is then derived with the help of the hydrodynamic description of quantum mechanics. We begin by specifying the full many-body Hamiltonian and will follow along similarly to density functional theory (DFT) [93]. The potential corresponds to the nuclear motion of a collection of atoms with pair-wise interaction potentials.

$$H = -\sum_{i=1}^{N} \frac{1}{2m_i} \vec{\nabla}_i^2 + \sum_{i \neq j} V(ij), \tag{5.39}$$

where the first is the sum of the kinetic energies of the individual atoms and the second is the sum of the potential energy contributions. $\vec{r}_i$ is the vector location of atom i, and n_i is the corresponding density. We also have an arbitrary N-body trial density given by:

$$n(\vec{r}) = \sum_i n_i(\vec{r}_i). \tag{5.40}$$

The energy functional corresponding to this density and Hamiltonian is given by:

$$E[n] = T[n] + \sum_{i \neq j} \int \int n_i(\vec{r}_i) n_j(\vec{r}_j) V(ij) d\vec{r}_i d\vec{r}_j. \tag{5.41}$$

The kinetic energy operator is separable since we have assumed distinguishability amongst the constituent atoms. Therefore, the kinetic energy term is the sum of the individual kinetic energy functionals.

$$T[n(1 \cdots N)] = \sum_{i=1}^{N} T_i[n_i(\vec{r}_i)] \tag{5.42}$$

As in electronic structure DFT, evaluating the kinetic energy functionals is problematic since evaluating the quantum kinetic energy is a nonlocal operator and the density is a local function [93].

If instead we write the quantum wave function in polar form, as in the hydrodynamic formulation of QM [31, 33, 83] and also in the time-dependent DFT formulation [6, 7]:

$$\Psi(\vec{r}) = \sqrt{n(\vec{r})} e^{i\phi(\vec{r})}, \tag{5.43}$$

we can arrive at a stationary condition that if $\vec{\nabla}\phi = 0$ [52]:

$$V(1\cdots N) - \sum_i \frac{1}{2m_i}\frac{1}{\sqrt{n_i(\vec{r}_i)}}\nabla_i^2\sqrt{n_i(\vec{r}_i)} = const, \qquad (5.44)$$

at all points in space. The constant is the energy of the system. We note here the similarity of the second term in the previous equation with the quantum force from diffusion Monte Carlo, $\vec{\nabla}\psi/\psi$. This term is also known as the quantum potential in the Bohm formulation. By inspection, then, we can define our kinetic energy functional as:

$$T[n(\vec{r}_i)] = -\frac{1}{2m_i}\int \sqrt{n_i(\vec{r}_i)}\nabla_i^2\sqrt{n_i(\vec{r}_i)}d\vec{r}_i. \qquad (5.45)$$

Integrating by parts and taking $n(i) \rightarrow 0$ at $\pm\infty$ produces the familiar von Weizsacker kinetic energy functional [102]:

$$T_W[n(\vec{r}_i)] = +\frac{1}{8m}\int \frac{1}{n_i(\vec{r}_i)}\vec{\nabla}_i n_i(\vec{r}_i)\cdot\vec{\nabla}_i n_i(\vec{r}_i)d\vec{r}_i. \qquad (5.46)$$

Thus, the total energy functional is given in terms of the single-particle densities.

$$E[n] = \sum_{i=1}^{N} T_W[n_i(\vec{r}_i)] + \sum_{i\neq j}\int\int n_i(\vec{r}_i)n_i(\vec{r}_j)V(ij)d\vec{r}_i d\vec{r}_j. \qquad (5.47)$$

Taking the variation of $E[n]$ with respect to the single-particle densities with the constraint that $\sum_i \int n_i(\vec{r}_i)di = N$:

$$\delta\left\{\sum_{i=1}^{N}\left[T_W[n_i(\vec{r}_i)] + \sum_{j\neq i}\int\int n_i(\vec{r}_i)n_j(\vec{r}_j)V(ij)d\vec{r}_i d\vec{r}_j \right.\right.$$
$$\left.\left. - \mu\left(\int n_i(\vec{r}_i)d\vec{r}_i - 1\right)\right]\right\} = 0, \qquad (5.48)$$

leads to the following Euler–Lagrange equations:

$$\frac{\delta T_W[n_i(\vec{r}_i)]}{\delta n_i(\vec{r}_i)} + \sum_{j\neq i}\int V(ij)n_j(\vec{r}_j)d\vec{r}_j - \mu = 0. \qquad (5.49)$$

When satisfied, μ is the vibrational ground state energy, and the $n_i(\vec{r}_i) = |\phi_i(i)|^2$ are the probability densities of the individual nuclei. This leads to an effective mean field potential for each atom of the form:

$$V_i^e = Q(\vec{r}_i) + V_e(\vec{r}_i) + \sum_{j=1}^{N} V_p(\mathbf{r}_i, \vec{r}_j). \qquad (5.50)$$

Here, $Q(\vec{r})$ is the quantum potential, $V_e(\vec{r}_i)$ is an external potential, which corresponds to any external driving field ($V_e = 0$ in the present study), and $V_p(\vec{r}_i, \vec{r}_j)$ is the pair-wise interatomic interaction potential.

5.4.2 Finite Temperature Theory

For a system at finite temperature under the conditions of a grand ensemble, $Z(T, V, \mu) = \mathrm{Tr}\{e^{-\beta(H-\mu N)}\}$, an equilibrium state density matrix will minimize Ω, the grand potential. This is given by:

$$\Omega = -\frac{1}{\beta}\ln(\mathrm{Tr}\{e^{-\beta(H-\mu N)}\}), \tag{5.51}$$

where H is the Hamiltonian, N is the number operator, and μ, the chemical potential. Nearly 40 years ago, Mermin [88] showed that by writing Ω as a functional of an arbitrary trial density matrix:

$$\Omega[\hat{\rho}_T] = \mathrm{Tr}\{\hat{\rho}_T(K + V - \mu N + \frac{1}{\beta}\ln\hat{\rho}_T)\}, \tag{5.52}$$

$\delta\Omega = 0$ only if the correct density matrix is used. Thus, for any trial density matrix $\hat{\rho}_T \neq \hat{\rho}$ then $\Omega[\hat{\rho}_T] \geq \Omega[\hat{\rho}]$. Mermin also shows that there is a unique density associated with the equilibrium density matrix, $n(\vec{r}) = tr\{\hat{\rho}|\psi(\vec{r})|^2\}$. This implies then that one can write the grand potential as a functional of the density:

$$\Omega[n(\vec{r})] = F[n(\vec{r})] - \mu \int n(\vec{r})d\vec{r}, \tag{5.53}$$

where our free energy functional $F[n]$ is given by:

$$F[n(\vec{r})] = \mathrm{Tr}\{\hat{\rho}[n(\vec{r})](K + V + \frac{1}{\beta}\ln\hat{\rho}[n(\vec{r})])\}. \tag{5.54}$$

$$= T_w[n(\vec{r})] + V[n(\vec{r})] + \frac{1}{\beta}S[n(\vec{r})]. \tag{5.55}$$

Here we have made the substitution of $S[n(\vec{r})]$ for the entropic term. The T_w functional is the Weizsacker functional, which along with the potential functional, is identical to our previous work [37] as well as Section 5.4.1. We have excluded exchange and correlation terms since we have assumed noninteracting particles. The inverse temperature, β is a Lagrange multiplier used in the determination of the ground state density. This is similar to the chemical potential used previously in the determination of the ground state at $T = 0$. The minimum of the Ω functional will correspond to the atomic density profile of the system at a given temperature. Note that the free energy functional, $F[n]$, contains the kinetic energy and external potential operators as well as an entropy/temperature term, so that now the stationary equilibrium state will be an energetic compromise between the quantum and the entropic potentials, both of which tend to destabilize the clusters, and a mean field interaction potential that tends to stabilize the clusters. The net effect is that as the cluster temperature increases, the clusters will be increasingly unstable and undergo transitions from ordered to disordered states.

The thermodynamic justification for the form of the entropic functional defined by Mermin can be seen from the form of the entropy in Boltzmann's eulogistic equation

$$S = k_B \ln(\Omega_{mc}), \tag{5.56}$$

where Ω_{mc} is the microcanonical density of states. Since we can write the microcanonical density of states in terms of the density matrix in the von Neumann definition of entropy, we can write the above equation as:

$$S = -k_B \mathrm{Tr}\{\hat{\rho} \ln(\hat{\rho})\}. \tag{5.57}$$

This is also sometimes called the differential entropy. The entropic functional takes into account the contribution from temperature-entropy work into our energy functional. β problem of path integral based approaches is avoided.

We can minimize Ω to obtain the chemical potential:

$$\mu = \frac{1}{\beta}\frac{\delta S[n]}{\delta n(\vec{r})} + Q(\vec{r}) + V_{ext}(\vec{r}). \tag{5.58}$$

Again Q is the quantum potential derived from the functional derivative of the Weizsacker term, $Q = \delta T[n]/\delta n$, and in a similar manner V is simply the mean field potential of a given atom in terms of all the other atoms. Now all that remains is to calculate this iteratively as before with a temperature correction related to $S[n(\vec{r})]$.

We assume that the entropic contribution is additive and can be derived using the von Neumann entropy:

$$S[n(\vec{r})] = \frac{1}{\beta}\sum_i \int n_i(\vec{r}_i) \ln(n_i(\vec{r}_i)) d\vec{r}_i = \sum_i S[n(i)]. \tag{5.59}$$

where the sum is over individual atoms. Taking the functional derivative with respect to the density, needed in the equations of motion of the particles, we define an "entropic force" as:

$$\frac{\delta S[n(\vec{r})]}{\delta n_i} = \frac{1}{\beta}(\ln(n_i(\vec{r}_i)) + 1). \tag{5.60}$$

In order to test our assumption and verify its range of applicability, we consider a simple harmonic system with a normalized Gaussian density function:

$$n(x) = \sqrt{\frac{a}{\pi}} e^{-x^2/2\langle x^2 \rangle}. \tag{5.61}$$

As such, the free energy is given by (with $\hbar = 1$):

$$\langle F \rangle = \frac{1}{8m\langle x^2 \rangle} + \frac{m\omega^2}{2}\langle x^2 \rangle - \frac{1}{2\beta}\left(\ln(2\pi\langle x^2 \rangle) + 1\right) \tag{5.62}$$

The first two terms are simply the average kinetic and potential energies. The last term is temperature dependent and arises from the entropy contribution. Minimizing $\langle F \rangle$ with respect to a yields an optimal width parameter of:

$$\langle x^2 \rangle_{opt} = \frac{k_B T}{2m\omega^2}\left(1 + \sqrt{1 + (\hbar\omega/k_B T)^2}\right) \tag{5.63}$$

For comparison the exact expression for the width of a harmonic oscillator at finite temperature is [40]:

$$\langle x^2 \rangle_{exact} = \frac{\hbar}{2m\omega} \coth\left(\frac{\Theta_v}{2T}\right) \tag{5.64}$$

where $\Theta_v = \hbar\omega/k_B$ is the vibrational temperature. In Fig. 5.8 we compare the Mermin functional versus the exact result for a system with $\hbar/m\omega = 1$. Clearly, the delocalization obtained using the Mermin functional approaches the exact result asymptotically at high temperatures. It also agrees perfectly at $T = 0$. However, for intermediate temperatures $T \lesssim 2\Theta_v$, $\langle x^2 \rangle_{\text{approx}} > \langle x^2 \rangle_{\text{exact}}$. Consequently, we anticipate that the Mermin approximation to the entropy functional will systematically over estimate quantum delocalization effects for $T < \Theta_v$.

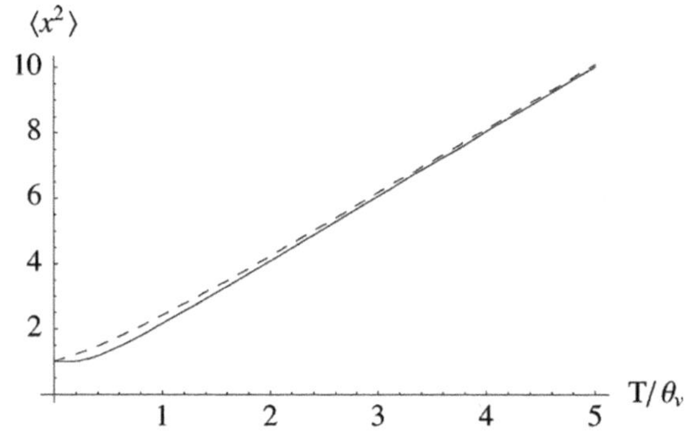

Figure 5.8 $\langle x^2 \rangle$ versus T comparing the approximate entropy functional of Eq. 5.59 (- - -) to the exact (—) value for a harmonic system at finite temperature. Reprinted with permission from Derrickson and Bittner [36]. Copyright 2007 American Chemical Society.

5.4.2.1 Computational approach: The mixture model

In order to utilize the hydrodynamic description one needs a quantitative description of the density. This will be done directly from an ensemble of points sampled from the initial quantum density in the following way. To begin, the single-particle probability distribution functions (PDF) can be represented by a mixture model [45, 87] by summing a finite number M of density approximates:

$$n(\vec{r}) = \sum_{m}^{M} p(\vec{r}, c_m), \tag{5.65}$$

where $p(\vec{r}, cm)$ is the probability that a randomly chosen member of the ensemble has the configuration $\vec{r}$ and is a variant of the mth approximate designated by c_m. These approximates may be Gaussians or any other integrable multidimensional function, which can be parameterized by its moments. For Gaussian clusters, we have a weight $p(c_m)$, a mean position vector μ_m, and a covariance matrix C_m.

By definition, each joint probability in Eq. 5.65 is related to a pair of conditional probabilities according to the relation:

$$p(\vec{r}, c_m) = p(c_m)p(\vec{r}|c_m) = n(\vec{r})p(c_m|\vec{r}). \tag{5.66}$$

The forward conditional probability $p(\vec{r}|c_m)$ refers to the probability that a randomly chosen variant of c_m has the configuration $\vec{r}$, and the posterior probability $p(c_m|\mathbf{r})$ refers to the probability that the configuration point $\vec{r}$ is a variant of the approximate c_m. Notice, $n(\vec{r})$ and $p(c_m)$ are the quantum density and weight of the mth approximate, respectively.

As shown in our previous works [37, 82], this formulation can be used to define a multidimensional quantum density with user-defined amounts of correlation between the particles. Briefly we can outline our procedure as follows. With a Gaussian model representing the full $3N$ dimensional system:

$$p(\vec{r}|c_m) = \sqrt{\frac{\|\mathbf{C}^{-1}\|}{(2\pi)^{N_d}}} e^{(\vec{r}_d-\mu_{m,d}).\mathbf{C}_{\mathbf{m}}^{-1}.(\mathbf{r}_\mathbf{d}-\mu_{\mathbf{m},\mathbf{d}})}. \tag{5.67}$$

the covariance matrix, **C**, can be used to eliminate (or maintain) coupling between various degrees of freedom. Then one must determine the Gaussian parameters $p(c_m)$, μ_m, and C_m, that define the density. This is facilitated using an iterative expectation maximization (EM) algorithm. In each case, these are readily approximated by summing over an ensemble of points $\{\vec{r}_n\}$ sampled from the $n(\vec{r})$ or probability distribution function. For instance, the mean positions are approximated with:

$$\vec{\mu}_m \approx \frac{1}{Np(c_m)} \sum_n^N \vec{r}_n p(c_m|\vec{r}_n). \tag{5.68}$$

The updated Gaussian parameters are then used to update the posterior terms $p(c_m|\vec{r}_n)$ for each $\vec{r}_n$ sample point by inserting this back into Eq. 5.67 and using Bayes's equation:

$$p(c_m|\vec{r}_n) = \frac{p(c_m)p(\vec{r}_n|c_m)}{\sum_m p(c_m)p(\vec{r}_n|c_m)}. \tag{5.69}$$

This procedure progressively solves for the best set of parameters, given a distribution of sample points.

The EM algorithm described above allows us to generate an approximate analytical functional form for the single-particle density via statistical sampling over an ensemble of points. The next step is to adjust the single-particle densities themselves to produce a lower total energy. We do this by deriving the quantum hydrodynamic equations of motion for the sample points, $\vec{r}_{in}$ where i labels a given atom and n labels a given sample point associated with density $n_i(\vec{r})$.

5.4.2.2 Computational approach: Equations of motion for the sample points

The quantum Hamilton–Jacobi equation generates the equations of motion for the ray lines of a time-dependent solution to the Schrödinger equation [15, 16, 18, 111]. This allows convergence to the ground state by relaxing along an action field determined for each atom. This gives a set of time-dependent, self-consistent field equations whereby the motion of atom i is determined by the average potential interaction between atom i and the rest of the atoms in the system.

$$\dot{S}_i(\vec{r}) + \frac{|\vec{\nabla}_i S|^2}{2m_i} + \sum_{j\neq i} \int V(ij) n_j(\vec{r}) d\vec{r} - \frac{1}{2m_i} \frac{1}{\sqrt{n_i(\vec{r})}} \nabla_i^2 \sqrt{n_i(\vec{r})} + \frac{1}{\beta} \int n_i(\vec{r}) \ln(n_i(\vec{r})) = 0. \quad (5.70)$$

Taking $\vec{\nabla} S = \mathbf{p}$ as a momentum of a particle, the equations of motion along a given ray line or sample particle $\mathbf{r_{in}}(t)$ of the quantum wave function are given by:

$$m_i \ddot{\vec{r}}_{in} = -\sum_{j\neq i} \int (\vec{\nabla}_i V(ij)) n(\vec{r}_j) d\vec{r}_j - \vec{\nabla}_i Q[n(\vec{r}_i)] + \vec{\nabla}_i \frac{\delta S[n(i)]}{\delta n_i} \quad (5.71)$$

where $Q[n(i)]$ is the Bohmian quantum potential specified by the last term in Eq. 5.70. Stationary solutions of the time-dependent Schrödinger equation are obtained whenever $m_i \ddot{\vec{r}}_n = 0$. Consequently, we reach the ground state by relaxing the sample points in a direction along the energy gradient:

$$\vec{\nabla}_i E = -\sum_{j\neq i} \int (\vec{\nabla}_i V(ij)) n_j(\vec{r}_j) d\vec{r}_j - \vec{\nabla}_i Q[n_i(\vec{r}_i)] + \vec{\nabla}_i \frac{\delta S[n(i)]}{\delta n_i}. \quad (5.72)$$

keeping $n(\vec{r}_j)$ fixed. This generates a *new* statistical sampling that we then use to determine a new set of approximates and the process is repeated.

The algorithm can be summarized as follows:

(1) For each atom, generate and sample a normalized trial density $n_i(\vec{r}_i)$.
(2) Using the EM routines and the given sample of points, compute the coefficients for the density approximates.
(3) Compute the forces on each point using Eq. 5.70 and advance each point along the energy gradient for one "time" step, either discarding or dampening the velocity of each point. This generates a new sample of points describing the single-particle density for each atom. The new distribution should have a lower total energy since we moved the sample points in the direction towards lower energy.

Iterating through these last two steps, we rapidly converge towards the global quantum energy minimum of the system.

5.4.3 Computational Studies

5.4.3.1 Zero temperature results

In all the calculations presented here, we used 100 statistical points to represent the density of each atom and propagated the SCF equations described above until the energy and the density were sufficiently converged. To reach convergence, this typically required a few hundred thousand iterations. The LJ parameters used for the neon atoms are $\epsilon = 0.3059$ kJ/mole and $\sigma = 2.79$Å [76]. The initial centers of the Gaussian approximates correspond to the position of the global energy minima for each cluster with initial widths taken from a harmonic oscillator approximation.

The primary motivation for continuing the study of these clusters at zero temperature is the desire to be able to simulate the quantum dynamics of bulk systems. In our previous work [37], we were limited to systems with less than 20 atoms. Subsequent improvement of our algorithms have allowed us to substantially advance past this limit. Furthermore, it is recognized that roughly 110 atoms per unit cell are required to approximate the bulk behavior of many-cluster systems [98]. Even so, our current

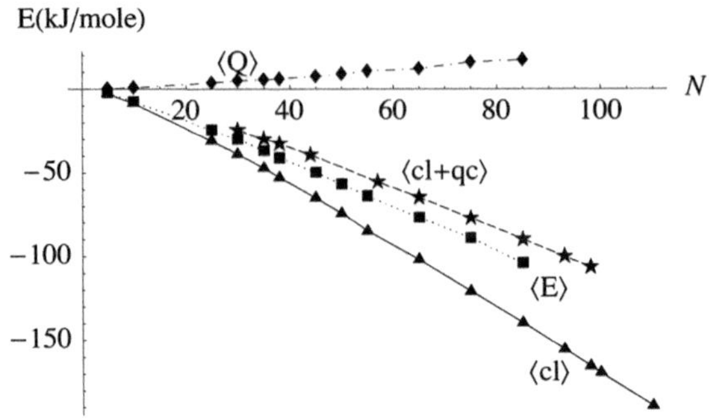

Figure 5.9 Various energetic contributions for quantum and classical neon clusters versus cluster size, N. Key: ⟨cl⟩ = classical global potential minimum energy, ⟨cl+qc⟩ = zero-point energy corrections from Calvo et al. [24], ⟨E⟩ = total energy, ⟨Q⟩ = quantum kinetic energy (from quantum potential). Reprinted with permission from Derrickson and Bittner [36]. Copyright 2007 American Chemical Society.

computational resources limited us to clusters with up to 85 atoms at $T = 0\ K$ and roughly half this at higher temperatures.

Our $T = 0K$ results are summarized in Fig. 5.9, which shows the various contributions to the total energy. First, we note that the contribution from the quantum potential (average kinetic energy) increases monotonically with system size. Moreover, the total energy $\langle E \rangle$ decreases monotonically. This is to be expected since the larger clusters have increasingly more nearest-neighbor interactions as the size of the system increases.

In Fig. 5.9 we also compare the present results to a similar semiclassical study by Calvo et al. [24]. In their results the zero-point energy of the static structure of the global minimum was calculated and then added in an ad hoc fashion to the pair-potential interaction. Generally, our results lie somewhat lower in total energy than the semiclassical estimates but above the classical global energy minimum for each cluster. We do note, however, that the inclusion of the quantum potential alters the total energy surface. Consequently, in some cases, the system could relax to a different minimum or in a superposition of close-lying minima due to tunneling. The clustering model can handle this situation through the inclusion of multiple Gaussian approximates for each atom. However, in each case examined here, we did not observe serious deviations or tunneling between nearly degenerate structures.

However, it is possible that the inclusion of quantum delocalization can influence the energetic ordering of nearly equivalent structures. Calvo et al. also investigated changes in ground state structure as a result of quantum delocalization. They did this using a basin-hopping Monte Carlo optimization algorithm to explore the energy landscape of small Ne clusters with less than 100 atoms. In this study the zero point energy contributions were again approximated in an ad hoc fashion similar to that shown in Fig. 5.9. This can be summarized with the following: an initial Monte Carlo search over the potential energy hypersurface is performed to determine a test configuration. The zero-point energy of this test configuration is determined using the static atomic positions. The calculated zero-point energy is then added to the classical potential energy and this sum is used for the Metropolis acceptance criteria. This process is repeated until the lowest energy configuration is determined, now including both the pair-potential and the zero-point energy.

In our study as well as that from Calvo et al. [24], the starting configurations were based upon the global classical minimum on the potential energy hypersurface of the cluster. In Calvo et al.'s semiclassical results, quantum effects produced a different global minimum for 35 out 99 cases for Ne_n in the range of $n \leq 100$. For example, the 17-atom cluster has three nearly equivalent minima (17A, 17B, and 17C) with energies $E_C < E_B < E_A$ separated by substantial potential barriers. Likewise, $n = 27$ and $n = 28$ each have two energetically similar minima. The energies (from Calvo et al. [24]) of these are given in Table 5.1. The geometries of representative clusters discussed in this section are shown in Fig. 5.10. Remarkably, our results show a different ordering of the energies of these structures compared to the semiclassical results. The difference between the two results is consistent with the general trend shown in Fig. 5.9 and corresponds to the different levels of theory used in each study. In the semiclassical approach, zero-point contributions are estimated from the curvature of the potential, *after energy "relaxation"* on the *potential* energy hypersurface. However, in our approach the quantum delocalization self-consistently *alters* the $3N$-dimensional *total* energy hypersurface being sampled.

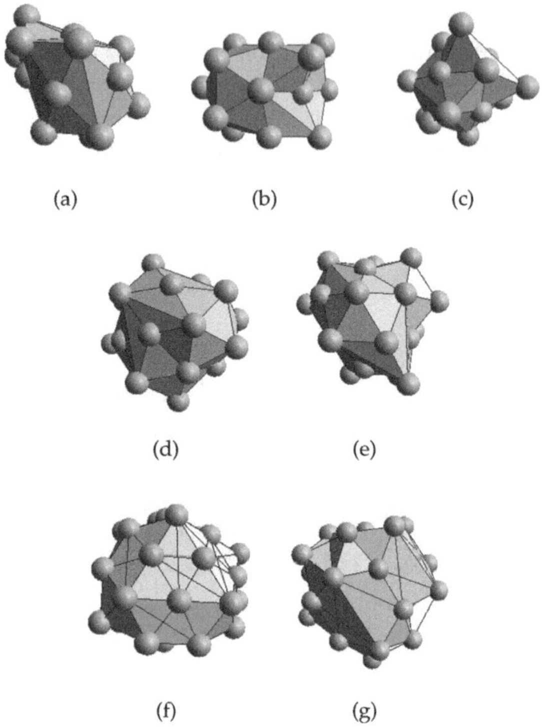

Figure 5.10 Minimal energy Lennard–Jones clusters from the Cambridge Cluster database. (http://www-wales.ch.cam.ac.uk) Point group is given in parenthesis. a: 17a (C_2), b: 17b (C_1), c: 17c (C_{3v}), d: 27a (C_{2v}), e: 27b (C_s), f: 37a (C_1) g: 37b (C_1).

The effects from quantum delocalization were approximated by using the structure's pair-potential value as well as the zero-point energy contribution inserted in an ad hoc manner to the Metropolis acceptance probability of the sampling from the potential surface. The initial structures used were based upon the global classical minimum. We note that there could be slight differences in the ground state structures of the clusters compared in Fig. 5.9. This is because, although the initial atom positions used for the hydrodynamic simulations were also taken from the global classical minimum, the hydrodynamic approach has a slightly altered energy surface and could possibly relax to a new minimum. This is not expected to result in serious deviations in the structure for the sizes listed.

In Calvo et al. [24] also analyzed the lowest-energy structures using a global optimization algorithm but with zero-point energy added in an ad hoc fashion to see the changes the quantum delocalization would have on the different ground state structures compared to classical results. According to their results neon was heavily influenced by quantum affects and showed a different global minimum for 35 out 99 cases for Ne_n in the range of $n \leq 100$. They state

> Although challenging, it would be interesting to verify these results (the structures and their energies) with more accurate quantum Monte Carlo calculations at $T = 0$. A reasonable test case would be the size $n = 17\ldots$ [24].

We have done this for all three structures of 17 atoms as well as the two structures they identified for 27- and 28-atom clusters of neon. We summarize these results in Table 5.1. The different structures are identified with 17A, 17B, etc., and the orderings of the energies are identified in the columns, i.e., the lowest-energy structure is labeled with (1) and so on. The important point shown in Table 5.1 is that the orderings of the energies are different. Note that the difference in energies between columns 2 and 3 is consistent with the general trend shown in Fig. 5.9 and corresponds to the different levels of theory used. We think that our calculations are

Table 5.1 Ground state vibrational energies for Ne_n clusters for our results compared to the results tabulated by Calvo et al. [24]

Cluster	Order	Energy (From [24])	Order	This work
17C	(1)	−11.0853 kJ/mol	(2)	−16.6336 kJ/mol
17B	(2)	−11.0814	(3)	−16.3188
17A	(3)	−11.0633	(1)	−16.6699
27B	(1)	−21.5483	(2)	−27.6994
27A	(2)	−21.5099	(1)	−28.2823
28B	(1)	−22.5892	(2)	−28.7459
28A	(2)	−22.5496	(1)	−29.3524

Adapted with permission from Derrickson and Bittner [36]. Copyright 2007 American Chemical Society.

consistent with the general statement that ground state structures will be altered depending on the quantum delocalization, but our results indicate that ad hoc techniques of correcting for quantum delocalization are insufficient for global optimization algorithms since the zero-point energy is added after the "relaxation" portion. This is because the quantum delocalization actually changes the $3N$-dimensional energy surface. The hydrodynamic method we have outlined might be used in conjunction with these global optimization algorithms for a more accurate picture of the quantum effects.

5.4.3.2 Finite temperature results

The thermodynamics of small mesoscale systems is of considerable interest since what are typically extensive variables (e.g., total energy, entropy, etc.) that scale monotonically with system size can exhibit anomalous behavior as the system size becomes small. Add to this the influences of quantum delocalization, and one anticipates the predicted thermodynamics of these system to exhibit behavior quite different from the bulk or even from a purely classical prediction.

One attractive way to introduce quantum corrections into an otherwise classical molecular dynamics or Monte Carlo simulation is through the use of an effective "quantum potential." Typically such effective potentials are expansions of the quantum partition function in powers of $\hbar$. The Feynman–Hibbs potential is derived by characterizing a quantum particle with a Gaussian that has a width equal to the thermal de Broglie length centered about the particle and accounts for the spread in density expected for quantum particles. Under these assumptions the partition function can be simplified, and with a Gaussian density the pair potential term would be evaluated as:

$$V(\vec{r}_{ij}) = \left(\frac{2\mu}{\pi\beta\hbar^2}\right)^{3/2} \int dR V(|r+R|) e^{-\frac{2\mu}{\beta\hbar^2}R^2}, \quad (5.73)$$

with some reduced mass, μ. The effective potential can then be found by expanding about r and truncating at some convenient order.

$$V_{\text{eff}}(r) = V(r) + \frac{\hbar^2 \beta}{24m} V''(r). \qquad (5.74)$$

Such an approach was used by Calvo et al. [24] in their very comprehensive survey of how quantum delocalization affects the structure and energetics of rare-gas clusters and as such provides a highly useful point of comparison for our approach. We do note that these expansions assume λ to be small (compared to the local variation in the potential), as per the semiclassical WKB criteria. Consequently, for lower temperatures and higher degrees of quantum delocalization such effective quantum corrections are not applicable.

Here we focus on three clusters, Ne_{13}, Ne_{17}, and Ne_{37} over a temperature range spanning the solid to liquid transition for bulk Ne. In the figures which display the thermodynamic data the temperature is given in terms of reduced unit which is the temperature in Kelvin multiplied by Boltzmann's constant and divided by the well depth of the LJ potential, $T' = T k_B/\epsilon$. Figure 5.11 shows the total free energy (scaled to a common $T = 0$K origin) versus temperature for the three clusters. Figure 5.12 shows the various contributions to the total free energy for the 13-atom cluster with similar behavior for the other clusters. First, the contribution from the quantum potential increases, as it should as T increases. The averaged quantum potential is simply the average quantum kinetic energy and as such is approximately inversely proportional to the de Broglie wavelength squared, $\langle Q \rangle \propto \lambda^{-2}$. Hence, $\langle Q \rangle$ increases as the system becomes more localized, corresponding to an increasingly shorter thermal de Broglie wavelength as T increases.

At higher temperatures, though, the quantum effects will be washed out as the de Broglie wavelength goes to zero. So, we expect that these factors will only be apparent at lower temperatures. The de Broglie wavelength is decreasing because the entropic potential causes an increase in the effective well depth that the atom feels with increasing T. As this happens, the cohesive forces increase in response to the decreased delocalization. This is a counterintuitive result since the cohesive forces are expected to decrease at higher temperatures. This results from the ability of atomic clusters to store

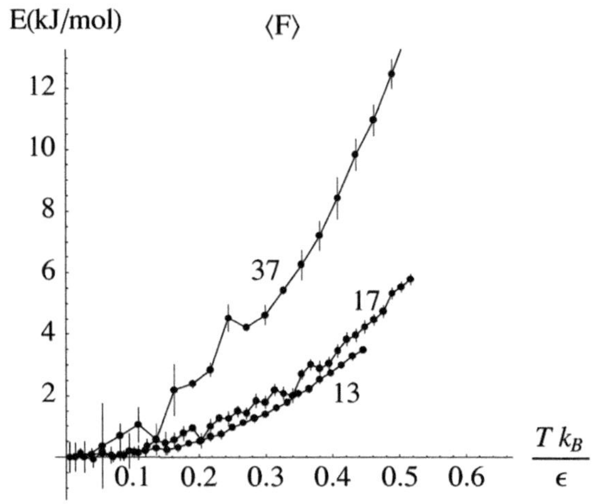

Figure 5.11 The free energy of the different clusters versus reduced temperature $k_B T/\epsilon$. Error bars indicate numerical/statistical precision of each computed free energy value. Note the $T = 0$ values are offset to a common origin for comparison. The energies at $T = 0$ K for the three clusters are as follows: $F_0^{13} = -11.21$ kJ/mol, $F_0^{17} = -15.216$ kJ/mol, and $F_0^{37} = -39.03$ kJ/mol. Reprinted with permission from Derrickson and Bittner [36]. Copyright 2007 American Chemical Society.

energy as internal interaction energy rather than as kinetic energy. This aspect of mesoscopic clusters is discussed later.

It is useful to compare the results we have obtained with the analytical results obtained using the Debye model which is known to have the correct low-temperature behavior for the heat capacity in the bulk material. The Debye model has a single adjustable parameter, the Debye temperature, defined by:

$$T_D = \frac{hc_s}{2k_B}\left(\frac{6N}{\pi V}\right)^{-3}. \tag{5.75}$$

where $N/V = \rho$ is the bulk density and c_s is the speed of sound in the medium. From this we can derive the internal energy as:

$$U = 9Nk_B T(T/T_D)^3 \int_0^{T_D/T} \frac{x^3}{e^x - 1}dx. \tag{5.76}$$

In general, T_D is determined by fitting the model to experimental thermodynamic data. For bulk Ne, $T_D = 75$K.

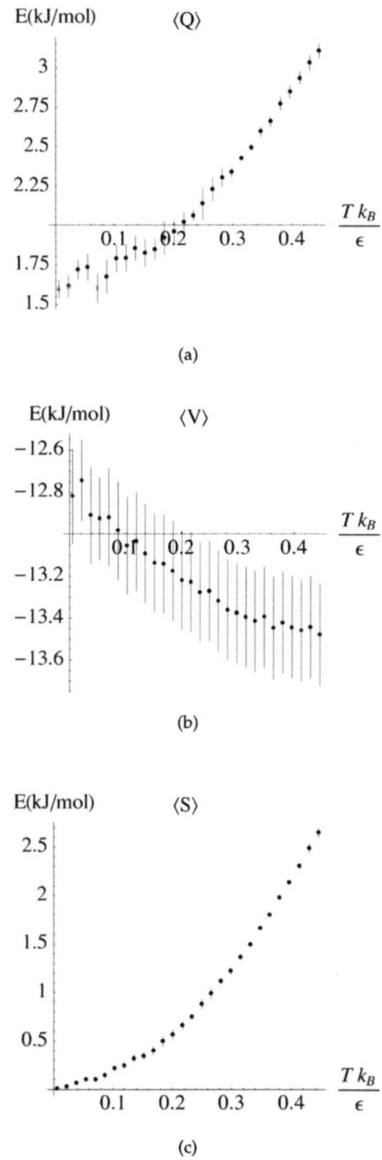

Figure 5.12 Plots of the quantum (a), total potential (b), and entropic (c) contributions to the total free energy versus reduced temperature $k_B T/\epsilon$ for Ne_{13}. Reprinted with permission from Derrickson and Bittner [36]. Copyright 2007 American Chemical Society.

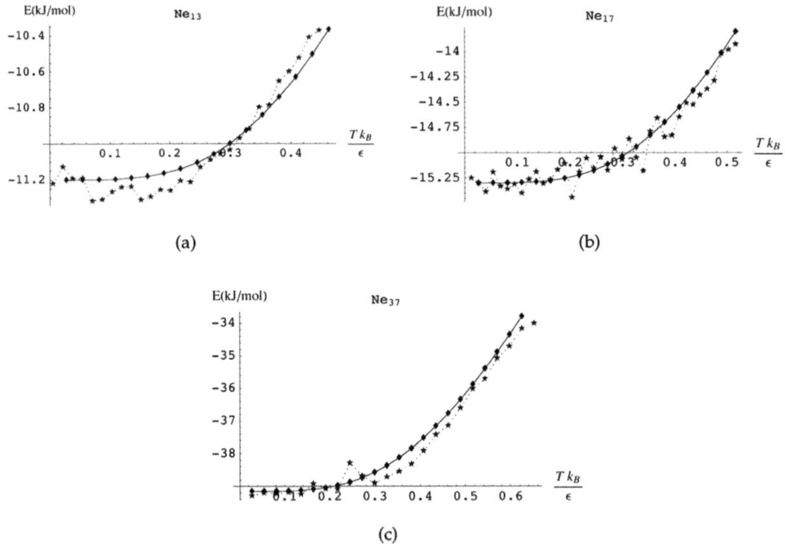

Figure 5.13 Internal energy for for various sized clusters (··★··) compared with the Debye model (—): (a) Ne_{13}, (b) Ne_{17}, (c) Ne_{37}. Reprinted with permission from Derrickson and Bittner [36]. Copyright 2007 American Chemical Society.

Figure 5.13 compares the internal energy from our results to the Debye model with the Debye energy shifted so that it corresponds at $T = 0\,\text{K}$ with our results. By comparing the curves it is evident our results for the 17- and 37-atom systems will give similar Debye temperatures to the bulk limit. The melting region can be identified as the nonlinear regions of the internal energy curves. In all three clusters, similar melting regions are observed for both approaches. It may seem remarkable that the Debye model is still useful given the fact that these clusters are far from the bulk limit. However, the Debye model was constructed to account for both the high and low temperature caloric curves in condensed phase systems, and there is no fundamental problem with it as an approximation in this case.

Closer inspection of the internal energy curve for Ne_{13} indicates that tor temperatures, $0 < T \leq 0.2T'$, the internal energy *decreases* to some extent. This corresponds to a *negative heat capacity*. Even given a computational error estimate of ± 0.1 kJ/mole in the internal energy, the dip is clearly present in our results. This is not entirely unreasonable or unprecedented as several recent studies have predicted negative heat capacities for atomic clusters [14, 24, 66]. However, by Calvo et al. [24] it was dismissed as an unphysical result. In addition negative heat capacities have also been observed recently for sodium clusters of 147 atoms [96] and they have been predicted in astrophysics, where energy can be added to a star whose temperature subsequently cools down [78]. Schmidt et al. [96] explain this for small atomic systems as a purely microscopic phenomenon. That is, for larger systems at a phase transition, energy is added as potential energy rather than kinetic energy so that the temperature remains constant over the course of the transition. For mesoscopic-scale atomic systems, on the other hand, it can be entropically favorable to avoid a partially melted state so that some energy is actually transferred from kinetic to potential energy causing a negative heat capacity near phase transitions. Since this is not observed in classical simulations nor in the Debye model, it is possibly due to anharmonic quantum delocalization effects in the system. Another factor is that $N = 13$ and 147 clusters form complete icosohedral structures in their lowest energy state. These are called magic-number clusters because of the stability of these highly symmetric forms. Since negative heat capacities have only been observed and/or predicted for magic-number clusters, this suggests that the negative heat capacity may be related to the symmetry of the system.

In all instances of negative heat capacity the common factor is that the energy is not an extensive quantity and the interactions *between* subsystems must be taken into account. In the clusters we are examining the temperature is raised but the atoms adjust themselves to store energy in the pair-potential interaction between atoms rather than increase the kinetic energy. Although this explanation offered by Schmidt et al. [96] appears to indicate that this is a purely classical effect; to our knowledge, no classical molecular dynamic methods have predicted negative heat capacities. Hence, we attribute the negative heat capacity to purely quantum mechanical effects in this system.

The reason for the failure of the Debye model for the smaller clusters is the continuous density of states for the phonons assumed by the model as seen in Eq. 5.76. As the clusters get smaller this becomes substantially different than the actual 3N-6 vibrational levels that the cluster contains. Though this is a powerful approximation, since the levels very quickly become tightly packed approach the continuous limit. If we loosely think of the portions of the density which are not adequately accounted for in this assumption as anharmonic regions, then our results indicate that these are important for describing the thermodynamics of these systems. Though we assume a Gaussian form for the statistical approximates used in the present work, which amounts to a harmonic approximation, we can sample the anharmonic regions of the density. This is because the statistical points themselves are free to sample the anharmonic regions. It happens that the anharmonic regions are more important for smaller clusters since they have more surface atoms and consequently experience more delocalization relative to the larger systems [37].

Based upon the above discussions we can say the thermodynamics of these clusters is influenced greatly by their relative ability to store energy, preferentially in the potential energy. This aspect of these systems can be studied by introducing a virial like parameter consisting of the ratio of the quantum potential with the total internal energy, $\langle Q\rangle/\langle U\rangle$. This parameter essentially measures the percentage of energy contained in the kinetic energy. This will be given by:

$$\tau_m = -\frac{\langle Q\rangle}{\langle Q\rangle + \langle V\rangle}.$$

τ_m should approach 1 as the temperature is raised, since the averaged quantum potential value is increasing with temperature and the averaged potential interaction energy should remain about constant, although the cluster will dissociate into a disordered state long before this point is reached. The averaged quantum potential value is a monotonically increasing function of the temperature because it is inversely proportional to the delocalization, or the de Broglie wavelength, $\langle Q\rangle \approx 1/\lambda^2 \approx T$. τ_m is shown for the three clusters in Fig. 5.14 and the curves clearly show that the smaller clusters must increase the amount of kinetic energy at a greater rate with temperature. Essentially the different rates of increase for τ_m are due to the larger clusters increased ability to store energy

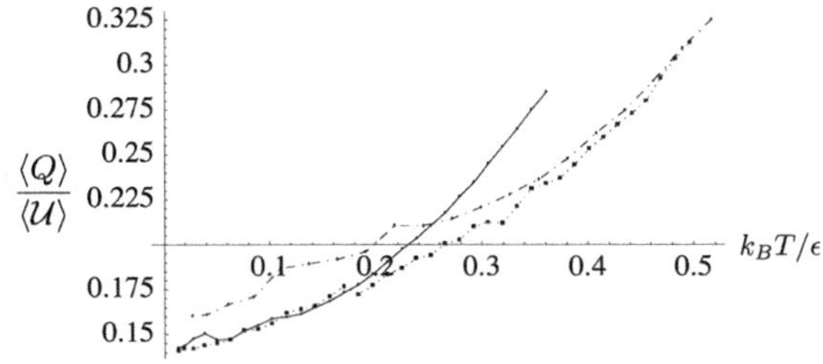

Figure 5.14 $(\langle Q\rangle/\langle U\rangle)$ versus reduced temperature k_BT/ϵ . (Key: —: 13 atoms, $\cdots$:17 atoms, $- \cdot -$: 37 atoms). Reprinted with permission from Derrickson and Bittner [36]. Copyright 2007 American Chemical Society.

in the pair-potential. This explains the marked decrease in the temperatures of phase transitions as the size of the clusters drops.

Locating the melting point for mesoscopic systems is often a difficult task, even for systems with negligible quantum effects, because of the characteristic melting regions often seen for these systems. Some factors that influence the melting characteristics of these clusters are the well depth for the lowest-energy configuration, the depths of the nearest neighbor wells around the ground state, and the time scales of the transitions [8]. A simple means of examining when melting has actually occurred is from the relative percentage of energy contained in the kinetic energy. In the bulk this will show a clear discontinuity at a phase transition, and for microscopic or mesoscopic systems. To analyze this possibility we introduce a virial like term consisting of the ratio of the quantum potential with the total internal energy, $\langle Q\rangle/\langle U\rangle$, to help identify when melting has occured. This is given by:

$$\tau_m = -\frac{\langle Q\rangle}{\langle Q\rangle + \langle V\rangle},$$

where we have taken the negative in order to have positive values for the parameter. τ_m should approach 1 as the temperature is raised, since the averaged quantum potential value is increasing with temperature, and the averaged potential interaction energy should remain about constant, but the cluster will dissociate into a disordered state long before this point is reached. The averaged quantum potential value is a monotonically increasing function of the temperature because it is inversely proportional to the delocalization, or the de Broglie wavelength, $\langle Q \rangle \approx 1/\lambda^2 \approx T$. τ_m is shown for all three cluster sizes examined at finite temperature in Fig. 5.14. These plots reveal some interesting aspects of these clusters. They show that the rate of increase of τ_m versus temperature is dependent on the cluster size, which explains the significant decrease in the phase transition temperatures observed for mesoscopic clusters compared to bulk values. The different rates of increase in τ_m for the different cluster sizes is due to the larger clusters increased ability to store energy in the pair-potential.

5.5 Overcoming the Node Problem

One of the difficulties encountered in a quantum density-based scheme is that the quantum density is a positive quantity that vanishes whenever there is a node in the quantum wave function. The "node problem" has been a bugbear in the development of time-dependent quantum trajectory approaches and plagues Monte Carlo approaches. The difficulty within the quantum trajectory scheme is that when $\rho(x) \to 0$, the quantum potential becomes singular giving rise to a sharply repulsive force that pushes sample points away from the node. One can adopt a "fixed-node" scheme keep separate ensembles of sample points for each nodal region; however, this approach is not very flexible for high-dimensional systems, which may have nodal hypersurfaces.

We present here an idea we developed that uses supersymmetric (SUSY) QM within the context of a quantum Monte Carlo scheme to get around the node issue. By using the fact that SUSY QM gives rises to a series of isospectral Hamiltonians, we show that Monte Carlo ground state calculations in the SUSY partners can be used to reconstruct accurately both the spectrum and the states of an arbitrary Schrödinger equation. Since the ground state of each partner potential is nodeless, we avoid any node problem typically associated with the Monte Carlo technique. While we provide an example of using this approach to determine the tunneling states in a double-well potential, the method is applicable to any one-dimensional (1D) potential problem [13, 63]. We conclude by discussing the extension to higher dimensions.

The variational Monte Carlo (VMC) technique is a powerful way to estimate the ground state of a quantum mechanical system. The basic idea is that one can use the variational principle to minimize the energy expectation value with respect to a set of parameters $\{\alpha\}$

$$E(\alpha) = \frac{\int |\psi(x,\alpha)|^2 (H\psi)/\psi(x,\alpha))dx}{\int |\psi(x,\alpha)|^2 dx}. \tag{5.77}$$

Following the Monte Carlo method for evaluating integrals, one interprets:

$$p(x)dx = \frac{|\psi(x,\alpha)|^2 dx}{\int |\psi(x,\alpha)|^2 dx} \tag{5.78}$$

as a probability distribution function. Typically, one assumes a functional form for the trial wave function, $\psi(x,\alpha)$ and the numerical advantage is that one can evaluate the energy integral by simply evaluating $\psi(x,\alpha)$. The method becomes variational when one then adjusts the parameters to optimize the trial wave function. Since the spectrum of H is bounded from below, the optimized trial wave function provides a best approximation to the true ground state of the system. However, since $p(x) = |\psi(x,\alpha)|^2$ is a positive definite function, this procedure fails if the system has nodes or if the position of the nodes is determined by the parameters. One can in principle obtain excitation energies by constraining the trial function to have a fixed set of nodes perhaps determined by symmetry.

Given that VMC is a robust technique for ground states, it would be highly desirable if the technique could be extended to facilitate the calculation of excited states. In this section, we present such an extension (albeit in one dimension) using SUSY QM. The underlying mathematical idea behind SUSY QM is that every Hamiltonian $H_1 = T + V_1$ has a partner Hamiltonian, $H_2 = T + V_2$ (T being the kinetic energy operator) in which the spectrum of H_1 and H_2 are identical for all states above the ground state of H_1. That is to say, the ground state of H_2 has the same energy as the first excited state of H_1 and so on. This hierarchy of related Hamiltonians and the algebra associated with the SUSY operators present a powerful formal approach to determine the energy spectra for a wide number of systems [5, 9, 26, 34, 39, 49, 55, 57, 72, 84, 90]. To date, little has been done exploiting SUSY QM as a way to develop new numerical techniques.

We shall first use the ideas of SUSY QM to develop a Monte Carlo-like scheme for computing the tunneling splittings in a symmetric double-well potential. While the model can be solved using other techniques, this provides a useful proof of principle for our approach. We find that the SUSY/VMC combination provides a useful and accurate way to obtain the tunneling splitting and excited state wave function for this system. While our current focus is on a 1D system, we conclude by commenting upon how the technique can be extended to multiparticle systems and to higher dimension. In short, our results strongly suggest that this approach can be brought to bear on a more general class of problems involving multiple degrees of freedom. Surprisingly, the connection between the Monte Carlo technique and the SUSY hierarchy has not been exploited until recently [13, 63, 64].

5.5.1 Supersymmetric Quantum Mechanics

SUSY QM is obtained by factoring the Schrödinger equation into the form [29, 105, 106]:

$$H\psi = A^{\dagger}A\psi_o^{(1)} - 0 \tag{5.79}$$

using the operators:

$$A = \frac{\hbar}{\sqrt{2m}}\partial_x + W \tag{5.80}$$

$$A^{\dagger} = -\frac{\hbar}{\sqrt{2m}}\partial_x + W. \tag{5.81}$$

Since we can impose $A\psi_o^{(1)} = 0$, we can immediately write that:

$$W(x) = -\frac{\hbar}{\sqrt{2m}}\partial_x \ln\psi_o. \tag{5.82}$$

$W(x)$ is the *superpotential* which is related to the physical potential by a Riccati equation.

$$V(x) = W^2(x) - \frac{\hbar}{\sqrt{2m}}W'(x). \tag{5.83}$$

The SUSY factorization of the Schrödinger equation can always be applied in one dimension.

From this point on we label the original Hamiltonian operator and its associated potential, states, and energies as H_1, V_1, $\psi_n^{(1)}$ and $E_n^{(1)}$. One can also define a partner Hamiltonian, $H_2 = AA^\dagger$ with a corresponding potential

$$V_2 = W^2 + \frac{\hbar}{\sqrt{2m}} W'(x). \tag{5.84}$$

All of this seems rather circular and pointless until one recognizes that V_1 and its partner potential, V_2, give rise to a common set of energy eigenvalues. This principle result of SUSY can be seen by first considering an arbitrary stationary solution of H_1:

$$H_1 \psi_n^{((1)} = A^\dagger A \psi_n = E_n^{(1)} \psi_n^{(1)}. \tag{5.85}$$

This implies that $(A\psi_n^{(1)})$ is an eigenstate of H_2 with energy $E_n^{(1)}$ since:

$$H_2(A\psi_n^{(1)}) = AA^\dagger A\psi_n^{(1)} = E_n^{(1)}(A\psi_n^{(1)}). \tag{5.86}$$

Likewise, the Schrödinger equation involving the partner potential $H_2\psi_n^{(2)} = E_n^{(2)}\psi_n^{(2)}$ implies that:

$$A^\dagger AA^\dagger \psi_n^{(2)} = H_1(A^\dagger \psi_n^{(2)}) = E_n^{(2)}(A^\dagger \psi_n^{(2)}). \tag{5.87}$$

This (along with $E_o^{(1)} = 0$) allows one to conclude that the eigenenergies and eigenfunctions of H_1 and H_2 are related in the following way: $E_{n+1}^{(1)} = E_n^{(2)}$,

$$\psi_n^{(2)} = \frac{1}{\sqrt{E_{n+1}^{(1)}}} A\psi_{n+1}^{(1)}, \text{ and } \psi_{n+1}^{(1)} = \frac{1}{\sqrt{E_n^{(2)}}} A^\dagger \psi_n^{(2)} \tag{5.88}$$

for $n > 0$.[a] Thus, the *ground state of H_2 has the same energy as the first excited state of H_1*. If this state $\psi_o^{(2)}$ is assumed to be nodeless, then $\psi_1^{(1)} \propto A^\dagger \psi_o^{(2)}$ will have a single node. We can repeat this analysis and show that H_2 is partnered with another Hamiltonian, H_3 whose ground state is isoenergetic with the first excited state of H_2 and thus isoenergetic with the second excited state of the original H_1. This hierarchy of partners persists until all of the bound states of H_1 are exhausted.

[a]Our notation from here on is that $\psi_n^{(m)}$ denotes the nth state associated with the mth partner Hamiltonian with similar notation for related quantities such as energies and superpotentials.

5.5.2 Implementation of SUSY QM in an Adaptive Monte Carlo Scheme

Having defined the basic terms of SUSY QM, let us presume that one can determine an accurate approximation to the ground state density $\rho_o^{(1)}(x)$ of Hamiltonian H_1. One can then use this to determine the superpotential using the Riccati transform:

$$W_o^{(1)} = -\frac{1}{2}\frac{\hbar}{\sqrt{2m}}\frac{\partial \ln \rho_o^{(1)}}{\partial x} \tag{5.89}$$

and the partner potential:

$$V_2 = V_1 - \frac{\hbar^2}{2m}\frac{\partial^2 \ln \rho_o^{(1)}}{\partial x^2}. \tag{5.90}$$

Certainly, our ability to compute the energy of the ground state of the partner potential V_2 depends on having first obtained an accurate estimate of the ground state density associated with the original V_1.

For this we turn to an adaptive VMC approach developed by Maddox and Bittner [82], as discussed earlier in this chapter. To recapitulate this approach, we assume we can write the trial density as a sum over N Gaussian approximate functions:

$$\rho_T(x) = \sum_n G_n(x, \mathbf{c}_n). \tag{5.91}$$

parameterized by their amplitude, center, and width.

$$G_n(x, \{\mathbf{c}_n\}) = c_{no} e^{-c_{n2}(x-c_{n3})^2} \tag{5.92}$$

This trial density then is used to compute the energy:

$$E[\rho_T] = \langle V_1 \rangle + \langle Q[\rho_T] \rangle \tag{5.93}$$

where $Q[\rho_T]$ is the Bohm quantum potential:

$$Q[\rho_T] = -\frac{\hbar^2}{2m}\frac{1}{\sqrt{\rho_T}}\frac{\partial^2}{\partial x^2}\sqrt{\rho_T}. \tag{5.94}$$

The energy average is computed by sampling $\rho_T(x)$ over a set of trial points $\{x_i\}$ and then moving the trial points along the conjugate gradient of:

$$E(x) = V_1(x) + Q[\rho_T](x). \tag{5.95}$$

After each conjugate gradient step, a new set of $\mathbf{c_n}$ coefficients are determined according to an EM criteria such that the new trial density provides the best N-Gaussian approximation to the actual probability distribution function sampled by the new set of trial points. The procedure is repeated until $\delta\langle E\rangle = 0$. In doing so, we simultaneously minimize the energy and optimize the trial function. Since the ground state is assumed to be nodeless, we will not encounter the singularities and numerical instabilities associated with other Bohmian equations of motion based approaches [12, 16, 52, 77, 82, 108].

5.5.3 Test Case: Tunneling in a Double-Well Potential

As a nontrivial test case, consider the tunneling of a particle between two minima of a symmetric double-well potential. One can estimate the tunneling splitting using semiclassical techniques by assuming that the ground and excited states are given by the approximate form:

$$\psi_{\pm} = \frac{1}{\sqrt{2}}(\phi_o(x) \pm \phi_o(-x)) \tag{5.96}$$

where ϕ_o is the lowest-energy state in the right-hand well in the limit the wells are infinitely far apart. From this, one can easily estimate the splitting as [67]

$$\delta = 4\frac{\hbar^2}{m}\phi_o(0)\phi_o'(0) \tag{5.97}$$

If we assume the localized state (ϕ_o) to be gaussian, then:

$$\psi_{\pm} \propto \frac{1}{\sqrt{2}}(e^{-\beta(x-x_o)^2} \pm e^{-\beta(x+x_o)^2}) \tag{5.98}$$

and we can write the superpotential as:

$$W = \sqrt{\frac{2}{m}}\hbar\beta\,(x - x_o\tanh(2xx_o\beta))\,. \tag{5.99}$$

From this, one can easily determine both the original potential and the partner potential as:

$$V_{1,2} = W^2 \pm \frac{\hbar}{\sqrt{2m}}W' \tag{5.100}$$

$$= \frac{\beta^2\hbar^2}{m}\,(2(x - x_o\tanh(2xx_o\beta))^2 \pm (2x_o^2\mathrm{sech}^2(2xx_o\beta) - 1) \tag{5.101}$$

While the V_1 potential has the characteristic double minima giving rise to a tunneling doublet, the SUSY partner potential V_2 has a central dimple, which in the limit of $x_o \to \infty$ becomes a δ-function, which produces an unpaired and nodeless ground state [29]. Using Eq. 5.87, one obtains $\psi_1^{(1)} = \psi_- \propto A^\dagger \psi_o^{(2)}$, which now has a single node at $x = 0$.

For a computational example, we take the double-well potential to be of the form:

$$V_1(x) = ax^4 + bx^2 + E_o. \tag{5.102}$$

with $a = 438.9\ \mathrm{cm}^{-1}/(bohr^2)$, $b = 877.8\ \mathrm{cm}^{-1}/(bohr)^4$, and $E_o = -181.1\ \mathrm{cm}^{-1}$, which (for $m = m_H$) gives rise to exactly two states below the barrier separating the two minima with a tunneling splitting of 59.32 cm^{-1} as computed using a DVR approach [74]. For the calculations reported here, we used $n_p = 1000$ sample points and $N = 15$ Gaussians and in the expansion of $\rho_T(x)$ to converge the ground state. This converged the ground state to $1 : 10^{-8}$ in terms of the energy. This is certainly a bit of an overkill in the number of points and number of Gaussians since far fewer DVR points were required to achieve comparable accuracy (and a manifold of excited states). The numerical results, however, are encouraging since the accuracy of generic Monte Carlo evaluation would be $1/\sqrt{n_p} \approx 3\%$ in terms of the energy.[a] Plots of V_1 and the converged ground state is shown in Fig. 5.15.

The partner potential $V_2 = W^2 + \hbar W'/\sqrt{2m}$, can be constructed once we know the superpotential, $W(x)$. Here, we require an accurate evaluation of the ground state density and its first two log-derivatives. The advantage of our computational scheme is that one can evaluate these analytically for a given set of coefficients. In Fig. 5.15a we show the partner potential derived from the ground state density. Whereas the original V_1 potential exhibits the double-well structure with minima near $x_o = \pm 1$, the V_2 partner potential has a pronounced dip about $x = 0$. Consequently, its ground state should have a simple Gaussian-like form peaked about the origin.

[a] In our implementation, the sampling points are only used to evaluate the requisite integrals and they themselves are adjusted along a conjugate gradient rather than by resampling. One could in principle forego this step entirely and optimize the parameters describing the Gaussians directly.

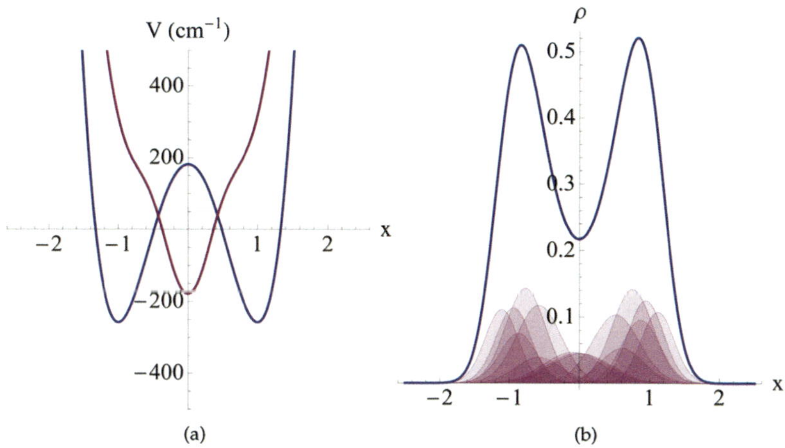

Figure 5.15 (a) Model double well potential (blue) and partner potential (purple). The energies of the tunneling doublets are indicated by the horizontal lines at $V = 0$ cm^{-1} and $V = 59.32$ cm^{-1} indicating the positions of the sub-barrier tunneling doublet. (b) Final ground state density (blue) superimposed over the Gaussians used in its expansion (purple). Reprinted with permission from Bittner et al. [13]. Copyright 2009 American Chemical Society.

Once we determined an accurate representation of the partner potential, it is now a trivial matter to reintroduce the partner potential into the optimization routines. The ground state converges easily and is shown in Fig. 5.16b along with its Gaussians. After 1000 CG steps, the converged energy is within 0.1% of the exact tunneling splitting for this model system. Again, this is an order of magnitude better than the $1/\sqrt{n_p}$ error associated with a simple Monte Carlo sampling. Furthermore, Fig. 5.16b shows $\psi_1^{(1)} \propto A^\dagger \psi_0^{(2)}$ computed using the converged $\rho_0^{(2)}$ density. As anticipated, it shows the proper symmetry and nodal position.

By symmetry, one expects the node to lie precisely at the origin. However, since we have not imposed any symmetry restriction or bias on our numerical method, the position of the node provides a sensitive test of the convergence of the trial density for $\rho_0^{(2)}$. In the example shown in Fig. 5.17, the location of the node oscillates about the origin and appears to converge exponentially with the number of CG steps. This is remarkably good considering that this is ultimately

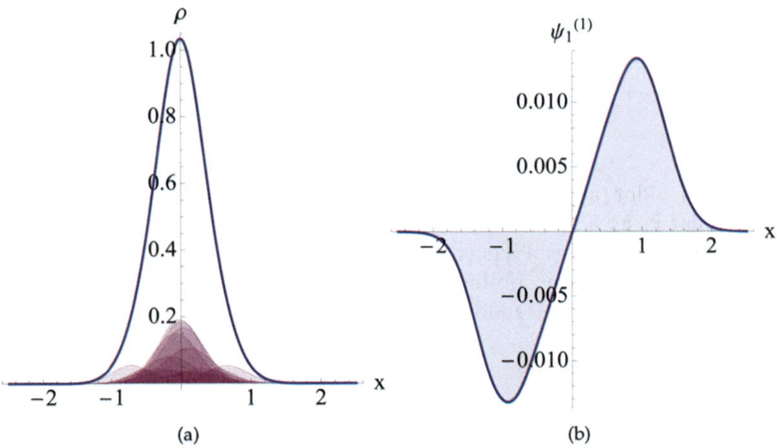

Figure 5.16 (a) Ground state density of the partner Hamiltonian H_2 (blue) superimposed over its individual Gaussian components. (b) Excited state $\psi_1^{(1)}$ derived from the ground state of the partner potential, $\psi_o^{(2)}$. Reprinted with permission from Bittner et al. [13]. Copyright 2009 American Chemical Society.

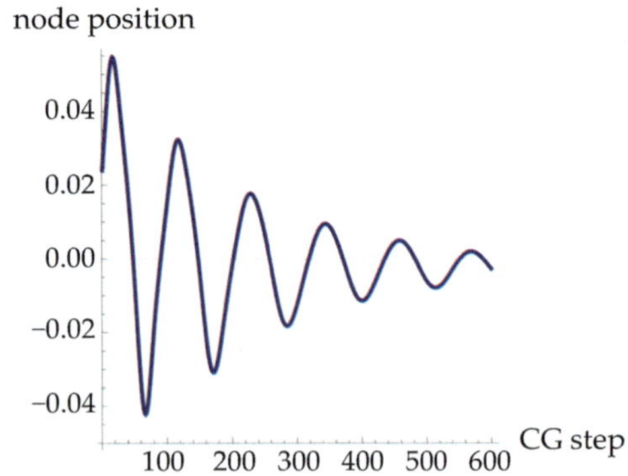

Figure 5.17 Location of excited state node for the last 600 CG steps. Reprinted with permission from Bittner et al. [13]. Copyright 2009 American Chemical Society.

determined by the quality of the third and fourth derivatives of $\rho_o^{(1)}$ since these appear when computing the conjugate gradient of V_2. We have tested this approach on a number of other 1D bound-state problems with similar success.

5.5.4 Extension to Higher Dimensions

Having demonstrated that the SUSY approach can be used to compute excitation energies and wave functions starting from a Monte Carlo approach, the immediate next step is to extend this to arbitrarily higher dimensions. To move beyond 1D SUSY, Ioffe and coworkers have explored the use of higher-order charge operators [1–3, 25], and Kravchenko has explored the use of Clifford algebras [65]. Unfortunately, this is difficult to do in general. The reason being that the Riccati factorization of the 1D Schrödinger equation does not extend easily to higher dimensions. One remedy is write the charge operators as vectors $\vec{A} = (+\vec{\partial} + \vec{W})$ and with $\vec{A}^\dagger = (-\vec{\partial} + \vec{W})^T$ as the adjoint charge operator. The original Schrödinger operator is then constructed as an inner-product:

$$H_1 = \vec{A}^\dagger \cdot \vec{A}. \tag{5.103}$$

Working through the vector product produces the Schrödinger equation:

$$H_1\phi = (-\nabla^2 + W^2 - (\vec{\nabla} \cdot \vec{W}))\phi = 0 \tag{5.104}$$

and a Riccati equation of the form:

$$U(x) = W^2 - \vec{\nabla} \cdot \vec{W}. \tag{5.105}$$

For a 2D harmonic oscillator, we would obtain a vector superpotential of the form:

$$\vec{W} = -\frac{1}{\psi_0^{(1)}}\vec{\nabla}\psi_0^{(1)} = (x, y) = (W_x, W_y) \tag{5.106}$$

Let us look more closely at the $\vec{\nabla}\cdot\vec{W}$ part. If we use the form that $\vec{W} = -\vec{\nabla}\ln\psi$, then $-\vec{\nabla}\cdot\vec{\nabla}\ln\psi = -\nabla^2\ln\psi$ which for the 2D oscillator results in $\vec{\nabla}\cdot\vec{W} = 2$. Thus:

$$W^2 - \vec{\nabla}\cdot\vec{W} = (x^2 + y^2) - 2 \tag{5.107}$$

which agrees with the original symmetric harmonic potential. Now, we write the scaled partner potential as:

$$U_2 = W^2 + \vec{\nabla}\cdot\vec{W} = (x^2 + y^2) + 2. \tag{5.108}$$

This is equivalent to the original potential shifted by a constant amount.

$$U_2 = U_1 + 4. \tag{5.109}$$

The ground state in this potential would be have the same energy as the states of the original potential with quantum numbers $n+m=2$. Consequently, even with this naïve factorization, one can in principle obtain excitation energies for higher-dimensional systems, but there is no assurance that one can reproduce the entire spectrum of states.

The problem lies in the fact that neither Hamiltonian H_2 nor its associated potential U_2 is given in its most general form by the expression implied by Eq. 5.104 and Eq. 5.108. Rather, the correct approach is to write the H_2 Hamiltonian as a *tensor* by taking the outer product of the charges $\overline{H}_2 = \vec{A}\vec{A}^\dagger$ rather than as a scalar $\vec{A}\cdot\vec{A}^\dagger$. At first this seems unwieldy and unlikely to lead anywhere since the wave function solutions of:

$$\overline{H}_2 \cdot \vec{\psi} = E\vec{\psi} \tag{5.110}$$

are now vectors rather than scalars. However, rather than adding an undue complexity to the problem, it actually simplifies matters considerably. As we demonstrate in a recent paper, this tensor factorization preserves the SUSY algebraic structure and produces excitation energies for any $n-$dimensional SUSY system [62].

5.5.4.1 Discussion

In brief, we have used the ideas of SUSY QM to obtain excitation energies and excited state wave functions within the context of a Monte Carlo scheme. This was accomplished without prespecifying the location of nodes or restriction to a specific symmetry. While it is clear that one could continue to determine the complete spectrum of H_1, the real challenge is to extend this technique to higher dimensions. Furthermore, the extension to multi-Fermion systems may be accomplished through the use of the Gaussian Monte Carlo method in which any quantum state can be expressed as a real probability distribution [4, 30].

5.6 Summary

In this chapter, we have explored an efficient strategy for determining the quantum density associated with a statistical ensemble of space-time trajectories. Given a statistical ensemble of probability elements, we can estimate the quantum force in terms of a set of Gaussian fitting parameters. Our methodology incorporates Bayesian probabilities and a mixture model approximation to calculate a parameterized estimate of Bohm's quantum force. The EM procedure used to fit the density is not sensitive to trajectory crossings because the error associated with an individual rogue trajectory is essentially washed out by the statistical ensemble. After a sufficient equilibration time the ensemble is representative of the ground state distribution and can be used to gather statistics on ground state properties such as the zero-point energy and other expectation values. Moreover, because the density fitting is formulated in terms of simple sums over data points, our method is easily extended to high dimensions and can be conveniently implemented on parallel computers.

We applied this approach to investigate the ground vibrational state energies at zero temperature and the Low-temperature thermodynamics of mesoscopic rare gas clusters. Improvements in the algorithm allowed the calculation of the ground state structure at zero temperature approaching the size necessary to simulate bulk systems. Our method compares favorably against path-integral Monte Carlo results on these systems and offers a systematic improvement over semiclassical treatments.

Finally, we present some of our most recent work in using the SUSY approach to compute quantum excited states using our adaptive approach. This work holds considerable promise in avoiding the node problem that has plagued both Monte Carlo and Bohmian trajectory approaches for years.

Most implementations of the Bohmian theory use trajectories as a way to gain deeper insight into a given problem or as a way to compute properties *given* a quantum density, we take more of *synthetic* approach to construct the quantum density from an ensemble of particles which obey the Bohm quantum equations of motion. As discussed above, presents a very difficult challenge especially when extended to high-dimensional systems. The methods and results presented in this chapter represent the *first* successful implementations of a *synthetic* Bohmian approach that is robust, stable, and computationally efficient enough to study large numbers of atoms using realistic interatomic potentials. While we have not discussed it here, our approach can be used in conjunction with standard classical molecular dynamics algorithms to impart quantum-like behavior (such as tunneling) to specific atoms within the simulation.

Acknowledgements

The work at the University of Houston was funded in part by the National Science Foundation (CHE-1011894) and the Robert A. Welch Foundation (ERB: E-1334, DJK: E-0608). We also acknowledge fruitful discussions with Prof. R. E. Wyatt, and Prof. Irene Burghardt concerning various aspects of the work presented in this chapter.

References

1. A. A. Andrianov, M. V. Ioffe, and V. P. Spiridonov, *Higher-Derivative Supersymmetry and the Witten Index*, Physics Letters A, **174**, 273 (1993).
2. A. A. Andrianov, M. V. Ioffe, and D. N. Nishnianidze, *Polynomial Supersymmetry and Dynamical Symmetries in Quantum Mechanics*, Theoretical and Mathematical Physics **104**, no. 3, 1129–1140 (1995).
3. A. A. Andrianov, M. V. Ioffe, and D. N. Nishnianidze, *Polynomial SUSY in Quantum Mechanics and Second Derivative Darboux Transformations*, Physics Letters A, **201**, 103–110 (2002).
4. F. F. Assaad, P. Werner, P. Corboz, E. Gull, and M. Troyer, *Symmetry projection Schemes for Gaussian Monte Carlo Methods*, Physics Review B, **72**, no. 22, 224518 (2005).
5. L. Balents and M. P. A. Fisher, *Delocalization Transition via Supersymmetry in One Dimension*, Physical Review B (Condensed Matter) **56**, no. 20, 12970–12991 (1997).
6. A. Banerjee and M. K. Harbola, *Hydrodynamic Approach to Time-Dependent Density Functional Theory; Response Properties of Metal Clusters*, Journal of Chemical Physics, **113**, no. 14, 5614–5623 (2000).
7. A. Banerjee and M. K. Harbola, *Calculation of van der waals Coefficients in Hydrodynamic Approach to Time-Dependent Density Functional*

Theory, The Journal of Chemical Physics **117**, no. 17, 7845–7851 (2002).

8. T. L. Beck and R. S. Berry, *The Interplay of Structure and Dynamics in the Melting of Small Clusters*, Journal of Chemical Physics, **88**, no. 6, 3910–3922 (1988).
9. V. Berezinsky and M. Kachelriess, *Monte Carlo Simulation for Jet Fragmentation in SUSY QCD*, Physical Review D (Particles and Fields) **63**, no. 3, 034007 (2001).
10. R. S. Berry, *Clusters: Tools for Studying Potential Surfaces and Their Connection to Molecular Dynamics*, The Journal of Physical Chemistry **98**, no. 28, 6910–6918 (1994).
11. E. R. Bittner, *Quantum Tunneling Dynamics Using Hydrodynamic Trajectories*, Journal of Chemical Physics, **112**, no. 22, 9703 (2000).
12. E. R. Bittner and R. E. Wyatt, *Integrating the Quantum Hamilton-Jacobi Equations by Wavefront Expansion and Phase Space Analysis*, Journal of Chemical Physics, **113**, no. 20, 8888 (2000).
13. E. R. Bittner, J. B. Maddox, and D. J. Kouri, *Supersymmetric Approach to Excited States*, The Journal of Physical Chemistry A **113**, no. 52, 15276–15280 (2009).
14. M. Bixon and J. Jortner, *Energetic and Thermodynamic Size Effects in Molecular Clusters*, Journal of Chemical Physics, **91**, no. 3, 1631–1642 (1989).
15. D. Bohm, *A Suggested Interpretation of the Quantum Theory in Terms of "Hidden" Variables. I*, Physical Review, **85**, 166 (1952).
16. D. Bohm, *A Suggeted Interpretation of the Quantum Theory in Terms of "Hidden" Variables. II*, Physical Review, **85**, 180 (1952).
17. D. Bohm and B. Hiley, *The Undivided Universe: An Ontological Interpretation of Quantum Theory*, (Routlegde, 1993).
18. D. Bohm, B. J. Hiley, and P. N. Kaloyerou, *An Ontological Basis for the Quantum Theory*, Physics Reports, **144** (19878), 321.
19. D. F. R. Brown, M. N. Gibbs, and D. C. Clary, *Combining ab initio Computations, Neural Networks, and Diffusion Monte Carlo: An Efficient Method to Treat Weakly Bound Moleclues*, Journal of Chemical Physics, **105**, no. 17, 7597 (1996).
20. J. C. Burant and J. C. Tully, *Nonadiabatic Dynamics via the Classical Limit Schrödinger Equation*, Journal of Chemical Physics, **112**, no. 14, 6097–6103 (2000).

21. I. Burghardt and L. S. Cederbaum, *Hydrodynamic Equation for Mixed Quantum States. I. General Formulation*, Journal of Chemical Physics, **115**, no. 22, 10303 (2001).
22. I. Burghardt and L. S. Cederbaum, *Hydrodynamic Equations for Mixed Quantum States. II. Coupled Electronic States*, Journal of Chemical Physics, **115**, no. 22, 10312 (2001).
23. W. Butine, *A Guide to the Literature on Learning Probabilistic Networks from Data*, IEEE Transactions Knowledge Data Engineering, **8**, no. 2, 195 (1996).
24. F. Calvo, J. P. K. Doye, and D. J. Wales, *Quantum Partition Functions from Classical Distributions: Application to Rare-Gas Clusters*, Journal of Chemical Physics, **114**, no. 17, 7312–7329 (2001).
25. F. Cannata, M. V. Ioffe, and D. N. Nishnianidze, *New Methods for the Two-Dimensional Schrödinger Equation: SUSY-Separation of Variables and Shape Invariance*, Journal of Physics A: Mathematical and General, **35**, no. 6, 1389–1404 (2002).
26. F. Cannata, G. Junker, and J. Trost, *Solvable Potentials, Non-Linear Algebras, and Associated Coherent States*, J. Rembielinski and K. A. Smolinski, (eds.), vol. 453, pp. 209–218 (AIP, 1998).
27. C. M. Caves, C. A. Fuchs, and R. Schack, *Quantum Probabilities as Bayesian Probabilities*, Physical Review A **65**, 022305 (2002).
28. D. M. Ceperley and E. L. Pollock, *Path-Integral Computation of the Low-Temperature Properties of Liquid ^4He*, Physical Review Letters, **56**, 351 (1986).
29. F. Cooper, A. Khare, and Uday Sukhatme, *Supersymmetry and Quantum Mechanics*, Physics Reports, **251**, 267–385 (1995).
30. J. F. Corney and P. D. Drummond, *Gaussian Quantum Monte Carlo Methods for Fermions and Bosons*, Physical Review Letters, **93**, no. 26, 260401 (2004).
31. L. de Broglie, C. R. Academy of Sciences Paris, **183**, 24 (1926).
32. L. de Broglie, *Interference and Corpuscular Light*, Nature, **118**, 441–442 (1926).
33. L. de Broglie, C. R. Academy of Sciences Paris, **184**, 273 (1927).
34. R. de Lima Rodrigues, P. B. da Silva Filho, and A. N. Vaidya, *SUSY QM and Solitons from Two Coupled Scalar Fields in Two Dimensions*, Physical Review D (Particles, Fields, Gravitation, and Cosmology) **58**, no. 12, 125023 (1998).

35. A. P. Dempster, N. M. Laird, and D. B. Rubin, *Maximum Likelihood from Incomplete Data via the EM Algorithm*, Journal of the Royal Statistical Society, Series B **29** (1977).
36. S. W. Derrickson and E. R. Bittner, *Thermodynamics of Atomic Clusters Using Variational Quantum Hydrodynamics*, Journal of Physical Chemistry A **111**, no. 41, 10345–10352 (2007).
37. S. W. Derrickson and E. R. Bittner, *A Self-Consistent Field Quantum Hydrodynamic Approach for Molecular Clusters*, Journal of Physical Chemistry A **110**, no. 16, 5333–5341 (2006).
38. B. K. Dey, A. Askar, and H. Rabitz, *Multidimensional Wave Packet Dynamics Within the Fluid Dynamical Formulations of the Schrödinger Equation*, Journal of Physical Chemistry, **109**, no. 20, 8770 (1998).
39. D. J. Fernández-C. and N. Fernández-García, *Higher-Order Supersymmetric Quantum Mechanics*, vol. 744, pp. 236–273 (AIP, 2004).
40. R. P. Feynman, *Statistical Mechanics: A Set of Lectures*, (Addison-Wesley, 1972).
41. R. P. Feynman and A. R. Hibbs, *Quantum Mechanics and Path Integrals*, (McGraw-Hill, 1965).
42. S. Garashchuk and V. A. Rassolov, *Semiclassical Dynamics Based on Quantum Trajectories*, Chemical Physics Letters, **364**, 562 (2002).
43. S. Garashchuk and V. A. Rassolov, *Semiclassical Dynamics with Quantum Trajectories: Formulation and Comparison with the Semiclassical Initial Value Representation Propagator*, Journal of Chemical Physics, **118**, no. 6, 2482–2490 (2003).
44. C. W. Gardiner, *Handbook of Stochastic Methods for Phyics, Chemistry, and the Natural Sciences*, 2nd ed., Springer Series in Synergetics, vol. 13, (Springer-Verlag, New York, 1985).
45. N. Gershenfeld, *The Nature of Mathematical Modeling*, (Cambridge University Press, Cambridge, 1999).
46. E. Gindensperger, C. Meier, and J. A. Beswick, *Mixing Quantum and Classical Dynamics Using Bohmian Trajectories*, Journal of Chemical Physics, **113**, no. 21, 9369 (2000).
47. E. Gindensperger, C. Meier, and J. A. Beswick, *Quantum-Classical Dynamics Including Continuum States Using Quantum Trajectories*, Journal of Chemical Physics, **116** (2002).
48. E. Gindensperger, C. Meier, J. A. Beswick, and M. C. Heitz, *Quantum-Classical Description of Rotational Diffractive Scattering Using Bohmian Trajectories: Comparison with Full Quantum Wave Packet Results*, Journal of Chemical Physics, **116**, no. 23, 10051 (2002).

49. M. Günther, J. Hellmig, G. Heusser, M. Hirsch, H. V. Klapdor-Kleingrothaus, B. Maier, H. Päs, F. Petry, Y. Ramachers, H. Strecker, M. Völlinger, A. Balysh, S. T. Belyaev, A. Demehin, A. Gurov, I. Kondratenko, D. Kotel'nikov, V. I. Lebedev, and A. Müller, *Heidelberg-Moscow $\beta\beta$ Experiment with ^{76}Ge: Full Setup with Five Detectors*, Physical Review D (Particles and Fields) **55**, no. 1, 54–67 (1997).
50. B. L. Hammond, W. A. Lester, and P. J. Reynolds, *Monte Carlo methods in* ab initio *quantum chemistry*, in Lecture and Course Notes in Chemistry, vol. 1 (World Scientific, River Edge, NJ, 1994).
51. E. J. Heller, *Frozen Gaussians: A Very Simple Semiclassical Approximation*, Journal of Chemical Physics, **75**, no. 6, 2923 (1981).
52. P. R. Holland, *The Quantum Theory of Motion: An Account of the de Broglie-Bohm Causal Interpretation of Quantum Mechanics*, (Cambridge University Press, 1993).
53. X. Hu, T. Ho, H. Rabitz, and A. Askar, *Solution of the Quantum Fluid Dynamical Equations with Radial Basis Function Interpolation*, Physical Review E **61**, 5967 (2000).
54. K. E. Hughes and R. E. Wyatt, *Wavepacket Dynamics on Dynamically Adapting Grids: Application of the Equidistribution Principle*, Chemical Physics Letters, **366**, 336 (2002).
55. M. Humi, *Darboux Transformations for Schrödinger Equations in Two Variables*, Journal of Mathematical Physics **46**, no. 8, 083515 (2005).
56. C. Huscroft, R. Gross, and M. Jarrell, *Maximum Entropy Method of Obtaining Thermodynamic Properties from Quantum Monte Carlo Simulations*, Physical Review B **61**, no. 14, 9300 (2000).
57. A. Inomata and G. Junker, *Quasiclassical Path-Integral Approach to Supersymmetric Quantum Mechanics*, Physical Review A (Atomic, Molecular, and Optical Physics) **50**, no. 5, 3638–3649 (1994).
58. J. Jellinek, T. L. Beck, and R. S. Berry, *Solid-Liquid Phase Changes in Simulated Isoenergetic Ar_{13}*, Journal of Chemical Physics **84**, no. 5, 2783–2794 (1986).
59. A. N. Korotkov, *Output Spectrum of a Detector Measuring Quantum Oscillations*, Physical Review B **63**, 085312 (2001).
60. A. N. Korotkov, *Selective Quantum Evolution of a Qubit due to Continuous Measurement*, Physical Review B **63**, 115403 (2001).
61. A. N. Korotkov, *Continuous Measurement of Entangled Qubits*, Physical Review A **65**, 052304 (2002).
62. D. J. Kouri, K. Maji, T. Markovich, and E. R. Bittner, *New Generalization of Supersymmetric Quantum Mechanics to Arbitrary Dimensionality or*

Number of Distinguishable Particles, Journal of Physical Chemistry A **114**, no. 32, 8202–8216 (2010).

63. D. J. Kouri, T. Markovich, N. Maxwell, and E. R. Bittner, *Supersymmetric Quantum Mechanics, Excited State Energies and Wave Functions, and the Rayleigh-Ritz Variational Principle: A Proof of Principle Study*, Journal of Physical Chemistry A **113**, no. 52, 15257–15264 (2009).
64. D. J. Kouri, T. Markovich, N. Maxwell, and B. G. Bodman, *The Heisenberg-Weyl Algebra on the Circle and a Related Quantum Mechanical Model for Hindered Rotation*, Journal of Physical Chemistry A **113**, 7698–7705 (2009).
65. V. V. Kravchenko, *On the Reduction of the Multidimensional Stationary Schrödinger Equation to a First-Order Equation and its Relation to the Pseudoanalytic Function Theory*, Journal of Physics A: Mathematical and General **38**, 851 (2005).
66. P. Labastie and R. L. Whetten, *Statistical Thermodynamics of the Cluster Solid-Liquid Transition*, Physical Review Letters, **65**, no. 13, 1567–1570 (1990).
67. L. D. Landau and E. M. Lifshitz, *Quantum Mechanics (non-relativistic theory)*, in Course of Theoretical Physics, 3rd ed., vol. 3 (Pergammon, Oxford, 1974).
68. P. M. Lee, *Bayesian Statistics: An Introduction*, (Wiley, New York, 1997).
69. C. Leforestier, *A Comparison of Different Propagation Schemes for the Time Dependent Schrödinger Equation*, Journal of Computational Physics, **94**, 59 (1991).
70. J. C. Lemm and J. Uhlig, *Hartree-Fock Approximation for Inverse Many-Body Problems*, Physical Review Letters, **84**, no. 20, 4517 (2000).
71. J. C. Lemm, J. Uhlig, and A. Weiguny, *Bayesian Approach to Inverse Quantum Statistics*, Physical Review Letters, **84**, no. 10, 2068 (2000).
72. P. T. Leung, A. M. van den Brink, W. M. Suen, C. W. Wong, and K. Young, *SUSY Transformations for Quasinormal Modes of Open Systems*, Journal of Mathematical Physics **42**, no. 10, 4802–4820 (2001).
73. J. C. Light, *Discrete variable representations in quantum dynamics*, In Time-Dependent Quantum Molecular Dynamics, J. Broeckhove and L. Lathouwers, (eds.), p. 185, (Plenum Press, New York, 1992).
74. J. C. Light, I. P. Hamilton, and J. V. Lill, *Generalized Discrete Variable Approximation in Quantum Mechanics*, Journal of Chemical Physics, **82**, no. 3, 1400–1409 (1985).

75. T. J. Liszka, C. A. M. Duarte, and W. W. Tworzydlo, *hp-Meshless Cloud Method*, Computational Methods in Applied Mechanical Engineering, **139**, 263 (1996).

76. D. M. Liveley, *Phase Separation in fFreezing ^{3}He-Ne and ^{4}He-Ne Mixtures*, Journal of Physics C: Solid State Physics, **16**, 2889 (1983).

77. C. L. Lopreore and R. E. Wyatt, *Quantum Wavepacket Dynamics with Trajectories*, Physical Review Letters, **82**, 5190 (1999).

78. D. Lynden-Bell, *Negative Specific Heat in Astronomy, Physics and Chemistry*, Physica A: Statistical Mechanics and Its Applications, **263**, no. 1–4, 293–304, Proceedings of the 20th IUPAP International Conference on Statistical Physics (1999).

79. R. M. Lynden-Bell and D. J. Wales, *Free Energy Barriers to Melting in Atomic Clusters*, Journal of Chemical Physics, **101**, no. 2, 1460–1476 (1994).

80. J. B. Maddox and E. R. Bittner, *Quantum Relaxation Dynamics with Bohmian Trajectories*, Journal of Chemical Physics, **115**, no. 14, 6309 (2001).

81. J. B. Maddox and E. R. Bittner, *Quantum Dissipation in Unbounded Systems*, Physical Review E **65**, 026143 (2002).

82. J. B. Maddox and E. R. Bittner, *Estimating Bohm's Quantum Force Using Bayesian Statistics*, Journal of Chemical Physics, **119**, no. 13, 6465–6474 (2003).

83. E. Madelung, Zeitschrift für Physik, **40**, 322 (1926).

84. J. Margueron and P. Chomaz, *Supersymmetry Transformation for Excitation Processes*, Physical Review C (Nuclear Physics) **71**, no. 2, 024318 (2005).

85. F. S. Mayor, A. Askar, and H. Rabitz, *Quantum Fluid Dynamics in the Lagrangian Representation and Applications to Photodissociation Problems*, Journal of Physical Chemistry **111**, 2423 (1999).

86. G. J. McLachlan and K. E. Basford, *Mixture Models: Inference and Applications to Clustering*, (Marcel Dekker, New York, 1988).

87. G. J. McLachlan and K. E. Basford, *Mixture Models: Inference and Applications to Clustering*, (Dekker, New York, 1998).

88. N. David Mermin, *Thermal Properties of the Inhomogeneous Electron Gas*, Physical Review, **137**, no. 5A, A1441–A1443 (1965).

89. T. Biben and D. Frenkel, *Density Functional Approach to Helium at Finite Temperature*, Journal of Physics: Condensed Matter, **14**, 9077–9088 (2002).

90. J. Niederle and A. G. Nikitin, *Extended Supersymmetries for the Schrödinger–Pauli Equation*, Journal of Mathematical Physics **40**, no. 3, 1280–1293 (1999).
91. X. Oriols, J. J. García, F. Martín, J. Suñé, T. Gonzàlez, J. Mateos, and D. Pardo, *Bohm Trajectories for the Monte Carlo Simulation of Quantum-Based Devices*, Applied Physics Letters, **72**, 806 (1998).
92. X. Oriols, F. Martín, and J. Suñé, *High Frequency Components of Current Fluctuations in Semiconductor Tunneling Barriers*, Applied Physics Letters, **80**, no. 21, 4048–4050 (2002).
93. R. G. Parr and W. Yang, *Density Functional Theory of Atoms and Molecules*, (Clarendon Press, Oxford, England; Oxford University Press, New York, 1989).
94. C. Predescu, P. A. Frantsuzov, and V. A. Mandelshtam, *Thermodynamics and Equilibrium Structure of Ne_{38} Cluster: Quantum Mechanics versus Classical*, Journal of Chemical Physics, **122**, no. 15, 154305 (2005).
95. O. V. Prezhdo and C. Brooksby, *Quantum Backreaction through the Bohmian Particle*, Physical Review Letters, **86**, no. 15, 3215 (2001).
96. M. Schmidt, R. Kusche, T. Hippler, J. Donges, W. Kronmüller, B. von Issendorff, and H. Haberland, *Negative Heat Capacity for a Cluster of 147 Sodium Atoms*, Physical Review Letters, **86**, no. 7, 1191–1194 (2001).
97. M. Shapiro and R. Bersohn, *Vibrational Energy Distribution of the CH_3 Radical Phostodissociated from CH_3I*, Journal of Chemical Physics, **73**, no. 8, 3810 (1980).
98. J. Solca, A. J. Dyson, G. Steinebrunner, B. Kirchner, and H. Huber, *Melting Curves for Neon Calculated from Pure Theory*, Journal of Chemical Physics, **108**, no. 10, 4107–4111 (1998).
99. A. M. Steinberg, *How Much Time Does a Tunneling Particle Spend in the Barrier Region?*, Physical Review Letters, **74**, no. 13, 2405 (1995).
100. M. Toda, R. Kubo, and N. Saito, *Statistical Physics I: Equilibrium Statistical Mechanics*, (Springer, Berlin, 1992).
101. C. Trahan and R. E. Wyatt, *An Arbitrary Lagrangian-Eulerian Approach to Solving the Quantum Hydrodynamic Equations of Motion: Implementation with Smart Springs*, Journal of Chemical Physics, **118**, 4784 (2003).
102. C. F. von Weizsacker, Zeitschrift für Physik, **96**, 431 (1935).
103. Z. S. Wang, G. R. Darling, and S. Holloway, *Dissociation Dynamics from the de Broglie-Bohm Perspective*, Journal of Chemical Physics, **115**, no. 22, 10373 (2001).

104. H. M. Wiseman, S. Mancini, and J. Wang, *Bayesian Feedback versus Markovian Feedback in a Two-Level Atom*, Physical Review A **66**, 013807 (2002).

105. E. Witten, *Dynamical Breaking of Supersymmetry*, Nuclear Physics B (Proceedings Supplements), **188**, 513 (1981).

106. E. Witten, *Supersymmetry and Morse Theory*, Journal of Differential Geometry **17**, 661–692 (1982).

107. R. E. Wyatt, *Quantum Wave Packet Dynamics with Trajectories: Application to Reactive Scattering*, Journal of Chemical Physics, **111**, no. 10, 4406 (1999).

108. R. E. Wyatt, C. L. Lopreore, and G. Parlant, *Electronic Transitions with Quantum Trajectories*, Journal of Physical Chemistry, **114**, no. 12, 5113 (2001).

109. R. E. Wyatt and K. Na, *Electronic Transitions with Quantum Trajectories. II*, Journal of Physical Chemistry, **116**, 1228 (2002).

110. R. E. Wyatt and K. Na, *Quantum Trajectory Analysis of Multinode Subsystem-Bath Dynamics*, Physical Review E, **65**, 016702 (2002).

111. R. E. Wyatt, *Quantum Wavepacket Dynamics with Trajectories: Wavefunction Synthesis along Quantum Paths*, Chemical Physics Letters, **313**, no. 1–2, 189–197 (1999).

Chapter 6

Nanoelectronics: Quantum Electron Transport

Enrique Colomés,[a] Guillermo Albareda,[b] Zhen Zhan,[c] Devashish Pandey,[a] Alfonso Alarcón,[a] Fabio Traversa,[d] and Xavier Oriols[a]

[a]*Departament d'Enginyeria Electrònica, Universitat Autonòma de Barcelona (UAB), 08193 Bellaterra, Spain*
[b]*Theory Department, Max Planck Institute for the Structure and Dynamics of Matter, Luruper Chaussee 149, 22761 Hamburg, Germany*
[c]*School of Physics and Technology, Wuhan University, Wuhan 430072, China*
[d]*Department of Physics, University of California, San Diego, California, USA*
xavier.oriols@uab.es

Nowadays, the need for faster and smaller devices is pushing the electronics industry to develop electron devices with solid-state structures of a few nanometers, with lower dimensions, and driven by picosecond signals. Electron dynamics in such scenarios is in general governed by quantum mechanical laws. This chapter is devoted to discuss how Bohmian mechanics can help us to understand and model the behavior of novel electron device prototypes at the nanometer and picosecond scales. The adaptation of Bohmian mechanics for describing electron transport in open systems leads to a quantum Monte Carlo algorithm, where

Applied Bohmian Mechanics: From Nanoscale Systems to Cosmology (Second Edition)
Edited by Xavier Oriols and Jordi Mompart

ISBN 978-981-4800-10-5 (Hardcover), 978-0-429-29474-7 (eBook)
www.jennystanford.com

additional randomness appears because of the uncertainties in the number of electrons, their energies, and the initial positions of the (Bohmian) trajectories. A general, versatile, and time-dependent three-dimensional (3D) electron transport simulator for nanoelectronic devices, named BITLLES (Bohmian Interacting Transport for nonequiLibrium eLEctronic Structures), is presented, showing its ability for a full prediction (direct current [DC], alternating current [AC], fluctuations, etc.) of the electrical characteristics of any nanoelectronic device. Numerical examples for a resonant tunneling diode and two-dimensional (2D) graphene transitors are presented.

6.1 Introduction: From Electronics to Nanoelectronics

In order to introduce the reader to the topic of this chapter, we explain briefly the birth of the electronics and its evolution towards nanoelectronics. Electronics as a scientific discipline was born around 1897, when Thomson proved that cathode rays were composed of negatively charged particles, named *electron* [1]. Later, in 1904, Fleming discovered that placing an electrode (a metallic material with free electrons inside) close to the filament of an incandescent bulb was enough to establish a net flow of electrons from the filament to the new electrode. Next, De Forest improved Fleming's original invention by creating the triode with an additional third terminal, the grid. Thus, the flux of electrons from the filament to the electrode was controlled by the voltage applied in the grid [2]. A particular voltage was used to define an *On* state with a net current through the triode and another voltage for the *Off* state without current.

During half a century, spectacular electronic applications were developed with these vacuum valve tubes (diodes and triodes). However, the short life and high power consumption of the vacuum tubes made the Bell Laboratories to establish a research group focused on investigating the possibility of using semiconductor solid-state electron devices. In 1947, Bardeen, Brattain, and Shockley created the first solid-state transistor at Bell Laboratories. Although the functionality of the solid-state transistor was quite similar to that of the triode, the former was much smaller, faster, cheaper, and more reliable. Thus, it became the fundamental element of the electronic technology in the second half of the 20th century. In the 60's, the previous solid-state transistor was improved by using a metal-oxide-semiconductor (MOS) solid-state capacitor. A third terminal, the (metal) gate, separated from the (semiconductor) channel by an (oxide) dielectric, controlled the *On* and *Off* states of the electrical current through the transistor by means of a simple electrostatic force between gate and channel electrons. Because of the importance of the electric field in defining the behavior of such MOS transistor, it was also called field-effect-transistor (FET).[a]

Nowadays, the International Roadmap for Devices and Systems (IRDS), previously called the International Technology Roadmap for Semiconductors (ITRS), is focused on the improvement of the FET transistor as the best strategy to be followed [3]. Nonetheless, the scientific community is looking for completely different alternatives to the FET transistors because the midterm scaling required by Moore's law[b] (transistors with less than 10 nm channel length [3]) will be technologically and mainly economically unattainable with the present FET technology. It is, however, still not clear which proposals will replace the present FET transistors in the midterm future. Some works suggest that a *revolution* (similar to the substitution of vacuum tubes by solid-state transistor) is awaiting for the electronic industry. Other affirm that such *revolution* will not take place, but we will see just an *evolution* of present FET transistors into smaller structures. The novel 2D materials, such as graphene [4], are clear example of the trade off between *revolution* and *evolution*.

---

[a] It was also named MOSFET by combining both previous acronyms.

[b] In 1958, with the invention of the integrated circuit (the chip) a race for chip miniaturization started that lead to an empirical law known as Moore's Law: *The number of transistors that can be implemented in a chip doubles approximately every 2 years*. The increase in the number of transistors in a chip, offers more functions per chip with much lower cost per function, which gives as a result smaller electron devices, higher performance, and greater energy efficiency.

In any case, what is unquestionable is that the dimensions of the new commercial electron devices will attain few nanometers, so that we are entering into the new fascinating nanoelectronic world. Electrons can no longer be described by classical particles and must be understood as quantum particles. The aim of this chapter is to show how Bohmian mechanics can help us in the understanding and designing the upcoming electron device prototypes. Other helpful reviews can be found in Refs. [5–8].

6.2 Evaluation of the Electrical Current and Its Fluctuations

In this section, we will see that the Bohmian perspective provides useful tools when dealing with the continuous measurement of the electrical current in nanodevices. The functionality of any electronic device is determined by the relationship between the current measured by an ammeter and the voltage applied at the external battery. See Fig. 6.1 for a description of a typical setup of an electron device and its associated circuit. The device-active region is connected to a battery (modeled by an ideal voltage source plus an internal resistance R_{IN}) and to the ground by metallic wires. The ammeter located far from the device-active region measures the current. A load resistance R_L is also depicted in Fig. 6.1.

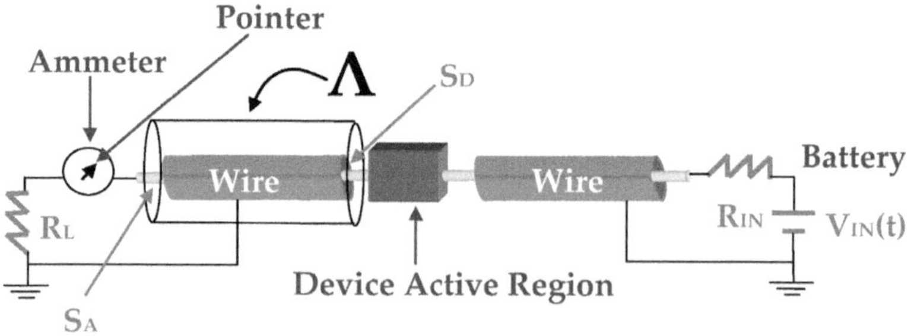

Figure 6.1 Schematic representation of a typical electric circuit used in this chapter for studying the current measurement in an electron device. Device simulators compute the current on the surface, S_D, of the device-active region, while the ammeter measures it on the surface, S_A.

6.2.1 Bohmian Measurement of the Current as a Function of the Particle Positions

The ammeter in Fig. 6.1 provides a relationship between the value of the electrical current in the active region $I(t)$ and the value of the electrical current $I_{\text{ammeter}}(t)$ *observed* in the (position of a pointer inside the) ammeter.[a] The Bohmian explanation of the *measurement* process discussed in the Chapter 1 tells us what we have ultimately *measured* is the position of the ammeter pointer. Therefore, the pointer (Bohmian) positions, $\{\vec{r}_{p1}[t], \cdots, \vec{r}_{p_N}[t]\}$, at time t, have to be included as a part of the simulated degrees of freedom (the subindex p refers to the pointer Bohmian positions). In this sense the whole Hamiltonian of the system (including the degrees of freedom of the measuring apparatus, the cables and the active device region) in Fig. 6.1 is:

$$\begin{aligned} H_{\text{circuit}} = &\sum_{k=1}^{M_T} \left\{ K(\vec{p}_k) + \frac{1}{2} \sum_{\substack{j=1 \\ j \neq k}}^{M_T} q V_0(\vec{r}_k, \vec{r}_j) \right\} \\ &+ \sum_{k=M_T+1}^{M_P} \left\{ K(\vec{p}_k) + \frac{1}{2} \sum_{\substack{j=M_T+1 \\ j \neq k}}^{M_P} q Z_k Z_j V_0(\vec{r}_k, \vec{r}_j) \right\} \\ &- \sum_{k=1}^{M_T} \sum_{j=M_T+1}^{M_P} q Z_j V_0(\vec{r}_k, \vec{r}_j). \end{aligned} \tag{6.1}$$

Such Hamiltonian will contain $i = \{1, 2, \cdots, M_T\}$ electrons. Each one with momentum $\vec{p}_i = \{p_{x,i}, p_{y,i}, p_{z,i}\}$, electron charge $-q$, and position $\vec{r}_i = \{x_i, y_i, z_i\}$). In addition, it contains $i = \{M_T + 1, \ldots, M_P\}$ atomic nuclei, (with Z_i the atomic number of the i-th atom). The term $K(\vec{p}_k)$ is the kinetic energy of the k-particle and $V_0(\vec{r}_k, \vec{r}_j) = \frac{q}{4\pi\,\varepsilon_0\,|\vec{r}_k - \vec{r}_j|}$ the Coulomb interaction between the k and j particle (with ε_0 the vacuum permittivity). We remind

[a]The relationship between the current $I_{\text{ammeter}}(t)$ and the pointer positions can be established, for example, by the magnetic deflection of a pointer: a current passing through a coil in a magnetic field causes the coil to move. The position of a *pointer* fixed to this coil will indicate the final value of the current.

that a small set of those degrees of freedom, that we labeled as $\{\vec{r}_{p1}[t], \cdots, \vec{r}_{p_N}[t]\}$, belongs to the pointer of the ammeter.

In principle, if it were possible to know the total *circuit* wave function $\psi_T(\vec{r}_1, \cdots, \vec{r}_{M_P}, t)$ by solving the many-particle Schrödinger equation with the Hamiltonian of Eq. 6.1, then, we would be able to compute the Bohmian trajectories of the *pointer* and know the measured current $I_{\text{ammeter}}(t)$ at any time. However, the previous procedure has so many computational difficulties (the so-called many-body problem [9, 10]) that the current can only be obtained with strong approximations. In any case, from a Bohmian conceptual point of view, it has no difficulties at all and we can define a relationship between $I(t)$ and $I_{\text{ammeter}}(t)$. Since $I_{\text{ammeter}}(t)$ is directly related to the M_P Bohmian trajectories (among them, those belonging to the pointer), we can establish the following relationship f_B between all trajectories and the value of the current on the active region $I_g(t)$ as:

$$I_g(t) = f_B\left(\vec{r}_{1,g}[t_0], \cdots \vec{r}_{p1,g}[t_0], \cdots, \vec{r}_{p_N,g}[t_0], \cdots, \vec{r}_{M_P,g}[t_0], t\right). \tag{6.2}$$

The detailed definition of such function f_B is the main goal of this section. Bohmian mechanics assures that f_B exists because it allows a discussion about the ontological reality of electrical current in the active region (independently of the fact that it is measured or not).

The probability to find a particular value of the current in a particular experiment depends on the probabilities of $\{\vec{r}_{1,g}[t], \cdots, \vec{r}_{M_P,g}[t]\}$, which in turn depend on the initial positions. Thus, we use the subindex g in $I_g(t)$ to recall which are the initial positions $\{\vec{r}_{1,g}[t_0], \cdots, \vec{r}_{M_P,g}[t_0]\}$ that determines the current value $I_g(t)$ at any later time. From the quantum equilibrium hypothesis of Bohmian mechanics mentioned in the postulates of Chapter 1, we define the g-distribution as the infinite ensemble of all possible Bohmian initial positions, $g = \{1, 2, \cdots, N_g\}$, defined by the condition:

$$|\psi_T(\vec{r}_{1,g}[t_0], \cdots, \vec{r}_{M_P,g}[t_0], t_0)|^2 = \lim_{N_g \to \infty} \frac{1}{N_g}\left(\sum_{g=1}^{N_g}\prod_{i=1}^{M_P}\delta(\vec{r}_i - \vec{r}_{i,g}[t_0])\right). \tag{6.3}$$

The probability of each element of the g-distribution of initial positions (i.e., of each element $I_g(t)$) is, by construction, equal to $1/N_g$.

6.2.1.1 Relationship between current in the ammeter $I_{\text{ammeter},g}(t)$ and the current in the device-active region $I_g(t)$

Before computing f_B, it will be interesting to establish the well-known relationship between the current $I_{S_A,g}(t)$ crossing the surface S_A of the ammeter shown in Fig. 6.1, and the current $I_{S_D,g}(t)$ on the surface of the device-active region, S_D. Here, to simplify the notation, we have re-defined $I_g(t) \equiv I_{S_D,g}(t)$ and $I_{\text{ammeter},g}(t) \equiv I_{S_A,g}(t)$. Let us assume that we deal with a particular g-element of the set of Bohmian trajectories. We will keep the subindex g to remind this point. We rewrite the current conservation at any point along the wire:

$$\vec{\nabla}\vec{j}_{c,g}(\vec{r},t) + \frac{\partial \rho_g(\vec{r},t)}{\partial t} = 0, \tag{6.4}$$

where $\vec{j}_{c,g}(\vec{r},t)$ is the particle or conduction current density and $\rho_g(\vec{r},t) = -q\sum_{i=1}^{M_P}\delta(\vec{r}-\vec{r}_{i,g}[t])$ is the density of Bohmian particles at time t in the whole circuit. The second term, $\rho_g(\vec{r},t)$, can be related to the electric field, $\vec{E}_g(\vec{r},t)$, by using the Poisson (i.e., first Maxwell) equation:

$$\vec{\nabla}\left(\varepsilon(\vec{r})\vec{E}_g(\vec{r},t)\right) = \rho_g(\vec{r},t), \tag{6.5}$$

where the electric permittivity, $\varepsilon(\vec{r})$, is assumed to be a time-independent scalar function. Thus, we can rewrite Eq. 6.4 as:

$$\vec{\nabla}\vec{j}_{c,g}(\vec{r},t) + \vec{\nabla}\left(\varepsilon(\vec{r})\frac{\partial \vec{E}_g(\vec{r},t)}{\partial t}\right) = \vec{\nabla}\left(\vec{j}_{c,g}(\vec{r},t) + \vec{j}_{d,g}(\vec{r},t)\right) = 0, \tag{6.6}$$

where $\vec{j}_{d,g}(\vec{r},t) = \varepsilon(\vec{r})\frac{\partial \vec{E}_g(\vec{r},t)}{\partial t}$ is the displacement current density. From Equation (6.6), we can define the total current, $\vec{j}_{T,g}(\vec{r},t) = \vec{j}_{c,g}(\vec{r},t) + \vec{j}_{d,g}(\vec{r},t)$, as a divergence-free vector. Finally, by integrating Eq. 6.6 on the wire volume Λ of Fig. 6.1, we arrive to the following identity for the total currents:

$$\int_{\Lambda}\vec{\nabla}\vec{j}_{T,g}(\vec{r},t)dv = \int_{S}\vec{j}_{T,g}(\vec{r},t)d\vec{s} = 0, \tag{6.7}$$

where the first integral is evaluated inside the volume Λ and the second integral over the closed surface S limiting this volume Λ. In detail, the surface S is composed of the ammeter surface, S_A, the device surface S_D and a lateral cylindrical surface drawn in Fig. 6.1. We assume that this lateral surface is so far away from the metallic wire that the electric field there is almost zero and there are no particles crossing it.[a] Thus, the integral surface of the right-hand side of expression (6.7) can be rewritten as:

$$\int_{S_D} \vec{j}_{T,g}(\vec{r}, t) d\vec{s} + \int_{S_A} \vec{j}_{T,g}(\vec{r}, t) d\vec{s} = 0. \tag{6.8}$$

Expression (6.8) tells us that $I_{S_A,g}(t) = -I(t)_{S_D,g}$. There is an irrelevant sign related with the direction of the vector $d\vec{s}$. In conclusion, we have established a (trivial) relationship between the current in the ammeter $I_{\text{ammeter},g}(t)$ (evaluated on the surface S_A) and the current in the device-active region $I_g(t)$ (evaluated on the surface of the simulation box S_D). This is the first element to construct f_B.

6.2.1.2 Relationship between the current on the device-active region $I_g(t)$ and the Bohmian trajectories $\{\vec{r}_{1,g}[t], \cdots, \vec{r}_{M_P,g}[t]\}$

Now, we determine the function f_B of Eq. 6.2 between the positions of the Bohmian trajectories in the whole circuit and the value of the current $I_g(t) \equiv I_{S_D,g}(t)$ measured on the surface S_D.

First, we will discuss the charge associated with the particle (conduction) and displacement components of the total current. From the continuity equation of Eq. 6.4, which is accomplished by Bohmian trajectories, we conclude that:

$$\int_{S_D} \vec{j}_{c,g}(x_D, y, z, t)\, d\vec{s} = \frac{d}{dt} \int_{S_D} \int_{x=-\infty}^{x=x_D} \rho_g(x, y, z, t)\, dx\, dy\, dz. \tag{6.9}$$

[a] In fact, the relevant point is not only that the lateral surface is far away from the wire, but also that the lines of the electromagnetic field leaving the lateral surface of a two-terminal device will always return to one of the two terminals. An ideal two-terminal device assumes that there is no (positive/negative) charge anywhere outside the device, except along the two terminals that connect the device to the battery.

where we have defined x_D as the x-position on the surface S_D. We define the *particle (or conduction)* charge as:

$$Q_{c,g}(t) = -q \int_{S_D} d\vec{s} \int_{x=-\infty}^{x_D} dx \sum_{i=1}^{M_P} \delta(\vec{r} - \vec{r}_{i,g}[t])$$
$$= -q \left(\sum_{i=1}^{M_P} u(x_{i,g}[t] - x_D) \right), \tag{6.10}$$

where $u(\xi)$ is the unit step (Heaviside) function, it is $u(\xi) = 1$ when $\xi > 0$ and zero elsewhere.[a]

Now, we can interpret the charge associated to the *displacement* current, as:

$$Q_{d,g}(t) = \int_{S_D} d\vec{s}\varepsilon(\vec{r})\vec{E}_g(x_D, y, z, t). \tag{6.11}$$

The electric field, $\vec{E}_g(x_D, y, z, t)$ depends directly on the position of the Bohmian trajectories $\{\vec{r}_{1,g}[t], \ldots, \vec{r}_{M_P,g}[t]\}$.[b] Finally, we can interpret one particular outcome of the total current measured by an ammeter, $I_g(t)$, as the time derivative of the *particle* plus *displacement* charges:

$$I_g(t) = \frac{d\left(Q_{c,g}(t) + Q_{d,g}(t)\right)}{dt}. \tag{6.12}$$

This Eq. 6.12 (together with Eqs. 6.10 and 6.11) provides the explicit definition of the function f_B mentioned in Eq. 6.2.

In summary, we have found a relationship between the measured current (i.e., the positions of the Bohmian particles that constitutes the pointer) and the positions of the Bohmian particles inside the device-active region. This is a trivial result for classical trajectories, but we have shown explicitly that it is also true for quantum (Bohmian) trajectories. Classical mechanics and Bohmian mechanics share a common ontological characteristic. Both theories assume the reality of some magnitudes (for example, the electrical current

[a] In order to compute $Q_{c,g}(t)$, we do only have to know at each time step how many particles are in each side of the boundary $x = x_D$. In practical computations, an appropriate injection model is necessary at the borders of the simulation box.

[b] In practical computations, $\vec{E}_g(x_D, y, z, t)$ depends on the simulated Bohmian particles inside the simulation box and on the boundary conditions on the borders of our simulation box

defined from the dynamics of the trajectories) independently of the fact of they being measured or not. This is the only ingredient that we have used to establish the function f_B.

From a practical point of view, getting f_B has a limited utility unless we are able to reduce the degrees of freedom of the Hamiltonian in Eq. 6.1. However, in making the approximations, we have to take into account that (contrarily to classical mechanics) the velocity of each Bohmian trajectories in the device-active region is computed from the many-particle wave function defined in the huge configuration space associated to the degrees of freedom of the whole system of Fig. 6.1.

6.2.1.3 Reducing the number of degrees of freedom of the whole circuit

When dealing with quantum electron transport in solid-state structures, there are some standard simplifying assumptions to reduce the Hamiltonian of Eq. 6.1 into a solvable equation:

(1) Since we are mainly interested in conducting electrons, we assume that the dynamics of the conducting electrons can be understood by neglecting the dynamics of the atomic nuclei and fixing their positions to their equilibrium value. This is the standard Born–Oppenheimer approximation that can be justified because the mass of atomic nuclei is much higher than the electron mass. In many cases, ballistic transport cannot be justified and the dissipation of energy through the interaction with atomic nuclei can be included through stochastic collision phenomena, as it will be shown in Section 6.4.

(2) From the M_T electrons, we consider explicitly only $N(t)$ (free) electrons inside the 3D device-active region. We named this limited spatial region the *simulation box*. We neglect those electrons outside the simulation box and also those electrons inside which are so strongly coupled with the atomic nuclei that they do not contribute to the current (i.e., core electrons).

(3) The interaction of the $N(t)$ electrons with the (fixed) atomic nuclei in the device-active region can be considered, in materials

with a parabolic band structure, through the effective electron mass approximation [11–14]. Hence, we can neglect the terms with electron-atomic nuclei interaction by just modifying the value of the electron mass to an *effective* value. For materials with a linear band structure, a similar strategy can be followed.

With these standard approximations, we arrive to the following Hamiltonian for $\{\vec{r}_1, \ldots, \vec{r}_{N(t)}\}$ electrons in the device-active region:

$$H = \sum_{k=1}^{N(t)} \left\{ K(\vec{p}_k) + \frac{1}{2} \sum_{\substack{j=1 \\ j \neq k}}^{N(t)} q V_0(\vec{r}_k, \vec{r}_j) \right\} = \sum_{k=1}^{N(t)} K(\vec{p}_k) + U(\vec{r}_1, \ldots, \vec{r}_N, t). \tag{6.13}$$

where $U(\vec{r}_1, \ldots, \vec{r}_N, t)$ is now the potential energy due to the Coulomb interaction among those electrons inside the simulation box. At each time, the number of electrons explicitly simulated inside the simulation box is $N(t)$ which changes with time because there are electrons entering and leaving.[a]

Let us discuss what are the physical consequences of reducing the degrees of freedom. Bohmian mechanics provides an interesting path to deal with such a reduction. Instead of working with the whole many-particle wave function $\psi_T(\vec{r}_1, \ldots, \vec{r}_{M_P}, t)$ of the Hamiltonian in Eq. 6.1, we can work with the conditional wave function defined in the introductory Chapter 1 of this book:

$$\Psi_a(\vec{r}_a, t) \equiv \psi_T(\vec{r}_1[t], \cdots \vec{r}_{a-1}[t], \vec{r}_a, \vec{r}_{a+1}[t], \cdots, \vec{r}_{M_P}[t], t). \tag{6.14}$$

In such conditional wave function, only $\vec{r}_a$ is considered as the variable. The rest of the degrees of freedom are substituted by Bohmian trajectories (which introduce an additional time-dependence on the conditional wave function). We know that additional real and imaginary potentials have to be included into the Hamiltonian that describes the time evolution of the conditional wave function.

[a]Let us notice that, in practical simulations, we need to keep a large (infinite) number of additional electrons waiting to enter into the simulation box (i.e., the injected electrons). The important point is that only a small fraction, $N(t)$, of such *infinite* electrons is explicitly simulated in the Hamiltonian of Eq. 6.13.

Another consequence of the reduction of the degrees of freedom is that we cannot completely specify the initial $N(t)$-particle wave function inside the simulation box (we do not know with certainty the number of electrons $N(t)$, their energies...). We are dealing with an open system. There is no specific preparation for the wave packets associated to electrons entering from the reservoirs to the active region. We can only know the characteristics of injected electrons in a statistical manner. For example, we can assume that the mean energy of the injected electrons follows Fermi–Dirac statistics. In other words, we can no longer deal with a single *pure* $N(t)$-particle state, but with a *mixed quantum system* prepared by statistically combining different pure states. We take into account the probabilities of these states by an additional stochastic h-distribution. Such a statistical ensemble of pure states takes into account our ignorance on the number of electrons inside the simulation box, $N(t)$, the energy of the wave packets associated to these electrons, the injection time of each electron, etc. From a practical point of view, this means that apart from the distribution in the initial position of Bohmian particles (g-distribution, mentioned at the beginning of Section 6.2.1), we do also have an additional distribution of these other properties of injected electrons, $h = \{1, \ldots, N_h\}$, due to the lack of information that appears when we cut (open) our (initially closed) circuit.

For readers interested in a more technical discussion on what type of quantum measurement can model the electrical current flowing through an electron device, we refer them to Refs. [15–17]. The measurement of the electrical current corresponds to a weak measurement, where a noisy value of the electrical current is obtained at the price of a small perturbation of the (conditional) wave function. This *measurement* noise can be related to plasmonic dynamics of electrons in the contacts and it tends to disappear when the observation frequency of the electrical current is reduced [15]. For DC results, with a temporal average of the electrical current, such *measurement* noise simply disappears. However, at higher frequencies, these noise can be relevant implying that the *instantaneous* value of the measured current at the ammeter I_{ammeter} in a particular experiment is different from the value on the active region.

6.2.2 Practical Computation of DC, AC, and Transient Currents

As discussed above, the measured value of the current can be associated to two sources of uncertainty, g and h, in order to compute $I_{g,h}(t)$. Thus, we determine the average value of the current at time t_1 (or the expectation value, or the mean, or the first moment) from the following ensemble average [5, 6]:

$$\langle I(t_1)\rangle = \lim_{N_g, N_h \to \infty} \frac{1}{N_g N_h} \sum_{g=1}^{N_g} \sum_{h=1}^{N_h} I_{g,h}(t_1). \qquad (6.15)$$

As we have discussed, the probability of each element of the g-distribution is $1/N_g$. Identically, we can always assume that the h-distribution is selected so that the probability of each element of the ensemble is $1/N_h$.[a]

Therefore, the procedure to compute the average current from Eq. 6.15 will be the following:

(1) At $t = 0$, we select a particular realization of the h-distribution and a particular realization of the g-distribution (see Fig. 6.2).
(2) We solve the (many-particle) Schrödinger equation from time $t = 0$ till $t = t_1$ as we will be later described in next Section 6.3.
(3) From Eq. 6.12, we compute the value $I_{g,h}(t_1)$.
(4) We repeat the (1)–(3) steps for the whole ensemble $g = \{1, \ldots, \infty\}$ and $h = \{1, \ldots, \infty\}$ to evaluate Eq. 6.15.

When the battery in Fig. 6.1 is fixed to a constant value, then the whole circuit becomes stationary. For a stationary process, the mean current in Eq. 6.15 is independent of time. Then, if the process is ergodic, we can compute the mean current from the following (first-order) time average expression:

$$\overline{I_{g,h}(t)} = \lim_{T \to \infty} \frac{1}{T} \int_{-T/2}^{T/2} I_{g,h}(t)dt. \qquad (6.16)$$

In this case, the practical procedure for the computation of the mean current is simpler. Before beginning the simulation, we select

[a] In fact, the g-element depends on the particular value of the h-element.

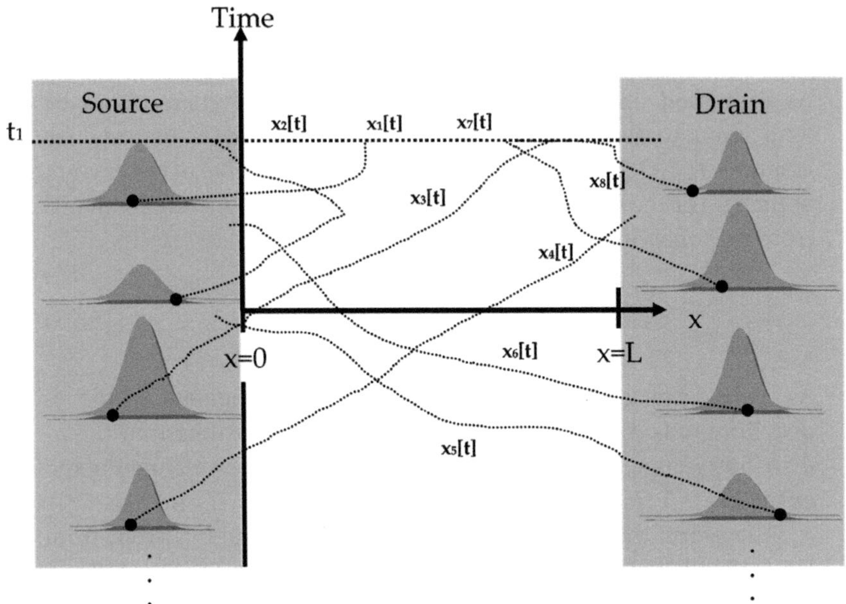

Figure 6.2 Schematic representation of the simulation box for a particular selection of energies and injecting times of wave packets (h-distribution) and the initial position of Bohmian trajectories (g-distribution) during an infinite time.

only one particular realization of the h-distribution for an *infinite*[a] number of electrons. Simultaneously, we fix the g-distribution of the initial positions for the previous (infinite) realization of wave packets. See Fig. 6.2 where we have represented a scheme of a single h- and g-element of the distributions in the simulation box in position and time.

A single sample function often provides little information about the statistics of the process. However, if the process is ergodic, that is, time average equal ensemble averages, then all statistical information can be derived from just one sample element of the h- and g-distributions.

Electrical engineers are not only interested in the DC behavior of nanoelectronic devices, but also in their transient or AC

[a]The practical procedure for the *infinite* number is selecting a number large enough so that the mean current remains practically unchanged for successive times.

performance. For these time-dependent scenarios, the circuit is no longer stationary nor ergodic. Then, we cannot use ergodicity and we have to compute the mean value of the current at each particular time, t_1, only from the ensemble average of Eq. 6.15.

6.2.3 Practical Computation of Current Fluctuations and Higher Moments

When we measure the current experimentally, we do not find a fixed (DC) value but a randomly varying function of time. These fluctuations of the current are simply called noise. The characterization of the noise is an important issue in electronics in order to understand how to avoid it in practical circuit applications. On the other hand, from a physical point of view, there is a lot of useful information in the noise that is missing (because of the average) in the mean values discussed above. Roughly speaking, the noise can be characterized by the *variance* of the probability distribution of the current. The average current is known as the first moment of the current probability distribution, while the variance is the second moment. In fact, once we know all the values $I_{g,h}(t)$ from the infinite g- and h-distributions, one can compute the probability of each current value and, from them, any higher moments of the current distribution. For example, the third moment measures the *skewness*[a] of the current distribution. In this sense, our Bohmian simulation algorithm can be compared to the *orthodox* full-counting statistics technique. Both provide full information about the current distribution for quantum transport. See Refs. [18–20] for an explanation of the latter technique.

In the next subsection, we will only discuss the *variance* of the current distribution, i.e., the second moment which is the standard method to characterize how noisy is the current. Before writing the mathematical expression for the practical evaluation of the variance of the noise, we provide a brief summary of the typical noise features in electronics that will be useful latter to discuss the numerical results.

---

[a]A positive (negative) skew indicates that the tail on the right (left) side of the distribution is longer than the left (right) side and the bulk of the values lie to the left (right) of the mean.

6.2.3.1 Thermal and shot noise

Noise in electronic devices can be understood as a consequence of the uncertainties (related to the g- and h-distributions) present in the electron transport process. Theoretical approaches to analyze and model noise in semiconductor devices are carried out using stochastic calculus and microscopic noise sources (see, for example, Ref. [21] for a semiclassical treatment of noise in semiconductor devices).

There are basically two fundamental sources of noise in nanostructures, thermal and shot noise. The former appears at nonzero temperature, where the thermal agitation causes the occupation of energy states to fluctuate. Typically, we can characterize the presence or absence of an electron in a particular region of phase space by the occupation number n, which is either 0 or 1, at each particular time.[a] In simple scenarios, the average of the occupation number $\langle n \rangle$ is determined by the Fermi distribution function $f(E)$. At equilibrium, the probability that a state is empty is given by $1 - f(E)$, and that it is occupied is given by $f(E)$. Taking into account that $n^2 = n$, we find immediately that the fluctuations of the occupation number are given by:

$$\left\langle (n - \langle n \rangle)^2 \right\rangle = f(E)\,(1 - f(E)) \,. \tag{6.17}$$

At high energies and low temperatures, $f(E)$ is much smaller than 1 and thus the factor $1 - f(E)$ in Eq. 6.17 can be replaced by 1. Thus the fluctuations are determined by the Boltzmann distribution.

The other noise source, namely, shot noise, can be explained with the simple case of a constant flux of particle incident (only from the left) upon a barrier.[b] At the barrier, the particle is either transmitted with probability T or reflected with probability $R = 1 - T$ (not both!). We now only consider the occupation numbers n_T associated to a (right) region of the phases space where electrons have crossed the barrier. Its average value is simply given by $\langle n_T \rangle = T$. To find

[a] Alternatively, the parameter n can be visualized as the number of Bohmian particles in a particular point $\vec{r}$ and $\vec{k}$ of the phase space. We will avoid the dependence of n on such variables to simplify the notation in the present section.

[b] At zero temperature, the injection of electrons with energies below the Fermi level is constant without fluctuations (i.e., without thermal noise). To simplify our discussion of shot noise, we will assume such zero temperature in our present argument

the mean squared fluctuations in n_T, we consider the average of the product of the occupation numbers of the transmitted and reflected beam $\langle n_T (1 - n_T) \rangle$. Since in each event the particle is either transmitted or reflected, the product $n_T (1 - n_T)$ vanishes. Notice that this mechanism is essentially the particle nature of the charge as anticipated above. Hence the average $\langle n_T (1 - n_T) \rangle$ vanishes also, and we find easily:

$$\left\langle (n_T - \langle n_T \rangle)^2 \right\rangle = T(1 - T). \qquad (6.18)$$

Such fluctuations are also called *partition noise* since the scatterer divides the incident carrier stream into two. In electron devices out of equilibrium at room temperature, the electronic noise is a combination of thermal noise in the reservoirs, that affects the injection into the active region, and the partition noise due to the presence of tunneling barriers. Such combination of noises in electron devices is the so-called shot noise. The thermal noise refers mainly to the noise present in systems in equilibrium (without partition noise).

Both the thermal and shot noise at low frequencies and low voltages can be studied from in many situations independent quasi-particle transport models. Electrons are, however, interacting particles, and both the fluctuations at finite frequencies and the fluctuation properties far from equilibrium require, in general, a discussion on the role of the Coulomb interaction among electrons. A quasi-particle picture is no longer sufficient, and collective properties of the electron system come into play. Moreover, for finite frequencies, the electrodynamics (Maxwell equations) must be included. See the review of Blanter and Büttiker [18] as a guide for theoretical approach to noise in mesoscopic devices.

6.2.3.2 Practical computation of current fluctuations

The practical expressions for the computation of the current fluctuations within Bohmian mechanics are quite simple. In fact, once the simulations are performed, we directly know all possible measurable current values $I_{g,h}(t)$ and their probabilities $1/(N_g N_h)$, thus, we can use standard statistical techniques to characterize the fluctuations of the current.

The fluctuating signal of the current can be defined as $\Delta I_{g,h}(t) = I_{g,h}(t) - \langle I_{g,h}(t)\rangle$. We can obtain information of the noise from the variance (or the mean square or the second moment) defined as $\langle \Delta I(t)^2\rangle = \langle I(t)^2\rangle - \langle I(t)\rangle^2$. However, experimentalists are interested in having information on how the noise is distributed along the different frequencies.[a] Therefore, the characterization of fluctuations of the current are computed from the covariance:

$$\langle \Delta I(t_1)\Delta I(t_2)\rangle = \lim_{N_g, N_h\to\infty} \frac{1}{N_g N_h}\sum_{g=1}^{N_g}\sum_{h=1}^{N_h} \Delta I_{g,h}(t_1)\Delta I_{g,h}(t_2). \quad (6.19)$$

If the process is ergodic, i.e., $\langle \Delta I_{g,h}(t)\Delta I_{g,h}(t+\tau)\rangle = \overline{\Delta I(t)\Delta I(t+\tau)}$, we can compute the noise equivalently from the autocorrelation function:

$$\overline{\Delta I(t)\Delta I(t+\tau)} = \lim_{T\to\infty}\frac{1}{T}\int_{-T/2}^{T/2} \Delta I_{g,h}(t)\Delta I_{g,h}(t+\tau)dt. \quad (6.20)$$

In addition, a process is called wide-sense (or weakly) stationary if its mean value is constant and its autocorrelation function depends only on $\tau = t_2 - t_1$. Then, we define the autocorrelation function $R(\tau)$ as:

$$R(\tau) = \overline{\Delta I(t)\Delta I(t+\tau)}, \quad (6.21)$$

because it depends only on $\tau = t_2 - t_1$. Wide-sense stationary processes are important because the autocorrelation function of Eq. 6.20 and the power spectral density function $S(f)$ (measured by experimentalists) are related by a Fourier transform:

$$S(f) = \int_{-\infty}^{\infty} R(\tau)e^{-j2\pi f\tau}\, d\tau. \quad (6.22)$$

This is known as the Wiener–Khinchine theorem. In many systems, one obtains the well-known Schottky's result [22] for the shot noise:

$$S_{\text{schot}}(f) = 2q\,\langle I\rangle\,, \quad (6.23)$$

which is also referred to in literature as *Poissonian value* of shot noise.

---

[a]Most of electronic apparatuses, and the ammeter itself, have to be interpreted as low-pass filters. Therefore, they are not able to measure *all* noise of the spectrum, but only up to a maximum frequency.

6.3 Solving Many-Particle Systems with Bohmian Trajectories

In Section 6.2, we have discussed the measurement of the current and its fluctuations in nanoelectronic devices. Certainly, the simplicity for computing the first and second (or even higher) moments of the current distribution is the first important advantage provided by Bohmian mechanics when we model nanoelectronic transport. In this section, we will discuss a second important advantage of Bohmian mechanics: its ability to treat many-particle systems. Let us notice that although we have been able to reduce the number of electrons of the whole circuit to only a few, present inside the simulating box, there are still strong correlations among the simulated (free) electrons, so that an approximate solution of the many-body problem is still needed. In this section, we will discuss how Bohmian mechanics can help us to study many-particle quantum electron transport phenomena in nanoelectronic devices.

In a typical nanoscale structure, the device-active region contains hundreds of (conducting) electrons [23]. Thus, time-dependent quantum simulators have to provide reasonable approximations for handling the many-particle (electron-electron interaction) problem. In a system of $N(t)$ particles the evolution of the positions of these particles is guided by the wave function, which evolves according to the many-particle Schrödinger equation, using the Hamiltonian in Eq. 6.13. However, it is well-known that the many-particle Schrödinger equation can be solved for very few (two, three,...) degrees of freedom. To surpass this computational problem, we present next a novel algorithm to solve the many-particle Schrödinger equation with Bohmian mechanics, taking into account the Coulomb interaction. Next, in Section 6.3.2, we extend the previous algorithm to take into account Coulomb and exchange interaction together.

6.3.1 Coulomb Interaction Among Electrons

As indicated in Chapter 1 and as it can be seen in Ref. [24], a many-particle Bohmian trajectory $\vec{r}_a[t]$ associated to a-electron can be computed from a (conditional) single-particle wave function, $\Psi_a(\vec{r}_a, t)$, solution of the next pseudo single-particle dynamical equation:

$$i\hbar\frac{\partial\Psi_a(\vec{r}_a, t)}{\partial t} = \{K_a(\vec{r}_a) + U_a(\vec{r}_a, \vec{R}_a[t], t) + G_a(\vec{r}_a, \vec{R}_a[t], t) + i \cdot J_a(\vec{r}_a, \vec{R}_a[t], t)\}\Psi_a(\vec{r}_a, t), \quad (6.24)$$

where we define, $\vec{r}_a$, as the vector position of the a-electron and $\vec{R}_a[t] = \{\vec{r}_1[t], \vec{r}_{a-1}[t], \vec{r}_{a+1}[t], \vec{r}_N[t], t\}$ as a vector that contains all Bohmian trajectories except $\vec{r}_a[t]$. $K_a(\vec{r}_a)$ is the kinetic energy of the a-th particle, which explicitly is either $-\frac{\hbar^2}{2m}\nabla^2_{\vec{r}_a}$ when we study parabolic band structures or $-i\hbar v_f\nabla_{\vec{r}_a}\cdot\vec{\sigma}$ when studying linear band structures . The explicit expressions of the potentials $G_a(\vec{r}_a, \vec{R}_a[t], t)$ and $J_a(\vec{r}_a, \vec{R}_a[t], t)$ that appear in Eq. 6.24 can be found in reference [24]. In fact, their numerical values are unknown and need some educated guesses. In contrast, the term $U_a(\vec{r}_a, \vec{R}_a[t], t)$ can take into account (short and long) range Coulomb interaction without any approximation. In fact, the potential $U_a(\vec{r}_a, \vec{R}_a[t], t)$ can be any type of many-particle potential defined in the position-representation. Similar strategies have also been adapted to describe the many-body electron-nuclei interaction [25, 26], and it has been shown that usually the potentials G_a and J_a take discontinuous steps [27].

The relevant point in our quantum trajectory model is that, in order to find the trajectories, $\vec{r}_a[t]$, from Eq. 6.24, we do not have to evaluate the wave function and potential energies everywhere in the configuration space, $\{\vec{r}_1, \vec{r}_{a-1}, \vec{r}_a, \vec{r}_{a+1}, \vec{r}_N\}$, but only over a smaller number of configurations points, $\{\vec{r}_a, \vec{R}_a[t]\} = \{\vec{r}_1[t], \vec{r}_{a-1}[t], \vec{r}_a, \vec{r}_{a+1}[t], \vec{r}_N[t]\}$, because all positions of electrons are fixed at $\vec{R}_a[t]$ in Eq. 6.24, except $\vec{r}_a$.

In summary, we have been able to decompose an irresolvable single $N(t)$-particle Schrödinger equation into a set of $N(t)$ single-particle Schrödinger equation with time-dependent potentials. From a practical point of view, all Bohmian trajectories $\vec{r}_a[t]$ have to be computed simultaneously. The steps of the algorithm are the following:

(1) At the initial time t_0, we fix the initial position of all a-particles and their associated single-particle wave function $\Psi_a(\vec{r}_a, t_0)$.

(2) From all particle positions, we compute the exact value of the potential $U_a(\vec{r}_a, \vec{R}_a[t_0], t)$ for each particle. An approximation for the other terms $G_a(\vec{r}_a, \vec{R}_a[t_0], t_0)$ and $J_a(\vec{r}_a, \vec{R}_a[t_0], t_o)$ is needed.
(3) We use a finite-difference numerical method to solve each single-particle Schrödinger equation, Eq. 6.24, from t_0 till $t_0 + dt$.
(4) From the knowledge of the single-particle wave function $\Psi_a(\vec{r}_a, t_0 + dt)$, we can compute the new Bohmian velocity $\vec{v}_a(t_0 + dt)$ of each a-particles.
(5) From the velocity, we compute the new position of each a-particle as $\vec{r}_a[t_0 + dt] = \vec{r}_a[t_0] + \vec{v}_a(t_0 + dt)dt$.
(6) Finally, with the set of new positions and wave functions, we repeat the whole procedure, steps (1) to (5), for another infinitesimal time dt till the simulation time is finished.

The algorithm explained above, can be applied to many different many-particle quantum problems (not only to study quantum electrons transport in nanoelectronics). More details and numerical results discussing the accuracy and numerical viability of this *many-particle* procedure can be found in Ref. [24] and in Chapter 1.

6.3.2 Exchange and Coulomb Interaction Among Electrons

As explained in many textbooks, the exchange interaction for fermionic particles is described by the Pauli exclusion principle [28]. Electrons are fermions (spin 1/2) and, therefore, suffer from this exchange interaction. Electrons with identical spin repel each other when they try to occupy the same regions of the phase space. This interaction is not classical, and we cannot find a term in the potential energies of the many-particle Schrödinger equation that accounts for it. Alternatively, this new interaction is introduced in the *shape* of the wave function through the requirement of providing antisymmetrical wave functions. We say that a many-particle wave function is antisymmetrical when the interchange of the degrees of freedom associated to two different electrons (positions and spin) provides only a global change of the sign of the wave function. The fact that Schrödinger equation does not take into account the spin of particles does not invalidate this equation or the results obtained by means of it. Since the Hamiltonian of a system of electrically interacting particles (in the absence of a magnetic field) does not contain a spin operator, the wave function of the system of particles can be written in the form of a product of a function that depends only on the coordinate of the particles (that we call *coordinate* or *orbital* wave function and it is solution of the Schrödinger equation) and a function of their spins. Thus, if we are only interested in the orbital wave function, we can ignore the spin evolution. However, in general, the spin is still needed to determine the exchange properties of the orbital wave function.

In this subsection, we provide the explicit form of the anti-symmetrical wave function in two different cases. First, assuming that the spin of electrons have no effect on the consideration of antisymmetrical wave functions (spinless electrons) and, second, considering that the spin of electrons determines the exchange properties of the many-particle wave function.

6.3.2.1 Algorithm for spinless electrons

First, let us discuss what we mean by spinless electrons. If we consider that all electrons have the same spin, then the interchange of spins is irrelevant and we know for sure that the orbital wave function must be antisymmetrical when we interchange the positions of the electrons. The orbital wave function will be identical to the one constructed by considering that electrons have no spin (i.e., spinless electrons). Thus, let us focus on constructing an antisymmetrical (orbital) wave function.

For spinless electrons, it has been shown in the introductory chapter that the exchange interaction is effectively introduced into Eq. 6.24 through the terms $G_a(\vec{r}_a, \vec{R}_a[t], t)$ and $J_a(\vec{r}_a, \vec{R}_a[t], t)$. Due to the Pauli exclusion principle, the modulus of the wave function tends to zero, $R(\vec{r}_a, \vec{R}_a[t], t) \to 0$, whenever the position $\vec{r}_a \to \vec{r}_k[t]$. Thus, the term $G_a(\vec{r}_a, \vec{R}_a[t], t)$ and $J_a(\vec{r}_a, \vec{R}_a[t], t)$ has asymptotes at $\vec{r}_a \to \vec{r}_k[t]$ that *repel* other electrons. Therefore, when the exchange interaction is present, the simplification done in these potential terms has to be considered with care. In this section we extend the algorithm presented in Section 6.3 to take into account the

Coulomb and the exchange interactions together. The trick will be to decompose the many-particle antisymmetrical wave function as a sum of many-particles wave functions without any special symmetry. Because of the linearity of the Schrödinger equation, such sum of terms without symmetry is maintained during the time evolution of the wave function. So, the solution of the Schrödinger equation in Eq. 6.24 is:

$$\Psi_a(\vec{r}_a, t) = \sum_{j=1}^{N!} \Psi_{1,p(j)_1}(\vec{r}_1[t], t) \ldots \times \Psi_{a,p(j)_a}(\vec{r}_a, t) \ldots \Psi_{N,p(j)_N}(\vec{r}_N[t], t) s(\vec{p}(j)). \quad (6.25)$$

The first label i of the subindex of the wave functions $\Psi_{i,j}(\vec{r}_i, t)$ that appear in Eq. 6.25 fixes the potential energy $U_i(\vec{r}_i, \vec{R}_i[t], t)$ in Eq. 6.24. The second label j fixes the initial wave function, i.e., $\Psi_{i,j}(\vec{r}_i, t_0) = \Psi_j(\vec{r}_i, t_0)$. In other words, identical initial wave functions $\Psi_{i,j}(\vec{r}_i, t_0) = \Psi_{k,j}(\vec{r}_k, t_0)$ can evolve differently when using $U_i(\vec{r}_i, \vec{R}_i[t], t)$ or $U_k(\vec{r}_k, \vec{R}_k[t], t)$.

The sum is performed over all $N!$ permutations $\vec{p}(l) = \{p(l)_1, p(l)_2, \ldots, p(l)_N\}$ and $s(\vec{p}(l)) = \pm 1$ is the sign of the permutations. Equation (6.25) can be simply computed from a complex matrix (Slater) determinant. The practical solution of a system of N electrons with Coulomb and exchange interaction will require $N \cdot N$ equation wave functions. See Chapter 1 for more details.

6.3.2.2 Algorithm for electrons with spins in arbitrary directions

As mentioned above we cannot neglect the degrees of freedom of electron spins because the exchange of particles must interexchange the particle coordinates and the spins in order to build an antisymmetric wave function. Thus, the symmetry of the function of spins is very relevant to determine the exchange properties of the orbital part. This is the only effect of spin that we will consider in this chapter.

Following Ref. [29], it is also possible to extend the formalism mentioned above and explained in Ref. [24] to study many-particle wave functions where the electrons have different spin orientations. The main difficulty to treat N different electron spin orientations with time-dependent wave packets is that one must study all possible $N!N!$ products of permutations among spin states in the construction of the antisymmetrical wave function. Therefore, for N electrons, the explicit evaluation of $N!N!$ products of permutations is intractable for more than very few electrons.[a] To overcome this computationally inaccessible problem, the total wave function can be treated as a separated product of two many-particle wave functions, the first with spin up and the second with spin down. For this purpose, we assume that the many-particle wave function can be separated into a product of spin-up ($\uparrow$) and spin-down ($\downarrow$) many-particle wave functions:

$$\Psi\left(\vec{r}_1, \vec{r}_2, \vec{r}_3 \ldots \uparrow_1, \downarrow_2, \downarrow_3 \ldots,\right) \approx \phi_\uparrow\left(\vec{r}_1, \vec{r}_4 \ldots \uparrow_1, \uparrow_4 \ldots,\right) \cdot \phi_\downarrow\left(\vec{r}_2, \vec{r}_3 \ldots \downarrow_2, \downarrow_3 \ldots\right). \quad (6.26)$$

Using the approximation written in the right-hand side of Eq. 6.26, the numerical difficulties in the computation of the many-particle wave function disappear because, $\phi_\uparrow\left(\vec{r}_1, \vec{r}_4 \ldots \uparrow_1, \uparrow_4 \ldots\right)$, and $\phi_\downarrow\left(\vec{r}_2, \vec{r}_3 \ldots \downarrow_2, \downarrow_3 \ldots\right)$ can be computed again from a complex matrix (Slater) determinant [29].

6.4 Dissipation with Bohmian Mechanics

In the previous section, we have shown how the many-particle problem can be tackled within the Bohmian framework. In this section, we show how Bohmian mechanics faces dissipation, which is another intrinsic phenomenon present in all nanodevices (in the active region or in the reservoirs due to the interaction of electrons with phonons, impurities, defects, etc.). In principle, one could just compute the dynamics of all particles, including atoms, doping particles, etc., but obviously this is computationally unaffordable. The inclusion of dissipation in the equation of motion of nanodevices is faced in the literature with different approaches. We can split them into completely positive (CP) ones and the non CP ones [30]. CP methods are the ones which always give positive dynamical maps of the system which, for example, ensures that the charge density is always positive. Obviously CP is a desirable property, since then

[a]Note that for instance $8!^2 = 40320^2$.

no unphysical results in the charge [31] appear. In Ref. [32], it was proved that through the use of the conditional wave function, Eq. 6.24, Bohmian mechanics provides a CP method to deal with dissipation.

Here, we show how Bohmian mechanics allows to account for the interaction of electrons with ions (phonons). We remark that any other interaction (for example, those interactions with photons) can be treated similarly. A general Hamiltonian which considers electron-ion interaction is the following:

$$H(\vec{r}, \vec{R}) = K_e(\vec{r}) + K_h(\vec{R}) + V_{ee}(\vec{r}) + V_{hh}(\vec{R}) + H_{ep}(\vec{r}, \vec{R}), \quad (6.27)$$

with $K_e(\vec{r})$ the electron kinetic energies, $K_h(\vec{R})$ the ion kinetic energies, $V_{ee}(\vec{r})$ the electron-electron interactions, $V_{hh}(\vec{R})$ the ion-ion interactions, and $H_{ep}(\vec{r}, \vec{R})$ the total electron-lattice interaction. If we condition the system to a configuration of all the particles (ions and electrons) in the system except the a-th electron, we find (for more details see Ref. [32]) that the modification of the $J_a(\vec{r}_a, \vec{R}_a[t], t)$ and $G_a(\vec{r}_a, \vec{R}_a[t], t)$ terms must change the momentum of the conditional wave function.

For future purposes, we describe the pseudo single-particle Schrödinger equation and also the pseudo single-particle Dirac equation when including the $J_a(\vec{r}_a, \vec{R}_a[t], t)$ and $G_a(\vec{r}_a, \vec{R}_a[t], t)$ terms to tackle dissipative quantum transport.

6.4.1 Parabolic Band Structures: Pseudo Schrödinger Equation

For parabolic band structures, assuming that the degrees of freedom of the a-electron are collectively described in a 3D spatial space by $\vec{r}_a = \{x_a, y_a, z_a\}$, the pseudo single-particle Schrödinger equation of Eq. 6.24 including the dissipation [32] takes the form:

$$i\hbar\frac{\partial\Psi_a(\vec{r}_a, t)}{\partial t} = \left[\frac{1}{2m^*}\left(\vec{p}_a + \vec{\lambda}_a\Theta(t_1, t_2)\right)^2 + U_a(\vec{r}_a, \vec{R}_a[t], t)\right]\Psi_a(\vec{r}_a, t), \quad (6.28)$$

where $\vec{p}_a = -i\hbar\vec{\nabla}_a$ is the momentum related to a 3D wave vector $\vec{k}_a = \{k_{x,a}, k_{y,a}, k_{z,a}\}$. We have defined $m^* = 0.067m_e$ as the electron effective mass typical for the AlGaAs materials used later (m_e is the free electron mass), $\vec{\lambda}_a$ is the momentum of the phonon and $\Theta(t_1, t_2)$ is a function which describes the time interaction of the phonon with the electron, being zero before the interaction and 1 when it finishes [32]. We will use this equation of motion for the discussion of the energy dissipation in AlGaAs heterostructure devices.

6.4.2 Linear Band Structures: Pseudo Dirac Equation

We will also discuss dissipation in the so-called 2D materials, where the degrees of freedom of the electrons are described in a 2D spatial space by $\vec{r}_a = \{x_a, y_a\}$. In particular, graphene is one of these 2D materials whose band structure has a linear behavior. The wave nature of electrons with dissipation in graphene is predicted by the pseudo single-particle Dirac equation:

$$i\hbar\frac{\partial\Psi_a(\vec{r}_a, t)}{\partial t} = v_f\begin{pmatrix} U_a(\vec{r}_a, \vec{R}_a[t], t)/v_f & p_a^- + \lambda_a^-\Theta(t_1, t_2) \\ (p_a^+ + \lambda_a^+\Theta(t_1, t_2)\chi_t & U_a(\vec{r}_a, \vec{R}_a[t], t)\chi_t/v_f \end{pmatrix}\Psi_a(\vec{r}_a, t). \tag{6.29}$$

As we will discuss later in Section 6.7, there is no gap between positive and negative energies in the graphene band structure (the zero of energy is fixed in the Dirac point). The Dirac equation deals with a wave function defined as a bispinor: $\Psi_a(\vec{r}_a, t) \equiv (\Psi_{a,1}(\vec{r}_a, t), \Psi_{a,1}(\vec{r}_a, t))^T$. The parameter v_f is the Fermi velocity. We define $p_a^\pm = -i\hbar\partial_{x_a} \pm \hbar\partial_{y_a}$ and $\lambda_a^\pm = \lambda_{ax} \pm i\lambda_{ay}$. The a-electron momentum is described by a two dimensional vector $\vec{p}_a = \{p_{x,a}, p_{y,a}\}$ and, accordingly, the electron wave vector by $\vec{k}_a = \{k_{x,a}, k_{y,a}\}$. When the interaction occurs, the term $\chi_t = e^{i(m\pi+\beta_{\vec{k}_{fa}}-\beta_{\vec{k}_{0a}})\Lambda}$ makes sure that the final state is either in the conduction band (positive energy branch) or in the valence band (negative energy states). If the electron moves from the conduction to the valence band (or vice versa), then $m = 1$ and if there is no change of band $m = 0$, with $e^{i\beta_{\vec{k}_{0a}}} = (k_{0ax} + ik_{0ay})/|\vec{k}_{0a}|$, where $\vec{k}_{0a}$ ($\vec{k}_{fa}$) is the central initial (final) wave vector of the wave packet and $e^{i\beta_{\vec{k}_{fa}}}$ having the same definition. χ_t is only relevant when the interaction is occurring, i.e., $\Lambda(t) = 0$ except $\Lambda(t_1 < t < t_2) = 1$. See Ref. [32].

6.5 The BITLLES Simulator

In the three preceding sections, we have discussed the ability of Bohmian mechanics to provide, on one hand, simple mathematical expressions for predicting the DC, AC, transient currents, and fluctuations variance (or even higher moments) for general nanoelectronic devices and, on the other hand, an algorithm to solve the many-body electron problem with or without dissipation. Based on these three important contributions of Bohmian mechanics for modeling electron quantum transport, we develop a general, versatile, and time-dependent 3D electron transport simulator for nanoelectronic devices. As indicated in the title of this section, the commercial name of this simulator is **BITLLES** (Bohmian Interacting Transport for nonequiLibrium eLEctronic Structures).[a] Visit the website http://europe.uab.es/bitlles for more information. By construction, as discussed in Section 6.3, the numerical algorithm is implemented as a fully quantum Monte Carlo simulator in order to account for the g- and h-uncertainties. Here, we discuss the general algorithm of the simulator and some meritorious technical details. Next, in Section 6.6, we present some numerical examples of this simulator.

6.5.1 Overall Charge Neutrality and Current Conservation

In general, the coulomb interaction in the whole circuit introduces screening among electrons implying that the total charge in the whole circuit is zero, i.e., *overall charge neutrality*. Thus, quantum electron transport simulators have to include, somehow, the Coulomb correlations among electrons.

However, in order to achieve charge neutrality, one has to deal with a very large simulation box (that includes not only the device-active region but also the leads and part of the contacts). The strategy of the BITLLES simulator is to use a smaller simulation box (tens of nanometers) where the Poisson equation is solved directly

[a]The acronym BITLLES (Bohmian Interacting Transport for nonequiLibrium eLEctronic Structures) is also the catalan name of the bowling pins, which are solid pieces of plastic or wood situated in a *periodic* structure (similar to a solid-state structure) waiting for a ball (an electron) to impinge on them.

and introduce the rest of the lead-sample Coulomb correlation into appropriate boundary conditions at the boundaries of the simulation box. The total electrostatic potential energy among the M_T electrons that appears in Eq. 6.24 can be divided into two parts (one including the degrees of freedom of the a-th electron and other which does not):

$$U(\vec{r}_a, \vec{R}_a[t], t) = U_a(\vec{r}_a, \vec{R}_a[t], t) + U_b(\vec{R}_a[t], t), \tag{6.30}$$

where the term $U_a(\vec{r}_a, \vec{R}_a[t], t)$ can be evaluated as:

$$U_a(\vec{r}_a, \vec{R}_a[t], t) = \sum_{\substack{j=1 \\ j\neq a}}^{N(t)} \frac{q^2}{4\pi\varepsilon|\vec{r}_a - \vec{r}_j[t]|} + \sum_{j=N(t)+1}^{M_T} \frac{q^2}{4\pi\varepsilon|\vec{r}_a - \vec{r}_j[t]|}. \tag{6.31}$$

We use the nomenclature of Section 6.3, where the first ensemble $\{1, \ldots, N(t)\}$ correspond to the electrons that are explicitly simulated in the device-active region and the second ensemble $\{N(t) + 1, \ldots, M_T\}$ correspond to those that are outside. Since we ignore the dynamics of outside electrons, the scalar potential inside the simulation box can be introduced through the solution of the Poisson equation with proper boundary condition. Let us first explain how we solve the Poisson equation and then, how we specify the appropriate boundary conditions.

6.5.1.1 The Poisson equation in the simulation box

Since we deal with well-defined trajectories, instead of using Eq. 6.31, the term $U_a(\vec{r}_a, \vec{R}_a[t], t)$ can be alternatively computed from the following 3D Poisson equation:

$$\nabla^2_{\vec{r}_a}\left(\varepsilon(\vec{r}_a)U_a(\vec{r}_a, \vec{R}_a[t], t)\right) = \rho_a(\vec{r}_a, \vec{R}_a[t], t). \tag{6.32}$$

The charge density that appears in Eq. 6.32 only considers the first ensemble of $N(t)$ electrons inside the simulate box:

$$\rho_a(\vec{r}_a, \vec{R}_a[t], t) = \sum_{\substack{j=1 \\ j\neq a}}^{N} -q\delta(\vec{r}_a - \vec{r}_j[t]). \tag{6.33}$$

To solve Eq. 6.32 we have to specify the appropriate boundary conditions at the border of the 3D simulation box, which determine the role of the second term of the right-hand side of Eq. 6.31, i.e., the interaction of the first ensemble $\{1, \ldots, N(t)\}$ with the second $\{N(t) + 1, \ldots, M_T\}$. Interestingly, the use of the Poisson equation allows one to use spatial dependent permittivities, $\varepsilon(\vec{r})$ that accounts for the interaction between the (simulated) free electrons and the (nonsimulated) atomic nuclei and core electrons. We notice that the restriction $j \neq a$ in Eq. 6.33 eliminates the Coulomb self-interaction.

In summary, there is a Coulomb potential (or electric field) for each a-electron. The Poisson equation, Eq. 6.32, provides the electron-electron Coulomb interaction that reproduces accurately the electrostatics of the system. Then, at each simulation time step, dt, we solve $N(t)$ Poisson equation with $N(t)$ different charge densities. These potential energies solution of the $N(t)$ Poisson equation are, then, introduced into the $N(t)$ Schrödinger equations defined from Eq. 6.24.

This procedure provides a self-consistent solution of the Poisson and the many-particle Schrödinger equations beyond the mean field approximation [33, 34].

6.5.1.2 Time-dependent boundary conditions for the Poisson equation

As already mentioned, we need proper boundary conditions of the Poisson equation, Eq. 6.32, to model correctly the effect of the Coulomb interaction with other electrons outside the simulation box, which are responsible for overall charge neutrality.

All boundary conditions of electron transport simulators are based on specifying the value of the scalar potential (or the electric field) and the charge density on the borders of the simulating box. However, it is very difficult to anticipate an educated guess for such quantities on the borders of a small simulation box that excludes the leads and the reservoirs mainly because of the far-from-equilibrium conditions present there [35–38]. Nonetheless, in order to avoid such a complexity, we can develop analytical expressions for the charge density, the electric field, and the scalar potential along the leads, and reservoirs [35, 36, 38] letting us to transfer the specification of the boundary conditions at the borders of the small simulation box into a much simpler ones deep inside the reservoirs [35, 36, 38]. In particular, the two boundary conditions are:

(1) Noticing that the total charge in a large volume, Λ (see Fig. 6.1), including the device device-active, the leads, and the reservoirs, tends to zero within the dielectric relaxation time, $\tau_c = \varepsilon/\sigma$. It can be easily proved that the electric field deep inside the reservoirs, $E^C_{S/D}(t)$, tends to its drift value $E^C_{S/D}(t) \rightarrow E^{\text{drift}}_{S/D}(t)$ within the same time $\tau_c = \varepsilon/\sigma$.

(2) The scalar potentials deep inside the reservoir are fixed by the external bias $V^C_S(t) = 0$ and $V^C_D(t) = V_{\text{external}}(t)$.

The temporal and spatial relations for the charge density, the electric field, and the scalar potential must be, however, ultimately coupled to the injection model [35, 36] controlling the amount of charge on the borders of the simulation box (this electron injection model will, in fact, introduce the g- and h-uncertainties discussed in Section 6.2.1). Such a coupling process constitutes the last piece of the puzzle providing *overall charge neutrality* and *current conservation* without any fitting parameter. See Refs. [35, 36] for a detailed explanation of this algorithm. Let us clarify that, with the consideration of the overall charge neutrality, we introduce somehow correlations between the simulated Bohmian trajectories and those in the ammeter. However, such correlations do only affect the boundary conditions that appear in the potential profile in Eq. 6.13. In this sense, one can consider that this is just part of the effects that the ammeter has on the simulated particles in case the whole system were fully quantum coherent.

6.5.2 Practical Computation of Time-Dependent Electrical Currents

In Section 6.5.1, we have emphasized the importance of the Coulomb interaction among electrons to assure *overall charge neutrality*. In fact, there is an additional argument that justifies, even more, the

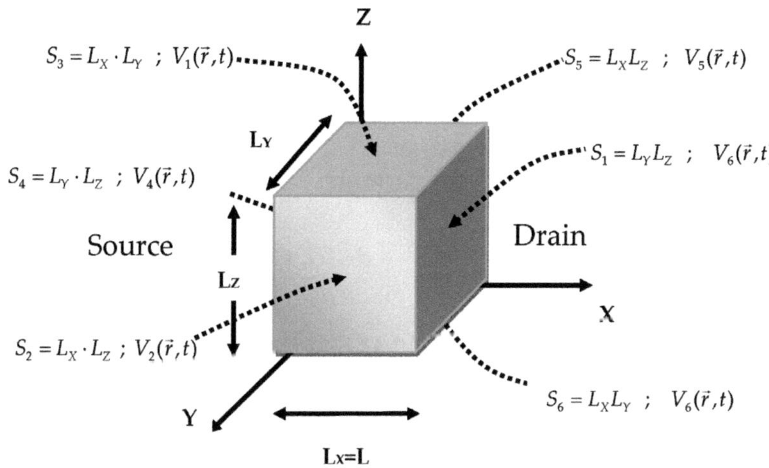

Figure 6.3 Volume Ω: Schematic representation of the arbitrary 3D geometry considered in this chapter as simulation box (i.e., device-active region) for the computation of quantum transport with local current conservation.

importance of properly introducing the Coulomb interaction. As discussed in Section 6.2.1.1, one has to compute time-dependent variations of the electric field (i.e., the displacement current) to assure that the total time-dependent current computed in a surface of the simulate box is equal to that measured by an ammeter, i.e., *current conservation*.

By means of Eq. 6.12, we can compute the current from the Bohmian trajectories crossing the surface S_D. Identically, an algorithm based on the Ramo-Shockley-Pellegrini theorem can be implemented [39–44] to compute the total current.

We will briefly describe the direct computation of the total current and the Ramo-Shockley method. We will use the volume Ω drawn in Fig. 6.3 which contains the whole device-active region. In particular, we assume that one of the surface of the volume Ω of Fig. 6.3 is in contact with the surface S_A of the volume Λ of Fig. 6.1. Again, we will discuss one particular element of the g- and h-distributions, but we will omit the subindexes to simplify the notation.

6.5.2.1 The direct method for the computation of the total current

According to Eq. 6.2, the total current flowing through one surface S_i (of the volume Ω of Fig. 6.3) has two different components: The particle and the displacement currents. The displacement current can be written as:

$$I_d(t) = \int_{S_i} \varepsilon(\vec{r}_i)\left(\frac{d\vec{E}(\vec{r}_i, t)}{dt}\right) d\vec{s}. \qquad (6.34)$$

We use the vector $\vec{r}_i$ to remind us that Eq. 6.34 is evaluated only on the surface S_i. It is important to specify that $\vec{E}(\vec{r}_i, t)$ is the electric field seen by an additional electron probe located at the observation position $\vec{r}_i$ at time t due to the presence of the rest of the electrons.

Since our algorithms deal with Bohmian trajectories $\vec{r}_a[t]$, the particle (conduction) current density on a particular point $\vec{r}_i$ of the surface S_i at time t has to be computed from:

$$I_c(t) = \int_{S_i} \vec{j}_i(\vec{r}_i, t) d\vec{s} = \lim_{dt\to 0} \sum_{i=1}^{N'(t)} \frac{-q}{dt}\text{sgn(i)}. \qquad (6.35)$$

In Eq. 6.35 the sum $N'(t)$ is over the number of Bohmian trajectories that have crossed the surface S_i during the temporal step dt. The function $sgn(i) = 1$ when one trajectory leaves the volume Ω through the surface S_i, while $sgn(i) = -1$ when the trajectory enters. In this equation the simulation temporal step, dt, provides the temporal resolution of our approach. We observe that the current density of Eq. 6.35 tends to infinity (when dt tends to zero) each time that a Bohmian trajectory crosses S_i. In principle, this infinite value of the particle current and the infinity of the displacement current have opposite signs and therefore both compensates. See Refs. [45–47] for a compact computation of both terms with the direct method.

6.5.2.2 The Ramo-Shockley-Pellegrini method for the computation of the total current

Here, we directly follow the development of Refs. [44, 48–50]. A vector function $\vec{F}_i(\vec{r})$ inside the volume Ω is defined through the expression $\vec{F}_i(\vec{r}) = -\vec{\nabla}\Phi_i(\vec{r})$ where the scalar function $\Phi_i(\vec{r})$ is the solution of the Laplace equation for the particular boundary condition at the surface $\Phi_i(\vec{r}) = 1$ when $\vec{r} \epsilon S_i$, and zero elsewhere, i.e., $\Phi_i(\vec{r}) = 0$ when $\vec{r} \epsilon S_{h \neq i}$:

$$\vec{\nabla}\left(\varepsilon(\vec{r})\vec{F}_i(\vec{r})\right) = -\vec{\nabla}\left(\varepsilon(\vec{r})\Phi_i(\vec{r})\right) = 0. \tag{6.36}$$

The total time-dependent current through the surface S_i can be then decomposed into three terms [48]:

$$I_i(t) = \Gamma_i^q(t) + \Gamma_i^e(t) + \Gamma_i^m(t), \tag{6.37}$$

where:

$$\Gamma_i^q(t) = -\int_\Omega \vec{F}_i(\vec{r})\vec{j}_c(\vec{r}_i, t)d^3\vec{r}, \tag{6.38}$$

$$\Gamma_i^e(t) = \int_S \vec{F}_i(\vec{r})\varepsilon(\vec{r})\frac{\partial}{\partial t}V(\vec{r}, t)d\vec{s}, \tag{6.39}$$

$$\Gamma_i^m(t) = \int_\Omega \vec{F}_i(\vec{r})\varepsilon(\vec{r})\frac{\partial^2}{\partial t^2}\vec{A}(\vec{r}, t)d^3\vec{r}. \tag{6.40}$$

We use the subindex, i, in Eqs. 6.37–6.40 because the current through a surface different from S_i leads to a different definition of $\vec{F}_i(\vec{r})$. Let us remark that the first method provides a computation of the total current through the integration on the surface S_i. Contrarily, this second method computes the total current in a surface S_i through a volume integral in Eq. 6.38 and an integral over all surfaces of the volume Ω in Eq. 6.39. Let us notice that the evaluation of the third term $\Gamma_i^m(t)$ can be omitted because it is related to the radiation properties of the electromagnetic field generated inside Ω that are negligible for the system studied here [48].

The evaluation of $\Gamma_i^q(t)$, Eq. 6.38 can be simplified by [49]:

$$\Gamma_i^q(t) = \sum_{i=1}^{N(t)} \vec{F}_i(\vec{r}_i[t])q\vec{v}_i(\vec{r}_i[t]). \tag{6.41}$$

Let us mention that Eq. 6.41 does not only contains the particle (conduction) current, but also contains partially the displacement current. The evaluation of $\Gamma_i^e(t)$ follows directly the Eq. 6.39. We rewrite it here for convenience:

$$\Gamma_i^e(t) = \int_S \vec{F}_i(\vec{r})\varepsilon(\vec{r})\frac{\partial V(\vec{r}, t)}{\partial t}d\vec{s}. \tag{6.42}$$

The time derivative is obtained with a simple finite-difference evaluation, $\partial V(\vec{r},t)/\partial t = (V(\vec{r},t) - V(\vec{r},t-dt))/dt$.

In summary, the numerical evaluation of the total current through a particular surface S_l due to a set of $N(t)$ Bohmian trajectories can be computed from Eqs. 6.34 and 6.35 with the direct method, and from Eqs. 6.41 and 6.42 with Ramo-Shockley-Pellegrini theorem. In principle, both provide identical results. Depending on our interest we can use one or the other.

6.6 Application of the BITLLES Simulator to Resonant Tunneling Diodes

In previous sections we have demonstrated the ability of the Bohmian trajectories to model the DC, AC, transient, and noise features for state-of-the-art nanoelectronic devices. Here we will discuss numerical results for a Resonant Tunneling Diodes (RTDs) that have played a crucial role in understanding many of the electronic transport features belonging to the quantum world. The use of the BITLLES simulator to evaluate the performance of other typical electron devices can be found, among others, in Refs. [51–56].

6.6.1 Device Characteristics and Available Simulation Models

From the original electronic transport analysis of structures exploiting a finite sequence of potential barriers spaced by a distance shorter than the electron mean free path, Esaki, in 1970, introduced for the first time the RTDs [57–59]. The simple case of two barriers, useful to study the relevant quantum features generally involved in superlattices, is also the basic physical picture of the RTD active zone, see Fig. 6.4.

To understand electronic transport in this device, it can be assumed that the only permitted energies in the well for the electrons are those very close to the resonant energies[a] E_{R_n} depicted

[a]Roughly speaking, the value E_{R_n} are the energies of the single-particle stationary quantum states of the well with infinite thickness barriers.

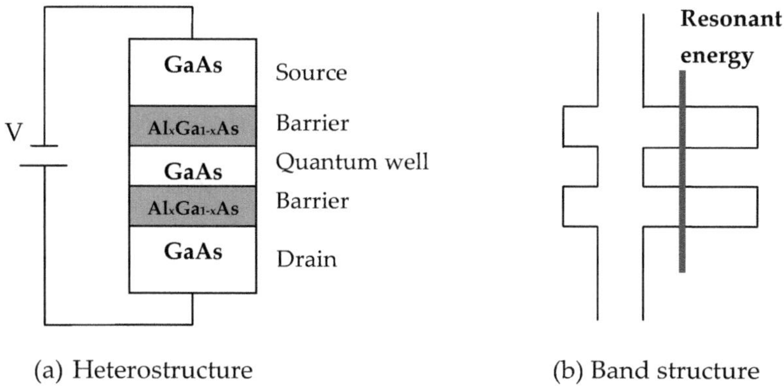

Figure 6.4 (a) Basic configuration for a typical RTD and (b) related conduction and valence band structure.

in Fig. 6.4. If the transport is assumed ballistic, tunneling is only possible for electrons with energies close to E_{R_n}. This is the *resonant tunneling transport* mechanics. Varying the applied voltage V, the whole potential profile is modified. Let us assume that we keep the source[a] contact grounded and the drain contact at V bias. Then, the resonant energies shift down when we increase bias. The electrons from the source will be able to reach the drain until the resonant energy is lower than the bottom of the source conduction band. If the resonant energy overcomes this limit, resonant tunneling is no longer permitted and the RTD differential conductance rapidly decreases, see Fig. 6.5. This transport behavior results in a Negative Differential Conductance (NDC) region of the current-voltage (I–V) characteristic. For more details about features of RTDs, please see [60, 61].

The resonant tunneling is of general interest in many applications of quantum mechanics (see references therein [62]), the particular case of RTDs is very intriguing not only for its peculiar properties [60] but also for its potential applications in both analog [63, 64] and digital [65] electronics. Nevertheless

[a]In the literature, the source and drain contacts are usually referred as emitter and collector, respectively, when dealing with RTD. Here, we will keep the names source and drain mentioned in the previous sections for general devices.

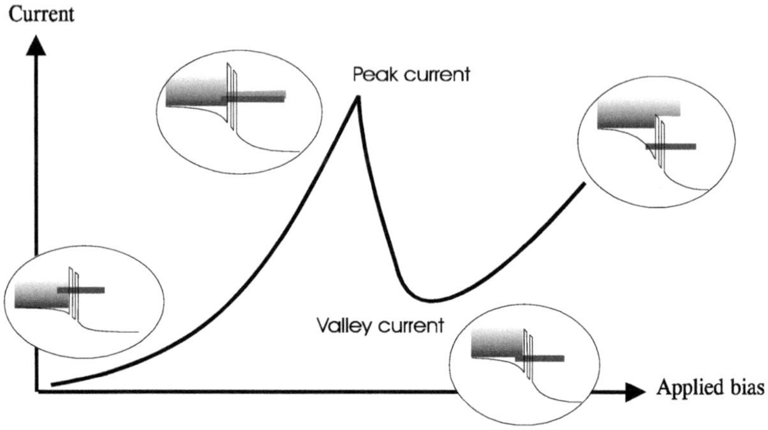

Figure 6.5 Schematic representation of the I–V curve of a typical RTD. The resonant energy inside the quantum well acts like an energetic filter that lets the electrons from the source to arrive at the drain.

technology solutions to integrate RTDs in electronic circuits are still under investigation. Requiring deposition of few-nanometers-thick layers, the epitaxial technologies were the most used in the last years [66] allowing RTDs to be designed with III–V materials and technologies. However the difficulty to integrate these materials in silicon processes has forced the search of possible solutions for its implementation with silicon-compatible materials. Recent results of Gilbert et al. [67] established a value of the mean free path for Si to a few nanometers. Currently, it is possible to think of Si based RTDs as good candidates for the same applications that were previously implemented only with III–V materials. As a matter of fact, evidence of resonant tunneling effects has been recently observed in Si structures [68–70]. However, performances obtained in these works in terms of RTD-relevant parameters are still not satisfactory.

From a theoretical point of view RTDs have been widely studied. Their peculiar properties have attracted the interest either of important corporations such as IBM (with the works of Büttiker [18, 62] and Fischetti [71]) and Texas Instruments (Frensley [72] and Klimeck [73]) or leading universities like MIT (Brown [74]) and Purdue University (Datta [75]). Singular transmission coefficient shape and Negative Differential Conductance (NDC) region of the I–V characteristics [60, 61], are only some of the most useful RTD properties. Unfortunately single-particle theory for mesoscopic structures is not adequate to describe the totality of the typical behavior of these devices. Also in the most idealized case of RTDs, the inclusion of the Coulomb correlation between electrons is enough to spoil the results of the single-particle theory. Many-body based theories and simulations, confirmed by experimental measurements show for example, different current patterns [33, 73, 75] or very enhanced noise spectrum in NDC region [76, 77]. However, as already discussed in this chapter, theories involving the solution of the many-body Schrödinger equation are in general very hard to treat. The widely used approximation to partially overcome this problem is the Breit-Wigner formalism [17, 78]. This picture includes many of the effects arising from the quantum correlations by introducing a phenomenological scattering parameter based on the concept of decoherence [18, 80]. It furnishes a powerful tool to analyze RTDs theoretically and reproduce their characteristics in an analytical way [61]. Nevertheless, some other tools such as second quantization ([76] and references therein) or Hartree approximation and Bardeen approach ([81] and references therein), must be adopted in order to treat more complicated phenomena such as the noise. Other approaches devoted to the numerical simulation are the non-equilibrium Green function combined with the Landauer formalism [75], and the Wigner distribution function [72, 73, 82].

6.6.2 Numerical Results

In this subsection, the BITLLES simulator will be used to predict DC, AC, and noise characteristics of an RTD. An enlarged explanation of the results discussed here can be found in Ref. [83].

6.6.2.1 Coulomb interaction in DC scenarios

As a first example, we consider the importance of the Coulomb interaction in the prediction of the current-voltage characteristics of a typical RTD. We consider two highly doped drain-source $GaAs$

regions (the leads), two $AlGaAs$ barriers, and a quantum well (the device-active region). Transport takes place from source to drain in the x direction. The lateral dimensions are $L_y = L_z = 48.6$ nm. Room temperature is assumed. As previously discussed in Section 6.3, the practical quantum algorithm for the RTD implies solving numerically $N(t)$ time-dependent single-particle $1D$ Schrödinger equations. All Schrödinger equations are coupled to the Poisson equations with the boundary conditions given by our boundary condition algorithm. More technical details can be found in Refs. [24, 33–35].

As discussed in Section 6.5.1, we distinguish between (i) the Coulomb interaction among electrons inside the simulation box plus (ii) the Coulomb correlations among these electrons and those outside the simulation box. Therefore, we will compare the results obtained by means of the previously described model, with those obtained, in one hand by eliminating the Coulomb correlations among the device-active region and the leads (i.e., assuming standard Dirichlet external bias at the borders of the device-active region of the RTD (see Section 6.5.1.2 for a detailed description), and on the other hand, with those obtained by switching off the Coulomb correlations (i.e., assuming simple single-particle electron transport). In Fig. 6.6, we present the current-voltage curves of the simulated RTD using our boundary conditions algorithm (solid circles), standard Dirichlet external bias boundary conditions (open circles), and switching off the Coulomb correlations (open triangles). As it can be observed, the differences between these three approaches appear not only in the magnitude of the current but also in the position of the resonant region. More details are explained in Ref. [36].

6.6.2.2 Coulomb interaction in high-frequency scenarios

Next, we provide an example of the computation of the total current in time-dependent scenarios with the BITLLES simulator. First, we will consider a single electron crossing an RTD to show the accuracy of our quantum electron transport approach in providing local current conservation, that is the sum of the particle (conduction) plus the displacement currents to be zero when integrated on any

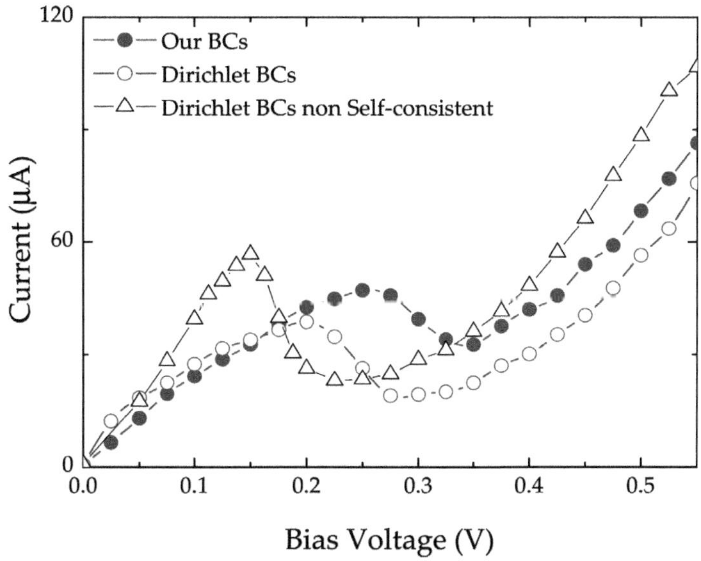

Figure 6.6 RTD current-voltage characteristic. Results taking into account the Coulomb correlations between the leads and the device-active region are presented in solid circles. Open circles refer to the same results neglecting the lead-device-active region interaction. Open triangles refer to a wholly noninteracting scenario, i.e., both coulomb interaction between the leads and the device-active region and coulomb interaction among electrons within the device-active region are neglected.

closed surface, and to discuss some technical details announced in Section 6.3. Second, as an enlightening example, we will compute transient currents with our BITLLES simulator.

For the first example, we consider an RTD diode composed of two highly doped source-drain *GaAs* regions, two *AlGaAs* barriers with the length of 1.6 nm and height 0.5 eV, and a quantum well with a length of 2.4 nm. The lateral dimensions y and z are small enough for considering electron confinement. The Bohmian velocities in the lateral directions are zero because we consider that the wave function involves only one quantized level. In the x direction, the time evolution of an initial Gaussian wave packet with initial kinetic energy of 0.25 eV coming from the source contact is computed. The computation of the trajectory requires the algorithm explained in the Section 6.3. At each simulation time step dt, we solve the Poisson (in this single-electron case we solve the Laplace version) equation with the appropriate boundary conditions. From this potential profile, we solve Eq. 6.24 to find $\Psi_a(\vec{r}_a, t)$ following the numerical method explained in Section 6.5 with a temporal step $dt = 10^{-17}$ s and a spatial resolutions in the x direction $dx = 0.2$ nm. Finally, we compute the next position of the Bohmian trajectory.

The scalar electrostatic potential energy is obtained from the numerical solution of the $3D$ Poisson equation. The boundary conditions at the source and drain surfaces are equal to zero volts. In the rest of the surfaces, we assume an electric field equal to zero, but no explicit restriction is imposed on the value of the potential. For simplicity, we assume a homogeneous dielectric constant $\varepsilon = 13\varepsilon_0$ (with ε_0 the vacuum permittivity) in the whole device-active region. The time-dependent electric field, $\vec{E}(\vec{r}, t) = -\vec{\nabla} V(\vec{r}, t)$, changes with time due to electron dynamics, and it is different from zero even in the source and drain surfaces.

The computation of the total current following the expressions developed in Section 6.5.2 needs the knowledge of (i) the Bohmian trajectory depicted in Fig. 6.7, (ii) the time derivative of the potential profile (and the associated electric field), and (iii) the function $\vec{F}_i(\vec{r})$ defined as the solution of the Laplace equation, see Eq. 6.35. In Fig. 6.8, we show the total time-dependent current on the six surfaces of volume Ω of Fig. 6.3. The numerical evaluation of the total current through each of the six surfaces is computed from the Eqs. 6.34 and 6.35 with the direct method, and from Eqs. 6.41 and 6.42 for Ramo-Shockley-Pellegrini theorem. It is interesting to notice that there is a correlation between the three accelerations of the Bohmian trajectory in Fig. 6.7 and the three oscillatory behaviors of the currents in Fig. 6.8. In general, the results obtained for the total current are identical for the two methods. However, we observe in the plots of the surfaces 1 and 4 two peaks when the total current is computed from the first method. As mentioned in Section 6.5.2, these spurious peaks are a consequence of the infinities generated in Eq. 6.35 when a Bohmian trajectory crosses these surfaces. The computation of the current using the second method is free from these spurious numerical peaks because Eq. 6.41 contains not only the particle(conduction) current, but also part of the displacement current. A similar solution to avoid the spurious peaks

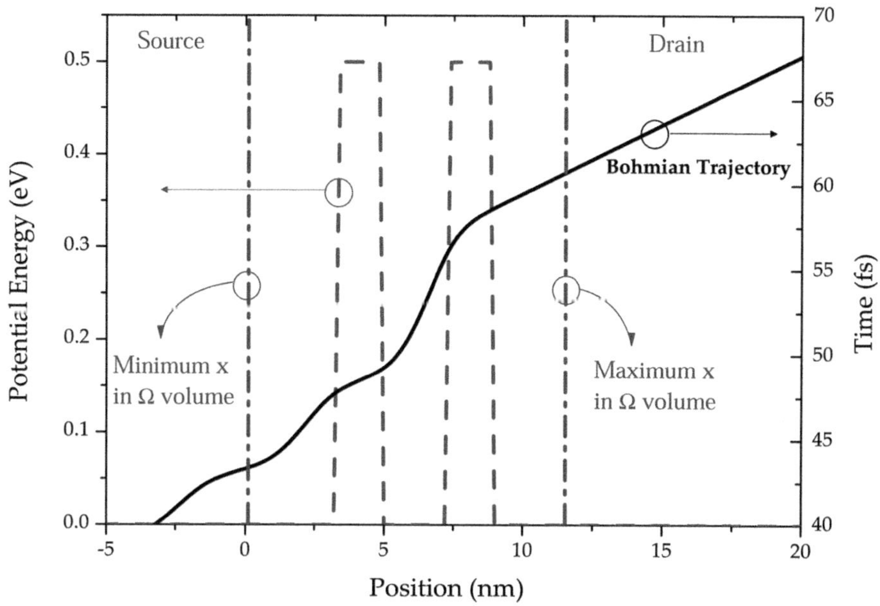

Figure 6.7 Dashed line: Potential energy profile for a double-barrier structure; solid line: Bohmian trajectory of an electron crossing the heterostructure; dashed dotted line: Schematic representation of the limits of the volume Ω in the *x* direction.

is possible also for the first method when particle (conduction) and displacement currents are computed with a unique expression.

Once we have discussed the ability of the BITLLES simulator to compute time-dependent currents with current conservation, we show here, the current response to the step input voltage in the NDC region.

The input signal is the step voltage $V(t) = V_1 u(t) + V_2 [1 - u(t)]$, where $u(t)$ is the Heaviside (step function) function mentioned in Eq. 6.10. The voltages V_1 and V_2 are constant. Then the current response can be expressed as $I(t) = I_{\text{tran}}(t) + I_1 u(t) + I_2 [1 - u(t)]$ where I_1 and I_2 are the stationary currents corresponding to V_1 and V_2, respectively, and I_{tran} is the *intrinsic* transient current. This formalism permits to focus on the current component carrying the dynamics involved in the RTD without performing any time

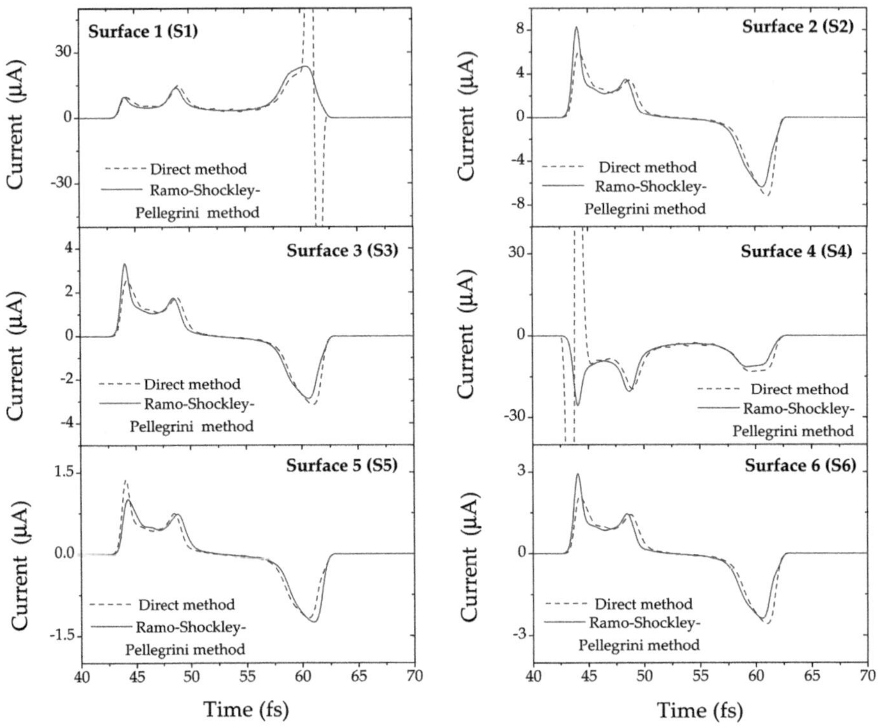

Figure 6.8 Time-dependent total current computed on the six surfaces that form the volume Ω of Fig. 6.3. The computation of the current within the first method (dashed lines) has spurious effects that are not present when the displacement and particle components are computed with a compact formula including both terms [45–47] or when the second method (solid line) is used.

derivative that from a numerical point of view could be very inaccurate.

The results reported in Fig. 6.9 are for an RTD including Coulomb correlation among the $N(t)$ electrons and among those in the leads. As pointed out in the inset of Fig. 6.9, $I_{\text{tran}}(t)$ manifests a delay of about $0.1\,ps$ with respect to the step input voltage, due to the dynamical adjustment of the electric field in the leads. After the delay, the current response becomes an RLC-like response (inset 1a, solid line) i.e., purely exponential. Performing the Fourier transform

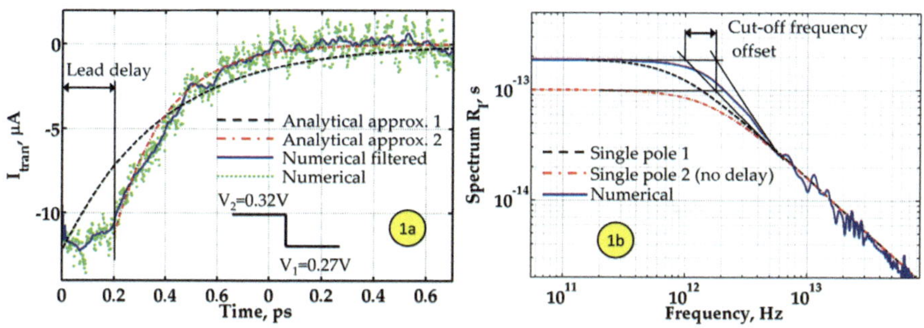

Figure 6.9 $I_{\text{tran}}(t)$ an its Fourier transform in inset a and b, respectively.

of $I_{\text{tran}}(t)$ (inset 1b solid line) and comparing with the single pole spectra (Fourier transform of RLC-like responses, inset 1b, dashed and dashed dotted lines) as depicted in the inset 1b, we are able to estimate the cut-off frequency (about 1.6 THz for this device) and the frequency offset (about 0.76 THz) due to the delay.

6.6.2.3 Current-current correlations

In the following example, we show how the BITLLES simulator can compute noise features. As discussed in Section 6.2.3, each Bohmian trajectory used to evaluate the current is selected according to the g- and h-distribution giving rise to the noise. Thus, in the BITLLES simulator the noise in the current can be easily obtained evaluating the autocorrelation function $R(\tau)$ of the current $I_{g,h}(t)$ from Eq. 6.21. Taking the Fourier transform, it gives the two-sided power spectrum $S(f)$ of the fluctuations of Eq. 6.21. In the following we will refer to the Fano factor as the ratio $\gamma = S(0)/S_{\text{schot}}$ with S_{schot} given by Eq. 6.23.

We now briefly discuss on how the many-body Coulomb interaction might affect noise in RTDs. Specifically, we investigate on the correlation between an electron trapped in the resonant state during a dwell time τ_d and the ones remaining in the left reservoir. This correlation occurs essentially because the trapped electron perturbs the potential energy felt by the electrons in the reservoir. In the limit of noninteracting electrons and mean field

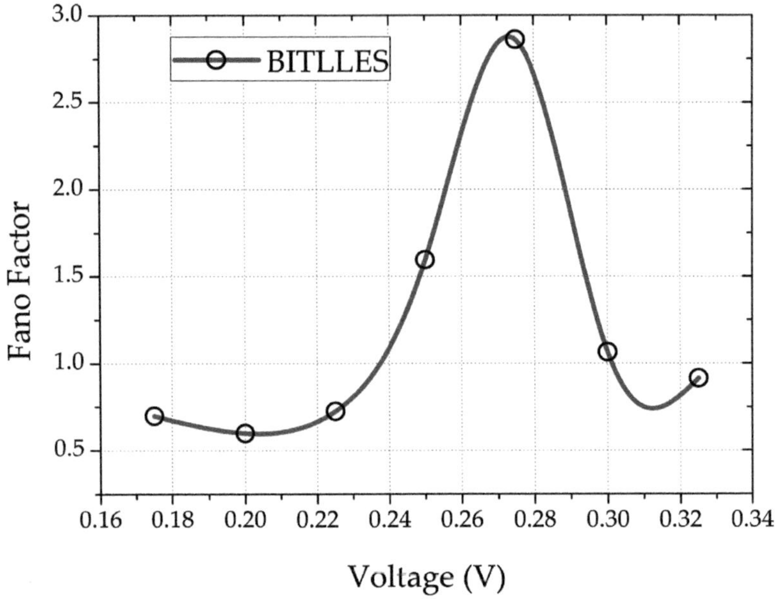

Figure 6.10 Fano factor evaluated using the current fluctuations directly available from BITLLES.

approximation, the Fano factor will be essentially proportional to the partition noise as discussed in Section 6.2.3 (see Fig. 6.10). However if the Coulomb correlation is self-consistently included in the simulations (see Section 6.5) this result is no longer reached. Roughly speaking, as depicted in Fig. 6.11, an electron tunneling into the well from the Source, raises the potential energy of the well by an amount of e/C_{equ}, where e is the electron charge and C_{eq} the structure equivalent capacitance. As a consequence, the density of state in the well is shifted upwards by the same amount. This can affect the noise in the following ways: if the resonant energy E_{R_1} is over the bottom of the conduction band in the source, when an electron enters into the well, the density of state inside the well is raised but it does not change much the transmittance of the sample for other electrons. The noise remains in the sub-Poissonian regime ($\gamma < 1$) already present in the limit of partition noise only. Conversely if the resonant energy is under the conduction band at the source, the density of states is raised because of the presence of

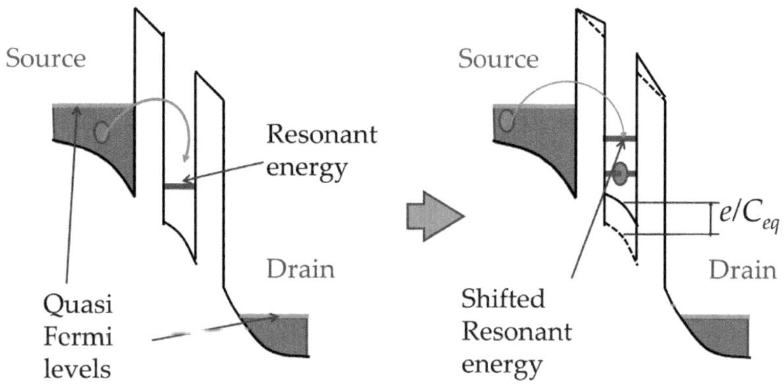

Figure 6.11 RTD band diagram deformation caused by particle tunneling in the well.

an electron in the well makes the resonant energy accessible to other electrons staying near the bottom of the conduction band. Therefore, many electrons in the source can tunnel into the well thanks to the first transmitted electron. Thus the Coulomb interaction tries to regroup the electrons providing a super-Poissonian noise ($\gamma > 1$) (see Fig. 6.11).

Because of the inclusion of the time dependence of self-consistent solution of the potential, with our BITLLES simulator we are not only able to reproduce Fano factor for 0 frequency but we can also evaluate the high frequency spectrum $S(f)$ given by Eq. 6.22 revealing information about internal energy scale of RTDs not available from DC transport, as reported in Fig. 6.12.

6.6.2.4 RTD with dissipation

We present here the last example performed with RTDs. As discussed previously, the inclusion of dissipation in quantum devices is unavoidable once we cannot justify the ballistic regime. For that purpose Section 6.4 was devoted to see how Bohmian mechanics provides a CP method to account for scattering. With the use of Eq. 6.28 we include in an RTD the scattering with impurities, and acoustic and optical (absorption and emission) phonons. The characteristics of the GaAs/AlGaAs RTD simulated are the following: The barrier height is 0.5 eV, its width is 1.6 nm and the well

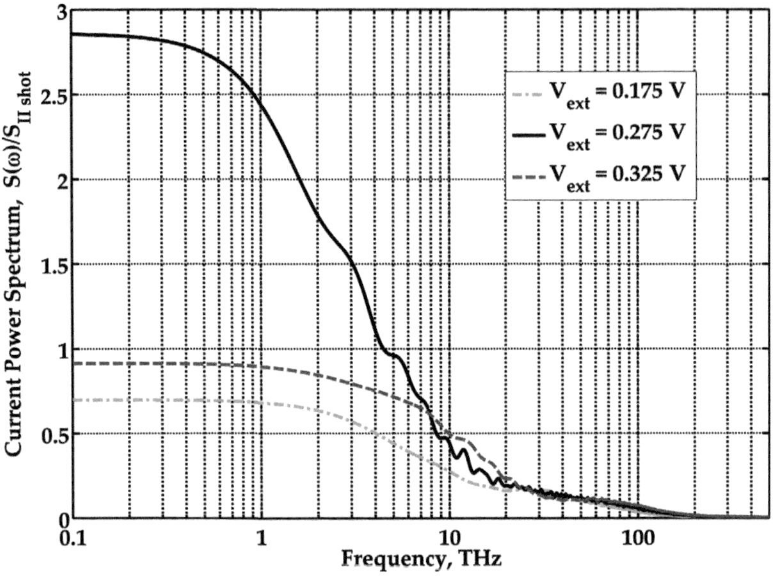

Figure 6.12 Current noise power spectrum referred to Poissonian shot noise at different biases.

width is 2.4 nm. There is n-type doping and the Fermi level is $E_f = 0.15$ eV above the conduction band. The number and type of collisions are obtained from the Fermi Golden Rule for GaAs materials [84].

We compute the current as the net number of trajectories $\vec{r}_a[t]$ transmitted from one side to the other, divided by the total simulation time (5 ps). Identically the DC current is also computed as the time average of the total (particle plus displacement) current. Both types of DC computations provide the same value at each bias point, validating the accuracy of the simulation.

The expected reduction of the transmission seen in Fig. 6.13a is because of the randomization of the momentum due to acoustic phonons and to the energy dissipation due to the emission of optical phonons. We see in Fig. 6.13b that the number of collisions at resonance is three times larger than out of resonance, showing that the ballisticity of tunneling devices also depends on the electron transit time that varies from one voltage to another.

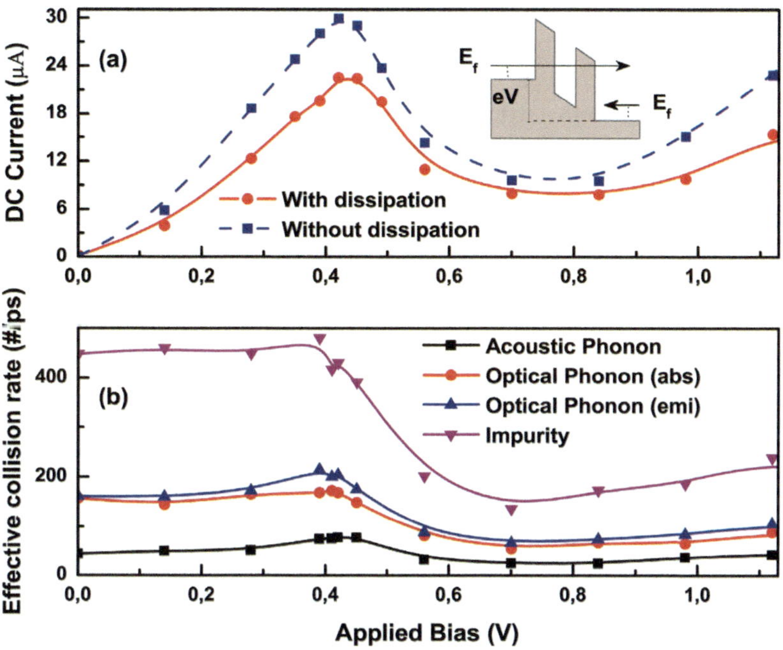

Figure 6.13 (a) Current-voltage characteristic for a RTD with (solid red curve) and without (blue dashed curve) dissipation due to acoustic and optical phonons and impurities. (b) Effective collision rate as a function of the applied bias. Optical phonons lead to an inelastic change of the electron energy of ± 0.036 eV. Reprinted with permission from Colomés et al. [32]. Copyright 2017 American Physical Society.

6.7 Application of the BITLLES Simulator to Graphene and 2D Linear Band Structures

As it was pointed out in the introduction of this chapter, a new *revolution* (or *evolution*) of the current materials and techniques are mandatory to reduce the scale of electronic devices. Although many new materials have been proposed as candidates to substitute the old-fashioned Silicon FETs, a recent article concluded that: *"many such saviours have come and gone, yet the reliable silicon CMOS continues to be scaled and to reach even higher performance levels"* [85]. Among these new materials, graphene was viewed by the scientific community as a new material with a great potential impact in our society, in general, and in electronics, in particular [4, 86]. Historically, in 1947, graphene [87] (a one atom thick carbon honeycomb lattice) was just a 2D crystal whose stability was considered impossible because it was thermodynamically unstable [88]. However, since its experimental discovery in 2004 [89] much of the attention has been attracted specifically to this material and in general, to other 2D materials.

Graphene is a single 2D layer of carbon atoms with a hexagonal lattice [86]. The energy dispersion of graphene mimics the energy dispersion of massless relativistic particles but with velocity $v_f = 10^6$ m/s, much smaller than the speed of light $c = 3 \times 10^8$ m/s. Apart from the fact that it exhibits relativistic effects -such as the so called Klein paradox- because of its size and its mobility (at room temperature around 15000 $cm^2V^{-1}s^{-1}$), graphene (or other 2D similar materials) is expected to occupy a relevant role in future nanoelectronic devices. It has a linear energy-momentum dispersion (see Fig. 6.14), an extraordinary elasticity (allowing flexible electronics) and extremely large electrical conductivity (with electron velocities of 10^6 m/s) [90]. However, graphene has a zero

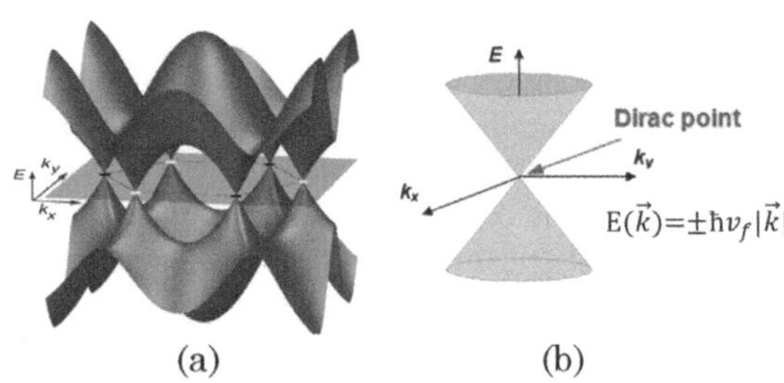

Figure 6.14 (a) General graphene band structure in the reciprocal space and (b) Linear band structure of graphene around the Dirac point where electron transport is relevant. A linear relation between the kinetic energy E and the electron wave vector $\vec{k} = k_x, k_y$ is given by $E = \pm\hbar v_f|\vec{k}|$. The positive sign is associated to electrons defined by states with energies above the Dirac point and the negatives sign with electrons with states associated to energies below the Dirac point. In some scenarios, for example those involving Klein tunneling, where due to a potential barrier an electron from the CV moves to the BV, a single electron involves states with positive and negative energies.

bandgap implying a small on-off ratio for graphene digital FETs. It is important to notice that reliable techniques to create a sizeable gap degrade a lot the properties mentioned above. Therefore, it seems that successful logic applications with graphene are not currently feasible. On the contrary, the large conductivity of graphene is very suitable for (small-signal) radio frequency applications (such as amplifiers or mixers) [4, 91, 92] which are not required to be switched off.

Although the performance of the contemporary radio-frequency graphene transistors is still quite below the one obtained from Silicon and III–V HEMTs [4], significant progress has been made since the experimental demonstration of the first GigaHertz graphene transistors in 2008 [93]. Most notably, a research group reported graphene FETs breaking the 100-GHz cut-off frequency (f_T) mark in 2010 [94]. Furthermore, only a few months later, researchers demonstrated [95] a graphene FET that has an f_T of 300 GHz. In 2012, graphene FETs with cut-off frequency exceeding 400 GHz has been reported [96]. Undoubtedly, graphene FETs are a very young class of devices compared to the traditional FETs. After carrying out an the enthusiastic span of research, the electronic community has realized that graphene would not be able to fulfill these high expectations owing to its zero bandgap since a reasonable bandgap is mandatory for proper device operation [97].

In 2011, a paper on the fabrication of single-layer MoS_2 attracted the attention to the research on other 2D materials [98]. The IRDS and the ITRS has mentioned other 2D amterials as candidates for future electronics [3]. In fact, in 2005, the Novoselov-Geim group reported the preparation of single-layer materials (for instance, MoS_2, $MoSe_2$, $MoTe_2$ and WS_2) other than graphene [99]. The experimental MoS_2 FETs show reasonable mobilities and an excellent ON-OFF ratio [98, 100]. On the theoretical work, simulation results qualitatively confirm the experimental results and predict very high ON-OFF ratios with monolayer dichalcogenide channels [101, 102].

In order to deal with graphene material in the BITLLES simulator and other 2D linear band structure materials, the Dirac equation needs to be implemented and velocities for the trajectories needs to be obtained in a similar way as in Chapter 1, which we briefly describe in the next subsection.

6.7.1 Bohmian Trajectories for Linear Band Structures

In order to obtain the Bohmian trajectories in graphene, the starting point is the single-particle Dirac equation:

$$i\hbar\frac{\partial\psi(\bar{r},t)}{\partial t} = -i\hbar v_f(\bar{\sigma}\cdot\bar{\nabla})\psi(\bar{r},t) + V(\bar{r})\psi(\bar{r},t), \qquad (6.43)$$

where $\bar{\sigma}$ are the Pauli matrices and $V(\bar{r})$ a potential. One must follow the same procedure as is done with the Schrödinger Hamiltonian in Chapter 1, but now using the Dirac Hamiltonian (Eq. 6.43). By carrying out some mathematical manipulations, we end up straightforwardly to the following continuity equation:

$$\frac{\partial|\psi(\bar{r},t)|^2}{\partial t} + \bar{\nabla}\cdot\left(v_f\psi(\bar{r},t)^*\bar{\sigma}\psi(\bar{r},t)\right) = 0. \qquad (6.44)$$

From Eq. 6.44 we can associate a density current to the term $\vec{J}(\bar{r},t) = v_f\psi(\bar{r},t)^*\bar{\sigma}\psi(\bar{r},t)$. From here, we can easily identify the Bohmian velocities, because the density current can be expressed as $\vec{J}(r,\bar{t}) = \rho\bar{v} = |\psi(\bar{r},t)|^2\bar{v}$. Then, the mentioned Bohmian velocities correspond to:

$$\bar{v}(\bar{r},t) = \frac{\vec{J}(\bar{r},t)}{|\psi(\bar{r},t)|^2} = \frac{v_f\psi(\bar{r},t)^*\bar{\sigma}\psi(\bar{r},t)}{|\psi(\bar{r},t)|^2} \qquad (6.45)$$

Differently from the Bohmian velocities obtained from the Schrödinger equation, where $\bar{v}(\bar{r},t) = \frac{\hbar}{m^*}Im\left(\frac{\bar{\nabla}\psi(\bar{r},t)}{\psi(\bar{r},t)}\right)$, we observe that no derivatives appear in Eq. 6.45. Finally, trajectories are trivially obtained through $\bar{r}(t) = \bar{r_0}(t) + \int_{t_0}^{t}\bar{v}(\bar{r},t')dt'$.

6.7.2 Numerical Results

In this subsection, we present different current-voltage curves for a graphene transistor (GFET) for both ballistic and dissipative quantum transport. The GFET simulated channel length is L_x = 40 nm, its width L_y = 250 nm and the Fermi energy is 0.15 eV above the Dirac point. It has bottom and top gates, whose voltages are set equal to zero, $V_{bg} = V_{tg} = 0$ V. Electrons are injected described by conditional Gaussian wave packets, following Fermi statistics

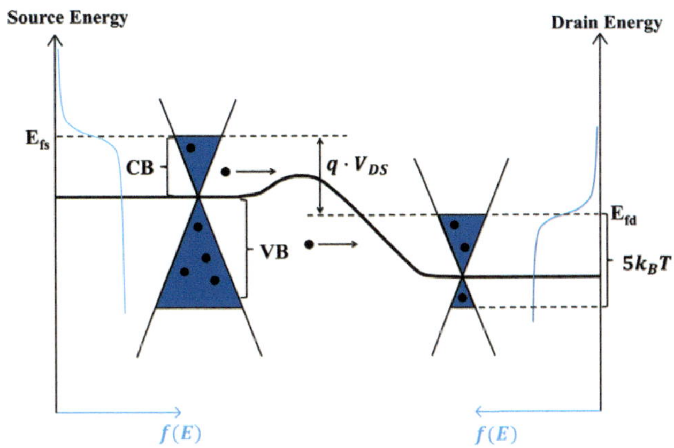

Figure 6.15 Schematic representation of an energy profile in the transport direction of devices in BITLLES simulator. The applied drain-source bias, V_{DS}, provides a different source, E_{fs}, and drain, E_{fd}, Fermi levels, the Fermi–Dirac distribution function at each contact are indicated. Relevant electrons for computing the quantum current are the ones belonging to the blue shaded energy region.

at room temperature as indicated in Fig. 6.15. All pseudo Dirac equations (one for each injected electron) are coupled to the Poisson equation and solved self-consistently.

In Fig. 6.16a, we see four different current-voltage characteristics. The dashed lines correspond to ballistic transport, whereas the solid ones takes into account energy dissipation. The insets of Fig. 6.16a are related to the one plotted in Fig. 6.15 indicating the relevant presence of electrons with energy above the Dirac point, conduction band (CB), and below, valence band (VB). In particular we consider acoustic and optical phonons with emission and absorption from, both, zone edge (ZE) and zone center (ZC) with energy interchange of ± 0.160 and ± 0.196 eV, respectively. The scattering rates are obtained from Ref. [103]. Since typical graphene mean free path values are low, the dark blue (without dissipation) and orange (with dissipation) lines are very similar. For that reason, we added the red line where the scattering rates are enhanced. We see in Fig. 6.16b and 6.16c the effective collision rate as a function of the applied bias for the orange and red lines, respectively. In the treatment of scattering in graphene structures one has to remind

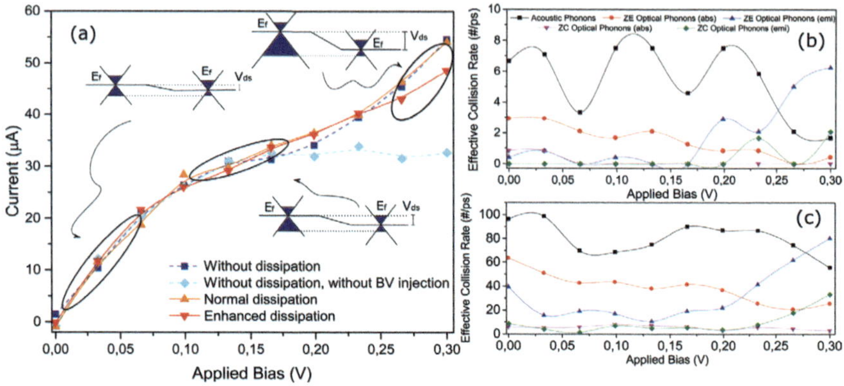

Figure 6.16 (a) Current-voltage characteristic for the GFET described in the text for ballistic transport (dashed lines) and transport with dissipation (solid ones). The dark blue (square) line corresponds to normal graphene injection (electrons injected from the CB and the VB), while in the light blue (diamond) line only electrons from the CB are injected. The orange (up triangle) line takes into account dissipation due to acoustic and optical phonons. The red (down triangle) line corresponds to enhanced dissipation. The insets sketch different energy profiles for applied bias. Similarly to Fig. 6.15, only electrons in the blue energy (above and below the Dirac point) are relevant. (b) Effective collision rate as a function of the applied bias for the orange (up triangle) line in (a) for acoustic and optical phonons. (c) Same as (b) but for the red (down triangle) line in (a), where dissipation is enhanced.

that graphene is gapless. Thus, in principle, there is no minimal value of energy required for an electron to suffer phonon emission. However, in practice, the Pauli principle avoids collision where the final energy of the electron is lower than the Fermi energy.[a] For that reason, we observe in Fig. 6.16b that phonon emission is only relevant for applied bias (V_{ds}) higher than the optical phonon energy for ZE and ZC (± 0.160 and ± 0.196 eV respectively). In such situations, an electron injected with an energy roughly equal to the

[a]The Fermi energy is not well defined inside the device because it is a system outside of thermodynamical equilibrium when an external bias is applied. However, a reasonable estimation of a quasi-Fermi energy inside the device can be easily approximated to take into account such many-body (exchange interaction) effect in a manageable way. We defined such quasi-Fermi energy so that the difference between the electrostatic potential and the Fermi energy level in the reservoir is constant along the device.

Fermi energy will acquire enough kinetic energy during its ballistic transport along the device to thermalize. If this condition is fulfilled, emission is higher than absorption, obtaining current values smaller than in the ballistic scenario. We remind also that forward scattering is prohibited and backscattering is preferred for ZE phonons and just the opposite for acoustic phonon scattering, while the ZC phonons are isotropic [103].

To successfully explain the shape of the current-voltage characteristic of Fig. 6.16a, it is necessary to (i) understand how electrons are injected in graphene and which ones participate in the current and (ii) the density of states of graphene. Contrary to standard semiconductors with energy gap, apart from electrons in the conduction band, electrons belonging to the valence band also contribute to the current. As sketched in Fig. 6.15, relevant electrons for transport are the ones whose energy is in between the Fermi energies of the drain and source, i.e., electrons from the blue area in Fig. 6.15 (strictly speaking, Fermi statistics are taken into account since we are at room temperature). This means that, contrary to normal FETs, there is no saturation current, since the more the voltage is applied between source and drain, the more number of electrons are transmitted from the source to the drain (from valence band in the source to conduction band in the drain). In Fig. 6.16, we can appreciate a saturation region when the applied voltage is close to the Fermi energy level. To understand this behaviour it is important to be aware that the density of states is proportional to the energy, i.e., $D_{gr}(E) \propto E$. Then, if due to the dynamics of the system, electrons in an energy level E_1 try to move to a lower energy level E_2 not all electrons will be able to arrive, since in this scenario $D_{gr}(E_2) < D_{gr}(E_1)$. Thus, some of the electrons will be rejected. This is sketched in the different insets of Fig. 6.16a. In these insets, similarly to Fig. 6.15, the blue area depicts the relevant electrons contributing to the current and explains the saturation when the voltage is close to the Fermi energy level.

The dark blue current-voltage characteristic behavior can be easily understood with the above considerations. When we increase the drain voltage, more electron flows from the source to the drain and current increases. However, when the voltage corresponds to the Fermi energy, the density of states vanishes in the drain and no more electrons can reach the drain, thus current saturates. However, if the drain voltage keeps on increasing, density of states again is non zero, and more electrons from the source in the valence band can reach the drain in the conduction band. In the case of the light blue curve, when no electrons are injected from the valence band, current saturates because after the voltage reaches the Fermi energy value, the same amount of electrons from the conduction band are injected independently of the applied voltage. This is similar to typical FETs with semiconductors with energy gap where typically only electrons from the conduction band (or only electrons from the valence band) are considered. Since graphene mean free path is of the order of a micron and our device is much smaller, when scattering is considered (orange line), the current-voltage line is very similar to the ballistic one. On the contrary, if we increase the scattering rates (red line), we observe how dissipation occurs and then current decreases compared to the ballistic case.

All these results emphasizes the success of the BITLLES simulator to study graphene. In fact, all AC tools seen for the RTD can be used for analyzing GFETs, converting the BITLLES an excellent tool to study transients, high-frequency scenarios in graphene and other exotic linear band structure materials.

6.8 Conclusions

As the size of the electronic devices shrink to the nanoscale range, electron transport has to take into account the quantum nature of the electrons. In this chapter we have presented an algorithm to study quantum electron transport using Bohmian trajectories for either parabolic band structure materials or linear ones. The adaptation of Bohmian mechanics to electron transport leads to a quantum Monte Carlo algorithm, where randomness appears because of the uncertainties in (i) the number of electrons in the simulation box, their initial energies, initial times and (ii) initial positions of (Bohmian) trajectories. In particular, we have presented a general, versatile, and time-dependent 3D electron transport simulator for nanoelectronic devices. We named such simulator as BITLLES (Bohmian Interacting Transport for nonequiLibrium eLEctronic Structures). Two of its most relevant features emphasized in this chapter are:

(1) *Its ability to deal with many-particle (Coulomb and exchange) correlations:* The many-particle Schrödinger equation can only be solved for very few degrees of freedom. This is at the heart of most of the unsolved problems in quantum transport. We have recently shown that Bohmian trajectories allows a direct treatment of the many-particle interaction among electrons. The ability of our simulator to deal with strongly correlated systems is demonstrated for the many-particle tunneling phenomena present in an RTD leading to a super-Poissonian value of the Fano factor. In addition, continuity in the total (particle plus displacement) current is guaranteed in the BITLLES simulator because of the self-consistent time-dependent solution of the Poisson equation. Dissipation to tackle electron-phonon interaction has also been discussed.

(2) *Its ability to provide full information about the current probability distribution:* The continuous measurement of the current is not a delicate issue with Bohmian mechanics because the measured quantities depend, ultimately, on the distribution of positions of a set of Bohmian particles whose position is (ontologically) well-defined at any time, independently of being measured or not (i.e., the orthodox collapse of the wave function can be avoided in the modeling). Only a reasonable approximation of the role of the ammeter on the dynamics of simulated electrons is needed. We have shown that we can compute the probabilities of all available current moments. From this result, we can compute the mean value (first moment), the variance (second moment), and any other higher moment for the distribution of the current. Such abilities are proved by computing the power spectral density of the current fluctuations for typical RTDs and graphene transistors.

The ultimate reason that determines the success of different quantum transport models is their practical ability to improve our predicting capabilities on electronic devices. We believe that the present Bohmian formulation of quantum transport has some important computational advantages over previous *orthodox* simulation tools. Therefore, since electron devices are present in almost every moment of our lives, we hope the BITLLES simulator to be one such example where we can assert that Bohmian mechanics does really help in improving our daily life.

Acknowledgments

This work has been partially supported by the "Ministerio de Ciencia e Innovación" through the Spanish Project TEC2015-67462-C2-1-R, the Generalitat de Catalunya (2014 SGR-384), the European Unions Horizon 2020 grant agreement no: 604391 of the Flagship initiative "Graphene-Based Revolutions in ICT and Beyond" and the European Unions Horizon 2020 research and innovation programme under the Marie Skodowska-Curie grant agreement No 765426 (TeraApps). G. A. acknowledges funding support from the Spanish "Ministerio Economía y Competividad" through the Project CTQ2016-76423-P and the European Commission Project 752822-BEBOP. We thank Xavier Cartoixà, David Jiménez and Damiano Marian for fruitful discussions.

References

1. R. K. Wangsness, *Electromagnetic Fields* (John Wiley and Sons, 1996).
2. A. Corbin, *The Third Element: A Brief History of Electronics* (Author-House, Bloomington, USA, 2006).
3. International Roadmap for Devices and Systems (IRDS), https://irds.ieee.org/; International Technology Roadmap for Semiconductors (ITRS), http://www.itrs2.net/.
4. F. Schwierz, *Graphene Transistors*, Nature Nanotechnology, **5**, 487 (2010).
5. F. Triozon and P. Dollfus, *Simulation of Transport in Nanodevices* (John Wiley and Sons, 2016).
6. G. Albareda, D. Marian, A. Benali, S. Yaro, N. Zanghì, and X. Oriols, *Time-Resolved Electron Transport with Quantum Trajectories*, Journal of Computational Electronics, **12**, 405 (2013).
7. X. Oriols and D. K. Ferry, *Quantum Transport Beyond DC*, Journal of Computational Electronics, **12**, 317 (2013).

8. A. Benseny, G. Albareda, A. Sanz, J. Mompart, and X. Oriols, Applied Bohmian Mechanics, The European Physical Journal D, **68** (10), 286 (2014).
9. M. Born, *It would indeed be remarkable if Nature fortified herself against further advances in knowledge behind the analytical difficulties of the many-body problem* 1960.
10. P. A. M. Dirac, *The general theory of quantum mechanics is now almost complete. The underlying physical laws necessary for the mathematical theory of a large part of physics and the whole of chemistry are thus completely known, and the difficulty is only that the exact application of these laws leads to equations much too complicated to be soluble* 1929.
11. R. Landauer, *Electrical Resistance of Disordered One-Dimensional Lattices*, The Philosophical Magazine, **21**, 863 (1970).
12. S. Datta, *Nanoscale Device Modeling: The Green's Function Method*, Superlattices and Microstructures, **28**, 253 (2000).
13. W. R. Frensley, *Boundary-Conditions for Open Quantum-Systems Driven Far from Equilibrium*, Reviews of Modern Physics, **62**, 745 (1990).
14. M. Büttiker, *Scattering-Theory of Thermal and Excess Noise in Open Conductors*, Physical Review Letters, **65**, 2901 (1990).
15. D. Marian, N. Zanghì, and X. Oriols, *Weak Values from Displacement Currents in Multiterminal Electron Devices*, Physical Review Letters, **116**, 110404 (2016).
16. F. L. Traversa, G. Albareda, M. Ventra, and X. Oriols, *Robust Weak-Measurement Protocol for Bohmian Velocities*, Physical Review A, **87**, 052124 (2013).
17. E. Colomés, Z. Zhan, and X. Oriols *Comparing Wigner, Husimi and Bohmian Distributions: Which One is a True Probability Distribution in Phase Space?*, Journal of Computational Electronics, **14**, 894 (2015) and also in arXiv1507.07828.
18. Y. M. Blanter and M. Büttiker, *Shot Noise in Mesoscopic Conductors*, Physics Reports, **336**, 1 (2000).
19. L. S. Levitov and G. B. Lesovik, *Charge-Distribution in Quantum Shot Noise*, JETP Letters, **58**, 230 (1993).
20. L. S. Levitov, H. W. Lee, and G. B. Lesovik, *Electron Counting Statistics and Coherent States of Electric Current*, Journal of Mathematical Physics, **37**, 4845 (1996).
21. F. Bonani and G. Ghione, *Noise in Semiconductors Devices, Modeling and Simulation* (Spriger-Verlag, Berlin, Heidelberg, 2001).

22. W. Schottky, *Über spontane stromschwankungen in Verschiedenen Elektrizitätsleitern*, Annalen der Physik (Leipzig), **57**, 541 (1918).
23. R. Landauer, *Conductance from Transmission: Common Sense Points*, Physica Scripta, **T42**, 10 (1992).
24. X. Oriols, *Quantum Trajectory Approach to Time Dependent Transport in Mesoscopic Systems with Electron-Electron Interactions*, Physical Review Letters, **98**, 066803 (2007).
25. G. Albareda, J. M. Bofill, I. Tavernelli, F. Huarte-Larraaga, F. Illas, and A. Rubio, *Conditional Born-Oppenheimer Dynamics: Quantum Dynamics Simulations for the Model Porphine*, Journal of Physical Chemistry Letters, **6**, 1529 (2015).
26. G. Albareda, H. Appel, I. Franco, A. Abedi, and A. Rubio, *Correlated Electron-Nuclear Dynamics with Conditional Wavefunctions*, Physical Review Letters, **113**, 083003 (2014).
27. G. Albareda, A. Abedi, I. Tavernelli, and A. Rubio, *Universal Steps in Quantum Dynamics with Time-Dependent Potential-Energy Surfaces: Beyond the Born-Oppenheimer Picture*, Physical Review A, **94**, 062511 (2016).
28. J. J. Sakurai, *Modern Quantum Mechanics* (Addison Wesley, 1994).
29. A. Alarcón, X. Cartoixà, and X. Oriols, *Towards the Explicit Computation of Bohm Velocities Associated to n-Electron Wavefunctions with Arbitrary Spin-Orientations*, Physica Status Solidi **7**, 2636 (2010).
30. I. Vega and D. Alonso, *Dynamics of non-Markovian Open Quantum Systems*, Reviews of Modern Physics, **89**, 015001 (2017).
31. Z. Zhan, E. Colomés, and X. Oriols, *Unphysical Features in the Application of the Boltzmann Collision Operator in the Time-Dependent Modeling of Quantum Transport*, Journal of Computational Electronics, **15**, 1206 (2016).
32. E. Colomés, Z. Zhan, D. Marian, and X. Oriols, *Quantum Dissipation with Conditional Wave Functions: Application to the Realistic Simulation of Nanoscale Electron Devices*, Physical Review B, **96**, 075135 (2017).
33. G. Albareda, J. Suñé, and X. Oriols, *Many-Particle Hamiltonian for Open Systems with Full Coulomb Interaction: Application to Classical and Quantum Time-Dependent Simulations of Nanoscale Electron Devices*, Physical Review B, **79**, 075315 (2009).
34. G. Albareda, J. Suñé, and X. Oriols, *Monte Carlo Simulations of Nanometric Devices Beyond the "Mean-Field" Approximation*, Journal of Computational Electronics, **7**, (3) 197 (2008).

35. G. Albareda, A. Benali, and X. Oriols, *Self-Consistent Time-Dependent Boundary Conditions for Static and Dynamic Simulations of Small Electron Devices*, Journal of Computational Electronics, **12**, 730 (2013).
36. G. Albareda, H. Lopez, X. Cartoixà, J. J. Suñé, and X. Oriols, *Time-Dependent Boundary Conditions with Lead-Sample Coulomb Correlations: Application to Classical and Quantum Nanoscale Electron Device Simulators*, Physical Review B, **82**, 085301 (2010).
37. F. L. Traversa, Z. Zhan, and X. Oriols, *Absorption and Injection Models for Open Time-Dependent Quantum Systems*, Physical Review E, **90**, 023304 (2014) and also in arXiv:1405.7074.
38. H. López, G. Albareda, X. Cartoixà, J. Suñé, and X. Oriols, *Boundary Conditions with Pauli Exclusion and Charge Neutrality: Application to the Monte Carlo Simulation of Ballistic Nanoscale Devices*, Journal of Computational Electronics, **7**, 213 (2008).
39. W. Shockley, *Currents to Conductors Induced by a Moving Point Charge*, Journal of Applied Physics, **9**, 635 (1938).
40. S. Ramo, *Currents Induced by Electron Motion*, Proceedings of IRE, **27**, 584 (1939).
41. B. Pellegrini, *Electric Charge Motion, Induced Current, Energy Balance, and Noise*, Physical Review B, **34**, 5921 (1986).
42. B. Pellegrini, *Elementary Applications of Quantum-Electrokinematics Theorem*, Il Nuovo Cimento D, **15**, 881 (1993).
43. B. Pellegrini, *Extension of the Electrokinematics Theorem to the Electromagnetic Field and Quantum Mechanics*, Il Nuovo Cimento D, **15**, 855 (1993).
44. G. Albareda, F. L. Traversa, A. Benali, and X. Oriols, *Computations of Quantum Electrical Transport through the Ramo-Shockley-Pellegrini Theorem with Trajectories*, Fluctuations and Noise Letters, **11**, 1242008 (2012).
45. Z. Zhan, *On the Displacement Current in THz Quantum Nanodevices: Application to the Simulation of Graphene Transistors* (PhD Thesis, Barcelona, Spain, 2017).
46. Z. Zhan, E. Colomés, and X. Oriols, *Limitations of the Intrinsic Cutoff Frequency to Correctly Quantify the Speed of Nanoscale Transistors*, IEEE Transactions on Electron Devices, **64**, 2617 (2017) and also in arXiv:1611.03856.
47. Z. Zhan, E. Colomés, A. Benali, D. Marian, and X. Oriols, *Time-Dependent Simulation of Particle and Displacement Currents in THz Graphene*

Transistors, Journal of Statistical Mechanics: Theory and Experiment, **5**, 054019 (2016) and also in arXiv:1511.05515.

48. X. Oriols, A. Alarcón, and E. Fernández-Díaz, *Time Dependent Quantum Current for Independent Electrons Driven Under Non-Periodic Conditions*, Physical Review B, **71**, 245322 (2005).
49. A. Alarcón and X. Oriols, *Computation of Quantum Electron Transport with Local Current Conservation Using Quantum Trajectories*, Journal of Statistical Mechanics: Theory and Experiment, **2009**, 01051 (2009).
50. B. Eisenberg, X. Oriols, and D. K. Ferry, *Dynamics of Current, Charge and Mass*, Molecular Based Mathematical Biology, **5**, 78 (2017) and in arXiv:1708.07400.
51. A. Benali, F. L. Traversa, G. Albareda, M. Aghoutane, and X. Oriols, *Improving the Intrinsic Cut-Off Frequency of Gate-All-Around Quantum-Wire Transistors Without Channel Length Scaling*, Applied Physics Letters, **102**, 173506 (2013).
52. G. Albareda, X. Saura, X. Oriols, and J. Suñé, *Many-Particle Transport in the Channel of Quantum Wire Double-Gate Field-Effect Transistors with Charged Atomistic Impurities*, Journal of Applied Physics, **108**, 043706 (2010).
53. G. Albareda, D. Jiménez, and X. Oriols, *Intrinsic Noise in Aggressively Scaled Field-Effect Transistors*, Journal of Statistical Mechanics: Theory and Experiment, **2009**, 01044 (2009).
54. G. Albareda, F. L. Traversa, and X. Oriols, *Local Energy and Power in Many-Particle Quantum Systems Driven by an External Electrical Field*, Journal of Statistical Mechanics: Theory and Experiment, **5**, 054026 (2016).
55. G. Albareda, A. Alarcón, and X. Oriols, *Electric Power in Nanoscale Devices with Full Coulomb Interaction*, International Journal of Numerical Modelling, **23**, 354 (2010).
56. A. Benali, F. L. Traversa, G. Albareda, A. Alarcón, M. Aghoutane, and X. Oriols, *Effect of Gate-All-Around Transistor Geometry on the High-Frequency Noise: Analytical Discussion*, Fluctuations and Noise Letters, **11**, 1241002 (2012).
57. L. Esaki and R. Tsu, *Superlattice and Negative Differential Conductivity in Semiconductors*, IBM Journal of Research and Development, **14**, 61 (1970).
58. R. Tsu and L. Esaki, *Tunneling in a Finite Superlattice*, Applied Physics Letters, **22**, 562 (1973).

59. L. L. Chang, R. Tsu, and L. Esaki, *Resonant Tunneling in Semiconductor Double Barriers*, Applied Physics Letters, **24**, 593 (1974).

60. J. P. Sun, G. I. Haddad, P. Mazumder, and J. N. Schulman, *Resonant Tunneling Diodes: Models and Properties*, Proceedings of the IEEE, **86**, 641 (1998).

61. J. B. Ferry and S. M. Goodnick, *Transport in Nanostructures* (Cambridge University Press, 2009).

62. M. Büttiker, *Coherent and Sequential Tunneling in Series Barriers*, IBM Journal of Research and Development, **32**, 63 (1998).

63. H. Mizuta and T. Tanoue, *The Physics and Applications of Resonant Tunnelling Diodes* (Cambridge Studies in Semiconductor Physics and Microelectronic Engineering, 1995).

64. H. J. De Los Santos, K. K. Chui, D. H. Chow, and H. L. Dunlap, *An Efficient HBT/RTD Oscillator for Wireless Applications*, IEEE Microwave and Wireless Components Letters, **11**, 193 (2001).

65. P. Mazumder, S. Kulkarni, M. Bhattacharya, J. P. Sun, and G. I. Haddad, *Digital Circuit Applications of Resonant Tunneling Devices*, Proceedings of the IEEE, **86**, 664 (1998).

66. M. T. Björk, B. J. Ohlsson, C. Thelander, A. I. Persson, K. Deppert, L. R. Wallenberg, and L. Samuelson, *Nanowire Resonant Tunneling Diodes*, Applied Physics Letters, **81**, 4458 (2002).

67. M. J. Gilbert, R. Akis, and D. K. Ferry, *Phonon-Assisted Ballistic to Diffusive Crossover in Silicon Nanowire Transistors*, Journal of Applied Physics, **98**, 094303 (2005).

68. S. Luryi and A. Zaslavsky, *Nonclassical Devices in SOI: Genuine or Copyright from III–V*, Solid-State Electronics, **51**, 212 (2007).

69. G. K. Celler and S. Cristoloveanu, *Frontiers of Silicon-on-Insulator*, Journal of Applied Physics, **93**, 4955 (2003).

70. C. H. Choi, Z. Yu, and R. W. Dutton, *Resonant Gate Tunneling Current in Double-Gate SOI: A Simulation Study*, IEEE Transactions on Electron Devices, **50**, 2579 (2003).

71. M. V. Fischetti, *Theory of Electron Transport in Small Semiconductor Devices Using the Pauli Master Equation*, Journal of Applied Physics, **83**, 6202 (1998).

72. W. R. Frensley, *Wigner-Function Model of a Resonant-Tunneling Semiconductor Device*, Physical Review B **36**, 1570 (1987).

73. R. C. Bowen, G. Klimeck, R. K. Lake, W. R. Frensley, and T. Moise, *Quantitative Simulation of a Resonant Tunneling Diode*, Journal of Applied Physics, **81**, 3207 (1997).

74. E. R. Brown and C. D. Parker, *Resonant Tunnel Diodes as Submillimetre-Wave Sources*, Philosophical Transactions: Mathematical, Physical and Engineering Sciences, **354**, 2365 (1996).
75. R. Lake and S. Datta, *Nonequilibrium Green's-Function Method Applied to Double-Barrier Resonant-Tunneling Diodes*, Physical Review B, **45**, 6670 (1991).
76. Y. M. Blanter and M. Büttiker, *Transition from Sub-Poissonian to Super-Poissonian Shot Noise in Resonant Quantum Wells*, Physical Review B, **59**, 10217 (1999).
77. G. Iannaccone, G. Lombardi, M. Macucci, and B. Pellegrini, *Enhanced Shot Noise in Resonant Tunneling: Theory and Experiment*, Physical Review Letters, **80**, 1054 (1998).
78. G. Breit and E. Wigner, *Capture of Slow Neutrons*, Physical Review, **49**, 519 (1936).
79. L. D. Landau and L. M. Lifshitz, *Quantum Mechanics Non-Relativistic Theory* (Butterworth-Heinemann, 1977).
80. M. Büttiker, *Role of Quantum Coherence in Series Resistors*, Physical Review B, **33**, 3020 (1986).
81. G. Iannaccone, M. Macucci, and B. Pellegrini, *Shot Noise in Resonant-Tunneling Structures*, Physical Review B, **55**, 4539 (1997).
82. K. Blanks, G. Klimeck, R. Lake, R. C. Bowen, M. Leng, C. Fernando, W. R. Frensley, and D. Jovanovic, *Nemo Quantum Device Simulator*, Government Microcircuit Applications Conference Digest of Papers (GOMAC), 218 (1998).
83. F. L. Traversa, E. Buccafurri, A. Alarcón, G. Albareda, R. Clerc, F. Calmon, A. Poncet, and X. Oriols, *Time-Dependent Many-Particle Simulation for Resonant Tunneling Diodes: Interpretation of an Analytical Small-Signal Equivalent Circuit*, IEEE Transaction on Electron Devices **58**, 2104 (2011).
84. C. Jacoboni and L. Paolo, *The Monte Carlo Method for Semiconductor Device Simulation* (Springer, Berlin, 1989).
85. D. K. Ferry, M. J. Gilbert, and R. Akis, *Some Considerations on Nanowires in Nanoelectronics*, IEEE Transactions on Electron Devices, **55**, (2008).
86. A. H. C. Neto, F. Guinea, N. M. R. Peres, K. S. Novoselov, and A. K. Geim, *The Electronic Properties of Graphene*, Reviews of Modern Physics, **81**, 109 (2009).
87. P. R. Wallace, *The Band Theory of Graphite*, Physical Review, **71**, 622 (1947).

88. R. E. Peierls, *Quelques proprietes typiques des corpses solides*, Annales de l'institut Henri Poincare, **5**, 177 (1935).
89. K. S. Novoselov, A. K. Geim, S. V. Morozov, D. Jiang, Y. Zhang, S. V. Dubonos, I. V. Grigorieva, and A. A. Firsov, *Electric Field Effect in Atomically Thin Carbon Films*, Science, **306**, 666 (2004).
90. A. K. Geim, *Graphene: Status and Prospects*, Science, **324**, 1530 (2009).
91. F. Schwierz, *Electronics: Industry-Compatible Graphene Transistors*, Nature, **472**, 41 (2011).
92. R. Sordan and A. C. Ferrari, *Gigahertz Multi-Transistor Graphene Integrated Circuits*, IEEE International Electron Devices Meeting, 1 (2013).
93. I. Meric, N. Baklitskaya, P. Kim, and K. L. Shepard, *RF Performance of Top-Gated, Zero-Bandgap Graphene Field-Effect Transistors*, IEEE International Electron Devices Meeting, 1 (2008).
94. Y. M. Lin, C. Dimitrakopoulos, K. A. Jenkins, D. B. Farmer, H. Y. Chiu, A. Grill, and Ph. Avouris, *100-GHz Transistors from Wafer-Scale Epitaxial Graphene*, Science, **327**, 662 (2010).
95. L. Liao, Y.-C. Lin, M. Bao, R. Cheng, J. Bai, Y. Liu, Y. Qu, K. L. Wang, Y. Huang, and X. Duan, *High-Speed Graphene Transistors with a Self-Aligned Nanowire Gate*, Nature, **467**, 305 (2010).
96. R. Cheng, J. Bai, L. Liao, H. Zhou, Y. Chen, L. Liu, Y.-C. Lin, S. Jiang, Y. Huang, and X. Duan, *High-Frequency Self-Aligned Graphene Transistors with Transferred Gate Stacks*, Proceedings of the National Academy of Sciences of the United States of America, **109**, 11588 (2012).
97. F. Schwierz, *Graphene Transistors: Status, Prospects, and Problems*, Proceedings of the IEEE, **101**, 1567 (2013).
98. B. Radisavljevic, A. Radenovic, J. Brivio, V. Giacometti, and A. Kis, *Single-Layer* MoS_2 *Transistors*, Nature Nanotechnology, **6**, 147 (2011).
99. K. S. Novoselov, D. Jiang, F. Schedin, T. J. Booth, V. V. Khotkevich, S. V. Morozov, and A. K. Geim, *Two-Dimensional Atomic Crystals*, Proceedings of the National Academy of Sciences of the United States of America, **102**, (2005).
100. H. Liu and D. Y. Peide, MoS_2 *Dual-Gate MOSFET with Atomic-Layer-Deposited* Al_2O_3 *as Top-Gate Dielectric*, IEEE Electron Device Letters, **33**, 546 (2012).
101. L. Liu, S. B. Kumar, Y. Ouyang, and J. Guo, *Performance Limits of Mono-layer Transition Metal Dichalcogenide Transistors*, IEEE Transactions on Electron Devices, **58**, 3042 (2011).

102. Y. Yoon, K. Ganapathi, and S. Salahuddin, *How Good Can Monolayer* MoS_2 *Transistors Be?*, Nano Letters, **11**, 3768 (2011).
103. T. Fang, A. Konar, H. Xing, and D. Jena, *High-Field Transport in Two-Dimensional Graphene*, Physical Review B, **84**, 125450 (2011).

Chapter 7

Beyond the Eikonal Approximation in Classical Optics and Quantum Physics

Adriano Orefice, Raffaele Giovanelli, and Domenico Ditto
Università degli Studi di Milano,
Via Giovanni Celoria 2, 20133 Milano, Italy
adriano.orefice@unimi.it

The analysis of the Helmholtz equation is shown to lead to an exact Hamiltonian system describing in terms of ray trajectories a very wide family of wave-like phenomena (including diffraction and interference) containing the eikonal approximation as a simple limiting case. Since, moreover, the time-independent Schrödinger equation is itself a Helmholtz-like equation, the same mathematics holding for a classical optical beam turns out to apply to a quantum particle beam moving in a stationary force field, and the so-called quantum potential of the Bohmian theory turns out to be a particular case of a general wave-like property. A suitable set of Hamiltonian equations is shown to provide both the exact and deterministic trajectories of the particles and their dynamical laws, reducing to the laws of classical mechanics in the eikonal limit. Wave-like features in Hamiltonian form are presented in some typical numerical cases.

Applied Bohmian Mechanics: From Nanoscale Systems to Cosmology (Second Edition)
Edited by Xavier Oriols and Jordi Mompart

ISBN 978-981-4800-10-5 (Hardcover), 978-0-429-29474-7 (eBook)
www.jennystanford.com

7.1 Introduction

It is often believed that the "naïve" concept of optical *rays* applies only to a very limited set of physical cases, which may be ascribed to the so-called *geometrical optics approximation*, while more general and complex phenomena (such as diffraction and interference) would necessarily require a fully wave-like treatment.

In the first (classical) part of the present chapter (Sections 7.2 and 7.3), we show that this commonplace is not correct. Starting from the Helmholtz equation, we obtain, *without any omission or approximation*, a Hamiltonian set of ray-tracing equations providing (in stationary media) the *exact* description *in term of rays* of a family of wave phenomena much wider than that allowed by the standard geometrical optics, which is contained as a simple limiting case. Such a description includes both the ray *geometry* and their *motion laws*.

The rays of a classical electromagnetic wave beam are shown to be mutually coupled (in a kind of self-refractive behavior, strongly dependent on the space amplitude distribution of the beam) by a term acting *perpendicularly* to the rays themselves and affecting therefore their geometry without altering the amplitude of their velocity. This property is shown to provide, moreover, a basic tool for the numerical solution of the Hamiltonian ray-tracing system. An important consequence of this mutual coupling is the necessity of taking simultaneously into account the propagation of all the rays of the beam.

And here begins the second (quantum) part (Sections 7.4–7.8) of the present chapter.

We recall, in fact, that the particle trajectories encountered in classical mechanics were often suggested to constitute the *geometrical optics approximation* of the wave-like quantum behavior of the particles themselves. This suggestion was aimed, of course, to enforce the necessity of abandoning, in general, the very concept of particle trajectories, just as the ray picture was supposed to collapse in standard optics. The basic consequence of this philosophy is *quantum uncertainty*, according to which no position *and* momentum can be exactly and simultaneously associated to a particle.

Since, however—thanks to our previous extension of the concept of "ray" much beyond the limits of the geometrical optics approximation—no collapse of the optical ray description does occur, and since, after all, the time-independent Schrödinger equation is itself a Helmholtz-like equation, it is quite natural to expect, now, that the same peculiar properties holding in classical optics may be extended to quantum mechanics.

This expectation, indeed, is immediately satisfied, and the trajectories and dynamical laws of a quantum particle beam turn out to be provided by an *exact* Hamiltonian system *mathematically coïnciding* (in suitable dimensionless variables) with the system found in the previous optical case, and involving therefore—in correspondence with the same boundary conditions—the same geometrical trajectories, with the same transverse correlation property discovered in the classical case, requiring to take simultaneously into account the motion of all the particles of the beam.

This Hamiltonian system provides a complete description of the motion of a quantum particle beam, showing no trace of probabilistic features and containing the laws of classical dynamics as a limiting case (just as the geometrical optics approximation turns out to be a particular case of our exact ray approach).

Although, of course, neither the Helmholtz equation nor the time-independent (Helmholtz-like) Schrödinger equation can directly describe any propagation phenomenon, they provide (for any assigned stationary medium and for any set of boundary conditions) a fixed frame of trajectories, which is determined at the very outset (in a way somewhat reminding Fermat's and Maupertuis' variational principles) by the boundary conditions, and along which each particle (or ray) moves according to well-defined motion laws, in agreement with the time-independent character of the underlying Hamiltonian. These trajectory patterns and motion laws, in their turn, are shown to strongly depend on the beam amplitude distribution but not on the number of particles (or rays), which could even be reduced to a single particle at a time: a peculiar feature that has sometimes induced to speak (in the quantum context) of single-particle self-diffraction.

Within the mathematical coïncidence observed here between classical optical rays and quantum trajectories, an important feature is represented by the fact that the same mathematical term omitted in the standard geometrical optics approximation and taken into account in the present work under the name of "wave potential" turns out to give rise, in the quantum case, to the so-called quantum potential of the Bohmian theory [1–5]. Such a term is simply due, therefore, to the structure itself of Helmholtz-like equations [6], both in classical and in quantum waves.

Beyond the physical and even philosophical insight given by these observations into the nature of quantum potential, a practical consequence is the possibility of exchanging all numerical techniques between the quantum and classical cases.

7.2 Helmholtz Equation and Geometrical Optics

In order to establish our mathematical formalism, let us start from a *classical* case of wave-like behavior.

Although many kinds of physical waves would lend themselves to the considerations we have in mind here, we shall refer, in order to fix ideas, to a monochromatic electromagnetic wave beam, with a time dependence $\propto \exp(-i\omega t)$, travelling through a time-independent, isotropic, and (generally) inhomogeneous dielectric medium. Its basic features are accounted for by the Helmholtz equation:

$$\nabla^2\psi + (nk_0)^2\psi = 0, \tag{7.1}$$

where

$$\nabla^2 = \frac{\partial^2}{\partial x^2} + \frac{\partial^2}{\partial y^2} + \frac{\partial^2}{\partial z^2}; \tag{7.2}$$

$\psi(x, y, z)$ represents any component of the electric or magnetic field; $n(x, y, z)$ is the refractive index of the medium, and

$$k_0 \equiv \frac{2\pi}{\lambda_0} = \frac{\omega}{c}, \tag{7.3}$$

with obvious meaning of λ_0 and c. The phase velocity is given, in its turn, by

$$v_{\mathrm{ph}}(x, y, z) = \frac{c}{n(x, y, z)}. \tag{7.4}$$

Because of its time independence, Eq. 7.1 cannot directly describe propagation phenomena: it determines, however, as we shall see, both the fixed trajectory frame where propagation occurs and the ray motion laws along such a frame. By performing, in fact, the quite general replacement

$$\psi(x, y, z) = R(x, y, z)e^{i\varphi(x,y,z)}, \tag{7.5}$$

with real $R(x, y, z)$ and $\varphi(x, y, z)$ and separating the real from the imaginary part, Eq. 7.1 splits into the well-known [7] and strictly equivalent system of coupled equations

$$(\vec{\nabla}\varphi)^2 - (nk_0)^2 = \frac{\nabla^2 R}{R} \tag{7.6a}$$

$$\vec{\nabla} \cdot (R^2\vec{\nabla}\varphi) = 0 \tag{7.6b}$$

where $\vec{\nabla} \equiv \partial/\partial\vec{r} \equiv (\partial/\partial x, \partial/\partial y, \partial/\partial z)$, and Eq. 7.6b expresses the constancy of the flux of the vector $R^2\vec{\nabla}\varphi$ along any tube formed by the lines of $\vec{\nabla}\varphi$ itself.

When the space variation length, L, of the beam amplitude $R(x, y, z)$ may be assumed to satisfy the condition $k_0 L \gg 1$, Eq. 7.6a is well approximated by the *eikonal equation*

$$(\vec{\nabla}\varphi)^2 \simeq (nk_0)^2, \tag{7.7}$$

decoupled from Eq. 7.6b (whose presence is generally neglected) and allowing the so-called *geometrical optics approximation*, which describes the wave propagation in terms of "rays" travelling along the field lines of the *wave vector*

$$\vec{k} = \vec{\nabla}\varphi \tag{7.8}$$

independently from one another and from the amplitude distribution $R(x, y, z)$ of the beam. To be sure, by multiplying Eq. 7.7, for convenience, by the constant factor $c/2k_0$, we obtain the relation

$$D(\vec{r}, \vec{k}) \equiv \frac{c}{2k_0}\left[k^2 - (nk_0)^2\right] \simeq 0, \tag{7.9}$$

(where $\vec{r} = (x, y, z)$), whose differentiation

$$\frac{\partial D}{\partial\vec{r}} \cdot d\vec{r} + \frac{\partial D}{\partial\vec{k}} \cdot d\vec{k} = 0, \tag{7.10}$$

with $\partial/\partial\vec{k} \equiv (\partial/\partial k_x, \partial/\partial k_y, \partial/\partial k_z)$, directly provides, for any assigned refractive function $n(x, y, z)$, both the geometrical form of the rays and their motion law in the simple Hamiltonian form

$$\frac{d\vec{r}}{dt} = \frac{\partial D}{\partial \vec{k}} = \frac{c\vec{k}}{k_0} \tag{7.11a}$$

$$\frac{d\vec{k}}{dt} = -\frac{\partial D}{\partial \vec{r}} = \frac{c}{2k_0}\frac{\partial}{\partial \vec{r}}(nk_0)^2 \tag{7.11b}$$

where a ray velocity $\vec{v}_{\text{ray}} = c\vec{k}/k_0$ is implicitly defined. We may observe that $v_{\text{ray}} = |\vec{v}_{\text{ray}}| = c$ when $k = k_0$, and that $v_{\text{ray}}v_{\text{ph}} = c^2$. The rays of a wave beam are described, within the eikonal approximation 7.11, *independently* from one another, requiring only the knowledge of the (time-independent) refraction index $n(x, y, z)$ of the medium. The generalization of the eikonal approach of Eq. 7.11 to the anisotropic case of relativistic electron cyclotron waves travelling through a magnetoactive plasma in steady-state thermonuclear devices was treated, for instance, in Refs. [8, 9].

We conclude the present section by recalling Fermat's variational principle, according to which any optical ray travelling between two points A, B shall follow a trajectory satisfying the condition

$$\delta \int_A^B k ds = 0, \tag{7.12}$$

where $k = |\vec{k}|$ and ds is an element of a (virtual) line connecting A and B.

The relevance of this variational principle in the context of the present work is limited to the single *but fundamental* fact that it finds its quantum mechanical counterpart, as we shall see, in Maupertuis' principle.

7.3 Beyond the Geometrical Optics Approximation

Let us consider now Eq. 7.6a in its *complete form*, arriving therefore at the *exact* relation, generalizing the function $D(\vec{r}, \vec{k})$ of Eq. 7.9,

$$D(\vec{r}, \vec{k}) \equiv \frac{c}{2k_0}\left[k^2 - (nk_0)^2 - \frac{\nabla^2 R}{R}\right] = 0, \tag{7.13}$$

whose differentiation, formally coinciding with Eq. 7.10, leads to the exact Hamiltonian ray-tracing system

$$\frac{d\vec{r}}{dt} = \frac{\partial D}{\partial \vec{k}} = \frac{c\vec{k}}{k_0} \tag{7.14a}$$

$$\frac{d\vec{k}}{dt} = -\frac{\partial D}{\partial \vec{r}} = \frac{c}{2k_0}\frac{\partial}{\partial \vec{r}}\left[(nk_0)^2 + \frac{\nabla^2 R}{R}\right]. \tag{7.14b}$$

Such a system lends itself, as we shall see, to a practicable numerical solution even in physical cases where the standard geometrical optics approximation is completely inapplicable. It avoids the approximations of the "eikonal" approach 7.11, although fully retaining the idea of optical rays travelling along the field lines of $\vec{k} = \vec{\nabla}\varphi$. The right-hand side (r.h.s.) of Eq. 7.14b contains, with respect to the eikonal case, an additional term, which may be written in the form $-\vec{\nabla}W \equiv -\partial W/\partial\vec{r}$, thus defining a function

$$W(x, y, z) = -\frac{c}{2k_0}\frac{\nabla^2 R}{R}, \tag{7.15}$$

which we shall call "wave potential," coupling all the rays of the beam. Such a function (having the dimensions of a *frequency*) is an intrinsic, stationary property of the beam, entirely determined in the whole space by the launching conditions imposed to the beam amplitude $R(x, y, z)$, and therefore by the experimental setup.

For any set of boundary conditions imposed to the function $R(x, y, z)$ on the launching surface of the beam, and for any assigned refractive stationary medium, the system 7.14 provides both a fixed "weft" of "rails" along which the rays are channelled and the motion laws along these rails.

In order to analyze the properties of the wave potential, we make use of Eq. 7.6b, which may be written in the form

$$\vec{\nabla}\cdot(R^2\vec{\nabla}\varphi) \equiv 2R\vec{\nabla}R\cdot\vec{\nabla}\varphi + R^2\vec{\nabla}\cdot\vec{\nabla}\varphi = 0. \tag{7.16}$$

Since no new ray trajectory may suddenly arise in the space region spanned by the beam, we must have, of course, $\vec{\nabla}\cdot\vec{\nabla}\varphi = 0$, so that Eq. 7.16 splits into the system

$$\vec{\nabla}\cdot\vec{\nabla}\varphi = 0, \tag{7.17a}$$

$$\vec{\nabla}R\cdot\vec{\nabla}\varphi = 0. \tag{7.17b}$$

where the second equation is automatically entailed by the first one. The term $\vec{\nabla}R$ is therefore perpendicular to the field lines of $\vec{k} \equiv \vec{\nabla}\varphi$, and this transverse character is shared by the gradient $\vec{\nabla}W$. The absolute value v_{ray} of the ray velocity remains, *in vacuo*, equal to c all along the ray trajectories because such a gradient may only modify the *direction*, but not the *amplitude*, of the wave vector $\vec{k}$: the only possible changes of $k \equiv |\vec{k}|$ could be due, in a medium different from vacuum, to its refractive function $n(x, y, z)$.

Thanks to the constancy of the flux of $R^2\vec{\nabla}\varphi$ along any tube formed by the field lines of $\vec{k} \equiv \vec{\nabla}\varphi$, the function $R(x, y, z)$, once assigned on the launching surface from which the ray beam is assumed to start, may be numerically built up step by step, together with its derivatives and therefore with the *wave potential* $W(x, y, z)$, in the whole region crossed by the beam.

The beam amplitude distribution $R(x, y, z)$ is entirely determined, in the whole space, by the form assumed on the launching surface in the experimental setup. The relevant energy input of the beam, on the other hand, is quite arbitrary and could even be reduced to a single ray at a time: a ray, however, that belongs to a pattern fixed from the very start by the launching conditions and that is coupled to all the other rays of the beam. The virtual presence of the other trajectories of the beam must be kept into account step by step, in any numerical computation, in order to build up (wave front after wave front, since every front may be considered as the launching surface of the beam) the proper form of the wave potential.

7.4 The Time-Independent Schrödinger Equation

The *classical* motion of a monoenergetic beam of noninteracting particles of mass m in a force field deriving from a potential energy $V(x, y, z)$ not explicitly depending on time may be described for each particle of the beam, as is well known, by means of the so-called reduced (or time-independent) Hamilton–Jacobi equation [7]

$$(\vec{\nabla}S)^2 = 2m(E - V), \tag{7.18}$$

where E is the total energy and one of the main properties of the function $S(x, y, z)$ is that the particle momentum is given by

$$\vec{p} = \vec{\nabla} S. \tag{7.19}$$

Recalling Maupertuis' variational principle

$$\delta \int_A^B p ds = 0, \tag{7.20}$$

with $p = |\vec{p}|$, the formal analogy between Eqs. 7.7, 7.8, and 7.12 on the one hand, and Eqs. 7.18, 7.19, and 7.20, on the other hand, suggests, as is well known, that the *classical* particle trajectories could constitute the *geometrical optics approximation* of an equation analogous to the Helmholtz Eq. 7.1, from which it is immediately obtained by means of the substitutions

$$\varphi = \frac{S}{a} \tag{7.21a}$$

$$\vec{k} = \vec{\nabla}\varphi = \frac{\vec{\nabla} S}{a} = \frac{\vec{p}}{a} \tag{7.21b}$$

$$k_0 \equiv \frac{2\pi}{\lambda_0} = \frac{\sqrt{2mE}}{a} \equiv \frac{p_0}{a} \tag{7.21c}$$

$$n^2(x, y, z) = 1 - \frac{V(x, y, z)}{E} \tag{7.21d}$$

where the parameter "a" represents a constant *action* whose value is *a priori* arbitrary—as far as the relations 7.21 are concerned—but is imposed by the history itself of quantum mechanics:

$$a = \hbar \simeq 1.0546 \times 10^{-27} \text{ erg} \times \text{s}. \tag{7.22}$$

The equation obtained from Eq. 7.1 by means of the substitutions Eq. 7.21 and Eq. 7.22 takes up the form

$$\nabla^2\psi + \frac{2m}{\hbar^2}(E - V)\psi = 0, \tag{7.23}$$

which is the standard *time-independent* Schrödinger equation.

By applying now to Eq. 7.23 the same procedure leading from Eq. 7.1 to Eqs. 7.6, and assuming therefore

$$\psi(x, y, z) = R(x, y, z)e^{iS(x,y,z)/\hbar}, \tag{7.24}$$

Eq. 7.23 splits [10] into the well-known coupled system

$$(\vec{\nabla} S)^2 - 2m(E - V) = \hbar^2 \frac{\nabla^2 R}{R}, \tag{7.25a}$$

$$\vec{\nabla} \cdot (R^2 \vec{\nabla} S) = 0. \tag{7.25b}$$

By taking the gradient of Eq. 7.25a we get moreover

$$\left(\frac{\vec{\nabla} S}{m} \cdot \vec{\nabla}\right) \frac{\vec{\nabla} S}{m} + \frac{\vec{\nabla} V}{m} = \frac{\hbar^2}{2m^2} \vec{\nabla} \left(\frac{\nabla^2 R}{R}\right). \tag{7.26}$$

Equation 7.26, together with Eq. 7.25b, is often interpreted as describing, in the "classical limit" $\hbar \to 0$ (whatever such a limit may mean), a "fluid" of particles with mass m and velocity $\vec{\nabla} S/m$, moving in an external potential $V(x, y, z)$: an interpretation consistent with the probabilistic character usually ascribed to the Schrödinger equation.

7.5 Hamiltonian Description of Quantum Particle Motion

Let us now observe that, by simply maintaining Eq. 7.19, Eq. 7.25a may be written in the form of a generalized, time-independent Hamiltonian:

$$H(\vec{r}, \vec{p}) \equiv \frac{p^2}{2m} + V - \frac{\hbar^2}{2m} \frac{\nabla^2 R}{R} = E \tag{7.27}$$

including the crucial term $-(\hbar^2/2m)\nabla^2 R/R$, coïnciding with the well-known *quantum potential* of the Bohmian theory [1–5]. By differentiating Eq. 7.27 we get the relation

$$\frac{\partial H}{\partial \vec{r}} \cdot d\vec{r} + \frac{\partial H}{\partial \vec{p}} \cdot d\vec{p} = 0 \tag{7.28}$$

with $\partial/\partial\vec{p} \equiv (\partial/\partial p_x, \partial/\partial p_y, \partial/\partial p_z)$, leading to a Hamiltonian dynamical system of the form

$$\frac{d\vec{r}}{dt} = \frac{\partial H}{\partial \vec{p}} = \frac{\vec{p}}{m} \tag{7.29a}$$

$$\frac{d\vec{p}}{dt} = -\frac{\partial H}{\partial \vec{r}} = -\frac{\partial}{\partial \vec{r}} \left[V(x, y, z) - \frac{\hbar^2}{2m} \frac{\nabla^2 R}{R}\right] \tag{7.29b}$$

strictly similar to the classical ray-tracing system 7.14. The quantum potential

$$Q(x, y, z) = -\frac{\hbar^2}{2m}\frac{\nabla^2 R}{R} \tag{7.30}$$

is analogous to the wave potential $W(x, y, z)$ of Eq. 7.15, and is therefore due to the wave-like nature of quantum particles. Its presence causes the interconnection between the trajectories of the whole beam, and its absence would reduce the system 7.29 to the standard classical set of Hamiltonian dynamical equations, which constitute its eikonal approximation.

If we envisage the Hamiltonian system 7.29 for what it appears to be, without superimposing any interpretative prejudice, it is quite evident that its mathematical treatment is the same employed in the classical ray-tracing case, including the fact that $\vec{\nabla} R$ turns out to be perpendicular the field lines of $\vec{p} \equiv \vec{\nabla} S$. The gradient $\vec{\nabla} Q(x, y, z)$, in its turn, is tangent to the wave front and cannot act on the *amplitude* of the particle momentum (while modifying, in general, its *direction*). The only possible amplitude changes could be due to the presence of an external potential $V(x, y, z)$, but no energy exchange may occur between particles and quantum potential. Once more the function $R(x, y, z)$ may be assigned on the launching surface from where the beam is assumed to start, and numerically built up step by step (i.e. wave front after wave front) in the whole region spanned by the motion of the beam; and once more the numerical treatment of each trajectory requires the simultaneous treatment of all the (real or virtual) trajectories of the beam consistent with the experimental setup.

7.6 The Unique Dimensionless Hamiltonian System

A quite expedient step is now the passage to the new, *dimensionless* variables $(\tilde{x}, \tilde{y}, \tilde{z})$, $(\tilde{p}_x, \tilde{p}_y, \tilde{p}_z)$, and $\tilde{t}$, defined as the ratios between the *dimensioned* quantities (x, y, z), (p_x, p_y, p_z), and t, respectively, and $\lambda_0 \equiv 2\pi\hbar/p_0$ for the space variables, $p_0 = \sqrt{2mE}$ for the momentum variables (so that $\tilde{p}_0 = 1$), and $m\lambda_0/p_0$ for the time variable. The equation system 7.29 takes up, therefore, the form

$$\frac{d\tilde{x}}{d\tilde{t}} = \tilde{p}_x \tag{7.31a}$$

$$\frac{d\tilde{y}}{d\tilde{t}} = \tilde{p}_y \tag{7.31b}$$

$$\frac{d\tilde{z}}{d\tilde{t}} = \tilde{p}_z \tag{7.31c}$$

$$\frac{d\tilde{p}_x}{d\tilde{t}} = -\frac{\partial}{\partial\tilde{x}}\left[\frac{V(\tilde{x},\tilde{y},\tilde{z})}{2E} - \frac{1}{8\pi^2}G(\tilde{x},\tilde{y},\tilde{z})\right] \tag{7.31d}$$

$$\frac{d\tilde{p}_y}{d\tilde{t}} = -\frac{\partial}{\partial\tilde{y}}\left[\frac{V(\tilde{x},\tilde{y},\tilde{z})}{2E} - \frac{1}{8\pi^2}G(\tilde{x},\tilde{y},\tilde{z})\right] \tag{7.31e}$$

$$\frac{d\tilde{p}_z}{d\tilde{t}} = -\frac{\partial}{\partial\tilde{z}}\left[\frac{V(\tilde{x},\tilde{y},\tilde{z})}{2E} - \frac{1}{8\pi^2}G(\tilde{x},\tilde{y},\tilde{z})\right] \tag{7.31f}$$

with

$$G(\tilde{x},\tilde{y},\tilde{z}) = \frac{1}{R}\left(\frac{\partial^2 R}{\partial\tilde{x}^2} + \frac{\partial^2 R}{\partial\tilde{y}^2} + \frac{\partial^2 R}{\partial\tilde{z}^2}\right). \tag{7.32}$$

It may be observed that no direct reference is present, in the dimensionless form 7.31 assumed by the quantum dynamical system 7.29, to the mass of the moving particles and not even to $\hbar$.

Let us also observe that *the same dimensionless form* 7.31 is taken up by the ray-tracing system 7.14—relevant to the *classical* electromagnetic case—by simply assuming

$$\tilde{\vec{r}} = \frac{k_0\vec{r}}{2\pi}, \tag{7.33a}$$

$$\tilde{\vec{p}} = \frac{\vec{k}}{k_0} \equiv \frac{\vec{v}_{\text{ray}}}{c}, \tag{7.33b}$$

$$\tilde{t} = \frac{ck_0 t}{2\pi}, \tag{7.33c}$$

and replacing $V(\tilde{x},\tilde{y},\tilde{z})/E$ with $\left[1 - n^2(\tilde{x},\tilde{y},\tilde{z})\right]$, in agreement with the relations 7.21.

The basic Hamiltonian equations underlying the *classical* as well as the *quantum* cases are therefore *exactly the same*.

Once assigned on the launching surface of the beam, the function $G(\tilde{x},\tilde{y},\tilde{z})$ may be numerically determined step by step, in principle, together with its derivatives, along the full set of trajectories of the beam. This dimensionless function (to which both the wave potential W and the quantum potential Q reduce, with opposite signs) determines, together with the external refractive index or force field, the trajectory geometry and the motion laws both of *classical* electromagnetic rays and of *quantum* particles. In its absence the system 7.31 would simply describe either the behavior of the rays of an electromagnetic beam in the *eikonal* approximation or the *classical* motion of the particles of a material beam, and both rays and particles would move along trajectories independent from one another.

Due to the small coefficient $1/8\pi^2$, the transverse gradient $\vec{\nabla}G$ acts along the ray (or particle) trajectories in a soft and cumulative way: a fact granting the main justification for omitting such a term, as is done both in classical dynamics and in the standard geometrical optics approximation.

The trajectory pattern, in its turn, is a stationary structure determined at the very outset along which rays and particles move in a way strongly reminding the spirit of classical variational principles, such as the ones of Fermat and Maupertuis, in time-independent media. For any set of boundary conditions imposed to the function $R(\tilde{x}, \tilde{y}, \tilde{z})$ on the launching surface of the beam, and for any assigned stationary refractive medium (or force field), the system 7.31 provides both a fixed weft of rails and a motion law (in agreement with the underlying time-independent Hamiltonian) to which particles (or rays) are deterministically bound, showing no trace of probabilistic features. Each quantum particle (as well as each classical electromagnetic ray) turns out to be conceivable indeed, on the basis of the present analysis, as *starting and remaining* on a well-defined trajectory belonging to a pattern which is *a priori* fixed, as a whole, by the properties of the medium and by the beam amplitude distribution on the launching surface and therefore by the experimental setup.

The equation system 7.31, derived from a time-independent Hamiltonian, provides in conclusion a complete set of motion laws that replace (and contain as an approximation, when the wave- or quantum-potential may be neglected) both geometrical optics and classical dynamics in stationary media. In striking divergence from these approximated limiting cases, however, the new set of equations does not represent an ensemble of mutually independent rays (or particles) but requires, in general, for its solution, the full set of boundary conditions and the simultaneous treatment of all the rays (or particles) of the beam, which are mutually coupled by the wave (or quantum) potential.

7.7 Wave-Like Features in Hamiltonian Form

We propose now a simple application of the Hamiltonian system 7.31 to the propagation of a collimated beam injected at $\tilde{z} = 0$ parallel to the $\tilde{z}$-axis and centered at $\tilde{x} = \tilde{y} = 0$ in order to simulate *wave diffraction* through a single slit.

The problem may be faced by taking into account for simplicity sake (but with no substantial loss of generality) either a (quantum) particle beam in the absence of external fields ($V = 0$) or a (classical) electromagnetic beam *in vacuo* ($n^2 = 1$), with a geometry allowing to limit the computation to the trajectories lying on the ($\tilde{x}$, $\tilde{z}$)-plane.

The system 7.31 takes up, therefore, the form

$$\frac{d\tilde{x}}{d\tilde{t}} = \tilde{p}_x \tag{7.34a}$$

$$\frac{d\tilde{z}}{d\tilde{t}} = \tilde{p}_z \tag{7.34b}$$

$$\frac{d\tilde{p}_x}{d\tilde{t}} = \frac{1}{8\pi^2}\frac{\partial}{\partial\tilde{x}}G(\tilde{x}, \tilde{z}) \tag{7.34c}$$

$$\frac{d\tilde{p}_z}{d\tilde{t}} = \frac{1}{8\pi^2}\frac{\partial}{\partial\tilde{z}}G(\tilde{x}, \tilde{z}) \tag{7.34d}$$

with

$$G(\tilde{x}, \tilde{z}) = \frac{1}{R}\left(\frac{\partial^2 R}{\partial\tilde{x}^2} + \frac{\partial^2 R}{\partial\tilde{z}^2}\right) \tag{7.35a}$$

$$\tilde{p}_x(\tilde{z} = 0) = 0 \tag{7.35b}$$

$$\tilde{p}_z(\tilde{z} = 0) = \tilde{p}_0 = 1 \tag{7.35c}$$

and with a suitable amplitude distribution $R(\tilde{x}; \tilde{z} = 0)$ (from whose normalization the function G is obviously independent) assigned on the launching plane.

Because of the transverse nature of the gradient of $G(\tilde{x}, \tilde{z})$, the *amplitude* of the vector φ remains unchanged (in the absence of external fields and/or refractive effects) along each trajectory, leading to the relation

$$\tilde{p}_z = \sqrt{\tilde{p}_0^2 - \tilde{p}_x^2} \equiv \sqrt{1 - \tilde{p}_x^2} \tag{7.36}$$

which may be employed to improve the second (and to suppress the fourth) equation of the Hamiltonian system 7.34. Three possible models of the amplitude distribution $R(\tilde{x}; \tilde{z} = 0)$ and, therefore of the beam-launching conditions are considered here:

(1) A simple Gaussian distribution centered at $\tilde{x} \equiv x/\lambda_0 = 0$, in the form

$$R_0(\tilde{x}; \tilde{z} = 0) \propto e^{-\frac{\lambda_0^2}{w_0^2}\tilde{x}^2} \equiv e^{-\varepsilon^2 \tilde{x}^2} \tag{7.37}$$

(where the length w_0 describes the half-width of the Gaussian and $\varepsilon = \lambda_0/w_0 \leq 1$);

(2) A superposition of Gaussian functions, each one with half-width w_0, to be defined in Eq. 7.40;

(3) An algebraic distribution of the form

$$R_N(\tilde{x}; \tilde{z} = 0) \propto \frac{1}{1 + (\varepsilon\tilde{x})^{2N}} \tag{7.38}$$

(with integer $N \geq 1$), allowing the presence of a flat central region, widening with increasing values of N.

In our numerical procedure we found it convenient to adopt a different set of dimensionless space and time variables, suggested by Eqs. 7.37 and 7.38 themselves. We assumed in fact $\hat{x} = \varepsilon\tilde{x}$, $\hat{z} = \varepsilon\tilde{z}$, $\hat{t} = \varepsilon\tilde{t}$. This amounts to measuring lengths in terms of w_0, and time in terms of the wave travel along a length equal to w_0 itself.

Our three amplitude distribution models take up, therefore, respectively, the simple forms

(1) $$R_0(\hat{x}; \hat{z} = 0) \propto e^{-\hat{x}^2}, \tag{7.39}$$

(2) $$R_S(\hat{x}; \hat{z} = 0) \propto \sum_{n=-m}^{m} e^{-(\hat{x} - n\hat{x}_0)^2}, \tag{7.40}$$

with suitable values of m and $\hat{x}_0$, and

(3)
$$R_N(\hat{x};\hat{z}=0) \propto \frac{1}{1+\hat{x}^{2N}}, \tag{7.41}$$

while the system 7.34, keeping Eq. 7.36 into account, reduces to the form

$$\frac{d\hat{x}}{d\hat{t}} = \tilde{p}_x \tag{7.42a}$$
$$\frac{d\hat{z}}{d\hat{t}} = \sqrt{1-\tilde{p}_x^2} \tag{7.42b}$$
$$\frac{d\tilde{p}_x}{d\hat{t}} = \frac{\varepsilon^2}{8\pi^2}\frac{\partial G(\hat{x},\hat{z})}{\partial \hat{x}} \tag{7.42c}$$

with

$$G(\hat{x},\hat{z}) = \frac{1}{R}\left(\frac{\partial^2 R}{\partial \hat{x}^2} + \frac{\partial^2 R}{\partial \hat{z}^2}\right) \tag{7.43a}$$
$$\tilde{p}_x(\hat{z}=0) = 0 \tag{7.43b}$$
$$\tilde{p}_z(\hat{z}=0) = \tilde{p}_0 = 1. \tag{7.43c}$$

We show in Figs. 7.1, 7.2, and 7.3, respectively, in arbitrary units, the launching distributions $R_0(\hat{x},\hat{z}=0)$, $R_S(\hat{x},\hat{z}=0)$ (with $m=2$

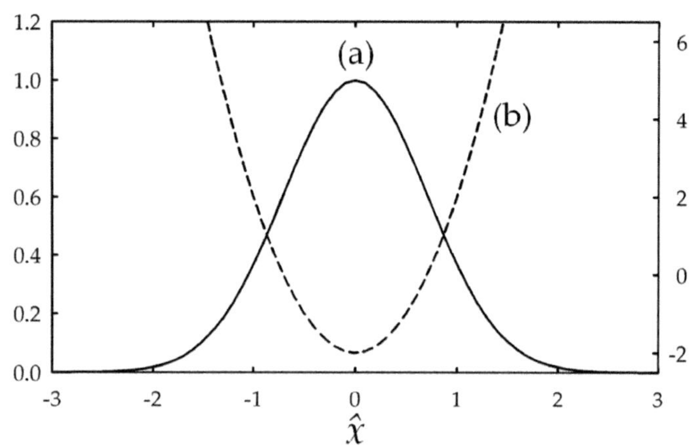

Figure 7.1 (a) Left scale: Transverse amplitude distribution R_0 assigned to the beam at $\hat{z}=0$ according to Eq. 7.39; (b) right scale: Its potential function G_0 given by Eq. 7.44.

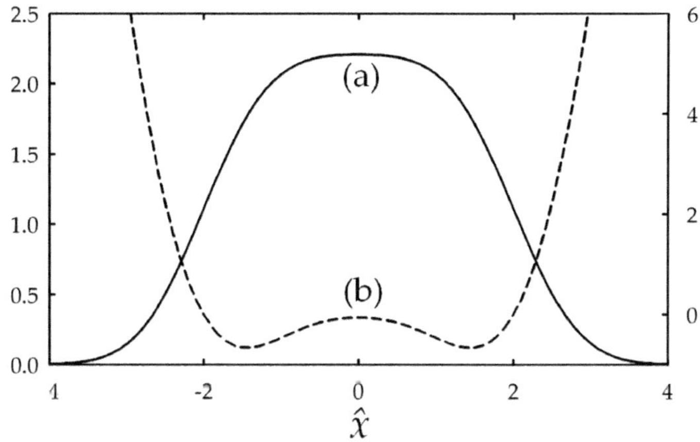

Figure 7.2 (a) Left scale: Transverse amplitude distribution R_S at $\hat{z} = 0$ according to Eq. 7.40, with $m = 2$ and $\hat{x}_0 = 0.8$; (b) right scale: Its potential function G_S given by Eq. 7.44.

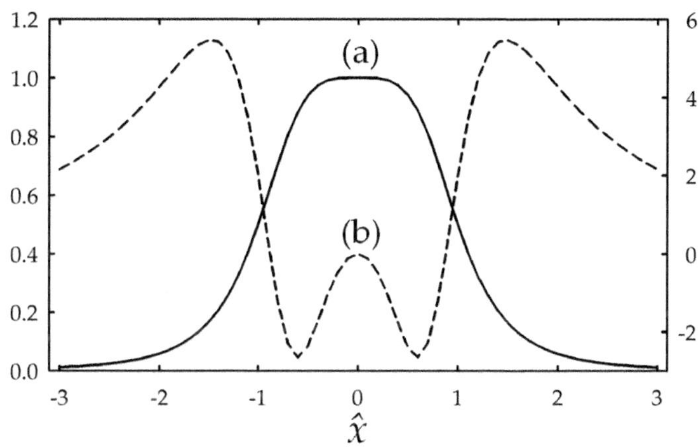

Figure 7.3 (a) Left scale: Algebraic transverse amplitude distribution R_N at $\hat{z} = 0$ according to Eq. 7.41, with $N = 2$; (b) right scale: Its potential function G_N given by Eq. 7.44.

and $\hat{x}_0 = 0.8$) and $R_N(\hat{x}, \hat{z} = 0)$ (with $N = 2$), each one represented together with its own "potential function":

$$G_{0,S,N}(\hat{x}, \hat{z} = 0) = \frac{1}{R_{0,S,N}} \frac{d^2 R_{0,S,N}}{d\hat{x}^2} \tag{7.44}$$

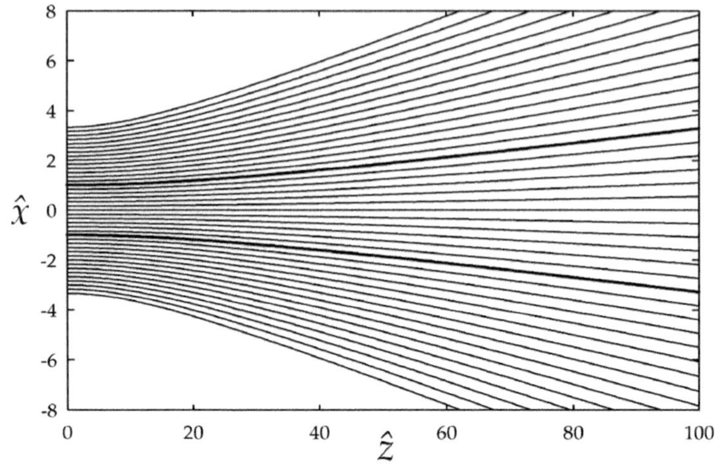

Figure 7.4 Trajectory pattern on the ($\hat{x}$, $\hat{z}$)-plane in the Gaussian case 7.39. The heavy lines are the waist rays of the beam given by Eq. 7.45.

Let us recall that the functions $G(\hat{x}, \hat{z} = 0)$ are dimensionless forms (with opposite sign) of the wave (or quantum) potential on the launching plane.

The ray (or particle) trajectories were obtained in our computations (for $\varepsilon \equiv \lambda_0/w_0 = 0.1$) by means of a *symplectic* numerical method (which is particularly apt to face Hamiltonian ray-tracing problems), and the potential functions $G(\hat{z} > 0)$ were built up step by step, together with their derivatives, by means of Lagrange interpolations and of Eq. 7.6b.

Figure 7.4 presents the trajectories obtained in the Gaussian case of Eq. 7.39. As predicted by the standard optical diffraction theory [11], no diffraction "fringe" is found in this case, due to the fact that the Fourier transform of the distribution R_0 consists of another Gaussian function. An excellent agreement is observed with the well-known relation

$$\hat{x}(\hat{z}) = \pm\sqrt{1 + \left(\frac{\varepsilon}{\pi}\right)^2 \hat{z}^2} \qquad (7.45)$$

holding, *in vacuo*, for the rays starting (at $\hat{z} = 0$) from $\hat{x} = \pm 1$, in correspondence with the (dimensioned) "waist" positions $x = \pm w_0$ which delimit the so-called paraxial part of the Gaussian beam [12].

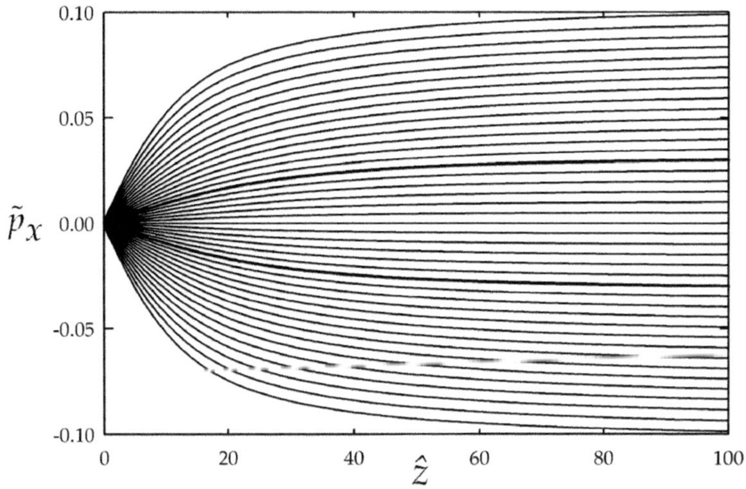

Figure 7.5 Plots versus $\hat{z}$ of the values assumed by the variable $\tilde{p}_x$ along the trajectories of Fig. 7.4. The heavy lines correspond to the paraxial trajectories.

These rays are represented in Fig. 7.4 (for $\varepsilon = 0.1$) by the heavy lines. A discussion of Eq. 7.45 is given in the appendix.

Figure 7.5 plots versus $\hat{z}$ the values of the variable $\tilde{p}_x$ corresponding to the Gaussian trajectories of Fig. 7.4. Each line is seen to be limited by a horizontal asymptote. Figures 7.6 and 7.7 present, respectively, in their turn, in arbitrary units, the transverse profiles of the beam intensity $\propto R^2$ and of the potential function G (computed at different times) corresponding to Fig. 7.4.

Figure 7.8 shows the beam trajectories starting with the amplitude distribution $R_S(\hat{x}; \hat{z} = 0)$ of Eq. 7.40, with $m = 2$ and $\hat{x}_0 = 0.8$. Figures 7.9 and 7.10 show the evolution, respectively, of the transverse profiles of the beam intensity and of the potential function corresponding to Fig. 7.8. A progressive formation of fringes is observed, occurring both later (in time) and further (in space) when the value of ε ($\equiv \lambda_0/w_0$) is decreased, as could be "a priori" expected from Eq. 7.42c.

Figures 7.11, 7.12, and 7.13 present the ray trajectories, the evolution of beam intensity and potential function, and the

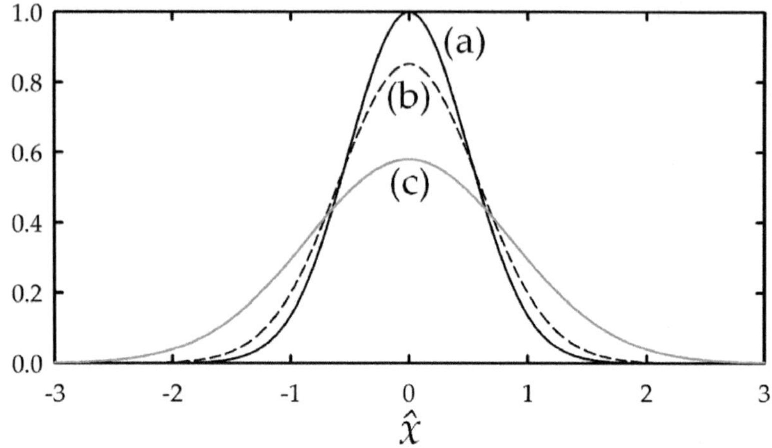

Figure 7.6 (a)–(c): Transverse beam intensity distributions of the beam corresponding to Fig. 7.4, at three successive times.

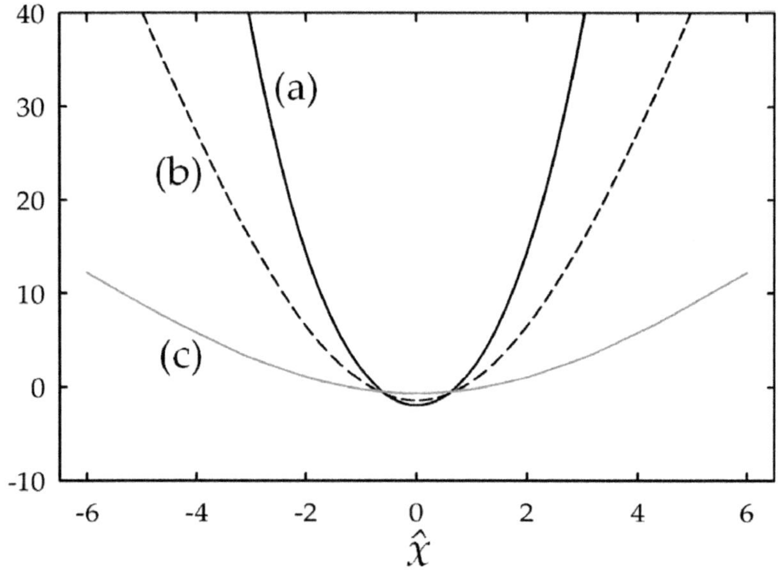

Figure 7.7 (a)–(c): Transverse profiles of the potential function G corresponding to Fig. 7.4, at three successive times.

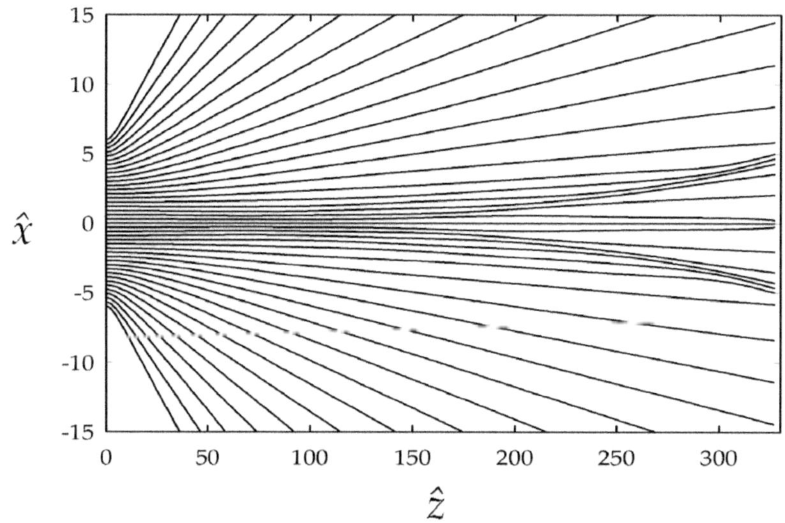

Figure 7.8 Trajectory pattern on the $(\hat{x}, \hat{z})$-plane in the case of Eq. 7.40, with $m = 2$ and $\hat{x}_0 = 0.8$.

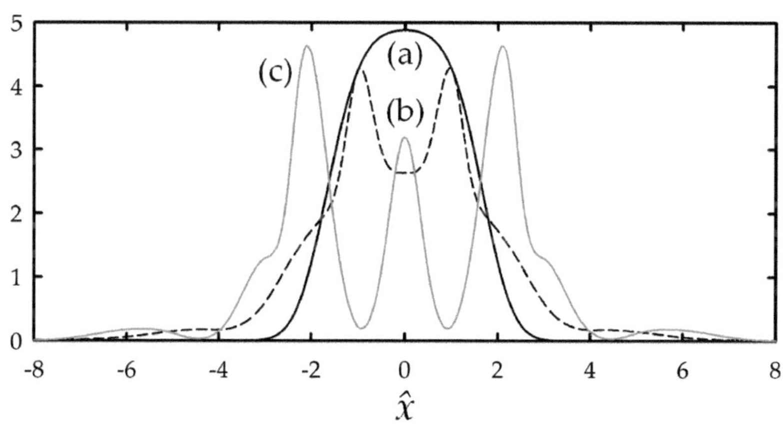

Figure 7.9 (a)–(c): Transverse intensity distributions of the beam corresponding to Fig. 7.8, at three successive times.

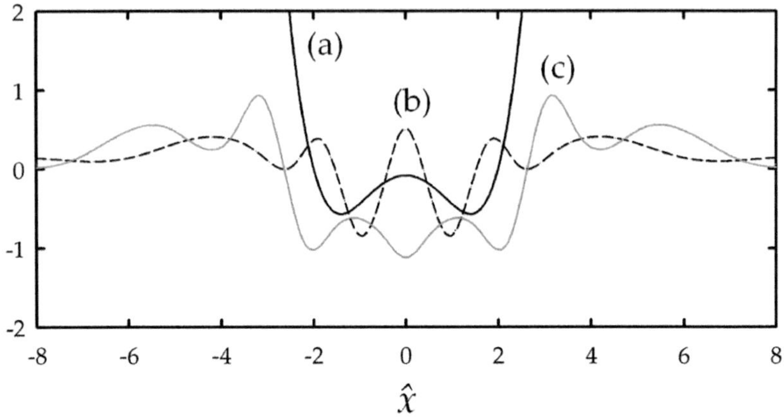

Figure 7.10 (a)–(c): Transverse plots of the potential function G corresponding to Fig. 7.8, at three successive times.

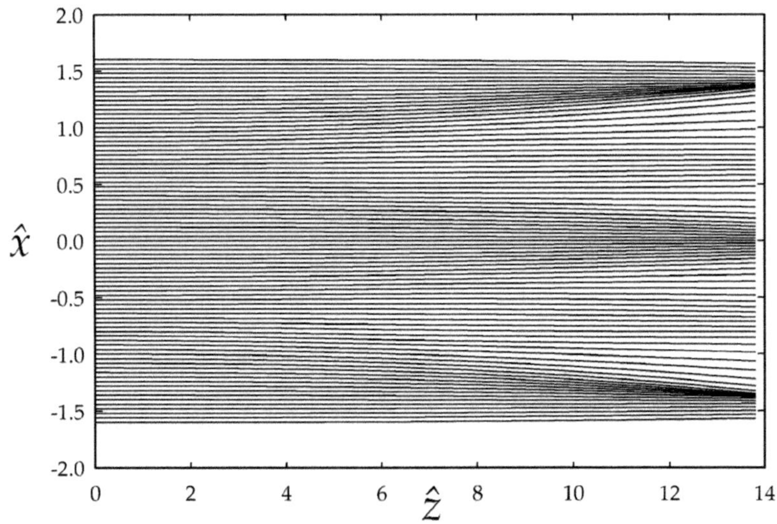

Figure 7.11 Trajectory pattern on the $(\hat{x}, \hat{z})$-plane in the algebraic case of Eq. 7.41, with $N = 2$.

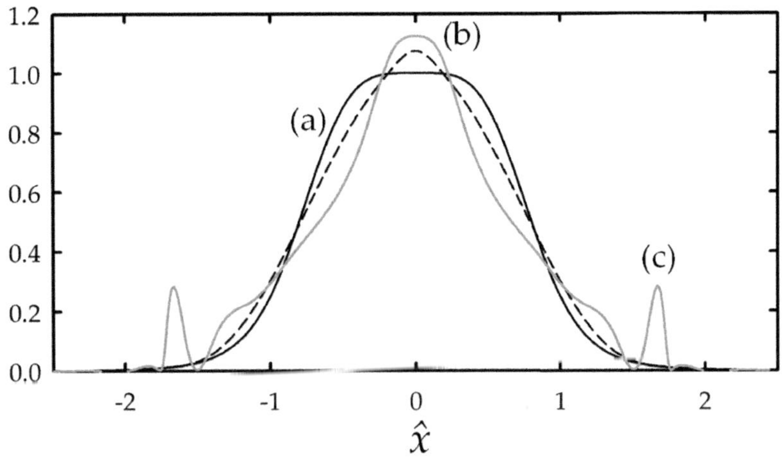

Figure 7.12 (a)–(c): Transverse intensity distributions of the beam corresponding to Fig. 7.11, at three successive times.

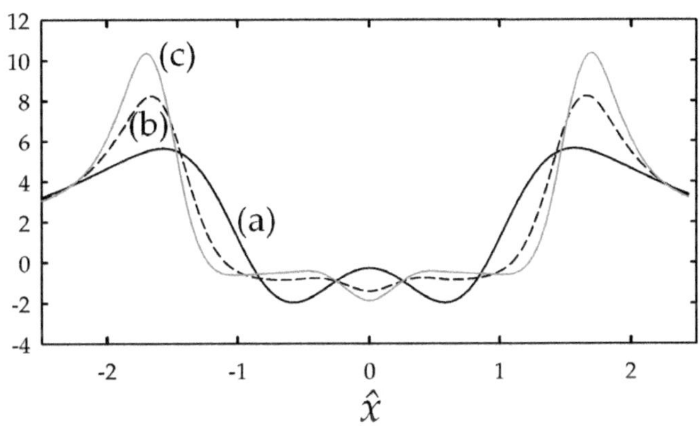

Figure 7.13 (a)–(c): Transverse plots of the potential function G corresponding to Fig. 7.11, at three successive times.

fringe formation obtained with the algebraic launching distribution $R_N(\hat{x}; \hat{z} = 0)$ of Eq. 7.41, with $N = 2$.

As an introduction to the problem of two-slits interference (to be thoroughly treated in a successive work), we finally present in Figs. 7.14, 7.15, and 7.16 the trajectory pattern and the evolution of

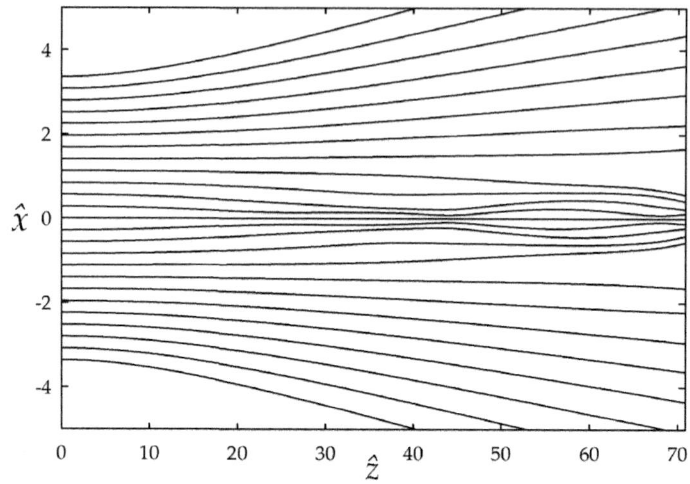

Figure 7.14 (a)–(c): Trajectory pattern on the ($\hat{x}$, $\hat{z}$)-plane, with $\hat{x}_0 = 1.2$, in the double-Gaussian case of Eq. 7.46.

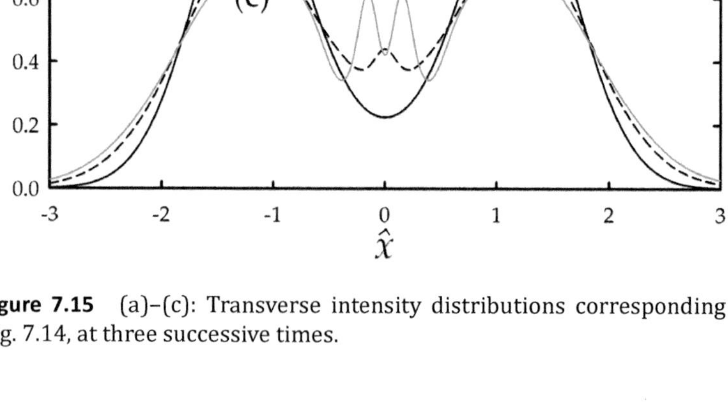

Figure 7.15 (a)–(c): Transverse intensity distributions corresponding to Fig. 7.14, at three successive times.

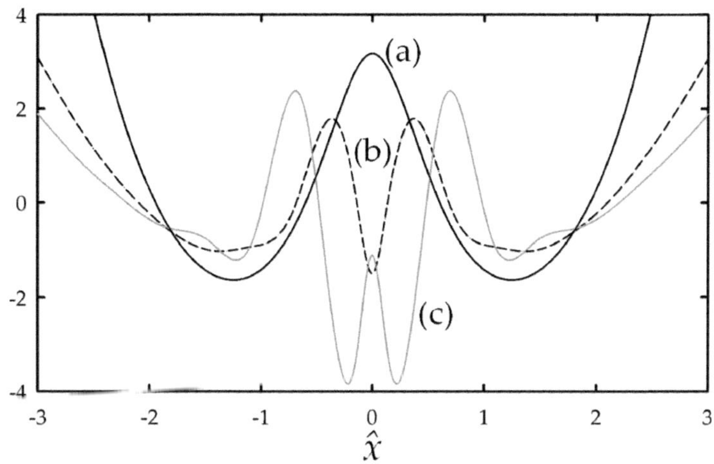

Figure 7.16 (a)–(c): Transverse plots of the potential function G corresponding to Fig. 7.14, at three successive times.

the wave intensity and potential, respectively, for a beam launched in the double-Gaussian form.

$$R(\hat{x}; \hat{z} = 0) \propto \left[e^{-(\hat{x}-\hat{x}_0)^2} + e^{-(\hat{x}+\hat{x}_0)^2} \right], \tag{7.46}$$

(with $\hat{x}_0 = 1.2$, small enough to allow a partial overlapping of the two Gaussians), showing the gradual formation of a central fringe. We may observe, once more, the striking mutual feedback between wave potential and wave intensity distribution: while, for instance, no trajectory may cross the symmetry axis, $\hat{z}$, of the system, a typical "sausage" shape around this axis is assumed by the trajectory set.

7.8 Discussion and Conclusions

As we already said, the main purpose of the present chapter is to provide an *exact, ray-based* Hamiltonian treatment of wave-like features, general enough to hold both for classical and quantum waves. Referring, in particular, to the quantum case, we remind that the subject of quantum trajectories was faced by de Broglie, Bohm and co-workers (see Refs. [1–5] and [13–16]) by representing particles, as is usual in Quantum Mechanics, as probabilistic *wave-packets* solving the *time-dependent* Schrödinger equation. An analogous line of thought, which appears to minimize difficulties, is developed in Refs. [17–19], where individual trajectories are suggested by the fluid-like stream-lines of the probability current density. A different point of view is adopted in Refs. [20, 21], where a time-independent eikonal approximation is employed for nuclear trajectories, coupling a classical molecular motion with quantum electron transitions. Finally, a ray-based *quasi-optical approximation* (based on a complex eikonal equation) was employed in classical, respectively, and quantum contexts, in Refs. [22, 23] and [24].

In the present work we avoid *any kind of approximation* by exploiting the general properties of Helmholtz-like equations, which are shown to encode a Hamiltonian system providing, for classical waves, the full *kinematics* (both rails and timetable) of exact ray-trajectories, mutually coupled by a Wave Potential function acting *orthogonally* to the ray velocities. The theory is extended (thanks to the Helmholtz-like nature of the *time-independent* Schrödinger equation) to de Broglie's waves, and provides an *exact* Hamiltonian *point-particle dynamics*, where a *mono-energetic* Wave Potential, acting orthogonally to the relevant particle trajectories, is seen to pilot them *without any wave-particle energy exchange*. This *mono-energetic* property cannot be enjoyed by the Bohmian "Quantum Potential" function, which involves *the entire energy spectrum* of the wave-packets *assumed* to represent the moving particles. It's important to notice, for comparison, that in our own approach *no further assumption* is added to the *original* de Broglie-Schrödinger discovery of the wave-particle duality: it's simply based on the application of exact, *standard* mathematics to *standard* Schrödinger equations. The main difference between our approach and the Bohmian one is due to the difference between our Wave Potential, involving classical-looking *point-particles*, and Bohm's "Quantum Potential," based on a *wave-packet* particle model. This difference is made dramatic by the fact that the use of wave-packets is both responsible for most "quantum paradoxes" [25] and subject to Borns well-known objection [26] that *"it's tempting, for us, to try to interpret a particle of matter as a wave-packet due to the superposition of a number of wave trains. This tentative interpretation, however, comes up against insurmountable difficulties, since a wave-packet of this kind is in general very soon dissipated".*

Referring the reader to Refs. [27–31] for a discussion of the point-particle and wave-packet models, we remind, concerning one of the most popular quantum paradoxes, that when, during a quantum diffraction experiment, the launching conditions are modified, the motion of each particle is declared, *in the Bohmian approach*, to be instantaneously (i.e.non-locally) affected by this change, thanks to the ubiquitous properties of the *Quantum Potential* [25]. According to *our approach*, on the contrary, one would simply say that the any information is carried along by the rays, with their characteristic velocity: every wave-front may be viewed, in fact, as a new launching surface, where the transverse distribution of the *Wave Potential* provides the necessary and sufficient boundary conditions determining—in a quite local way—the subsequent course of the ray trajectories. Recalling, therefore, that (as we have shown) the trajectories are exactly the same for classical and matter waves, we are induced to think that *the same interpretation* should hold in both cases, so that a non-local behavior should be either a property characterizing classical as well as quantum cases, or (to say the least) an avoidable concept.

Acknowledgements

Thanks are due to Prof. M. Cocucci, director of the Dipartimento di Produzione Vegetale (Di.Pro.Ve.) of Milan University, for his kind interest, generous hospitality, and logistic support.

A.1 Appendix: The Paraxial Approach

Although, as we have seen, any kind of approximation may be avoided, in principle, by means of our exact Hamiltonian treatment, we mention here the remarkable approximation allowed by the so-called "paraxial approach" [12] for a collimated beam launched, *in vacuo*, along the z-axis with $k_x = 0$ in the Gaussian form

$$R(x; y; z = 0) \propto e^{-\frac{x^2+y^2}{w_0^2}} \tag{A.1}$$

with constant w_0. The circle $x^2 + y^2 = w_0^2$ represents, on the plane $z = 0$, the so-called waist line of the beam. The rays starting from the points $x = \pm w_0$ of the waist line turn out to be well described, on the (x, z)-plane, by the (*dimensioned*) relation:

$$x(z) = \pm\sqrt{w_0^2 + \frac{4z^2}{k_0^2 w_0^2}}. \tag{A.2}$$

Equation 7.45 of the main text represents Eq. A.2 in *dimensionless* form, expressing lengths in terms of w_0. Eq. A.2 is well approximated, for $z^2 \gg k_0^2 w_0^4$, by the two straight lines (starting from the centre of the diffracting slit):

$$x(z) \simeq \pm\frac{2z}{k_0 w_0}. \tag{A.3}$$

It is easily verified that Eq. A.3 leads to the relation

$$\left|2k_0\frac{x}{z}\right|(2w_0) \simeq 8, \tag{A.4}$$

showing that a ray of the beam, starting with $k_x = 0$ from an unknown position $x(z = 0)$ in the launching slit, and therefore with an *uncertainty* on the x-coordinate given by $\Delta x \simeq 2w_0$, shall develop, along its trajectory, a k_x component of the wave vector endowed with an uncertainty $\Delta k_x \simeq |2k_0 x/z|$, such that

$$\Delta x \Delta k_x \simeq 8. \tag{A.5}$$

Equation A.5 leads, in its turn, by means of the replacements 7.21 of the main text, to the *quantum* relation

$$\Delta x \Delta p_x \simeq 8\hbar, \tag{A.6}$$

whose interpretation is obviously the same required by Eq. A.5 itself, i.e., that of a simple lack of information and not of an intrinsic uncertainty of the physical reality.

References

1. L. de Broglie, *La mécanique ondulatoire et la structure atomique de la matière et du rayonnement*, Journal de Physique et Le Radium, **8**, 225 (1927).
2. L. de Broglie, *Sur l'interprtation de la mécanique ondulatoire à l'aide d'ondes à régions singuliéres*, Comptes rendus de l'Academie des Sciences, **236**, 1459 (1953).
3. L. de Broglie, *Une tentative d'interprétation causale et non-linéaire de la mécanique ondulatoire* (Gauthier-Villars, 1956).
4. D. J. Bohm, *A Suggested interpretation of the Quantum Theory in Terms of "hidden variables," I and II*, Physical Review, **85**, 166 and 180 (1952).
5. D. J. Bohm, *Reply to a Criticism of a Causal Re-Interpretation of the Quantum Theory*, Physical Review, **87**, 389 (1952).
6. A. Orefice, R. Giovanelli, and D. Ditto, *Complete Hamiltonian Description of Wave-Like Features in Classical and Quantum Physics*, Foundations of Physics, **39**, 256 (2009).
7. H. Goldstein, *Classical Mechanics* (Addison-Wesley, 1965).
8. A. Airoldi, A. Orefice, and G. Ramponi, *E. C. W. Propagation in Axisymmetric Plasmas with General Equilibrium Surfaces*, Il Nuovo Cimento D, **6** , 527 (1985).
9. A. Airoldi, A. Orefice, and G. Ramponi, *Polarization and Energy Evolution of Electromagnetic Waves is Sheared Toroidal Plasmas*, Physics of Fluids B, **11**, 2143 (1989).
10. A. Messiah, *Mécanique Quantique* (Dunod, 1959).
11. G. R. Fowles, *Introduction to Modern Optics* (Dover, 1975).
12. A. T. Friberg, T. Jaakkola, and J. Tuovinen, *Electromagnetic Gaussian Beams Beyond the Paraxial Regime*, IEEE Transactions on Antennas Propagation, **40**, 984 (1992).
13. C. Philippidis, D. Dewdney, and B. J. Hiley , *Quantum Interference and the Quantum Potential*, Il Nuovo Cimento B, **52**, 15 (1979).
14. D. J. Bohm and B. J. Hiley, *The de Broglie Pilot Wave Theory and the Further Development of New Insights Arising Out of It*, Foundations of Physics, **12**, 1001 (1982).
15. P. R. Holland, *The Quantum Theory of Motion* (Cambridge University Press, 1993).
16. P. R. Holland, *Computing the Wavefunction from rtajectories: Particle and Wave Pictures in Quantum Mechanics and Their Relation*, Annals of Physics, **315**, 505 (2005).

17. A. S. Sanz, F. Borondo, and S. Miret-Artés, *Causal Trajectories Description of Atom Diffraction by Surfaces*, Physical Review B, **61**, 7743 (2000).
18. A. S. Sanz, F. Borondo, and S. Miret-Artés, *Quantum Trajectories in Atom-Surface Scattering with Single Adsorbates: The Role of Quantum Vortices*, Journal of Chemical Physics, **120**, 8794 (2004).
19. A. S. Sanz and S. Miret-Artés, *Aspects of Nonlocality from a Quantum Trajectory Perspective: A WKB Approach to Bohmian Mechanics*, Chemical Physics Letters, **445**, 350 (2007).
20. D. A. Micha, *A Self-Consistent Eikonal Treatment of Electronic Transitions in Molecular Collisions*, Journal of Chemical Physics, **78**, 7138 (1983).
21. J. M. Cohen and D. A. Micha, *Angular Distributions in Electronically Adiabatic Hyperthermal Collisions. An Eikonal Approach*, Journal of Chemical Physics, **98**, 2023 (1993).
22. S. Nowak and A. Orefice, *Quasioptical Treatment of Electromagnetic Gaussian Beams in Inhomogeneous and Anisotropic Plasmas*, Physics of Fluids B, **5**, 1945 (1993).
23. S. Nowak and A. Orefice, *Three-Dimensional Propagation and Absorption of High Frequency Gaussian Beams in Magnetoactive Plasmas*, Physics of Plasmas, **1**, 1242 (1994).
24. A. Orefice and S. Nowak, *Hamiltonian Approach to Electron Diffraction*, Physics Essays, **10**, 364 (1997).
25. D. J. Bohm, *Wholeness and the Implicate Order* (Routledge & Kegan Paul, 1980).
26. M. Born, *Atomic Physics* (Blackie & Son Ltd., 1935, VI edition,1958).
27. A. Orefice, R. Giovanelli, and D. Ditto, *A Non-probabilistic Insight into Wave Mechanics*, Annales de la Fondation L. de Broglie, **38**, 7 (2013)
28. A. Orefice, R. Giovanelli, and D. Ditto, *From Classical to Wave-Mechanical Dynamics*, Annales de la Fondation L. de Broglie, **40**, 95 (2015)
29. A. Orefice, R. Giovanelli, and D. Ditto, *Is Wave Mechanics Consistent with Classical Logic?* Physics Essays, **28**, 515 (2015).
30. A. Orefice, R. Giovanelli, and D. Ditto, *The Dynamics of Wave-Particle Duality*, Journal of Applied Mathematics and Physics, **6**, 1840–1859 (2018).
31. A. Orefice, R. Giovanelli, and D. Ditto, *Primary Assumptions and Guidance Laws in Wave Mechanics*, Journal of Applied Mathematics and Physics, **6**, 2621 (2018).

Chapter 8

Relativistic Quantum Mechanics and Quantum Field Theory

Hrvoje Nikolić
Theoretical Physics Division, Rudjer Bošković Institute,
P.O. Box 180, HR-10002 Zagreb, Croatia
hnikolic@irb.hr

A general formulation of classical relativistic particle mechanics is presented, with an emphasis on the fact that superluminal velocities and nonlocal interactions are compatible with relativity. Then a manifestly relativistic-covariant formulation of relativistic quantum mechanics (QM) of fixed number of particles (with or without spin) is presented, based on many-time wave functions and the space-time probabilistic interpretation. These results are used to formulate the Bohmian interpretation of relativistic QM in a manifestly relativistic-covariant form. The results are also generalized to quantum field theory (QFT), where quantum states are represented by wave functions depending on an infinite number of space-time coordinates. The corresponding Bohmian interpretation of QFT describes an infinite number of particle trajectories. Even though the particle trajectories are continuous, the appearance of creation and destruction of a finite number of particles results from the

Applied Bohmian Mechanics: From Nanoscale Systems to Cosmology (Second Edition)
Edited by Xavier Oriols and Jordi Mompart

ISBN 978-981-4800-10-5 (Hardcover), 978-0-429-29474-7 (eBook)
www.jennystanford.com

quantum theory of measurements describing entanglement with particle detectors.

8.1 Introduction

The following chapter somewhat differs from the previous ones, in the sense that this chapter does not deal with an application to a specific practical physical problem. Instead, the main goal of this chapter is to develop a *generalized formulation* of Bohmian mechanics such that effects of relativistic QM and QFT can also be incorporated into it.

Since this chapter deals with a general formulation of the theory, the practical utility of it may not be obvious. Nevertheless, we believe that the results of this chapter may lead to practical applications as well. For example, many physicists argue that the most practical result that emerged from the original Bohm reformulation of QM was the famous Bell theorem that revealed fundamental nonlocal nature of QM.[a] Bell's result is valid independently on validity of the Bohm reformulation, but to obtain this result, Bell was significantly guided and inspired by the Bohm reformulation in which nonlocality of QM is particularly manifest.[b] In a similar way, although our primary motivation lying behind the results of this chapter is to make the Bohmian formulation compatible with relativity and QFT, this motivation led us to some new fundamental results on relativistic QM and QFT valid even without the Bohmian interpretation. In our quest towards relativistic Bohmian mechanics, as a by-product we realize that even non-Bohmian relativistic QM

---

[a]Some even argue that the Bell theorem is the most important discovery in physics of the 20th century.

[b]Today many physicists still misinterpret Bell theorem as a proof that the Bohmian interpretation cannot be right. It cannot be overemphasized that just the opposite is true. Bell theorem proves that no *local* hidden variable theory can be compatible with QM, so that any hidden variable theory compatible with QM must necessarily be *nonlocal*. Since the Bohmian interpretation is a nonlocal hidden variable theory, Bell theorem gives further credit to it. Indeed, Bell himself had a very positive opinion on the Bohmian interpretation and significantly contributed to the popularization of it.

and QFT should be first made "more relativistic" than they are in the usual formulation, i.e., that time and space should be treated more symmetrically. First, the usual single-time wave function should be generalized to the many-time wave function, such that each particle has its own space-time coordinate. Second, $|\psi|^2$ should be reinterpreted as a probability density in space-time, rather than that in space. Eventually, this by-product may turn out to be even more useful than the relativistic Bohmian formulation itself.[a]

The primary motivation lying behind this chapter has very much to do with nonlocality of QM. One of the most frequent questions related to nonlocality is, how can it be compatible with relativity? If entangled particles communicate instantaneously, is it in contradiction with the relativistic rule that no information can travel faster than light? Since communication instantaneous in one Lorentz frame cannot be instantaneous in any other Lorentz frame, does it mean that there exists a preferred Lorentz frame which violates (the spirit of) relativity? In this chapter we offer an answer to these and many other questions regarding relativity, nonlocality, QM, and Bohmian mechanics. In particular, by developing the Bohmian interpretation of QFT, we also explain how continuous particle trajectories can be made compatible with phenomena of particle creation and destruction.

Of course, due to the lack of space, this chapter is not intended to be a general pedagogic introduction to relativistic QM and QFT. Instead, we assume that the reader is already familiar with some basics of those, as well as with some basics of special relativity in classical mechanics. (A knowledge of some basics of general relativity may also be useful, but is not necessary.) With these assumptions, we can pay more attention to aspects that are not widely known to experts in relativistic QM and QFT.

Our presentation is to a large extent based on the papers [3–6]. Some results from the papers [7–11] are also used.

---

[a]For example, the many-time formalism with space-time probability density can be used to avoid the black-hole information paradox [1, 2].

8.2 Classical Relativistic Mechanics

8.2.1 Kinematics

Our point of departure is a four-dimensional space-time with coordinates x^μ, $\mu = 0, 1, 2, 3$, and the Minkowski metric $\eta_{\mu\nu}$, where $\eta_{00} = 1, \eta_{ij} = -\delta_{ij}, \eta_{0i} = 0$, for $i = 1, 2, 3$. We work in units in which the velocity of light is $c \equiv 1$. At some places we also use the notation $x^\mu = (t, \mathbf{x})$, where $t = x^0$ is the time coordinate and $\mathbf{x} = (x^1, x^2, x^3)$ represents the space coordinates.

The physical objects that we study are particles living in space-time. By a particle we mean a material point in space. More precisely, since the concept of space is not a well-defined entity in relativity, a better definition of a particle is a curve in space-time. Thus, the particle is a one-dimensional object living in the four-dimensional space-time.

The simplest way to specify a curve is through a set of four equations:

$$x^\mu = X^\mu(s), \tag{8.1}$$

where s is an auxiliary real parameter and $X^\mu(s)$ are some specified functions of s. Each s defines one point on the curve and the set of all values of s defines the whole curve. In this sense, the curve can be identified with the functions $X^\mu(s)$.

The parameter s is a scalar with respect to Lorentz transformations or any other transformations of the space-time coordinates x^μ. In this sense, the parametric definition of the curve (8.1) is covariant. However, noncovariant definitions are also possible. For example, if the function $X^0(s)$ can be inverted, then the inverse $s(X^0)$ can be plugged into the space components $X^i(s(X^0)) \equiv \tilde{X}^i(X^0)$. This leads to the usual nonrelativistic view of the particle as an object with the trajectory $x^i = \tilde{X}^i(X^0)$, where X^0 is time.

A priori, the auxiliary parameter s does not have any physical interpretation. It is merely a mathematical parameter that cannot be measured. In fact, a transformation of the form

$$s \to s' = f(s) \tag{8.2}$$

does not change the curve in space-time.[a] This means that the functions $X^\mu(s)$ and $\tilde{X}^\mu(s) \equiv X^\mu(f(s))$ represent the same curve.

---

[a]The only restriction on the function $f(s)$ is that $df(s)/ds > 0$.

Since the curve is a one-dimensional manifold, the parameter s can be viewed as a coordinate on that manifold. The transformation (8.2) is a coodinate transformation on that manifold. One can also define the metric tensor $h(s)$ on that manifold, such that $h(s)ds^2$ is the (squared) infinitesimal length of the curve. Since the manifold is one-dimensional, the metric tensor h has only 1 component. It is important to stress that this is an intrinsic definition of the length of the curve that may be defined completely independently on the space-time metric $\eta_{\mu\nu}$. This intrinsic length is not measurable, so one can freely choose the metric $h(s)$. However, once $h(s)$ is chosen, the metric in any other coordinate s' is defined through

$$h(s)ds^2 = h'(s')ds'^2. \tag{8.3}$$

We say that the curve at a point s is timelike if the space-time vector tangent to the curve at this point is timelike. Spacelike and lightlike parts of the curve are defined analogously. Thus, the part of the curve is timelike if $\dot{X}^\mu \dot{X}_\mu > 0$, spacelike if $\dot{X}^\mu \dot{X}_\mu < 0$, and lightlike if $\dot{X}^\mu \dot{X}_\mu = 0$, where $\dot{X}^\mu = dX^\mu(s)/ds$.[a] A timelike trajectory describes a particle that moves slower than light, a lightlike trajectory describes a particle that moves with the velocity of light, and a spacelike trajectory describes a particle that moves faster than light. Contrary to what one might expect, we see that relativistic kinematics allows particles to move even faster than light. As we shall see in the next subsection, it is relativistic dynamics that may (or may not!) forbid motions faster than light, depending on details of the dynamics.

For a timelike trajectory, there exists one special choice of the parameter s. Namely, one can choose it to be equal to the proper time τ defined by

$$d\tau^2 = dX^\mu dX_\mu. \tag{8.4}$$

For such a choice, we see that

$$\dot{X}^\mu \dot{X}_\mu = 1. \tag{8.5}$$

In this case it is convenient to choose the metric on the trajectory such that $h(\tau) = 1$, so that the intrinsic length of the curve coincides with the proper time, which, by definition, is equal to the extrinsic

---

[a]Here $A^\mu B_\mu \equiv \eta_{\mu\nu} A^\mu B^\nu$ and the summation over repeated vector indices μ, ν is understood.

length defined by the space-time metric $\eta_{\mu\nu}$. Yet, such a choice is by no means necessary.

Finally, let us briefly generalize the results above to the case of many particles. If there are n particles, then they are described by n trajectories $X_a^{\mu}(s_a)$, $a = 1, \ldots, n$. Note that each trajectory is parameterized by its own parameter s_a. However, since the parameterization of each curve is arbitrary, one may parameterize all trajectories by the same parameter s, so that the trajectories are described by the functions $X_a^{\mu}(s)$. In fact, the functions $X_a^{\mu}(s)$, which describe n curves in the four-dimensional space-time, can also be viewed as *one* curve on a $4n$-dimensional manifold with coordinates x_a^{μ}.

8.2.2 Dynamics

8.2.2.1 Action and equations of motion

Dynamics of a relativistic particle is described by an action of the form

$$A = \int ds\, L(X(s), \dot{X}(s), s), \tag{8.6}$$

where $X \equiv \{X^{\mu}\}$, $\dot{X} \equiv \{\dot{X}^{\mu}\}$. We require that the Lagrangian L should be a scalar with respect to space-time coordinate transformations. This means that all space-time indices μ must be contracted. We also require that the action should be invariant with respect to reparameterizations of the form of (8.2). From (8.3), we see that this implies that any ds should by multiplied by $\sqrt{h(s)}$, because such a product is invariant with respect to (8.2). To restrict the dependence on s as much as possible, we assume that there is no other explicit dependence on s except through the dependence on $h(s)$. To further restrict the possible forms of the action, we require that L should be at most quadratic in the velocities $\dot{X}^{\mu}(s)$. With these requirements, the most general action can be written in the form

$$A = -\int ds\, \sqrt{h(s)} \left[\frac{1}{2h(s)} \frac{dX^{\mu}}{ds} \frac{dX^{\nu}}{ds} C_{\mu\nu}(X) + \frac{1}{\sqrt{h(s)}} \frac{dX^{\mu}}{ds} C_{\mu}(X) + C(X) \right]. \tag{8.7}$$

The functions $C(X)$, $C_\mu(X)$, and $C_{\mu\nu}(X)$ are referred to as scalar potential, vector potential, and tensor potential, respectively.

What is the dynamical role of the function $h(s)$? Requiring that $h(s)$ is a dynamical variable, the dynamical equation of motion $\delta A/\delta h(s) = 0$ leads to

$$h^{-1}C_{\mu\nu}(X)\dot{X}^\mu\dot{X}^\nu = 2C(X). \tag{8.8}$$

Viewed as an equation for h, it can be trivially solved as $h = C_{\mu\nu}\dot{X}^\mu\dot{X}^\nu/2C$. However, since h is not a physical quantity, this solution does not bring an important physical information. Nevertheless, Eq. (8.8) does play an important physical role, as we shall see soon.

Equation (8.8) determines $h(s)$ only when the coordinate s is chosen. Thus, $h(s)$ can still be changed by changing the coordinate. In particular, from (8.3) we see that the coordinate transformation of the form $s'(s) = \text{const} \int ds\, \sqrt{h(s)}$ makes $h'(s')$ a constant. Thus, omitting the prime, we can fix $\sqrt{h(s)} = m^{-1}$, where m is a positive constant. For convenience, we choose s to have the dimension of time and $C_{\mu\nu}$ to be dimensionless. Then the action (8.7) implies that m has the dimension of mass (recall that we work in units $c = 1$). Hence, we can rewrite (8.7) as

$$A = -\int ds \left[\frac{m}{2}C_{\mu\nu}(X)\dot{X}^\mu\dot{X}^\nu + C_\mu(X)\dot{X}^\mu + \frac{C(X)}{m}\right]. \tag{8.9}$$

Now m is no longer a dynamical quantity, but Eq. (8.8) rewritten as

$$C_{\mu\nu}(X)\dot{X}^\mu\dot{X}^\nu = \frac{2C(X)}{m^2} \tag{8.10}$$

should be added to (8.9) as an additional constraint.

Now we are ready to study the physical role of the potentials C, C_μ and $C_{\mu\nu}$. By writing $C_\mu(x) \equiv eA_\mu(x)$, one recognizes that the second term in (8.9) looks just like the action for the particle with the charge e moving under the influence of the external electromagnetic potential $A_\mu(x)$ (see, e.g., [12]). Similarly, by writing $C_{\mu\nu}(x) \equiv g_{\mu\nu}(x)$, one recognizes that the first term in (8.9) looks just like the action for the particle moving in a gravitational background described by the curved metric tensor $g_{\mu\nu}(x)$ (see, e.g., [13]). Since the physical properties of electromagnetic and gravitational forces are well known, we shall not study them in further discussions. Instead, from now on we assume $C_\mu(x) = 0$, $C_{\mu\nu}(x) = \eta_{\mu\nu}$.

Therefore, introducing the notation $U(X) \equiv C(X)/m$, Eqs. (8.9) and (8.10) reduce to

$$A = -\int ds \left[\frac{m}{2}\dot{X}^{\mu}\dot{X}_{\mu} + U(X)\right], \tag{8.11}$$

$$\dot{X}^{\mu}\dot{X}_{\mu} = \frac{2U(X)}{m}. \tag{8.12}$$

We see that the scalar potential $U(X)$ has the dimension of energy. The dynamical equation of motion for $X^{\mu}(s)$ is $\delta A/\delta X^{\mu}(s) = 0$. Applying this to (8.11), one obtains a relativistic Newton equation:

$$m\frac{d^2X^{\mu}(s)}{ds^2} = \partial^{\mu}U(X(s)), \tag{8.13}$$

where $\partial^{\mu} \equiv \eta^{\mu\nu}\partial/\partial X^{\nu}$. The constraint (8.12) is compatible with (8.13). Indeed, by applying the derivative d/ds on (8.12), one obtains

$$[m\ddot{X}^{\mu} - \partial^{\mu}U(X)]\dot{X}_{\mu} = 0, \tag{8.14}$$

which is consistent because the expression in the square bracket trivially vanishes when (8.13) is satisfied.

The constraint (8.12) implies that the sign of $\dot{X}^{\mu}\dot{X}_{\mu}$ is equal to the sign of U. Thus, we see that the particle moves slower than light if $U > 0$, with the velocity of light if $U = 0$, and faster than light if $U < 0$. Since $U(X)$ may change sign as X varies, we see that the particle may, e.g., start motion with a velocity slower than light and *accelerate to a velocity faster than light.*

At first sight, one may think that acceleration to velocities faster than light is in contradiction with the well-known "fact" that the principle of relativity does not allow particles to accelerate to velocities faster than light. However, there is no contradiction because this well-known "fact" is valid only if some additional assumptions are fulfilled. In particular, if all forces on particles are either of the electromagnetic type (vector potential) or of the gravitational type (tensor potential), then acceleration to velocities faster than light is forbidden. Indeed, as far as we know, all relativistic classical forces on particles that exist in nature are of those two types. Nevertheless, the principle of relativity allows also relativistic forces based on the scalar potential, which, as we have seen, does allow acceleration to velocities faster than light. Such classical forces have not yet been found in nature, but it does not imply that they are forbidden. More precisely, they may be forbidden by some additional physical principle taken together with the principle of relativity, but they are *not forbidden by the principle of relativity alone.*

8.2.2.2 Canonical momentum and the Hamilton–Jacobi formulation

Physics defined by (8.11)–(8.12) can also be described by introducing the canonical momentum:

$$P_\mu = \frac{\partial L}{\partial \dot{X}^\mu}, \tag{8.15}$$

where

$$L(X, \dot{X}) = -\frac{m}{2}\dot{X}^\mu \dot{X}_\mu - U(X). \tag{8.16}$$

This leads to

$$P^\mu = -m\dot{X}^\mu. \tag{8.17}$$

The canonical Hamiltonian is

$$H(P, X) = P_\mu \dot{X}^\mu - L = -\frac{P^\mu P_\mu}{2m} + U(X). \tag{8.18}$$

Note that this Hamiltonian is *not* the energy of the particle. In particular, while particle energy transforms as a time component of a space-time vector, the Hamiltonian above transforms as a scalar. This is a consequence of the fact $\dot{X}^\mu$ is not a derivative with respect to time x^0, but a derivative with respect to the scalar s.

The constraint (8.12) now can be written as

$$P^\mu P_\mu = 2mU(X). \tag{8.19}$$

In relativity, it is customary to *define* the invariant mass M through the identity $P^\mu P_\mu \equiv M^2$. This shows that the mass depends on X as

$$M^2(X) = 2mU(X). \tag{8.20}$$

Since $U(X)$ may change sign as X varies, we see that the particle may, e.g., start motion as an "ordinary" massive particle ($M^2 > 0$) and later evolve into a tachyon ($M^2 < 0$). The usual proof that an ordinary particle cannot reach (or exceed) the velocity of light involves an assumption that the mass is a constant. When mass is not a constant, or more precisely when M^2 can change sign, then the particle can reach and exceed the velocity of light.

The existence of the Hamiltonian allows us to formulate classical relativistic mechanics with the relativistic Hamilton–Jacobi formalism. One introduces the scalar Hamilton–Jacobi function $S(x, s)$ satisfying the Hamilton–Jacobi equation:

$$H(\partial S, x) = -\frac{\partial S}{\partial s}. \tag{8.21}$$

Comparing (8.19) with (8.18), we see that the constraint (8.19) can be written as

$$H(P, X) = 0. \tag{8.22}$$

The constraint (8.22) implies that the right-hand side of (8.21) must vanish, i.e., that $S(x, s) = S(x)$. Hence (8.21) reduces to $H(\partial S, x) = 0$, which in an explicit form reads

$$-\frac{(\partial^\mu S)(\partial_\mu S)}{2m} + U(x) = 0. \tag{8.23}$$

The solution $S(x)$ determines the particle momentum

$$P^\mu = \partial^\mu S(X), \tag{8.24}$$

which, through (8.17), determines the particle trajectory

$$\frac{dX^\mu(s)}{ds} = -\frac{\partial^\mu S(X(s))}{m}. \tag{8.25}$$

8.2.2.3 Generalization to many particles

Now, let us briefly generalize all this to the case of many particles. We study the dynamics of n trajectories $X_a^\mu(s)$, $a = 1, \ldots, n$, parameterized by a single parameter s. In the general action (8.7), the velocity-dependent terms generalize as follows:

$$\dot{X}^\mu C_\mu \to \sum_{a=1}^{n} \dot{X}_a^\mu C_{a\mu}, \tag{8.26}$$

$$\dot{X}^\mu \dot{X}^\nu C_{\mu\nu} \to \sum_{a=1}^{n} \sum_{b=1}^{n} \dot{X}_a^\mu \dot{X}_b^\nu C_{ab\mu\nu}. \tag{8.27}$$

Since the scalar potential is our main concern, we consider trivial vector and tensor potentials $C_{a\mu} = 0$ and $C_{ab\mu\nu} = c_a \delta_{ab} \eta_{\mu\nu}$, respectively, where c_a are constants. Thus, Eqs. (8.11)–(8.12) generalize to

$$A = -\int ds \left[\sum_{a=1}^{n} \frac{m_a}{2} \dot{X}_a^{\mu} \dot{X}_{a\mu} + U(X_1, \ldots, X_n) \right], \tag{8.28}$$

$$\sum_{a=1}^{n} m_a \dot{X}_a^{\mu} \dot{X}_{a\mu} = 2U(X_1, \ldots, X_n), \tag{8.29}$$

where c_a are dimensionless and $m_a = mc_a$. The relativistic Newton equation (8.13) generalizes to

$$m_a \frac{d^2 X_a^{\mu}(s)}{ds^2} = \partial_a^{\mu} U(X_1(s), \ldots, X_n(s)). \tag{8.30}$$

In general, from (8.30) we see that the force on the particle a at the space-time position $X_a(s)$ depends on positions of all other particles for the same s. In other words, the forces on particles are nonlocal. Nevertheless, since s is a scalar, such nonlocal forces are compatible with the principle of relativity; the nonlocal equation of motion (8.30) is relativistic covariant. Thus we see that *relativity and nonlocality are compatible with each other*. Even though for each s there may exist a particular (s-dependent) Lorentz frame with respect to which the force between two particles is instantaneous, such a Lorentz frame is by no means special or "preferred." Instead, such a particular Lorentz frame is determined by covariant equations of motion supplemented by a particular choice of initial conditions $X_a^{\mu}(0)$. (Of course, the initial velocities $\dot{X}_a^{\mu}(0)$ also need to be chosen for a solution of (8.30), but the initial velocities can be specified in a covariant manner through the equation (8.34) below.)

Note also that the phenomena of *nonlocal forces between particles* and *particle motions faster than light* are independent of each other. The force (8.30) becomes local when

$$U(X_1, \ldots, X_n) = U_1(X_1) + \cdots + U_n(X_n), \tag{8.31}$$

in which case (8.30) reduces to

$$m_a \frac{d^2 X_a^{\mu}(s)}{ds^2} = \partial_a^{\mu} U_a(X_a(s)). \tag{8.32}$$

Thus we see that particle motions faster than light ($U_a < 0$) are possible even when the forces are local. Similarly, $U(X_1, \ldots, X_n)$

may be such that particles move only slower than light, but that the forces are still nonlocal.

The Hamilton–Jacobi formalism can also be generalized to the many-particle case. In particular, Eqs. (8.23) and (8.25) generalize to

$$-\sum_{a=1}^{n} \frac{(\partial_a^\mu S)(\partial_{a\mu} S)}{2m_a} + U(x_1, \ldots, x_n) = 0, \tag{8.33}$$

$$\frac{dX_a^\mu(s)}{ds} = -\frac{\partial_a^\mu S(X_1(s), \ldots, X_n(s))}{m_a}, \tag{8.34}$$

respectively. In the local case (8.31), the solution of (8.33) can be written in the form

$$S(x_1, \ldots, x_n) = S_1(x_1) + \cdots + S_n(x_n), \tag{8.35}$$

so (8.34) reduces to

$$\frac{dX_a^\mu(s)}{ds} = -\frac{\partial_a^\mu S_a(X_a(s))}{m_a}. \tag{8.36}$$

8.2.2.4 Absolute time

Finally, let us give a few conceptual remarks on the physical meaning of the parameter s. As discussed in more detail in [6, 11], its role in the equations above is formally analogous to the role of the Newton absolute time t in nonrelativistic Newtonian mechanics. In particular, even though s cannot be measured directly, it can be measured indirectly in the same sense as t is measured indirectly in Newtonian mechanics. Namely, one measures time by a "clock," where "clock" is nothing but a physical process periodic in time. Hence, if at least one of the $4n$ functions $X_a^\mu(s)$ is periodic in s, then the number of cycles (which is a measurable quantity) can be interpreted as a measure of elapsed s. Thus, it is justified to think of s as an absolute time in relativistic mechanics.

The parameter s is also related to the more familiar relativistic notion of proper time τ. As discussed in more detail in [6, 11], s can be thought of as a generalization of the notion of proper time.

8.3 Relativistic Quantum Mechanics

8.3.1 Wave Functions and Their Relativistic Probabilistic Interpretation

Let us start with QM of a single particle without spin. The basic object describing the properties of the particle is the wave function $\psi(x)$. We normalize the wave function such that

$$\int d^4x\, \psi^*(x)\psi(x) = 1. \tag{8.37}$$

More precisely, to avoid a divergence, the integral $\int d^4x$ is taken over some very large but not necessarily infinite four-dimensional region. (For most practical purposes it is more than sufficient to take a region of the astronomical size.) If the integral (8.37) happens to converge even when the boundary of the region is at infinity, then an infinite four-dimensional region is also allowed.

The probability of finding the particle in the (infinitesimal) four-volume d^4x is postulated to be

$$dP = |\psi(x)|^2 d^4x, \tag{8.38}$$

which is compatible with the normalization (8.37), as $|\psi|^2 \equiv \psi^*\psi$. At first sight, (8.38) may seem to be incompatible with the usual probabilistic interpretation in three-space[a]

$$dP_{(3)} \propto |\psi(\mathbf{x}, t)|^2 d^3x. \tag{8.39}$$

Nevertheless, (8.38) is compatible with (8.39). If (8.38) is the fundamental *a priori* probability, then (8.39) is naturally interpreted as the conditional probability corresponding to the case in which one knows that the particle is detected at time t. More precisely, the conditional probability is

$$dP_{(3)} = \frac{|\psi(\mathbf{x}, t)|^2 d^3x}{N_t}, \tag{8.40}$$

where

$$N_t = \int d^3x |\psi(\mathbf{x}, t)|^2 \tag{8.41}$$

---

[a]To our knowledge, the first version of probabilistic interpretation based on (8.38) rather than (8.39) was proposed in [14].

is the normalization factor. If ψ is normalized such that (8.38) is valid, then (8.41) is also the marginal probability that the particle will be found at t. Of course, in practice a measurement always lasts a finite time Δt and the detection time t cannot be determined with perfect accuracy. Thus, (8.40) should be viewed as a limiting case in which the fundamental probability (8.38) is averaged over a very small Δt. More precisely, if the particle is detected between $t - \Delta t/2$ and $t + \Delta t/2$, then (8.40) is the probability of different three-space positions of the particle detected during this small Δt.

Can the probabilistic interpretation (8.38) be verified experimentally? In fact, it already is! In practice one often measures cross sections associated with scattering experiments or decay widths and lifetimes associated with spontaneous decays of unstable quantum systems. These experiments agree with standard theoretical predictions. Our point is that these standard theoretical predictions actually use (8.38), although not explicitly. Let us briefly explain it. The basic theoretical tool in these predictions is the transition amplitude A. Essentially, the transition amplitude is the wave function (usually Fourier transformed to the three-momentum space) at $t \to \infty$, calculated by assuming that the wave function at $t \to -\infty$ is known. Due to energy conservation one obtains

$$A \propto \delta(E_{\text{in}} - E_{\text{fin}}), \tag{8.42}$$

where E_{in} and E_{fin} are the initial and final energy, respectively. Thus, the transition probability is proportional to

$$|A|^2 \propto [\delta(E_{\text{in}} - E_{\text{fin}})]^2 = \frac{T}{2\pi}\delta(E_{\text{in}} - E_{\text{fin}}), \tag{8.43}$$

where $T = \int dt = 2\pi\delta(E = 0)$ and we work in units $\hbar = 1$. Since T is infinite, this transition probability is physically meaningless. The standard interpretation (see, e.g., [15] for the nonrelativistic case or [16, 17] for the relativistic case), which agrees with experiments, is that the physical quantity is $|A|^2/T$ and that this quantity is (proportional to) the transition probability *per unit time*. But this is essentially the same as our equation (8.38), which says that $\int d^3x|\psi|^2$ is not probability itself, but probability *per unit time*. Although the interpretation of $|A|^2/T$ as probability per unit time may seem plausible even without explicitly postulating (8.38), without this postulate, such an interpretation of $|A|^2/T$ is at best heuristic and cannot be strictly derived from other basic postulates of QM, including (8.39). In this sense, the standard interpretation of transition amplitudes in terms of transition probabilities per unit time is better founded in basic axioms of QM if (8.38) is also adopted as one of its axioms.

Now let us generalize it to the case of n particles. Each particle has its own space position $\mathbf{x}_a$, $a = 1, \ldots, n$, as well as its own time coordinate t_a. Therefore, the wave function is of the form $\psi(x_1, \ldots, x_n)$, which is a many-time wave function. (For an early use of many-time wave functions in QM see [18]). Then (8.38) generalizes to

$$dP = |\psi(x_1, \ldots, x_n)|^2 d^4x_1 \cdots d^4x_n. \tag{8.44}$$

Hence, if the first particle is detected at t_1, the second particle at t_2, etc., then Eq. (8.40) generalizes to

$$dP_{(3n)} = \frac{|\psi(\mathbf{x}_1, t_1, \ldots, \mathbf{x}_n, t_n)|^2 d^3x_1 \cdots d^3x_n}{N_{t_1,\ldots,t_n}}, \tag{8.45}$$

where

$$N_{t_1,\ldots,t_n} = \int |\psi(\mathbf{x}_1, t_1, \ldots, \mathbf{x}_n, t_n)|^2 d^3x_1 \cdots d^3x_n. \tag{8.46}$$

The many-time wave function contains also the familiar single-time wave function as a special case:

$$\psi(\mathbf{x}_1, \ldots, \mathbf{x}_n; t) = \psi(\mathbf{x}_1, t_1, \ldots, \mathbf{x}_n, t_n)|_{t_1=\cdots=t_n=t}. \tag{8.47}$$

In this case (8.45) reduces to the familiar expression:

$$dP_{(3n)} = \frac{|\psi(\mathbf{x}_1, \ldots, \mathbf{x}_n; t)|^2 d^3x_1 \cdots d^3x_n}{N_t}, \tag{8.48}$$

where N_t is given by (8.46) calculated at $t_1 = \cdots = t_n = t$.

Finally, let us generalize all this to particles that carry spin or some other additional discrete degree of freedom. For one particle, instead of one wave function $\psi(x)$, one deals with a collection of wave functions $\psi_l(x)$, where l is a discrete label. Similarly, for n particles with discrete degrees of freedom we have a collection of wave functions of the form $\psi_{l_1\ldots l_n}(x_1, \ldots, x_n)$. To simplify the notation, it is convenient to introduce a collective label $L = (l_1, \ldots, l_n)$, which means that the wave function for n particles can be written as $\psi_L(x_1, \ldots, x_n)$. Now all equations above can be easily generalized through the replacement:

$$\psi^*\psi \to \sum_L \psi_L^*\psi_L. \tag{8.49}$$

In particular, the joint probability for finding particles at the positions $x_1, \ldots, x_n$ is given by a generalization of (8.44)

$$: dP = \sum_L \psi_L^*(x_1, \ldots, x_n)\psi_L(x_1, \ldots, x_n)d^4x_1 \cdots d^4x_n. \tag{8.50}$$

Another useful notation is to introduce the column $\psi = \{\psi_L\}$ and the row $\psi^\dagger = \{\psi_L^*\}$, i.e.,

$$\psi = \begin{pmatrix} \psi_1 \\ \psi_2 \\ \vdots \end{pmatrix}, \quad \psi^\dagger = \left(\psi_1^* \; \psi_2^* \cdots \right). \tag{8.51}$$

With this notation, (8.50) can also be written as

$$dP = \psi^\dagger(x_1, \ldots, x_n)\psi(x_1, \ldots, x_n)d^4x_1 \cdots d^4x_n. \tag{8.52}$$

8.3.2 Theory of Quantum Measurements

Let $\psi(x)$ be expanded as

$$\psi(x) = \sum_b c_b\psi_b(x), \tag{8.53}$$

where $\psi_b(x)$ are eigenstates of some hermitian operator $\hat{B}$ on the Hilbert space of functions of x. Let $\psi_b(x)$ be normalized such that $\int d^4x\, \psi_b^*(x)\psi_b(x) = 1$. Assume that one measures the value of the observable B described by the hermitian operator $\hat{B}$. In a conventional approach to QM, one would postulate that $|c_b|^2$ is the probability that B will take the value b. Nevertheless, there is no need for such a postulate because, whatever the operator $\hat{B}$ is, this probabilistic rule can be derived from the probabilistic interpretation in the position space discussed in Section 8.3.1.

To understand this, one needs to understand how a typical measuring apparatus works, i.e., how the wave function of the measured system described by the coordinate x interacts with the wave function of the measuring apparatus described by the coordinate y. (For simplicity, we assume that y is a coordinate of a single particle, but essentially the same analysis can be given by considering a more realistic case in which y is replaced by a macroscopically large number N of particles $y_1, \ldots, y_N$ describing the macroscopic measuring apparatus. Similarly, the same analysis can also be generalized to the case in which x is replaced by $x_1, \ldots, x_n$.) Let the wave function of the measuring apparatus for times before the interaction be $E_0(y)$. Thus, for times x^0 and y^0 before the interaction, the total wave function is $\psi(x)E_0(y)$. But what happens after the interaction? If $\psi(x) = \psi_b(x)$ before the interaction, then the interaction must be such that after the interaction, the total wave function takes the form $\psi_b(x)E_b(y)$, where $E_b(y)$ is a macroscopic state of the measuring apparatus, normalized so that $\int d^4y\, E_b^*(y)E_b(y) = 1$. The state $E_b(y)$ is such that one can say that "the measuring apparatus shows that the result of measurement is b" when the measuring apparatus is found in that state. Schematically, the result of interaction described above can be written as

$$\psi_b(x)E_0(y) \to \psi_b(x)E_b(y). \tag{8.54}$$

Of course, most interactions do not have the form (8.54), but only those that do can be regarded as measurements of the observable $\hat{B}$. The transition (8.54) is guided by some *linear* differential equation (we study the explicit linear dynamical equations for wave functions in the subsequent sections), which means that the superposition principle is valid. Therefore, (8.54) implies that for a general superposition (8.53) we have

$$\sum_b c_b\psi_b(x)E_0(y) \to \sum_b c_b\psi_b(x)E_b(y) \equiv \psi(x,y). \tag{8.55}$$

The states $E_b(y)$ must be macroscopically distinguishable. In practice, it means that they do not overlap (or more realistically that their overlap is negligible), i.e.,

$$E_b(y)E_{b'}(y) \simeq 0 \quad \text{for} \quad b \neq b', \tag{8.56}$$

for all values of y. Instead of asking "what is the probability that the measured particle is in the state $\psi_b(x)$," the operationally more meaningful question is, what is the probability that the measuring apparatus will be found in the state $E_b(y)$? The (marginal) probability density for finding the particle describing the measuring apparatus at the position y is

$$\rho(y) = \int d^4x\, \psi^*(x, y)\psi(x, y). \tag{8.57}$$

Using (8.55) and (8.56), this becomes

$$\rho(y) \simeq \sum_b |c_b|^2 |E_b(y)|^2. \tag{8.58}$$

Now let supp E_b be the support of $E_b(y)$, i.e., the region of y-space on which $E_b(y)$ is not negligible. Then, from (8.58), the probability that y will take a value from the support of $E_b(y)$ is

$$p_b = \int_{\text{supp } E_b} d^4y\, \rho(y) \simeq |c_b|^2. \tag{8.59}$$

In other words, the probability that the measuring apparatus will be found in the state $E_b(y)$ is (approximately) equal to $|c_b|^2$.

8.3.3 Relativistic Wave Equations

In this subsection we consider particles which are free on the classical level, i.e., particles classically described by the action (8.11) with a constant scalar potential

$$U(X) = \frac{m}{2}. \tag{8.60}$$

The constraint (8.19) becomes

$$P^\mu P_\mu - m^2 = 0, \tag{8.61}$$

implying that m is the mass of the particle.

In QM, the momentum P_μ becomes the operator $\hat{P}_\mu$ satisfying the canonical commutation relations

$$[x^\mu, \hat{P}_\nu] = -i\eta^\mu_\nu, \tag{8.62}$$

where we work in units $\hbar = 1$. These commutation relations are satisfied by taking

$$\hat{P}_\nu = i\partial_\nu. \tag{8.63}$$

8.3.3.1 Single particle without spin

Let us start with a particle without spin. The quantum analog of the classical constraint (8.61) is

$$[\hat{P}^{\mu}\hat{P}_{\mu} - m^2]\psi(x) = 0, \tag{8.64}$$

which is nothing but the Klein–Gordon equation

$$[\partial^{\mu}\partial_{\mu} + m^2]\psi(x) = 0. \tag{8.65}$$

From a solution of (8.65), one can construct the real current

$$j_{\mu} = \frac{i}{2}\psi^{*}\overleftrightarrow{\partial_{\mu}}\psi, \tag{8.66}$$

where

$$\psi_1\overleftrightarrow{\partial_{\mu}}\psi_2 \equiv \psi_1(\partial_{\mu}\psi_2) - (\partial_{\mu}\psi_1)\psi_2. \tag{8.67}$$

Using (8.65), one can show that this current is conserved:

$$\partial_{\mu}j^{\mu} = 0. \tag{8.68}$$

By writing $\psi = Re^{iS}$, where R and S are real functions, the complex Klein–Gordon equation (8.65) is equivalent to a set of two real equations,

$$\partial^{\mu}(R^2\partial_{\mu}S) = 0, \tag{8.69}$$

$$-\frac{(\partial^{\mu}S)(\partial_{\mu}S)}{2m} + \frac{m}{2} + Q = 0, \tag{8.70}$$

where (8.69) is the conservation equation (8.68) and

$$Q = \frac{1}{2m}\frac{\partial^{\mu}\partial_{\mu}R}{R}. \tag{8.71}$$

It is easy to show that the equations above have the correct nonrelativistic limit. In particular, by writing

$$\psi = \frac{e^{-imt}}{\sqrt{m}}\psi_{\mathrm{NR}} \tag{8.72}$$

and using $|\partial_t\psi_{\mathrm{NR}}| \ll m|\psi_{\mathrm{NR}}|$, $|\partial_t^2\psi_{\mathrm{NR}}| \ll m|\partial_t\psi_{\mathrm{NR}}|$, from (8.66) and (8.65) we find the approximate equations:

$$j_0 = \psi_{\mathrm{NR}}^{*}\psi_{\mathrm{NR}}, \tag{8.73}$$

$$-\frac{\nabla^2}{2m}\psi_{\mathrm{NR}} = i\partial_t\psi_{\mathrm{NR}}, \tag{8.74}$$

which are the nonrelativistic probability density and the nonrelativistic Schrödinger equation for the evolution of the wave function $\psi_{\rm NR}$, respectively.

Note that (8.72) contains a positive-frequency oscillatory function e^{-imt} and not a negative-frequency oscillatory function e^{imt}. If we took e^{imt} in (8.72) instead, then we would obtain $-i\partial_t\psi_{\rm NR}$ on the right-hand side of (8.74), which would be a Schrödinger equation with the wrong sign of the time derivative. In other words, even though (8.65) contains solutions with both positive and negative frequencies, only positive frequencies lead to the correct nonrelativistic limit. This means that only solutions with positive frequencies are physical, i.e., that the most general physical solution of (8.65) is

$$\psi(x) = \int d^3k\, a(\mathbf{k}) e^{-i[\omega(\mathbf{k})x^0 - \mathbf{k}\mathbf{x}]}, \tag{8.75}$$

where $a(\mathbf{k})$ is an arbitrary function and

$$\omega(\mathbf{k}) = \sqrt{\mathbf{k}^2 + m^2} \tag{8.76}$$

is positive. More precisely, this is so if the particle is not charged, i.e., if the particle is its own antiparticle. When particles are charged, then ψ with positive frequencies describes a particle, while ψ with negative frequencies describes an antiparticle.

8.3.3.2 Many particles without spin

Now let us generalize it to the case of n identical particles without spin, with equal masses $m_a = m$. The wave function ψ satisfies n Klein–Gordon equations:

$$(\partial_a^\mu \partial_{a\mu} + m^2)\psi(x_1, \ldots, x_n) = 0, \tag{8.77}$$

one for each x_a. Therefore, one can introduce n real four-currents:

$$j_a^\mu = \frac{i}{2}\psi^* \overleftrightarrow{\partial_a^\mu} \psi, \tag{8.78}$$

each of which is separately conserved

$$\partial_a^\mu j_{a\mu} = 0. \tag{8.79}$$

Equation (8.77) also implies

$$\left(\sum_a \partial_a^\mu \partial_{a\mu} + nm^2\right) \psi(x_1, \ldots, x_n) = 0, \tag{8.80}$$

while (8.79) implies

$$\sum_a \partial_a^\mu j_{a\mu} = 0. \tag{8.81}$$

Next we write $\psi = Re^{iS}$, where R and S are real functions. Equation (8.80) is then equivalent to a set of two real equations:

$$\sum_a \partial_a^\mu (R^2 \partial_{a\mu} S) = 0, \tag{8.82}$$

$$-\frac{\sum_a (\partial_a^\mu S)(\partial_{a\mu} S)}{2m} + \frac{nm}{2} + Q = 0, \tag{8.83}$$

where

$$Q = \frac{1}{2m} \frac{\sum_a \partial_a^\mu \partial_{a\mu} R}{R}. \tag{8.84}$$

Equation (8.82) is equivalent to (8.81).

In the nonrelativistic limit we have n equations of the form of (8.74)

$$-\frac{\nabla_a^2}{2m} \psi_{\mathrm{NR}} = i\partial_{t_a} \psi_{\mathrm{NR}}, \tag{8.85}$$

where $\psi_{\mathrm{NR}} = \psi_{\mathrm{NR}}(\mathbf{x}_1, t_1, \ldots, \mathbf{x}_n, t_n)$ is the nonrelativistic many-time wave function. The single-time wave function is defined as in (8.47), so we see that

$$\sum_a \partial_{t_a} \psi_{\mathrm{NR}}(\mathbf{x}_1, t_1, \ldots, \mathbf{x}_n, t_n)|_{t_1=\cdots=t_n=t} = \partial_t \psi_{\mathrm{NR}}(\mathbf{x}_1, \ldots, \mathbf{x}_n; t). \tag{8.86}$$

Therefore (8.85) implies the usual many-particle single-time Schrödinger equation:

$$\left[\sum_a -\frac{\nabla_a^2}{2m}\right] \psi_{\mathrm{NR}}(\mathbf{x}_1, \ldots, \mathbf{x}_n; t) = i\partial_t \psi_{\mathrm{NR}}(\mathbf{x}_1, \ldots, \mathbf{x}_n; t). \tag{8.87}$$

8.3.3.3 Single particle with spin $\frac{1}{2}$

A relativistic particle with spin $\frac{1}{2}$ is described by a four-component wave function $\psi_l(x)$, $l = 1, 2, 3, 4$ (see, e.g., [17]). Each component satisfies the Klein–Gordon equation:

$$[\partial^\mu \partial_\mu + m^2]\psi_l(x) = 0. \tag{8.88}$$

Introducing the column

$$\psi = \begin{pmatrix} \psi_1 \\ \psi_2 \\ \psi_3 \\ \psi_4 \end{pmatrix}, \tag{8.89}$$

known as Dirac spinor, (8.88) can also be written as

$$[\partial^\mu \partial_\mu + m^2]\psi(x) = 0. \tag{8.90}$$

However, the four components of (8.90) are not completely independent. They also satisfy an additional constraint linear in the space-time derivatives, known as the Dirac equation

$$[i\gamma^{\bar{\mu}}\partial_\mu - m]\psi(x) = 0. \tag{8.91}$$

Here each $\gamma^{\bar{\mu}}$ is a 4×4 matrix in the spinor space. These matrices satisfy the anticommutation relations:

$$\gamma^{\bar{\mu}}\gamma^{\bar{\nu}} + \gamma^{\bar{\nu}}\gamma^{\bar{\mu}} = 2\eta^{\bar{\mu}\bar{\nu}}. \tag{8.92}$$

In fact, by multiplying (8.91) from the left with the operator $[-i\gamma^{\bar{\mu}}\partial_\mu - m]$ and using (8.92), one obtains (8.90). This means that the Klein–Gordon equation (8.90) is a consequence of the Dirac equation (8.91). Note, however, that the opposite is not true; one cannot derive (8.91) from (8.90).

The matrices $\gamma^{\bar{\mu}}$ are known as Dirac matrices. Even though they carry the index $\bar{\mu}$, they do *not* transform as vectors under space-time transformations. In fact, this is why $\bar{\mu}$ has a bar over it, to remind us that it is not a space-time vector index.[a] Instead, $\bar{\mu}$ is only a label. More precisely, since $\gamma^{\bar{\mu}}$ does not carry any space-time index like μ, it is a scalar with respect to space-time transformations. Similarly, the spinor ψ also does not carry space-time indices, so it is also a scalar with respect to space-time transformations.[b]

---

[a]In most literature, like [17], the bar is omitted and the Dirac matrices are denoted by γ^μ. In our opinion, such a notation without a bar causes a lot of confusion.

[b]In most literature, like [17], the spinor ψ transforms in a rather complicated and unintuitive way under Lorentz transformations of space-time coordinates. Even worse, it turns out that such a complicated transformation of spinors cannot be generalized to arbitrary transformations of space-time coordinates. This is why it is more convenient to adopt a more intuitive formalism in which ψ is a scalar with respect to space-time transformations [13, 19]. Nevertheless, as long as only Lorentz transformations of physically measurable quantities are concerned, the two formalisms turn out to be physically equivalent.

Nevertheless, there is a way to introduce a matrix γ^μ that transforms as a true vector [13, 19]. At each point of space-time, one introduces the tetrad $e^\mu_{\bar{\alpha}}(x)$, which is a collection of four space-time vectors, one for each $\bar{\alpha} = 0, 1, 2, 3$. The tetrad is chosen so that

$$\eta^{\bar{\alpha}\bar{\beta}} e^\mu_{\bar{\alpha}}(x) e^\nu_{\bar{\beta}}(x) = g^{\mu\nu}(x), \tag{8.93}$$

where $g^{\mu\nu}(x)$ is the space-time metric (which, in general, may depend on x) and $\eta_{\bar{\alpha}\bar{\beta}}$ are components of a matrix equal to the Minkowski metric. The space-time-vector indices are raised and lowered by $g^{\mu\nu}(x)$ and $g_{\mu\nu}(x)$, respectively, while $\bar{\alpha}$-labels are raised and lowered by $\eta^{\bar{\alpha}\bar{\beta}}$ and $\eta_{\bar{\alpha}\bar{\beta}}$, respectively. Thus, (8.93) can also be inverted as

$$g^{\mu\nu}(x) e^{\bar{\alpha}}_\mu(x) e^{\bar{\beta}}_\nu(x) = \eta^{\bar{\alpha}\bar{\beta}}. \tag{8.94}$$

Now from the constant Dirac matrices $\gamma^{\bar{\alpha}}$ we define

$$\gamma^\mu(x) = e^\mu_{\bar{\alpha}}(x)\gamma^{\bar{\alpha}}. \tag{8.95}$$

The spinor indices carried by matrices $\gamma^{\bar{\alpha}}$ and $\gamma^\mu(x)$ are interpreted as indices of the spinor representation of the *internal* group SO(1,3). Just like $\psi(x)$, $\psi^\dagger(x)$ is also a scalar with respect to space-time coordinate transformations. It is also convenient to define the quantity

$$\bar{\psi}(x) = \psi^\dagger(x)\gamma^{\bar{0}}, \tag{8.96}$$

which is also a scalar with respect to space-time coordinate transformations. Thus we see that the quantities

$$\bar{\psi}(x)\psi(x), \quad \psi^\dagger(x)\psi(x), \tag{8.97}$$

are both scalars with respect to space-time coordinate transformations and that the quantities

$$\bar{\psi}(x)\gamma^\mu(x)\psi(x), \quad \frac{i}{2}\psi^\dagger(x)\overleftrightarrow{\partial^\mu}\psi(x), \tag{8.98}$$

are both vectors with respect to space-time coordinate transformations.

Note that in the flat Minkowski space-time, there is a particular global Lorentz frame of coordinates in which

$$\gamma^\mu(x) = \gamma^{\bar{\mu}}. \tag{8.99}$$

Indeed, this is why Eq. (8.91) makes sense. However, (8.99) is not a covariant expression, but is only valid in one special system of coordinates. In other global Lorentz frames we have

$$\gamma'^{\mu} = \Lambda^{\mu}{}_{\nu}\gamma^{\nu}, \tag{8.100}$$

where $\Lambda^{\mu}{}_{\nu}$ are the matrix elements of the Lorentz transformation. Since $\Lambda^{\mu}{}_{\nu}$ do not depend on x, it follows that the vector γ^{μ} is x-independent in *any* Lorentz frame. Therefore, in an arbitrary Lorenz frame, (8.91) should be replaced by a truly Lorentz-covariant equation:

$$[i\gamma^{\mu}\partial_{\mu} - m]\psi(x) = 0. \tag{8.101}$$

The two quantities in (8.98)

$$j^{\mu}_{\text{Dirac}} = \bar{\psi}(x)\gamma^{\mu}\psi(x), \tag{8.102}$$

$$j^{\mu} = \frac{i}{2}\psi^{\dagger}(x)\overleftrightarrow{\partial^{\mu}}\psi(x), \tag{8.103}$$

are referred to as Dirac current and Klein–Gordon current, respectively. They are both conserved:

$$\partial_{\mu}j^{\mu}_{\text{Dirac}} = 0, \quad \partial_{\mu}j^{\mu} = 0. \tag{8.104}$$

The first conservation is a consequence of (8.101), while the second conservation is a consequence of (8.90).

8.3.3.4 Many particles with spin $\frac{1}{2}$

The wave function for n particles with spin $\frac{1}{2}$ has the form $\psi_{l_1\ldots l_n}(x_1, \ldots, x_n)$, where each l_a is a spinor index. It satisfies n Dirac equations. A convenient way to write them is

$$[i\gamma_a^{\mu}\partial_{a\mu} - m]\psi = 0, \tag{8.105}$$

where γ_a^{μ} is a "matrix" with $2n$ indices:

$$(\gamma_a^{\mu})_{l_1\ldots l_n l'_1\ldots l'_n} = \delta_{l_1 l'_1}\cdots(\gamma^{\mu})_{l_a l'_a}\cdots\delta_{l_n l'_n}. \tag{8.106}$$

In the more abstract language of direct products, we can also write (8.106) as

$$\gamma_a^{\mu} = 1\otimes\cdots\otimes\gamma^{\mu}\otimes\cdots\otimes 1. \tag{8.107}$$

Similarly, the wave function satisfies also n Klein–Gordon equations:

$$[\partial_a^\mu \partial_{a\mu} + m^2]\psi = 0. \tag{8.108}$$

Consequently, there are n conserved Klein–Gordon currents:

$$j_a^\mu = \frac{i}{2}\psi^\dagger \overleftrightarrow{\partial_a^\mu} \psi, \tag{8.109}$$

$$\partial_a^\mu j_{a\mu} = 0, \tag{8.110}$$

which imply a single conservation equation:

$$\sum_a \partial_a^\mu j_{a\mu} = 0. \tag{8.111}$$

A similar generalization of the Dirac current also exists, but we shall not need it.

8.3.3.5 Particles with spin 1

The case of spin 1 is much simpler than the case of spin $\frac{1}{2}$. Consequently, we shall only briefly outline how spin 1 particles are described.

A one-particle wave function is $\psi_\alpha(x)$ and carries one vector index α. It satisfies 4 equations (see, e.g., [20]):

$$\partial^\alpha F_{\alpha\beta} + m^2 \psi_\beta = 0, \tag{8.112}$$

where

$$F_{\alpha\beta} = \partial_\alpha \psi_\beta - \partial_\beta \psi_\alpha. \tag{8.113}$$

By applying the derivative ∂^β on (8.112), one finds

$$\partial_\beta \psi^\beta = 0. \tag{8.114}$$

Therefore, (8.112) implies four Klein–Gordon equations:

$$[\partial^\mu \partial_\mu + m^2]\psi_\alpha(x) = 0. \tag{8.115}$$

However, (8.112) implies that not all four components ψ_α are independent. For example, the time-component can be expressed in terms of other components as $\psi_0 = -\partial^\alpha F_{\alpha 0}/m^2$. Therefore, the most general positive-frequency solution of (8.112) can be written in the form

$$\psi^\alpha(x) = \int d^3k \sum_{l=1}^{3} \epsilon_l^\alpha(\mathbf{k}) a_l(\mathbf{k}) e^{-i[\omega(\mathbf{k})x^0 - \mathbf{kx}]}, \tag{8.116}$$

which can be thought of as a generalization of (8.75). Here $a_l(\mathbf{k})$ are arbitrary functions, while $\epsilon_l^\alpha(\mathbf{k})$ are fixed polarization vectors [20]. Thus, a wave function is completely determined by three independent functions $a_l(\mathbf{k}), l = 1, 2, 3$. This implies that the system can also be described by a three-component wave function:

$$\psi_l(x) = \int d^3k\, a_l(\mathbf{k}) e^{-i[\omega(\mathbf{k})x^0 - \mathbf{kx}]}, \tag{8.117}$$

where all three components are independent. Since each component of (8.117) also satisfies the Klein–Gordon equation, the Klein–Gordon current

$$j^\mu = \frac{i}{2} \sum_l \psi_l^* \overleftrightarrow{\partial^\mu} \psi_l \tag{8.118}$$

is conserved:

$$\partial_\mu j^\mu = 0. \tag{8.119}$$

In the case on n particles the wave function $\psi_{l_1 \ldots l_n}(x_1, \ldots, x_n)$ carries n polarization labels. It satisfies n Klein–Gordon equations:

$$[\partial_a^\mu \partial_{a\mu} + m^2] \psi_{l_1 \ldots l_n}(x_1, \ldots, x_n) = 0, \tag{8.120}$$

so (8.118) and (8.119) generalize to

$$j_a^\mu = \frac{i}{2} \sum_{l_1, \ldots, l_n} \psi_{l_1 \ldots l_n}^* \overleftrightarrow{\partial_a^\mu} \psi_{l_1 \ldots l_n}, \tag{8.121}$$

$$\partial_\mu j_a^\mu = 0, \tag{8.122}$$

which implies

$$\sum_a \partial_{a\mu} j_a^\mu = 0. \tag{8.123}$$

The case $m = 0$ is special because this case describes a photon, the wave function of which contains also a gauge symmetry. Namely, the (one-particle) wave function satisfies the free Maxwell equation:

$$\partial^\alpha F_{\alpha\beta} = 0, \tag{8.124}$$

which is invariant with respect to gauge transformations:

$$\psi_\alpha(x) \to \psi'_\alpha(x) = \psi_\alpha(x) + \partial_\alpha \Lambda(x), \tag{8.125}$$

where $\Lambda(x)$ is an arbitrary function. This gauge freedom can be partially removed by imposing the Lorentz-gauge condition (8.114).

However, when the gauge freedom is removed completely, then only two independent physical (transverse) polarizations remain. Consequently, the equations above involving l-labels modify such that l takes only two values $l = 1, 2$. A gauge transformation can be reduced to a change of the polarization vectors $\epsilon_l^{\alpha}(\mathbf{k})$. Thus, unlike $\psi_{\alpha}(x)$, the wave function $\psi_l(x)$ is gauge invariant.

Finally note that, in the massless case, the wave function $\psi_{\alpha}(x)$ is *not* the electromagnetic vector potential $A_{\alpha}(x)$. The latter is real (not complex), so is represented by a superposition of positive and negative frequencies. The former is a superposition of positive frequencies only, so it cannot be real at all x.

8.3.4 Bohmian Interpretation

Now we are finally ready to deal with the Bohmian interpretation of relativistic QM. Of course, the Bohmian interpretation could also be introduced without a lot of the background described in the preceding sections, but with this background the Bohmian interpretation is very natural and almost trivial.

We start from the observation that the quantum equation (8.70) has the same form as the classical equation (8.23), provided that we make the replacement:

$$U(x) \to \frac{m}{2} + Q(x). \tag{8.126}$$

The first term on the right-hand side of (8.126) is the classical potential (8.60), while the second term is the quantum potential.[a] This suggests the Bohmian interpretation, according to which (8.70) is the quantum Hamilton–Jacobi equation and the particle has the trajectory given by (8.25):

$$\frac{dX^{\mu}(s)}{ds} = -\frac{\partial^{\mu} S(X(s))}{m}. \tag{8.127}$$

From (8.127), (8.70), and the identity,

$$\frac{d}{ds} = \frac{dX^{\mu}}{ds}\partial_{\mu}, \tag{8.128}$$

[a]Recall that we work in units $\hbar = 1$. In units in which $\hbar \neq 1$, it is easy to show that (8.71) attains an additional factor $\hbar^2$, showing that the quantum potential Q vanishes in the classical limit.

one finds a quantum variant of (8.13)

$$m\frac{d^2X^\mu(s)}{ds^2} = \partial^\mu Q(X(s)). \tag{8.129}$$

But is such motion of quantum particles consistent with the probabilistic predictions studied in Sections 8.3.1 and 8.3.2? We first observe that (8.127) can be written as

$$\frac{dX^\mu}{ds} = \frac{j^\mu}{m\psi^*\psi}, \tag{8.130}$$

where j^μ is given by (8.66). It is convenient to eliminate the factor $1/m$ by rescaling the parameter s, so that (8.130) becomes

$$\frac{dX^\mu}{ds} = V^\mu, \tag{8.131}$$

where

$$V^\mu = \frac{j^\mu}{\psi^*\psi}. \tag{8.132}$$

Second, we observe that (8.68) can be written as

$$\partial_\mu(|\psi|^2 V^\mu) = 0. \tag{8.133}$$

Since $\psi(x)$ does not explicitly depend on s, we also have a trivial identity $\partial|\psi|^2/\partial s = 0$. Therefore (8.133) can be written as

$$\frac{\partial|\psi|^2}{\partial s} + \partial_\mu(|\psi|^2 V^\mu) = 0. \tag{8.134}$$

This implies that the trajectories satisfying (8.131) are consistent with the probabilistic interpretation (8.38). Namely, if a statistical ensemble of particles has the distribution (8.38) of space-time particle positions for some "initial" s, then (8.134) guarantees that this statistical ensemble has the distribution (8.38) for *any* s.

This shows that particles have the same distribution of space-time positions as predicted by the purely probabilistic interpretation of QM. But what about other measurable quantities? For example, what about the space distribution of particles described in purely probabilistic QM by (8.40)? Or what about the statistical distribution of particle velocities? In general, in the Bohmian interpretation all these other quantities may have a distribution totally different from those predicted by purely probabilistic QM. In particular, the Bohmian velocities of particles may exceed the velocity of light (which occurs when the right-hand side of (8.126) becomes negative[a]), while purely probabilistic QM does not allow such velocities because the eigenstates $e^{-ip_\mu x^\mu}$ of the velocity operator $\hat{p}_\mu/m$ are not solutions of (8.65) for $p^\mu p_\mu < 0$. Yet, when a quantity is *measured*, then the two theories have *the same* measurable predictions. Namely, since the Bohmian interpretation is compatible with (8.38), the probability that the measuring apparatus will be found in the state $E_b(y)$ in (8.55) is given by (8.59), which is the same as that in the purely probabilistic interpretation.

Now the generalization to n particles without spin is straightforward. Essentially, all equations above are rewritten such that each quantity having the index μ receives an additional index a. In particular, Eqs. (8.127), (8.129), (8.131), (8.132), (8.134) generalize to

$$\frac{dX_a^\mu(s)}{ds} = -\frac{\partial_a^\mu S(X_1(s), \ldots, X_n(s))}{m}, \tag{8.135}$$

$$m\frac{d^2X_a^\mu(s)}{ds^2} = \partial_a^\mu Q(X_1(s), \ldots, X_n(s)), \tag{8.136}$$

$$\frac{dX_a^\mu}{ds} = V_a^\mu, \tag{8.137}$$

$$V_a^\mu = \frac{j_a^\mu}{\psi^*\psi}, \tag{8.138}$$

$$\frac{\partial|\psi|^2}{\partial s} + \sum_{a=1}^{n} \partial_{a\mu}(|\psi|^2 V_a^\mu) = 0, \tag{8.139}$$

respectively. In general, particles have nonlocal influences on each other, in exactly the same way as in classical relativistic mechanics studied in Section 8.2.2.

Now let us generalize these results to particles with spin. When spin is present, the analogy with the classical Hamilton–Jacobi equation is less useful. The crucial requirement is the consistency with the purely probabilistic interpretation (8.52). This is achieved by generalizing (8.137) and (8.138) to

$$\frac{dX_a^\mu}{ds} = V_a^\mu, \tag{8.140}$$

[a]Chapter 9 studies a possible cosmological relevance of such faster-than-light velocities.

$$V_a^\mu = \frac{j_a^\mu}{\psi^\dagger \psi}, \tag{8.141}$$

where j_a^μ is a conserved current given by (8.109) for spin $\frac{1}{2}$ particles and (8.121) for spin 1 particles. The compatibility with (8.52) is provided by the generalization of (8.139):

$$\frac{\partial \psi^\dagger \psi}{\partial s} + \sum_{a=1}^{n} \partial_{a\mu}(\psi^\dagger \psi V_a^\mu) = 0. \tag{8.142}$$

8.4 Quantum Field Theory

8.4.1 Main Ideas of QFT and Its Bohmian Interpretation

So far, we have been considering systems with a fixed number n of particles. However, in many physical systems the number of particles is not fixed. Instead, particles may be created or destroyed. To describe such processes, a more general formalism is needed. This formalism is known as QFT.

The simplest way to understand the kinematics of QFT is as follows. Let $\mathcal{H}^{(n)}$ denote the Hilbert space associated with QM of a fixed number n of particles, where $n \geq 1$. An element of this Hilbert space is a quantum state of n particles, denoted abstractly by $|n\rangle$. In fact, the case $n = 0$ can also be included by defining a new trivial one-dimensional Hilbert space $\mathcal{H}^{(0)}$. This trivial space has only one linearly independent element denoted by $|0\rangle$, which represents the vacuum, i.e., the state with no particles. From all these Hilbert spaces, one can construct a single Hilbert space $\mathcal{H}$ containing all of them as subspaces, through a direct sum:

$$\mathcal{H} = \bigoplus_{n=0}^{\infty} \mathcal{H}^{(n)} \equiv \mathcal{H}^{(0)} \oplus \mathcal{H}^{(1)} \oplus \mathcal{H}^{(2)} \oplus \cdots . \tag{8.143}$$

QFT is nothing but the theory of states in the Hilbert space $\mathcal{H}$. A general state in this space is a linear combination of the form

$$|\Psi\rangle = \sum_{n=0}^{\infty} c_n |n\rangle. \tag{8.144}$$

QFT is the theory of states (8.144).[a]

As a simple example, consider a QFT state of the form

$$|\Psi\rangle = |1\rangle + |2\rangle, \qquad (8.145)$$

which is a superposition of a one-particle state $|1\rangle$ and a two-particle state $|2\rangle$. For example, it may represent an unstable particle for which we do not know if it has already decayed into two new particles (in which case it is described by $|2\rangle$) or has not decayed yet (in which case it is described by $|1\rangle$). However, it is known that one always observes either one unstable particle (the state $|1\rangle$) or two decay products (the state $|2\rangle$). One never observes the superposition (8.145). Why?

To answer this question, let us try with a Bohmian approach. One can associate a one-particle wave function $\Psi_1(x_1)$ with the state $|1\rangle$ and a two-particle wave function $\Psi_2(x_2, x_3)$ with the state $|2\rangle$, where x_A is the space-time position of the particle labeled by $A = 1, 2, 3$. Then the state (8.145) is represented by a superposition:

$$\Psi(x_1, x_2, x_3) = \Psi_1(x_1) + \Psi_2(x_2, x_3). \qquad (8.146)$$

However, the Bohmian interpretation of such a superposition will describe *three* particle trajectories. On the other hand, we should observe either one or two particles, not three particles. How to explain that?

The key is to take into account the properties of the *measuring apparatus*. If the number of particles is measured, then instead of (8.146) we actually have a wave function of the form

$$\Psi(x_1, x_2, x_3, y) = \Psi_1(x_1)E_1(y) + \Psi_2(x_2, x_3)E_2(y). \qquad (8.147)$$

The detector wave functions $E_1(y)$ and $E_2(y)$ do not overlap. Hence, if y takes a value Y in the support of E_2, then this value is not in

[a]In such a view of QFT, the fundamental physical objects are particles, while fields only play an auxiliary role. There is also a different view of QFT in which fields play a more fundamental role than particles. An example of such a different view is presented in Chapter 9. However, in the context of Bohmian interpretation, there are at least two problems when fields are viewed as being more fundamental. First, it is not known how to make the Bohmian equations of motion for bosonic fields relativistic covariant. Second, it is not known how to include the fermionic fields. Various proposals for solving these two problems exist, but none of them seems completely satisfying. On the other hand, we shall see that such problems can be solved in a simple and natural way when the Bohmian interpretation is based on particles.

the support of E_1, i.e., $E_1(Y) = 0$. Consequently, the motion of the measured particles is described by the conditional wave function $\Psi_2(x_2, x_3)E_2(Y)$. The effect is the same as if (8.146) collapsed to $\Psi_2(x^2, x^3)$.

Now, what happens with the particle having the space-time position x_1? In general, its motion in space-time may be expected to be described by the relativistic Bohmian equation of motion:

$$\frac{dX_1^\mu(s)}{ds} = \frac{\frac{i}{2}\Psi^* \overleftrightarrow{\partial_1^\mu} \Psi}{\Psi^*\Psi}. \tag{8.148}$$

However, if the absence of the overlap between $E_1(y)$ and $E_2(y)$ is exact, then the effective wave function does not depend on x_1, i.e., the derivatives in (8.148) vanish. Consequently, all four components of the four-velocity (8.148) are zero. The particle does not change its space-time position X_1^μ. It is an object without an extension not only in space, but also in time. It can be thought of as a pointlike particle that exists only at one instant of time X_1^0. It lives too short to be detected. Effectively, this particle behaves as if it did not exist at all.

Now consider a more realistic variation of the measuring procedure, taking into account the fact that the measured particles become entangled with the measuring apparatus at some finite time T. Before that, the wave function of the measured particles is really well described by (8.146). Thus, before the interaction with the measuring apparatus, all three particles described by (8.146) have continuous trajectories in space-time. All three particles exist. But at time T, the total wave function significantly changes. Either (i) y takes a value from the support of E_2 in which case dX_1^μ/ds becomes zero, or (ii) y takes a value from the support of E_1 in which case dX_2^μ/ds and dX_3^μ/ds become zero. After time T, either the particle 1 does not longer change its space-time position, or the particles 2 and 3 do not longer change their space-time positions. The trajectory of the particle 1 or the trajectories of the particles 2 and 3 terminate at T, i.e., they do not exist for times $t > T$. This is how relativistic Bohmian interpretation describes the particle destruction.

Unfortunately, the mechanism above works only in a very special case in which the absence of the overlap between $E_1(y)$ and $E_2(y)$ is exact. In a more realistic situation this overlap is negligibly small but not exactly zero. In such a situation neither of the particles will have exactly zero four-velocity. Consequently, neither of the particles will be really destroyed. Nevertheless, the measuring apparatus will still behave as if some particles have been destroyed. For example, if y takes value Y for which $E_1(Y) \ll E_2(Y)$, then for all practical purposes the measuring apparatus behaves as if the wave function collapsed to the second term in (8.147). The particles with positions X_2 and X_3 also behave in that way. Therefore, even though the particle with the position X_1 is not really destroyed, an effective wave function collapse still takes place. The influence of the particle with the position X_1 on the measuring apparatus described by Y is negligible, which is effectively the same as if this particle has been destroyed.

Of course, the interaction with the measuring apparatus is not the only mechanism that may induce destruction of particles. Any interaction with the environment may do that. Or more generally, any interactions among particles may induce not only particle destruction, but also particle creation. Whenever the wave function $\Psi(x_1, x_2, x_3, x_4, \ldots)$ does not really vary (or when this variation is negligible) with some of x_A for some range of values of x_A, then at the edge of this range, a trajectory of the particle A may exhibit true (or apparent) creation or destruction.

In general, a QFT state may be a superposition of n-particle states with n ranging from 0 to ∞. Thus, $\Psi(x_1, x_2, x_3, x_4, \ldots)$ should be viewed as a function that lives in the space of infinitely many coordinates x_A, $A = 1, 2, 3, 4, \ldots, \infty$. In particular, the one-particle wave function $\Psi_1(x_1)$ should be viewed as a function $\Psi_1(x_1, x_2, \ldots)$ with the property $\partial_A^\mu \Psi_1 = 0$ for $A = 2, 3, \ldots, \infty$. It means that any wave function in QFT describes an infinite number of particles, even if most of them have zero four-velocity. As we have already explained, particles with zero four-velocity are dots in space-time. The initial space-time position of any particle may take any value, with the probability proportional to $|\Psi_1(x_1, x_2, \ldots)|^2$. In addition to one continuous particle trajectory, there is also an infinite number of "vacuum" particles, which live for an infinitesimally short time.

The purpose of the remaining subsections of this section is to further elaborate the ideas presented in this subsection and to put them into a more precise framework.

8.4.2 Measurement in QFT as Entanglement with the Environment

Let $\{|b\rangle\}$ be some orthonormal basis of one-particle states. A general normalized one-particle state is

$$|\Psi_1\rangle = \sum_b c_b|b\rangle, \tag{8.149}$$

where the normalization condition implies $\sum_b |c_b|^2 = 1$. From the basis $\{|b\rangle\}$ one can construct the n-particle basis $\{|b_1, \ldots, b_n\rangle\}$, where

$$|b_1, \ldots, b_n\rangle = S_{\{b_1,\ldots,b_n\}}|b_1\rangle \cdots |b_n\rangle. \tag{8.150}$$

Here $S_{\{b_1,\ldots,b_n\}}$ denotes the symmetrization over all $\{b_1, \ldots, b_n\}$ for bosons, or antisymmetrization for fermions. The most general state in QFT describing these particles can be written as

$$|\Psi\rangle = c_0|0\rangle + \sum_{n=1}^{\infty} \sum_{b_1,\ldots,b_n} c_{n;b_1,\ldots,b_n}|b_1, \ldots, b_n\rangle, \tag{8.151}$$

where the vacuum $|0\rangle$ is also introduced. Now the normalization condition implies $|c_0|^2 + \sum_{n=1}^{\infty} \sum_{b_1,\ldots,b_n} |c_{n;b_1,\ldots,b_n}|^2 = 1$.

Now let as assume that the number of particles is measured. It implies that the particles become entangled with the environment such that the total state describing both the measured particles and the environment takes the form

$$|\Psi\rangle_{\text{total}} = c_0|0\rangle|E_0\rangle + \sum_{n=1}^{\infty} \sum_{b_1,\ldots,b_n} c_{n;b_1,\ldots,b_n}|b_1, \ldots, b_n\rangle|E_{n;b_1,\ldots,b_n}\rangle. \tag{8.152}$$

The environment states $|E_0\rangle$, $|E_{n;b_1,\ldots,b_n}\rangle$ are macroscopically distinct. They describe what the observers really observe. When an observer observes that the environment is in the state $|E_0\rangle$ or $|E_{n;b_1,\ldots,b_n}\rangle$, then one says that the original measured QFT state is in the state $|0\rangle$ or $|b_1, \ldots, b_n\rangle$, respectively. In particular, this is how the number of particles is measured in a state (8.151) with an uncertain number of particles. The probability that the environment will be found in the state $|E_0\rangle$ or $|E_{n;b_1,\ldots,b_n}\rangle$ is equal to $|c_0|^2$ or $|c_{n;b_1,\ldots,b_n}|^2$, respectively.

Of course, (8.151) is not the only way the state $|\Psi\rangle$ can be expanded. In general, it can be expanded as

$$|\Psi\rangle = \sum_{\xi} c_{\xi} |\xi\rangle, \tag{8.153}$$

where $|\xi\rangle$ are some normalized (not necessarily orthogonal) states that do not need to have a definite number of particles. A particularly important example are coherent states (see, e.g., [21]), which minimize the products of uncertainties of fields and their canonical momenta. Each coherent state is a superposition of states with all possible numbers of particles, including zero. The coherent states are overcomplete and not orthogonal. Yet, the expansion (8.153) may be an expansion in terms of coherent states $|\xi\rangle$ as well.

Furthermore, the entanglement with the environment does not necessarily need to take the form (8.152). Instead, it may take a more general form:

$$|\Psi\rangle_{\text{total}} = \sum_{\xi} c_{\xi} |\xi\rangle |E_{\xi}\rangle, \tag{8.154}$$

where $|E_{\xi}\rangle$ are macroscopically distinct. In principle, the interaction with the environment may create the entanglement (8.154) with respect to any set of states $\{|\xi\rangle\}$. In practice, however, some types of expansions are preferred. This fact can be explained by the theory of decoherence [22], which explains why states of the form of (8.154) are stable only for some particular sets $\{|\xi\rangle\}$. In fact, depending on details of the interactions with the environment, in most real situations the entanglement takes either the form (8.152) or the form (8.154) with coherent states $|\xi\rangle$. Since coherent states minimize the uncertainties of fields and their canonical momenta, they behave very much like classical fields. This explains why experiments in quantum optics can often be better described in terms of fields rather than particles (see, e.g., [21]). In fact, the theory of decoherence can explain under what conditions the coherent-state basis becomes preferred over basis with definite numbers of particles [23, 24].

Thus, decoherence induced by interaction with the environment can explain why do we observe either a definite number of particles or coherent states that behave very much like classical fields. However, decoherence alone cannot explain why do we observe some particular state of definite number of particles and not some other, or why do we observe some particular coherent state and not some other. Instead, a possible explanation is provided by the Bohmian interpretation.

8.4.3 Free Scalar QFT in the Particle-Position Picture

The purpose of this subsection is to see in detail how states of free QFT without spin can be represented by wave functions. They include wave functions with definite number of particles (discussed in Section 8.3), as well as their superpositions.

Consider a free scalar hermitian field operator $\hat{\phi}(x)$ satisfying the Klein–Gordon equation:

$$\partial^{\mu}\partial_{\mu}\hat{\phi}(x) + m^2\hat{\phi}(x) = 0. \tag{8.155}$$

The field can be decomposed as

$$\hat{\phi}(x) = \hat{\psi}(x) + \hat{\psi}^{\dagger}(x), \tag{8.156}$$

where $\hat{\psi}$ and $\hat{\psi}^{\dagger}$ can be expanded as

$$\begin{aligned}\hat{\psi}(x) &= \int d^3k\, f(\mathbf{k})\, \hat{a}(\mathbf{k}) e^{-i[\omega(\mathbf{k})x^0 - \mathbf{kx}]},\\ \hat{\psi}^{\dagger}(x) &= \int d^3k\, f(\mathbf{k})\, \hat{a}^{\dagger}(\mathbf{k}) e^{i[\omega(\mathbf{k})x^0 - \mathbf{kx}]}.\end{aligned} \tag{8.157}$$

Here

$$\omega(\mathbf{k}) = \sqrt{\mathbf{k}^2 + m^2} \tag{8.158}$$

is the k_0 component of the four-vector $k = \{k_{\mu}\}$, and $\hat{a}^{\dagger}(\mathbf{k})$ and $\hat{a}(\mathbf{k})$ are the creation and destruction operators, respectively (see, e.g., [25]), satisfying the commutation relations $[\hat{a}(\mathbf{k}), \hat{a}(\mathbf{k}')] = [\hat{a}^{\dagger}(\mathbf{k}), \hat{a}^{\dagger}(\mathbf{k}')] = 0$, $[\hat{a}(\mathbf{k}), \hat{a}^{\dagger}(\mathbf{k}')] \propto \delta^3(\mathbf{k} - \mathbf{k}')$. The function $f(\mathbf{k})$ is a real positive function which we do not specify explicitly, because several different choices appear in the literature, corresponding to several different choices of normalization. All subsequent equations will be written in forms that do not explicitly depend on this choice.

We define the operator

$$\hat{\psi}_n(x_{n,1}, \ldots, x_{n,n}) = d_n S_{\{x_{n,1},\ldots,x_{n,n}\}} \hat{\psi}(x_{n,1}) \cdots \hat{\psi}(x_{n,n}). \tag{8.159}$$

The symbol $S_{\{x_{n,1},\ldots,x_{n,n}\}}$ denotes the symmetrization, reminding us that the expression is symmetric under the exchange of coordinates

$\{x_{n,1}, \ldots, x_{n,n}\}$. (Note, however, that the product of operators on the right-hand side of (8.159) is in fact automatically symmetric because the operators $\hat{\psi}(x)$ commute, i.e., $[\hat{\psi}(x), \hat{\psi}(x')] = 0$.) The parameter d_n is a normalization constant determined by the normalization condition that will be specified below. The operator (8.159) allows us to define n-particle states in the basis of particle space-time positions as

$$|x_{n,1}, \ldots, x_{n,n}\rangle = \hat{\psi}_n^\dagger(x_{n,1}, \ldots, x_{n,n})|0\rangle. \tag{8.160}$$

The normalization function $f(\mathbf{k})$ in (8.157) can be chosen such that all states of the form (8.160) at a fixed common time $x_{n,1}^0 = \cdots = x_{n,n}^0 = t$, together with the vacuum $|0\rangle$, form a complete and orthogonal basis in the Hilbert space of physical states. For example, for one-particle states the orthogonality relation reads $\langle \mathbf{x}; t|\mathbf{x}'; t\rangle = \delta^3(\mathbf{x} - \mathbf{x}')$, and similarly for n-particle states. However, for such a choice of $f(\mathbf{k})$, the operators (8.157) are not Lorentz invariant. Thus, it is more appropriate to sacrifice orthogonality by choosing $f(\mathbf{k})$ such that (8.157) are Lorentz invariant. In the rest of the analysis we assume such a Lorentz-invariant normalization of (8.157).

If $|\Psi_n\rangle$ is an arbitrary (but normalized) n-particle state, then this state can be represented by the n-particle wave function:

$$\psi_n(x_{n,1}, \ldots, x_{n,n}) = \langle x_{n,1}, \ldots, x_{n,n}|\Psi_n\rangle. \tag{8.161}$$

We also have

$$\langle x_{n,1}, \ldots, x_{n,n}|\Psi_{n'}\rangle = 0 \quad \text{for} \quad n \neq n'. \tag{8.162}$$

We choose the normalization constant d_n in (8.159) such that the following normalization condition is satisfied:

$$\int d^4x_{n,1} \cdots \int d^4x_{n,n}\, |\psi_n(x_{n,1}, \ldots, x_{n,n})|^2 = 1. \tag{8.163}$$

However, this implies that the wave functions $\psi_n(x_{n,1}, \ldots, x_{n,n})$ and $\psi_{n'}(x_{n',1}, \ldots, x_{n',n'})$, with different values of n and n', are normalized in different spaces. On the other hand, we want these wave functions to live in the same space such that we can form superpositions of wave functions describing different numbers of particles. To accomplish this, we define

$$\Psi_n(x_{n,1}, \ldots, x_{n,n}) = \sqrt{\frac{\mathcal{V}^{(n)}}{\mathcal{V}}}\, \psi_n(x_{n,1}, \ldots, x_{n,n}), \tag{8.164}$$

where

$$\mathcal{V}^{(n)} = \int d^4x_{n,1} \cdots \int d^4x_{n,n}, \tag{8.165}$$

$$\mathcal{V} = \prod_{n=1}^{\infty} \mathcal{V}^{(n)}, \tag{8.166}$$

are volumes of the corresponding configuration spaces. In particular, the wave function of the vacuum is

$$\Psi_0 = \frac{1}{\sqrt{\mathcal{V}}}. \tag{8.167}$$

This provides that all wave functions are normalized in the same configuration space as

$$\int \mathcal{D}\vec{x}\, |\Psi_n(x_{n,1}, \ldots, x_{n,n})|^2 = 1, \tag{8.168}$$

where we use the notation

$$\vec{x} = (x_{1,1}, x_{2,1}, x_{2,2}, \ldots), \tag{8.169}$$

$$\mathcal{D}\vec{x} = \prod_{n=1}^{\infty} \prod_{a_n=1}^{n} d^4x_{n,a_n}. \tag{8.170}$$

Note that the physical Hilbert space does not contain nonsymmetrized states, such as a three-particle state $|x_{1,1}\rangle|x_{2,1}, x_{2,2}\rangle$. It also does not contain states that do not satisfy (8.158). Nevertheless, the notation can be further simplified by introducing an extended kinematic Hilbert space that contains such unphysical states as well. Every physical state can be viewed as a state in such an extended Hilbert space, although most of the states in the extended Hilbert space are not physical. In this extended space it is convenient to denote the pair of labels (n, a_n) by a single label A. Hence, (8.169) and (8.170) are now written as

$$\vec{x} = (x_1, x_2, x_3, \ldots), \tag{8.171}$$

$$\mathcal{D}\vec{x} = \prod_{A=1}^{\infty} d^4x_A. \tag{8.172}$$

Similarly, (8.166) with (8.165) is now written as

$$\mathcal{V} = \int \prod_{A=1}^{\infty} d^4x_A. \tag{8.173}$$

The particle-position basis of this extended space is denoted by $|\vec{x})$ (which should be distinguished from $|\vec{x}\rangle$ which would denote a symmetrized state of an infinite number of physical particles). Such a basis allows us to write the physical wave function (8.164) as a wave function on the extended space:

$$\Psi_n(\vec{x}) = (\vec{x}|\Psi_n\rangle. \tag{8.174}$$

Now (8.168) takes a simpler form:

$$\int \mathcal{D}\vec{x}\, |\Psi_n(\vec{x})|^2 = 1. \tag{8.175}$$

The unit operator on the extended space is

$$1 = \int \mathcal{D}\vec{x}\, |\vec{x})(\vec{x}|, \tag{8.176}$$

while the scalar product is

$$(\vec{x}|\vec{x}') = \delta(\vec{x} - \vec{x}'), \tag{8.177}$$

with $\delta(\vec{x} - \vec{x}') \equiv \prod_{A=1}^{\infty} \delta^4(x_A - x'_A)$. A general physical state can be written as

$$\Psi(\vec{x}) = (\vec{x}|\Psi\rangle = \sum_{n=0}^{\infty} c_n \Psi_n(\vec{x}). \tag{8.178}$$

It is also convenient to write this as

$$\Psi(\vec{x}) = \sum_{n=0}^{\infty} \tilde{\Psi}_n(\vec{x}), \tag{8.179}$$

where the tilde denotes a wave function that is not necessarily normalized. The total wave function is normalized, in the sense that

$$\int \mathcal{D}\vec{x}\, |\Psi(\vec{x})|^2 = 1, \tag{8.180}$$

implying

$$\sum_{n=0}^{\infty} |c_n|^2 = 1. \tag{8.181}$$

Next, we introduce the operator

$$\Box = \sum_{A=1}^{\infty} \partial_A^{\mu} \partial_{A\mu}. \tag{8.182}$$

From the equations above (see, in particular, (8.155)–(8.161)), it is easy to show that $\Psi_n(\vec{x})$ satisfies

$$\Box\Psi_n(\vec{x}) + nm^2\Psi_n(\vec{x}) = 0. \tag{8.183}$$

Introducing a hermitian number-operator $\hat{N}$ with the property

$$\hat{N}\Psi_n(\vec{x}) = n\Psi_n(\vec{x}), \tag{8.184}$$

one finds that a general physical state (8.178) satisfies the generalized Klein–Gordon equation

$$\Box\Psi(\vec{x}) + m^2\hat{N}\Psi(\vec{x}) = 0. \tag{8.185}$$

We also introduce the generalized Klein–Gordon current

$$J_A^\mu(\vec{x}) = \frac{i}{2}\Psi^*(\vec{x})\overset{\leftrightarrow}{\partial_A^\mu}\Psi(\vec{x}). \tag{8.186}$$

From (8.185) one finds that, in general, this current is not conserved

$$\sum_{A=1}^{\infty}\partial_{A\mu}J_A^\mu(\vec{x}) = J(\vec{x}), \tag{8.187}$$

where

$$J(\vec{x}) = -\frac{i}{2}m^2\Psi^*(\vec{x})\overset{\leftrightarrow}{\hat{N}}\Psi(\vec{x}), \tag{8.188}$$

and $\Psi'\overset{\leftrightarrow}{\hat{N}}\Psi \equiv \Psi'(\hat{N}\Psi) - (\hat{N}\Psi')\Psi$. From (8.188) we see that the current is conserved in two special cases: (i) when $\Psi = \Psi_n$ (a state with a definite number of physical particles), or (ii) when $m^2 = 0$ (any physical state of massless particles).

Finally, let us rewrite some of the main results of this (somewhat lengthy) subsection in a form that will be suitable for a generalization in the next subsection. A general physical state can be written in the form

$$|\Psi\rangle = \sum_{n=0}^{\infty}c_n|\Psi_n\rangle = \sum_{n=0}^{\infty}|\tilde{\Psi}_n\rangle. \tag{8.189}$$

The corresponding unnormalized n-particle wave functions are

$$\tilde{\psi}_n(x_{n,1},\ldots,x_{n,n}) = \langle 0|\hat{\psi}_n(x_{n,1},\ldots,x_{n,n})|\Psi\rangle. \tag{8.190}$$

There is a well-defined transformation

$$\tilde{\psi}_n(x_{n,1},\ldots,x_{n,n}) \to \tilde{\Psi}_n(\vec{x}) \tag{8.191}$$

from the physical Hilbert space to the extended Hilbert space, so that the general state (8.189) can be represented by a single wave function

$$\Psi(\vec{x}) = \sum_{n=0}^{\infty}c_n\Psi_n(\vec{x}) = \sum_{n=0}^{\infty}\tilde{\Psi}_n(\vec{x}). \tag{8.192}$$

8.4.4 Generalization to Interacting QFT

In this subsection we discuss the generalization of the results of the preceding subsection to the case in which the field operator $\hat{\phi}$ does not satisfy the free Klein–Gordon equation (8.155). For example, if the classical action for the field is

$$S = \int d^4x \left[\frac{1}{2}(\partial^\mu \phi)(\partial_\mu \phi) - \frac{m^2}{2}\phi^2 - \frac{\lambda}{4}\phi^4 \right], \qquad (8.193)$$

then (8.155) generalizes to

$$\partial^\mu \partial_\mu \hat{\phi}_H(x) + m^2 \hat{\phi}_H(x) + \lambda \hat{\phi}_H^3(x) = 0, \qquad (8.194)$$

where $\hat{\phi}_H(x)$ is the field operator in the Heisenberg picture. (From this point of view, the operator $\hat{\phi}(x)$ defined by (8.156) and (8.157) and satisfying the free Klein–Gordon equation (8.155) is the field operator in the interaction (Dirac) picture.) Thus, instead of (8.190) now we have

$$\tilde{\psi}_n(x_{n,1}, \ldots, x_{n,n}) = \langle 0|\hat{\psi}_{nH}(x_{n,1}, \ldots, x_{n,n})|\Psi\rangle, \qquad (8.195)$$

where $|\Psi\rangle$ and $|0\rangle$ are states in the Heisenberg picture. Assuming that (8.195) has been calculated (we shall see below how in practice it can be done), the rest of the job is straightforward. One needs to make the transformation (8.191) in the same way as in the free case, which leads to an interacting variant of (8.192):

$$\Psi(\vec{x}) = \sum_{n=0}^{\infty} \tilde{\Psi}_n(\vec{x}). \qquad (8.196)$$

The wave function (8.196) encodes the complete information about the properties of the interacting system.

Now let us see how (8.195) can be calculated in practice. Any operator $\hat{O}_H(t)$ in the Heisenberg picture depending on a single time-variable t can be written in terms of operators in the interaction picture as

$$\hat{O}_H(t) = \hat{U}^\dagger(t)\hat{O}(t)\hat{U}(t), \qquad (8.197)$$

where

$$\hat{U}(t) = Te^{-i\int_{t_0}^{t} dt' \hat{H}_{\text{int}}(t')}, \qquad (8.198)$$

t_0 is some appropriately chosen "initial" time, T denotes the time ordering, and $\hat{H}_{\text{int}}$ is the interaction part of the Hamiltonian expressed as a functional of field operators in the interaction picture (see, e.g., [26]). For example, for the action (8.193), we have

$$\hat{H}_{\text{int}}(t) = \frac{\lambda}{4} \int d^3x \; :\hat{\phi}^4(\mathbf{x}, t):, \tag{8.199}$$

where : : denotes the normal ordering. The relation (8.197) can be inverted, leading to

$$\hat{O}(t) = \hat{U}(t)\hat{O}_H(t)\hat{U}^\dagger(t). \tag{8.200}$$

Thus, the relation (8.159), which is now valid in the interaction picture, allows us to write an analogous relation in the Heisenberg picture:

$$\hat{\psi}_{nH}(x_{n,1}, \ldots, x_{n,n}) = d_n S_{\{x_{n,1},\ldots,x_{n,n}\}} \hat{\psi}_H(x_{n,1}) \cdots \hat{\psi}_H(x_{n,n}), \tag{8.201}$$

where

$$\hat{\psi}_H(x_{n,a_n}) = \hat{U}^\dagger(x^0_{n,a_n})\hat{\psi}(x_{n,a_n})\hat{U}(x^0_{n,a_n}). \tag{8.202}$$

By expanding (8.198) in powers of $\int_{t_0}^{t} dt' \hat{H}_{\text{int}}$, this allows us to calculate (8.201) and (8.195) perturbatively. In (8.195), the states in the Heisenberg picture $|\Psi\rangle$ and $|0\rangle$ are identified with the states in the interaction picture at the initial time $|\Psi(t_0)\rangle$ and $|0(t_0)\rangle$, respectively.

To demonstrate that such a procedure leads to a physically sensible result, let us see how it works in the special (and more familiar) case of the equal-time wave function. It is given by $\tilde{\psi}_n(x_{n,1}, \ldots, x_{n,n})$ calculated at $x^0_{n,1} = \cdots = x^0_{n,n} \equiv t$. Thus, (8.195) reduces to

$$\begin{aligned}\tilde{\psi}_n(\mathbf{x}_{n,1}, \ldots, \mathbf{x}_{n,n}; t) = d_n \langle 0(t_0)|\hat{U}^\dagger(t)\hat{\psi}(\mathbf{x}_{n,1}, t)\hat{U}(t) \\ \cdots \hat{U}^\dagger(t)\hat{\psi}(\mathbf{x}_{n,n}, t)\hat{U}(t)|\Psi(t_0)\rangle.\end{aligned} \tag{8.203}$$

Using $\hat{U}(t)\hat{U}^\dagger(t) = 1$ and

$$\hat{U}(t)|\Psi(t_0)\rangle = |\Psi(t)\rangle, \quad \hat{U}(t)|0(t_0)\rangle = |0(t)\rangle, \tag{8.204}$$

the expression further simplifies:

$$\tilde{\psi}_n(\mathbf{x}_{n,1}, \ldots, \mathbf{x}_{n,n}; t) = d_n \langle 0(t)|\hat{\psi}(\mathbf{x}_{n,1}, t) \cdots \hat{\psi}(\mathbf{x}_{n,n}, t)|\Psi(t)\rangle. \tag{8.205}$$

In practical applications of QFT in particle physics, one usually calculates the S-matrix, corresponding to the limit $t_0 \to -\infty$,

$t \to \infty$. For Hamiltonians that conserve energy (such as (8.199)), this limit provides the stability of the vacuum, i.e., obeys

$$\lim_{t_0 \to -\infty,\ t \to \infty} \hat{U}(t)|0(t_0)\rangle = e^{-i\varphi_0}|0(t_0)\rangle, \tag{8.206}$$

where φ_0 is some physically irrelevant phase [25]. Essentially, this is because the integrals of the type $\int_{-\infty}^{\infty} dt' \cdots$ produce δ-functions that correspond to energy conservation, so the vacuum remains stable because particle creation from the vacuum would violate energy conservation. Thus we have

$$|0(\infty)\rangle = e^{-i\varphi_0}|0(-\infty)\rangle \equiv e^{-i\varphi_0}|0\rangle. \tag{8.207}$$

The state

$$|\Psi(\infty)\rangle = \hat{U}(\infty)|\Psi(-\infty)\rangle \tag{8.208}$$

is not trivial, but whatever it is, it has some expansion of the form

$$|\Psi(\infty)\rangle = \sum_{n=0}^{\infty} c_n(\infty)|\Psi_n\rangle, \tag{8.209}$$

where $c_n(\infty)$ are some coefficients. Plugging (8.207) and (8.209) into (8.205) and recalling (8.159)–(8.162), we finally obtain

$$\tilde{\psi}_n(\mathbf{x}_{n,1}, \ldots, \mathbf{x}_{n,n}; \infty) = e^{i\varphi_0} c_n(\infty)\psi_n(\mathbf{x}_{n,1}, \ldots, \mathbf{x}_{n,n}; \infty). \tag{8.210}$$

This demonstrates the consistency of (8.195), because (8.208) should be recognized as the standard description of evolution from $t_0 \to -\infty$ to $t \to \infty$ (see, e.g., [25, 26]), showing that the coefficients $c_n(\infty)$ are the same as those described by standard S-matrix theory in QFT. In other words, (8.195) is a natural many-time generalization of the concept of single-time evolution in interacting QFT.

8.4.5 Generalization to Other Types of Particles

In Sections 8.4.3 and 8.4.4, we have discussed in detail scalar hermitian fields, corresponding to spinless uncharged particles. In this subsection we briefly discuss how these results can be generalized to any type of fields and the corresponding particles.

In general, fields ϕ carry some additional labels, which we collectively denote by l, so we deal with fields ϕ_l. For example, spin 1 field carries a polarization label (see Section 8.3.3.5), fermionic spin

$\frac{1}{2}$ field carries a spinor index, non-Abelian gauge fields carry internal indices of the gauge group, etc. Thus Eq. (8.159) generalizes to

$$\hat{\psi}_{n,L_n}(x_{n,1},\ldots,x_{n,n}) = d_n S_{\{x_{n,1},\ldots,x_{n,n}\}} \hat{\psi}_{l_{n,1}}(x_{n,1}) \cdots \hat{\psi}_{l_{n,n}}(x_{n,n}), \tag{8.211}$$

where L_n is a collective label $L_n = (l_{n,1},\ldots,l_{n,n})$. The symbol $S_{\{x_{n,1},\ldots,x_{n,n}\}}$ denotes symmetrization (antisymmetrization) over bosonic (fermionic) fields describing the same type of particles. Hence, it is straightforward to make the appropriate generalizations of all results of Sections 8.4.3 and 8.4.4. For example, (8.179) generalizes to

$$\Psi_{\vec{L}}(\vec{x}) = \sum_{n=0}^{\infty} \sum_{L_n} \tilde{\Psi}_{n,L_n}(\vec{x}), \tag{8.212}$$

with self-explaining notation.

To further simplify the notation, we introduce the column $\Psi \equiv \{\Psi_{\vec{L}}\}$ and the row $\Psi^\dagger \equiv \{\Psi^*_{\vec{L}}\}$. With this notation, the appropriate generalization of (8.180) can be written as

$$\int \mathcal{D}\vec{x} \sum_{\vec{L}} \Psi^*_{\vec{L}}(\vec{x}) \Psi_{\vec{L}}(\vec{x}) \equiv \int \mathcal{D}\vec{x}\, \Psi^\dagger(\vec{x}) \Psi(\vec{x}) = 1. \tag{8.213}$$

8.4.6 Probabilistic Interpretation

The quantity

$$\mathcal{D}P = \Psi^\dagger(\vec{x}) \Psi(\vec{x})\, \mathcal{D}\vec{x} \tag{8.214}$$

is naturally interpreted as the probability of finding the system in the (infinitesimal) configuration-space volume $\mathcal{D}\vec{x}$ around a point $\vec{x}$ in the configuration space. Indeed, such an interpretation is consistent with our normalization conditions such as (8.180) and (8.213). In more physical terms, (8.214) gives the joint probability that the particle 1 is found at the space-time position x_1, particle 2 at the space-time position x_2, etc.

As a special case, consider an n-particle state $\Psi(\vec{x}) = \Psi_n(\vec{x})$. It really depends only on n space-time positions $x_{n,1}, \ldots x_{n,n}$. With respect to all other positions x_B, Ψ is a constant. Thus, the probability of various positions x_B does not depend on

x_B; such a particle can be found anywhere and anytime with equal probabilities. There is an infinite number of such particles. Nevertheless, the Fourier transform of such a wave function reveals that the four-momentum k_B of these particles is necessarily zero; they have neither three-momentum nor energy. For that reason, such particles can be thought of as "vacuum" particles. In this picture, an n-particle state Ψ_n is thought of as a state describing n "real" particles and an infinite number of "vacuum" particles.

To avoid a possible confusion with the usual notions of vacuum and real particles in QFT, in the rest of the paper we refer to "vacuum" particles as *dead* particles and "real" particles as *live* particles. Or let us be slightly more precise: We say that the particle A is dead if the wave function in the momentum space $\Psi(\vec{k})$ vanishes for all values of k_A except $k_A = 0$. Similarly, we say that the particle A is live if it is not dead.

The properties of live particles associated with the state $\Psi_n(\vec{x})$ can also be represented by the wave function $\psi_n(x_{n,1}, \ldots, x_{n,n})$. By averaging over physically uninteresting dead particles, (8.214) reduces to

$$dP = \psi_n^\dagger(x_{n,1}, \ldots, x_{n,n})\psi_n(x_{n,1}, \ldots, x_{n,n})\, d^4x_{n,1} \cdots d^4x_{n,n}, \tag{8.215}$$

which involves only live particles. In this way, the probabilistic interpretation is reduced to the probabilistic interpretation of relativistic QM with a fixed number of particles, which is studied in Section 8.3.1.

Now let us see how the wave functions representing the states in interacting QFT are interpreted probabilistically. Consider the wave function $\tilde{\psi}_n(x_{n,1}, \ldots, x_{n,n})$ given by (8.195). For example, it may vanish for small values of $x^0_{n,1}, \ldots, x^0_{n,n}$, but it may not vanish for their large values. Physically, it means that these particles cannot be detected in the far past (the probability is zero), but that they can be detected in the far future. This is nothing but a probabilistic description of the creation of n particles that have not existed in the far past. Indeed, the results obtained in Section 8.4.4 (see, in particular, (8.210)) show that such probabilities are consistent with the probabilities of particle creation obtained by the standard S-matrix methods in QFT.

Having developed the probabilistic interpretation, we can also calculate the average values of various quantities. In particular, the average value of the four-momentum P_A^μ is

$$\langle P_A^\mu \rangle = \int \mathcal{D}\vec{x}\, \Psi^\dagger(\vec{x}) \hat{P}_A^\mu \Psi(\vec{x}), \tag{8.216}$$

where $\hat{P}_A^\mu = i\partial_A^\mu$ is the four-momentum operator. Equation (8.216) can also be written as

$$\langle P_A^\mu \rangle = \int \mathcal{D}\vec{x}\, \rho(\vec{x}) U_A^\mu(\vec{x}), \tag{8.217}$$

where

$$\rho(\vec{x}) = \Psi^\dagger(\vec{x})\Psi(\vec{x}) \tag{8.218}$$

is the probability density and

$$U_A^\mu(\vec{x}) = \frac{J_A^\mu(\vec{x})}{\Psi^\dagger(\vec{x})\Psi(\vec{x})}. \tag{8.219}$$

Here J_A^μ is given by an obvious generalization of (8.186):

$$J_A^\mu(\vec{x}) = \frac{i}{2}\Psi^\dagger(\vec{x}) \overleftrightarrow{\partial_A^\mu} \Psi(\vec{x}). \tag{8.220}$$

The expression (8.217) will play an important role in the next subsection.

8.4.7 Bohmian Interpretation

In the Bohmian interpretation, each particle has some trajectory $X_A^\mu(s)$. Such trajectories must be consistent with the probabilistic interpretation (8.214). Thus, we need a velocity function $V_A^\mu(\vec{x})$, so that the trajectories satisfy

$$\frac{dX_A^\mu(s)}{ds} = V_A^\mu(\vec{X}(s)), \tag{8.221}$$

where the velocity function must be such that the following conservation equation is obeyed:

$$\frac{\partial\rho(\vec{x})}{\partial s} + \sum_{A=1}^{\infty} \partial_{A\mu}[\rho(\vec{x}) V_A^\mu(\vec{x})] = 0. \tag{8.222}$$

Namely, if a statistical ensemble of particle positions in space-time has the distribution (8.218) for some initial s, then (8.221) and

(8.222) will provide that this statistical ensemble will also have the distribution (8.218) for *any* s, making the trajectories consistent with (8.214). The first term in (8.222) trivially vanishes: $\partial\rho(\vec{x})/\partial s = 0$. Thus, the condition (8.222) reduces to the requirement

$$\sum_{A=1}^{\infty} \partial_{A\mu}[\rho(\vec{x})V_A^{\mu}(\vec{x})] = 0. \tag{8.223}$$

In addition, we require that the average velocity should be proportional to the average momentum (8.217), i.e.,

$$\int \mathcal{D}\vec{x}\,\rho(\vec{x})V_A^{\mu}(\vec{x}) = \text{const} \times \int \mathcal{D}\vec{x}\,\rho(\vec{x})U_A^{\mu}(\vec{x}). \tag{8.224}$$

In fact, the constant in (8.224) is physically irrelevant, because it can always be absorbed into a rescaling of the parameter s in (8.221). Thus we fix const $= 1$.

As a first guess, Eq. (8.224) with const $= 1$ suggests that one could take $V_A^{\mu} = U_A^{\mu}$. However, it does not work in general. Namely, from (8.218) and (8.219), we see that $\rho U_A^{\mu} = J_A^{\mu}$, and we have seen in (8.187) that J_A^{μ} does not need to be conserved. Instead, we have

$$\sum_{A=1}^{\infty} \partial_{A\mu}[\rho(\vec{x})U_A^{\mu}(\vec{x})] = J(\vec{x}), \tag{8.225}$$

where $J(\vec{x})$ is some function that can be calculated explicitly whenever $\Psi(\vec{x})$ is known. Therefore, instead of $V_A^{\mu} = U_A^{\mu}$ we must take

$$V_A^{\mu}(\vec{x}) = U_A^{\mu}(\vec{x}) + \rho^{-1}(\vec{x})[e_A^{\mu} + E_A^{\mu}(\vec{x})], \tag{8.226}$$

where

$$e_A^{\mu} = -\mathcal{V}^{-1}\int \mathcal{D}\vec{x}\,E_A^{\mu}(\vec{x}), \tag{8.227}$$

$$E_A^{\mu}(\vec{x}) = \partial_A^{\mu}\int \mathcal{D}\vec{x}'\,G(\vec{x},\vec{x}')J(\vec{x}'), \tag{8.228}$$

$$G(\vec{x},\vec{x}') = \int \frac{\mathcal{D}\vec{k}}{(2\pi)^{4\aleph_0}}\frac{e^{i\vec{k}(\vec{x}-\vec{x}')}}{\vec{k}^2}, \tag{8.229}$$

and $\aleph_0 = \infty$ is the cardinal number of the set of natural numbers. It is straightforward to show that Eqs. (8.228)–(8.229) provide that

(8.226) obeys (8.223), while (8.227) provides that (8.226) obeys (8.224) with const = 1.

We note two important properties of (8.226). First, if $J = 0$ in (8.225), then $V_A^\mu = U_A^\mu$. In particular, since $J = 0$ for free fields in states with a definite number of particles (it can be derived for any type of particles analogously to the derivation of (8.188) for spinless uncharged particles), it follows that $V_A^\mu = U_A^\mu$ for such states. Second, if $\Psi(\vec{x})$ does not depend on some coordinate x_B^μ, then both $U_B^\mu = 0$ and $V_B^\mu = 0$. [To show that $V_B^\mu = 0$, note first that $J(\vec{x})$ defined by (8.225) does not depend on x_B^μ when $\Psi(\vec{x})$ does not depend on x_B^μ. Then the integration over $dx_B'^\mu$ in (8.228) produces $\delta(k_B^\mu)$, which kills the dependence on x_B^μ carried by (8.229)]. This implies that dead particles have zero four-velocity.

Having established the general theory of particle trajectories by the results above, now we can discuss particular consequences.

The trajectories are determined uniquely if the initial space-time positions $X_A^\mu(0)$ in (8.221), for all $\mu = 0, 1, 2, 3$, $A = 1, \ldots, \infty$, are specified. In particular, since dead particles have zero four-velocity, such particles do not really have trajectories in space-time. Instead, they are represented by dots in space-time. The space-time positions of these dots are specified by their initial space-time positions.

Since $\rho(\vec{x})$ describes probabilities for particle creation and destruction, and since (8.222) provides that particle trajectories are such that space-time positions of particles are distributed according to $\rho(\vec{x})$, it implies that particle trajectories are also consistent with particle creation and destruction. In particular, the trajectories in space-time may have beginning and ending points, which correspond to points at which their four-velocities vanish. For example, the four-velocity of the particle A vanishes if the conditional wave function $\Psi(x_A, \vec{X}')$ does not depend on x_A (where $\vec{X}'$ denotes the actual space-time positions of all particles except the particle A).

One very efficient mechanism of destroying particles is through the interaction with the environment such that the total quantum state takes the form (8.152). The environment wave functions $(\vec{x}|E_0\rangle$, $(\vec{x}|E_{n;b_1,\ldots,b_n}\rangle$ do not overlap, so the particles describing the environment can be in the support of only one of these environment wave functions. Consequently, the conditional wave function is described by only one of the terms in the sum (8.152), which effectively collapses the wave function to only one of the terms in (8.151). For example, if the latter wave function is $(\vec{x}|b_1, \ldots, b_n\rangle$, then it depends on only n coordinates among all x_A. All other live particles from sectors with $n' \neq n$ become dead, i.e., their four-velocities become zero which appears as their destruction in spacetime. More generally, if the overlap between the environment wave functions is negligible but not exactly zero, then particles from sectors with $n' \neq n$ will not become dead, but their influence on the environment will still be negligible, which still provides an effective collapse to $(\vec{x}|b_1, \ldots, b_n\rangle$.

Another physically interesting situation is when the entanglement with the environment takes the form (8.154), where $|\xi\rangle$ are coherent states. In this case, the behavior of the environment can very well be described in terms of an environment that responds to a presence of classical fields. This explains how classical fields may appear at the macroscopic level, even though the microscopic ontology is described in terms of particles. Since $|\xi\rangle$ is a superposition of states with all possible numbers of particles, trajectories of particles from sectors with different numbers of particles coexist; there is an infinite number of live particle trajectories in that case.

8.5 Conclusion

The usual formulation of Bohmian mechanics is not relativistic covariant because it is based on standard QM, which is also not relativistic covariant. Thus, to make Bohmian mechanics covariant, one needs first to reformulate the standard QM in a covariant way such that time is treated on an equal footing with space. More specifically, it means the following. First, the space probability density should be generalized to the space-time probability density. Second, the single-time wave function should be generalized to the many-time wave function. When standard QM is generalized in that way, then the construction of a relativistic-covariant version of Bohmian mechanics is straightforward.

To make the Bohmian mechanics of particles compatible with QFT and particle creation and destruction, one needs to do the following. First, QFT states should be represented by wave functions that depend on an infinite number of coordinates. Second, one needs to use the quantum theory of measurements, which then leads to an effective collapse into states of definite number of particles.

Acknowledgments

This work was supported by the Ministry of Science of the Republic of Croatia under Contract No. 098-0982930-2864.

References

1. H. Nikolić, *Resolving the Black-Hole Information Paradox by Treating Time on an Equal Footing With Space*, Physics Letters B, **678**, 218 (2009).
2. H. Nikolić, *The Spacetime View of the Information Paradox*, International Journal of Quantum Information, **10**, 1250024 (2012).
3. H. Nikolić, *Time in Relativistic and Nonrelativistic Quantum Mechanics*, International Journal of Quantum Information, **7**, 595 (2009).
4. H. Nikolić, *QFT as Pilot-Wave Theory of Particle Creation and Destruction*, International Journal of Modern Physics A, **25**, 1477 (2010).
5. H. Nikolić, *Making Nonlocal Reality Compatible with Relativity*, International Journal of Quantum Information, **9**, 367 (2011).
6. H. Nikolić, *Superluminal Velocities and Nonlocality in Relativistic Mechanics with Scalar Potential*, arXiv:1006.1986.
7. H. Nikolić, *Bohmian Particle Trajectories in Relativistic Bosonic Quantum Field Theory*, Foundations of Physics Letters, **17**, 363 (2004).
8. H. Nikolić, *Bohmian Particle Trajectories in Relativistic Fermionic Quantum Field Theory*, Foundations of Physics Letters, **18**, 123 (2005).
9. H. Nikolić, *Relativistic Quantum Mechanics and the Bohmian Interpretation*, Foundations of Physics Letters, **18**, 549 (2005).
10. H. Nikolić, *Relativistic Bohmian Interpretation of Quantum Mechanics*, AIP Conference Proceedings, **844**, 272 (2006) [quant-ph/0512065].
11. H. Nikolić, *Time and Probability: From Classical Mechanics to Relativistic Bohmian Mechanics*, arXiv:1309.0400.
12. J. D. Jackson, *Classical Electrodynamics*, John Wiley & Sons, New York (1962).

13. S. Weinberg, *Gravitation and Cosmology*, John Wiley & Sons, New York (1972).
14. E. C. G. Stückelberg, *La signification du temps propre en mecanique ondulatoire*, Helvetica Physica Acta, **14**, 322 (1941); *Remarque a propos de la creation de paires de particules en theorie de relativite*, Helvetica Physica Acta, **14**, 588 (1941).
15. L. I. Schiff, *Quantum Mechanics*, McGraw-Hill, Singapore (1968).
16. F. Halzen and A. D. Martin, *Quarks and Leptons*, John Willey & Sons, New York (1984).
17. J. D. Bjorken and S. D. Drell, *Relativistic Quantum Mechanics*, McGraw-Hill, New York (1964).
18. S. Tomonaga, *On a Relativistically Invariant Formulation of the Quantum Theory of Wave Fields*, Progress of Theoretical Physics, **1**, 27 (1946).
19. N. D. Birrell and P. C. W. Davies, *Quantum Fields in Curved Space*, Cambridge Press, New York (1982).
20. L. H. Ryder, *Quantum Field Theory*, Cambridge University Press, Cambridge (1984).
21. L. E. Ballentine, *Quantum Mechanics: A Modern Development*, World Scientific Publishing, Singapore (2000).
22. M. Schlosshauer, *Decoherence and the Quantum-to-Classical Transition*, Springer, Berlin (2007).
23. O. Kübler and H. D. Zeh, *Dynamics of Quantum Correlations*, Annals of Physics, **76**, 405 (1973).
24. J. R. Anglin and W. H. Zurek, *Decoherence of Quantum Fields: Pointer States and Predictability*, Physics Review D, **53**, 7327 (1996).
25. J. D. Bjorken and S. D. Drell, *Relativistic Quantum Fields*, McGraw-Hill Book Company, New York, (1965).
26. T.-P. Cheng and L.-F. Li, *Gauge Theory of Elementary Particle Physics*, Clarendon Press, Oxford (1984).

Chapter 9

Quantum Accelerating Universe

Pedro F. González-Díaz[a] **and Alberto Rozas-Fernández**[b]
[a]*Colina de los Chopos, Instituto de Física Fundamental,*
Consejo Superior de Investigaciones Científicas, Serrano 121, 28006 Madrid, Spain
[b]*Instituto de Astrofísica e Ciências do Espaço, Faculdade de Ciências da Universidade de Lisboa, Edifício C8, Campo Grande, P-1749-016 Lisbon, Portugal*
a.rozas@fc.ul.pt

Starting with the original quantum dark-energy model, the current accelerating phase of the evolution of the universe is considered by constructing most economical cosmic models that use just general relativity and some dominating quantum effects associated with the probabilistic description of quantum physics. Two of such models are explicitly analyzed. They are based on the existence of a quantum potential and correspond to a generalization of the spatially flat exponential model of de Sitter space. The thermodynamics of these two cosmic solutions is discussed, using the second principle as a guide to choose which among the two is more feasible. This chapter also discusses the relativistic physics on which the models are based, their holographic description, some implications from the classical energy conditions, and an interpretation of dark energy in terms of the entangled energy of the universe. Also contained in this

Alberto Rozas-Fernández wishes to dedicate the revision of this chapter to the memory of Prof. Pedro F. González-Díaz.

Applied Bohmian Mechanics: From Nanoscale Systems to Cosmology (Second Edition)
Edited by Xavier Oriols and Jordi Mompart

ISBN 978-981-4800-10-5 (Hardcover), 978-0-429-29474-7 (eBook)
www.jennystanford.com

chapter is a discussion on the quantum cosmic models that result from the existence of a nonzero entropy of entanglement. In such a realm, we obtain new cosmic solutions for any arbitrary number of spatial dimensions, studying the stability of these solutions, as well as the emergence of gravitational waves in the realm of the most general models. The occurrence of the scaling accelerated phase after matter dominance has been shown to be rather associated with the existence of quantum potentials, which make the effective mass of the matter particles to vanish at the coincidence time so that a cosmic system where the matter dominance phase is followed by accelerating expansion can be allowed.

9.1 Introduction

For the time being, the concept of dark energy continues to pose one of the biggest problems of all physics, which, in spite of many attempts and theories intended to solve or at least ameliorate it, has hitherto not found a conclusive outcome. Among such attempts and tentative theories, without trying to be at all exhaustive, we may count what has been dubbed as quintessence, a scalar-field theory satisfying a equation of state $p = w\rho$, where p and ρ are the pressure and the energy density of the quintessence field, respectively. The parameter w is bounded in such a way that $-1 \leq w \leq -1/3$, or its phantom-energy extension for which $w < -1$. Also very popular have been the so-called cosmic generalized Chaplygin gas theories, where the equation of state adopts a more exotic structure, or the tachyonic models for dark energy that describe suitable generalizations from the quintessential scalar fields (see [1] for a recent review). Besides some rather serious difficulties in trying to fix the observational data, all of the above theories appear to be artificial, such as inflation theories are within the inflationary paradigm. In the last few years, there have become quite fashionable some forms of modified-gravity theories in which one does not include any vacuum field but changes instead the gravitational Lagrangian by adding some convenient extra terms that are able to match inflation for large values of the Ricci curvature and describe an accelerating behavior at the smallest curvatures. Some of such theories are mathematically equivalent to the introduction of quintessence and phantom fields, but all of them suffer from the typical problems associated with having a non-Hilbert–Einstein action and may violate some solar system tests.

From the observational standpoint, the rapidly accumulating data coming from supernova Ia luminosity distance measurements, quasar statistics determinations, or studies of the fluctuations in the cosmic microwave background (CMB) radiation seems to imply a value for the parameter of the equation of state, which becomes, each time, closer and closer to $w = -1$, which corresponds to a typical cosmological constant, and it is at 1σ level in the following range

$$-1.5 < w < -0.7 \tag{9.1}$$

in the light of some recent cosmological observations [2–5]. Therefore, there is a significant probability that $w < -1$, i.e., that w could be in the phantom regime. From another point of view, there is currently a tension between the value of the Hubble constant provided by Planck [3] and the HST value given by Riess et al. [6]. This discrepancy could be solved by appealing to an equation of state with $w < -1$ [7]. Thus, the realm where our accelerating universe appears to approximately lie on is one that can be expressed as a phantom-like small perturbation of the de Sitter space. Even though one could eventually accommodate the above dark energy and modified-gravity models to account for such an observational scenario, that would ultimately appear rather unnatural. Moreover, none of such models can be shown to simultaneously satisfy the following two requirements: (i) exactly predicting what observational data points out in a natural way and (ii) an economic principle according to which one should not include unnecessary ingredients, such as mysterious cosmic fluids or fields, nor modifications of the very-well-tested background theories such as general relativity. The use of scalar fields in quintessence or k-essence scenarios is notwithstanding quite similar to including an inflation in inflationary theories for the early universe [8]. Even though, owing to the success of the inflationary paradigm, which actually shares its main characteristics with those of the present universal acceleration, many could take this similarity to be a reason enough to justify the presence of a scalar field also pervading the current universe. It could well be that a cosmic Occam's Razor principle would turn out to be over and above the nice coincidence between predictions of usual models for inflation and what has been found in cosmic observations such as the measurement of background anisotropies. After all, the medieval opinion that the simplest explanation must be the correct one has proved to be extremely fruitful so far, and on the other hand, the paradigm of inflation, by itself still raises some deep criticisms. Occam's Razor is also against the idea of modifying gravity by adding to the relativistic Lagrangian some convenient extra terms.

Besides general relativity, quantum theory is the other building block which can never be ignored while constructing a predicting model for any physical system. Although it is true that a quantum behavior must in general be expected to manifest for small-size systems, cosmology is providing us with situations where the opposite really holds. In fact, fashionable phantom models for the current universe are all characterized by an energy density that increases with time, making in this way the curvature larger as the size of the universe becomes greater. In such models quantum effects should be expected to more clearly manifest at the latest times where the universe becomes the largest. Thus, it appears that quantum theory should necessarily be another ingredient in our task to build up an economical theory of current cosmology without contravening the Occam's Razor philosophy.

A cosmological model satisfying all the above requirements has been advanced [9]. It was in fact constructed using just a gravitational Hilbert–Einstein action without any extra terms and taking into account the probabilistic quantum effects on the trajectories of the particles but not the dynamical properties of any cosmic field, such as quintessence or *k*-essence. The resulting most interesting cosmic model describes an accelerating universe with an expansion rate that goes beyond that of the de Sitter universe into the phantom regime where the tracked parameter of the universal state equation becomes slightly less than -1 and the future is free from any singularity. Such a model will thus describe what can be dubbed a *benigner* phantom universe because, besides being regular along its entire evolution, it does not show the violent instabilities driven by a noncanonical scalar-field kinetic term as by construction the model does not have a negative kinetic term nor it classically violate the null energy condition which guarantees the stability of the theory, contrary to what the customary phantom models do. Another cosmic model was also obtained, which describes an initially accelerating universe with equation of state parameter always greater than -1, that eventually becomes decelerating for a while, to finally contract down to a vanishing size asymptotically at infinity. The latter model seems to be less adjustable to current observational data.

9.2 The Original Quantum Dark-Energy Model

In this section we shall review the new interpretation for dark energy based on a Bohmian quantum potential which was first suggested in Ref. [9]. Keeping in mind the idea that dark energy should somehow reflect the otherwise unobservable existence of a cosmological substance with an essentially quantum-mechanical nature, and promoting the so-called Bohm's interpretation of quantum mechanics [10] to the status of a field theory in a similar way to how it is made from classical relativistic mechanics to finally produce the model of tachyonic dark energy [11], we will thus be able to finally propose simple "classical" models for dark energy that do not necessarily depend on the existence of any potential for the vacuum scalar field, and bring the imprint of their truly quantum origin, formally in much the same way as Bohm's interpretation of quantum mechanics does. From the real part of the Klein–Gordon wave equation applied to a quasi-classical wave function $R\exp(iS/\hbar)$, where the probability amplitude R ($P = |R|^2$) and the action S are real functions of the relativistic coordinates, if the classical energy $E = \partial S/\partial t$ and momentum $\vec{p} = \vec{\nabla} S$ are defined, one can write [12]

$$E^2 - p^2 + V_Q^2 = m_0^2, \tag{9.2}$$

where m_0 is the rest mass of the involved particle and V_Q is a relativistic quantum potential,

$$V_Q^2 = \frac{\hbar^2}{R}\left(\nabla^2 R - \frac{\partial^2 R}{\partial t^2}\right), \tag{9.3}$$

which should be interpreted according to the Bohm's idea [10] as the hidden quantum potential that accounts for precisely defined unobservable relativistic variables whose effects would physically manifest in terms of the indeterministic behavior shown by the given particles. From Eq. (9.1) it immediately follows that $p = \sqrt{E^2 + V_Q^2 - m_0^2}$. Thus, since classically $p = \partial L/\partial[q(\dot{t})]$ (with L being the Lagrangian of the system and q the spatial coordinates, which depends only on time $t, q \equiv q(t)$), we have for the Lagrangian[a]

$$L = \int d\dot{q}\, p = \int dv \sqrt{\frac{m_0^2}{1 - v^2} + M^2}, \tag{9.4}$$

in which $v = \dot{q}$ and $M^2 = V_Q^2 - m_0^2$. In the classical limit $\hbar \to 0$, $V_Q \to 0$, and hence we are left with just the classical relativistic Lagrangian for a particle with rest mass m_0. As shown by Bagla, Jassal, and Padmanabhan [11], promoting the quantities entering this simple Lagrangian to their field-theory counter-parts allows us to get a cosmological model with tachyonic dark energy. In what follows we shall explore the question of what kind of cosmological models can be derived if we apply a similar upgrading-to-field procedure starting with Lagrangian (9.3). Two limiting situations will be considered. First of all, we shall look at the case of most cosmological interest which corresponds to the limit of small values of the rest mass, $m_0 \to 0$, for which the Lagrangian becomes

$$L \simeq \sqrt{V_Q^2 - m_0^2}\left(\int dv \left(1 + \frac{m_0^2}{2\left(V_Q^2 - m_0^2\right)\left(1 - v^2\right)}\right)\right.$$

$$= \sqrt{V_Q^2 - m_0^2} v + \frac{m_0^2}{\sqrt{V_Q^2 - m_0^2}} \ln\left[\left(\frac{1+v}{1-v}\right)^{1/4}\right]. \tag{9.5}$$

[a] A somehow related approach can be found in Section 1.2.2 of Chapter 8 in this volume. We notice that the difficulties to be expected when applying a Bohmian procedure to a gravitational field are circumvented in these formalisms.

This Lagrangian is positive definite whenever $V_Q > 0$. For nonzero values of the quantum potential, we can have physical systems with nonzero Lagrangian even for the massless case where $v = 1$ and $m_0 = 0$ simultaneously. This is made possible because the existence of the quantum potential allows us to consider an effective rest mass given by $M \equiv \sqrt{V_Q^2 - m_0^2}$. On the other hand, since the quantum potential V_Q can take on both positive and negative values, the associated field theory can lead to positive or negative pressure, respectively. Choosing $V_Q < 0$ and hence $L < 0$, in the massless case $m_0 = 0$, $v = 1$, we have

$$L = -|V_Q|. \tag{9.6}$$

Generalizing to a field theory in the general case $m_0 \neq 0$, $v < 1$ requires the upgrading $q(t) \to \phi$, a field which will thereby depend on both space and time, $\phi(\vec{r}, t)$, replacing $v^2 \equiv \dot{q}^2$ for $\partial_i \phi \partial^i \phi$ and the rest mass m_0 for a generic potential $V(\phi)$). In the extreme massless case however the Lagrangian (9.6) does not contain any quantity which can be upgraded to depend on ϕ, so that the Lagrangian for the field theory in the massless case is no longer zero, but it is also given by Eq. (9.6).

In what follows we shall regard Lagrangian (9.6) as containing all the cosmological information that corresponds to a universe whose dark energy is given by a positive cosmological constant, provided the field ϕ is homogeneously and isotropically distributed. This can be accomplished if, e.g., the quantum potential is interpreted as that potential associated to the hidden dynamics of the particles making of the CMB radiation. Assuming next a perfect fluid form for the equation of state of the cosmic field ϕ, i.e., introducing a stress-energy tensor

$$T^i_k = (\rho + P)u^i u_k - P\delta^i_k, \tag{9.7}$$

where the energy density ρ and the pressure P that correspond to Lagrangian (9.4) are given by

$$\rho = |V_Q|, \quad P = -|V_Q|, \tag{9.8}$$

and the four-velocity is

$$u_k = \frac{\partial_k \phi}{\sqrt{\partial_i \phi \partial^i \phi}}. \tag{9.9}$$

From Eqs. (9.8) and the conservation equation for cosmic energy, $d\rho = -3(\rho + P)da/a$, it again follows that $\rho = \kappa^2 = |V_Q| = \text{const.}$, so that the resulting Friedmann equation, $\dot{a} = \kappa a/m_P$ (m_P being the Planck mass), yields the expected solution for the scale factor $a = a_0 \exp[\kappa(t - t_0)/m_P]$. Equations (9.8) immediately lead, moreover to a characteristic parameter for the perfect fluid-state equation, which turns out to be constant and given by $w = P/\rho = -1$. We can conclude, therefore, that if $m_0 = 0$, $v = 1$ (i.e., $V(\phi) = 0$ and $\partial_i\phi\partial^i\phi = 1$ in the field theory), and $V_Q < 0$, the observable CMB radiation makes to appear a quantum potential inducing the presence of a pure cosmological constant given by $\Lambda = \kappa = \sqrt{V_Q}$. In case that the rest mass is $m_0 \neq 0$ and very small, there would be a nonzero field-theory potential $V(\phi) \to m_0$ and the quantum medium would correspond to a cosmic dark energy, which would behave like some form of a "tracking" quintessential field [13]. In fact, for in such a case we had for negative V_Q and small but nonzero m_0,

$$L = P = -|M|\sqrt{\partial_i\phi\partial^i\phi} - \frac{V(\phi)^2}{4|M|}\ln\left(\frac{1+\sqrt{\partial_i\phi\partial^i\phi}}{1-\sqrt{\partial_i\phi\partial^i\phi}}\right), \qquad (9.10)$$

with M being now given by $M \equiv M[V(\phi)] = -\sqrt{V_Q^2 - V(\phi)^2}$. The pressure P is then a definite negative quantity such that $\partial_i\phi\partial^i\phi < 2V(\phi)$ only if $\partial_i\phi\partial^i\phi$ is sufficiently smaller than $\left(\partial_i\phi\partial^i\phi\right)_c$, with

$$\frac{\sqrt{\left(\partial_i\phi\partial^i\phi\right)_c}}{1-\left(\partial_i\phi\partial^i\phi\right)_c} = \ln\left[\frac{1+\sqrt{\left(\partial_i\phi\partial^i\phi\right)_c}}{\sqrt{1-\left(\partial_i\phi\partial^i\phi\right)_c}}\right].$$

The energy density which together with the pressure P enters the equation of state $P = w(\phi)\rho$ would then read

$$\rho = -\frac{V(\phi)^2}{2|M(\phi)|}\left[\frac{\sqrt{\partial_i\phi\partial^i\phi}}{1-\partial_i\phi\partial^i\phi} - \ln\left(\frac{1+\sqrt{\partial_i\phi\partial^i\phi}}{\sqrt{1-\partial_i\phi\partial^i\phi}}\right)\right]. \qquad (9.11)$$

We then note that for the considered range of the kinetic term, we always can, in fact, choose a range for the parameter entering the equation of state which satisfies $0 \geq w(\phi) \geq -1$.

In the limit that the rest mass and the quantum potential take on very similar values, which is the second situation we shall briefly consider, the Lagrangian can be approximated to

$$L \simeq m_0 \int \frac{dv}{\sqrt{1-v^2}} = \frac{1}{2} m_0 \ln \left(\frac{1-\sqrt{1-v^2}}{1+\sqrt{1-v^2}} \right)^{1/2} . \tag{9.12}$$

Such a Lagrangian is negative definite, and if we upgrade the quantities involved in it so that they become field-theory variables, $m_0 \to V(\phi)$, with $V(\phi)$ a classical potential for the scalar field ϕ, and $v^2 \to \partial_i \phi \partial^i \phi$, it would correspond to a negative pressure:

$$P = \frac{1}{2} V(\phi) \ln \left(\frac{1-\sqrt{1-\partial_i \phi \partial^i \phi}}{1+\sqrt{1-\partial_i \phi \partial^i \phi}} \right), \tag{9.13}$$

which is definite negative, and a positive energy density

$$\rho = \frac{V(\phi)}{\sqrt{1-\partial_i \phi \partial^i \phi}} - P. \tag{9.14}$$

Thus, for a perfect fluid equation of state $P = w(\phi)\rho$, this would again correspond to a tracking quintessence-like field.

9.3 Relativistic Bohmian Backgrounds

In this section we shall consider new fundamental aspects that strengthen the consistency and provide further physical motivation to the general model reviewed in Section 9.2. These new aspects concern both the use of a quantum potential model derived from the application of the Klein–Gordon equation and the background relativistic theory associated with the cosmic quantum models.

9.3.1 The Klein–Gordon Quantum Model

We note here that although for some time in the past, it was generally believed that the Klein–Gordon equation was unobtainable from the Bohm formalism [14], in recent years the Klein–Gordon equation has found satisfactory causal formulations. The solution presented in [15] by Horton et al. has to introduce the causal description of timelike flows in an Einstein–Riemann space (otherwise the probability current can assume negative values of its zeroth component and is not generally timelike). However, there exists a causal Klein–Gordon theory in Minkowski space [16] where this is achieved by introducing a cosmological constant as an additional assumption, which is justified in view of recent observations. Therefore, it makes perfect sense to use a Klein–Gordon equation in our model [9]. Moreover, the nonclassical character of the current whose continuity equation is derived from the purely imaginary part of the expression resulting from the application of the Klein–Gordon equation to the wave function is guaranteed by the fact that one can never obtain the classical limit by making $\hbar \to 0$. Thus, no classical verdict concerning that current of the kind pointed out by Holland [14] can be established. On the other hand, having a material object whose trajectory escapes out the light cone [14] cannot be used as an argument in favor of the physical unacceptability of the model. Quite the contrary, it expresses its actual essentially quantum content, much as the quite fashionable entangled states of sharp quantum theory seemed at first sight violate special relativity and then turned out to be universally accepted. In both cases, physics is preserved because we are not dealing with real signaling. Actually, we shall show later that our cosmic models can be also interpreted as being originated from the entanglement energy of the whole universe, without invoking any other cause.

9.3.2 Quantum Theory of Special Relativity

Consistent tachyonic theories for dark energy are grounded on the special theory of relativity in such a way that all the physics involved in them stems from Einstein relativity. Our cosmic quantum models actually come from a generalization from tachyonic theories for which the corresponding background relativistic description ought to contain the quantum probabilistic footprint. Thus, in order to check their consistency, viability and properly motivate the models reviewed in Section 9.2, one should investigate the characteristics of the quantum relativistic theory on which they are based. In what follows we shall consider in some detail the basic foundations of that background quantum relativity.

Actually, there are two ways of defining the action of a free system endowed with a rest mass m_0 [17]. The first one is by using the integral expression for the Lagrangian $L = \int p dv$, with the momentum $\vec{p}$ derived from the Hamilton–Jacobi equation, and inserting it in the expression $S = \int_{t_1}^{t_2} Ldt$. The second procedure stems from the definition $S = \beta \int_a^b ds$, where ds is the line element and the proportionality constant $\beta = m_0 c$ is obtained by going to the nonrelativistic limit. The strategy that we have followed here is to apply the first procedure to derive an integral expression for S in the case of a Hamilton–Jacobi equation containing an extra quantum term and then obtain the expression for ds by comparing the resulting expression for S with that is given by the second procedure.

As mentioned above, a Hamilton–Jacobi equation with the quantum extra term can be obtained by applying the Klein–Gordon equation to a quasiclassical wave function $\Psi = R(\vec{r}, t)\exp(iS(\vec{r}, t)/\hbar)$ [18], where $R(\vec{r}, t)$ is the quantum probability amplitude and $S(\vec{r}, t)$ is the classical action. By the second of the above procedures and $L_Q = -m_0 c^2 E(\varphi, k)$, we immediately get for the general space-time metric

$$ds = E(\phi, k)dt, \tag{9.15}$$

which consistently reduces to the metric of special relativity in the limit $\hbar \to 0$. If we take the above line element as invariant, then we obtain for time dilation

$$dt = \frac{E(k)dt_0}{E(\varphi, k)}, \tag{9.16}$$

in which $E(k)$ is the complete elliptic integral of the second kind [19].

A key question that arises now is, does the quantum relativistic description and, hence, our cosmic quantum models satisfy Lorentz invariance? What should be invariant in the present case is the quantity

$$I = ctE\left(\arcsin\sqrt{\frac{c^2t^2 - x^2}{c^2t^2}}.k\right) \tag{9.17}$$

If we chose a given transformation group in terms of hyperbolic or elliptic functions, which leaves invariant (such as it happens for Lorentz transformations) the usual relativistic combination $c^2t^2 - x^2 = c^2t'^2 - x'^2$, then we would obtain

$$I = cQ(t', x')E\left(\arcsin\frac{\sqrt{c^2t'^2 - x'^2}}{cQ(t', x')}, k\right), \tag{9.18}$$

where $Q(t', x') \equiv Q(t', x', \Psi)$ is the expression for the transformation of time t in terms of hyperbolic or elliptic functions. It would follow

$$\left(\frac{I}{cQ(t', x')}\right)^{-1} = \frac{\sqrt{c^2t'^2 - x'^2}}{cQ(t', x')}, \tag{9.19}$$

with $(\,)^{-1}$ denoting the inverted function associated to the elliptic integral of the second kind, generally one of the Jacobian elliptic functions or a given combination of them [19]. Thus, the quantity I can only be invariant under the chosen kind of transformations in the classical limit where $k = 1$. Therefore, a quantum relativity built up in this way would clearly violate Lorentz invariance, at least if we take usual classical values for the coordinates.

In order to obtain the desired transformation equations, we first notice that if we take the coordinate transformation formulas in terms of the usual hyperbolic or some elliptic functions of the rotation angle Φ, we can always reexpress the invariant quantity I of Einstein special relativity in the form

$$I = cQ(t', x')E\left(\arcsin\left(\frac{\sqrt{c^2t'^2 - x'^2}}{cQ(t', x')}\right)^{-1}, k\right). \tag{9.20}$$

From Eq. (9.20) one can write

$$\left(\frac{I}{cQ(t', x')}\right)^{-1} = \left(\frac{\sqrt{c^2t'^2 - x'^2}}{cQ(t', x')}\right)^{-1}$$

and hence

$$I = \sqrt{c^2t'^2 - x'^2} = ct'E\left(\arcsin\left(\frac{\sqrt{c^2t'^2 - x'^2}}{ct'}\right)^{-1}, k\right), \tag{9.21}$$

that is, I would in fact have the form of the Einstein relativistic invariant. If we interpret the coordinates entering Eq. (9.21) as quantum-mechanical coordinates, then our quantum expression for the invariant I given by Eq. (9.17) can be directly obtained from the last equality by making the replacement

$$\sqrt{1 - \frac{x^2}{c^2t^2}} = E\left(\arcsin\sqrt{1 - \frac{x_{clas}^2}{c^2t_{clas}^2}}, k\right) \tag{9.22}$$

or

$$\left(\sqrt{1-\frac{x^2}{c^2t^2}}\right)^{-1}=\sqrt{1-\frac{x_{clas}^2}{c^2t_{clas}^2}},\qquad(9.23)$$

where the notation $(\,)^{-1}$ again means inverted function of the elliptic integral of the second kind, and if the coordinates entering the right-hand-side are taken to be classical coordinates, then those on the left-hand-side must still in fact be considered to be quantum-mechanical coordinates. Classical coordinates are those coordinates used in Einstein special relativity and set the occurrence of a classical physical event in that theory. By quantum coordinates we mean those coordinates which are subject to quantum probabilistic uncertainties and would define what one may call a quantum physical event, i.e., that event which is quantum-mechanically spread throughout the whose existing space-time with a given probability distribution fixed by the boundaries specifying the extent and physical content of the system.

In what follows we will always express all equations in terms of classical coordinates, and therefore, for the sake of simplicity, we shall omit the subscript "clas" from them. The equivalence relation given by expressions (9.22) and (9.23) is equally valid for primed and non primed coordinates and should be ultimately related with the feature that for a given, unique time, t or t', the position coordinate, x or x', must be quantum-mechanically uncertain. From the equalities (9.22) and (9.23) for primed coordinates, we get then an expression for I' in terms of classical coordinates:

$$I'=ct'E\left(\arcsin\frac{\sqrt{c^2t'^2-x'^2}}{ct'},k\right),\qquad(9.24)$$

which shows the required invariance and, in fact, becomes the known relativistic result $I'=\sqrt{c^2t'^2-x'^2}$ in the classical limit $\hbar\to 0$.

From expressions (9.22) and (9.23), we also have

$$1-\frac{V^2}{c^2}=E(\varphi,k)^2\to\frac{V}{c}=\sqrt{1-E(\varphi,k)^2}=\tanh\Phi,\qquad(9.25)$$

where V is velocity, $\varphi=\arcsin\sqrt{1-\frac{x^2}{c^2t^2}}$ and we have specialized to using the usual hyperbolic functions. Whence $\cosh\Phi=1/E(\varphi,k)$,

$\sinh\Phi = \sqrt{1 - E(\varphi, k)^2}/E(\varphi, k)$, and from the customary hyperbolic transformation formulas for coordinates

$$x = x'\cosh\Phi + ct'\sinh\Phi, \quad ct = ct'\cosh\Phi + x'\sinh\Phi, \tag{9.26}$$

we derive the new quantum relativistic transformation equations

$$x = \frac{x' + ct'\sqrt{1 - E(\varphi, k)^2}}{E(\varphi, k)}, \quad ct = \frac{ct' + x'\sqrt{1 - E(\varphi, k)^2}}{E(\varphi, k)}. \tag{9.27}$$

Had we started with formulas expressed in terms of the Jacobian elliptic functions [19], such that

$$\frac{V}{c} = \mathrm{sn}(\Phi, k) = \sqrt{1 - E(\varphi, k)^2} \tag{9.28}$$

$$x = x'\mathrm{nc}(\Phi, k) + ct'\mathrm{sc}(\Phi, k), \quad ct = ct'\mathrm{nc}(\Phi, k) + x'\mathrm{sc}(\Phi, k), \tag{9.29}$$

then we have again obtained Eqs. (9.27), so confirming the quantum-mechanical character of the coordinates entering the left-hand-side of Eqs. (9.22) and (9.23). The above-derived expressions are not yet the wanted expressions as they still contain an unnecessary element of classicality due to the feature that when using quantum-mechanical coordinates for the derivation of the velocity V setting $x = 0$, the unity of the left-hand-side of Eq. (9.22) would correspond to the complete elliptic integral of the second kind $E(k)$ [19]. Thus, we finally get for the transformation equations

$$x = \frac{\left(x' + ct'\sqrt{1 - E(\varphi, k)^2}\right) E(k)}{E(\varphi, k)}$$

$$ct = \frac{\left(ct' + x'\sqrt{1 - E(\varphi, k)^2}\right) E(k)}{E(\varphi, k)}, \tag{9.30}$$

that are the wanted final expressions in terms of classical coordinates, which, in fact, reduce to the known Lorentz transformations in the classical limit $\hbar \to 0$. From the formula for time transformation, we, in fact, get time dilation to be the same as that (Eq. 9.16) directly obtained from the metric when referring to two events occurring at one and the same point x', i.e.,

$$\Delta t = \frac{E(k)\Delta t_0}{E(\varphi, k)}, \tag{9.31}$$

and from that for space transformation the formula for length contraction referred to one and the same time t'

$$\Delta\ell = \frac{E(\varphi, k)\Delta\ell_0}{E(k)}. \tag{9.32}$$

In any case, the quantum effects would be expected to be very small, that is, usually k is generally very close to unity for sufficiently large rest masses of the particles.

For the sake of completeness we shall derive, in what follows, the transformation of velocity components one can also derive from the coordinate transformations (9.30) that if space and time themselves are subject to the quantum-mechanical uncertainties, they should be now given as

$$\begin{aligned} v_x &= \frac{v'_x + c\sqrt{1 - E(\varphi, k)^2}}{1 + \frac{v'_x}{c}\sqrt{1 - E(\varphi, k)^2}} \\ v_y &= \frac{v'_y E(\varphi, k)}{E(k)\left(1 + \frac{v'_x}{c}\sqrt{1 - E(\varphi, k)^2}\right)} \\ v_z &= \frac{v'_z E(\varphi, k)}{E(k)\left(1 + \frac{v'_x}{c}\sqrt{1 - E(\varphi, k)^2}\right)}, \end{aligned} \tag{9.33}$$

which reduce once again to the well-known velocity transformation law of Einstein special relativity. Even though they are quantitatively distinct of the latter transformation law, Eqs. (9.33) behave qualitatively in a similar fashion and produce the analogous general velocity addition law as in Einstein special relativity.

We finally turn to the essentials of the relativistic mechanics and find the formulas for momentum and energy that must be satisfied by the cosmic quantum models to be given by

$$p = \frac{\partial L}{\partial v} = \frac{m_0 c\sqrt{1 - k^2\left(1 - \frac{v^2}{c^2}\right)}}{\sqrt{1 - \frac{v^2}{c^2}}} \tag{9.34}$$

$$E = pv - L = \frac{m_0 c^2}{\sqrt{1 - \frac{v^2}{c^2}}} \times$$
$$\left[\frac{v}{c}\sqrt{1 - k^2\left(1 - \frac{v^2}{c^2}\right)} + \sqrt{1 - \frac{v^2}{c^2}}E(\varphi, k)\right]. \tag{9.35}$$

Obviously, these expressions reduce to $p = m_0 v/\sqrt{1 - v^2/c^2}$ and $E = m_0 c^2/\sqrt{1 - v^2/c^2}$, respectively, in the limit $\hbar \to 0$. Moreover, if we set $v = 0$ then $p = V_Q/c$ and $E = m_0 c^2 E(k)$ which become, respectively, 0 and $m_0 c^2$ when $\hbar \to 0$. It follows then that our quantum special relativistic model has the expected good limiting behavior.

Except for rather extreme cases, the value of parameter k is very close to unity and, therefore, the corrections to the customary expressions induced by the present model should be expected to be very small locally. However, they could perhaps be detectable in specially designed experiments using extremely light particles.

The main conclusion that can be drawn from the above discussion is that whereas Lorentz invariance appears to be violated in our quantum description if classical coordinates are considered, such an invariance is preserved when one uses quantum coordinates in that description.

9.4 Dark Energy Without Dark Energy

In this section we shall look at current acceleration by using the same general economical philosophy as in the previous model, even without invoking, moreover, any effects induced at the primordial inflationary period. The sole ingredients which we shall explicitly include, besides general relativity, are the quantum effects on the trajectories of the particles that make up the background radiation. Such effects will be modeled through the relativistic generalization of the original quantum potential formalism by Bohm [10] and lead by themselves to an accelerating expansion, which, consistently, goes slightly beyond what is predicted by a cosmological constant. Thus, we use a version of the quantum model for dark energy [12] stemming from the analogy with the classically interpreted Hamilton–Jacobi equation derived from the Klein–Gordon wave equation for a quasi-classical wave function $\Psi = R\exp(iS/\hbar)$, i.e.,

$$E^2 - p(v)^2 + \tilde{V}_Q^2 = m_0^2, \qquad (9.36)$$

where

$$\tilde{V}_Q = \hbar\sqrt{\frac{\nabla^2 R - \ddot{R}}{R}} \tag{9.37}$$

is the quantum potential, $v = \dot{q}(t)$ and $p = \partial\tilde{L}/\partial\dot{q}$, with $\dot{} = d/dt$ and $\tilde{L}$ being the Lagrangian

$$\tilde{L} = \int d\dot{q}\, p = \int dv\sqrt{\tilde{V}_Q^2 - m_0^2 + \frac{m_0^2}{1 - v^2}}. \tag{9.38}$$

The dynamical structure of this expression is once again able to circumvent the difficulties stemming from dealing with a Bohmian description of a gravitational field (see footnote on page 550). As shown first by Bagla, Jassal, and Padmanabhan [11] for the fully classical case and later on by one of the present authors [12] for the case that the Lagrangian contains a quantum potential, upgrading the quantities entering this simple Lagrangian to their field-theory counterparts actually leads to a cosmological tachyonic model which can be used to predict cosmic acceleration. Following Ref. [12] we shall replace then the quantity q for a scalar field ϕ, the quantity $\dot{q}^2 \equiv v^2$ for $\partial_i\phi\partial^i\phi \equiv \dot{\phi}^2$, and the rest mass m_0 for the potential $\tilde{V}(\phi)$. With these replacements and leaving $\tilde{V}_Q$ constant for the moment, we can then integrate Eq. (9.38) to have for the field Lagrangian $\tilde{L} = -\tilde{V}(\phi)E\,(x(\phi), k(\phi))$, with $E(x, k)$ the elliptic integral of the second kind, $x(\phi) = \arcsin\sqrt{1 - \dot{\phi}^2}$ and $k = \sqrt{1 - \tilde{V}_Q^2/\tilde{V}(\phi)^2}$. At first sight one should also upgrade $\tilde{V}_Q$ to depend on ϕ. However, it will be seen later that such a upgrading would lead to a final expression for $\tilde{V}_Q$, which depends only on $\dot{\phi}$, a dependence that disappears because for the present model to avoid divergences, it is necessary that $\dot{\phi}$ be constant.

The idea now is that we have either pure background radiation or background radiation plus a cosmological constant Λ and interpret that dark energy is nothing but the cosmic footprint left in the classical universe from the set of quantum characteristics associated with the background radiation when such characteristics are taken to be given by the quantum potential. Thus, the condition that we have to impose to the scalar field theory derived in the quantum model [12] to satisfy the requirement that dark energy disappears once we erase any trace of the background quantum effects is that the Lagrangian, energy density, and pressure will all vanish in the limit where the cosmological constant and the quantum potential are both zero, i.e., $\Lambda \rightarrow 0$, $\tilde{V}_Q \rightarrow 0$. Using a more appropriate vector field instead of the scalar field ϕ does not make any difference relative to the final results of our model, which turns out to be finally independent of the precise characteristics of the field other than being characterized by the speed of light and a zero rest mass. It will be seen in what follows that such a condition is fulfilled, provided that we start with the Lagrangian density:

$$L = -V\left(E(x, k) - \sqrt{1 - \dot{\phi}^2}\right), \tag{9.39}$$

where again $x = \arcsin\sqrt{1 - \dot{\phi}^2}$ and now $k = \sqrt{1 - V_Q^2/V^2}$, with $V \equiv V(\phi)$ the density of potential energy associated to the field ϕ. We do not expect $\tilde{V}_Q$ to remain constant along the universal expansion but to increase like the volume of the universe V does. It is the quantum potential density $V_Q = \tilde{V}_Q/\mathrm{V}$ appearing in Eq. (9.39) that should be expected to remain constant at all cosmic times. In fact, from the imaginary part of the Klein–Gordon equation applied to the wave function Ψ we can get $v \cdot \vec{\nabla} R - \dot{R}$ and hence the continuity equation for the probability flux $\vec{J} = \hbar\ \mathrm{Im}(\Psi^* \vec{\nabla}\Psi)/(m\mathrm{V})$, $\vec{\nabla} \cdot J - \dot{\mathrm{P}} = 0$, where P is the probability density $\mathrm{P} = \mathrm{Probability/V}$. This continuity equation is the mathematical equivalent of a probability conservation law. Upgrading then the velocity v to $\dot{\phi}$ and noting that $\dot{\phi} = \pm 1$ (see later) it follows that $(\nabla^2 R - \ddot{R})/R = (\nabla^2 P - \ddot{P})/(2P)$, with $P = R^2$. Assuming that the particles move locally according to some causal law [10], one can now average Eq. (9.36) with the Probability-weighting function $P = R^2$, so that one obtains for the averaged quantum potential squared, $\langle \tilde{V}_Q^2 \rangle_{av} = \iiint dx^3 P \tilde{V}_Q^2 = \hbar^2 \iiint dx^3 (\nabla^2 P - \ddot{P}) \equiv \hbar^2 \left(\langle \nabla^2 P \rangle_{av} - \langle \ddot{P} \rangle_{av}\right)$. Since the universe is isotropic and homogeneous, the corresponding conserved quantity can then be obtained by simply taking $\langle \tilde{V}_Q^2 \rangle_{av}^{1/2}/\mathrm{V} = \langle V_Q^2 \rangle_{av}^{1/2}$, that is, renaming for the aim of simplicity all the quantities $\langle f^2 \rangle_{av}^{1/2}$ involved in the averaged version of Eq. (9.36) as f, we can again derive Eq. (9.39), now with V_Q a constant conserved quantity when referred to the whole volume V of the isotropic and homogeneous universe.

It is easy to see that in the limit of vanishing V_Q, $V E(x, k)$ reduces to $\sqrt{1 - \dot{\phi}^2}$ so that the Lagrangian (9.39) vanishes, as required. The pressure and energy density are then obtained from Eq. (9.39) to read

$$p_\phi = -V\left(E(x,k) - \sqrt{1-\dot{\phi}^2}\right) \tag{9.40}$$

$$\rho_\phi = V\left(\frac{\sqrt{\dot{\phi}^2 + \frac{V_Q^2}{V^2}(1-\dot{\phi}^2)\dot{\phi}}}{\sqrt{1-\dot{\phi}^2}} + E(x,k) - \frac{1}{\sqrt{1-\dot{\phi}^2}}\right), \tag{9.41}$$

where we have considered $V \equiv V(\phi)$. In any case, for a source with parameter $w(t) = p_\phi/\rho_\phi$, we must always have

$$\frac{\dot{\rho}_\phi}{\rho_\phi} = -3H\,(1+w(t)) = \frac{2\dot{H}}{H}. \tag{9.42}$$

By itself this expression can generally determine the solution for the scale factor $a(t)$, provided $w = \text{const}$. In such a case, we obtain after integrating Eq. (9.42) for the scale factor

$$a = \left(a_0^{3(1+w_0)/2} + \frac{3}{2}(1+w_0)\kappa t\right)^{2/[3(1+w_0)]},$$

in which a_0 is the initial value of the scale factor and κ is a constant. However, we shall not restrict ourselves in this chapter to a constant value for the parameter w of the equation of state but leave it as a time-dependent parameter whose precise expression will be determined later on. Combining now Eq. (9.42) with the expression for $w(t)$, we can then obtain an expression for $d(H^{-1})/dt$ by using Eqs. (9.40) and (9.41) as well. Moreover, multiplying Eqs. (9.40) and (9.41) and using Eq. (9.42), a relation between the potential density V and the elliptic integral E can be derived from the Friedmann equation $H^2 = 8\pi G\rho_\phi/3$. These manipulations allow us to finally obtain

$$E = -\left[\frac{A(\dot{\phi}, V, V_Q)\dot{\phi}\left(1+\frac{3H^2}{2\dot{H}}\right) - 1 - \frac{3H^2\dot{\phi}^2}{2\dot{H}}}{\sqrt{1-\dot{\phi}^2}}\right]$$

$$= -\left\{\frac{\frac{3H^2\dot{\phi}^4V_Q^2}{\dot{H}} - \left(\frac{2\dot{H}}{G}\right)^2(1-\dot{\phi}^2) + \dot{\phi}^2V_Q^2(1+\dot{\phi}^2)}{\sqrt{1-\dot{\phi}^2}\left[\left(\frac{\dot{H}}{4\pi G}\right)^2 - \dot{\phi}^2V_Q^2\right]}\right\} \tag{9.43}$$

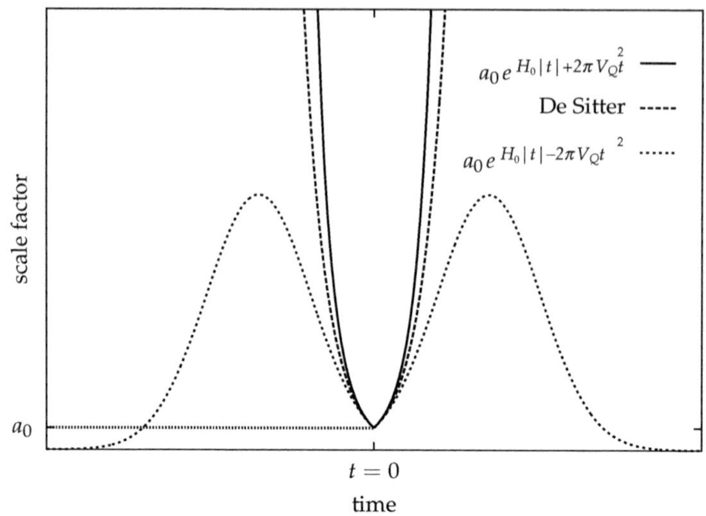

Figure 9.1 Cosmic solutions that result from the introduction of a quantum potential density V_Q when $\dot{\phi}^2 = 1$. Solution (a) goes like in de Sitter space with the same H_0 but with higher acceleration. Solution (b) corresponds to the case where $H_0^2 > 4\pi V_Q$, and represents a universe that is initially expanding in an accelerated way (at a rate slower than in de Sitter space with the same H_0), then expands in a decelerated way for a while, to finally contract toward a zero radius as $t \to \infty$. In the figure we have used units such that $\hbar = c = G = 1$

with $A(\dot{\phi}, V, V_Q) = \sqrt{\dot{\phi}^2 + \frac{V_Q^2}{V^2}(1 - \dot{\phi}^2)}$, and

$$V = -\frac{2\pi G\sqrt{1-\dot{\phi}^2}}{\dot{H}\dot{\phi}^2}\left[\left(\frac{\dot{H}}{4\pi G}\right)^2 - \dot{\phi}^2 V_Q^2\right]. \tag{9.44}$$

Thus, simple general expressions for the energy density and pressure can be finally derived to be

$$\rho_\phi = 6\pi G\left(\dot{H}^{-1} H\dot{\phi} V_Q\right)^2 \tag{9.45}$$

$$p_\phi = -4\pi G\dot{H}^{-1}\dot{\phi}^2 V_Q^2\left(1 + \frac{3H^2}{2\dot{H}}\right) = w(t)\rho_\phi, \tag{9.46}$$

where

$$w(t) = -\left(1 + \frac{2\dot{H}}{3H^2}\right). \tag{9.47}$$

The Friedmann equation $H^2 = 8\pi G\rho_\phi/3$, derived from the action integral with the Lagrangian (9.39), corresponds to a universe dominated by quantum energy. Using Eq. (9.45) this Friedmann equation leads to

$$\dot{H} = \pm 4\pi G\dot{\phi}V_Q, \tag{9.48}$$

with a slowly varying $w(t)$ that should be quite close to but still less than -1 (that is, the case that current observations each time more clearly are pointing to [20]). We have also

$$H = \pm 4\pi G\phi V_Q + C_1, \tag{9.49}$$

with C_1 an integration constant. Moreover, if we assume that $\dot{\phi}$ is constant (an assumption which would indeed be demanded by the fact that $v^2 = 1$ for radiation), then from the equation of motion that corresponds to the Lagrangian for the field ϕ alone [11] $\ddot{\phi} + (1 - \dot{\phi}^2)(3H\dot{\phi} + dV/Vd\phi) = 0$, we have $\dot{\phi}^2 = +1$. Actually, from the Lagrangian density $L_{SQ} = -V(\phi)E(x,k)$, we can also obtain

$$\dot{\phi}\ddot{\phi} = (1-\dot{\phi}^2)\left\{-3H\left[\dot{\phi}^2 + \frac{V_Q^2}{V(\phi)^2}(1-\dot{\phi}^2)\right]\right. \tag{9.50}$$
$$\left. + \sqrt{1-\dot{\phi}^2}\sqrt{\dot{\phi}^2 + \frac{V_Q^2}{V(\phi)^2}(1-\dot{\phi}^2)}\frac{\partial L_Q}{V(\phi)\partial\phi} - \frac{\partial V}{V\partial\phi}\dot{\phi}^3\right\},$$

from which we again derive the conclusion that $\ddot{\phi} = 0$ implies $\dot{\phi}^2 = 1$. Indeed, the assumption that $\dot{\phi}^2 = 1$ can be really regarded as a regularity requirement for $\ddot{\phi}$ at very large ratios $t/\sqrt{4\pi G V_Q}$ or $H_0/\sqrt{4\pi G V_Q}$, because if $\dot{\phi}^2 \neq 1$ then $\ddot{\phi}$ would necessarily tend to diverge at these extreme ratios since $V(\phi)$, by itself, would then tend to vanish even when $\dot{\phi}^2 \neq 1$, as it can be checked from Eqs. (9.44) and (9.48). Hence a vanishing $\ddot{\phi}$ implies that strictly $\dot{\phi}^2 = 1$ and $V = 0$, which, once we have downgraded to the original relativistic formalism, means that the present model describes the cosmic quantum effects necessarily associated with an isotropic and homogeneous sea of bosonic particles with zero rest mass which move at the speed of light, i.e., photons-identifying that photon sea with the CMB is just a reasonable assumption.

It follows then

$$H = \pm 4\pi G V_Q t + C_0, \tag{9.51}$$

in which C_0 is another integration constant, and for the scale factor

$$a_{\pm} = a_0 e^{\pm 2\pi G V_Q t^2 + C_0 t}. \tag{9.52}$$

Both solutions are depicted in Fig. 9.1. The solution a_- would predict a universe which initially expands but that immediately starts to contract, tending to vanish as $t \to \infty$. An always accelerating solution slightly beyond the speeding-up predicted by a de Sitter universe is given by the scale factor a_+. In what follows we shall consider the latter solution as that representing the evolution of our current universe and restrict ourselves to deal with that solution only for the branch $t > 0$, denoting $a_+ \equiv a$ and taking then H and $\dot{H}$ to be definite positive.

Thus, the time-dependent parameter of the equation of state will be given by

$$w(t) = -1 - \frac{8\pi G V_Q}{3(4\pi G V_Q t + C_0)^2}, \tag{9.53}$$

which takes on values very close, though slightly less than -1 on the regimes considered in this chapter.

Notice that in the limit $V_Q \to 0$, H becomes a constant $H_0 = C_0$, and hence $\rho_\phi \to 3C_0/(8\pi G)$ and $w \to -1$. Clearly, $H_0^2 = \Lambda$ must be interpreted as the cosmological constant associated with the de Sitter solution $a = a_0 e^{H_0 t}$. When we set $C_0 = 0$ instead, then all remaining quantities have the following limiting values:

$$\rho_\phi = \frac{p_\phi}{w(t)} = 6\pi G V_Q^2 t^2 \to 0, \tag{9.54}$$

$$w(t) = -1 - \frac{1}{6\pi G V_Q t^2} \to -\infty \tag{9.55}$$

and

$$a = a_0 e^{2\pi G V_Q t^2} \to a_0, \tag{9.56}$$

as $V_Q \to 0$. That is precisely the result we wanted to have and means that all the cosmic speed-up effects currently observed in the universe should be attributed to the purely quantum dynamics that one can associate to the background radiation, rather than to the presence of a dark-energy component or any modifications of Hilbert–Einstein gravity. In fact, it can be readily checked that the obtained expression for $\dot{H}$ inexorably leads to a vanishing value for the potential $V(\phi)$, and hence to $\dot{\phi}^2 = 1$, which correspond to pure radiation.

The result that, if there is not constant cosmological term, then it is the considered quantum effects associated with the background radiation which are responsible for a current accelerating expansion of the universe that goes beyond the cosmological constant limit, implies, on the other hand, that (i) the parameter of the equation of state is necessarily less than -1, though very close to it, (ii) the energy density increases with time, (iii) $\rho_\phi + p_\phi < 0$, that is, the dominant energy condition (DEC) is violated, and (iv) the kinetic term $\dot{\phi}^2 > 0$. Whereas the first three properties are shared by the so-called phantom models [21], unlike such models, the fourth one guarantees stability of the resulting universe because $V(\phi) = 0$. Also unlike the usual phantom scenarios, the present model does not predict, moreover, any big rip singularity in the future. Finally, the considered quantum effects may justify violation of the DEC.

On the other hand, if we place a Schwarzschild black hole with initial mass M_0 in the universe described by the suggested model, the mechanism advanced by Babichev, Dokuchaev, and Eroshenko [22] would imply that the black hole will accrete this quantum phantom energy so that it would progressively lose mass down to finally vanish at $t = \infty$, according to the equation

$$M = \frac{M_0}{1 + \pi^2 D V_Q M_0 t}, \tag{9.57}$$

with D a constant. If we place a Morris-Thorne wormhole with initial throat radius b_0 instead, the corresponding accretion mechanism [23] leads now to a progressive increase of the wormhole size governed by

$$b = \frac{b_0}{1 - \pi^2 D' V_Q b_0 t}, \tag{9.58}$$

with D' another constant, bringing us to consider the existence of a big trip process [23] by which, relative to an asymptotic observer at $r = \infty$, the wormhole will quickly grow up to engulf the universe itself, blowing up at a finite time in the future given by

$$\tilde{t} = \frac{1}{\pi^2 D' V_Q b_0}. \tag{9.59}$$

In this case, on times $t > \tilde{t}$, the wormhole converts to an Einstein-Rosen bridge, which decays into a black hole plus a white hole that will, in this case, progressively lose mass to vanish at $t = \infty$ [23]. This result holds both for a static wormhole metric and when the throat radius is allowed to be time dependent [23].

Before closing up we shall briefly consider solution a_-. As it has already been pointed out before, if $C_0 = H_0 = \Lambda^{1/2} >> \sqrt{4\pi G V_Q}$, then this solution corresponds to an initial period of accelerating expansion with an equation-of-state parameter w greater than, though very close to -1. This situation would stand until a time

$$t_a = \frac{H_0 - \sqrt{4\pi G V_Q}}{4\pi G V_Q}, \qquad (9.60)$$

which corresponds to $w = -1/3$. After t_a the universe would keep expanding but now in a decelerating way until a time

$$t_c = \frac{H_0}{4\pi G V_Q}, \qquad (9.61)$$

after which the universe would enter a contracting phase, which would be maintained until $t = \infty$. If $H_0 \leq \sqrt{4\pi G V_Q}$, then the present model would no longer be valid.

It could be at first sight thought that the universe might now be in the phase $t < t_a$ of solution a_-, but current constraints on w [20] seem to preclude that it can be greater than -1. Perhaps another argument against solution a_- be the fact that for this kind of solution, while the accretion of the quantum energy onto a Morris-Thorne wormhole leads to a progressive decrease of the wormhole size according to the law $b = b_0/(1 + \pi^2 D' V_Q t)$, the size of a black hole of initial mass M_0 will progressively increase with quantum energy accretion so that $M = M_0/(1 - \pi^2 D V_Q M_0 t)$. In this way, at a time $t_* = 1/(\pi^2 D V_Q M_0)$, the black hole would blow up. Clearly, for a supermassive black hole at a galactic center, one would then expect that by the present time, the black hole had grown up so big that its astronomical effects would probably be observable.

All the above results have been obtained in the case that the energy density associated with the quantum potential would dominate over any other type of energy. More realistic models where contributions from dark and observable matter are taken into account as well will be considered elsewhere.

9.5 Benigner Phantom Cosmology

9.5.1 Thermodynamics

The thermodynamical description of dark energy has offered an alternative route to investigate the evolution of the current universe [18–24]. However, whereas well-defined expressions can be obtained for dark energy models with equations of state $p = w\rho$ where $w > -1$, in the phantom regime characterized by $w < -1$, there are violent instabilities and a future singularity, the Lagrangian contains a nonphysical kinetic term, and either the temperature or the entropy must be definite negative. In what follows we shall discuss the thermodynamical properties of what we can call the benign phantom cosmic models, in which it will be seen that these problems are largely alleviated. By using the equations for the pressure and the energy density given in the previous section, we proceed now to derive expressions for the thermodynamical functions according to the distinct models implied by the sign ambiguity in Eq. (9.52) and the possibility that the cosmological term be zero or not, only for the solution branches that correspond to a positive time $t > 0$. On the one hand, the translational energy that can be associated with the scalar field would be proportional to [25] $a^3\dot{\phi}^2$, and therefore, because $\dot{\phi}^2 = 1$ [9], the essentially quantum temperature associated with the quantum models must be generally given by

$$T_Q = \kappa a^3, \tag{9.62}$$

with κ a given positive constant whose value will be determined later. It is worth noting that, unlike for phantom energy models [31], in this case the temperature is definite positive even though the value of the state equation parameter w be less than -1. Moreover, this temperature is an increasing function of the scale factor and hence it will generally increase with time. It must be also stressed that T_Q must be a quantum temperature as it comes solely from the existence of a quantum potential.

On the other hand, one can define the entropy and the enthalpy. If, since the universe evolves along an irreversible way, following the general thermodynamic description for dark energy [24, 25], one defines the total entropy of the quantum medium as $S_Q(a) = \rho \mathrm{V}/T_Q$, with $\mathrm{V} = a^3$ the volume of the universe, then in the case that we choose for the scale factor the simplest expanding solution (without cosmological constant) $a_+ = a_0 \exp(2\pi G V_Q t^2)$, with V_Q the quantum potential density, we obtain the increasing, positive quantity:

$$S_Q(a_+) = \frac{V_Q}{\kappa} \ln\left[\left(\frac{a_+}{a_0}\right)^3\right]. \tag{9.63}$$

This definition of entropy satisfies the second law of thermodynamics.

For the kind of systems we are dealing with, one may always define a quantity which can be interpreted as the total enthalpy of the universe by using the same expression as for entropy but referred to the internal energy, which, in the present case, is given by $\rho + p$, instead of just ρ. Thus, we can write for the enthalpy $H_Q = (\rho + p)\mathrm{V}/T_Q$, which leads to the same cosmic solution to the constant, negative-definite quantity

$$H_Q(a_+) = -\frac{V_Q}{\kappa}, \tag{9.64}$$

whose negative sign actually implies a quantum violation of the DEC and indicates that we are in the phantom regime.

The consistency of the above definitions of entropy and enthalpy will be guaranteed in what follows, because the expressions that we obtain from them in the limit $V_Q \to 0$ are the same as for the de Sitter space.

Since the third power of the ratio a_+/a_0 must be proportional to the number of states in the whole universe, the mathematical expression of the entropy given by Eq. (9.63) could still be interpreted to be just the statistical classical Boltzmann's formula, provided we take the constant V_Q/κ to play the role of the Boltzmann's constant k_B or, in other words, k_B is taken to be given by $k_B = V_Q/\kappa$ in such a way that the temperature becomes $T_Q(a_+) = V_Q a^3/k_B$, which consistently vanishes at the classical limit $\hbar \to 0$. If we let $\hbar \to 0$, then it would be $T_Q(a_+)$ but not $S_Q(a_+)$ that vanishes. In this way, Eq. (9.64) becomes

$$H_Q(a_+) = -k_B. \tag{9.65}$$

The negative value of this enthalpy can be at first sight taken as a proof of an unphysical character. However, one could also interpret $H_Q(a_+)$ the way Schrödinger did [32] with the so-called "negentropy" as a measure of the information available in the given system, which in the present case is the universe itself.

The above results correspond to the case in which the universe is endowed with a vanishing cosmological constant. If we allow now a nonzero cosmological term H_0 to exist, i.e., if we first choose the solution $a_- = a_0 \exp(H_0 t - 2\pi G V_Q t^2)$, then we have for the expressions of the entropy and enthalpy that correspond to a universe, which, if $H_0 > \sqrt{4\pi G V_Q}$, first expands in an accelerated way with $w > -1$, then expands in a decelerating way, to finally progressively contracts all the way down until it fades out at an infinite time

$$S_Q(a_-, H_0) = \frac{3H_0^2}{8\pi G\kappa} - \frac{V_Q}{\kappa} \ln\left[\left(\frac{a_-}{a_0}\right)^3\right], \tag{9.66}$$

and again for this case

$$H_Q(a_-, H_0) = \frac{V_Q}{\kappa} = k_B, \tag{9.67}$$

which is now positive definite.

Equation (9.66) contains two different terms. The first term, $S_{dS} = 3H_0^2 k_B/(8\pi G V_Q)$, corresponds to a de Sitter quantum entropy, which diverges in the classical limit $\hbar \to 0$. The second one is the same as the statistical-mechanic entropy in Eq. (9.63) but with the sign reversed. It would be worth comparing the first entropy term with the Hawking formula for the de Sitter space-time, which is given by the horizon area in Planck units, $S_H \propto H_0^{-2} k_B/(\ell_P^2)$ [37]. At first sight the entropy term S_{dS} appears to be proportional to just the inverse of the Hawking's formula. However, one can rewrite S_{dS} as $S_{dS} = k_B/(2G H_0 \bar{V}_Q)$, where $\bar{V}_Q = V_Q \mathrm{V}_{dS}$, with V_{dS} the equivalent volume occupied by the de Sitter space-time with horizon at $r = H_0^{-1}$. Now, $\bar{V}_Q$ is the amount of quantum energy contained in that equivalent de Sitter volume so that we must have $\bar{V}_Q = \hbar H_0$. It follows that S_{dS} actually becomes given by the horizon area in Planck units, too. It is worth noticing that the temperature $T_Q(a_-, H_0)$ can similarly be decomposed into two parts, one of which is given by the Gibbons–Hawking expression [37] $\hbar H_0/k_B$ and the other corresponds to the negative volume deficit that the factor $\exp(-2\pi G V_Q t^2)$ introduces in the de Sitter space-time volume.

We note that also for this kind of solution, a universe with $T_Q(a_-, H_0) = V_Q a_0^3/k_B$ and $S_Q(a_-, H_0) = S_{dS}$ is left when we set $t = 0$. If we let $\hbar \to 0$, then $T_{SQ}(a_-, H_0) \to 0$ and $S_Q(a_-, H_0) \to \infty$. On the other hand, it follows from Eq. (9.66) that as the universe evolves from the initial size a_0, the initially positive entropy $S_Q(a_-, H_0)$ progressively decreases until it vanishes at time $t = t_* = H_0/(4\pi G V_Q)$, after which the entropy becomes negative. This would mean a violation of the second law of thermodynamics even on the current evolution of the universe, which is induced by quantum effects. Therefore the model that corresponds to Eqs. (9.66) and (9.67) appears to be prevented by the second law.

Finally, we consider the remaining solution $a_+ = a_0 \exp(H_0 t + 2\pi G V_Q t^2)$, which predicts a universe expanding in a superaccelerated fashion all the time up to infinity with $w < -1$. In this case we obtain

$$S_Q(a_+, H_0) = \frac{3H_0^2}{8\pi G\kappa} + \frac{V_Q}{\kappa}\ln\left[\left(\frac{a_+}{a_0}\right)^3\right], \tag{9.68}$$

with $3H_0^2/(8\pi G\kappa) = 3H_0^2 k_B/(8\pi G V_Q) \propto S_H$, and

$$H_Q(a_+, H_0) = -\frac{V_Q}{\kappa} = -k_B. \tag{9.69}$$

All the above discussion on the relation of the quantum thermodynamical functions with the Hawking temperature and entropy holds also in this case, with the sole difference that now $S_Q(a_+, H_0)$ and $T_Q(a_+, H_0)$ are larger than their corresponding Hawking counterparts. Again for this solution a universe with $T_Q(a_+, H_0) = \kappa a_0^3$ and $S_Q(a_+, H_0) = S_{dS}$ is left when we set $t = 0$ whereas $T_Q(a_+, H_0) \to 0$ and $S_Q(a_+, H_0) \to \infty$ in the classical limit $\hbar \to 0$. Moreover, such as it happens when $H_0 = 0$, there is here no violation of the second law for $S_Q(a_+, H_0)$, but $H_Q(a_+, H_0)$ is again a negative constant interpretable like a negative entropy that would mark the onset of existing structures in the universe which are capable to store and process information [15].

In any case, we have shown that the thermodynamical laws derived in this section appear to preclude any model with $w > -1$ and so leave only a kind of phantom universe with $w < -1$ as the only possible cosmological alternative compatible with such laws. That kind of model does not show, however, the sort of shortcomings, including instabilities, negative kinetic field terms, or future singularities named big rips, that the usual phantom models have [17]. Since we have dealt with an essentially quantum system, the violation of the DEC that leads to the negative values of the enthalpy H_Q in the thermodynamically allowed models appears to be a rather benign problem from which one could even get some interpretational advantages. In fact, from Eqs. (9.45–9.49), we notice that the violation of the null energy condition (NEC)

$$\rho + p = -V_Q, \tag{9.70}$$

has an essentially quantum nature so that such a violation vanishes in the classical limit where $\hbar \to 0$. In fact, it is currently believed that even though classical general relativity cannot be accommodated to a violation of the DEC [34], such a violation can be admitted quantum mechanically, at least temporarily. Moreover, since the violating term $-V_Q$ is directly related to the negentropy $H_Q = -k_B$, it is really tempting to establish a link between that violation and the emergence of life in the universe. After all, one cannot forget that if living beings are fed on with negative entropy [32], then we ought to initially have some amount of negentropy to make the very emergence of life a more natural process, which, by itself, satisfies the second law.

9.5.2 Violation of Classical NEC

Thus, the quantum violation of the NEC has not any classical counterpart and therefore is physically allowable. We shall investigate in what follows the sense in which that violation would permit the formation of Lorentzian wormholes. Choosing the simplest mixed-energy-momentum tensor components and the ansatz that correspond to a static, spherically symmetric wormhole space-time with vanishing shift function, $ds^2 = -dt^2 + e^{\lambda}dr^2 + r^2 d\Omega_2^2$ (where $d\Omega_2^2$ is the metric on the unit two-sphere), we can obtain a wormhole space-time solution from the corresponding Einstein equations containing the extra quantum energy density and pressure, that is,

$$-\frac{\lambda'}{r}e^{-\lambda} - \frac{1}{r^2}\left(e^{-\lambda} - 1\right) = -\frac{8\pi G}{3}\left(\frac{9r_0^2}{8\pi G r^4} + \rho\right)$$

$$-\frac{1}{r^2}\left(e^{-\lambda}-1\right)=\frac{8\pi G}{3}\left(\frac{3r_0^2}{8\pi G r^4}+p\right)$$

$$-\frac{1}{2}e^{-\lambda}\frac{\lambda'}{r}=\frac{8\pi G}{3}\left(\frac{3r_0^2}{8\pi G r^4}+p\right),$$

supplemented by the condition $\rho + p = -V_Q$, to obtain

$$ds^2 = -dt^2 + \frac{dr^2}{1-\frac{r_0^2}{r^2}+\ell_P^2 V_Q r^2} + r^2 d\Omega_2^2, \tag{9.71}$$

with r_0 the radius of the spherical wormhole throat and ℓ_P the Planck length. Note that if $\rho + p$ was positive, then no cosmic wormhole could be obtained, such as it happens for the de Sitter space. The metric (9.71) is by itself nevertheless an actual cosmic wormhole because, if that metric is written as

$$ds^2 = -dt^2 + d\ell^2 + r^2 d\Omega_2^2, \tag{9.72}$$

then the new parameter [35]

$$\ell = \pm\int_{r_0}^{r}\frac{r'dr'}{\sqrt{r'^2 - r_0^2 + \ell_P^2 V_Q r'^4}} \tag{9.73}$$

$$=\pm\frac{1}{2\ell_P\sqrt{V_Q}}ln\left(\frac{2\ell_P\sqrt{V_Q}\sqrt{r^2-r_0^2+\ell_P^2 V_Q r^4}+2\ell_P^2 V_Q r^2+1}{1+4\ell_P^2 V_Q r_0^2}\right)$$

goes from $-\infty$ (when $r = +\infty$) to zero (at $r = r_0$) and finally to $+\infty$ (when $r = \infty$ again), such as it is expected for a wormhole with a throat at $r = r_0$, which is traversable and can be converted into a time machine. It can be readily checked that for $\rho + p > 0$, there is no metric like (9.73) which can show these properties.

9.5.3 Holographic Models

Holographic models which are related with the entropy of a dark-energy universe have been extensively considered [36, 38]. We shall discuss now the main equation that would govern the holographic model for the quantum cosmic scenario. If we try to adjust that model to the Li's holographic description for dark energy [36], then we had to define the holographic quantum model by the relation

$$H^2 = \frac{8\pi G\rho}{3} = 4\pi G V_Q \mu(t)^2 \ln\left(8G V_Q R_h^2\right), \tag{9.74}$$

where the future event horizon $R_h = a(t)\int_t^{\infty} dt'/a(t')$ is given by

$$R_h = \frac{e^{x^2}}{\sqrt{8GV_Q}}\left[1 - \Phi(x)\right], \tag{9.75}$$

with $\Phi(x)$ the probability integral [19],

$$x = \frac{H_0}{\sqrt{8\pi GV_Q}} + \sqrt{2\pi GV_Q}t, \tag{9.76}$$

and

$$\mu(t)^2 = \frac{1}{1 + 3(1 + w(t))\ln\left[1 - \Phi\left(-\frac{1}{1+w(t)}\right)\right]}. \tag{9.77}$$

Note that (1) $R_h \rightarrow \infty$ as $t \rightarrow \infty$ or $V_Q \rightarrow 0$, (2) in the latter limit $H^2 \rightarrow 0$, (3) $\mu(t)^2$ is no longer a constant because we are dealing with a tracking model where the parameter w depends on time, and (4) the holographic model does not have the problems posed by the usual holographic phantom-energy models. However, this formulation does not satisfy the general holographic equation originally introduced by Li, which reads [36] $\rho \propto H^2 \propto c^2/R^2$, where R is the proper radius of the holographic surface and c is a parameter of order unity that depends on w according to the relation $w = -(1+2/c)/3$, and therefore does not seem satisfactory enough. A better and quite simpler holographic description which comes from saturating the original bound on entropy [39] and conforms the general holographic equation stems directly from the very definitions of the energy density (9.45) and the entropy (9.68). Such a definition would read

$$\rho = \kappa S_Q(a_+, H_0) = \frac{3H^2}{8\pi G} = \frac{3}{8\pi G R_H^2}. \tag{9.78}$$

It appears that if the last equality in Eq. (9.79) holds, then the holographic screen is related to the Hubble horizon rather than the future event horizon or the particle horizon. In order to confirm that identification, we derive now the vacuum metric that can be associated to our ever-accelerating cosmic quantum model with the ansatz $ds^2 = -e^{\nu}dt^2 + e^{\lambda}dr^2 + r^2 d\Omega_2^2$. For an equation of state $p = w\rho$ the Einstein equations then are

$$e^{-\lambda}\left(\frac{\lambda'}{r} - \frac{1}{r^2}\right) + \frac{1}{2r^2} = 8\pi G\rho \tag{9.79}$$

$$e^{-\lambda}\left(\frac{\nu'}{r}+\frac{1}{r^2}\right)-\frac{1}{2r^2}=8\pi G w\rho. \tag{9.80}$$

We get finally the nonstatic metric:

$$ds^2=-\left(1-H^2r^2\right)^{-(1+3w)/2}dt^2+\frac{dr^2}{1-H^2r^2}+r^2d\Omega_2^2, \tag{9.81}$$

which consistently reduces to the de Sitter static metric for $w=-1$. It follows that there exists a time-dependent horizon at $r = H^{-1}$, which is always apparent for $w < -1/3$, playing, in Fact, the role of a Hubble horizon, like in the de Sitter case. Thus, for example, for the case that $w=-5/3$, the above metric reduces to

$$ds^2=\left(1-H^2r^2\right)^2dt^2+\frac{dr^2}{1-H^2r^2}+r^2d\Omega_2^2, \tag{9.82}$$

so that we can introduce a tortoise coordinate

$$r^*=\int\frac{dr}{\left(1-H^2r^2\right)^{3/2}}=-\frac{r}{\sqrt{1-H^2r^2}},\quad 0\leq r^*\leq\infty. \tag{9.83}$$

Using then advanced and retarded coordinates so that $U = t+r^*$ and $V = t-r^*$, we always can rewrite the above metric as a line Element, which, in fact, is no longer singular at $r=H^{-1}$, that is,

$$ds^2=-\left(1-H^2r^2\right)^2dU\,dV+r^2d\Omega_2^2, \tag{9.84}$$

in which r is implicitly determined from r^*. We can follow now the procedure described in Ref. [37] in order to obtain the maximally extended metric and from it the known expression for temperature and entropy of the de Sitter space.

This holographic model has several advantages over the previous Li's model [36] and other models [38], including its naturalness (it has been many times stressed that choosing the Hubble horizon is quite more natural than using, for the sake of mathematical consistency, particle or future event horizons), simplicity (no ad hoc assumption has been made), implication of an infrared cutoff depending on time, formal equivalence with Barrow's hyperinflationary model [40] (but here respecting the thermodynamical second law as, in this case, $S_Q(a_+, H_0)$ increases with time), and allowance of a unification between the present model and that for dark energy from vacuum entanglement [41].

9.5.4 Quantum Cosmic Models and Entanglement Entropy

The latter property deserves some further comments. In fact, if we interpret $a^3 V_Q$ as the total entanglement energy of the universe, due to the additiveness of the entanglement entropy, one can then add up [41] the contributions from all existing individual fields in the observable universe so that the entropy of entanglement $S_{\rm Ent} = \beta R_H^2$ (see comment after Eq. (9.69), with β a constant including the spin degrees of freedom of quantum fields in the observable volume of radius R_H and a numerical constant of order unity. On the other hand, the presence of a boundary at the horizon leads us to infer that the entanglement energy ought to be proportional to the radius of the associated spherical volume, i.e., $E_{\rm Ent} = \alpha R_H$ [41], with α a given constant. We have then

$$E_{\rm Ent} = a^3 V_Q = \alpha R_H \tag{9.85}$$

$$S_{\rm Ent} = \beta R_H^2. \tag{9.86}$$

It is worth noticing that one can then interpret the used temperature as the entanglement temperature so that $E_{\rm Ent} = k_B T(a_+)$. Now, integrating over R_H the expression for $dE_{\rm Ent}$ derived by Lee, Lee, and Kim [41] from the saturated black hole energy-bound [42],

$$dE_{\rm Ent} = T_{\rm Ent} dS_{\rm Ent} \tag{9.87}$$

(where $T_{\rm Ent} = (2\pi R_H)^{-1}$ is the Gibbons–Hawking temperature), we consistently recover expression (9.86) for $\alpha = \beta/\pi$. This result is also consistent with the holographic expression introduced before. It follows therefore that the quantum cosmic holographic model considered in the present paper can be consistently interpreted as an entangled dark-energy holographic model, similar to the one discussed in Ref. [41], with the quantum potential V_Q playing the role of the entanglement energy density.

Before closing up this section, it would be worth mentioning that the recent data [46] seem to point to a value $w < -1$, with $\dot{w}$ small and positive, just the result predicted in the present chapter. We, in fact, note that from Eq. (9.46) we obtain that $\dot{w} = 4\dot{H}^2/(3H^3) \propto t^{-3}$, at sufficiently large time.

9.6 Generalized Cosmic Solutions

It looks like an obvious empirical fact that the universe is four-dimensional at large scales. However, many extensions of the standard model involve theories which inhabit in a higher dimensional space-time. More than century ago Nordstrom [43] already formulated a unified theory based on extra dimensions. Later, Kaluza and Klein unified gravity and electromagnetism resorting to additional spatial dimensions [44, 45]. Extra dimensions are a theoretical necessity for superstring theory since it requires ten space-time dimensions to be consistent from the quantum mechanical point of view and M-theory requires eleven dimensions. Other important reasons to study extra dimensions are the Higgs mass hierarchy problem and the cosmological constant problem. Given that we have not yet been able to detect these extra dimensions, we usually assume that they are compactified on a small radius or have a strong curvature. The physical motivation for exploring cosmology in the quantum cosmic models with extra dimensions is to search for characteristic features that could not be explained in the four-dimensional case, in particular, the spectrum of the gravitational waves that are generated. For this reason, we expect that the future gravitational waves observations will be a crucial tool in detecting positive signatures of extra dimensions. As we have already discussed, the quantum cosmic solutions can be regarded as either some generalizations from the flat version of de Sitter space or, if V_Q is sufficiently small, such as it appears to actually be the case, as perturbations of that de Sitter space. Since most of such models correspond to equations of state whose parameter is less than -1, such as it was mentioned before, they are also known as *benigner* phantom cosmic models. In this section we shall derive even more general expressions for these quantum cosmic solutions by (i) considering the similar generalizations or perturbations of the hyperbolic version of the de Sitter space and (ii) using a d-dimensional manifold. Actually, some observational data have implied that our universe is not perfectly flat, and recent works [47, 48] contemplate the possibility of the universe having spatial curvature. Thus, although the Wilkinson Microwave

Anisotropy Probe (WMAP) alone abhors open models, requiring $\Omega_{\text{total}} \equiv \Omega_m + \Omega_\Lambda = 1 - \Omega_k \geq 0.9$ (95%), closed model with Ω_{total} as large as 1.4 are still marginally allowed provided that the Hubble parameter $h \sim 0.3$ and the age of the Universe $t_0 \sim 20Gyr$. The combinations of the WMAP plus the type Ia supernova data or the Hubble constant data also imply the possibility of the closed universe, giving curvature parameters $k = -0.011 \pm 0.012$ and $k = -0.014 \pm 0.017$, respectively [47], although the estimated values are still consistent with the flat Friedmann-Robertson-Walker (FRW) world model. Moreover, in Ref. [49], it is said that the best-fit closed universe model has $\Omega_m = 0.415$, $\Omega_\Lambda = 0.630$, and $H_0 = 55\ \text{kms}^{-1}\text{Mpc}^{-1}$ and is a better fit to the WMAP data alone than the flat universe model ($\Delta\chi^2_{\text{eff}} = 2$. However, the combination of WMAP data with either supernovae data, large-scale structure data, or measurements of H_0 favors models with Ω_K close to 0.

The d-dimensional de Sitter space has already been considered elsewhere [88]. Here we shall extend it to the also maximally symmetric space whose space-time curvature is still negative (positive Ricci scalar) but no longer constant. Our space-time will be solution of the Einstein equation:

$$R_{ab} = t_{ab}, \quad a, b = 0, 1, \ldots (d-1), \tag{9.88}$$

with

$$t_{ab} = (H \pm \hbar\xi t)^2 g_{ab}, \tag{9.89}$$

where $H^2 = \Lambda/(d-1)$ is a cosmological constant and the constant $\hbar\xi$ generalizes the quantum potential considered in the quantum cosmic models described in Section 9.4. We notice that in the classical limit $\hbar \to 0$, the above definition becomes that of the usual d-dimensional de Sitter space. We shall restrict ourselves in this paper to the case in which our generalized d-dimensional de Sitter space can still be visualized as a $d+1$ hyperboloid defined as [50]:

$$-x_0^2 + \sum_{j=1}^{d} x_j^2 = H^{-2}. \tag{9.90}$$

This $(d+1)$-dimensional hyperboloid is embedded in E^{d+1} so that the most general expression of the metric for our extended quantum-corrected solutions is provided by the metric induced in this embedding, that is,

$$ds^2 = -dx_0^2 + \sum_{j=1}^{d} dx_j^2, \tag{9.91}$$

which has the same topology and invariance group as the d-dimensional de Sitter space [88].

This metric can now be exhibited in coordinates $\Theta_\pm = t(1 \pm \hbar\xi t/H)\epsilon(\mp H_0/(4\hbar\xi), \pm\infty)$-notice that our solutions only then cover a portion of the de Sitter time, while $t\epsilon(-\infty, +\infty)$), $\psi_{d-1}, \psi_{d-2}, \ldots \psi_2\epsilon(0, \pi)$, $\psi_1\epsilon(0, 2\pi)$, defined by

$$\begin{aligned}
x_d &= H^{-1}\cosh(H\Theta)\sin\psi_{d-1}\sin\psi_{d-2}\ldots\sin\psi_2\cos\psi_1\\
x_{d-1} &= H^{-1}\cosh(H\Theta)\sin\psi_{d-1}\sin\psi_{d-2}\ldots\sin\psi_2\sin\psi_1\\
x_{d-2} &= H^{-1}\cosh(H\Theta)\sin\psi_{d-1}\sin\psi_{d-2}\ldots\cos\psi_2\\
x_1 &= H^{-1}\cosh(H\Theta)\cos\psi_{d-1}\\
x_0 &= H^{-1}\sinh(H\Theta),
\end{aligned} \tag{9.92}$$

which should be referred to as either time Θ_+ or time Θ_-. In terms of these coordinates metric (9.91) splits into

$$ds_\pm^2 = -\left(1 \pm \frac{2\hbar\xi t}{H}\right)^2 dt^2 + H^{-2}\cosh^2\left[t\left(H \pm \hbar\xi t\right)\right] d\Omega_{d-1}^2, \tag{9.93}$$

where $d\Omega_{d-1}^2$ is the metric on the $(d-1)$-sphere. The metric (9.97) is a closed $(d-1)$-dimensional FRW metric whose spatial sections are $(d-1)$-spheres of radius $H^{-1}\cosh(H\Theta)$. The coordinates defined by Eq. (9.92) describe two closed quantum cosmic spaces, $B_\pm$, which interconvert into each other at $t = 0$. B_+ first steadily contracts until $t = 0$, where it converts into B_- to first expand up to a finite local maximum value at $t = H/(2\hbar\xi)$, then contract down to a_0 at $t = H/(\hbar\xi)$, expanding thereafter to infinite. B_- would first contract until $t = -H/(\hbar\xi)$, then expand up to reach a local maximum at $t = -H/(2\hbar\xi)$, to contract again until $t = 0$, where it converts into a_+, which will steadily expand thereafter to infinite.

In terms of the conformal times $\eta_\pm = \int d\Theta_\pm/a_\pm$, which is given by

$$\tan\eta_\pm = \sinh\left(t \pm \hbar\xi t^2/H\right), \tag{9.94}$$

with $\pi/2 \geq \eta_+ \geq 0$ and $3\pi/2 \geq \eta_- \geq \pi$, the metrics can be reexpressed in a unitary form as

$$ds_\pm^2 = \frac{a_0^2}{\cos^2\eta_\pm}\left(-d\eta_\pm^2 + \gamma_{\alpha\beta}dx^\alpha dx^\beta\right), \alpha,\ \beta = 1, 2, \ldots (d-1), \tag{9.95}$$

where $\gamma_{\alpha\beta}$ is the metric for a unit $(d-1)$-sphere.

We shall consider in what follows the equivalent in our quantum cosmic scenarios of the static $(d-1)$-dimensional metric. Using the new coordinates

$$\begin{aligned}
x_d &= H^{-1}\sin\psi_{d-1}\sin\psi_{d-2}\ldots\sin\psi_2\cos\psi_1 \\
x_{d-1} &= H^{-1}\sin\psi_{d-1}\sin\psi_{d-2}\ldots\sin\psi_2\sin\psi_1 \\
x_{d-2} &= H^{-1}\sin\psi_{d-1}\sin\psi_{d-2}\ldots\cos\psi_2 \\
x_3 &= H^{-1}\sin\psi_{d-1}\sin\psi_{d-2}cos\psi_{d-3} \\
x_2 &= H^{-1}\sin\psi_{d-1}\cos\psi_{d-2} \\
x_0 &= H^{-1}\cos\psi_{d-1}\sinh(H\Theta')
\end{aligned} \tag{9.96}$$

where the coordinates are defined by $t'\epsilon(-\infty, +\infty))$, $r\epsilon(0, H^{-1})$, $\psi_{d-1}, \psi_{d-2}, \ldots \psi_2\epsilon(0, \pi)$, $\psi_1\epsilon(0, 2\pi)$. These coordinates will again be referred to either time Θ'_+ or time Θ'_-. Setting $r = H^{-1}\sin\psi_{d-1}$, we then find the metrics

$$ds_\pm^2 = -\left(1 \pm \frac{\hbar\xi t'}{H}\right)^2 dt'^2\left(1 - H^2r^2\right) + \frac{dr^2}{1 - H^2r^2} + r^2 d\Omega_{d-2}^2, \tag{9.97}$$

where $d\Omega_{d-2}^2$ is the metric on the $(d-2)$-sphere. We immediately note that this metric is no longer static. The coordinates defined by that metric cover only the portion of the spaces with $x_1 > 0$ and $\sum_{j=2}^{d} x_j^2 < H^{-2}$, i.e., the region inside the particle and event horizons of an observer moving along $r = 0$.

Respective instantons can now be obtained by analytically continuing $\Theta_\pm \to i\mathrm{T}_\pm$ (where we have taken $\Theta' \equiv \Theta$ for the sake of simplicity in the expressions), that is, $t' \to i\tau$ and $\xi \to -i\chi$, which contain singularities at $r = H^{-1}$, which are only apparent singularities if $\mathrm{T}_\pm$ are identified with periods $\pm 2\pi H^{-1}$, or, in other words, if τ is respectively identified with periods $H(\sqrt{1 + 8\pi\hbar\chi H^{-2}} \pm 1)/(2\hbar\chi)$. It follows then that the two spaces under consideration would respectively behave as though if they would emit a bath of thermal radiation at the intrinsic temperatures given by

$$T_{\pm}^{th} = \frac{2\hbar\chi}{H\left(\sqrt{1+8\pi\hbar\chi H^{-2}} \pm 1\right)}. \tag{9.98}$$

It must be remarked that in the limit when $\chi \to 0$, both temperatures $T_{\pm}^{th}$ consistently reduce to the unique value $H/(2\pi) = \sqrt{\Lambda(d-1)^{-1}}/(2\pi)$, that is, the temperature of a d-dimensional de Sitter space [88], even though T_{-}^{th} does it more rapidly than T_{+}^{th} (in fact, for sufficiently small χ, we can check that $T_{-}^{th} \simeq H/(2\pi)$ and $T_{+}^{th} \simeq \hbar\chi H/(H^2 + 2\pi\hbar\chi)$). Note that while we keep $\hbar$ in all definitions concerning the quantum cosmic spaces, natural units so that $\hbar = G = c = k_B = 1$ are otherwise used when such definitions are used. Now, one can estimate the entropy of these spaces by taking the inverse to their temperature. Thus, it can be seen that the entropy of the universe with scale factor a_+ will always be larger than that for a universe with scale factor a_-. It follows then that whereas the transition from a_+ to a_- at $t = 0$ would violate the second law of thermodynamics, the transition from a_- to a_+ at $t = 0$ would satisfy it, so making the model with scale factor a_+ evolving along positive time more likely to happen.

The time variables t and t' in Eqs. (9.89), (9.92), and (9.96) do not admit any bounds other than $(-\infty, +\infty)$, so that the involved models can be related with the Barrow's hyper inflationary model [40], albeit the solution a_+ here always respect the second law of thermodynamics because for such a solution the entropy is an ever-increasing function of time [51].

Before closing up this section, we shall briefly consider the static Schwarzschild-quantum mechanically perturbed solutions. It can be shown that in that case the line element is again not properly static as they depend on time in their g_{tt} component, that is,

$$ds_{\pm}^2 = -\left(1 \pm \frac{\hbar\xi t'}{H}\right)^2 dt'^2 \left(1 - \frac{2M}{r} - H^2 r^2\right) + \frac{dr^2}{1 - \frac{2M}{r} - H^2 r^2} + r^2 d\Omega_{d-2}^2, \tag{9.99}$$

Instantons for such solutions can also be similarly constructed. One readily may show that again such instantons describe thermal baths at given temperatures given now by

$$T_{\pm}^{th} = \frac{2\hbar\chi}{H\left(\sqrt{1+8\pi\hbar\chi H^{-2}} \pm 1\right)\left(1 \mp \frac{2}{3}\epsilon\right) + O\left(\epsilon^2\right)}. \tag{9.100}$$

where the second sign ambiguity in the denominator refers to the cosmological (upper) and black hole (lower) horizons and, according to Ginsparg and Perry [52], $9M^2\Lambda = 1 - 3\epsilon^2$, with $0 \leq \epsilon \ll 1$, the degenerate case corresponding just to $\epsilon \to 0$.

9.7 Gravitational Waves and Semiclassical Instability

The recent detection of gravitational waves by the LIGO Scientific Collaboration and the Virgo Collaboration [53] marks the birth of gravitational wave astronomy and provides a new window that can be used to test cosmological models [54–72]. Gravitational waves of cosmological origin can be produced in the early epochs of the universe, such as inflation and reheating, and are called primordial gravitational waves. These waves will leave a unique imprint on the B-mode of the cosmic microwave background. Different regions of the gravitational wave spectrum may be probed by current and future gravitational wave detectors. Ground-based, such as the advanced LIGO [73], advanced Virgo [74] and KAGRA [75], or space-borne such as evolved LISA [76] and DECIGO [77]. Also by pulsar timing experiments: Parkes [78], European pulsar timing arrays [79], NANOGRAV [80] and the Square Kilometer Array [81] as well as small-scale fluctuations and B-mode polarization of the CMB [82] and big bang nucleosynthesis [83].

We shall now make use of the solutions derived in the previous section, considering the generation of gravitational waves and some semiclassical instabilities that arise when the higher-dimensional solutions are Euclideanised ($t \to i\tau$).

According to Lifshitz [84, 85], the perturbations in the d metric are of the form $g_{\mu\nu} = g_{\mu\nu} + h_{\mu\nu}$. Because the constant-time spatial sections of our general metric are maximally symmetric $(d-1)$-spheres, an arbitrary initial perturbation of the metric tensor component $h_{\mu\nu}$ can be decomposed into d-dimensional spherical harmonics, i.e., the representations of SO(d). The most general metric perturbation can be written as $h_{\mu\nu} = \lambda(\eta)P_{\alpha\beta}) + \mu(\eta)Q_{\alpha\beta} + \sigma(\eta)S_{\alpha\beta} + \nu(\eta)H_{\alpha\beta}$, where the coefficients λ, μ, σ and ν depend only on the η time parameter, and $P_{\alpha\beta}$, $Q_{\alpha\beta}$, $S_{\alpha\beta}$ and $H_{\alpha\beta}$ are d-dimensional tensor harmonics. The generalisation of the Lifshitz-Khalatnikov perturbative formalism [84, 85] to the case of a closed FRW higher dimensional Universe has already been studied in [86] (see also [87]), where the following differential equation for ν is derived

$$\nu'' + (d-2)\tan\eta\nu' + \nu[\ell(\ell+d-2)] = 0\,. \tag{9.101}$$

where η and $' = d/d\eta$ refer to the conformal time, either η_+ or η_-, defined in Eq. (9.94). We shall consider d-dimensional gravitational waves, that is to say, the perturbations generated by the tensor spherical harmonics $H_{\alpha\beta}$. There are two cases to be studied (a) $\ell = 0$. The solution to Eq. (9.101) in this situation is the following

$$\nu = C_0 + C_1\left[2^{2-d}\frac{(d-2)!\eta}{\left[\left(\frac{d-2}{2}\right)!\right]^2} + 2^{3-d}\sum_{k=0}^{(d-4)/2}\frac{(d-2)!}{(d-2-k)!k!}\frac{\sin[(d-2-2k)\eta}{d-2-2k}\right] \tag{9.102}$$

for any even d. When the dimension d is odd, the solution becomes

$$\nu = C_0 + C_1\left[2^{3-d}\sum_{k=0}^{(d-3)/2}\frac{(d-2)!}{(d-2-k)!k!}\frac{\sin[(d-2-2k)\eta}{d-2-2k}\right]. \tag{9.103}$$

In both Eqs. (9.102) and (9.103), C_0 and C_1 are integration constants.

(b) $\ell \neq 0$. In this case the solution is

$$\nu = \cos^{(d-1)}\eta C_{\ell-1}^{(d/2)}(\sin\eta)\,, \tag{9.104}$$

where $C_{\ell-1}^{(d/2)}(\sin\eta)$ are the ultraspherical (Gegenbauer) polynomials.

If the modes (9.104) were excited during the inflationary epoch, before or at the time at which any possible compactification of the extra dimensions may have occurred, they could perhaps survive the phase transition and any compactification process to our current universe before being exponentially damped. They would then produce inhomogeneities for which there are strict observational bounds. These bounds could in principle allow us to constrain the initial dimensionality of space-time.

On the other hand, it was shown in [86] that the qualitative behaviour of semiclassical and tensorial perturbations in the d-dimensional de Sitter space is the same as in the four-dimensional case [87] (provided that the number of dimensions is finite). For this reason, and given that the quantum solutions presented in the previous section can be considered as scalar perturbations on the de Sitter space, we shall only discuss the evolution of the gravitational waves that arise from these quantum solutions in the usual four-dimensional scenario where any initial d-dimensional de Sitter symmetry must have broken long ago. We have then that Eq. (9.101) when $d = 4$ and $\ell = 0$ gives

$$\nu'' + 2\tan\eta\nu' = 0\,. \tag{9.105}$$

This differential equation has as general solution

$$\nu = C_0 + C_1\left(\eta + \frac{1}{2}\sin(2\eta)\right)\,, \tag{9.106}$$

where C_0 and C_1 are given integration constants. We must now particularize the above solution to be referred to $\eta_\pm$. In the case η_+ we see that the conformal time runs from 0 ($t = 0$) to $\pi/2$ ($t = \infty$). These waves do not destabilize the space because as, though their amplitude does not vanish at the limit where $\eta_+ \to \pi/2$, neither does it grow with time t. For η_- the conformal time runs from π ($t = 0$ or $t = H^2/(\hbar\xi)$) to $3\pi/2$ ($t = \infty$). It can be easily seen that neither these waves can destabilize the space.

For the general case $\ell \neq 0$, we have the general differential equation, likewise referred to either η_+ or η_-:

$$\nu'' + 2\tan\eta\nu' + \ell(\ell+2)\nu = 0\,. \tag{9.107}$$

The solution to this differential equation as

$$\begin{aligned}\nu(\eta) &= c_1\left[\sin(\ell\eta) + \ell\cos\eta\sin((\ell+1)\eta)\right]\\ &\quad + c_2\left[\cos(\ell\eta) + \ell\cos\eta\cos((\ell+1)\eta)\right],\end{aligned} \tag{9.108}$$

where c_1 and c_2 are arbitrary constants. This solution can also be expressed as

$$\nu = \cos^3\eta C^{(2)}_{\ell-1}(\sin\eta)\,, \tag{9.109}$$

with $C_n^{(\alpha)}$ the ultraspherical (Gegenbauer) polynomials of degree 2. Now, for $\eta_+ = 0$ or $\eta_- = \pi$, the amplitude vanishes for even $\ell = 2, 4, 6, \ldots$, and becomes

$$\nu = (-1)^{(\ell-1)/2} \frac{\Gamma\left(2 + \frac{\ell-1}{2}\right)}{\Gamma(2)\left(\frac{\ell-1}{2}\right)!}, \tag{9.110}$$

for odd $\ell = 1, 3, 5, \ldots$. For $\eta_+ = \pi/2$, $\nu = (\ell+2)!/[6(l-1)!]$ and for $\eta_- = 3\pi/2$, $\nu = (-1)^{\ell-1}(\ell+2)!/[6(l-1)!]$. Once the considered spaces are, therefore, stable to tensorial perturbations for nonzero ℓ. It is worth mentioning that for the solution corresponding to η_- and even ℓ, the absolute value of the amplitude of the gravitational waves would first increase from zero (at $t = 0$) to reach a maximum value at $t = H/(2\hbar\xi)$, to then decrease down to zero at $t = H/(\hbar\xi)$, and finally steadily increase all the time to reach its final finite value of unit order as $t \to \infty$. Clearly, a distinctive observational effect predicted by that cosmic model would be the generation of gravitational waves whose amplitude adjusted to the given pattern. The same, of course, can be said for the gravitational wave spectrum amplitude in the d-dimensional case given by Eqs. (9.102),(9.103) and (9.104) for $\eta_\pm$, which if confirmed by future observations would support the existence of extra dimensions.

A general derivation of Eqs. (9.105) and (9.107) from a general traceless rank-two tensor harmonics, which is an eigenfunction of the Laplace operator on S^3 and satisfies the eigenvalue equation $\nabla_a \nabla^a H_{cd}^{(n)} = -(n^2-3)H_{cd}^{(n)}$, can be found in Refs. [52, 88].

We add finally some comments to the possibility that our closed spaces develop a semiclassical instability. We shall use the Euclidean approach. In order to see if our Euclideanized solutions are stable or correspond to semiclassical instabilities, it will suffice to determine the eigenvalues of the differential operator [52, 89]:

$$G_{abcd}\phi^{ab} \simeq -\Box\phi_{cd} - 2R_{acbd}\phi^{ab} \simeq \lambda\phi_{cd}, \tag{9.111}$$

where ϕ_{ab} is a metric perturbation. Now, if all $\lambda \geq 0$, the Euclideanized spaces are stable, showing a semiclassical instability otherwise. Stability can most readily be shown if, by analytically continuing metrics (9.97), the metric on the $(d-2)$-sphere, $d\Omega_{d-2}$, turns out to be expressible as the Kahler metric associated to a two-sphere. Thus, let us introduce the complex transformation:

$$Z = 2\tan(\psi_{d-2}/2)\exp\left(i\int d\Omega_{d-3}\right), \tag{9.112}$$

and, hence, in fact, we can derive

$$d\Omega_{d-2} = \frac{d\bar{Z}dZ}{\left(1+\frac{1}{4}\bar{Z}Z\right)^2} \tag{9.113}$$

and the Kahler potential

$$K = 2\log\left(1+\frac{1}{4}\bar{Z}Z\right), \tag{9.114}$$

so showing that, quite similarly to what it happens in the d-dimensional de Sitter space, the instantons constructed from metrics (9.97) are stable. Whether or not a space-time corresponding to Schwarzschild generalized de Sitter metric would show a semiclassical instability is a question that required further developments and calculations.

9.8 On the Onset of the Cosmic Accelerating Phase

A recent paper by Amendola, Quartin, Tsujikawa, and Waga (hereafter denoted as AQTW) [90] has put all existing models for dark energy in an apparent very serious trouble. Actually, if the result obtained by AQTW was confirmed with full generality, then according to these authors the whole paradigm of dark energy ought to be abandoned (see, however, the results in Ref. [91], e.g.). Such as it happens with other aspects of the current accelerating cosmology, the problem is to some extend reminiscent of the difficulty initially confronted by earliest inflationary accelerating models [8], which could not smoothly connect with the following FRW decelerating evolution [92]. As is well known, such a difficulty was solved by invoking the new inflationary scenario [93]. In fact, the problem recently posed for dark energy can be formulated by saying that a previous decelerating matter-dominated era cannot be followed by an accelerating universe dominated by dark energy, and it is in this sense that it can be somehow regarded as the time-reversed version of the early inflationary exit difficulty. In more technical terms what AQTW have shown is that it is impossible to find a sequence of matter and scaling acceleration for any scaling Lagrangian that can be approximated as a polynomial, because a scaling Lagrangian is always singular in the phase space, so either the matter-dominated era is prevented or the region with a viable matter is isolated from that where the scaling acceleration occurs. Ways out from this problem required assuming either a sudden emergence of dark-energy domination or a cyclic occurrence of dark energy, both assumptions being quite hard to explain and implement. However, we consider here a dark-energy model where such problems are no longer present due to some sort of quantum characteristics, which can be assigned to particles and radiation in that model.

We start with an action integral that contains all the ingredients of our model. Such an action is a generalization of the one used by AQTW, which contains a time-dependent coupling between dark energy and matter and leads to a general Lagrangian that admits scaling solutions formally the same as those derived in Ref. [90]. Setting the Planck mass unity, our Lorentzian action reads

$$S = \int d^4x\sqrt{-g}\,[R + p(X,\phi)] + S_m\left[\psi_i, \xi, m_i(V_Q), \phi, g_{\mu\nu}\right] + ST\,(K, \psi_i, \xi)\,, \qquad (9.115)$$

where g is the determinant of the four-metric; p is a generically noncanonical general Lagrangian for the dark-energy scalar field ϕ with kinetic term $X = g^{\mu\nu}\partial_\mu\phi\partial_\nu\phi$, formally the same as the one used in Ref. [90]; and S_m corresponds to the Lagrangian for the matter fields ψ_i, each with mass m_i, which is going to depend on a quantum potential V_Q in a way that will be made clear in what follows, so as on the time-dependent coupling ξ of the matter field to the dark-energy field ϕ. The term ST denotes the surface term that generally depends on the trace on the second fundamental form K, the matter fields ψ_i, and the time-dependent coupling $\xi(t)$ between ψ_i and ϕ for the following reasons.

We first of all point out that in the theory being considered, the coupling between the matter and the scalar fields can generally be regarded to be equivalent to a coupling between the matter fields and gravity plus a set of potential energy terms for the matter fields. In fact, if we restrict ourselves to this kind of theories, a scalar field ϕ can always be mathematically expressed in terms of the scalar curvature R [94]. More precisely, for the scaling accelerating phase, we shall consider a quantum dark-energy model (see Refs. [10] and [9, 12, 51]) in which the Lagrangian for the field ϕ vanishes in the classical limit where the quantum potential is made zero; i.e., we take $p = L = -V(\phi)\left(E(x,k) - \sqrt{1-\dot{\phi}^2}\right)$, where $V(\phi)$ is the potential energy and $E(x,k)$ is the elliptic integral of the second kind, with $x = \arcsin\sqrt{1-\dot{\phi}^2}$ and $k = \sqrt{1 - V_Q^2/V(\phi)^2}$, and the overhead dot $\dot{}$ means derivative with respect to time. Using then a potential energy density for ϕ and the quantum medium (note that the quantum potential energy density becomes constant [9, 12, 51][see later on]), we have for the energy density and pressure, $\rho \propto X(HV_Q/\dot{H})^2 = p(X)/w(t)$, with $H \propto \phi V_Q + H_0$, $\dot{H} \propto \sqrt{2X}V_Q$, where H_0 is constant. For the resulting field theory to be finite, the condition that $2X = 1$ (i.e. $\phi = C_1 + t$) had to be satisfied [9, 12, 51], and from the Friedmann equation, the scale factor ought to be given by $a(t) \propto \exp\left(C_2 t + C_3 t^2\right)$, with C_1, C_2 and C_3 being constants. It follows then that for at least a flat space-time, we generally have $R \propto 1 + \alpha\phi^2$ (where α is another constant and we have rescaled time) in that type of theories, and hence the matter fields-scalar field couplings, which can be generally taken to be proportional to $\phi^2\psi_i^2$-turn out to yield $\xi R\psi_i^2 - K_0\psi_i^2$, with K_0 again a given constant. The first term of this expression corresponds to a coupling between matter fields and gravity, which requires an extra surface term, and the second one ought to be interpreted as a potential energy term for the matter fields $V_i \equiv V(\psi_i) \propto \psi_i^2$. In this way, for a general theory that satisfied the latter requirement, the action integral (9.115) should be rewritten as

$$S = \int d^4x\sqrt{-g}\left[R(1-\xi\psi_i^2) + p(X,\phi)\right] + S_m\left[\psi_i, V_i, m_i(V_Q), g_{\mu\nu}\right] - 2\int d^3x\sqrt{-h}\,\mathrm{Tr}K(1-\xi\psi_i^2), \tag{9.116}$$

in which h is the determinant of the three-metric induced on the boundary surface and it can be noticed that the scalar field ϕ is no longer involved at the matter Lagrangian. We specialize now in the minisuperspace that corresponds to a flat FRW metric in conformal time $\eta = \int dt/a(t)$

$$ds^2 = -a(\eta)\left(-d\eta^2 + a(\eta)^2 d\mathrm{x}^2\right), \tag{9.117}$$

with $a(\eta)$ the scale factor. In this case, if we assume a time dependence of the coupling such that it reached the value $\xi(\eta_c) = 1/6$ at the coincidence time η_c and choose suitable values for the arbitrary constants entering the above definition of R in terms of ϕ^2, then the action at that coincidence time would reduce to

$$S = \frac{1}{2}\int d\eta\left[a'^2 - \sum_i(\chi_i'^2 - \chi_i^2) + a^4\left(p(X,\phi) + \sum_i m_i(V_Q)^2\right)\right], \tag{9.118}$$

where the prime $'$ denotes derivative with respect to conformal time η and $X = \frac{1}{2a^2}(\phi')^2$. Clearly, the fields χ_i would then behave as if they formed a collection of conformal radiation fields were it not by the presence of the nonzero mass terms m_i^2 also at the coincidence time. If for some physical cause the latter mass terms could all be made to vanish at the coincidence time, then all matter fields would behave like though they were a collection of radiation fields filling the universe at around the coincidence time and there would not be the disruption of the evolution from a matter-dominated era to a stable accelerated scaling solution of the kind pointed out by AQTW, but the system smoothly entered the accelerated regime after a given brief interlude where the matter fields behave like pure radiation. In what follows we shall show that in the quantum scenario considered above, such a possibility can actually be implemented.

At the end of the day, any physical system always shows the actual quantum nature of its own. One of the most surprising implications tough by dark-energy and phantom-energy scenarios is that the universal system is not exception on that at any time or value of the scale factor. Thus, we shall look at the particles making up the matter fields in the universe as satisfying the Klein–Gordon wave equation [95] for a Bohmian quasi-classical wave function [10] $\Psi_i = R_i \exp(iS_i/\hbar)$, where we have restored an explicit Planck constant, R_i is the probability amplitude for the given particle to occupy a certain position within the whole homogeneous and isotropic space-time of the universe, as expressed in terms of relativistic coordinates, and S_i is the corresponding classical action also defined in terms of relativistic coordinates.

Taking the real part of the expression resulting from applying the Klein–Gordon equation to the wave function Ψ_i, and defining the classical energy as $E_i = \partial_i S/\partial t$ and the classical momentum as $p_i = \vec{\nabla} S_i$, one can then derive the modified Hamilton–Jacobi equation:

$$E_i^2 - p_i^2 + V_{Qi}^2 = m_{0i}^2, \tag{9.119}$$

where V_{Qi} is the relativistic version of the so-called quantum potential [10], which is here given by

$$V_{Qi} = \hbar \sqrt{\frac{\nabla^2 R_i - \ddot{R}_i}{R_i}}, \tag{9.120}$$

that should also satisfy the continuity equation (i.e., the probability conservation law) for the probability flux, $\vec{J} = \hbar\ \mathrm{Im}(\Psi^* \vec{\nabla} \Psi)/(m\mathrm{V})$ (with $\mathrm{V} \propto a^3$ the volume), stemming from the imaginary part of the expression that results by applying the Klein–Gordon equation to the wave equation Ψ. Thus, if the particles are assumed to move locally according to some causal laws [10], then the classical expressions for E_i and p_i will be locally satisfied. Therefore we can now interpret the cosmology resulting from the above formulae as a classical description with an extra quantum potential and average Eq. (9.119) with a probability weighting function for which we take $P_i = |R_i|^2$, so that

$$\iiint dx^3 P_i \left(E_i^2 - p_i^2 + V_{Qi}^2\right) = \langle E_i^2 \rangle_{\mathrm{av}} - \langle p_i^2 \rangle_{\mathrm{av}} + \langle V_{Qi}^2 \rangle_{\mathrm{av}} = \langle m_{0i}^2 \rangle_{\mathrm{av}}, \tag{9.121}$$

with the averaged quantities coinciding with the corresponding classical quantities and the averaged total quantum potential squared being given by $\langle V_Q^2 \rangle_{\mathrm{av}} = \hbar^2 \left(\langle \nabla^2 P \rangle_{\mathrm{av}} - \langle \ddot{P} \rangle_{\mathrm{av}}\right)$.

It is worth noticing that in the above scenario, the velocity of the matter particles should be defined to be given by

$$\langle v_i \rangle_{\mathrm{av}} = \frac{\langle p_i^2 \rangle_{\mathrm{av}}^{1/2}}{\left(\langle p_i^2 \rangle_{\mathrm{av}} + \langle m_{0i}^2 \rangle_{\mathrm{av}} - \langle V_{Qi}^2 \rangle_{\mathrm{av}}\right)^{1/2}}. \tag{9.122}$$

It follows that in the presence of a quantum potential, a particle with nonzero rest mass $m_{0i} \neq 0$ can behave like though if was a particle moving at the speed of light (i.e., a radiation massless particle) provided $\langle m_{0i}^2 \rangle_{\mathrm{av}} = \langle V_{Qi}^2 \rangle_{\mathrm{av}}$. Thus, if we introduce an effective particle rest mass $m_{0i}^{\mathrm{eff}} = \sqrt{\langle m_{0i}^2 \rangle_{\mathrm{av}} - \langle V_{Qi}^2 \rangle_{\mathrm{av}}}$, then we get that the speed of light again corresponds to a zero effective rest mass. It has been noticed [9, 12, 51], moreover, that in the cosmological context the averaged quantum potential defined for all existing radiation in the universe should be regarded as the cosmic stuff expressible in terms of a scalar field ϕ that would actually make up our scaling dark-energy solution. At the coincidence time, that idea should actually extend in the present formalism to also encompass, in an incoherent way, together with the averaged quantum potential for CMB radiation, the averaged quantum potential for matter particles, as a source of dark energy. On the other hand, it has been pointed out as well [9, 12, 51] that the quantum potential ought to depend on the scale factor $a(t)$ in such a way that it steadily increases with time, being the quantum energy density satisfying the above continuity equation what keeps constant along the whole cosmic evolution.

Assuming the mass m_i appearing in the action (9.118) to be an effective particle mass, it turns out that the onset of dark energy dominance would then be precisely at the coincidence time when $\langle V_{Qi}^2 \rangle_{\rm av} \equiv \langle V_{Qi}(a)^2 \rangle_{\rm av}$ reached a value, which equals $\langle m_{0i}^2 \rangle_{\rm av}$, and all the matter fields behaved in this way like a collection of radiation fields which are actually irrelevant to the issue of the incompatibility of the previous eras with a posterior stable accelerated current regime. In this case, the era of matter dominance can be smoothly followed by the current accelerated expansion, where all matter fields would effectively behave like though if they cosmologically were tachyons. This interpretation would ultimately amount to the unification of dark matter and dark energy, as the dark-energy model being dealt here with is nothing but a somehow quantized version of tachyon dark energy [96], so that one should expect both effective tachyon matter and tachyon dark energy to finally decay to dark matter, so providing a consistent solution to the cosmic coincidence problem.

Now, from our action integral (9.118), one can derive the equation of motion for the field ϕ, that is, (See also Refs. [97] and [98]),

$$\ddot{\phi}\left(p_X + 2Xp_{XX}\right) + 3Hp_X\dot{\phi} + 2Xp_{X\rho} - p_\phi = \frac{\delta S}{a^3 \delta\phi}, \qquad (9.123)$$

where we have restored the cosmic time t, using the notation of Refs. [90, 97], and [98], so that a suffix X or ϕ denotes a partial derivative with respect to X or ϕ, respectively, and now the last coupling term is Time dependent. Note that if we confine ourselves to the theory where $a(t)$ accelerates in an exponential fashion and $\dot{\phi}^2 = 1$, then the first term of this equation would vanish. Anyway, in terms of the energy density ρ for the scalar field ϕ, the above general equation becomes formally the same as that which was derived in Ref. [90]:

$$\frac{d\rho}{dN} + 3(1+w)\rho = -Q\rho_m \frac{d\phi}{dN}, \tag{9.124}$$

with ρ_m the energy density for the matter field, $N = \ln a$, and $Q = -\frac{1}{a^3 \rho_m}\frac{\delta S_m}{\delta \phi}$. We can then derive the condition for the existence of scaling solutions for time-dependent coupling, which, as generally the latter two equations are formally identical to those derived by AQTW, is the same as that was obtained by these authors. Hence, we have the generalized master equation for p [90]:

$$\left[1 + \frac{2dQ(\phi)}{\lambda Q^2 d\phi}\right] \frac{\partial \ln p}{\partial \ln X} - \frac{\partial \ln p}{\lambda Q \partial \phi} = 1, \tag{9.125}$$

whose solution was already obtained by AQTW [90] to be

$$p(X, \phi) = X\, Q(\phi)^2 g\left(X\, Q(\phi)^2 e^{\lambda \kappa(\phi)}\right) \tag{9.126}$$

where g is an arbitrary function, λ is a given function of the parameters of the equations of state for matter and ϕ and the energy density for ϕ, being $\kappa = \int^{\phi} Q(\xi)d\xi$ (see Ref. [90]). In the phase space we then have an equation-of-state effective parameter for the system $w_{\text{eff}} = -1 - \frac{2\dot{H}}{3H^2} = gx^2 + z^2/3$, with H the Hubble parameter and x and z, respectively, being $x = \dot{\phi}/(\sqrt{6}H)$ and $z = \sqrt{\rho_{\text{rad}}/(3H^2)}$. At the coincidence time where we have just radiation ($z \neq 0$ and $\rho_m = \rho_{\text{rad}}$), the effective equation of state is [90] $w_{\text{eff}} = 1/3$. Hence at the coincidence time interval we can only have radiation, neither matter or accelerated expansion domination, just the unique condition that would allow the subsequent onset of the accelerated expansion era where conformal invariance of the field χ no longer holds.

Thus, it appears that in the considered model, a previous matter-dominated phase can be evolved first into a radiation phase at a physical regular coincidence short stage which is then destroyed to be finally followed by the required new, independent phase of current accelerating expansion. This conclusion can be more directly drawn if one notices that there is no way by which the general form of the Lagrangian (9.126) can accommodate the Lagrangian final form $L \equiv p = f(a, \dot{a})\dot{\phi}^2 V_Q^2$, which characterizes quantum dark-energy models whose pressure p vanishes in the limit $V_Q \to 0$. It thus appears that at least these models can be taken to be counterexamples to the general conclusion that current dark-energy and modified-gravity models (see, however, Ref. [91]) are incompatible with the existence of a previous matter-dominated phase, as suggested in Ref. [90].

We finally notice, moreover, that the kind of quantum dark-energy theory providing the above counterexample is one that shows no classical analog (i.e., the Lagrangian, energy density, and pressure are all zero in the classical limit $\hbar \to 0$) and is thereby most economical of all. Thus, the above conclusion can also be stated by saying that, classically, a previous phase of matter dominance is always compatible with the ulterior emergence of a dominating phase made up of "nothing." In this way, similarly to as the abrupt, unphysical exit of the old inflationary problem was circumvented by introducing [93] a scalar field potential with a flat plateau leading to a "slow-rollover" phase transition, the abrupt disruption of the scaling phase after matter dominance can be also avoided by simply considering a vanishing scalar field potential that smooths the transition and ultimately makes it to work.

9.9 Conclusions and Comments

This chapter deals with two new four-dimensional cosmological models describing an accelerating universe in the spatially flat case. The ingredients used for constructing these solutions are minimal as they only specify a cosmic relativistic field described by just Hilbert–Einstein gravity and the probabilistic quantum effects associated with particles in the universe. While one of the models is ruled out on general thermodynamical grounds as being unphysical, the other model corresponds to an equation of state $p = w\rho$ with parameter $w < -1$ for its entire evolution; that is to say, this solution is associated with the so-called phantom sector, showing, however, a future evolution of the universe that is free from most of the problems confronted by usual phantom scenarios, namely, violent instabilities, future singularities, and classical violations of energy conditions. We have shown furthermore that the considered phantom model implies a more consistent cosmic holographic description and the equivalence between the discussed models and the entangled dark-energy model of the universe. Therefore we name our phantom model a *benign* phantom model.

Indeed, if the ultimate cause for the current speeding-up of the universe is quantum entanglement associated with its matter and radiation contents, then one would expect that the very existence of the current universe implied violation of the Bell's inequalities and hence the quantum probabilistic description related to the quantum potential considered in this chapter, or the collapse of the superposed cosmic quantum state into the universe we are able to observe, or its associated complementarity between cosmological and microscopic laws, any other aspects that may characterize a quantum system. The current dominance of quantum repulsion over attractive gravity started at a given coincidence time would then mark the onset of a new *quantum* region along the cosmic evolution, other than that prevailed at the big bang and early primeval universe, this time referring to the quite macroscopic, apparently classical, large universe that we live in. Thus, quite the contrary to what is usually believed, quantum physics governs not just the microscopic aspects of nature but also the most macroscopic domain of it in such a way that we can say that current life is forming part and is a consequence of a true quantum system.

Observational data are being accumulated that each time more accurately point to an equation of state for the current universe, which corresponds to a parameter whose value is very close to that of the case of a cosmological constant but still being less than -1 [2–5]. It appears that one of the models considered in this chapter would adjust perfectly to such a requirement, while it does not show any of the shortcomings that the customary phantom or modified-gravity scenarios now at hand actually have. Therefore, one is tempted to call for more developments to be made on such benign cosmological model, aiming at trying to construct a final scenario which would consistently describe the current universe and could presumably shed some light on what really happened during the primordial inflationary period as well.

We have also dealt with new four-dimensional and d-dimensional cosmological models describing an accelerating universe in the spatially flat and closed cases. The ingredients used for constructing these solutions are minimal as they only specify a cosmic relativistic field described by just Hilbert–Einstein gravity and the notion of the quantum entanglement of the universe, that is, the probabilistic quantum effects associated with the general matter content existing in the universe or its generalization for the closed cases. Two of such models correspond to an equation of state $p = w\rho$ with parameter $w < -1$ for their entire evolution, and still another covers a period in its future also with $w < -1$; that is to say, these three solutions are associated with the so-called phantom sector, showing, however, a future evolution of the universe that is free of most of the problems confronted by usual phantom scenarios, namely, violent instabilities, future singularities, incompatibility with the previous existence of a matter-dominated phase, classical violations of energy conditions, and inadequacy of the holographic description. Therefore we also denote such quantum cosmic models as *benign* phantom models. All these models can be regarded as generalizations or perturbations of the either exponential or the hyperbolic form of the de Sitter space. The hyperbolic solution are given in a d-dimensional manifold, which is particularized in the four-dimensional case in the Euclideanized extension that allowed us to derive quantum formulas for the temperature that reduce to that of Gibbons–Hawking when the perturbation is made to vanish. Finally, the generation of gravitational waves in the spatially closed case for both d-dimensions and four dimensions, analysing the evolution of the gravitational waves in four dimensions. It is also shown that none of these waves destabilize the space-time, as neither the vector and scalar cosmological perturbations do in the spatially flat and closed cases.

Acknowledgments

This work was supported by MEC under research project no. FIS2008-06332. Alberto Rozas-Fernández acknowledges financial support by the Fundação para a Ciência e a Tecnologia (FCT) (Portugal) through the postdoctoral fellowship SFRH/BPD /96981/2013 and by Ministerio de Economía y Competitividad (Spain) through project nu. FIS2012-38816. The authors benefited from discussions with C. Sigüenza and G. Readman of the *Estación Ecológica de Biocosmología* de Medellín, Spain.

References

1. E. J. Copeland, M. Sami, and S. Tsujikawa, *Dynamics of Dark Energy*, International Journal of Modern Physics D, **15**, 1753 (2006).
2. N. Said, C. Baccigalupi, M. Martinelli, A. Melchiorri, and A. Silvestri, *New Constraints on the Dark Energy Equation of State*, Physical Review D, **88**, 043515 (2013) and arXiv:1303.4353 [astro-ph.CO].
3. P. A. R. Ade, et al. [Planck Collaboration], *Planck 2015 results. XIV. Dark Energy and Modified Gravity*, Astronomy & Astrophysics, **594**, A14 (2016) and arXiv:1502.01590 [astro-ph.CO].
4. E. Piedipalumbo, E. Della Moglie, M. De Laurentis, and P. Scudellaro, *High Redshift Investigation on the Dark Energy Equation of State*, Monthly Notices of the Royal Astronomical Society, **441**, 3643 (2014) and arXiv:1311.0995 [astro-ph.CO].
5. W. Zheng, S. Y. Li, H. Li, J. Q. Xia, M. Li, and T. Lu, *Constraints on Dark Energy from New Observations Including Pan-STARRS*, Journal of Cosmology and Astroparticle Physics, **1408**, 030 (2014); Erratum: [Journal of Cosmology and Astroparticle Physics, **1409**, E01 (2014)] and arXiv:1405.2724 [astro-ph.CO].
6. A. G. Riess, et al., *A 2.4 % Determination of the Local Value of the Hubble Constant*, Astrophysical Journal, **826**, 56 (2016) and arXiv:1604.01424 [astro-ph.CO].
7. E. Di Valentino, A. Melchiorri, and J. Silk, *Reconciling Planck with the Local Value of H_0 in Extended Parameter Space*, Physics Letters B, **761**, 242 (2016) and arXiv:1606.00634 [astro-ph.CO].

8. A. H. Guth, *Inflationary Universe: A Possible Solution to the Horizon and Flatness Problems*, Physical Review D, **23**, 347 (1981).
9. P. F. González-Díaz, *Dark Energy Without Dark Energy*, AIP Conference Proceedings, **878**, 277 (2006); P. F. González-Díaz and A. Rozas-Fernández, *Accelerating Hilbert-Einstein Universe Without Dynamic Dark Energy*, Physics Letters B, **641**, 134 (2006).
10. D. Bohm, *A Suggested Interpretation of the Quantum Theory in Terms of Hidden Variables. 1*, Physical Review **85** 166 (1952); D. Bohm, *A Suggested Interpretation of the Quantum Theory in Terms of Hidden Variables. 2*, Physical Review **85**, 180 (1952).
11. J. S. Bagla, H. K. Jassal, and T. Padmanabhan, *Cosmology with Tachyon Field as Dark Energy*, Physical Review D, **67**, 063504 (2003).
12. P. F. González-Díaz, *Sub-Quantum Dark Energy*, Physical Review D, **69**, 103512 (2004).
13. A. C. Baccigalupi, A. Balbi, S. Matarrase, F. Perrotta, and N. Vittorio, Physical Review D, **65**, 063520 (2002); M. Melchiorri, L. Mersini, C. J. Odman, and M. Tradden, Physical Review D, **68**, 043509 (2003); M. Doupis, A. Riazuelo, Y. Zolnierowski, and A. Blanchard, Astronomy & Astrophysics, **405**, 409 (2003).
14. P. R. Holland, *The Quantum Theory of Motion*, Cambridge University Press, Cambridge, UK (1993).
15. G. Horton, C. Dewdney, and A. Nesteruk, *Time-Like Flows of Energy Momentum and Particle Trajectories for the Klein–Gordon Equation*, Journal of Physics A, **33**, 7337 (2000).
16. G. Grössing, *Quantum Cybernetics: A New Perspective for Nelson's Stochastic Theory, Nonlocality, and the Klein–Gordon Equation*, Physical Letters A, **296**, 1 (2002).
17. L. D. Landau and E. M. Lifshitz, *The Classical Theory of Fields*, Pergamon Press, Oxford, UK (1975).
18. D. J. Bohm, *Quantum Theory*, Prentice Hall, Dover, New York, USA (1951); Physical Review, **85**, 180 (1952).
19. M. Abramowitz and I. A. Stegun, *Handbook of Mathematical Functions*, Dover, New York, USA (1972); I. S. Gradshteyn and I. M. Ryzhik, Tables and Integrals, Series and Products, Academic Press, New York (1980).
20. Hannestad and E. Mortsell, Physical Review D, **66**, 063508 (2002); M. Tegmark, et al. [Collaboration], Physical Review D, **69**, 103501 (2004); U. Seljak, et al., Physical Review D, **71**, 103 515 (2005); U. Alam, et al., Monthly Notices of the Royal Astronomical Society, **354**, 275 (2004); astro-ph/0406672; D. Huterer and A. Cooray, Physical

Review D, **71**, 023506 (2005); B. Feng, X. L. Wang, and X. M Zhang, Physics Letters B, **607**, 35 (2005); J. Jonsson, A. Goobar, R. Amanullah, and L. Bergstrom, Journal of Cosmology and Astroparticle Physics, **09**, 007 (2004). For most recent data on WMAP, see: http://lambda.gsfc.nasa.gov/outreach/recent_papers.cfm

21. R. R. Caldwell, Physics Letters B, **545**, 23 (2002); R. R. Caldwell, M. Kamionkowski, and N. N. Weinberg, Physical Review Letters, **91**, 071301 (2003); P. F. González-Díaz, Physics Letters B, **586**, 1 (2004); Physical Review D, **69**, 063522 (2004); S. M. Carroll, M. Hoffman, and M. Trodden, Physical Review D, **68**, 023509 (2003); S. Nojiri and S. D. Odintsov, Physical Review D, **70**, 103522 (2004).
22. E. Babichev, V. Dokuchaev, and Yu. Eroshenko, *Black Hole Mass Decreasing due to Phantom Energy Accretion*, Physical Review Letters, **93**, 021102 (2004).
23. P. F. González-Díaz, *Achronal Cosmic Future*, Physical Review Letters, **93**, 071301 (2004); *On the Accretion of Phantom Energy onto Wormholes*, Physics Letters B, **632**, 159 (2006); *Some Notes on the Big Trip*, Physics Letters B, **635**, 1 (2006).
24. J. A. S. Lima and J. S. Alcaniz, *Thermodynamics and Spectral Distribution of Dark Energy*, Physical Letters B, **600**, 191 (2004).
25. P. F. Gonzalez-Diaz and C. L. Siguenza, *Phantom Thermodynamics*, Nuclear Physics B, **697**, 363 (2004).
26. S. Nojiri and S. D. Odintsov, *The Final State and Thermodynamics of Dark Energy Universe*, Physical Review D, **70**, 103522 (2004).
27. G. Izquierdo and D. Pavon, *Dark Energy and the Generalized Second Law*, Physical Letters B, **633**, 420 (2006).
28. G. Izquierdo and D. Pavon, *The Generalized Second Law in Dark Energy Dominated Universes* and arXiv:[gr-qc/0612092].
29. Y. Gong, B. Wang, and A. Wang, *Thermodynamical Properties of the Universe with Dark Energy*, Journal of Cosmology and Astroparticle Physics, **0701**, 024 (2007).
30. Y. Gong, B. Wang, and A. Wang, *On Thermodynamical Properties of Dark Energy*, Physical Review D, **75**, 123516 (2007).
31. R. R. Caldwell, *A Phantom Menace? Cosmological Consequences of a Dark Energy Component with Super-Negative Equation of State*, Physics Letters B, **545**, 23 (2002). S. M. Carroll, M. Hoffman, and M. Trodden, *Can the Dark Energy Equation-of-State Parameter w be Less Than −1?*, Physical Review D, **68**, 023509 (2003).

32. E. Schrödinger, *What is Life?. The Physical Aspects of the Living Cell*, Cambridge University Press, Cambridge, UK (1959); S. A. Kauffman, in *"What is Life?". The Next Fifty Years*, Cambridge University Press, Cambridge, UK (1995); L. Smolin, *The Life of the Cosmos*, Phoenix, London, UK (1997).
33. G. W. Gibbons and S. W. Hawking, *Cosmological Event Horizons, Thermodynamics, and Particle Creation*, Physical Review D, **15**, 2738 (1977).
34. S. W. Hawking and G. F. R. Ellis, *The Large Scale Structure of Space-Time*, Cambridge University Press, Cambridge, UK (1973).
35. M. S. Morris and K. S. Thorne,*Wormholes in Space-Time and Their Use for Interstellar Travel: A Tool for Teaching General Relativity*, American Journal of Physics, **56**, 395 (1988).
36. M. Li, *A Model of Holographic Dark Energy*, Physics Letters B, **603**, 1 (2004).
37. G. W. Gibbons and S. W. Hawking, *Cosmological Event Horizons, Thermodynamics, and Particle Creation*, Physical Review D, **15**, 2738 (1977).
38. D. Pavón and W. Zimdahl, *Holographic Dark Energy and Cosmic Coincidence*, Physics Letters B, **628**, 206 (2005); A. G. Cohen, D. B. Kaplan, and A. E. Nelson, *Effective Field Theory, Black Holes, and the Cosmological Constant*, Physical Review Letters, **82**, 4971 (1999).
39. P. F. González-Díaz, *Bounds on the Entropy*, Physical Review D, **27** 3042 (1983).
40. J. D. Barrow, *The Deflationary Universe: An Instability of the de Sitter Universe*, Physics Letters B, **180**, 335 (1986); *Deflationary Universes with Quadratic Lagrangians* Physics Letters B, **183**, 285 (1987).
41. J.-W. Lee, J. Lee, and H.-C. Kim, *Quantum Informational Dark Energy: Dark Energy from Forgetting*, Journal of Cosmology and Astroparticle Physics, **0708**, 005 (2007) and arXiv: [hep-th/0709.0047].
42. S. Mukohyama, M. Seriu, and H. Kodama, Physical Review D, **55**, 7666 (1997).
43. G. Nordstrom, *On the Possibility of Unifying the Electromagnetic and the Gravitational Fields*, Zeitschrift für Physik, **15**, 504 (1914) and arXiv: [physics/0702221].
44. T. Kaluza, *On the Problem of Unity in Physics*, Sitzungsber. Preuss. Akad. Wiss. Berlin (Math. Phys.) **1921**, 966 (1921).

45. O. Klein, *Quantum Theory and Five-Dimensional Theory of Relativity. (In German and English)*, Zeitschrift für Physik, **37**, 895 (1926). [Surveys High Energy Physics, **5**, 241 (1986).].

46. D. J. Mortlock and R. L. Webster, *The Statistics of Wide-Separation Lensed Quasars*, Monthly Notices of the Royal Astronomical Society, **319**, 872 (2000) and arXiv:[astro-ph/0008081]; A. G. Riess, et al. [Supernova Search Team Collaboration], *Observational Evidence from Supernovae for an Accelerating Universe and a Cosmological Constant*, Astronomical Journal, **116**, 1009 (1998) and arXiv:[astro-ph/9805201]; S. Perlmutter, et al. [Supernova Cosmology Project Collaboration], *Measurements of Omega and Lambda from 42 High-Redshift Supernovae*, Astrophysical Journal, **517**, 565 (1999) and arXiv:[astro-ph/9812133]; J. L. Tonry, et al. [Supernova Search Team Collaboration], *Cosmological Results from High-z Supernovae*, Astrophysical Journal, **594**, 1 (2003). D. N. Spergel, et al. [WMAP Collaboration], *First Year Wilkinson Microwave Anisotropy Probe (WMAP) Observations: Determination of Cosmological Parameters*, Astrophysical Journal, Suppl. **148**, 175 (2003). C. L. Bennett, et al., *First Year Wilkinson Microwave Anisotropy Probe (WMAP) Observations: Preliminary Maps and Basic Results*, Astrophysical Journal, Suppl. **148**, 1 (2003).
M. Tegmark, et al. [SDSS Collaboration], *Cosmological Parameters from SDSS and WMAP*, Physical Review D, **69**, 103501 (2004).

47. C. L. Bennet, et al., *First Year Wilkinson Microwave Anisotropy Probe (WMAP) Observations: Preliminary Maps and Basic Results* Astrophysical Journal, Suppl. **148**, 1 (2003); D. N. Spergel, *First Year Wilkinson Microwave Anisotropy Probe (WMAP) Observations: Determination of Cosmological Parameters*, Astrophysical Journal, Suppl. **148**, 175 (2003).

48. M. Tegmark, et al. *Cosmological Parameters from SDSS and WMAP*, Physical Review D, **69**, 103501 (2004).

49. D. N. Spergel, et al., *Three-Year Wilkinson Microwave Anisotropy Probe (WMAP) Observations: Implications for Cosmology*, Astrophysical Journal, Suppl. Ser. **170**, 377 (2007).

50. E. Schrödinger, *Expanding Universe*, Cambridge University Press, Cambridge, UK (1965).

51. P. F. González-Díaz and A. Rozas-Fernández, *Quantum Cosmic Models and Thermodynamics*, Classical and Quantum Gravity, **25**, 175023 (2008).

52. P. Ginsparg and M. J. Perry, *Semiclassical Perdurance of de Sitter Space*, Nuclear Physics B, **222**, 245 (1983).

53. B. P. Abbott, et al. [LIGO Scientific and Virgo Collaborations], *Observation of Gravitational Waves from a Binary Black Hole Merger*, Physical Review Letters, **116**, 061102 (2016) and arXiv:1602.03837 [gr-qc].
54. B. F. Schutz, *Determining the Hubble Constant from Gravitational Wave Observations*, Nature, **323**, 310 (1986).
55. D. Markovic, *On the Possibility of Determining Cosmological Parameters from Measurements of Gravitational Waves Emitted by Coalescing, Compact Binaries*, Physical Review D, **48**, 4738 (1993).
56. L. S. Finn, *Binary Inspiral, Gravitational Radiation, and Cosmology*, Physical Review D, **53**, 2878 (1996) and arXiv: [gr-qc/9601048].
57. D. E. Holz and E. V. Linder, *Safety in Numbers: Gravitational Lensing Degradation of the Luminosity Distance-Redshift Relation*, Astrophysical Journal, **631**, 678 (2005) and arXiv [astro-ph/0412173].
58. D. E. Holz and S. A. Hughes, *Using Gravitational-Wave Standard Sirens*, Astrophysical Journal, **629**, 15 (2005) and arXiv:[astro-ph/0506616].
59. C. Cutler and D. E. Holz, *Ultra-high Precision Cosmology from Gravitational Waves*, Physical Review D, **80**, 104009 (2009) and arXiv:0906.3752 [astro-ph.CO].
60. S. R. Taylor and J. R. Gair, *Cosmology with the Lights Off: Standard Sirens in the Einstein Telescope Era*, Physical Review D, **86**, 023502 (2012) and arXiv:1204.6739 [astro-ph.CO].
61. S. Camera and A. Nishizawa, *Beyond Concordance Cosmology with Magnification of Gravitational-Wave Standard Sirens*, Physical Review Letters, **110**, 151103 (2013) and arXiv:1303.5446 [astro-ph.CO].
62. N. Yunes and X. Siemens, *Gravitational-Wave Tests of General Relativity with Ground-Based Detectors and Pulsar Timing-Arrays*, Living Reviews Relativity, **16**, 9 (2013) and arXiv:1304.3473 [gr-qc].
63. B. P. Abbott, et al. [LIGO Scientific and Virgo Collaborations], *Tests of General Relativity with GW150914*, Physical Review Letters, **116**, 221101 (2016) and arXiv:1602.03841 [gr-qc].
64. T. E. Collett and D. Bacon, *Testing the Speed Of Gravitational Waves over Cosmological Distances with Strong Gravitational Lensing*, Physical Review Letters, **118**, 091101 (2017) and arXiv:1602.05882 [astro-ph.HE].
65. M. Oguri, *Measuring the Distance-Redshift Relation with the Cross-correlation of Gravitational Wave Standard Sirens and Galaxies*, Physical Review D, **93**, 083511 (2016) and arXiv:1603.02356 [astro-ph.CO].

66. I. Ben-Dayan, *Gravitational Waves in Bouncing Cosmologies from Gauge Field Production*, Journal of Cosmology and Astroparticle Physics, **1609**, 017 (2016) and arXiv:1604.07899 [astro-ph.CO].
67. N. Yunes, K. Yagi, and F. Pretorius, *Theoretical Physics Implications of the Binary Black-Hole Mergers GW150914 and GW151226*, Physical Review D, **94**, 084002 (2016) and arXiv:1603.08955 [gr-qc].
68. C. Caprini and N. Tamanini, *Constraining Early and Interacting Dark Energy with Gravitational Wave Standard Sirens: The Potential of the eLISA Mission*, Journal of Cosmology and Astroparticle Physics, **1610**, 006 (2016) and arXiv:1607.08755 [astro-ph.CO].
69. I. Sawicki, I. D. Saltas, M. Motta, L. Amendola, and M. Kunz, *Nonstandard Gravitational Waves Imply Gravitational Slip: On the Difficulty of Partially Hiding New Gravitational Degrees of Freedom*, Physical Review D, **95**, 083520 (2017) and arXiv:1612.02002 [astro-ph.CO].
70. T. Baker and M. Trodden, *Multimessenger Time Delays from Lensed Gravitational Waves*, Physical Review D, **95**, 063512 (2017) and arXiv:1612.02004 [astro-ph.CO].
71. C. M. Will, *The Confrontation between General Relativity and Experiment*, Living Reviews Relativity, **17**, 4 (2014) and arXiv:1403.7377 [gr-qc].
72. E. Berti et al., *Testing General Relativity with Present and Future Astrophysical Observations*, Classical and Quantum Gravity, **32**, 243001 (2015) and arXiv:1501.07274 [gr-qc].
73. D. Sigg [LIGO Scientific Collaboration], *Status of the LIGO detectors*, Classical and Quantum Gravity, **25**, 114041 (2008).
74. T. Accadia, et al., *Status of the Virgo Project*, Classical and Quantum Gravity, **28**, 114002 (2011).
75. K. Kuroda [LCGT Collaboration], *Status of LCGT*, Classical and Quantum Gravity, **27**, 084004 (2010).
76. P. Amaro-Seoane, et al., *eLISA/NGO: Astrophysics and Cosmology in the Gravitational-Wave Millihertz Regime*, GW Notes **6**, 4 (2013) and arXiv:1201.3621 [astro-ph.CO].
77. S. Kawamura, et al., *The Japanese Space Gravitational Wave Antenna: DECIGO*, Classical and Quantum Gravity, **28**, 094011 (2011).
78. R. N. Manchester, *The Parkes Pulsar Timing Array*, AIP Conference Proceedings, **983**, 584 (2008) and arXiv:0710.5026 [astro-ph].
79. R. D. Ferdman, et al., *The European Pulsar Timing Array: Current Efforts and a LEAP Toward the Future*, Classical and Quantum Gravity, **27**, 084014 (2010) and arXiv:1003.3405 [astro-ph.HE].

80. P. B. Demorest, et al., *Limits on the Stochastic Gravitational Wave Background from the North American Nanohertz Observatory for Gravitational Waves*, Astrophysical Journal, **762**, 94 (2013) and arXiv: 1201.6641 [astro-ph.CO].
81. G. Janssen, et al., *Gravitational Wave Astronomy with the SKA*, PoS AASKA **14**, 037 (2015) and arXiv: 1501.00127 [astro-ph.IM].
82. T. L. Smith, E. Pierpaoli, and M. Kamionkowski, *A New Cosmic Microwave Background Constraint to Primordial Gravitational Waves*, Physical Review Letters, **97**, 021301 (2006) and arXiv: [astro-ph/0603144].
83. R. H. Cyburt, B. D. Fields, K. A. Olive, and E. Skillman, *New BBN limits on Physics beyond the Standard Model from* ^4He, Astroparticle Physics, **23**, 313 (2005) and arXiv: [astro-ph/0408033].
84. E. Lifshitz, *Republication of: On the Gravitational Stability of the Expanding Universe*, Journal of Physics (USSR), **10**, 116 (1946) [General Relativity and Gravitation, **49** no. 2, 18 (2017).
85. E. M. Lifshitz and I. M. Khalatnikov, *Investigations in Relativistic Cosmology*, Advances in Physics, **12**, 185 (1963).
86. P. F. Gonzalez-Diaz, *Perdurance of Higher Dimensional de Sitter Space*, Physical Review D, **36**, 3651 (1987).
87. P. H. Ginsparg and M. J. Perry, *Semiclassical Perdurance of de Sitter Space*, Nuclear Physics B, **222**, 245 (1983).
88. P. F. González-Díaz, *Perdurance of Higher-Dimensional de Sitter Space*, Physical Review D, **36**, 3651 (1987).
89. P. F. González-Díaz, *Instability of Higher Dimensional de Sitter Space*, Physics Letters B, **191**, 263 (1987).
90. L. Amendola, M. Quartin, S. Tsujikawa, and I. Waga, *Challenges for scaling cosmologies*, Physical Review D, **74**, 023325 (2006).
91. S. Capozzielo, S. Nojiri, S. D. Odintsov, and A. Troisi, *Cosmological Viability of f(R)-gravity as an Ideal Fluid and its Compatibility with a Matter Dominated Phase*, Physics Letters B, **639**, 135 (2006).
92. T. Padmanabhan, *Cosmological Constant - the Weight of the Vacuum*, Physical Reports, **380**, 235 (2003).
93. A. D. Linde, *A New Inflationary Universe Scenario: A Possible Solution of the Horizon, Flatness, Homogeneity, Isotropy and Primordial Monopole Problems*, Physics Letters B, **108**, 389 (1982); A. Albrecht and P. J. Steinhardt, *Reheating an Inflationary Universe*, Physical Review Letters, **48**, 1437 (1982).

94. S. Nojiri, S. D. Odintsov and O. G. Gorbunova, Journal of Physics A, **39**, 6627 (2006); S. Nojiri and S. D. Odintsov, *Modified Gauss–Bonnet Theory as Gravitational Alternative for Dark Energy*, Physics Letters B, **631**, 1 (2005).
95. Strictly speaking, observable baryonic matter fields should be described by the Dirac equation rather than the Klein–Gordon equation. However, for our present purposes the use of the latter equation will suffice. On the other hand, even before coincidence time, most of the matter content had to be in the form of dark matter, a stuff whose nature is still unknown.
96. G. W. Gibbons, *Cosmological Evolution of the Rolling Tachyon*, Physics Letters B, **537**, 1 (2002).
97. F. Piazza and S. Tsujikawa, *Dilatonic Ghost Condensate as Dark Energy*, Journal of Cosmology and Astroparticle Physics, **0407**, 004 (2004).
98. S. Tsujikawa and M. Sami, *A Unified Approach to Scaling Solutions in a General Cosmological Background*, Physics Letters B, **603**, 113 (2004).

Chapter 10

Bohmian Quantum Gravity and Cosmology

Nelson Pinto-Neto[a] **and Ward Struyve**[b]

[a]*COSMO - Centro Brasileiro de Pesquisas Físicas – CBPF, rua Xavier Sigaud, 150, Urca, CEP22290-180, Rio de Janeiro, Brazil*

[b]*Mathematisches Institut, Ludwig-Maximilians-Universität München, Theresienstr. 39, 80333 München, Germany*

npintoneto@gmail.com; ward.struyve@gmail.com

Quantum gravity aims to describe gravity in quantum mechanical terms. How exactly this needs to be done remains an open question. Various proposals have been put on the table, such as canonical quantum gravity, loop quantum gravity, string theory, etc. These proposals often encounter technical and conceptual problems. In this chapter, we focus on canonical quantum gravity and discuss how many conceptual problems, such as the measurement problem and the problem of time, can be overcome by adopting a Bohmian point of view. In a Bohmian theory (also called pilot-wave theory or de Broglie–Bohm theory, after its originators de Broglie and Bohm); a system is described by certain variables in space-time such as particles or fields or something else, whose dynamics depends on the wave function. In the context of quantum gravity, these variables are a space-time metric and suitable variables for the matter fields (e.g., particles or fields). In addition to solving the

Applied Bohmian Mechanics: From Nanoscale Systems to Cosmology (Second Edition)
Edited by Xavier Oriols and Jordi Mompart

ISBN 978-981-4800-10-5 (Hardcover), 978-0-429-29474-7 (eBook)
www.jennystanford.com

conceptual problems, the Bohmian approach yields new applications and predictions in quantum cosmology. These include space-time singularity resolution, new types of semiclassical approximations to quantum gravity, and approximations for quantum perturbations moving in a quantum background.

10.1 Introduction

Quantum theory arose as a description of the world on the smallest scales. It culminated in the standard model of high energy physics which describes the electro-weak and strong interaction. Our current best theory for gravity, on the other hand, is the general theory of relativity which is a classical theory about space-time and matter. While each theory is highly successful in its own domain, quantum theory on small scales and general relativity on large scales, it is as yet unknown how to harmonize both theories. Matter seems fundamentally quantum mechanical. So a theory of gravity should take this into account. One way to do this is by assuming gravity to be quantum as well. The most conservative approach to achieve this, is by applying the usual quantization techniques, which turn classical theories into quantum theories, to general relativity. This results in a theory called canonical quantum gravity, described by the Wheeler–DeWitt equation. While these quantization techniques have proved to be enormously successful in the context of the standard model, there is no guarantee for success in the gravitational case. It is therefore crucial to look for possible experimental tests. However, it has been hard to extract possible predictions from canonical quantum gravity because the theory is problematic, not only due to the many technical issues surrounding the Wheeler–DeWitt equation, such as dealing with its infinities, but also due to the conceptual problems, such as the measurement problem and the problem of time, when it is considered in the framework of orthodox quantum theory.

In this chapter we show that a significant progress can be made by considering the canonical quantization of gravity from the Bohmian point of view. In a Bohmian theory, a system is described by certain variables in space-time such as particles or fields or something else, whose dynamics depends on the wave function [1–4]. In the context of nonrelativistic Bohmian mechanics, these variables are particle positions. So in this case there are actual particles whose motion depends on the wave function. Bohmian mechanics can also be extended to quantum field theory [5], where the variables may be particles or fields, and to canonical quantum gravity [6–10]. In the context of quantum gravity, the extra variables are a space-time metric and whatever suitable variable for the matter fields. This Bohmian formulation of quantum gravity solves the aforementioned conceptual problems. As such it can make unambiguous predictions in, for example, quantum cosmology. The aim of this chapter is to give an introduction to Bohmian quantum gravity, explain how it solves the conceptual problems with the conventional approach, and give examples of practical applications and novel predictions.

The chapter is organized as follows. We start with an introduction to nonrelativistic Bohmian mechanics in Section 10.2, and highlight some important properties that will also be used in the context of quantum gravity. In Section 10.3, we introduce canonical quantum gravity and discuss the conceptual problems that appear when trying to interpret the theory in the context of orthodox quantum theory. In Section 10.4, we turn to Bohmian quantum gravity and explain how it solves these conceptual problems. In Section 10.5, we discuss minisuperspace models. These are simplified models of quantum gravity, which assume certain symmetries such as homogeneity and isotropy. In the context of such models, we consider the problem of space-time singularities in Section 10.6. In Section 10.7, the minisuperspace models are extended to include perturbations. These perturbations are important in the description of structure formation. Bohmian approximation techniques will be employed to obtain tractable equations of motion. Observational consequences will be discussed for a particular model with Bohmian matter bounces. In addition, we show how the problem of the quantum-to-classical limit in inflationary and bouncing models is solved. Finally, in Section 10.8, we discuss a new approach to semiclassical gravity, which treats gravity classically and matter quantum mechanically, based on Bohmian mechanics.

10.2 Nonrelativistic Bohmian Mechanics

Nonrelativistic Bohmian mechanics is a theory about point-particles in physical space moving under the influence of the wave function [1–4]. The equation of motion for the configuration $X = (\mathbf{X}_1, \ldots, \mathbf{X}_n)$ of the particles, called the *guidance equation*, is given by[a]

$$\dot{X}(t) = v^{\psi}(X(t), t)\,, \tag{10.1}$$

where $v^{\psi} = (\mathbf{v}_1^{\psi}, \ldots, \mathbf{v}_n^{\psi})$, with

$$\mathbf{v}_k^{\psi} = \frac{1}{m_k}\mathrm{Im}\left(\frac{\nabla_k \psi}{\psi}\right) = \frac{1}{m_k}\nabla_k S \tag{10.2}$$

and $\psi = |\psi|\mathrm{e}^{\mathrm{i}S}$. The wave function $\psi(x, t) = \psi(\mathbf{x}_1, \ldots, \mathbf{x}_n, t)$ itself satisfies the nonrelativistic Schrödinger equation

$$\mathrm{i}\partial_t \psi(x, t) = \left(-\sum_{k=1}^{n} \frac{1}{2m_k}\nabla_k^2 + V(x)\right)\psi(x, t)\,. \tag{10.3}$$

For an ensemble of systems all with the same wave function ψ, there is a distinguished distribution given by $|\psi|^2$, which is called the *quantum equilibrium distribution*. This distribution is *equivariant*. That is, it is preserved by the particle dynamics (10.1) in the sense that if the particle distribution is given by $|\psi(x, t_0)|^2$ at some time t_0, then it is given by $|\psi(x, t)|^2$ at all times t. This follows from the fact that any distribution ρ that is transported by the particle motion satisfies the continuity equation

$$\partial_t \rho + \sum_{k=1}^{n} \nabla_k \cdot (\mathbf{v}_k^{\psi} \rho) = 0 \tag{10.4}$$

and that $|\psi|^2$ satisfies the same equation, i.e.,

$$\partial_t |\psi|^2 + \sum_{k=1}^{n} \nabla_k \cdot (\mathbf{v}_k^{\psi} |\psi|^2) = 0\,, \tag{10.5}$$

as a consequence of the Schrödinger equation.

It can be shown that for a typical initial configuration of the universe, the (empirical) particle distribution for an actual ensemble of subsystems within the universe will be given by the

[a] Throughout the paper we assume units in which $\hbar = c = 1$.

quantum equilibrium distribution [3, 4, 11]. Therefore, for such a configuration Bohmian mechanics reproduces the usual quantum predictions.

Nonequilibrium distributions would lead to a deviation of the Born rule. While such distributions are atypical, they remain a logical possibility [12]. However, it remains to be seen whether they are physically relevant.

Note that the velocity field is of the form $j^{\psi}/|\psi|^2$, where $j^{\psi} = (\mathbf{j}_1^{\psi}, \ldots, \mathbf{j}_n^{\psi})$ with $\mathbf{j}_k^{\psi} = \text{Im}(\psi^* \nabla_k \psi)/m_k$ is the usual quantum current. In other quantum theories, such as for example quantum field theories and canonical quantum gravity, the velocity can be defined in a similar way by dividing the appropriate current by the density. In this way equivariance of the density will be ensured. (See [13] for a treatment of arbitrary Hamiltonians.)

One motivation to consider Bohmian mechanics is the measurement problem. Orthodox quantum mechanics works fine for practical purposes. However, the measurement problem implies that orthodox quantum mechanics cannot be regarded as a fundamental theory of nature. The problem arises from the fact that the wave function has two possible time evolutions. On the one hand there is the Schrödinger evolution, on the other hand there is wave function collapse. But it is unclear when exactly the collapse takes place. The standard statement is that collapse happens upon measurement. But which physical processes count as measurements? Which systems count as measurement devices? Only humans? Or rather humans with a PhD [14]? Bohmian mechanics solves this problem. In Bohmian mechanics the wave function never collapses; it always evolves according to the Schrödinger equation. There is no special role for measurement devices or observers. They are treated just as other physical systems.

There are two aspects of the theory whose analogue in the context of quantum gravity will play an important role. Firstly, Bohmian mechanics allows for an unambiguous analysis of the classical limit. Namely, the classical limit is obtained whenever the particles (or at least the relevant macroscopic variables, such as the center of mass) move classically, i.e., satisfy Newton's equation. By taking the time derivative of (10.1), it is found that

$$m_k \ddot{\mathbf{X}}_k(t) = -\nabla_k (V(x) + Q^{\psi}(x,t))\big|_{x=X(t)}, \tag{10.6}$$

where

$$Q^{\psi} = -\sum_{k=1}^{n} \frac{1}{2m_k} \frac{\nabla_k^2 |\psi|}{|\psi|} \tag{10.7}$$

is the quantum potential. Hence, if the quantum force $-\nabla_k Q^{\psi}$ is negligible compared to the classical force $-\nabla_k V$, then the k-th particle approximately moves along a classical trajectory.

Another aspect of the theory is that it allows for a simple and natural definition for the wave function of a subsystem [3, 11]. Namely, consider a system with wave function $\psi(x, y)$ where x is the configuration variable of the subsystem and y is the configuration variable of its environment. The actual configuration is (X, Y), where X is the configuration of the subsystem and Y is the configuration of the other particles. The wave function of the subsystem $\chi(x, t)$, called the *conditional wave function*, is then defined as

$$\chi(x,t) = \psi(x, Y(t), t). \tag{10.8}$$

This is a natural definition since the trajectory $X(t)$ of the subsystem satisfies

$$\dot{X}(t) = v^{\psi}(X(t), Y(t), t) = v^{\chi}(X(t), t). \tag{10.9}$$

That is, for the evolution of the subsystem's configuration we can either consider the conditional wave function or the total wave function (keeping the initial positions fixed). (The conditional wave function is also the wave function that would be found by a natural operationalist method for defining the wave function of a quantum mechanical subsystem [15].) The time evolution of the conditional wave function is completely determined by the time evolution of ψ and that of Y. The conditional wave function does not necessarily satisfy a Schrödinger equation, although in many cases it does. This wave function collapses during measurement situations. This explains the success of the collapse postulate in orthodox quantum mechanics. In the context of quantum gravity, the conditional wave function will be used to derive an effective time-dependent wave equation for a subsystem of the universe from a time-independent universal wave function.

10.3 Canonical Quantum Gravity

Canonical quantum gravity is the most conservative approach to quantum gravity. It is obtained by applying the usual quantization techniques, which were so successful in high energy physics, to Einstein's theory of general relativity. The quantization starts with passing from the Lagrangian to the Hamiltonian picture and then mapping Poisson brackets to commutation relations of operators. Let us start with an outline of this procedure.

In general relativity, gravity is described by a Lorentzian space-time metric $g_{\mu\nu}(x)$, which satisfies the Einstein field equations

$$G_{\mu\nu} = 8\pi G T_{\mu\nu}, \tag{10.10}$$

where G is the gravitational constant, $G_{\mu\nu}$ the Einstein tensor and $T_{\mu\nu}$ the energy-momentum tensor, whose form is determined by the type of matter. In order to pass to the Hamiltonian picture, a splitting of space and time is necessary. This is done by assuming a foliation of space-time into space-like hypersurfaces so that $\mathcal{M}$ is diffeomorphic to $\mathbb{R} \times \Sigma$, with Σ a three-surface. Coordinates $x^\mu = (t, \mathbf{x})$ can be chosen such that the time coordinate t labels the leaves of the foliation and $\mathbf{x}$ are coordinates on Σ. In terms of these coordinates the space-time metric and its inverse can be written as

$$g_{\mu\nu} = \begin{pmatrix} N^2 - N_i N^i & -N_i \\ -N_i & -h_{ij} \end{pmatrix}, \qquad g^{\mu\nu} = \begin{pmatrix} \frac{1}{N^2} & \frac{-N^i}{N^2} \\ \frac{-N^i}{N^2} & \frac{N^i N^j}{N^2} - h^{ij} \end{pmatrix}, \tag{10.11}$$

where $N > 0$ is the lapse function, $N_i = h_{ij} N^j$ are the shift functions, and h_{ij} is the induced Riemannian metric on the leaves of the foliation.

The geometrical meaning of the lapse and shift is the following [16]. The unit vector field normal to the leaves is $n^\mu = (1/N, -N^i/N)$. The lapse $N(t, \mathbf{x})$ is the rate of change with respect to coordinate time t of the proper time of an observer with four-velocity $n^\mu(t, \mathbf{x})$ at the point $(t, \mathbf{x})$. The lapse function also determines the foliation. Lapse functions that differ only by a factor $f(t)$ determine the same foliation. Lapse functions that differ by more than a factor $f(t)$ determine different foliations. $N^i(t, \mathbf{x})$ is the rate of change with respect to coordinate time t of the shift of the points with the same coordinates $\mathbf{x}$ when we go from one hypersurface to another. Different choices of N^i correspond to different choices of coordinates on the space-like hypersurfaces.

The Hamiltonian picture makes it manifest that the functions N and N^i are arbitrary functions of space-time. The spatial metric h_{ij} satisfies nontrivial dynamics, corresponding to how it changes along the succession of space-like hypersurfaces. The arbitrariness of N and N^i arises from the space-time diffeomorphism invariance of the theory (i.e., the invariance under space-time coordinate transformations). The motion of h_{ij} does not depend on the foliation. That is, the evolution of an initial three-metric on a certain space-like hypersurface to a future space-like hypersurface does not depend on the choice of intermediate hypersurfaces. Spatial metrics h_{ij} that differ only by spatial diffeormophisms determine the same physical three-geometry. The dynamics is therefore called *geometrodynamics*. A succession of three-metrics determines a four-geometry.

Canonical quantization introduces an operator $\widehat{h}_{ij}(\mathbf{x})$ which acts on wave functionals $\Psi(h_{ij})$, which are wave functionals of metrics on Σ. In the presence of matter, the wave functional also depends on the matter degrees of freedom. But we assume just gravity for now. The wave functional satisfies the functional Schrödinger equation

$$\mathrm{i}\partial_t \Psi = \int d^3x \left(N\widehat{\mathcal{H}} + N^i \widehat{\mathcal{H}}_i \right) \Psi \,, \tag{10.12}$$

where

$$\begin{aligned} \mathcal{H} &= -16\pi G G_{ijkl} \frac{\delta}{\delta h_{ij}} \frac{\delta}{\delta h_{kl}} + \frac{h^{1/2}}{16\pi G} (2\Lambda - R^{(3)}) \,, \\ \mathcal{H}_i &= -2h_{il} D_j \frac{\delta}{\delta h_{jl}} \,, \end{aligned} \tag{10.13}$$

with G_{ijkl} the DeWitt metric (which depends on the three-metric), h the determinant of h_{ij}, $R^{(3)}$ the three-curvature and Λ the cosmological constant. In addition to the Schrödinger equation (10.12), the wave functional has to satisfy the constraints:

$$\mathcal{H}\Psi = 0 \,, \tag{10.14}$$

$$\mathcal{H}_i \Psi = 0 \,, \quad i = 1, 2, 3 \,. \tag{10.15}$$

This immediately implies that

$$\partial_t \Psi = 0\,, \qquad (10.16)$$

i.e., the wave function does not depend on time. This is the source of the problem of time. The constraint (10.15), which is called the *diffeomorphism constraint*, implies invariance of Ψ under infinitesimal spatial coordinate transformations. The constraint (10.14) is called the Wheeler–DeWitt equation and is believed to somehow contain the time evolution.

There are technical problems with this theory. Namely the functional Schrödinger equation is merely formal and needs to be regularized. In addition, a proper Hilbert space needs to be found. The theory is also not renormalizable. The latter does not necessarily mean a failure of the theory, but merely that the usual perturbation techniques do not work.

In addition to the technical problems, there are also conceptual problems. First of all, there is the measurement problem, which carries over from nonrelativistic quantum mechanics. However, in this case the problem is even more severe. Namely, the aim is to describe the whole universe (albeit with simplified models), and then there are no outside observers or measurement devices that could collapse the wave function. In addition, the aim is also to describe, for example, the early universe, and there are no observers or measurement devices present even within the universe. Second, there is the problem of time [17–19]. The wave function is static. So how can time evolution be explained in terms of such a wave function? How can we tell from the theory whether the universe is expanding or contracting, or running into a singularity? Finally, there is the problem of what it means to have a space-time singularity. The universe is described solely by a wave function, but there is no actual metric. Various definitions of what a singularity could mean have been explored [20–25]: that the wave function has support on singular metrics, that the wave function is peaked around singular metrics, that the expectation value of the metric operator is singular, etc. Although these definitions may have something to say about the occurrence or nonoccurrence of singularities, neither of these is completely satisfactory. In fact, since there is merely the wave function, one might even consider the question about space-time singularities as off-target, since it is the dynamics of the wave function that needs to be well-defined (which in this case amounts to finding solutions to the constraints). The question of the meaning of singularities is important because loop quantum gravity is believed to eliminate singularities, while canonical quantum gravity is not.

10.4 Bohmian Canonical Quantum Gravity

In the Bohmian approach to canonical quantum gravity, there is an actual three-metric h_{ij}, whose motion is given by

$$\dot{h}_{ij} = 32\pi G N G_{ijkl} \frac{\delta S}{\delta h_{kl}} + D_i N_j + D_j N_i, \tag{10.17}$$

with $\Psi = |\Psi| \mathrm{e}^{\mathrm{i}S}$ and D_i the three-dimensional covariant derivative. This equation can be obtained by considering a suitable current as explained in Section 10.2 [8]. It can also be obtained by considering the classical Hamilton equation and replacing the momentum conjugate to h_{ij} by $\delta S/\delta h_{kl}$ [9]. So, just as in the case of general relativity the theory concerns geometrodynamics, i.e., it is about an evolving three-geometry.

Different choices of the shift vectors N_i correspond to different coordinates on the spatial hypersurfaces. The Bohmian dynamics does not depend on the choice of coordinates on the spatial hypersurfaces. That is, the dynamics is invariant under spatial diffeomorphisms. A convenient choice is to take the $N_i = 0$ [8]. The lapse function $N > 0$ determines the foliation. Lapse functions that differ only by a factor $f(t)$ (which only depends on the time t which labels the leaves of the foliation) determine the same foliation. The Bohmian dynamics does not depend on such different choices. Such a difference merely corresponds to a time-reparameterization. However, different lapse functions that differ by more than a factor $f(t)$ generically yield different Bohmian dynamics [7, 8, 26, 27]. That is, if we consider the motion of an initial three-metric along a certain space-like hypersurface and let it evolve according to the dynamics given in Eq. (10.17) to a future space-like hypersurface, then the final three-metric will depend on the choice of lapse function or, in other words, on the choice of intermediate hypersurfaces. This was shown in detail in [26]. This is unlike general relativity, where there is foliation-independence. So in the Bohmian theory a particular choice of lapse function or foliation needs to be made. As such the theory is not generally covariant. This is akin to the situation in special relativity where the nonlocality (which is unavoidable for any empirically adequate quantum theory, due to Bell's theorem) is hard to combine with Lorentz invariance. In that context, it is simpler to assume a preferred reference frame or foliation. The extent to which this extra space-time structure can be eliminated and the theory be made fully Lorentz invariant is discussed in [28]. One possibility is to let the foliation be determined in a covariant way by the wave function. Perhaps a similar approach can be taken in the case of quantum gravity [8], but so far no concrete examples have been considered.

This theory solves the aforementioned conceptual problems with interpreting canonical quantum gravity. First of all there is no measurement problem. Measurement devices or observers do not play a fundamental role in the theory. Second, even though the wave function is static, the evolution of the three-metric is generically time-dependent. This evolution will, for example, indicate whether the universe is expanding or contracting. Effective time-dependent Schrödinger equations can be derived for subsystems by considering the conditional wave function. Suppose, for example, that we are dealing with a scalar field in the presence of gravity, for which the wave function is $\Psi(h_{ij}, \varphi)$. Then for a solution $(h_{ij}(\mathbf{x}, t), \varphi(\mathbf{x}, t))$ of the guidance equations, one can consider the conditional wave function for the scalar field: $\chi(\varphi, t) = \Psi(h_{ij}(\mathbf{x}, t), \varphi)$. In certain cases χ will approximately satisfy a time-dependent Schrödinger equation. Explicit examples will be given in Sections 10.7 and 10.8. There is a similar procedure to derive the time-dependent Schrödinger equation in the context of orthodox quantum theory [19]. In this procedure, a classical trajectory $h_{ij}(\mathbf{x}, t)$ is plugged into the wave function, rather than a Bohmian one. While this procedure indeed gives a time-dependent wave function for the scalar field, it seems rather ad hoc. In Bohmian mechanics, the conditional wave function is motivated by the fact that the velocity, and hence the evolution of the actual scalar field φ, can be expressed in terms of either the conditional or the universal wave function, cf. Eq. (10.9). The same trajectory is obtained. In the context of orthodox quantum mechanics there seems to be no justification for conditionalizing the wave function on a classical trajectory. Moreover, Bohmian mechanics is broader in its scope because it not only allows to conditionalize on classical paths but also on nonclassical ones. This allows to go beyond the usual semiclassical analysis, as will be shown in Section 10.7.

Finally, the meaning of space-time singularities becomes unambiguous. Namely, it is the same meaning as in general relativity: there is a singularity whenever the actual metric becomes singular. We will discuss examples in Section 10.6.

Taking the time derivative of the guidance equation (10.17) and using the expression (10.11) for the metric $g_{\mu\nu}$, the modified Einstein equations are obtained:

$$G_{\mu\nu} = 8\pi G T_{Q\mu\nu}, \tag{10.18}$$

where $T_Q^{\mu\nu}$ is an energy-momentum tensor of purely quantum mechanical origin (there is no matter in this case). It is given by

$$T_Q^{\mu\nu}(x) = -\frac{2}{\sqrt{-g(x)}}\frac{\delta}{\delta g_{\mu\nu}(x)}\int d^4yN(y)Q(y), \tag{10.19}$$

with

$$Q = -16\pi G G_{ijkl}\frac{1}{|\Psi|}\frac{\delta^2|\Psi|}{\delta h_{ij}\delta h_{kl}} \tag{10.20}$$

the quantum potential. More explicitly:

$$\begin{aligned} T_Q^{ij}(\mathbf{x},t) = & \left(\frac{N^i(\mathbf{x},t)N^j(\mathbf{x},t)}{N^2(\mathbf{x},t)} - h^{ij}(\mathbf{x},t)\right)\frac{Q(\mathbf{x},t)}{\sqrt{h(\mathbf{x},t)}} \\ & - \frac{2}{N(\mathbf{x},t)\sqrt{h(\mathbf{x},t)}}\int d^3yN(\mathbf{y},t)\sqrt{h(\mathbf{y},t)}\frac{\delta}{\delta h_{ij}(\mathbf{x})} \\ & \times \left(\frac{Q(\mathbf{y},t)}{\sqrt{h(\mathbf{y},t)}}\right). \end{aligned} \tag{10.21}$$

While Eq. (10.18) was written in covariant form, it is not generally covariant due to the preferred choice of lapse function. When $T_{Q\mu\nu}$ vanishes, we obtain the classical equations which are generally covariant.

The classical limit is obtained whenever $T_{Q\mu\nu}$ is negligible. The deviation from classicality also often causes singularities to be avoided, as we will see in the next Section.

Note that if $T_{Q\mu\nu} \sim g_{\mu\nu}$, then it would act as a cosmological constant. It is interesting to speculate whether the observed cosmological constant can indeed be of quantum origin [29, 30].

In orthodox quantum mechanics, it is important to consider a particular Hilbert space (which is difficult in this case). However, from the Bohmian point of view, this is not necessary. It is necessary however that the Bohmian dynamics be well-defined. For example, in the context of nonrelativistic Bohmian mechanics, a plane wave is not in the Hilbert space but the corresponding trajectories are well-defined. They are just straight lines.

What about probabilities? The Bohmian dynamics preserves the density $|\Psi(h_{ij})|^2$. However, this density is not normalizable (with respect to some appropriate measure $\mathcal{D}h$)[a] due to the constraints, so it cannot immediately be used to make statistical predictions. For certain predictions, it is not required, because we only have a single universe. On the other hand, statistical predictions play an important role in case where one can identify subsystems within the universe. We will see later on how this can be accomplished.

10.5 Minisuperspace

The Wheeler–DeWitt equation (10.14) and the diffeomorphism constraint (10.15) are very complicated functional differential equations which are hard to solve. In order to make the equations tractable, one often assumes certain symmetries like translation and rotation invariance. This reduces the number of degrees of freedom to a finite one. It is physically justified because we are interested in applying this formalism to the primordial universe, and observations indicate that it was very homogeneous and isotropic at these early times. Even today, at large scales, the universe seems to be spatially homogeneous and isotropic.

Rather than deriving the quantum minisuperspace models from the full quantum theory, they are obtained from the canonical

[a]This is rather formal and requires some mathematical rigor to make precise, but similar statements can be made in the context of minisuperspace models, to be discussed in the next Section, which are mathematically precise.

quantization of the reduced classical theory, using the action obtained from the full action upon imposition of the considered symmetries. It is as yet unclear to what extent these reduced quantum theories follow from the full quantum theory. The starting point is to express the metric h_{ij} and the matter degrees of freedom in terms of a finite number of variables, say q^a, $a = 1, \ldots, n$. By moving from the Lagrangian to the Hamiltonian picture one obtains a Hamiltonian which generally has the form

$$H = N\mathcal{H} = N\left(\frac{1}{2}f^{ab}(q)p_a p_b + U(q)\right), \tag{10.22}$$

where p_a are the momenta conjugate to q^a, $f^{ab}(q)$ is a symmetric function of the q's whose inverse plays the role of a metric on q-space and U is a potential. $N(t) > 0$ is the lapse function, which is arbitrary. It does not depend on space, since we can choose a foliation where the fields are homogeneous. The dynamics is generated by H, but constrained to satisfy $\mathcal{H} = 0$. This yields the equations of motion

$$\dot{q}^a = Nf^{ab}p_b, \tag{10.23}$$

$$\dot{p}_a = -N\left(\frac{1}{2}\frac{\partial f^{bc}(q)}{\partial q^a}p_b p_c + \frac{\partial U(q)}{\partial q^a}\right), \tag{10.24}$$

together with the constraint

$$\frac{1}{2}f^{ab}(q)p_a p_b + U(q) = 0. \tag{10.25}$$

Since the lapse is arbitrary, the dynamics is time-reparameterization invariant. The time parameter t is itself unobservable. Physical clocks should be modeled in terms of one of the variables q, say q^a. Namely, if q^a changes monotonically with t, it can be treated as a clock variable, and t could be eliminated by inverting $q^a(t)$.

Quantization of this model is done by introducing an operator

$$\widehat{\mathcal{H}} = -\frac{1}{2}\frac{1}{\sqrt{f}}\frac{\partial}{\partial q^a}\left(f^{ab}(q)\sqrt{f}\frac{\partial}{\partial q^b}\right) + U(q), \tag{10.26}$$

where f is the determinant of the inverse of f^{ab}, and which acts on wave functions $\psi(q)$. Here, interpreting f^{ab} as a metric on q-space, the Laplace–Beltrami operator was chosen, as is usually done [31].

This imposes an operator ordering choice. The Wheeler–DeWitt equation now reads

$$\widehat{\mathcal{H}}\psi = 0 \tag{10.27}$$

and the guidance equations are

$$\dot{q}^a = N f^{ab} \frac{\partial S}{\partial q^b}, \tag{10.28}$$

where $\psi = |\psi|e^{iS}$. The function $N(t)$ is again the lapse function. It is arbitrary, just as in the classical case, which implies that the dynamics is time-reparameterization invariant.

The continuity equation implied by the Wheeler–DeWitt equation is

$$\frac{\partial}{\partial q^a}\left(f^{ab}\frac{\partial S}{\partial q^b}|\psi|^2\right) = 0, \tag{10.29}$$

which implies that the density $|\psi|^2$ is preserved by the Bohmian dynamics. As mentioned in Section 10.2, this motivates the choice of the guidance equations (10.28).

Denoting $p_a = \partial S/\partial q^a$, the Bohmian dynamics implies the classical equations (10.24), but with the potential U replaced by $U + Q$, with

$$Q = -\frac{1}{2\sqrt{f}|\psi|}\frac{\partial}{\partial q^a}\left(f^{ab}\sqrt{f}\frac{\partial}{\partial q^b}|\psi|\right) \tag{10.30}$$

the quantum potential.

In the next Section, when considering the question of space-time singularities, we will consider two types of minisuperspace models. A Friedmann-Lemaître-Robertson-Walker (FLRW) metric coupled to respectively a canonical scalar field and a perfect fluid.

10.6 Space-Time Singularities

According to general relativity, space-time singularities such as a big bang or big crunch are generically unavoidable. This is usually taken as signaling the limited validity of the theory and the hope is that a quantum theory for gravity will eliminate the singularities. In Bohmian quantum gravity, we can unambiguously analyse the question of singularities because there is an actual metric and the meaning of singularities is the same as in general relativity.

In this Section, the question of big bang or big crunch singularities is considered in the simple case of a homogeneous and isotropic metric respectively coupled to a homogeneous scalar field (with zero matter potential [32, 33] and with exponential matter potential [34, 35]) and to a perfect fluid, modelled also by a scalar field [10, 36–38]. After considering the Wheeler–DeWitt quantization, we also consider the loop quantization of the former model [33]. In the Wheeler–DeWitt case, there may be singularities depending on the wave function and the initial conditions. In the case of loop quantization there are no singularities.

Anisotropic models are discussed in [10].

10.6.1 Minisuperspace: Canonical Scalar Field

The simplest example of a minisuperspace model is that of a homogeneous and isotropic FLRW metric coupled to a homogeneous scalar field. The metric is

$$\mathrm{d}s^2 = N(t)^2\mathrm{d}t^2 - a(t)^2\mathrm{d}\Omega_k^2, \tag{10.31}$$

where N is the lapse function, $a = \mathrm{e}^{\alpha}$ is the scale factor, and $\mathrm{d}\Omega_k^2$ is the spatial line-element on three-space with constant curvature k. In the classical theory, the coupling to a homogeneous scalar field φ is described by the Lagrangian

$$L = N\mathrm{e}^{3\alpha}\left(\kappa^2\frac{\dot{\varphi}^2}{2N^2} - \kappa^2 V_M - \frac{\dot{\alpha}^2}{2N^2} - V_G\right), \tag{10.32}$$

where $\kappa = \sqrt{4\pi G/3}$, with G the gravitational constant, V_M is the potential for the scalar field, $V_G = -\frac{1}{2}k\mathrm{e}^{-2\alpha} + \frac{1}{6}\Lambda$, and Λ is the cosmological constant [39, 40]. The classical equations of motion are

$$\frac{d}{dt}\left(\frac{\mathrm{e}^{3\alpha}\dot{\varphi}}{N}\right) + N\mathrm{e}^{3\alpha}\partial_{\varphi}V_M = 0, \tag{10.33}$$

$$\frac{\dot{\alpha}^2}{N^2} = 2\kappa^2\left(\frac{\dot{\varphi}^2}{2N^2} + V_M\right) + 2V_G. \tag{10.34}$$

The latter equation is the Friedmann equation. The acceleration equation, which corresponds to the second-order equation for α, follows from (10.33) and (10.34).

There is a big bang or big crunch singularity when $a = 0$, i.e., $\alpha \to -\infty$. This singularity is obtained for generic solutions, as was shown by the Penrose–Hawking theorems.

Canonical quantization of the classical theory leads to the Wheeler–DeWitt equation

$$\left[-\frac{1}{2\mathrm{e}^{3\alpha}}\partial_\varphi^2 + \frac{\kappa^2}{2\mathrm{e}^{3\alpha}}\partial_\alpha^2 + \mathrm{e}^{3\alpha}\left(V_M + \frac{1}{\kappa^2}V_G\right)\right]\psi(\varphi, \alpha) = 0. \quad (10.35)$$

In the Bohmian theory [6, 32] there is an actual scalar field φ and an actual FLRW metric of the form (10.31), whose time evolution is determined by

$$\dot{\varphi} = \frac{N}{e^{3\alpha}}\partial_\varphi S, \quad \dot{\alpha} = -\frac{N}{e^{3\alpha}}\kappa^2\partial_\alpha S. \quad (10.36)$$

It follows from these equations that

$$\frac{d}{dt}\left(\frac{\mathrm{e}^{3\alpha}\dot{\varphi}}{N}\right) + N\mathrm{e}^{3\alpha}\partial_\varphi(V_M + Q_M) = 0, \quad (10.37)$$

$$\frac{\dot{\alpha}^2}{N^2} = 2\kappa^2\left(\frac{\dot{\varphi}^2}{2N^2} + (V_M + Q_M)\right) + 2(V_G + Q_G), \quad (10.38)$$

where

$$Q_M = -\frac{1}{2\mathrm{e}^{6\alpha}}\frac{\partial_\varphi^2|\psi|}{|\psi|}, \qquad Q_G = \frac{\kappa^4}{2\mathrm{e}^{6\alpha}}\frac{\partial_\alpha^2|\psi|}{|\psi|} \quad (10.39)$$

are respectively the matter and the gravitational quantum potential. These equations differ from the classical ones by the quantum potentials.

In order to discuss the singularities, we consider the case of a free massless scalar field and that of an exponential potential.

10.6.1.1 Free massless scalar field

In the case of $V_M = V_G = 0$, the classical equations lead to

$$\dot{\varphi} = \frac{N}{\mathrm{e}^{3\alpha}}c, \qquad \dot{\alpha} = \pm\frac{N}{\mathrm{e}^{3\alpha}}\kappa^2 c, \quad (10.40)$$

where c is an integration constant. In the case $c = 0$, the universe is static and described by the Minkowski metric. In this case there is no singularity. For $c \neq 0$, we have

$$\alpha = \pm\kappa^2\varphi + \bar{c}, \quad (10.41)$$

with $\bar{c}$ another integration constant. In terms of proper time τ for a co-moving observer (i.e., moving with the expansion of the universe), also called *cosmic proper time*, which is defined by $d\tau = Ndt$, integration of (10.40) yields $a = e^{\alpha} = [3(c\tau + \tilde{c})]^{1/3}$, where $\tilde{c}$ is an integration constant, so that $a = 0$ for $\tau = -\tilde{c}/c$ (and there is a big bang if $c > 0$ and a big crunch if $c < 0$). This means that the universe reaches the singularity in finite cosmic proper time.

In the usual quantum mechanical approach to the Wheeler–DeWitt theory, the complete description is given by the wave function and as such, as mentioned in the introduction, the notion of a singularity becomes ambiguous. Not so in the Bohmian theory. The Bohmian theory describes the evolution of an actual metric and hence there are singularities whenever this metric is singular, i.e., when $a = 0$. The question of singularities in the special case where $V_M = V_G = 0$ was considered in [32, 41]. In this case, the Wheeler–DeWitt equation is

$$\frac{1}{a^3}\partial_{\varphi}^2\psi - \kappa^2\frac{1}{a^2}\partial_a(a\partial_a\psi) = 0, \tag{10.42}$$

or in terms of α:

$$\partial_{\varphi}^2\psi - \kappa^2\partial_{\alpha}^2\psi = 0. \tag{10.43}$$

The solutions are

$$\psi = \psi_R(\kappa\varphi - \alpha) + \psi_L(\kappa\varphi + \alpha). \tag{10.44}$$

The actual metric might be singular; it depends on the wave function and on the initial conditions. For example, for a real wave function, $S = 0$, the universe is static, so that there is no singularity. On the other hand, for wave functions $\psi = \psi_{R,L}$ the solutions are always classical, i.e., they are either static (if $\partial_{\alpha}S(\kappa\varphi(0) - \alpha(0)) = 0$, with $(\varphi(0), \alpha(0))$ the initial configuration) or they reach a singularity in finite cosmic proper time (if $\partial_{\alpha}S(\kappa\varphi(0) - \alpha(0)) \neq 0$). Wave functions with $\psi_R = -\psi_L$ satisfy $\psi(\varphi, \alpha) = -\psi(\varphi, -\alpha)$ and lead to trajectories that do not cross the line $\alpha = 0$ in (φ, α)-space. As such, trajectories starting with $\alpha(0) > 0$ will not have singularities. In this way bouncing solutions can be obtained. These describe a universe that contracts at early times then reaches a minimal volume and then expands again. At early and late times the evolution is classical.

Wave functions that have no singularities are of the form

$$\psi(\varphi,\alpha) = |\psi_R(\kappa\varphi-\alpha)| + |\psi_L(\kappa\varphi+\alpha)|e^{i\theta} \tag{10.45}$$

(up to an irrelevant constant phase factor) with θ a constant. The product $|\psi_R(\kappa\varphi-\alpha)||\psi_L(\kappa\varphi+\alpha)|$ is a constant of the motion in this case. For example, in the case $\psi_R(x) = \psi_L(x) = e^{-x^2}$, then $\alpha^2+\varphi^2$ is constant and the solutions correspond to cyclic universes. In this case, we do not get classical behavior at early or late times.

In summary, there may or may not be singularities depending on the wave function and the initial conditions for the actual fields.

10.6.1.2 The exponential potential

Consider $V_G = 0$ and the exponential matter potential

$$V_M(\varphi) = V_0 e^{-\lambda\bar{\kappa}\varphi}, \tag{10.46}$$

where V_0 and λ are constant. $\bar{\kappa}^2 = 6\kappa^2 = 8\pi G$, so that λ is dimensionless. Such potentials have been widely explored in cosmology in order to describe in a simple way primordial inflation (which describes an exponential expansion of the universe driven by the inflaton field), the present acceleration of the universe, and matter bounces (which concern bouncing cosmologies with an initial matter-dominated phase of contraction). This is because they contain attractor solutions where the ratio between the pressure and the energy density is constant, $p/\rho = w$, with $w = (\lambda^2-3)/3$. In order to describe primordial or late accelerated expansion, one should have $-1/3 > w \geq -1$, and for matter bounces $w \approx 0$, or $\lambda \approx \sqrt{3}$. Here we will discuss in detail the latter case.

The classical dynamics of such models is very rich and simple to understand. Assuming the gauge $N = 1$ (so that the time is actually cosmic proper time) and defining the variables

$$x = \frac{\bar{\kappa}}{\sqrt{6}H}\dot{\varphi}, \qquad y = \frac{\bar{\kappa}\sqrt{V_M}}{\sqrt{3}H}, \tag{10.47}$$

where

$$H = \frac{\dot{a}}{a} = \dot{\alpha} \tag{10.48}$$

is the Hubble parameter, reduces the dynamical equations to

$$\frac{dx}{d\alpha} = -3\left(x - \frac{\lambda}{\sqrt{6}}\right)(1-x)(1+x) \tag{10.49}$$

Table 10.1 Critical points of the planar system defined by (10.49) and (10.50)

x	y	w
-1	0	1
1	0	1
$\frac{\lambda}{\sqrt{6}}$	$-\sqrt{1-\frac{\lambda^2}{6}}$	$\frac{1}{3}(\lambda^2-3)$
$\frac{\lambda}{\sqrt{6}}$	$\sqrt{1-\frac{\lambda^2}{6}}$	$\frac{1}{3}(\lambda^2-3)$

and the Friedmann equation to

$$x^2 + y^2 = 1. \tag{10.50}$$

The ratio $w = p/\rho$ reads

$$w = 2x^2 - 1. \tag{10.51}$$

As we are interested in investigating matter bounces, we will from now on set $\lambda = \sqrt{3}$.

The critical points are very easily identified from (10.49). They are listed in table (Section 10.6.1.2). The critical points are $x = \pm 1$ with $w = 1$ ($p = \rho$, stiff matter) and correspond to the space-time singularity $a = 0$. Around this region, the potential is negligible with respect to the kinetic term. The critical points $x = 1/\sqrt{2}$ with $w = 0$ ($p = 0$, dust matter) are attractors (repellers) in the expanding (contracting) phase. Asymptotically in the infinite future (past) they correspond to very large slowly expanding (contracting) universes, and the space-time is asymptotically flat. Note that at $x = 0$ the scalar field behaves like dark energy, $w = -1$, $p = -\rho$.

Hence we have four possible classical pictures:

(a) A universe undergoing a classical dust contraction from very large scales, the initial repeller of the model, and ending in a big crunch singularity around stiff matter domination with $x \approx 1$, without ever passing through a dark energy phase.
(b) A universe undergoing a classical dust contraction from very large scales, the initial repeller of the model, passing through a dark energy phase, and ending in a big crunch singularity around stiff matter domination with $x \approx -1$.

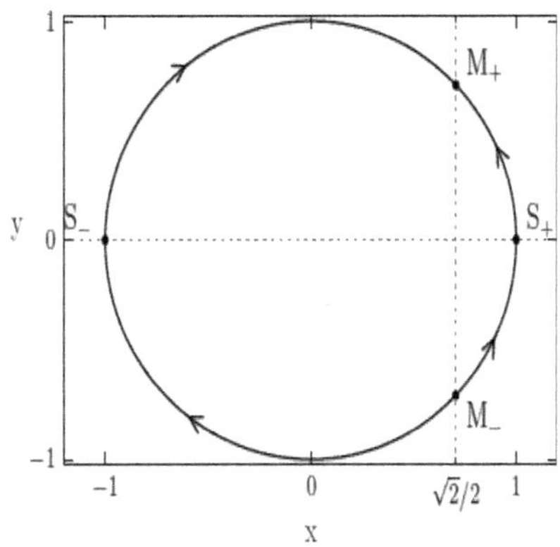

Figure 10.1 Phase space for the planar system defined by (10.49) and (10.50). The critical points are indicated by $M_{\pm}$ for a matter-type effective equation of state, and $S_{\pm}$ for a stiff-matter equation of state. For $y < 0$ we have the contracting phase, and for $y > 0$ the expanding phase. Lower and upper quadrants are not physically connected, because there is no classical mechanism that could drive a bounce between the contracting and expanding phases: there is a singularity in between.

(c) A universe emerging from a big bang singularity around stiff matter domination, with $x \approx 1$, and expanding to an asymptotically dust matter domination phase, without ever passing through a dark energy phase (which is the time-reversed of case a).

(d) A universe emerging from a big bang singularity around stiff matter domination, with $x \approx -1$, passing through a dark energy phase, and expanding to an asymptotically dust matter domination phase (which is the time-reversed of case b).

These classical possibilities are depicted in Fig. 10.1. The trajectories take place on a circle. The points $M_{\pm}$ are respectively the dust attractor and repeller, while $S_{\pm}$ are the singularities: the upper semicircle is disconnected from the down semicircle, and they respectively describe the expanding and contracting solutions.

In the quantum case, Bohmian bounce solutions were found. Exact solutions were given in [34] and with some approximation in [35], yielding the same qualitative picture. With these solutions, the quantum effects become important near the singularity. In

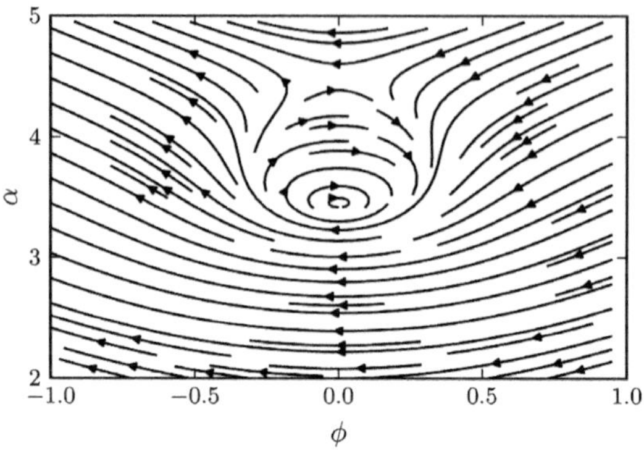

Figure 10.2 Phase space for the quantum bounce [35]. We can notice bounce and cyclic solutions. The bounces in the figure correspond to case B, where $\dot{\varphi} < 0$, and it connects regions around S_+ in the contracting phase with regions around S_- in the expanding phase.

this region, the potential is negligible and the quantum bounce is similar to the ones described in the preceding Section or as in [42]. The trajectories around the bounce are depicted in Fig. 10.2. For large scale factors, $\alpha \gg 1$, the classical stiff matter behavior is recovered, $x \approx \pm 1$, and from there on the Bohmian trajectories become classical, as described above.

One very important observation is that, looking at Fig. 10.2, the bounce can only connect $x \approx \pm 1$ classical stiff matter domination regions with $x \approx \mp 1$ classical stiff matter regions, respectively. In fact, a phase space analysis shows that such a connection of classical phases must happen for any bounce that might occur in the present model [34, 35]. This fact implies that there are only two possible bouncing scenarios, see Figs. 10.3 and 10.4:

(A) A universe undergoing a classical dust contraction from very large scales, which passes through a dark energy phase before reaching a stiff matter contracting phase with $x \approx -1$. In this regime, quantum effects become relevant and a bounce takes place, launching the universe to a classical stiff matter expanding phase with $x \approx 1$, which then evolves to an asymptotically dust matter expanding phase, without passing through a dark energy phase.

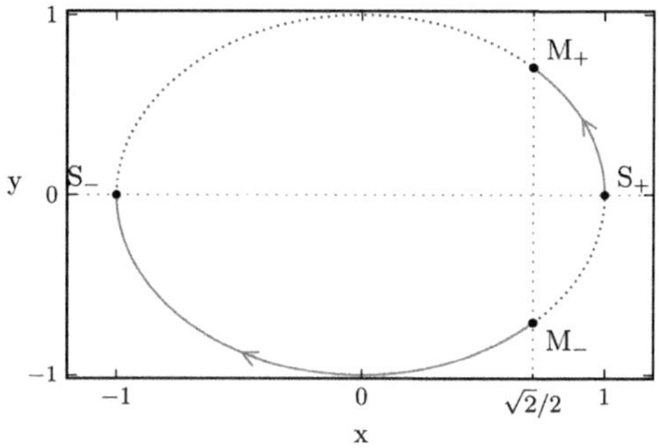

Figure 10.3 Case A: The scalar field has a dark energy type equation of state during contraction. By means of the quantum bounce, this system cannot address the dark energy in the future, since the matter attractor is reached before.

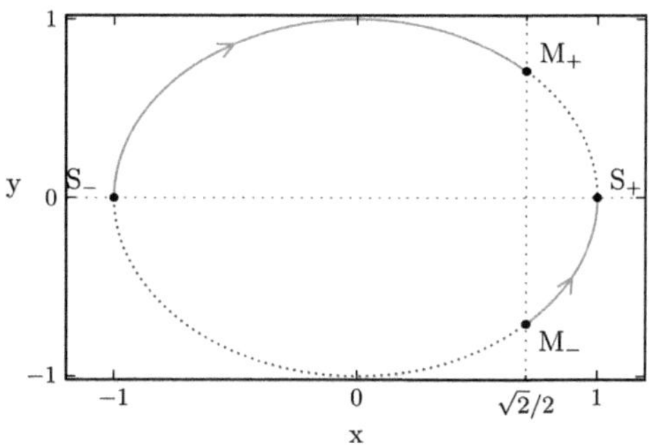

Figure 10.4 Case B: The contracting phase begins close to the unstable point M_+, in which the scalar field has a matter-type equation of state. After the quantum bounce, the system emerges from S_- and follows a dark energy phase until reaches the future attractor M_+.

(B) A universe undergoing a classical dust contraction from very large scales, which goes to a stiff matter contracting phase with $x \approx 1$, without passing through a dark energy phase. In this regime, quantum effects become relevant and a bounce takes place, launching the universe into a classical stiff matter expanding phase with $x \approx -1$, which passes through a dark energy phase before reaching an asymptotically dust matter expanding phase.

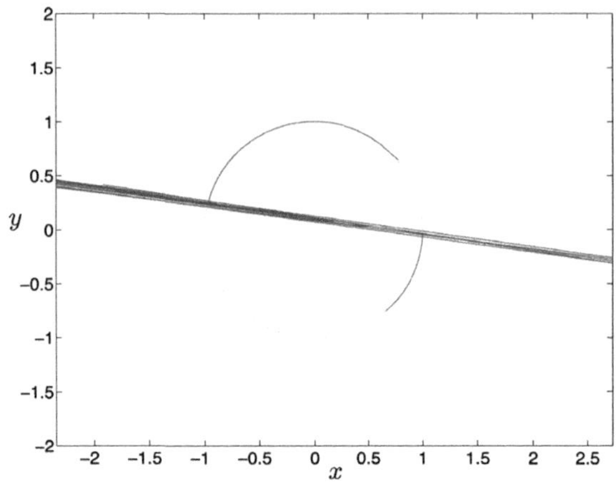

Figure 10.5 Bohmian trajectory corresponding to an exact solution [34]. It starts in the neighborhood of $(1/\sqrt{2}, -1/\sqrt{2})$ and ends in the neighborhood of $(1/\sqrt{2}, 1/\sqrt{2})$. The classical dynamics is valid almost everywhere, except near the singularity, where quantum effects become important and a bounce takes place, and the classical constraint $x^2 + y^2 = 1$ ceases to be satisfied.

Case B is the most physical one, because it can potentially describe the present observed acceleration of the universe as long as a dark energy era takes place in the expanding phase. Figure 10.5 shows an example of an exact Bohmian trajectory. Note that it satisfies almost everywhere the classical constraint $x^2 + y^2 = 1$, except near the singularity, where the quantum bounce takes place, and the trajectory goes from the region $x \approx -1$ to the region where $x \approx 1$.

In Section 10.7, we return to this bouncing model and we analyze the evolution of perturbations on these backgrounds. This leads to a promising alternative to inflation.

10.6.2 Minisuperspace: Perfect Fluid

Another example of a minisuperspace model is that of a FLRW spacetime with a perfect fluid, where the pressure is always proportional to the energy density, i.e., $p = w\rho$ with w constant. This kind of fluid may describe the hot universe. Namely, at high temperatures, fields and particles become highly relativistic, with a radiation equation of state $p \approx \rho/3$. We will see again that Bohmian mechanics gives rise to nonsingular solutions.

A perfect fluid can be modelled by a scalar field as follows. Consider the matter Lagrangian

$$L_M = \sqrt{-g}X^n, \tag{10.52}$$

where

$$X = \frac{1}{2}g^{\mu\nu}\partial_\mu\varphi\partial_\nu\varphi. \tag{10.53}$$

We will assume that $X \geq 0$ and we will interpret φ as the potential yielding the normalized four-velocity of the fluid

$$V_\mu = \frac{\partial_\mu\varphi}{(2X)^{1/2}}. \tag{10.54}$$

The energy-momentum tensor is given by

$$T_{\mu\nu} = \frac{2}{\sqrt{-g}}\frac{\partial L_M}{\partial g^{\mu\nu}} = 2nX^nV_\mu V_\nu - g_{\mu\nu}X^n. \tag{10.55}$$

Comparing with the usual expansion of the energy-momentum tensor in terms of energy density and pressure,

$$T_{\mu\nu} = (\rho + p)V_\mu V_\nu - pg_{\mu\nu}, \tag{10.56}$$

we get

$$p = X^n, \qquad p = \frac{1}{2n-1}\rho, \tag{10.57}$$

implying that $w = 1/(2n-1)$.

Assuming homogeneity, the scalar field depends only on time. The construction of the Hamiltonian is straightforward. The matter part reads

$$H_M = cN\frac{p_\varphi^{1+w}}{a^{3w}}, \tag{10.58}$$

where p_φ is the momentum conjugate to φ and $c = 1/w(\sqrt{2}n)^{1+w}$ is a constant. In the case of $w = 1$, the matter Hamiltonian is that of the previous Section. Before applying canonical quantization, the following canonical transformation is performed:

$$T = \frac{1}{c(1+w)}\frac{\varphi}{p_\varphi^w}, \qquad P_T = cp_\varphi^{1+w}, \tag{10.59}$$

so that

$$H_M = N\frac{P_T}{a^{3w}}. \tag{10.60}$$

An important property is that the momentum now appears linearly. Combining this perfect fluid Hamiltonian with the gravitational Hamiltonian for a FLRW geometry, the total minisuperspace Hamiltonian is obtained[a]:

$$H = N\left(-\frac{P_a^2}{4a} + \frac{P_T}{a^{3w}}\right). \tag{10.61}$$

It implies that $\dot{T} = N/a^{3w}$ or in terms of cosmic proper time τ, $dT/d\tau = 1/a^{3w}$ and hence T increases monotonically, so that it can be used as a clock variable. In terms of T, the scale factor evolves like $a \propto T^{2/3(1-w)}$ in the case $w \neq 1$, which is singular at $T = 0$ (if the proportionality constant is different from zero).

In the quantum case, because one momentum appears linearly in the Hamiltonian, the Wheeler–DeWitt equation assumes the Schrödinger form [36, 37, 43]

$$\mathrm{i}\frac{\partial}{\partial T}\Psi(a, T) = \frac{1}{4}\left\{a^{(3\omega-1)/2}\frac{\partial}{\partial a}\left[a^{(3\omega-1)/2}\frac{\partial}{\partial a}\right]\right\}\Psi(a, T). \tag{10.62}$$

Note that in the case $w = 1$, this equation differs from the Wheelder–DeWitt equation (10.42), due to the different pair of canonical variables which were quantized. In the rest of this Section, we will only consider $w \neq 1$ (for these cases we can apply the transformation (10.64)). The guidance equations are

$$\dot{T} = \frac{N}{a^{3w}}, \qquad \dot{a} = -\frac{N}{2a}\frac{\partial S}{\partial a}. \tag{10.63}$$

The dynamics can be simplified using the transformation

$$\chi = \frac{2}{3}(1-\omega)^{-1}a^{3(1-\omega)/2}, \tag{10.64}$$

to obtain

$$\mathrm{i}\frac{\partial\Psi(a, T)}{\partial T} = \frac{1}{4}\frac{\partial^2\Psi(a, T)}{\partial\chi^2}. \tag{10.65}$$

[a] In this Section, we follow the notation of [10], where units are such that $\kappa^2 = 1/2$. (Compared to the previous Section the total Lagrangian was also divided by κ^2.)

This is just the time-reversed Schrödinger equation for a one-dimensional free particle with mass 2 constrained to the positive axis.

In the context of orthodox quantum theory, the form of the Wheeler–DeWitt equation suggest to interpret T as time and to find a corresponding suitable Hilbert space. Since $\chi > 0$, the Hilbert space requires a boundary condition

$$\Psi\Big|_{\chi=0} = c\frac{\partial\Psi}{\partial\chi}\Bigg|_{\chi=0}, \tag{10.66}$$

with $c \in \mathbb{R} \cup \{\infty\}$ constant. $|\Psi^2|d\chi$ is then the probability measure for the scale factor. The boundary condition ensures that the total probability is preserved in time.

Note, however, that even though this form is suggestive, it is still rather ad hoc to interpret T as time. For example other variables could have been chosen (in particular if extra matter fields were considered). As explained before, in the Bohmian theory, the time t is unobservable and the physical clocks should be modeled by field or metric degrees of freedom. Since T increases monotonically with t, as long as the singularity $a = 0$ is not obtained, it can be used as a clock variable. But other monotonically increasing variables could also be used as clocks without ambiguities. The dynamics of the scale factor can be expressed in terms of T:

$$\frac{da}{dT} = -\frac{a^{3w-1}}{2}\frac{\partial S}{\partial a} \tag{10.67}$$

or

$$\frac{d\chi}{dT} = -\frac{1}{2}\frac{\partial S}{\partial\chi}. \tag{10.68}$$

In the Bohmian approach, the condition (10.66) implies that there are no singularities [38] (because the condition means that there is no probability flux $J_\chi \sim \operatorname{Im}\left(\Psi^* \frac{\partial\Psi}{\partial\chi}\right)$ through $\chi = 0$, so no trajectories will cross $a = 0$). However, for wave functions not satisfying the boundary condition (10.66), singularities will be obtained at least for some trajectories. For example, for a plane wave, the trajectories are the classical ones and hence a singularity is always obtained. From the Bohmian point of view this can motivate the consideration of a Hilbert space based on (10.66). It is then also natural to use $|\Psi^2|d\chi$ as the normalizable equilibrium distribution for the scale factor.

As an example of a wave function that satisfies the condition (10.66), consider the Gaussian

$$\Psi^{(\text{init})}(\chi) = \left(\frac{8}{T_0\pi}\right)^{1/4} \exp\left(-\frac{\chi^2}{T_0}\right), \tag{10.69}$$

where T_0 is an arbitrary constant. The wave solution for all times in terms of a is [36, 37]:

$$\Psi(a, T) = \left[\frac{8T_0}{\pi\left(T^2 + T_0^2\right)}\right]^{1/4} \exp\left[\frac{-4T_0 a^{3(1-\omega)}}{9(T^2 + T_0^2)(1-\omega)^2}\right]$$
$$\times \exp\left\{-\mathrm{i}\left[\frac{4T a^{3(1-\omega)}}{9(T^2 + T_0^2)(1-\omega)^2} + \frac{1}{2}\arctan\left(\frac{T_0}{T}\right) - \frac{\pi}{4}\right]\right\}.$$

The corresponding Bohmian trajectories are

$$a(T) = a_0\left[1 + \left(\frac{T}{T_0}\right)^2\right]^{\frac{1}{3(1-\omega)}}. \tag{10.70}$$

Note that this solution has no singularities for any initial value of $a_0 \neq 0$, and tends to the classical solution when $T \rightarrow \pm\infty$. The solution (10.70) can also be obtained for other initial wave functions [37].

For $w = 1/3$ (radiation fluid), and adjusting the free parameters, the solution (10.70) can reach the classical evolution before the nucleosynthesis era, where the standard cosmological model starts to be compared with observations. Hence, it can be a good candidate to describe a sensible cosmological model at the radiation dominated era which is free of singularities.

10.6.3 Loop Quantum Cosmology

Loop quantization is a different way to quantize general relativity [44, 45]. Application of this quantization method to the classical minisuperspace model defined by (10.32) results in the following theory. States are functions $\psi_\nu(\varphi)$ of a continuous variable φ and a discrete variable

$$\nu = \epsilon C a^3, \tag{10.71}$$

with

$$C = \frac{V_0}{2\pi G\gamma}, \tag{10.72}$$

where $\epsilon = \pm 1$ is the orientation of the triad (which is used in passing from the metric representation of general relativity to the connection representation), V_0 is the fiducial volume (which is introduced to make volume integrations finite) and γ is the Barbero-Immirzi parameter. ν is discrete as it is given by $\nu = 4n\lambda$ with $n \in \mathbb{Z}$ and $\lambda^2 = 2\sqrt{3}\pi\gamma G$. The value $\nu = 0$, which corresponds to the singularity, is included. One could also take $\nu = \epsilon + 4n\lambda$, with $\epsilon \in (0, 4\lambda)$. This does not include the value $\nu = 0$ and as such the singularity would automatically be avoided in the corresponding Bohmian theory (because, as will be discussed, in the Bohmian theory the possible values the scale factor can take are given by the discrete values of ν on which ψ has its support).

As usual, the quantization is not unique. Because of operator ordering ambiguities, different wave equations may be obtained. Different operator orderings are considered in the literature [23, 24, 46–48]. In all models, the wave equation is of the form

$$B_\nu \partial_\varphi^2 \psi_\nu(\varphi) + \sum_{\nu'} K_{\nu,\nu'} \psi_{\nu'}(\varphi) = 0, \tag{10.73}$$

with $\psi_\nu = \psi_{-\nu}$ and B_ν and $K_{\nu,\nu'} = K_{\nu',\nu}$ are real. The gravitational part, determined by K, is not a differential equation but a difference equation. For example, in the APS model [23, 24], the wave equation is

$$B_\nu \partial_\varphi^2 \psi_\nu(\varphi) - 9\kappa^2 D_{2\lambda}(|\nu| D_{2\lambda}\psi_\nu(\varphi)) = 0, \tag{10.74}$$

where

$$D_h \psi_\nu = \frac{\psi_{\nu+h} - \psi_{\nu-h}}{2h}, \tag{10.75}$$

so that

$$K_{\nu,\nu\pm 4\lambda} = -\frac{9\kappa^2}{16\lambda^2}|\nu \pm 2\lambda|, \qquad K_{\nu,\nu} = -K_{\nu,\nu+4\lambda} - K_{\nu,\nu-4\lambda} \tag{10.76}$$

and the other $K_{\nu,\nu'}$ are zero. Various choices for B_ν exist, again due to operator ordering ambiguities [49, 50]. One choice is [24]:

$$B_\nu = \left|\frac{3}{2} D_\lambda |\nu|^{2/3}\right|^3 = \left|\frac{3}{2}\frac{|\nu+\lambda|^{2/3} - |\nu-\lambda|^{2/3}}{2\lambda}\right|^3. \tag{10.77}$$

All choices of B_ν in all the models (except in the simplified APS model [24], called sLQC) share the important properties that $B_0 = 0$ and that for $|\nu| \gg \lambda$ (taking the limit $\lambda \to 0$, or equivalently, taking the Barbero-Immirzi parameter or the area gap to zero), $B_\nu \to 1/|\nu|$. For $|\nu| \gg \lambda$ (taking the limit $\lambda \to 0$), this wave equation reduces to the Wheeler–DeWitt equation

$$\frac{1}{|\nu|}\partial_\varphi^2\psi - 9\kappa^2\partial_\nu(|\nu|\partial_\nu\psi) = 0, \tag{10.78}$$

which is just the wave equation (10.43) in terms of ν.

Since the gravitational part of the wave equation (10.73) is now a difference operator, rather than a differential operator, the Bohmian dynamics now concerns a jump process rather than a deterministic process. Such processes have been introduced in the context of quantum field theory to account for particle creation and annihilation [51–53]. In the Bohmian theory, the scalar field evolves continuously, while the scale factor a, which will be expressed in terms of ν using (10.71), takes discrete values, determined by $\nu = 4n\lambda$ with $n \in \mathbb{Z}$. Since the evolution of the scale factor is no longer deterministic, but stochastic, the metric is no longer Lorentzian. Namely, once there is a jump, the metric becomes discontinuous. The metric is only "piece-wise" Lorentzian, i.e., Lorentzian in between two jumps.

The Bohmian dynamics can be found by considering the continuity equation, which follows from (10.73):

$$\partial_\varphi J_\nu(\varphi) = \sum_{\nu'} J_{\nu,\nu'}(\varphi), \tag{10.79}$$

where

$$J_\nu(\varphi) = B_\nu\partial_\varphi S_\nu(\varphi), \qquad J_{\nu,\nu'}(\varphi) = -K_{\nu,\nu'}\mathrm{Im}\left(\psi_\nu(\varphi)\psi^*_{\nu'}(\varphi)\right). \tag{10.80}$$

$J_{\nu,\nu'}$ is antisymmetric and nonzero only for $\nu' = \nu \pm 4\lambda$ for the LQC models considered above. Writing

$$\sum_{\nu'} J_{\nu,\nu'} = \sum_{\nu'}\left(\widetilde{T}_{\nu,\nu'}|\psi_{\nu'}|^2 - \widetilde{T}_{\nu',\nu}|\psi_\nu|^2\right), \tag{10.81}$$

where

$$\widetilde{T}_{\nu,\nu'}(\varphi) = \begin{cases} \frac{J_{\nu,\nu'}(\varphi)}{|\psi_{\nu'}(\varphi)|^2} & \text{if } J_{\nu,\nu'}(\varphi) > 0 \\ 0 & \text{otherwise} \end{cases}, \tag{10.82}$$

we can introduce the following Bohmian dynamics which preserves the quantum equilibrium distribution $|\psi_\nu(\varphi)|^2 d\varphi$. The scalar field satisfies the guidance equation

$$\dot{\varphi} = NC B_\nu \partial_\varphi S_\nu, \tag{10.83}$$

where $\psi_\nu = |\psi_\nu| e^{iS_\nu}$. The variable ν, which determines the scale factor, may jump $\nu' \to \nu$ with transition rates given by $T_{\nu,\nu'}(\varphi) = NC\widetilde{T}_{\nu,\nu'}(\varphi)$. That is, $T_{\nu,\nu'}(\varphi)$ is the probability to have a jump $\nu' \to \nu$ in the time interval $(t, t+dt)$. Note that the jump rates at a certain time depend on both the wave function and on the value of φ at that time. The properties of $J_{\nu,\nu'}$ imply that for a fixed ν either $T_{\nu,\nu+4\lambda}$ or $T_{\nu,\nu-4\lambda}$ may be nonzero (not both). The jump rates are "minimal", i.e., they correspond to the least frequent jump rates that preserve the quantum equilibrium distribution [53]. Just as in the classical case and the Bohmian Wheeler–DeWitt theory, the lapse function is arbitrary, which guarantees time-reparameterization invariance, just as in the case of Wheeler–DeWitt quantization. For $|\nu| \gg \lambda$ (taking the limit $\lambda \to 0$), this Bohmian theory reduces to the one of the Wheeler–DeWitt equation (using similar arguments as in [54]).

Let us now turn to the question of singularities. If $T_{0,\pm 4\lambda} = 0$, then the scale factor a (or ν) can never jump to zero, so a big crunch is not possible. If $T_{\pm 4\lambda,0} = 0$, then the scale factor cannot jump from zero to a nonzero value, so a big bang is not possible. Hence there are no singularities if $J_{0,\pm 4\lambda} = 0$. That this condition is satisfied can be seen as follows. Since $B_0 = 0$, we have

$$K_{0,4\lambda}\psi_{4\lambda} + K_{0,-4\lambda}\psi_{-4\lambda} + K_{0,0}\psi_0 = 0. \tag{10.84}$$

Using the properties $K_{0,\nu} = K_{0,-\nu}$ and $\psi_\nu = \psi_{-\nu}$, we obtain that

$$\mathrm{Im}\left(\psi_0^* K_{0,\pm 4\lambda}\psi_{\pm 4\lambda}\right) = 0 \tag{10.85}$$

and hence that $J_{0,\pm 4\lambda} = 0$. In summary, Bohmian loop quantum cosmology models for which the wave equation (10.73) has the properties that $B_0 = 0$, $K_{0,\nu} = K_{0,-\nu}$ and $\psi_\nu = \psi_{-\nu}$, do not have singularities. Importantly, no boundary conditions need to be assumed.

In the case that ψ is real, both φ and a are static. For other possible solutions, the wave equation needs to be solved first. This is rather hard, but can perhaps be done in the simplified model sLQC since the eigenstates of the gravitational part of the Hamiltonian are known in this case. Something can be said about the asymptotic behavior however. Since for large ν this Bohmian theory reduces to the Bohmian Wheeler–DeWitt theory, the trajectories will tend to be classical in this regime. Namely consider solutions (10.44) to the Wheeler–DeWitt equation for which the functions ψ_R and ψ_L go to zero at infinity. Then for $\alpha \to \infty$, the wave functions ψ_R and ψ_L become approximately nonoverlapping in (φ, α)-space. As such the Bohmian motion will approximately be determined by either ψ_R or ψ_L and hence classical motion is obtained. This implies an expanding or contracting (or static) universe. We expect that a bouncing universe will be the generic solution.

So far we assumed $k = \Lambda = 0$. In the case $k = \pm 1$ or $\Lambda \neq 0$ singularities are also eliminated [33].

In conclusion, in Bohmian loop quantum cosmology, there is no big bang or big crunch singularity regardless of the wave function. The result follows from a very simple dynamical analysis. It is in agreement with the results derived in the standard quantum mechanical framework [21–23]. However, in [21–23], φ is considered a time variable from the start, whereas in the Bohmian case, φ can only be used as a clock variable when it increases monotonically with t. In addition, often only a special class of wave functions is considered, namely the ones that behave classically at "late times."

10.7 Cosmological Perturbations

In Section 10.6, we have described Bohmian minisuperspace models. These simplified models of quantum gravity were obtained by assuming homogeneity and isotropy. In this Section, we consider deviations from homogeneity and isotropy by introducing perturbations. These perturbations are very important in current cosmological models, either inflationary or bouncing models, because they form the seeds of structure formation. Namely, according these models, in the far past the universe was so homogeneous that the only sources of inhomogeneities were quantum vacuum fluctuations. During the subsequent expansion of the universe the vacuum fluctuations result in classical fluctuations of the matter density. The classical fluctuations then grow through gravitational clumping and give rise to structures such as galaxies and clusters of galaxies we observe today. These vacuum fluctuations also leave an imprint on the cosmic microwave background radiation as temperature fluctuations.

There are a number of issues with this standard account that the Bohmian approach helps to solve. First of all, the conventional approach to deal with the cosmological perturbations is to consider a semiclassical treatment where only the first-order perturbations are quantized, while the background is treated classically (without back-reaction from perturbations onto fluctuations). This was largely explored in inflationary models in order to calculate the primordial power spectrum of scalar and tensor cosmological perturbations, and evaluate their observational consequences. However, the classical treatment of the background implies that there is a singularity, a point where no physics is possible, rendering the analysis incomplete. Using Bohmian mechanics, the usual approach to cosmological perturbations can be extended to include quantum corrections to the background evolution. This can then be used to infer consequences for the formation of structures in the universe, and for the anisotropies of the cosmic background radiation. Early attempts on this approach resulted in very complicated and intractable equations [31]. Using Bohmian mechanics, one is able to tremendously simplify the evolution equations of quantum cosmological perturbations in quantum backgrounds, rendering them into a simple and solvable form, suitable for the calculation of their observational consequences [55–62]. We start with illustrating the derivation of the motion of the quantum perturbations in a quantum background in Section 10.7.1 for the simple case of a canonical scalar with zero potential. Similar results can be obtained for nonzero potentials. Then, in Section 10.7.5, we will discuss the observational consequences in the case of an exponential matter potential, for which the background equations yield bouncing solutions as discussed in Section 10.6.1.2.

A second problem with the conventional approach is that of the quantum-to-classical transition of the perturbations [63, 64]. Namely, the quantum vacuum fluctuations somehow turn into classical fluctuations during the evolution of the universe.

But this is difficult to account for in the context of orthodox quantum theory. We will consider this in a bit more detail for the case of inflation theory in Section 10.7.4, for bounce theories see [65].

10.7.1 Cosmological Perturbations in a Quantum Cosmological Background

The minisuperspace bouncing nonsingular models described in Section 10.6 considered a hydrodynamical fluid or a scalar field as their matter contents. Here, we will present the main features for the quantum treatment of perturbations and background in the case of a canonical scalar field. We will consider a free scalar field, i.e., with zero field potential. The generalization to other potentials (like inflationary ones [61]) is straightforward. Hydrodynamical fluids are treated in [55–58].

The free massless scalar field is $\varphi(t, x) = \varphi_0(t) + \delta\varphi(t, x)$, where φ_0 is the background homogeneous scalar field and $\delta\varphi(t, x)$ is its linear perturbation. The FLRW metric together with its scalar perturbations is given by

$$g_{\mu\nu} = g^{(0)}_{\mu\nu} + h_{\mu\nu}, \tag{10.86}$$

where $g^{(0)}_{\mu\nu}$ represents a homogeneous and isotropic FLRW cosmological background,

$$\mathrm{d}s^2 = g^{(0)}_{\mu\nu}\mathrm{d}x^\mu \mathrm{d}x^\nu = N^2(t)\mathrm{d}t^2 - a^2(t)\delta_{ij}\mathrm{d}x^i \mathrm{d}x^j, \tag{10.87}$$

where we assumed a flat spatial metric, and $h_{\mu\nu}$ represents linear scalar perturbations around it, which we decompose into

$$\begin{aligned} h_{00} &= 2N^2\phi, \\ h_{0i} &= -NaB_{,i}, \\ h_{ij} &= 2a^2(\psi\gamma_{ij} - E_{,ij}), \end{aligned} \tag{10.88}$$

where we used the notation, $B_i = \partial_i B$. The case of tensor perturbations, i.e., gravitational waves, is very similar and actually easier [55, 56].

Starting from the classical action for this system, the Hamiltonian up to second-order can be brought into the following simple form (using a redefinition of N with terms which do not alter the equations of motion up to first order and performing canonical transformations), without ever using the background equations of motion [66] ($\kappa^2 = 1$):

$$H = \frac{N}{2e^{3\alpha}}\left[-P_\alpha^2 + P_\varphi^2 + \int d^3x\left(\frac{\pi^2}{\sqrt{\gamma}} + \sqrt{\gamma}e^{4\alpha}v^{,i}v_{,i}\right)\right], \quad (10.89)$$

where we dropped the subscript 0 from the background field and where again $a = e^\alpha$ and $v(\mathbf{x})$ is the usual Mukhanov–Sasaki variable [67], defined as

$$v = a\left(\delta\varphi + \frac{\varphi'\phi}{\mathcal{H}}\right), \quad (10.90)$$

with primes denoting derivatives with respect to conformal time η, defined by $d\eta = d\tau/a$, τ being cosmic proper time, and $\mathcal{H} = a'/a = \alpha'$.

This system is straightforwardly quantized and yields the Wheeler–DeWitt equation

$$(\hat{H}_0 + \hat{H}_2)\Psi = 0, \quad (10.91)$$

where

$$\hat{H}_0 = -\frac{\hat{P}_\alpha^2}{2} + \frac{\hat{P}_\varphi^2}{2}, \quad (10.92)$$

$$\hat{H}_2 = \frac{1}{2}\int d^3x\left(\frac{\hat{\pi}^2}{\sqrt{\gamma}} + \sqrt{\gamma}e^{4\hat{\alpha}}\hat{v}^{,i}\hat{v}_{,i}\right). \quad (10.93)$$

We now want to consider an approximation where the background evolves independently from the perturbations. The evolution of the background will be Bohmian rather than classical (as is usually considered). This approximation is obtained as follows. We write the wave function as

$$\Psi(\alpha, \varphi, v) = \Psi_0(\alpha, \varphi)\Psi_2(\alpha, \varphi, v) \quad (10.94)$$

and assume that $|\Psi_2| \ll |\Psi_0|$ and $|S_2| \ll |S_0|$, together with their derivatives with respect to the background variables. Then to lowest order we have

$$\hat{H}_0\Psi_0 = 0, \quad (10.95)$$

and the usual corresponding guidance equations. This is the minisuperspace model described in Section 10.6.1. As we have seen, quantum effects can eliminate the background singularity leading to bouncing models.

Using a Bohmian solution $(\alpha(\eta), \varphi(\eta))$ for the background, guided by Ψ_0, an approximate wave equation for the perturbations can now be obtained. It is found by considering the conditional wave function

$$\chi(v, \eta) = \Psi_2(\alpha(\eta), \varphi(\eta), v) \tag{10.96}$$

for the perturbations. It approximately satisfies (after suitable transformations)

$$\mathrm{i}\frac{\partial \chi(v,\eta)}{\partial \eta} = \frac{1}{2}\int d^3x \left(\hat{\pi}^2 + \hat{v}^{,i}\hat{v}_{,i} - \frac{a''}{a}\hat{v}^2\right)\chi(v,\eta). \tag{10.97}$$

This is the same wave equation for the perturbations known in the literature, in the absence of a scalar field potential [67]. When a scalar field potential is present, one just has to substitute a''/a by z''/z in this Hamiltonian, where $z = a\varphi'/\mathcal{H}$. The crucial difference with the standard account is that now the time-dependent potential a''/a or z''/z in Eq. (10.117) can be rather different from the semiclassical one because it is calculated from Bohmian trajectories, not from the classical ones. This can give rise to different effects in the region where the quantum effects on the background are important, which can propagate to the classical region, yielding different observations.

10.7.2 Bunch–Davies Vacuum and Power Spectrum

Having found the evolution equation for quantum perturbations in a quantum background, we recall the solution of interest in both inflationary and bouncing models, which is the Bunch–Davies vacuum.

Let us first apply the unitary transformation $e^{\mathrm{i}\frac{z'}{z}\hat{v}^2}$ to (10.97) (with a''/a replaced by z''/z to describe general potentials), which brings the Schrödinger equation into the form[a]

$$\mathrm{i}\frac{\partial \Psi(v,\eta)}{\partial \eta} = \frac{1}{2}\int d^3x \left[\hat{\pi}^2 + \hat{v}^{,i}\hat{v}_{,i} + \frac{z'}{z}\left(\hat{\pi}\hat{v} + \hat{v}\hat{\pi}\right)\right]\Psi(v,\eta). \tag{10.98}$$

[a]Both forms (10.97) and (10.98) are commonly used in the literature.

Introducing the Fourier modes $v_{\mathbf{k}}$ of the Mukhanov–Sasaki variable, defined by

$$v(\mathbf{x}) = \int \frac{d^3x}{(2\pi)^{3/2}} v_{\mathbf{k}} e^{i\mathbf{k}\cdot\mathbf{x}}, \tag{10.99}$$

and assuming a product wave function

$$\Psi = \Pi_{\mathbf{k}\in\mathbb{R}^{3+}} \Psi_{\mathbf{k}}(v_{\mathbf{k}}, v^*_{\mathbf{k}}, \eta), \tag{10.100}$$

each factor $\Psi_{\mathbf{k}}$ satisfies the Schrödinger equation

$$i\frac{\partial \Psi_{\mathbf{k}}}{\partial \eta} = \left[-\frac{\partial^2}{\partial v^*_{\mathbf{k}} \partial v_{\mathbf{k}}} + k^2 v^*_{\mathbf{k}} v_{\mathbf{k}} - i\frac{z'}{z}\left(\frac{\partial}{\partial v^*_{\mathbf{k}}} v^*_{\mathbf{k}} + v_{\mathbf{k}} \frac{\partial}{\partial v_{\mathbf{k}}} \right) \right] \Psi_{\mathbf{k}}. \tag{10.101}$$

The corresponding guidance equations are

$$v'_{\mathbf{k}} = \frac{\partial S_{\mathbf{k}}}{\partial v^*_{\mathbf{k}}} + \frac{z'}{z} v_{\mathbf{k}}. \tag{10.102}$$

The Bunch–Davies vacuum is of the form (10.100), with

$$\Psi_{\mathbf{k}} = \frac{1}{\sqrt{\pi}|f_k(\eta)|} \exp\left\{ -\frac{1}{2|f_k(\eta)|^2} |v_{\mathbf{k}}|^2 + i\left[\left(\frac{|f_k(\eta)|'}{|f_k(\eta)|} - \frac{z'}{z} \right) |v_{\mathbf{k}}|^2 - \int^{\eta} \frac{d\tilde{\eta}}{2|f_k(\tilde{\eta})|^2} \right] \right\}, \tag{10.103}$$

with f_k a solution to the classical mode equation

$$f''_{\mathbf{k}} + \left(k^2 - \frac{z''}{z} \right) f_{\mathbf{k}} = 0, \tag{10.104}$$

with initial conditions $f_k(\eta_i) = \exp(-ik\eta_i)/\sqrt{2k}$, at an early time $|\eta_i| \gg 1$ when the physical modes satisfy $k^2 \gg z''/z$. This state is homogeneous and isotropic. The guidance equations are easily integrated and yield

$$v_{\mathbf{k}}(\eta) = v_{\mathbf{k}}(\eta_i) \frac{|f_k(\eta)|}{|f_k(\eta_i)|}. \tag{10.105}$$

This result is independent of the precise form of $f_k(\eta)$ and hence is quite general.

The Bunch–Davies vacuum is motivated as follows. In inflationary models and bouncing models, $z'/z \propto 1/|\eta| \approx 0$ at early times, i.e., for $|\eta| \gg 1$. Hence, in this limit, the quantum perturbations behave like a bunch of quantum mechanical harmonic oscillators and the Bunch–Davies vacuum tends to the vacuum state of the quantum harmonic oscillator. In the case of inflation theory, the inflaton field drives the universe to a homogeneous state so that only vacuum fluctuations of these perturbations remain. Similarly, in the case of a bouncing model, in the far past in, the matter content of the universe was homogeneously and isotropically diluted in an immensely large space which was slowly contracting. In this very mild matter contraction, space-time was almost flat and empty, and the only source of inhomogeneities could only be small quantum vacuum fluctuations.

In the next Section, we discuss how this formalism connects to current cosmological observations. In Section 10.7.4, we discuss the quantum-to-classical transition of the perturbations and then finally, in Section 10.7.5, we discuss the cosmological perturbations for the matter bounce quantum background described in Section 10.6.1.2. This approach models the realistic situation where an accelerated era takes place in the expanding phase. In addition to the scalar perturbations, we will also discuss the results for the case of primordial gravitational waves. As we shall see, the quantum bounce solves important problems which cannot be addressed by classical bounces, and yield observational imprints on the cosmic microwave background radiation.

10.7.3 Power Spectrum and Cosmic Microwave Background

To make the connection between the early universe and present cosmological observations, in particular the anisotropies of the cosmic microwave background, the quantity of interest is the two-point correlation function

$$\langle \hat{v}(\mathbf{x}, \eta)\hat{v}(\mathbf{x}+\mathbf{r}, \eta)\rangle = \frac{1}{2\pi^2}\int d\ln k \frac{\sin kr}{kr} P(k), \tag{10.106}$$

which is written in terms of the Heisenberg picture, and

$$P(k, \eta) = k^3 |f_k(\eta)|^2 \tag{10.107}$$

is the power spectrum of v.

In Bohmian mechanics this quantity is obtained as follows. First, let us denote $v(\eta, \mathbf{x}; v_i)$, with v_i a field on space, a solution to the guidance equations such that $v(\eta_i, \mathbf{x}; v_i) = v_i(\mathbf{x})$. If the initial field v_i is distributed according to quantum equilibrium, i.e., $|\Psi(v_i, \eta_i)|^2$, then because of equivariance $v(\eta, \mathbf{x}; v_i)$ will be distributed according to $|\Psi(v, \eta)|^2$. For such an equilibrium ensemble, we can consider the two-point correlation function

$$\langle v(\eta, \mathbf{x}) v(\eta, \mathbf{x}+\mathbf{r})\rangle_{\mathrm{B}} \tag{10.108}$$

$$= \int \mathcal{D}v_i |\Psi(v_i, \eta_i)|^2 v(\eta, \mathbf{x}; v_i) v(\eta, \mathbf{x}+\mathbf{r}; v_i) \tag{10.109}$$

$$= \int \mathcal{D}v |\Psi(v, \eta)|^2 v(\mathbf{x}) v(\mathbf{x}+\mathbf{r}) \tag{10.110}$$

which leads to the usual expression (10.106) together with (10.107) for the correlation function and the power spectrum of v, respectively.

The power spectrum determines the temperature fluctuations of the cosmic microwave background. Let us consider this in a bit more detail. Let $T(\mathbf{n})$ denote the temperature of the cosmic microwave background in the direction $\mathbf{n}$, with $\bar{T}$ its average over the sky. The temperature anisotropy $\delta T(\mathbf{n})/\bar{T}$, where $\delta T(\mathbf{n}) = T(\mathbf{n}) - \bar{T}$, can be expanded in terms of spherical harmonics

$$\frac{\delta T(\mathbf{n})}{\bar{T}} = \sum_{l=2}^{\infty} \sum_{m=-l}^{m=l} a_{lm} Y_{lm}(\mathbf{n}). \tag{10.111}$$

The a_{lm} are determined by the Mukhanov–Sasaki variable. The main quantity used to study the temperature anisotropies is the angular power spectrum

$$C_l^0 = \frac{1}{2l+1} \sum_m |a_{lm}|^2. \tag{10.112}$$

In the standard treatments, one considers the operator $\widehat{C}_l^0$ and compares the observed value for the angular power spectrum with $C_l = \langle \Psi | \widehat{C}_l^0 | \Psi \rangle$, which is a function of the power spectrum (10.107). This is sometimes regarded as a puzzle, since the expectation value refers to an average over an ensemble of universes, while on the other hand we have only one sky to observe the microwave background radiation. It is sometimes claimed that this expectation value will agree with an average of the angular power spectrum seen for different observers at large spatial separations. While this may be true, it does not seem relevant, since we do not have observations from other places, just from earth.

What is relevant, however, is the variance. Since for the Bunch–Davies vacuum the modes are jointly Gaussian, then also the a_{lm} are jointly Gaussian, and the variance of C_l^0 is given by [68]

$$(\Delta C_l^0)^2 = \frac{2}{2l+1} C_l^2 \,. \tag{10.113}$$

This means that one would expect the observed value to deviate from C_l by an amount of the order $\sqrt{2/(2l+1)}C_l$. We will hence have a greater uncertainty for small l (which corresponds to large angles over the sky, since the angle is roughly π/l). For large l the observed value must lie closer to the expected value.

This kind of reasoning is justified in the Bohmian approach (while in the orthodox interpretation of quantum theory there remains a gap, viz., the measurement problem). Indeed, since in Bohmian mechanics the initial configuration Q_0 of the universe is a realization of (i.e., typical of) the $|\Psi|^2$ distribution, the a_{lm} obtained from Q_0 are a realization of the joint distribution that follows from $|\Psi|^2$ which, as we assumed, is Gaussian in the case at hand. And for a realization of a Gaussian random variable, its deviation from the expectation value is indeed of the order of the root mean square.

10.7.4 Quantum-to-Classical Transition in Inflation Theory

In both inflationary models and bouncing models, the seeds of structure are the vacuum fluctuations (usually) described by the Bunch–Davies vacuum. During the evolution of the universe, these vacuum fluctuations become classical fluctuations. It is problematic to explain this transition within the context of orthodox quantum mechanics. Namely, the fluctuations are initially described by a wave function that is homogeneous and isotropic. The transition to classical fluctuations implies that these symmetries are somehow broken. However, since the Schrödinger dynamics preserves these symmetries, the only way this can happen in the context of orthodox quantum theory is through the collapse of the wave function. But when exactly does the wave function collapse in this case? This is the measurement problem, as was mentioned in the introduction. In the early universe there were no observers or measurement devices. Moreover, observers and measurement devices are supposed to originate from these vacuum fluctuations. Bohmian mechanics provides an elegant and simple solution to the problem [65, 69, 70]. The key is that in Bohmian mechanics there is more than the wave function. There are actual field fluctuations whose motion is determined by the wave function. Even though the wave function may be homogeneous and isotropic, the actual fluctuations generically are not. Moreover the field fluctuations start to behave classically during the expansion, as expected. We will explain this in the context of inflation theory [69, 70]. For bouncing models, see [65].

According to the simplest inflationary models, the early universe has inflated almost exponentially driven by a single scalar field φ (the inflaton field). The homogeneous and isotropic background is assumed classical (rather than described by Bohmian mechanics), as is usually done in inflation theory. The scalar perturbations are described by the Bunch–Davies vacuum which satisfies (10.98). As discussed in the previous Section, this state is completely determined by the function $f_k(\eta)$ which satisfies the classical mode equation (10.104), which depends on z. In many inflationary models (like power-law or slow-roll inflation), the behavior of $f_k(\eta)$ for physical wave lengths at early times, i.e., $\eta \to \eta_i$, is given by

$$f_k(\eta) \sim \mathrm{e}^{-ik\eta}\left(1 + \frac{A_k}{\eta} + \ldots\right). \tag{10.114}$$

As such, as follows from (10.105), the Bohmian modes are given by

$$v_{\mathbf{k}}(\eta) \sim \left(1 + \frac{\mathrm{Re}A_k}{\eta} + \ldots\right) \tag{10.115}$$

(in many inflationary scenarios, $\mathrm{Re}A_k = 0$ and the first order term disappears). So $v_{\mathbf{k}}$ tends to be time independent for $|\eta| \gg 1$. (This is compatible with the fact that the Bohmian modes are stationary for the ground state of a quantized scalar field in Minkowski space-time [2].) Hence, the time dependence of the Bohmian modes is completely different from that of classical solutions, which oscillate for $|\eta| \gg 1$ and $k^2 \gg z''/z$, see Eq. (10.114).

The physical modes will grow larger during inflation and will eventually obtain wavelengths much bigger than the curvature scale, i.e., $k^2 \ll z''/z$. When this happens, the behavior of $f_k(\eta)$ is approximately given by the so-called growing mode, i.e.,

$$f_k(\eta) \sim A_k^g \eta^{\alpha_g}, \qquad (10.116)$$

where $\alpha_g < 0$, so that $|f_k|$ equals f_k, up to a time-independent complex factor. As such, the Bohmian modes approximately evolve according to the classical mode equation (10.104), so that the classical limit has been attained. The classical limit can also be investigated by examining the behavior of the quantum force and leads to the same result [70].

10.7.5 Observational Aspects for Matter Bounces

Matter bounces have been proposed as alternatives to inflation. First of all, they solve the horizon and flatness problems[a] of the big bang scenario (which also motivated inflation). Second, just as in inflation, they yield a viable causal mechanism for the origin of the seeds of all structures in the universe and for the anisotropies of the cosmic microwave background radiation. The idea is that, in the far past, the matter content of the universe was homogeneously and isotropically diluted in an immensely large space which was slowly contracting. In this very soft matter contraction, space-time was almost flat and empty and the only source of inhomogeneities are quantum vacuum fluctuations. It has been shown that perturbations originating from such quantum vacuum fluctuations during a slow matter dominated contracting phase become almost scale invariant in the expanding phase. This picture is called the matter bounce scenario. There are many proposals to cause the transition from contraction to the present expanding phase, most of them using another field (in addition to the regular matter field) which dominates at small scales and realizes the bounce. However, the

[a]The horizon problem refers to the question why the cosmic microwave background radiation had the same temperature in different regions of the primordial universe if the particle horizon of big bang models at that time was much smaller than the distances among these regions. The flatness problem concerns the question why the space-like hypersurface is so flat today if flatness is highly unstable under decelerated expansion. Both problems simply do not appear in bouncing models because particle horizons can be arbitrarily large in everlasting models and flatness is an attractor under decelerated contraction.

presence of another field yields entropy perturbations, the relative fluctuations between the individual energy densities of the different fields, which are not usually treated correctly (for the correct treatment, see e.g., [71]). Furthermore, in such classical bounce scenarios, primordial gravitational waves are also created, and they are as important as scalar perturbations, i.e., the ratio $r = T/S$ between the amplitude of primordial gravitational waves T and the amplitude of scalar perturbations S is approximately one. However, observations of the anisotropies of the cosmic microwave background indicate that the ratio r is small, i.e., $r < 0.1$. This is a shortcoming of these classical bouncing models.

In Section 10.6.1.2, we presented a quantum background model where, in the asymptotic past, the scalar field behaves like a matter fluid. Hence it is a matter bounce model, which leads to the observed almost scale invariant spectrum of scalar perturbations. Furthermore, this scalar field with exponential potential leads to a dark energy phase. As we have seen, the Bohmian quantum approach to this classical model opened a new possibility for bouncing scenarios with dark energy: the dark energy behavior may be a feature only of the expanding phase, where it can model the present observed accelerated expansion, and it was absent in the contracting phase. This is a very important property of this background because the presence of dark energy in the contracting phase of bouncing models makes problematic the imposition of vacuum initial conditions for cosmological perturbations in the far past of such models. For instance, if dark energy is a cosmological constant, all modes will eventually become bigger than the curvature scale in the far past and a vacuum prescription becomes quite contrived, possibly leading to divergences in the asymptotic past [72]. In the above cosmological model, however, the presence of dark energy in the universe does not make problematic the usual initial conditions prescription for cosmological perturbations in bouncing models, the universe will always be dust dominated in the far past (running back in time, the dust repeller becomes an attractor) and vacuum initial conditions can be easily imposed in this era. Consequently, we get a well-posed problem to calculate the observed spectrum and amplitude of scalar cosmological perturbations in bouncing models with dark energy.

Let us now investigate what are the amplitudes of scalar perturbations and primordial gravitational waves in this model, and whether the classical problem with their ratio $r = T/S$ is overcome [35]. The primordial gravitational waves are described by the variable μ which satisfies similar equations as the Mukhanov–Sasaki variable v, see Section 10.7.2, with the scale factor a playing the role of z. The Bunch–Davies vacuum state is also considered for these gravitational waves.

One can numerically calculate the modes of scalar perturbations and primordial gravitational waves up to the present epoch. The two-point correlation function (10.106) for scalar perturbations is expressed in terms of f_k, which satisfies Eq. (10.104). The analogous variable for gravitational waves, called g_k, satisfies

$$g''_{\mathbf{k}} + \left(k^2 - \frac{a''}{a}\right) g_{\mathbf{k}} = 0. \tag{10.117}$$

In order to understand qualitatively the final results, let us discuss what happens near the quantum bounce. When the universe contracts sufficiently and we get closer to the bounce in the contracting phase, the curvature of space-time increases a lot, and in this situation we arrive at the regime where $z''/z \gg k^2$ and $a''/a \gg k^2$ (the so-called super-Hubble behavior). The solutions for the scalar and tensor perturbations at leading order then read, respectively,

$$\zeta_k \equiv \frac{f_k}{z} \approx A_k^1 + A_k^2 \frac{1}{R_H} \int \frac{dt}{x^2 a^3}, \tag{10.118}$$

$$h_k \equiv \frac{g_k}{a} \approx B_k^1 + B_k^2 \frac{1}{R_H} \int \frac{dt}{a^3}, \tag{10.119}$$

where x was defined in Eq. (10.47).

In the classical contracting branch of case B, x varies between $1/\sqrt{2}$ and 1, while the scale factor goes through a large contraction. In other words, the value of this integral will be dominated by the values of a near the bounce phase, where a attains its smallest value. Hence, for classical bounces the amplitude of scalar and tensor perturbations will be similar, and their ratio $r \approx 1$, contrary to the observation limit $r < 0.1$. However, in a quantum bounce, when the quantum phase begins, the value of x is no longer restricted to $(1/\sqrt{2},\ 1)$, see Fig. 10.5, and it can increase the scalar perturbation amplitudes relative to the tensor perturbation amplitudes. Indeed,

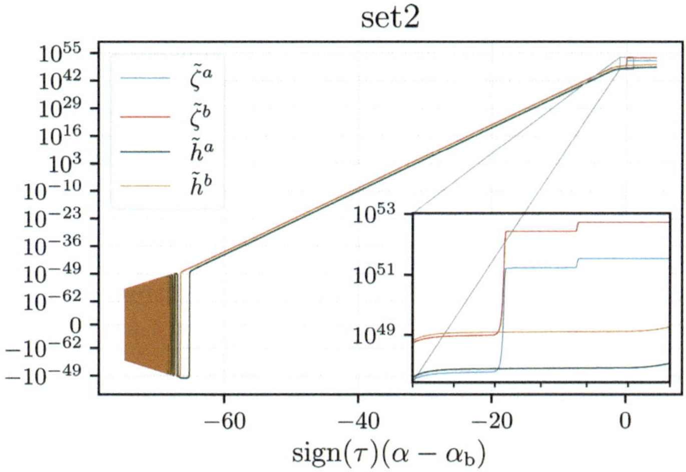

Figure 10.6 Evolution of scalar and tensor perturbations in the background of case B. The indices a and b refer to the real and imaginary parts of the perturbation amplitudes. Scalar and tensor perturbations grow almost at the same rate during classical contraction, but at the quantum bounce the scalar perturbations are hugely enhanced over the tensor perturbations due to the quantum effects (shown in the detail of the figure). After the bounce, the perturbations get frozen. The final amplitudes of both perturbations are compatible with observations.

during the classical matter phase, $1/x^2 \approx 2$, and during the classical stiff matter domination, $1/x^2 \approx 1$. Therefore, in the classical phase, the presence of $1/x^2$ in the integral (10.118) increases its value by a maximum factor of two. On the other hand, throughout the quantum phase, the $1/x$ factor can substantially increase the scalar perturbations amplitudes relative to the tensor perturbations amplitudes, as can be seen in Fig. 10.6. The curve shows that the presence of $1/x^2$ in the aforementioned integral will result in a sharp increase in the scalar perturbation amplitude around $|\alpha - \alpha_b| \approx 10^{-1}$, where α_b is the value of the scale factor at the bounce. This effect takes place close to the bounce.

This is a very important result: it shows that quantum cosmological effects may solve problems which plague classical bouncing models, namely, large ratios of tensor to scalar perturbations amplitudes. More than this, it shows that features of Bohmian trajectories can lead to observational consequences and explain involved cosmological issues. Whether these conclusions can also be reached under other quantum frameworks, is something yet to be verified.

The parameters of the theory can be adjusted in order to yield the right amplitudes and spectral indices of scalar and tensor perturbations. Hence, a single scalar field with a simple potential analyzed in the Bohmian framework yields a sensible bouncing model with dark energy behavior and correct perturbation amplitudes.

10.8 Semiclassical Gravity

Semiclassical gravity is an approximation to quantum gravity where gravity is treated classically and matter quantum mechanically. In the usual approach to semiclassical gravity, matter is described by quantum field theory on curved space-time. For example, in the case the matter is described by a quantized scalar field, the state vector can be considered to be a functional $\Psi(\varphi)$ on the space of fields, which satisfies a particular Schrödinger equation

$$\mathrm{i}\partial_t \Psi(\varphi, t) = \widehat{H}(\varphi, g)\Psi(\varphi, t)\,, \tag{10.120}$$

where the Hamiltonian operator $\widehat{H}$ depends on the classical space-time metric g. This metric satisfies Einstein's field equations

$$G_{\mu\nu}(g) = 8\pi G\langle\Psi|\widehat{T}_{\mu\nu}(\varphi, g)|\Psi\rangle\,, \tag{10.121}$$

where the source term is given by the expectation value of the energy-momentum tensor operator.

This will form a good approximation when the gravitational field and the matter field are behaving approximately classically, but will break down when the state of the matter enters in a macroscopic superposition. Namely, the Wheeler–DeWitt theory is linear, while the semiclassical approximation is not. So when matter is in a superposition of being in one location and another, then according to the Wheeler–DeWitt theory the state will be of the form $\Psi = (\psi_1(\varphi)\chi_1(g) + \psi_2(\varphi)\chi_1(g))/\sqrt{2}$, where ψ_i is the state of the matter at a particular location and χ_i the state of the corresponding gravitational field. However, according to the semiclassical theory the state will be of the form $\Psi = (\psi_1(\varphi) + \psi_2(\varphi))/\sqrt{2}$, so that $\langle\Psi|\widehat{T}_{\mu\nu}|\Psi\rangle \approx \left(\langle\psi_1|\widehat{T}_{\mu\nu}|\psi_1\rangle + \langle\psi_2|\widehat{T}_{\mu\nu}|\psi_2\rangle\right)/2$ and hence the gravitational field is affected by two matter sources, one coming from each term in the superposition. As experimentally shown by Page and Geilker, the semiclassical theory becomes inadequate in such situations [19, 73].

In the context of Bohmian mechanics it is natural to consider another type of semiclassical approximation [40, 74]. In this approximation, the back-reaction from the matter onto the gravitational field is by using the Bohmian energy-momentum tensor $T_{\mu\nu}(\varphi_B, g)$, where φ_B is the actual Bohmian field configuration, in Einstein's field equations:

$$G_{\mu\nu}(g) = 8\pi G T_{\mu\nu}(\varphi_B, g)\,. \tag{10.122}$$

While the resulting theory is still nonlinear, it solves the problem with the macroscopic superposition. Namely, even though the matter wave function is in a macroscopic superposition of being at two locations, the Bohmian field φ_B and the energy-momentum tensor $T_{\mu\nu}(\varphi_B, g)$ will correspond to a matter configuration at one of the locations. This means that according to Eq. (10.122) the gravitational field will correspond to that of matter localised at that location.

However, there is an immediate problem with this ansatz, namely that Eq. (10.122) is not consistent. The Einstein tensor $G_{\mu\nu}$ is identically conserved, i.e., $\nabla^\mu G_{\mu\nu} \equiv 0$. So the Bohmian energy-momentum tensor $T_{\mu\nu}(\varphi_B, g)$ must be conserved as well. However, the equation of motion for the scalar field does not guarantee this. (Similarly, in the Bohmian approach to nonrelativistic systems, the energy is generically not conserved.) The solution to the problem is that the usual expression for the Bohmian energy-momentum tensor is not the right source term in the Einstein field equations. The correct source term can in principle be derived by starting from the Bohmian Wheeler–DeWitt theory. However, in the derivation the gauge invariance, which in this case is the invariance under spatial diffeomorphisms (i.e., spatial coordinate transformations), should be dealt with, either by performing a gauge fixing or by working with gauge independent degrees of freedom. However, this is a notoriously difficult problem in the case of general relativity. In the case of minisuperspace model the spacial diffeomorphism invariance is eliminated and a consistent semiclassical approximation can be found straightforwardly.

A similar problem arises in the Bohmian semiclassical approach to scalar electrodynamics, which describes a scalar field interacting with an electromagnetic field. In this case, the wave equation for the scalar field is of the form

$$\mathrm{i}\partial_t \Psi(\varphi, t) = \widehat{H}(\varphi, A)\Psi(\varphi, t)\,, \tag{10.123}$$

where A is the vector potential. There is also a Bohmian scalar field φ_B and a charge current $j^\nu(\varphi_B, A)$ that could act as the source term in Maxwell's equations

$$\partial_\mu F^{\mu\nu}(A) = j^\nu(\varphi_B, A)\,, \tag{10.124}$$

where $F^{\mu\nu}$ is the electromagnetic field tensor. In this case, we have $\partial_\nu\partial_\mu F^{\mu\nu} \equiv 0$ due to the antisymmetry of $F^{\mu\nu}$. As such, the charge current must be conserved. However, the Bohmian equation of motion for the scalar field does not imply conservation. Hence, just as in the case of gravity, a consistency problem arises. In this case, the correct current can be derived by starting from the full scalar electrodynamics after eliminating the gauge freedom. For example, in the Coulomb gauge, there is an extra current j_Q^ν which appears in addition to the usual charge current and which depends on the quantum potential, so that Maxwell's equations read

$$\partial_\mu F^{\mu\nu}(A) = j^\nu(\varphi_B, A) + j_Q^\nu(\varphi_B, A)\,, \tag{10.125}$$

which is consistent since the total current is conserved, because of the equation of motion for the scalar field.

Let us consider now the derivation of the semiclassical approximation of the minisuperspace model defined by (10.35) and (10.36). For simplicity, we assume the gauge $N = 1$. Given a set of trajectories $(\alpha(t), \varphi(t))$, the conditional wave function for the scalar field is $\chi(\varphi, t) = \psi(\varphi, \alpha(t))$. Using

$$\partial_t \chi(\varphi, t) = \partial_\alpha \psi(\varphi, \alpha)\big|_{\alpha=\alpha(t)}\dot{\alpha}(t)\,, \tag{10.126}$$

and the Wheeler–DeWitt equation (10.35), we can write

$$\mathrm{i}\partial_t \chi = \widehat{H}_M \chi + I\,, \tag{10.127}$$

where

$$\widehat{H}_M = -\frac{1}{2\mathrm{e}^{3\alpha}}\partial_\varphi^2 + \mathrm{e}^{3\alpha} V_M \,. \tag{10.128}$$

and I a rest term [40]. When I is negligible (up to a real time-dependent function times χ), (10.127) becomes the Schrödinger equation for a homogeneous matter field in an external FLRW metric. If furthermore the quantum potential Q_G is negligible compared to other terms in Eq. (10.38), we are led to the semiclassical theory:

$$\mathrm{i}\partial_t \chi = \widehat{H}_M \chi \,, \tag{10.129}$$

$$\dot{\varphi} = \frac{1}{\mathrm{e}^{3\alpha}}\partial_\varphi S \,, \tag{10.130}$$

$$\frac{1}{2}\dot{\alpha}^2 = \kappa^2\left(\frac{1}{2}\dot{\varphi}^2 + V_M + Q_M\right) + V_G \equiv -\frac{\kappa^2}{\mathrm{e}^{3\alpha}}\partial_t S + V_G \,. \tag{10.131}$$

This whole procedure is very similar to what was done in Section 10.7.1.

In the usual semiclassical approximation, one has (10.129) and

$$\frac{1}{2}\dot{\alpha}^2 = \frac{\kappa^2}{\mathrm{e}^{3\alpha}}\langle \chi | \widehat{H}_M | \chi \rangle + V_G \,, \tag{10.132}$$

with χ normalized to one (which is the analogue of (10.120) and (10.121) for minisuperspace). In [40], an example was worked out showing that the Bohmian semiclassical approximation gives better results than the usual semiclassical approximation. As such the Bohmian semiclassical approximation may perhaps be used to find effects of quantum gravitational nature that are not present in the usual semiclassical approximation. Possible applications may be inflation theory (which will be discussed in the next Section), which is described by a minisuperspace model with fluctuations. Usually the fluctuations are described quantum mechanically while the homogeneous background is described by a classical minisuperspace theory. Including the back-reaction from the fluctuations onto the homogeneous background will lead to corrections which may be testable. Such an investigation has been carried out in [75]. A correction was found but is as yet unobservable. A Bohmian approach may perhaps improve on this result.

10.9 Conclusion

It is highly problematic to interpret canonical quantum gravity within the framework of orthodox quantum theory. The difficulty is that orthodox quantum theory merely makes predictions about outcomes of measurements and thereby relies on observers or measurement devices outside the quantum system of interest. When the quantum system of interest is the whole universe, there are no outside observers or measurement devices. As an alternative, there is Bohmian quantum gravity. Bohmian quantum gravity provides an objective description of the universe in terms of an actual space-time and matter fields (or particles), whose dynamics is determined by the universal wave function. This allows for a clear and unambiguous analysis of questions that are often rather ambiguous in the framework of orthodox quantum theory. Examples are the questions what the history of the universe is, whether it originated or will end up in a space-time singularity, how we can experimentally test the theory, etc.

In this chapter, we addressed various of these questions. We explained how Bohmian quantum gravity solves the problem of time and how it allows for the derivation of familiar time-dependent Schrödinger equations for subsystems in the universe.

We also analyzed the question of space-time singularities and described Bohmian solutions that are free of space-time singularities. This was done in the context of simplified models of quantum gravity which assume homogeneity and isotropy. In all the models we considered, bouncing solutions are possible which describe a contracting universe evolving into an expanding one. Such bouncing solutions may describe our actual universe. We also considered deviations from homogeneity and isotropy by introducing linear perturbations. Using the Bohmian approach, we described how effective equations can be found describing the motion of these quantum perturbations in an external homogeneous and isotropic quantum background. The study of these perturbations is important since they leave an imprint on, for example, the cosmic microwave background radiation. This allows to distinguish between different theories. We presented the results of a recent study that investigated the perturbations for a background model with a matter bounce, where the bounce follows from the Bohmian dynamics. The results are in good agreement with observation and hence such a model may provide a serious alternative to inflation theory, which is currently the prevailing approach to early universe cosmology. It is important to stress that the Bohmian dynamics for the background is crucial to obtain these results.

The Bohmian approach also deals in a natural way with the classical limit. The classical limit is obtained whenever the space-time or matter degrees of freedom behave approximately classically. We applied this to the problem of the quantum-to-classical transition in inflation theory and bouncing theories. This problem is solved very simply in Bohmian mechanics and gives the usual results (which are problematic to justify within the context of orthodox quantum theory).

In the final Section, we presented an approach to semiclassical gravity based on Bohmian mechanics. This approach goes beyond the usual semiclassical approach and might provide a new tool, unavailable in orthodox quantum theory, to probe quantum gravitational effects.

There are other applications of Bohmian quantum gravity that were not discussed here. One concerns the Boltzmann brain problem in cosmology. As was shown in [76], there is no such problem in the context of the Bohmian approach. Another application concerns the search for quantum nonequilibrium [77–79]. While a typical initial configuration gives the usual quantum predictions described by the Born rule, other initial configurations may lead to nonequilibrium distributions which yield deviations from the Born rule. Since nonequilibrium distributions tend to evolve very quickly to equilibrium distributions [80, 81], one needs to find special systems where this is not the case. Such special systems might be found in cosmological context, in particular with the primordial perturbations for which nonequilibrium distributions may leave observational imprints on the cosmic microwave background. In some of these investigations, it was conjectured that some anomalies found in the observation of the anisotropies of the cosmic background radiation might be explained by these deviations. However, there is still a large debate on whether such anomalies are indeed statistically significant, or whether they are produced by other causes. So far, there is no experimental evidence for a violation of the Born rule.

Acknowledgments

NPN and WS respectively acknowledge the CNPq of Brazil and the Deutsche Forschungsgemeinschaft for financial support.

References

1. D. Bohm and B. J. Hiley, *The Undivided Universe*, Routledge, New York (1993).
2. P. R. Holland, *The Quantum Theory of Motion*, Cambridge University Press, Cambridge (1993).
3. D. Dürr and S. Teufel, *Bohmian Mechanics*, Springer-Verlag, Berlin (2009).
4. D. Dürr, S. Goldstein, and N. Zanghì, *Quantum Physics Without Quantum Philosophy*, Springer-Verlag, Berlin (2012).
5. W. Struyve, *Pilot-Wave Approaches to Quantum Field Theory*, Journal of Physics: Conference Series, **306**, 012047 (2011) and arXiv: 1101.5819v1 [quant-ph].
6. J. C. Vink, *Quantum Potential Interpretation of the Wave Function of the Universe*, Nuclear Physics B, **369**, (1992).
7. Y. V. Shtanov, *Pilot Wave Quantum Cosmology*, Physical Review D, **54**, 2564 (1996) and arXiv:gr-qc/9503005.
8. S. Goldstein and S. Teufel, *Quantum Spacetime Without Observers: Ontological Clarity and the Conceptual Foundations of Quantum Gravity*, in *Physics Meets Philosophy at the Planck Scale*, eds. C. Callender and N. Huggett, Cambridge University Press, Cambridge (2004) and arXiv:quant-ph/9902018, pp. 275–289.
9. N. Pinto-Neto, *The Bohm Interpretation of Quantum Cosmology*, Foundations of Physics, **35**, 577 (2005) and arXiv:gr-qc/0410117.
10. N. Pinto-Neto and J. C. Fabris, *Quantum Cosmology from the de Broglie-Bohm Perspective*, Classical and Quantum Gravity, **30**, 143001 (2013) and arXiv:1306.0820 [gr-qc].

11. D. Dürr, S. Goldstein, and N. Zanghì, *Quantum Equilibrium and the Origin of Absolute Uncertainty*, Journal of Statistical Physics, **67**, 843 (1992) and arXiv:quant-ph/0308039. Reprinted in [4].
12. A. Valentini, *Signal-Locality, Uncertainty, and the Subquantum H-Theorem. I*, Physics Letters A, **156**, 5 (1991).
13. W. Struyve and A. Valentini, *de Broglie–Bohm Guidance Equations for Arbitrary Hamiltonians* Journal of Physics A, **42**, 035301 (2009) and arXiv:0808.0290v3 [quant-ph].
14. J. S. Bell, *Against Measurement*, Physics World, **3**, 33 (1990).
15. T. Norsen and W. Struyve, *Weak Measurement and Bohmian Conditional Wave Functions*, Annals of Physics, **350**, 178 (2014) and arXiv: 1305.2409 [quant-ph].
16. C. W. Misner, K. S. Thorne, and J. A. Wheeler, *Gravitation*, W. H. Freeman and Company, New York (1973).
17. C.J. Isham, *Canonical Quantum Gravity and the Problem of Time*, in *Integrable Systems, Quantum Groups, and Quantum Field Theories*, eds. L. A. Ibort and M. A. Rodriguez, Kluwer Academic Publishers, London (1993) and arXiv:gr-qc/9210011, p. 157.
18. K. V. Kuchař, *Time and Interpretations of Quantum Gravity*, in *Proceedings of the 4th Canadian Conference on General Relativity and Relativistic Astrophysics*, eds. G. Kunstatter, D. Vincent, and J. Williams, World Scientific, Singapore (1992), reprinted in International Journal of Modern Physics D, **20**, 3 (2011).
19. C. Kiefer, *Quantum Gravity*, International Series of Monographs on Physics, **124**, Clarendon Press, Oxford (2004).
20. B. S. DeWitt, *Quantum Theory of Gravity. I. The Canonical Theory*, Physical Review, **160**, 1113, (1967).
21. A. Ashtekar, T. Pawlowski, and P. Singh, *Quantum Nature of the Big Bang*, Physical Review Letters, **96**, 141301 (2006) and arXiv:gr-qc/0602086.
22. A. Ashtekar, T. Pawlowski, and P. Singh, *Quantum Nature of the Big Bang: An Analytical and Numerical Investigation*, Physical Review D, **73**, 124038 (2006) and arXiv:gr-qc/0604013.
23. A. Ashtekar, T. Pawlowski, and P. Singh, *Quantum Nature of the Big Bang: Improved Dynamics*, Physical Review D, **74**, 084003 (2006) and arXiv:gr-qc/0607039.
24. A. Ashtekar, A. Corichi, and P. Singh, *Robustness of Key Features of Loop Quantum Cosmology*, Physical Review D, **77**, 024046 (2008) and arXiv:0710.3565 [gr-qc].

25. A. Ashtekar and P. Singh, *Loop Quantum Cosmology: A Status Report*, Classical and Quantum Gravity, **28**, 213001 (2011) and arXiv:1108.0893 [gr-qc].
26. N. Pinto-Neto and E. Sergio Santini, *Must Quantum Spacetimes be Euclidean?*, Physical Review D, **59**, 123517 (1999) and arXiv:gr-qc/9811067.
27. A. Shojai and F. Shojai, *Constraint Algebra and Equations of Motion in the Bohmian Interpretation of Quantum Gravity*, Classical and Quantum Gravity, **21**, 1 (2004) and arXiv:gr-qc/0311076.
28. D. Dürr, S. Goldstein, T. Norsen, W. Struyve, and N. Zanghì, *Can Bohmian Mechanics be Made Relativistic?*, Proceedings of the Royal Society A, **470**, 20130699 (2014) and arXiv:1307.1714 [quant-ph].
29. E. J. Squires, *A Quantum Solution to a Cosmological Mystery*, Physics Letters A, **162**, 35 (1992).
30. N. Pinto-Neto and E. Sergio Santini, *The Accelerated Expansion of the Universe as a Quantum Cosmological Effect*, Physics Letters A, **315**, 36 (2003) and arXiv:gr-qc/0302112.
31. J. J. Halliwell and S. Hawking, *Origin of Structure in the Universe, Physical Review D,* **31**, 1777 (1985).
32. N. Pinto-Neto, F. T. Falciano, R. Pereira, and E. Sergio Santini, *Wheeler–DeWitt Quantization Can Solve the Singularity Problem, Physical Review D,* **86**, 063504 (2012) and arXiv:1206.4021 [gr-qc].
33. W. Struyve, *Loop Quantum Cosmology and Singularities,* Scientific Reports, **7**, 8161 (2017) and arXiv:1703.10274 [gr-qc].
34. S. Colin and N. Pinto-Neto, *Quantum Matter Bounce with a Dark Energy Expanding Phase*, Physical Review D, **96**, 063502 (2017) and arXiv:1706.03037 [gr-qc].
35. A. P. Bacalhau, N. Pinto-Neto, and S. D. P. Vitenti, *Consistent Scalar and Tensor Perturbation Power Spectra in Single Fluid Matter Bounce with Dark Energy Era*, arXiv:1706.08830 [gr-qc].
36. J. Acacio de Barros and N. Pinto-Neto, *The Causal Interpretation of Quantum Mechanics and the Singularity Problem and Time Issue in Quantum Cosmology*, International Journal of Modern Physics D, **07**, 201 (1998).
37. F. G. Alvarenga, J. C. Fabris, N. A. Lemos, and G. A. Monerat, *Quantum Cosmological Perfect Fluid Models*, General Relativity and Gravitation, **34**, 651 (2002) and arXiv:gr-qc/0106051 .

38. N. Pinto-Neto, E. Sergio Santini, and F. T. Falciano, *Quantization of Friedmann Cosmological Models with two Fluids: Dust Plus Radiation, Physics Letters A,* **A344**, 131 (2005) and arXiv:gr-qc/0505109.
39. J. J. Halliwell, *Introductory Lectures on Quantum Cosmology*, in *Quantum Cosmology and Baby Universes*, eds. S. Coleman, J. B. Hartle, T. Piran, and S. Weinberg, World Scientific, Singapore, 159 (1991) and arXiv:0909.2566 [gr-qc].
40. W. Struyve, *Semiclassical Approximations Based on Bohmian Mechanics*, arXiv:1507.04771 [quant-ph].
41. F. T. Falciano, N. Pinto-Neto, and W. Struyve, *Wheeler–DeWitt Quantization and Singularities*, Physical Review D, **91**, 043524 (2015) and arXiv:1501.04181 [gr-qc].
42. R. Colistete Jr., J. C. Fabris, and N. Pinto-Neto, *Gaussian Superpositions in Scalar-Tensor Quantum Cosmological Models*, Physical Review D, **62**, 083507 (2000) and arXiv:gr-qc/0005013.
43. F. G. Alvarenga and N. A. Lemos, *Dynamical Vacuum in Quantum Cosmology*, General Relativity and Gravitation, **30**, 681 (1998) and arXiv:gr-qc/9802029.
44. C. Rovelli, *Quantum Gravity*, Cambridge University Press, Cambridge (2004).
45. C. Rovelli and F. Vidotto, *Covariant Loop Quantum Gravity*, Cambridge University Press, Cambridge (2014).
46. M. Martín-Benito, G. A. Mena Marugán, and J. Olmedo, *Further Improvements in the Understanding of Isotropic Loop Quantum Cosmology*, Physical Review D, **80**, 104015 (2009).
47. G. A. Mena Marugán, J. Olmedo, and T. Pawłowski, *Prescriptions in Loop Quantum Cosmology: A Comparative Analysis*, Physical Review D **84**, 064012 (2011) and arXiv:1108.0829 [gr-qc].
48. K. Banerjee, G. Calcagni, and M. Martín-Benito, *Introduction to Loop Quantum Cosmology*, SIGMA, **8**, 016 (2012) and arXiv:1109.6801 [gr-qc].
49. M. Bojowald, *Quantization Ambiguities in Isotropic Quantum Geometry*, Classical and Quantum Gravity, **19**, 5113 (2002) and arXiv:gr-qc/0206053.
50. M. Bojowald, *Loop Quantum Cosmology*, Living Reviews in Relativity, **11**, 4 (2008) and arXiv:gr-qc/0601085.
51. J. S. Bell, *Beables for Quantum Field Theory*, in J. S. Bell, *Speakable and Unspeakable in Quantum Mechanics*, Cambridge University Press, Cambridge (1987).

52. D. Dürr, S. Goldstein, R. Tumulka, and N. Zanghì, *Quantum Hamiltonians and Stochastic Jumps*, Communications in Mathematical Physics, **254**, 129 (2005) and arXiv:quant-ph/0303056v3.
53. D. Dürr, S. Goldstein, R. Tumulka, and N. Zanghì, *Bell-Type Quantum Field Theories*, Journal of Physics A, **38**, R1-R43 (2005) and arXiv:quant-ph/0407116v1.
54. J. Vink, *Quantum Mechanics in Terms of Discrete Beables*, Physical Review A, **48**, 1808 (1993).
55. P. Peter, E. J. C. Pinho, and N. Pinto-Neto, *Tensor Perturbations in Quantum Cosmological Backgrounds*, Journal of Cosmology and Astroparticle Physics, **07**, 014 (2005) and arXiv:hep-th/0509232.
56. P. Peter, E. J. C. Pinho, and N. Pinto-Neto, *Gravitational Wave Background in Perfect Fluid Quantum Cosmologies*, Physical Review D, **73**, 104017 (2006) and arXiv:gr-qc/0605060.
57. P. Peter, E. J. C. Pinho, and N. Pinto-Neto, *Noninflationary Model with Scale Invariant Cosmological Perturbations*, Physical Review D, **75**, 023516 (2007) and arXiv:hep-th/0610205.
58. E. J. C. Pinho and N. Pinto-Neto, *Scalar and Vector Perturbations in Quantum Cosmological Backgrounds*, Physical Review D, **76**, 023506 (2007) and arXiv:hep-th/0610192.
59. S. D. P. Vitenti and N. Pinto-Neto, *Large Adiabatic Scalar Perturbations in a Regular Bouncing Universe*, Physical Review D, **85**, 023524 (2012) and arXiv:1111.0888 [astro-ph.CO].
60. S. D. P. Vitenti, F. T. Falciano, and N. Pinto-Neto, *Quantum Cosmological Perturbations of Generic Fluids in Quantum Universes*, Physical Review D **87**, 103503 (2013) and arXiv:1206.4374 [gr-qc].
61. F. T. Falciano, N. Pinto-Neto, and S. D. P. Vitenti, *Scalar Field Perturbations with Arbitrary Potentials in Quantum Backgrounds*, Physical Review D, **87**, 103514 (2013) and arXiv:1305.4664 [gr-qc].
62. N. Pinto-Neto and S. D. P. Vitenti, *Comment on "Growth of Covariant Perturbations in the Contracting Phase of a Bouncing Universe,"* Physical Review D, **89**, 028301 (2014) and arXiv:1312.7790 [astro-ph.CO].
63. A. Perez, H. Sahlmann, and D. Sudarsky, *On the Quantum Origin of the Seeds of Cosmic Structure*, Classical and Quantum Gravity, **23**, 2317 (2006) and arXiv:gr-qc/0508100v3.
64. D. Sudarsky, *Shortcomings in the Understanding of Why Cosmological Perturbations Look Classical*, International Journal of Modern Physics D, **20**, 509 (2011) and arXiv:0906.0315v3 [gr-qc].

65. N. Pinto-Neto, G. Santos, and W. Struyve, *Quantum-to-Classical Transition of Primordial Cosmological Perturbations in de Broglie-Bohm Quantum Theory: The Bouncing Scenario*, Physical Review D, **89**, 023517 (2014) and arXiv:1309.2670 [gr-qc].
66. F. T. Falciano and N. Pinto-Neto, *Scalar Perturbations in Scalar Field Quantum Cosmology*, Physical Review D, **79**, 023507 (2009) and arXiv:0810.3542v2 [gr-qc].
67. V. Mukhanov, H. A. Feldman, and R. H. Brandenberger, *Theory of Cosmological Perturbations*, Physics Reports, **215**, 203 (1992).
68. D. H. Lyth and A. R. Liddle, *The Primordial Density Perturbation*, Cambridge University Press, Cambridge (2009).
69. B. J. Hiley and A. H. Aziz Mufti, *The Ontological Interpretation of Quantum Field. Theory Applied in a Cosmological Context*, in *Fundamental Theories of Physics*, **73** eds. M. Ferrero and A. van der Merwe, Kluwer Dordrecht (1995), pp. 141–156.
70. N. Pinto-Neto, G. Santos, and W. Struyve, *The Quantum-to-Classical Transition of Primordial Cosmological Perturbations*, Physical Review D, **85**, 083506 (2012) and arXiv:1110.1339 [gr-qc].
71. P. Peter, N. Pinto-Neto, and S. D. P. Vitenti, *Quantum Cosmological Perturbations of Multiple Fluids*, Physical Review D, **93**, 023520 (2016) and arXiv:1510.06628 [gr-qc].
72. R. Maier, S. Pereira, N. Pinto-Neto, and B. B. Siffert, *Bouncing Models with a Cosmological Constant*, Physical Review D, **85**, 023508 (2012) and arXiv:1111.0946 [astro-ph.CO].
73. D. N. Page and C. D. Geilker, *Indirect Evidence for Quantum Gravity, Physical Review Letters,* **47**, 979 (1981).
74. W. Struyve, *Towards a Novel Approach to Semiclassical Gravity*, in *The Philosophy of Cosmology*, eds. K. Chamcham, J. Silk, J. D. Barrow, and S. Saunders, Cambridge University Press, Cambridge (2017), pp. 3566–375.
75. C. Kiefer and M. Krämer, *Quantum Gravitational Contributions to the Cosmic Microwave Background Anisotropy Spectrum*, Physical Review Letters, **108**, 021301 (2012) and arXiv:1103.4967 [gr-qc].
76. S. Goldstein, R. Tumulka, and W. Struyve, *The Bohmian Approach to the Problems of Cosmological Quantum Fluctuations*, to appear in *Guide to the Philosophy of Cosmology*, eds. A. Ijjas and B. Loewer, Oxford University Press, and arXiv:1508.01017 [gr-qc].

77. A. Valentini, *Astrophysical and Cosmological Tests of Quantum Theory*, Journal of Physics A, **40**, 3285 (2007) and arXiv:hep-th/0610032v2.

78. S. Colin and A. Valentini, *Primordial Quantum Nonequilibrium and Large-Scale Cosmic Anomalies*, Physical Review D, **92**, 043520 (2015) and arXiv:1407.8262 [astro-ph.CO].

79. S. Colin and A. Valentini, *Primordial Quantum Nonequilibrium and Large-Scale Cosmic Anomalies*, International Journal of Modern Physics D, **25**, 1650068 (2016) and arXiv:1510.03508 [gr-qc].

80. A. Valentini and H. Westman, *Dynamical Origin of Quantum Probabilities* Proceedings of the Royal Society A, **461**, 253 (2005) and arXiv:quant-ph/0403034.

81. M. D. Towler, N. J. Russell, and A. Valentini, *Time Scales for Dynamical Relaxation to the Born Rule*, Proceedings of the Royal Society A, **468**, 990 (2012) and arXiv:1103.1589 [quant-ph].

Index

For Product Safety Concerns and Information please contact our EU representative GPSR@taylorandfrancis.com Taylor & Francis Verlag GmbH, Kaufingerstraße 24, 80331 München, Germany

Batch number: 10726505

Printed by Printforce, the Netherlands